AF293812

Applied Fluorescence in Chemistry, Biology and Medicine

Springer

Berlin
Heidelberg
New York
Barcelona
Hong Kong
London
Milan
Paris
Singapore
Tokyo

Wolfgang Rettig · Bernd Strehmel · Sigurd Schrader ·
Holger Seifert

Applied Fluorescence in Chemistry, Biology and Medicine

With 235 Figures and 32 Tables

Springer

Prof. Dr. Wolfgang Rettig
Dipl.-Phys. Holger Seifert

Humboldt-Universität zu Berlin
Institut für Chemie
Fach-Institut für Physikalische und Theoretische Chemie
Bunsenstraße 1
D-10117 Berlin
Germany

Dr. Sigurd Schrader
Universität Potsdam
Institut für Physik
Lehrstuhl für Physik kondensierter Materie
An Neuen Palais 10
D-14469 Potsdam
Germany

Dr. Bernd Strehmel
Center for Photochemical Sciences
Department of Chemistry
Bowling Green State University
Bowling Green
OH 43403
USA

ISBN-13:978-642-64175-6   Springer-Verlag Berlin Heidelberg New York

Library of Congress Cataloging-in-Publication Data

Applied fluorescence in chemistry, biology and medicine / Wolfgang
Rettig ... [et al.].
    p. cm.
Includes bibliographical references and index.
ISBN-13:978-642-64175-6   e-ISBN-13:978-3-642-59903-3
DOI: 10.1007/978-3-642-59903-3

1. Fluorescence spectroscopy. I. Rettig, Wolfgang 1947.
QD96, F56 A67 1999
543'.08584-ddc21   98-40878

Typesetting: Fotosatz-Service Köhler GmbH, Würzburg
Coverdesign: Design & Production GmbH, Heidelberg
SPIN: 10634893   52/3020-5 4 3 2 1 0 – Printed on acid-free paper

Preface

The light of the sun – the source of life and a god in ancient times – was always associated with warmth or even burning heat. When man learnt to tame fire, he had another source of light, which was also hot. All the more puzzling for the amazed observer has been the dance of fireflies glimmering in the dark. How can an animal emit light? Or what about the faint glow of walls in some prehistoric tombs due to luminescent bacteria? This sort of "cold light" is known today as chemiluminescence. A related phenomenon is the aurora borealis with its swiftly moving curtains of light. This is a special sort of electroluminescence, another kind of "cold light".

The basis of "hot light" is the thermal generation of electronically excited states (normally of atoms or ions). The source of "cold light", on the other hand, is the chemical or electrochemical generation of excited states, possible also for larger molecules. By using light instead of chemical or electrical energy, we can generate yet another type of "cold light", the ordinary luminescence: fluorescence or phosphorescence. The possibilities here have increased enormously because we can color-tune the exciting light and gain specificity, and we can modulate it in time or even in polarization opening new dimensions of research and applications.

The present book tries to give a state-of-the-art of these developments in the chemical, biochemical, biological and medical sciences. Some of the areas most rapidly developing due to fluorescence are covered: Imaging methods in medicine, which complement NMR tomographic methods but can also be used for single cells; the easy and highly sensitive detection of antibodies in fluorescence immuno assays; the possibility for the human genome sequencing via single molecule fluorescence techniques; the development of simple analytical fluorescence methods in environmental and life sciences; fluorescence probes for microscopic properties of condensed matter, both liquid and solid. And last but not least, fluorescence allows us to follow and to understand excited-state processes and mechanisms.

Fluorescence as a science is young. The first (still classical) book on this topic was published in the 1950s by Theodor Förster [1], and since then, many other books covering both principle and applicational aspects have been written. A representative list covering both classical and also the more recent books are given in the bibliographic section [1–25]. The present book derives, to a large part, from the invited contributions to the 5th International Conference on Methods and Applications of Fluorescence Spectroscopy (MAFS5), held in

Berlin/Germany in September 1997, and includes a few further relevant contributions. The series of MAFS conferences led to the production of a book in 1991 [24] updating and summarizing the state-of-the-art in fluorescence spectroscopy and application. We felt that in this rapidly developing and important field a further overview is timely and we have tried to include some new aspects by introducing a section covering polymers.

The book is divided into the following sections focussing on both the methods and the applications of fluorescence in different fields:

I   Methods and Trends in Fluorescence Spectroscopy
II   Analytical Fluorescence Probes, and Environmental Research
III Fluorescence Probes in Polymers
IV   Applications of Fluorescence Spectroscopy in Biology
V   Fluorescence Techniques in Medicine – a Challenge for the Future

The first section provides a survey on some of the methods applied in fluorescence spectroscopy that have recently gained increased importance such as imaging techniques, single-molecule spectroscopy, time-resolved spectroscopy, fluorescence correlation spectroscopy, fluorescence microscopy on a submicron scale, three-photon excitation spectroscopy, and the evaluation of photophysical data at low temperatures. We are aware of the problem that not all methods of fluorescence spectroscopy could be included here. Some more will be found in subsequent chapters.

The second section focusses on fluorescence probes for the use in analytical chemistry and we included a contribution reporting about the use of fluorescence in environmental research. The chapter also contains an overview on crown-ether and photoinduced electron-transfer fluorescence probes. Both classes of probes have gained importance in analytical chemistry in recent years. A further contribution reports about the photostability of some selected dye chromophores important for single molecule spectroscopy. Photobleaching studies and the most important photochemical and photophysical properties are reported.

In the third section, we have collected some contributions dealing with fluorescence in polymers. The application of fluorescence in polymer sciences is relatively young. Unfortunately, it has received a lot of criticism because people have always asked about the relations between fluorescence and the properties of polymers. Two contributions treat the problem of how to explore polymeric materials by the use of stationary fluorescence spectroscopy. They focus on the examination of swollen materials and the thermodynamics of block copolymer micelle formation. A further contribution discusses the relation between fluorescence probes and liquid crystals including macromolecular chain dynamics. Another field of polymer fluorescence is highlighted in a contribution reporting on light-emitting polymers including a photochemical and photophysical study of the corresponding monomers and oligomers. There are further important applications which are not included in this section. Some of these concern polymer curing in real time which is important for many technical applications.

The fourth section includes contributions reporting on fluorescence microscopy and the examination of biological materials using single molecule spectroscopy, some quantitative relations about solvent relaxation in biomembranes, and the application of lanthanide chelates for improved resonance energy transfer including some interesting biological examples. This section is by no means comprehensive and only covers selected topics but is meant to give some ideas about the various application possibilities of fluorescence in biology.

The final section reports on the application of fluorescence in medical sciences focussing on application of fluorescence microscopy, fluorescence immuno assays, microfluorimetry of cellular and subcellular processes, and the fluorescence imaging of liver tumors. These contributions provide examples for medical scientists who may want to apply fluorescence spectroscopy in their fight against cancer.

We hope that this book can be used fruitfully by scientists who are interested in the further development and applications of fluorescence. It is intended to inspire scientists working at universities and in industry for their further research. The field is highly interdisciplinary. Therefore, this book addresses chemists, physicists, biologists, material and medical scientists who would like to apply the fluorescence method in their research or in application.

Berlin, October 1998

Wolfgang Rettig,
Bernd Strehmel,
Sigurd Schrader,
Holger Seifert

Selected Bibliography on Fluorescence

1. Förster T (1951 and 1982) Fluoreszenz Organischer Verbindungen. Vandenhoeck and Ruprecht, Göttingen
2. Hercules DM (ed) (1966) Fluorescence and Phosphorescence Analysis, Wiley Interscience, New York
3. Parker CA (1968) Photoluminescence of Solutions. Elsevier, Amsterdam
4. Becker RS (1969) Theory and Interpretation of Fluorescence and Phosphorescence. Wiley Interscience, New York
5. Udenfriend S (1962 and 1971) Fluorescence Assay in Biology and Medicine, vols 1 and 2. Academic, New York
6. Berlman IB (1971) Handbook of Fluorescence Spectra of Aromatic Molecules. Academic Press, New York
7. Guilbault CG (1973) Practical Fluorescence: Theory, Methods, and Techniques. Marcel Dekker, New York
8. Chen RF, Edelhoch H (1976) Biochemical Fluorescence: Concept, vols 1 and 2. Marcel Dekker, New, York
9. Wehry EL (ed) (1981) Modern Fluorescence Spectroscopy. Plenum, New York
10. Zander M (1981) Fluorimetrie. Springer, Berlin Heidelberg New York
11. Steiner RF (ed) (1983) Principles of Fluorescence Spectroscopy. Plenum, New York
12. Lakowicz JR (1983) Principles of Fluorescence Spectroscopy. Plenum, New York
13. O'Connor DV, Phillips D (1984) Time-Correlated Single Photon Counting, Academic Press, London

14. Schulman SG (ed) (1985, 1988, 1993) Molecular Luminescence Spectroscopy. Methods and Applications, vols 1, 2 and 3. John Wiley, New York
15. Rendell D (1987) Fluorescence and Phosphorescence (Analytical Chemistry by Open Learning). John Wiley, Chichester
16. Krasovitskii BM, Bolotin BM (1988) Organic Luminescent Materials. VCH, Weinheim
17. Goldberg MC (ed) (1989) Luminescence Applications in Biological, Chemical, Environmental and Hydrological Sciences. ACS Symposioum Ser. Vol. 383, Am. Chem. Soc. Washington, DC
18. Guilbault GG (ed) (1990) Practical Fluorescence. Marcel Dekker, New York.
19. Dewey TG (ed) (1991) Biophysical and Biochemical Aspects of Fluorescence Spectroscopy. Plenum, New York
20. Baeyens WRG, De Keukelaire D, Korkidis D (eds) (1991) Luminescence Techniques in Chemical and Biochemical Analysis. Marcel Dekker, New York
21. Hemmilä, IA (1991) Applications of Fluorescence in Immunoassays. John Wiley, New York.
22. Lakowicz JR (ed) (1991 and 1992) Topics in Fluorescence Spectroscopy, vol 1 (Techniques), vol 2 (Principles), vol 3 (Biochemical Applications). Plenum Press, New York
23. Czarnik AW (ed) (1992) Fluorescent Chemosensors for Ion and Molecule Recognition. ACS Symp. Series, Vol 538, Amer. Chem. Soc., Washington DC
24. Wolfbeis OS (ed) (1993) Fluorescence Spectroscopy. New Methods and Applications. Springer, Berlin Heidelberg
25. Slavik J (1993) Fluorescent Probes in Cellular and Molecular Biology. CRC Press, Boca Raton

Contents

Part 1
Methods and Trends in Fluorescence Spectroscopy

1 Fluorescence Lifetime Imaging and Spectroscopy in Random Media   3
T. L. Troy, E. M. Sevick-Muraca

2 Single-Molecule Detection in Biology with Multiplex Dyes
and Pulsed Semiconductor Lasers 39
M. Sauer, J. Wolfrum

3 Time-Resolved Fluorescence of Conjugated Polymers 59
H. Bässler, M. Hopmeier, R. F. Mahrt

4 Low-Temperature Photophysics of Permethylated n-Heptasilane:
The Borderline Between Excitation Localization and Delocalization
in a Conjugated Chain   79
M. K. Raymond, T. F. Magnera, I. Zharov, R. West, B. Dreczewski,
A. J. Nozik, J. R. Sprague, R. J. Ellingson, J. Michl

5 Characterization of Membrane Mimetic Systems
with Fluorescence Correlation Spectroscopy 101
M. Hink, A. J. W. G. Visser

6 Excited State Probing of Supramolecular Systems
on a Submicron Scale 119
P. Vanoppen, J. Hofkens, L. Latterini, K. Jeuris, H. Faes,
F. C. De Schryver, J. Kerimo, P. F. Barbara, A. E. Rowan, R. J. M. Nolte

7 Three-Photon Excitation of Fluorescence 137
J. R. Lakowicz, I. Gryczynski

Part 2
Analytical Fluorescence Probes, and Environmental Research

8 Fluorescence Properties of Crown-Containing Molecules 161
M. V. Alfimov, S. P. Gromov

9 Recent Developments in Luminescent PET
(Photoinduced Electron Transfer) Sensors and Switches 179
A. Prasanna de Silva, A. J. M. Huxley

10 Photostability of Fluorescent Dyes for Single-Molecule Spectroscopy: Mechanisms and Experimental Methods for Estimating Photobleaching in Aqueous Solution ... 193
C. Eggeling, J. Widengren, R. Rigler, C.A.M. Seidel

11 Analysis of Chemical Dynamics and Technical Combustion by Time-Resolved Laser-Induced Fluorescence ... 241
H.-R. Volpp, C. Schulz, J. Wolfrum

12 Fluorescence Techniques for Probing Molecular Interactions in Imprinted Polymers ... 277
O.S. Wolfbeis, E. Terpetschnig, S. Piletsky, E. Pringsheim

Part 3
Fluorescence Probes in Polymers

13 Advanced Light Emitting Dyes: Monomers, Oligomers, and Polymers   299
Y. Geerts, K. Müllen

14 Fluorescence Probes in Polymers and Liquid Crystals: Complex Macromolecular Chain Dynamics (Proposal from the Far East)   325
H. Ushiki

15 Fluorescence Method for Monitoring Gelation and Gel Swelling in Real Time ... 371
Ö. Pekcan, Y. Yılmaz

16 Photophysical Studies Provide Thermodynamic Insights into Block-Copolymer Micelle Formation in a Selective Solvent ... 389
C.W. Frank, D.Y. Ylitalo

Part 4
Applications of Fluorescence Spectroscopy in Biology

17 Fluorescence Microscopy and the Reactions of Single Molecules ... 417
G. Pilarczyk, S. Monajembashi, C. Hoyer, V. Uhl, K.O. Greulich

18 Solvent Relaxation in Biomembranes ... 439
M. Hof

19 Luminescent Lanthanide Chelates for Improved Resonance Energy Transfer and Application to Biology ... 457
P.R. Selvin

Part 5
Fluorescence Techniques in Medicine – a Challenge for the Future

20 Fluorescent Lifetime Imaging Microscopy 491
   B. Herman, X. F. Wang, P. Wodnicki, A. Periasamy, N. Mahajan,
   G. Berry, G. Gordon

21 Injection Based Heterogeneous Fluorescence Immunoassays 509
   J. N. Miller, M. Evans, M. T. French, D. A. Palmer

22 Microfluorometry of Cellular and Subcellular Processing in CNS Cells 521
   W. Müller, S. Schuchmann, V. A. Egorov, T. Gloveli, K. Bittner

23 Fluorescence Diagnosis in the Border Zone of Liver Tumors 537
   J. Beuthan, O. Minet

Subject Index 553

Contributors

Alfimov, M. V.
Photochemistry Center of Russian Academy of Sciences, Novatorov St. 7a,
117421 Moscow, Russia, e-mail: alf@mx.icp.rssi.ru

Bässler, H.
Institut für Physikalische Chemie, Kernchemie und Makromolekulare Chemie,
Fachbereich Chemie, Universität Marburg, Hans-Meerwein-Straße,
D-35032 Marburg, Germany, E-Mail:
akbaess@psl40s.phys-chemie.uni-marburg.de

Barbara, Paul F.
Department of Chemistry, University of Minnesota, Mineapolis,
MN USA 55455, U.S.A.

Berry, G.
Department of Cell Biology & Anatomy, CB # 7090,
University of North Carolina, Chapel Hill, NC 27599, U.S.A.,
E-mail: bhgf@med.unc.edu

Beuthan, J.
Freie Universität Berlin, Institut für Medizinische Physik und Laser-Medizin,
Krahmerstr. 6–10, D-12207 Berlin, Germany

Bittner, Katrin
Institut für Physiologie der Charite, Abt. Neurophysiologie,
AG Molekulare Zellphysiologie, Tucholskystr. 2, D-10117 Berlin, Germany

Dreczewski, Boguslaw
Department of Chemistry, University of Wisconsin, Madison, WI 53706, U.S.A.

Eggeling, C.
Max-Planck-Institut für Biophysikalische Chemie, Am Faßberg 11,
D-37077 Göttingen, Germany

Egorov, Alexei V.
Institut für Physiologie der Charite, Abt. Neurophysiologie,
AG Molekulare Zellphysiologie, Tucholskystr. 2, D-10117 Berlin, Germany

Ellingson, Randy J.
National Renewable Energy Laboratory, Golden, CO 80401, U.S.A.

Evans, Mark
Department of Chemistry, Loughborough University, Loughborough,
Leicestershire LE 11 3 TU, UK

Faes, Herman
Department of Chemistry, Katholieke Universiteit Leuven,
Celestijnenlaan 200 F, 3001 Heverlee-Leuven, Belgium

Frank, Curtis W.
Department of Chemical Engineering, Stanford University, Stanford,
CA 94305-5025, U.S.A

French, Martin T.
Kalibrant Ltd., Department of Chemistry, Loughborough University,
Loughborough, Leicestershire LE 11 3 TU, UK

Geerts, Yves
Université Libre de Bruxelles, Laboratoire de Chimie Macromoléculaire,
CP 206/1, Boulevard du Triomphe, B-1050 Bruxelles, Belgium,
E-mail: ygeerts@ulb.ac.be

Gloveli, Tengis
Institut für Physiologie der Charite, Abt. Neurophysiologie,
AG Molekulare Zellphysiologie, Tucholskystr. 2, D-10117 Berlin, Germany

Gordon, G.
Department of Cell Biology & Anatomy, CB #7090,
University of North Carolina, Chapel Hill, NC 27599, U.S.A.,
E-mail: bhgf@med.unc.edu

Greulich, K. O.
Inst. Mol. Biotech, Postfach 100813, D-07708 Jena, Germany,
E-mail: kog@imb-jena.de

Gromov, S. P.
Photochemistry Center of Russian Academy of Sciences, Novatorov St. 7 a,
117421 Moscow, Russia

Gryczynski, Ignacy
Center for Fluorescence Spectroscopy, Department of Biochemistry
and Molecular Biology, University of Maryland School of Medicine,
725 West Lombard Street, Baltimore, MD 21201, U.S.A

Herman, B.
Department of Cell Biology & Anatomy, CB #7090,
University of North Carolina, Chapel Hill, NC 27599, U.S.A.,
E-mail: bhgf@med.unc.edu

Hink, Mark
MicroSpectroscopy Centre, Department of Biomolecular Sciences,
Laboratory of Biochemistry, Wageningen Agricultural University,
Dreijenlaan 3, 6703 HA Wageningen, The Netherland

Hof, Martin
J. Heyrovský Institute of Physical Chemistry, Academy of Sciences
of the Czech Republic, Dolejskova 3, CZ-18223 Prague 8, Czech Republic
(extra address: Martin Hof, Institute for Physical Chemistry,
University of Würzburg, Am Hubland, D-97074 Würzburg, Germany)

Hofkens, Johan
Department of Chemistry, Katholieke Universiteit Leuven,
Celestijnenlaan 200F, 3001 Heverlee-Leuven, Belgium

Hopmeier, M.
Fachbereich Physik, Universität Marburg, Renthof 5–7,
D-35032 Marburg, Germany

Hoyer, C.
Inst. Mol. Biotech, Postfach 100813, D-07708 Jena, Germany,
E-mail: kog@imb-jena.de

Huxley, Allen J.M.
School of Chemistry, Queen's University Belfast BT9 5AG, Northern Ireland

Jeuris, Karin
Department of Chemistry, Katholieke Universiteit Leuven,
Celestijnenlaan 200F, 3001 Heverlee-Leuven, Belgium

Kerimo, Josef
Department of Chemistry, University of Minnesota, Mineapolis,
MN USA 55455, U.S.A.

Lakowicz, Joseph R.
Center for Fluorescence Spectroscopy, Department of Biochemistry
and Molecular Biology, University of Maryland School of Medicine,
725 West Lombard Street, Baltimore, MD 21201, U.S.A

Latterini, Loredana
Department of Chemistry, Katholieke Universiteit Leuven,
Celestijnenlaan 200 F, 3001 Heverlee-Leuven, Belgium

Magnera, Thomas F.
Department of Chemistry and Biochemistry, University of Colorado, Boulder,
CO 80309-0215, U.S.A.

Mahajan, N.
Department of Cell Biology & Anatomy, CB #7090,
University of North Carolina, Chapel Hill, NC 27599, U.S.A.,
E-mail: bhgf@med.unc.edu

Michl, Josef D
Department of Chemistry and Biochemistry, University of Colorado, Boulder,
CO 80309–0215, U.S.A.

Minet, O.
Freie Universität Berlin, Institut für Medizinische Physik und Laser Medizin,
Krahmerstr. 6–10, D-12207 Berlin, Germany

Miller, James N.
Department of Chemistry, Loughborough University, Loughborough,
Leicestershire LE11 3TU, UK

Monajembashi, S.
Inst. Mol. Biotech, Postfach 100813, D-07708 Jena, Germany,
E-mail: kog@imb-jena.de

Mahrt, R.F.
Institut für Physikalische Chemie, Kernchemie und Makromolekulare Chemie,
Fachbereich Chemie, Universität Marburg, Hans-Meerwein-Straße,
D-35032 Marburg, Germany

Klaus Müllen
Max-Planck-Institut für Polymerforschung, Ackermannweg 10,
55128 Mainz, Germany

Müller, Wolfgang
Institut für Physiologie der Charite, Abt. Neurophysiologie,
AG Molekulare Zellphysiologie, Tucholskystr. 2, D-10117 Berlin, Germany

Nolte, Roeland J.M.
Department of Organic Chemistry, NSR Center, University of Nijmegen,
Toernooiveld, 6525 ED Nijmegen, The Netherlands

Nozik, Arthur J.
National Renewable Energy Laboratory, Golden, CO 80401, U.S.A.

Palmer, Derek A.
Kalibrant Ltd., Department of Chemistry, Loughborough University,
Loughborough, Leicestershire LE11 3TU, UK

Pekcan, Ö.
Department of Physics, Istanbul Technical University, Maslak,
80626, Istanbul, Turkey

Periasamy, A.
Department of Biology, P 229 Gilmer Hall, University of Virginia,
Charlottesville, VA. 22903, U.S.A

Pilarczyk, G.
Inst. Mol. Biotech, Postfach 100813, D-07708 Jena, Germany,
E-mail: kog@imb-jena.de

Piletsky, Sergey
University of Regensburg, Institute of Analytical Chemistry,
Chemo- and Biosensors, 93040 Regensburg, Germany

Pringsheim, Erika
University of Regensburg, Institute of Analytical Chemistry,
Chemo- and Biosensors, 93040 Regensburg, Germany

Raymond, Mary Katherine
Department of Chemistry and Biochemistry, University of Colorado,
Boulder, CO 80309-0215, U.S.A.

Rigler, R.
Karolinska Institut, Department of Medical Biophysics,
S-10401 Stockholm, Sweden

Rowan, Alan E.
Department of Organic Chemistry, NSR Center, University of Nijmegen,
Toernooiveld, 6525 ED Nijmegen, The Netherlands

Sauer, M.
Physikalisch-Chemisches Institut, Universität Heidelberg,
Im Neuenheimer Feld 253, D-69120 Heidelberg, Germany

de Schryver, Frans C.
Department of Chemistry, Katholieke Universiteit Leuven, Celestijnenlaan
200 F, 3001 Heverlee-Leuven, Belgium

Schuchmann, Sebastian
Institut für Physiologie der Charite, Abt. Neurophysiologie,
AG Molekulare Zellphysiologie, Tucholskystr. 2, D-10117 Berlin, Germany

Schulz, C.
Physikalisch-Chemisches Institut, Universität Heidelberg,
Im Neuenheimer Feld 253, D-69120 Heidelberg, Germany

Seidel, C.A.M.
Max-Planck-Institut für Biophysikalische Chemie, Am Faßberg 11,
D-37077 Göttingen, Germany, E-mail: cseidel@gwdg.de

Selvin, Paul R.
Loomis Laboratory of Physics, 1110 W. Green St., University of Illinois,
Urbana, IL 61801, U.S.A., E-mail: selvin@uiuc.edu

Sevick-Muraca, Eva M.
The Photon Migration Laboratory, School of Chemical Engineering,
Purdue University, West Lafayette, IN 47907-1283, U.S.A.

de Silva, A. Prasanna
School of Chemistry, Queen's University, Belfast BT9 5AG, Northern Ireland

Sprague, Julian R.
National Renewable Energy Laboratory, Golden, CO 80401, U.S.A.

Terpetschnig, Ewald
University of Regensburg, Institute of Analytical Chemistry,
Chemo- and Biosensors, 93040 Regensburg, Germany

Troy, Tamara L.
The Photon Migration Laboratory, School of Chemical Engineering,
Purdue University, West Lafayette, IN 47907-1283, U.S.A.

Uhl, V.
Inst. Mol. Biotech, Postfach 100813, D-07708 Jena, Germany,
E-mail: kog@imb-jena.de

Ushiki, Hideharu
Laboratory of Transport and Transformation in Bio-Systems,
Department of Bio-Mechanics and Intelligent Systems (BMIS),
Graduate School of Bio-Applications and System Engineering (BASE),
Tokyo University of Agriculture and Technology, 3-5-8, Saiwai-cho, Fuchu-shi,
Tokyo 183, Japan

Vanoppen, Peter
Department of Chemistry, Katholieke Universiteit Leuven, Celestijnenlaan 200 F,
3001 Heverlee-Leuven, Belgium

Visser, Antonie J. W. G.
MicroSpectroscopy Centre, Department of Biomolecular Sciences,
Laboratory of Biochemistry, Wageningen Agricultural University, Dreijenlaan 3,
6703 HA Wageningen, The Netherlands, E-mail: Ton.Visser@laser.bc.wau.nl

Volpp, H.-R.
Physikalisch-Chemisches Institut, Universität Heidelberg,
Im Neuenheimer Feld 253, D-69120 Heidelberg, Germany

Wang, X. F.
Zevatech, Inc. 507 Airport Blvd., Morrisville, NC 27560, U.S.A.

West, Robert
Department of Chemistry, University of Wisconsin, Madison, WI 53706, U.S.A.

Widengren, J.
Karolinska Institut, Department of Medical Biophysics,
S-10401 Stockholm, Sweden

Wodnicki, P.
Zevatech, Inc. 507 Airport Blvd., Morrisville, NC 27560, U.S.A.

Wolfbeis, Otto S.
University of Regensburg, Institute of Analytical Chemistry,
Chemo- and Biosensors, 93040 Regensburg, Germany

Wolfrum, J.
Physikalisch-Chemisches Institut, Universität Heidelberg,
Im Neuenheimer Feld 253, D-69120 Heidelberg, Germany

Ylitalo, David Y.
Department of Chemical Engineering, Stanford University, Stanford,
CA 94305–5025, U.S.A

Yılmaz, Y.
Department of Physics, Istanbul Technical University, Maslak,
80626, Istanbul, Turkey

Zharov, Ilya
Department of Chemistry and Biochemistry, University of Colorado,
Boulder, CO 80309–0215, U.S.A.

**Part 1
Methods and Trends in Fluorescence**

CHAPTER 1

Fluorescence Lifetime Imaging and Spectroscopy in Random Media

T. L. Troy, E. M. Sevick-Muraca

Abstract. With the development of near-infrared (NIR) laser diodes, the synthesis of fluorescent dyes with excitation and emission spectra in the NIR wavelength regime has accelerated in the past decade for microscopy applications. Owing to the window of low absorbance in tissues in this wavelength regime, an opportunity also exists for the deployment of fluorescent dyes as in vivo diagnostic agents. Figure 1.1 illustrates the typical absorbance spectra of tissues showing that at wavelengths less than 650 nm, hemoglobin absorbance provides the predominant attenuation of light in tissues, and above 900 nm water absorbance provides the predominant attenuation of light in tissues. In the "therapeutic window" of 650–900 nm, a window of low absorption exists in which light will be preferentially scattered over being absorbed. Consequently, it is possible to transmit multiply scattered NIR light across several centimeters of tissue. In addition, it is possible to excite NIR fluorophores deep within tissues and to collect the fluorescence re-emitted from the air-tissue interface. Since fluorescence provides a sensitive means for assessing local biochemistry via changes in quantum efficiency and lifetime, the ability to diagnose tissues based on spectroscopic evaluation of lifetime and quantum efficiency of exogenous diagnostic agents is possible. Agents whose emission characteristics vary with tissue pH [1, 2] and pO_2 [3] have been employed to detect diseased tissues by the nature of differing fluorescent properties as well as to provide diagnostic information regarding the diseased tissue volume. While typical contrast agents employed for the detection of diseased tissues depend on and are limited by the poor preferential uptake, the alteration of fluorescent properties provides a unique mechanism for inducing additional contrast [4]. For example, using time-gating to collect the long-lived fluorescence from hematoporphyrin derivative (HpD), Cubbeddu and co-workers [5, 6] were able to differentiate normal tissues from intradermally or subcutaneously implanted murine tumors in mice. More recently, it has been reported that the fluorescent decay of HpD is appreciably slower in experimental mice tumors than in their surrounding normal healthy tissues. Consequently, the use of a fluorescent dye may provide contrast owing to changes in fluorescent properties within tissue compartments.

The difficulty, however, lies in understanding the use of multiply scattered excitation light to excite a fluorophore in the tissue, and secondly, to extract information of lifetime and quantum efficiency from the multiply scattered fluorescence detected at the air-tissue interface. If the optical properties of the tissue are spatially uniform and the fluorescent dye has constant fluorescent properties, then the problem is one of *fluorescent lifetime spectroscopy* in tissues. However, if the detection of diseased tissues is to be tackled, the problem becomes one of *fluorescent lifetime spectroscopic imaging*, since optical properties and fluorescent properties of diagnostic agents may vary with spatial location.

In this chapter, frequency domain photon migration fluorescence imaging is described as a method for generating an optical map or image of fluorescent lifetime and quantum efficiency from exterior measurements of modulation phase and modulation amplitude on tissues or highly scattering media. In Sect. 1, the theory behind the propagation of excitation light and generation of fluorescent light within scattering media such as tissues is presented. Section 2 describes experimental measurements which show that the delay in phase and decrease in amplitude of fluorescence measured in simulated tissue phantoms varies as a function of dye lifetime. Section 3 describes the general theory behind the derivation of an optical property map from measurements of modulation phase and amplitude of fluorescent light detected at the tissue surface, and Sect. 4 presents actual images generated from simulated measurements of modulation phase and amplitude. Finally, the prognosis for moving these theoretical and computational studies into an experimental demonstration is commented upon.

1
Principles of Photon Migration

1.1
Time and Frequency Domain Photon Migration

Photon migration techniques consist of measuring the time-dependent light propagation of multiply scattered light as it is transmitted through tissue. Two techniques are used to monitor the time dependence of photon migration in a scattering medium: (i) time domain, and (ii) frequency domain techniques. In the time domain technique, a picosecond impulse of light is launched into a scattering medium and the intensity of the detected light is recorded as a function of picosecond to nanosecond time. As time progresses, fewer photons are detected with longer "times-of-flight." Figure 1.2 shows the broadened pulse that is re-emitted and measured representing the distribution of photon "times-of-flight" traveled by the detected photons. In the frequency domain (Fig. 1.3), incident light whose intensity is sinusoidally modulated is continuously launched into a scattering sample. As the "photon density wave" of light propagates, it experiences a phase lag and an amplitude reduction relative to the incident light. The phase-shift and amplitude modulation are related to the optical properties of the medium. Both the modulation phase and amplitude are measured as a function of multiple frequencies. The modulation ratio is defined as the modulation amplitude divided by the average intensity.

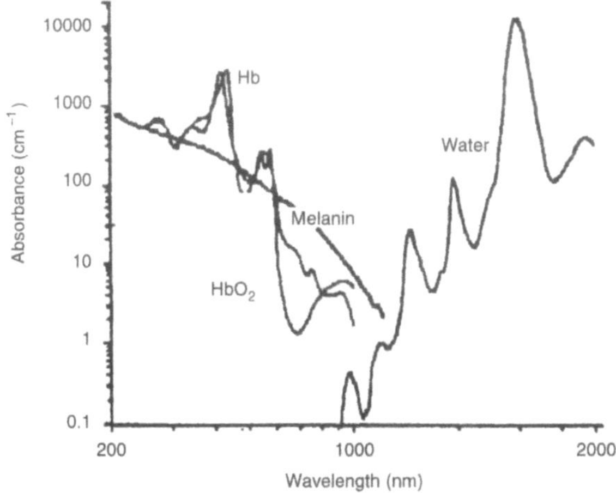

Fig. 1.1. Dependence of absorption on wavelength illustrating the "therapeutic window" in which the propagation of light through tissue is high. Reproduced from [7]

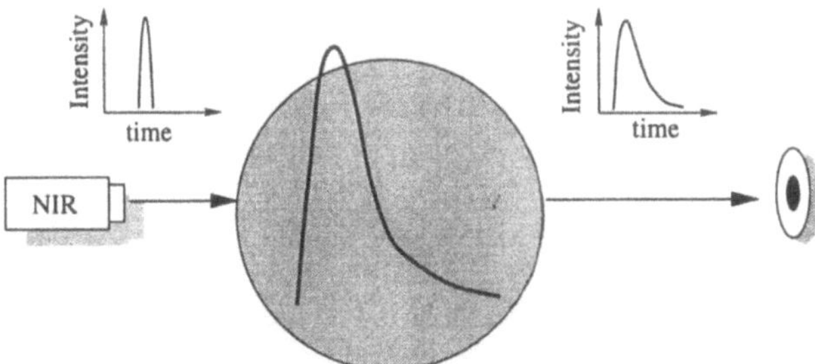

Fig. 1.2. In the time domain, an impulse of light is launched into a scattering medium. The detected pulse is broadened due to the increase in photon "times-of-flight"

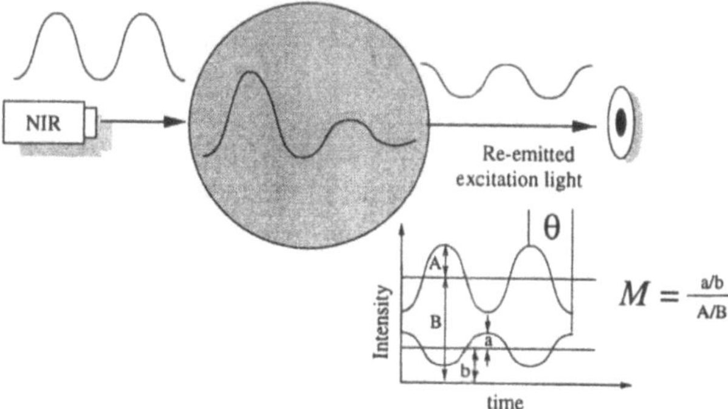

Fig. 1.3. In the frequency domain, a sinusoidally modulated source is launched into a scattering medium. The detected signal's phase is shifted and amplitude is decreased relative to the incident light

The intensity of the re-emitted light as a function of time (time domain), or the phase and amplitude of the modulated, re-emitted light as a function of the modulation frequency (frequency domain) contains identical information regarding the transport of light through the scattering medium. Time and frequency domain measurements are mathematically related to each other by the Fourier transform.

Photon migration imaging consists of (i) the forward problem of measuring the time-dependent characteristics of light propagation and (ii) the inverse problem of reconstructing an image of normal and diseased tissues based on their contrasting optical properties. In the following, the properties which impact the propagation of excitation light are reviewed.

1.2
Transport Property: Absorption Coefficient

The absorption coefficient, μ_a, is the inverse of the mean distance a photon will travel before being absorbed. In tissue, oxy- and deoxy-hemoglobin are the predominant endogenous absorbers in the therapeutic window (Fig. 1.1). In this wavelength range, most tissues have an absorption coefficient in the range of 10^{-2} to 10^{-1} cm^{-1} [8].

In the absence of scattering, the absorption coefficient can be found from the attenuation of light directly using the Beer-Lambert law:

$$\mu_a = -\frac{1}{L} \ln \frac{I}{I_0} = \varepsilon_\lambda [C] \tag{1.1}$$

where μ_a is the absorption coefficient of the medium (cm^{-1}); L is the optical path length light travels (cm); I is the intensity of the detected light [number of photons/(cm$^2 \cdot$ s)]; I_0 is the intensity of the incident light [number of photons/(cm$^2 \cdot$ s)]; ε_λ is the extinction coefficient at wavelength λ (cm^{-1} mM^{-1}); and [C] is the concentration of the light-absorbing species (mM). Note that the number of photons per second can be converted into units of power (Watts) by multiplying by the photon's energy.

Equation 1.1 describes an absorption coefficient that is inversely related to the optical path length, L. In the presence of a scattering medium such as tissue, there is no unique path length, but rather a distribution of path lengths (Fig. 1.4). Therefore, determination of the absorption coefficient in a scattering medium is not possible from the Beer-Lambert relationship.

1.3
Transport Property: Scattering Coefficient

The scattering coefficient, μ_s, describes the distance photons travel between scattering events. Scattering in tissue occurs due to the index of refraction mismatch between fluid and cellular organelles. Mitochondria are postulated to be the predominant scatterer in tissues [9, 10].

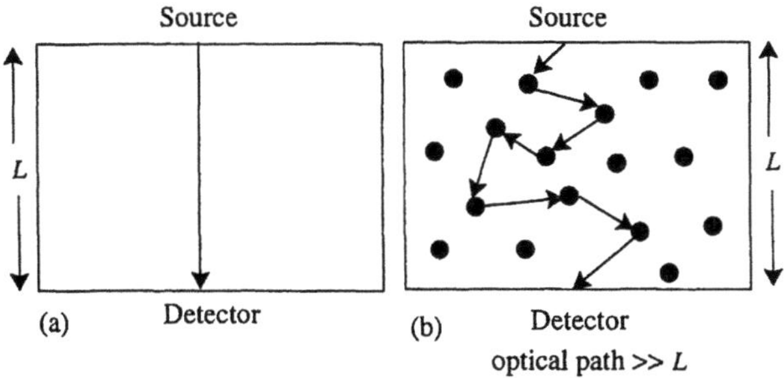

Fig. 1.4. Light propagation in **a** a non-scattering and **b** a scattering medium

When multiple scattering occurs, light no longer follows a straight path. Instead, photons travel a distribution of optical path lengths, rendering the Beer-Lambert law invalid. As shown in Fig. 1.4, the total optical path length, L, is now defined as the sum of all individual paths, ℓ_i's, between scatterers and the scattering coefficient is defined as the inverse of the mean scattering length ℓ^*:

$$\mu_s = \frac{1}{\ell^*} \tag{1.2}$$

Since "photon migration" describes a random walk of photons between scatterers, it is often more convenient to define the scattering parameter as the isotropic scattering coefficient $[\mu_s' = \mu_s(1-g)]$ where the anisotropy coefficient, g, is defined below. In the NIR wavelength range, the values of μ_s' for tissues typically range from 10 to 50 cm^{-1} [8].

1.4
Transport Property: Anisotropy Coefficient

The anisotropy coefficient, g, is defined as the mean cosine of the scattering angle at which a photon deflects from a scatterer. Flock et al. [11] have shown that the Henyey-Greenstein phase function describes the angular scatter of light in tissue. This phase function describes the relationship between g and the collimated transmittance, T_c, at a scattering angle θ where:

$$T_c(\theta) = \frac{(1-g^2)}{(1+g^2-2g\cos\theta)^{3/2}} \tag{1.3}$$

Collimated transmittance is a measure of the light intensity which is transmitted through the sample as a function of the scattering angle θ. Using a least-squares analysis of several measurements of T_c vs. θ, the anisotropy coefficient can be obtained from Eq. 1.3. The anisotropy coefficient varies between isotropic (g = 0) and forward (g = 1) scattering. In all tissues, light is known to be forward scattering and the anisotropy coefficient is reported to be around 0.9 [12]. The Henyey-Greenstein phase function is widely used in tissue optics studies not only because it provides a good representation of experimental tissue measurements but also because it is mathematically simple.

1.5
Governing Equations Describing the Transport of Light

The scattering, absorption, and anisotropy coefficients are parameters that govern the propagation of light in any medium. Analysis of light transport is analogous to that of the motion of neutrons in a reactor core where neutrons diffuse from regions of high to low density [13]. When photons rather than neutrons are employed, the one-speed radiative transport equation describes light propagation in a scattering medium. The radiative transport equation is written in terms of the integro-differential equation governing the local number of photons, $n(\vec{r}, E, \hat{\Omega}, t)$, at position $\vec{r}$ and time t, and traveling with energy E and angular direction $\hat{\Omega}$ [14]:

$$\frac{\partial n(\vec{r}, E, \hat{\Omega}, t)}{\partial t} = -c\hat{\Omega} \cdot \nabla n - c\mu_t n(\vec{r}, E, \hat{\Omega}, t) + s(\vec{r}, E, \hat{\Omega}, t) +$$

$$\int_{4\pi} d\hat{\Omega}' \int_0^\infty dE' c' \mu_s'(E' \to E, \hat{\Omega}' \to \hat{\Omega}) \, n(\vec{r}, E', \hat{\Omega}', t) \qquad (1.4)$$

The term on the left-hand side (LHS) represents the local accumulation of photons, while the first term on the right-hand side (RHS) represents the flux of photons into the system. The second term on the RHS represents the loss of photons traveling with energy E in direction $\hat{\Omega}$ due to absorption and scattering events. The total cross section for these extinction processes, μ_t, is given by the sum of the absorption and scattering cross sections, $\mu_t = \mu_a + \mu_s'$ (cm^{-1}). The third term on the RHS represents the source, and along with the second term accounts for the rate of addition of photons traveling in direction $\hat{\Omega}$ with energy E which have been scattered from direction $\hat{\Omega}'$ and energy E'. To account for scattering from photons traveling in all directions and energies, that last term on the RHS is integrated over all solid angles, $\int_{4\pi} = \int_0^{2\pi} \int_0^{\pi} \sin\theta \, d\theta \, d\phi$, and energies. The symbol c represents the speed of photons (cm/s).

1.5.1
One-Speed Approximation to the Transport Equation

When all the neutrons or photons travel at the same speed, the neutron transport equation can be reduced to the one-speed approximation by eliminating the dependence on energy, E. By writing the local concentration of photons in terms of an angular flux, φ [number of photons/($cm^2 \cdot$ steradians $\cdot$ s)], where $\varphi(\vec{r}, \hat{\Omega}, t) = cn(\vec{r}, \hat{\Omega}, t)$, the one-speed equation is given in Eq. 1.5 [13]:

$$\frac{1}{c}\frac{\partial \varphi}{\partial t} = -\hat{\Omega} \cdot \nabla\varphi - \mu_t(\vec{r}) \, \varphi(\vec{r}, \hat{\Omega}, t) + s(\vec{r}, \hat{\Omega}, t) \qquad (1.5)$$

$$+ \int_{4\pi} d\hat{\Omega}' \mu_s'(\hat{\Omega}' \to \hat{\Omega}) \, \varphi(\vec{r}, \hat{\Omega}', t)$$

The term $\hat{\Omega} \cdot \nabla\varphi$ represents the negative rate of addition of photons traveling in an angular direction $\hat{\Omega}$. The term $\mu_t(\vec{r})\,\varphi(\vec{r}, \Omega, t)$ accounts for the loss of photons traveling in direction $\hat{\Omega}$ due to scatter and absorption. The integral term represents photons scattering from direction $\hat{\Omega}'$ into the direction $\hat{\Omega}$. The term $s(\vec{r}, \hat{\Omega}, t)$ represents a source of photons [number of photons/(cm^3 · steradians · s)] traveling in a direction $\hat{\Omega}$ at a position $\vec{r}$ and time t [13].

1.5.2
Diffusion Equation Describing Light Propagation in Tissues

Following the derivation of Duderstadt and Hamilton [13], the one-speed equation can be reduced to the diffusion equation describing light propagation in tissues:

$$\frac{1}{c}\frac{\partial \Phi}{\partial t} = \nabla \cdot D(\vec{r})\,\nabla\Phi - \mu_a(\vec{r})\,\Phi(\vec{r}, t) + S(\vec{r}, t) \tag{1.6}$$

Equation 1.6 represents the time domain diffusion equation which can analogously be written in the frequency domain [14, 15] as:

$$\frac{i\omega}{c}\,\Phi^{AC}(\vec{r}, \omega) - \vec{\nabla}\cdot D\vec{\nabla}\,\Phi^{AC}(\vec{r}, \omega) + \mu_a \Phi^{AC} = M\delta(\vec{r}_s - \vec{r}) \tag{1.7}$$

where Φ^{AC} is the AC component of the fluence at position $\vec{r}$ and frequency ω; M is the modulation of the point source; and $\vec{r}_s$ denotes the source position.

1.5.3
Source Location and Boundary Conditions

A collimated light source incident on a scattering medium is slowly transformed into a diffuse source as it encounters scatterers. Patterson et al. [16] have analytically shown that in order to accurately represent an incident laser light source at the surface, the simulated source location has to be set to:

$$z_0 = \frac{1}{\mu_s'} \tag{1.8}$$

Therefore, in order to accurately simulate a laser source located on the surface, the placement of the simulated source needs to be one isotropic scattering length inside the physical boundary. A sinusoidally modulated light source can be represented by a complex number [17]:

$$S = S_0[\cos(\theta) + i\sin(\theta)] \tag{1.9}$$

where θ and S_0 are the phase and the amplitude of the source, respectively.

The three most common boundary conditions used to solve the diffusion equation at the air-scatterer interface of a random medium are the (i) zero fluence boundary condition, (ii) the partial current boundary condition, and

(iii) the extrapolated boundary condition. The zero fluence boundary condition defines $\Phi = 0$ on and outside the boundary. Although the zero fluence condition is physically inaccurate, it is mathematically simple and gives solutions to the diffusion equation which agree well with experimental data for biological systems [8].

1.6
Non-Radiative Mechanisms for Exogenous Contrast

As discussed above, the diffusion equation (Eqs. 1.6 and 1.7) describes the propagation of light in a highly scattering medium. This equation depends on three parameters: (i) the speed of light c, (ii) the isotropic scattering coefficient, μ_s', and (iii) the absorption coefficient μ_a. Since all these parameters govern the migration of photons in tissue, an exogenous compound which alters the characteristics of any of these parameters can be used to induce contrast for imaging and provide information for diagnostic information.

1.6.1
Contrast Due to Absorption

The propagation of a photon density wave through a medium containing a region of light absorbing dye is different from the propagation of a wave in a medium with uniform optical properties as illustrated in Fig. 1.5. The "homogeneous" propagation is similar to a wave formed by a pebble being dropped in a shallow pond. The propagating wave dampens as it travels across the surface but the wave maintains coherence. When the wave encounters an optical heterogeneity (or diseased tissue with different optical properties), it will reflect, diffract, and be absorbed [18]. The "heterogeneous" situation is similar to a pebble being dropped in the shallow pond with a solid obstruction in the wave's path. The solid obstruction will absorb and partially reflect a re-scattered wave. At any position, the diffusing wave will be the sum of the incident wave and the scattered wave.

Since these differences in light propagation enable contrast, photon migration measurements are conducted relative to an absence (homogeneous) condition. The phase contrast, $\Delta\theta$, and modulation contrast, ΔM, are defined as:

$$\Delta\theta = \theta_{\text{presence}} - \theta_{\text{absence}} \tag{1.10}$$

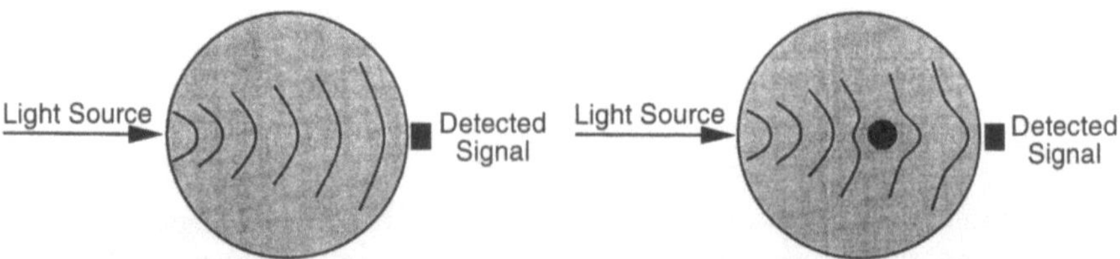

Fig. 1.5. Schematic describing the progagation of a photon density wave across a medium in the absence and presence of a heterogeneity with optical contrast

$$\Delta M = \frac{M_{presence}}{M_{absence}} \tag{1.11}$$

where $\theta_{presence}$ and $M_{presence}$ are, respectively, the measured modulation phase and modulation ratio in the presence of a heterogeneity. Likewise, $\theta_{absence}$ and $M_{absence}$ are, respectively, the measured modulation phase and modulation ratio in the absence of a heterogeneity. If the optical contrast relative to an absence condition is significant enough to cause $\Delta\theta \neq 0$ and $\Delta M \neq 1.0$, then it may be possible to reconstruct an interior optical property map from exterior measurements of modulation phase and amplitude as described in Sect. 3.

Since phase and amplitude contrast depend on differences caused by an absorbing heterogeneity, the physics which change the detected wave need to be examined. Figure 1.6a describes frequency domain photon migration imaging

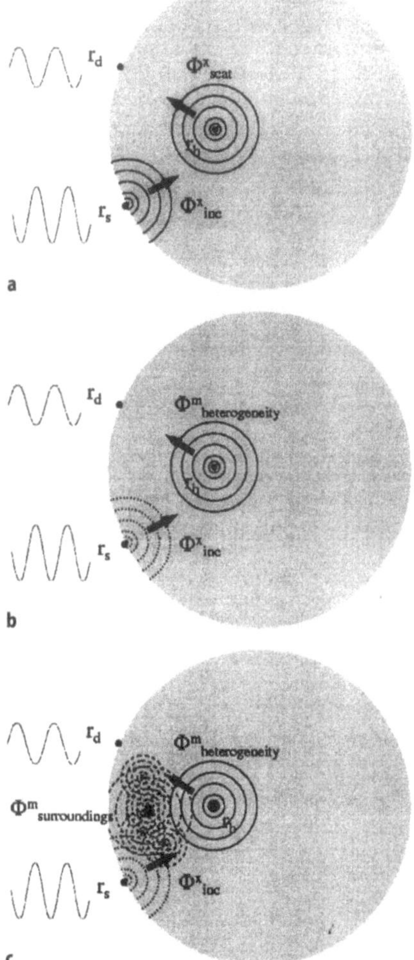

Fig. 1.6. Schematic illustrating frequency domain photon migration measurements of a sinusoidally modulated excitation light source at position $\vec{r}_s$ and the detected excitation and emission (fluorescent) signals at position $\vec{r}_d$ when contrast is due to **a** absorption, **b** perfect uptake of a fluorescent dye, and **c** imperfect uptake of fluorescent dye

which consists of launching a sinusoidally modulated light source at position $\vec{r}_s$ and measuring the detected signal at position $\vec{r}_d$. The propagation of the wave through the medium causes a decrease in modulation and an increase in phase at position $\vec{r}$ relative to the incident wave. The magnitude of a scattered wave depends on the heterogeneity's distance from the source, its dimensions, and the absorption difference between it and the surroundings. The magnitude of a scattered wave will be greatest closest to the source. The greatest magnitude possible will occur when there is no loss mechanism (no absorption) in the surroundings and the heterogeneity is a perfect absorber (a black body, or μ_a is infinity). Nonetheless, even for a perfect absorber, the incident propagated wave will constitute the predominant signal measured at the detector. Consequently, the scattered wave from an absorber will be small in comparison to the incident propagated wave and the contrast offered by absorption should be low. In other words, absorption contrast consists of the ability to detect a small scattered wave within a very large signal.

1.6.2
Contrast Due to Dye with Fluorescent Characteristics

Another means of inducing contrast is through fluorescence. Imagine an optical heterogeneity which contains a contrast agent that not only absorbs, but also fluoresces light (Fig. 1.6b). In essence, if an interference filter is placed at the detector so that only fluorescence light is seen, the heterogeneity would act like a beacon. Since there is no small signal to identify from a large background signal, the contrast should be significantly greater than that offered by absorption. Usually perfect uptake is not possible and one would imagine that while there is 10-fold or at most 20-fold more dye in a tissue volume of interest than in the surrounding tissues, the problem becomes one of picking the brightest beacon from a number of beacons uniformly distributed in the tissue (Fig. 1.6c). Under these conditions, the fluorescent properties (such as lifetime and quantum efficiency) of the compound may be used to impart additional contrast to that offered simply by a fluorophore concentration difference between the heterogeneity and its surroundings.

1.6.2.1
Physics of Photoluminence

Fluorescence is the emission of a photon which results when a molecule in an excited state relaxes to its original ground state (Fig. 1.7). The excited molecules can either relax non-radiatively to the ground state (without emitting fluorescence) or stay in the excited state for a period of time before returning to the ground state and emitting a fluorescent photon. Often when π-orbitals are close, an electron in the electronically activated level will experience a change in its spin state. Since relaxation to the ground state populated with same spin state is forbidden, the activated molecule will remain in the excited state for a longer period of time. In this case the emission is termed phosphorescence. Due to the loss of energy associated when the fluorescent or phosphorescent

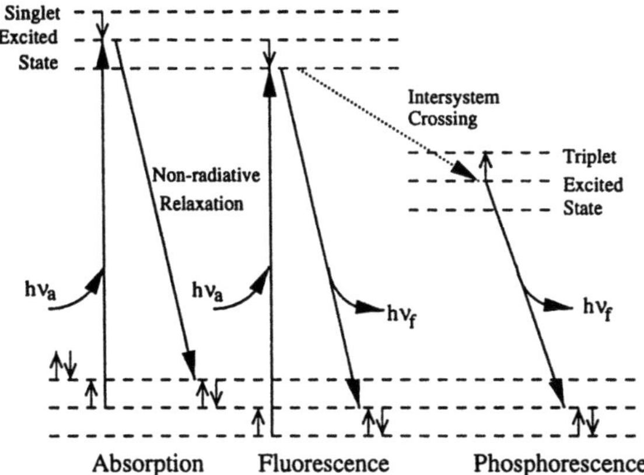

Fig. 1.7. Jablonski diagram illustrating the electronic transitions associated with absorption, fluorescence and phosphorescence

molecules return to the ground state, the emission light occurs at a longer wavelength than the excitation light [19].

The fluorescent properties of a compound are characterized by the quantum efficiency, ϕ, and the lifetime, τ, of the molecule. The quantum efficiency is the ratio of the number of fluorescent photons emitted to the number of excitation photons absorbed by the fluorophore. The lifetime of the molecule is the average time that the molecule stays in the activated state and is measured as the average time between absorption of an excitation photon and emission of a fluorescent photon. Both lifetime and quantum efficiency are sensitive to the local biochemical environment. For example, porphyrin fluorescence is quenched by oxygen enabling quantitative detection of oxygen from the shortening of its lifetime and reduction of its quantum efficiency [19].

1.6.2.2
Measurrement of Photolumination Properties in Dilute, Non-scattering Suspensions

The lifetime, τ, and quantum efficiency, ϕ, of a molecule can be determined using both a time and frequency domain analysis. In the time domain, a pulse of light excites a dilute non-scattering solution of fluorophores and the decay of the generated fluorescent intensity is recorded as a function of time after the initial impulse. The amount of fluorescent light, I_m, is defined by the absorption coefficient of the fluorophores, $\mu_{a_{xf}}$, the quantum efficiency, ϕ, and the lifetime, τ, of the molecule [19]:

$$I^m(t) = \frac{\phi\mu_{a_{xf}}}{\tau} e^{-1/\tau} = I_0 e^{-1/\tau} \tag{1.12}$$

where I_0 is the maximum fluorescent intensity. The lifetime of a fluorophore is equal to the time required for the intensity to decay to $1/e$ of its initial value.

Frequency domain measurements of lifetime, τ, and quantum efficiency, ϕ, are made by exciting a dilute, non-scattering solution of the fluorophore with a sinusoidally modulated source and measuring either the emitted modulation phase or modulation ratio of the fluorescent signal relative to the incident excitation light. The modulation phase of the fluorescent light is independent of the fluorophore concentration and the quantum efficiency:

$$\theta^m(\omega) = \tan^{-1}(\omega\tau) \tag{1.13}$$

where ω is the modulation frequency of the light source. The modulation ratio, however, is dependent on quantum efficiency, ϕ, and $\mu_{a_{xf}}$:

$$M^m(\omega) = \frac{\mu_{a_{xf}}}{\sqrt{1+(\omega\tau)^2}} \tag{1.14}$$

and can therefore be used to determine both lifetime, τ, and quantum efficiency, Φ, of a sample. If the modulation ratio is referenced to the modulation ratio obtained under continuous wave or constant illumination ($\omega = 0$), the resulting ratio is a function of modulation frequency and lifetime τ:

$$\frac{M^m(\omega)}{M^m(0)} = \frac{1}{\sqrt{1+(\omega\tau)^2}} \tag{1.15}$$

Equations 1.13–1.15 are obtained from the Fourier transform of Eq. 1.12.

The measurement of the emission properties of τ and Φ are determined in a dilute, non-scattering medium. Since the fluorescent lifetime, τ, is in the same order as photon "times-of-flight," the above equations are not valid in a scattering medium due to the additional time delay and amplitude decrease associated with photon migration. In order to accurately determine the emission properties in a scattering medium, photon migration times need to be deconvolved from time-dependent measurements.

As an alternative, long-lived (phosphorescent) probes have been suggested for photon migration measurements of the re-emission properties in a highly scattering medium since the phosphorescent lifetimes are longer than photon "times-of-flight". However, Hutchinson et al. [20, 21] found that the signal orginating from a medium containing phosphorescent agents is confined to surface or sub-surface regions. Since phosphorescent agents cannot be used to determine the lifetime deep within scattering media, fluorescent contrast agents are suggested.

1.6.2.3
Collisional Quenching Mechanisms

Collisional quenching is the mechanism in which the fluorescent intensity decreases due to collisional encounters between a fluorophore and a quencher. For this process to happen, the quencher must diffuse to the fluorophore while the fluorophore is in the excited state. Some examples of collisional quenchers are oxygen, chloride, chlorinated hydrocarbons, xenon, hydrogen peroxide, bromate, and iodide [19]. The lifetime of a fluorophore can be related to the concentration of a quencher using the Stern-Volmer equation [19]:

$$\frac{\tau_0}{\tau} = 1 + K_{SV}[Q]$$ (1.16)

where τ_0 is the lifetime of the probe in the absence of a quencher, τ is the lifetime of the probe in the presence of a quencher, K_{SV} is the Stern-Volmer constant and [Q] is the quencher concentration. As the concentration of the quencher increases, the excited state of the probe is quenched causing a reduction in the lifetime. Since the (bio)chemical environment (amount of quencher present) impacts the fluorescence lifetime, τ, fluorescent lifetime imaging may be used for metabolite sensing. It is important to note that Eq. 1.16 does not depend on the concentration of the fluorophore.

A class of lifetime sensitive fluorophores, called multiplex dyes [22] involve "tuning" cyanine dyes to exhibit different spectra lifetimes depending on local environments and binding characteristics. This ability to change the emission characteristics between normal and diseased tissue using dyes whose characteristics change by varying pH [1, 2] and O_2 [3] may not only enhance detection but may also contain diagnostic information regarding the disease.

1.6.3
Diffusion Equations Describing Excitation and Emission Photon Migration

To use fluorescent contrast agents in forward and inverse imaging problems, fluorescent generation and propagation must be accurately modeled. The light propagation due to fluorescence in a scattering medium can be described by the coupled diffusion equations for light propagation [20, 23, 24]:

$$-\nabla \cdot D_x(\vec{r})\,\nabla\Phi^x(\vec{r},\omega) + \left[\frac{i\omega}{c_x} + \mu_{a_{xi}}(\vec{r}) + \mu_{a_{xf}}(\vec{r})\right]\Phi^x(\vec{r},\omega) = S(\vec{r},\omega)$$ (1.17)

$$-\nabla \cdot D_m(\vec{r})\,\nabla\Phi^m(\vec{r},\omega) + \left[\frac{i\omega}{c_m} + \mu_{a_m}(\vec{r})\right]\Phi^m(\vec{r},\omega) = \phi\mu_{a_{xf}}\frac{1 - i\omega\tau}{1 + (\omega\tau)^2}\,\phi^x(\vec{r},\omega)$$ (1.18)

In the above equations, Φ^x and Φ^m are the AC components of fluence for excitation (superscript x) and emission (superscript m) light (photons/cm²); $\mu_{a_{xi}}$ is the absorption due to chromophores (cm⁻¹); $\mu_{a_{xf}}$ is the absorption due to fluorophores (cm⁻¹); μ_{a_m} represents the absorption of the emission light due to chromophores (cm⁻¹); μ_{s_x} and μ_{s_m} are the scattering coefficients of excitation and emission light (cm⁻¹), respectively; ϕ and τ are the quantum efficiency and lifetime (ns) of the fluorophore, respectively; and D_x and D_m are the optical diffusion coefficients for the excitation and emission light (cm) where

$$D_x = \frac{1}{3[\mu_{a_x} + (1 - g)\,\mu_{s_x}]}$$ (1.19)

and

$$D_m = \frac{1}{3\left[\mu_{a_m} + (1-g)\,\mu_{s_m}\right]} \tag{1.20}$$

Equations 1.17 and 1.18 are coupled by the excitation fluence, Φ^x, and the absorption due to fluorescence, $\mu_{a_{xf}}$. The solutions for these equations, Φ^x and Φ^m, are complex. It is important to note that fluorescence is assumed linear with the excitation fluence (e.g. photo-bleaching invalidates the model). The modulation phase, θ, and modulation amplitude, I_{AC}, of the excitation and emission light are represented by the real and imaginary components of the emission and excitation fluence:

$$\theta^{x,m} = \tan^{-1}\left[\frac{\text{Im}\,\Phi^{x,m}}{\text{Re}\,\Phi^{x,m}}\right] \tag{1.21}$$

$$I_{AC}^{x,m} = \sqrt{[\text{Re}\,\Phi^{x,m}]^2 + [\text{Im}\,\Phi^{x,m}]^2} \tag{1.22}$$

and the modulation ratio is obtained by dividing I_{AC} by the average intensity (or I_{DC}):

$$M^{x,m} = \frac{\sqrt{[\text{Re}\,\Phi^{x,m}]^2 + [\text{Im}\,\Phi^{x,m}]^2}}{I_{DC}} = \frac{I_{AC}}{I_{DC}}. \tag{1.23}$$

The mathematical framework for predicting modulation phase and modulation amplitude at emission wavelengths assumes an angle exponential decay kinetics, but can be easily modified for other kinetics.

1.6.4
Potential Fluorescent Contrast Agents for Photon Migration Imaging

An example of a class of potential contrast agents are porphyrin derivatives. Porphyrin molecules have characteristic ring structures derived from four pyrrole rings joined together by four methene bridges as shown in Fig. 1.8.

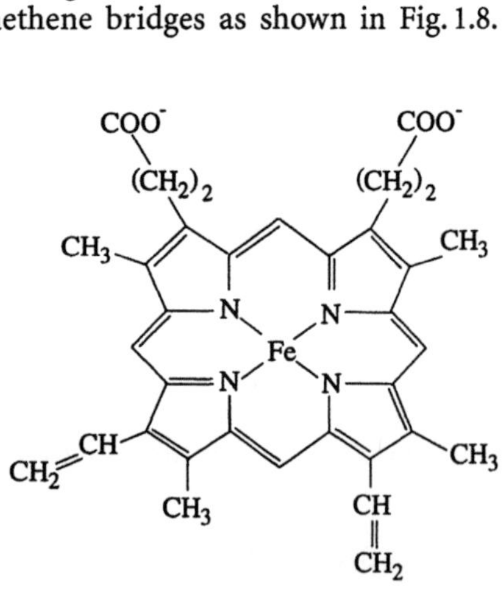

Fig. 1.8. Molecular structure of heme showing the characteristic heterocyclic rings that are common to porphyrin compounds

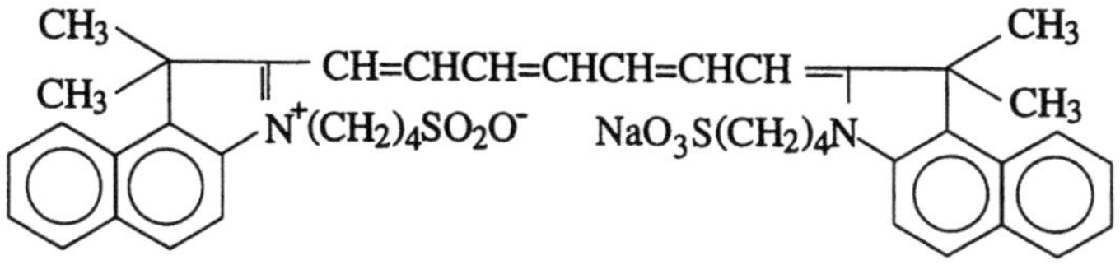

Fig. 1.9. Molecular structure of Indocyanine Green (ICG)

Porphyrin derivatives are typically employed in photodynamic therapy (PDT). PDT is a method in which a hematoporphyrin derivative drug is administered for photo-therapeutic destruction of cancer [25]. In this therapy, a photo-sensitizing drug, which has a high affinity for tumor tissue, is systematically administered to the patient. The tumor area is then irradiated with red light which photoactivates the dye to a triplet state to produce cytotoxic byproducts causing irreversible cellular damage. Many problems limit the efficacy of photo-dynamic therapy, such as damage to skin exposed to sunlight. Nonetheless, while PDT agents depend on triplet states for their therapeutic action, the singlet state may enable contrast for diagnostic imaging.

Indocyanine Green (ICG) (Fig. 1.9) is another optically active contrast agent that is approved by the Food and Drug Administration (FDA) for diagnostic studies. It is currently used to diagnose retinal and chorodial diseases by enhancing the images of the retinal vasculature of the eye [26]. In burn diagnostics, ICG is administered to the blood stream after which the burn area is illuminated to monitor the amount of blood flow to the area. Because blood flow is proportional to the depth of the burn, ICG is used to measure the severity of the wound [27]. ICG is also used to determine hepatic function by monitoring the amount of dye that clears the circulatory system of the liver as a function of time. Sevick et al. [4] measured the extinction coefficient, ε_λ, of ICG to be 1300 $(M\ cm)^{-1}$ at 780 nm and 22.000 $(M\ cm)^{-1}$ at 830 nm. They also report the quantum efficiency, ϕ, and lifetime, τ, of ICG to be 0.016 and 0.56 ns, respectively. Although their measurements were conducted with ICG diluted in water, the properties are assumed to be reflective of those in 0.5% Intralipid.

2
Contrast for Photon Migration Imaging: Fluorescence Lifetime

2.1
Single Pixel Measurements for Photon Migration Imaging

To experimentally assess the feasibility of external contrast agents, the impact of absorbing and fluorescing heterogeneities on light propagation in a tissue simulating phantom was examined. The influence of lifetime on photon migration measurements was also investigated. Single pixel measurements of modulation phase, θ, and modulation ratio, M, were measured as a function of modulation frequency, ω, in the presence and absence of: (i) light-absorbing, and (ii) fluorescing objects.

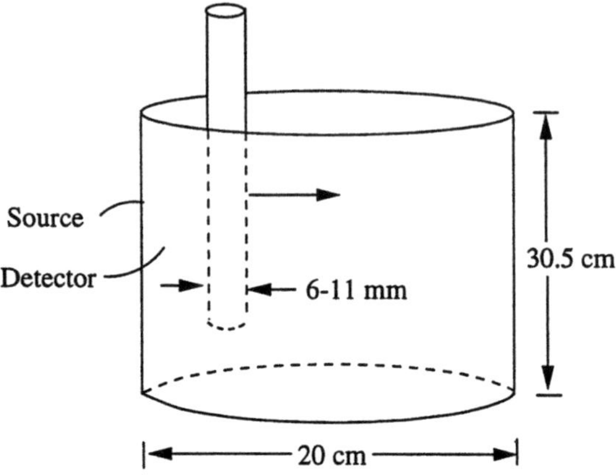

Fig. 2.1. Schematic of the cylindrical tank used in the single pixel measurements of modulation phase and modulation ratio for absorbing and fluorescing heterogeneities

2.2
Phantom Apparatus

The tissue phantom, illustrated in Fig. 2.1, consisted of a Plexiglas cylindrical tank, 20 cm in diameter and 30.5 cm in height, filled with 8 l of 0.5% by volume Intralipid solution (Kabi Pharmacia, Clayton, NC). This concentration of Intralipid mimics the scattering properties of tissue ($\mu_s' \approx 10\,\mathrm{cm}^{-1}$) [28]. Two small holes, an arc distance of 2.0 cm apart, were drilled half way up the side of the tank in order to couple the 1000 micron fiber optic (HCP-M1000T-08, Spectron Specialty Optics Co., Avon, CT) source and detector directly to the phantom. The fiber optics were placed approximately 1 mm inside the medium to avoid boundary effects caused by the wall of the tank. On the outside of the tank, silicon gel (Dow Corning Co., high vacuum grease, Midland, MI) was placed around the fibers to prevent leakage of Intralipid.

The perfect absorbing heterogeneity was a 6-mm diameter glass rod painted black. The fluorescent body consisted of a 9-mm inner diameter cylindrical glass container filled with micromolar solutions of either 3,3′-diethylthiatricarbocyanine iodide (DTTCI) (ACROS Organics, Fairlawn, NJ), a common laser dye, or ICG (Sigma Chemical Company, St. Louis, MO) diluted in 0.5% Intralipid solution. Dilution in Intralipid maintained a scattering coefficient inside the heterogeneity identical to the surrounding medium. Measurements of θ and M were conducted as the heterogeneity was moved various distances away from the wall midpoint between the source and detector. The object position was measured from the leading edge of the heterogeneity where a position of zero denotes contact between the heterogeneity and the wall of the tank.

2.3
Frequency Domain Instrumentation and Setup

The light propagation at both the excitation and emission wavelengths was measured using frequency domain measurements. The measurements were conducted at modulation frequencies between 80 and 240 MHz using a pulsed titanium/sapphire laser coupled to the detection system consisting of an ISS K2 phase fluorometer (ISS, Champaign, IL).

2.3.1
Heterodyne Detection Approach with Modulated Laser Source

To use simple electronics to detect the signal, a heterodyne technique was used. This detection method involves modulating the detectors at a frequency plus a small offset from the modulation frequency of the source. For example, the sinusoidal source is described by the expression:

$$A \cos(\omega t + \theta) \tag{2.1}$$

where A is the amplitude; ω is the modulation frequency; and θ is the phase of the source. The gain of the detector is modulated by $\omega + \Delta\omega$ so that the detector response is represented by:

$$\cos[(\omega + \Delta\omega)\,t] \tag{2.2}$$

where $\Delta\omega$ is the small offset frequency of the detector. The signal obtained by the detector is then:

$$A \cos(\omega t + \theta) \times \cos[(\omega + \Delta\omega)\,t] = \frac{A}{2} \cos(\Delta\omega t + \theta) +$$

$$+ \frac{A}{2} \cos[(2\omega + \Delta\omega)\,t + \theta] \tag{2.3}$$

A low pass filter is used to filter out frequencies above $\Delta\omega$, the signal received by the detector is:

$$\frac{A}{2} \cos(\Delta\omega t + \theta) \tag{2.4}$$

It is important to note that the phase and amplitude information of the high frequency signal is preserved.

2.3.2
Heterodyne Detection Approach with Pulsed Laser Source

When the light source is a mode locked pulsed laser, the signal can no longer be described by Eq. 2.1. Instead, the source is represented by:

$$\sum_{n=1}^{\infty} A_n \cos(n \omega t + \theta_n) \tag{2.5}$$

where the above equation is the frequency domain representation of a mode locked pulsed laser [29]. Again the detector is modulated at a frequency plus a small offset as described by Eq. 2.2, the signal obtained by the detector is:

$$\cos\left[(\omega + \Delta\omega)\,t\right] \times \left[A_1 \cos(\omega t + \theta_1) + A_2 \cos(2\omega t + \theta_2) + \ldots\right] =$$

$$+ \frac{A_1}{2}\left\{\cos(\Delta\omega t + \theta_1) + \cos\left[(2\omega + \Delta\omega)\,t + \theta_1\right]\right\} \qquad (2.6)$$

$$+ \frac{A_2}{2}\left\{\cos\left[(\omega + \Delta\omega)\,t + \theta_2\right] + \cos\left[(3\omega + \Delta\omega)\,t + \theta_2\right]\right\} + \ldots$$

Filtering out frequencies above $\Delta\omega$, the signal received by the detector is:

$$\frac{A_1}{2}\cos(\Delta\omega t + \theta_1) \qquad (2.7)$$

which is completely analogous to Eq. 2.4.

2.3.3
Instrumentation

The excitation light source consists of a titanium/sapphire laser pumped by a 10-W argon ion laser (Spectra Physics, CA), to produce a pulse train of 2 ps at a repetition rate of 80 MHz (Fig. 2.2). The output wavelength of the titanium/ sapphire laser is set to 780 nm. To establish a phase lock with the detection equipment, an 80 MHz signal origination from the titanium/sapphire laser is sent to a divide-by-8 circuit resulting in a 10-MHz signal. This 10-MHz signal is sent to a frequency synthesizer (Model 2022D, Marconi, Allendale, NJ) which

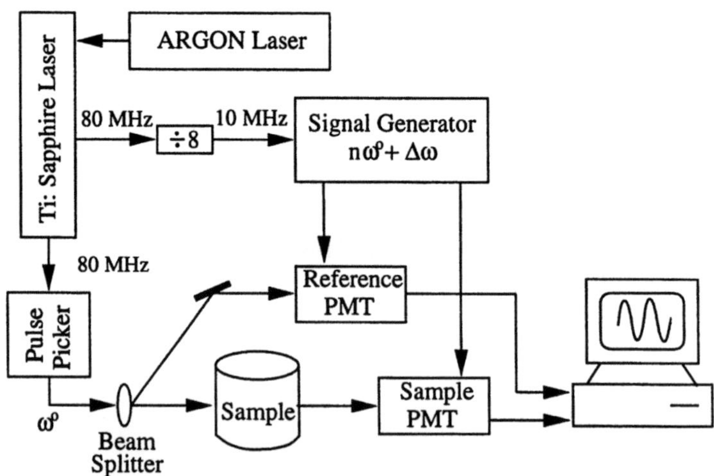

Fig. 2.2. Frequency domain system for single pixel measurements using a titanium/sapphire laser as the light source

controls the modulation of the detectors. The detectors are modulated at frequencies of 80, 160 and 240 MHz. The ligth from the laser has 1.3 W of average power.

The pulse train is focused onto a beam splitter. The unreflected portion of the pulse train (approximately 90 % of the signal) serves as the light source and is focused onto a fiber optic coupled to the side of the cylindrical tank. The reflected portion of the signal is focused onto another fiber optic cable which delivers the signal to a reference photomultiplier tube (PMT). This light serves as the reference signal. The scattered excitation and emission light detected at the periphery of the phantom is delivered to a sample PMT also using a fiber optic cable. An 830-nm interference filter (10-830-4-1.00, CVI Laser Co, Albuquerque, NM) is placed in front of the sample PMT and is used to make frequency domain measurements at the emission wavelengths.

The gain of the PMTs is modulated at a harmonic of the source frequency but offset by 100 Hz in order to produce a heterodyne signal. The output from both PMTs is sent to an ISS board for data acquisition. The modulated phase and amplitude are extracted from the signal at the 100-Hz cross correlation frequency through simple electronics and software (ISS, Champaign, IL).

2.4
Contrast Due to Perfect Absorption and Perfect Uptake of Fluorescence

The modulation phase and modulation ratio were measured to evaluate the contrast offered by, (i) a perfect absorber, and (ii) a fluorescent volume in a lossless scattering medium (no absorption in the surrounding). Since the fluorescent object had perfect uptake of dye and since the absorber was a black rod, they represent the best possible contrast owing to absorption and fluorescence. For this experiment DTTCI served as the fluorescent contrast agent.

Measurements of modulation phase, θ, and modulation ratio, M, were conducted as the heterogeneities were moved from the peripheral location towards the center of the phantom. Figure 2.3 shows the phase contrast, $\Delta\theta$ ($\theta_{presence} - \theta_{absence}$), due to the presence of both an absorbing and a fluorescing heterogeneity at 80-, 160-, and 240-MHz modulation frequencies. The largest phase contrast is measured at 240 MHz and the signal is only altered by approximately 20° for absorption; however, for fluorescence, an alteration of approximately 60° is observed. Therefore, these results show that the best contrast available from absorption (i.e., a perfect absorber) causes a smaller measurable disturbance in photon migration than a micromolar concentration of fluorescent dye. In addition, the fluorescent signal is detected approximately 1 cm deeper than the absorbance signal.

Measurements of modulation ratio, M, for the perfect absorber and the fluorescent volume were not performed. However, a similar experiment conducted by Lopez [30] using a 9-mm diameter perfect absorbing fluorescent object shows that the modulation contrast ($M_{presence}/M_{absence}$) of the fluorescent object at 240 MHz is approximately 0.1 while the absorber only offers 0.6 contrast. These results are consistent with those that would be expected.

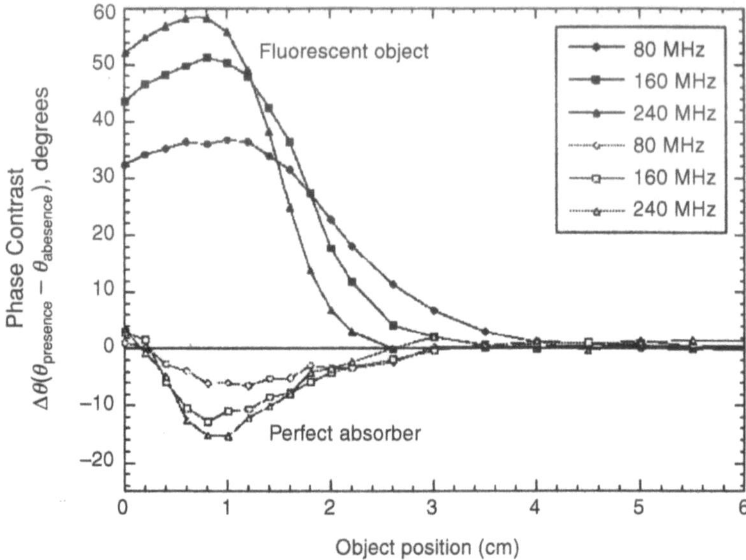

Fig. 2.3. Phase contrast, $\Delta\theta(\theta_{\text{presence}} - \theta_{\text{absence}})$, as a function of object position for both a perfect absorbing object (open symbols) and a fluorescent object (closed symbols) measured at 80, 160 and 240 MHz

In a realistic situation, the administration of contrast agents results in the imperfect uptake of dye into tissue regions of interest. Recently, Lopez et al. [4, 30] have conducted measurements of phase and modulation contrast with imperfect uptake ratios of 1:100 and 1:10 with a similar experimental apparatus as that described above. They found the resulting phase contrast to be dramatically smaller than the situation of perfect uptake. However, the magnitude of contrast offered by fluorescence phase contrast and modulation contrast still exceeds absorption contrast, which is in agreement with the above work. As discussed in Sect. 1, the additional time-delay associated with the lifetime of a fluorescent molecule during the emission process causes a greater phase and modulation contrast over that of absorption.

2.5
Contrast Due to the Influence of Lifetime

The next experiment assessed the influence of lifetime on contrast. In an analogous single pixel experiment to the one described above, Intralipid solutions with micromolar concentrations of DTTCI and ICG were investigated and compared with each other since DTTCI and ICG have different lifetimes of 1.18 and 0.56 ns, respectively [4].

Dilute non-scattering solutions of both DTTCI and ICG were separately analyzed in a small glass cuvette. Using the instrumentation discussed above, measurements of modulation phase and modulation ratio were conducted with

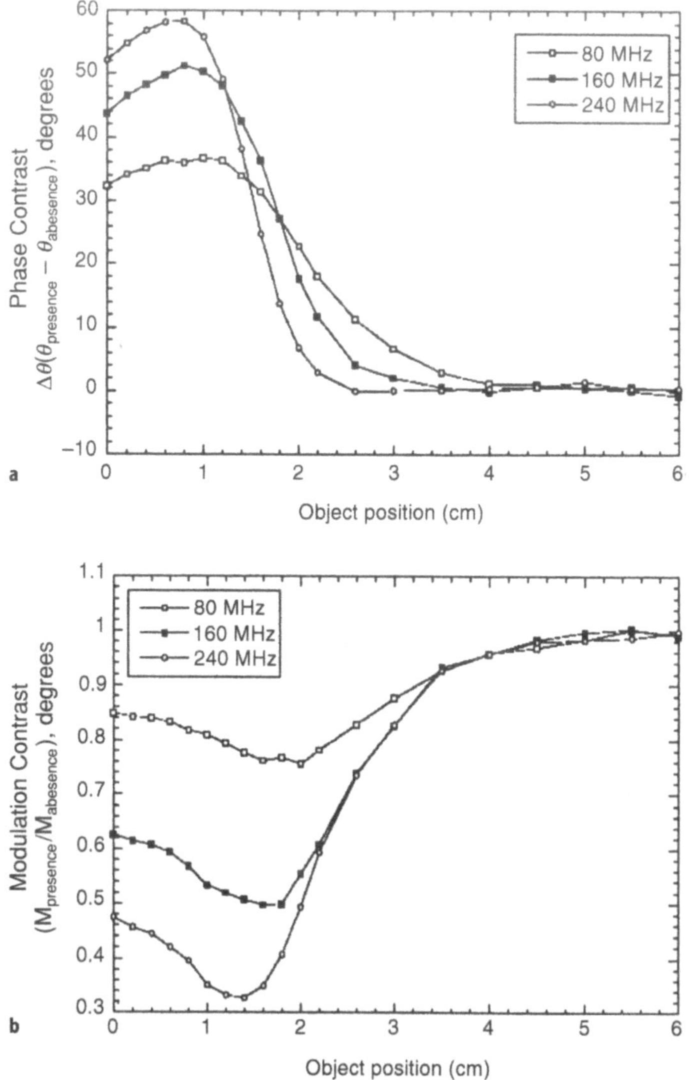

Fig. 2.4. Experimental measurements of **a** phase contrast, $\Delta\theta(\theta_{presence} - \theta_{absence})$, and **b** modulation contrast, ΔM ($M_{presence}/M_{absence}$) vs. object position for a micromolar solution of DTTCI at modulation frequencies of 80, 160 and 240 MHz. A 9.5-mm diameter heterogeneity was used to hold the dye solution inside the 20-cm diameter tank

the source at a right angle to the detector. Using a third dye, IR-140 (ACROS Organics, Fairlawn, NJ), as a standard, the lifetimes of DTTCI and ICG were obtained by evaluating the equations discussed in Sect. 1.6.2.1.

Both the modulation phase and modulation ratio were measured as a function of frequency as DTTCI and ICG scattering solutions were moved from the periphery to the center of the phantom. Figures 2.4 and 2.5 illustrate the phase contrast for DTTCI and ICG, respectively. As the lifetime of the dye increases, the phase contrast should also increase and the modulation ratio should decrease (see Sect. 1). Since the lifetimes of DTTCI and ICG are 1.05 and 0.58 ns, respectively, the phase contrast for DTTCI should be larger and the modulation contrast should be smaller. This is indeed the case since DTTCI

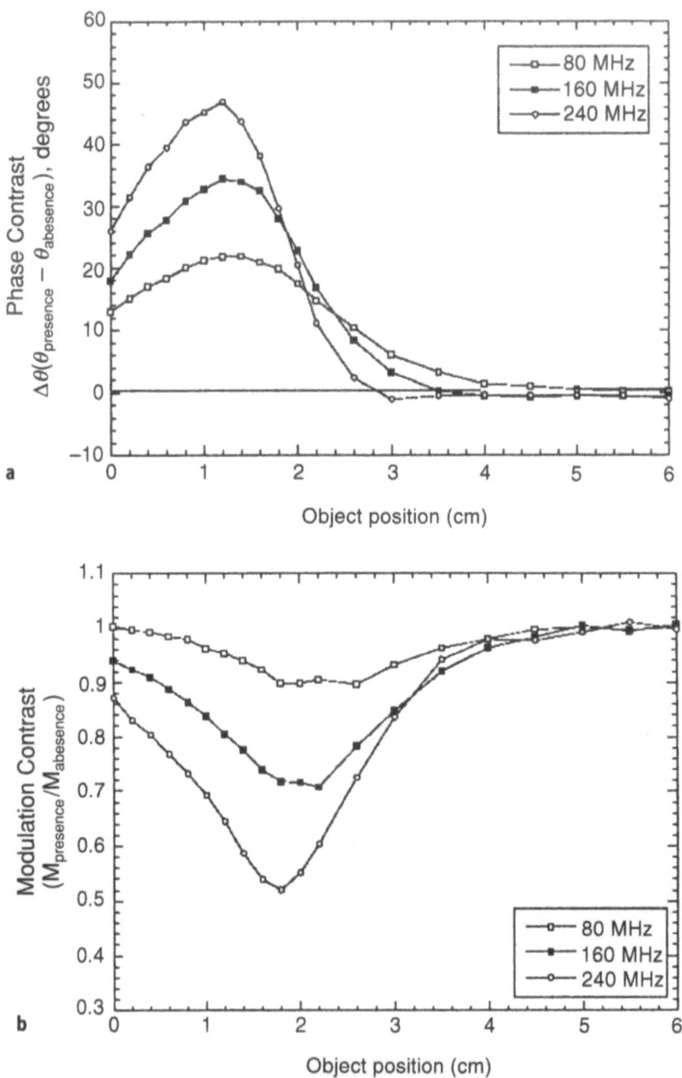

Fig. 2.5. Experimental measurements of **a** phase contrast, $\Delta\theta(\theta_{\text{presence}} - \theta_{\text{absence}})$, and **b** modulation contrast, ΔM ($M_{\text{presence}}/M_{\text{absence}}$) vs. object position for a micromolar solution of ICG at modulation frequencies of 80, 160 and 240 MHz. A 9.5-mm diameter heterogeneity was used to hold the dye solution inside the 20-cm diameter tank

exhibits values of $\Delta\theta = 60°$ and $\Delta M = 0.3$ and ICG shows values of $\Delta\theta = 46°$ and $\Delta M = 0.5$. These results confirm that frequency domain measurements contain information of fluorescent lifetime.

2.6
Discussion

The results in this section show that fluorescence offers improved contrast over absorption. Since measurements were conducted on a perfect absorber, the findings suggest that the endogenous contrast due to absorption from increased hemoglobin associated with angiogenesis may be too small for realistic measurements for optical tomography (Fig. 1.6). The second study shows that frequency domain measurements contain lifetime information which can provide diagnostic information about the local biochemical environment (see Eq. 1.16). However, these measurements need to be coupled to an inversion algorithm in order to extract the lifetime from the scattered signal (see Sect. 3 for more details). The single pixel measurements presented here show great promise for the detection of diseased tissues laden with contrast agents using photon migration measurements. But single pixel measurements only provide one dimensional information and therefore cannot be used to resolve an interior image of optical properties. To obtain enough information to locate a heterogeneity, measurements at multiple locations need to be obtained.

3
Reconstruction of Fluorescence Quantum Efficiency and Lifetime

3.1
Inverse Imaging Problem

Recently, we have developed multi-pixel measurements for rapid data acquisition of modulation phase and amplitude of re-emitted optical signals [31]. Using a gain-modulated image intensifier coupled to a CCD camera, acquisition of 128×128 measurements of modulation phase and amplitude has been possible in approximately 10 ms at excitation wavelengths and 1 s at emission wavelengths. In order to employ these measurements to reconstruct interior maps of lifetime and quantum efficiency from exterior frequency domain measurements made at the periphery, an inverse algorithm needs to be employed. Currently, two approaches for solving the inverse imaging problem have been demonstrated. Both approaches involve solutions to the diffusion equation where the first method uses perturbation expansions of the photon diffusion equation [32–35] and the second solves the diffusion equation directly using a finite difference or a finite element approach [36–40].

In this work, we employ the latter approach to reconstruct optical property maps of absorption, fluorescent quantum efficiency, and fluorescent lifetime of an optical heterogeneity without a priori information of position or volume. Specifically, the tissue volume of interest is mathematically represented by discretized volume elements in which the optical properties of absorption, fluorescence quantum efficiency and lifetime are constructed. The inversion schemes incorporate an iterative Newton-Raphson technique to update values of the optical properties at every volume element in order to minimize the least-squares difference between the experimental and calculated detector responses at the periphery. The method is similar to that used by Yorkey et al. [41] for electrical impedance tomography, Pogue et al. [37] for absorption and scattering reconstructions, and Paithankar et al. [40] who performed reconstructions on both fluorescent lifetime, τ, and fluorescent yield (the product of the absorption coefficient due to fluorophores and quantum efficiency, $\phi\mu_{a_{xf}}$). In this section, interested readers will find the numerical technique used in solving the forward problem along with the approach to solving the inverse imaging problem. Finally, applications of the inversion to simulated data are presented in Sect. 4 illustrating the feasibility of fluorescent quantum efficiency and lifetime imaging in tissues and other scattering media. Section 3 may be ignored by those readers not interested in the mathematics of solving the forward and inverse imaging problems.

3.1.1
Forward Problem

The coupled diffusion equations for excitation and emission light propagation (Eqs. 1.17 and 1.18) are complex linear elliptic equations which can be analytically solved as boundary value problems when the optical properties are independent of spatial position. The complex fluence can be used to predict modulation phase and amplitude from the real and imaginary components. However, when there is a spatial variation in optical properties (such is the case when an optical heterogeneity is present) numerical solutions are necessary. In this work, the finite difference method was used to solve both the excitation fluence, $\Phi^x(\vec{r}, \omega)$, and the emission fluence, $\Phi^m(\vec{r}, \omega)$. The method of finite differences involves discretizing the area of interest into a grid and a linear system is obtained by approximating derivatives in the diffusion equation by differences.

In this study, a two-dimensional simulated tissue phantom, as shown in Fig. 3.1, was employed. The phantom is modeled as a $4 \times 4\,\mathrm{cm}^2$ square surface with a 17×17 grid representing the area of potentially differing optical properties. The source and the detector locations were placed one pixel in from the edge of the phantom to overcome the effects of the zero fluence boundary condition. A source is centered on each side of the phantom for a total of 4 sources. The grid contains 56 detectors per source evenly spaced around the phantom. The placement of the source represents an isotropic excitation source located 0.25 cm or 2.5 scattering lengths from the surface. Since all the experiments discussed here are performed using simulated data, the placement of the source will not affect the solution to the inversion. However, when this algorithm is coupled to experimental measurements, the placement of the source needs to be correctly represented.

For solutions of the forward problem, the known optical properties of the background and the object are input as parameters into the program and then the fluence at each grid point is calculated. The fluence values are used at the detector positions and the modulation phase, $\theta^{x,m}$, and the log of the modula-

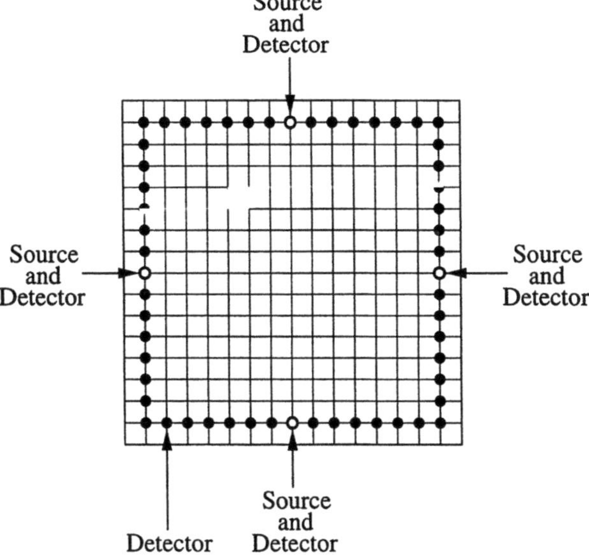

Fig. 3.1. Schematic of the two-dimensional simulated tissue phantom showing the location of the 4 sources and 56 detectors around the perimeter for the 17×17 mesh on a 16-cm^2 area. The small square object inside the phantom represents a simulated diseased tissue volume

tion amplitude, $I_{AC}^{x,m}$, are calculated from both the real and imaginary components of the excitation or emission fluence, $\Phi^{x,m}$, where:

$$\theta^{x,m} = \tan^{-1}\left[\frac{\operatorname{Im}\Phi^{x,m}}{\operatorname{Re}\Phi^{x,m}}\right] \times \frac{360}{2\pi} \tag{3.1}$$

$$I_{AC}^{x,m} = \log_{10}\sqrt{[\operatorname{Im}\Phi^{x,m}]^2 + [\operatorname{Re}\Phi^{x,m}]^2} \tag{3.2}$$

To run a single simulated experiment, only one source is active. For each forward solution, values of $\theta^{x,m}$ and $I_{AC}^{x,m}$ are calculated at all the detector positions. Once these detector values are obtained, the forward solution is repeated for the next source location. This procedure continues until all four sources have been investigated. A total of four simulated experiments constitute the data to be acquired experimentally. Consequently, the simulations will yield a total of 224 detector measurements. In all the simulated experiments, the sources are modulated at a frequency of 100 MHz. Typical forward calculations on a Ultra 2 Sun workstation took approximately 1.3 s.

After detector data have been evaluated for $\theta^{x,m}$ and $I_{AC}^{x,m}$ simulated Gaussian noise with standard deviations of 0.1° for phase and 1% for modulation amplitude are added using a MATLAB routine. The Gaussian noise is added to account for the noise which would be encountered in a real experiment. The simulated forward results with added noise are then input to the inversion algorithm.

3.1.2
Solution to the Inverse Problem

The solution to the inverse problem requires reconstructing images of the interior volume from exterior measurements of $\theta^{x,m}$ and $I_{AC}^{x,m}$. Since more information from exterior measurements enable better image reconstruction, more detectors and multiple modulation frequencies should improve image reconstructions. The inversion algorithm is conducted in two parts on the basis of, (i) absorption at the excitation wavelength due to fluorophores, $\mu_{a_{xf}}$, and (ii) quantum efficiency, ϕ, or lifetime, τ, at the emission wavelength. The two parts are similar except that the reconstruction based on $\mu_{a_{xf}}$ uses detector data at the excitation wavelength while the other part uses data at the emission wavelength.

Figure 3.2 depicts the flow diagram for these inversion algorithms. To begin the inverse calculation based on $\mu_{a_{xf}}$, an initial homogeneous guess for the

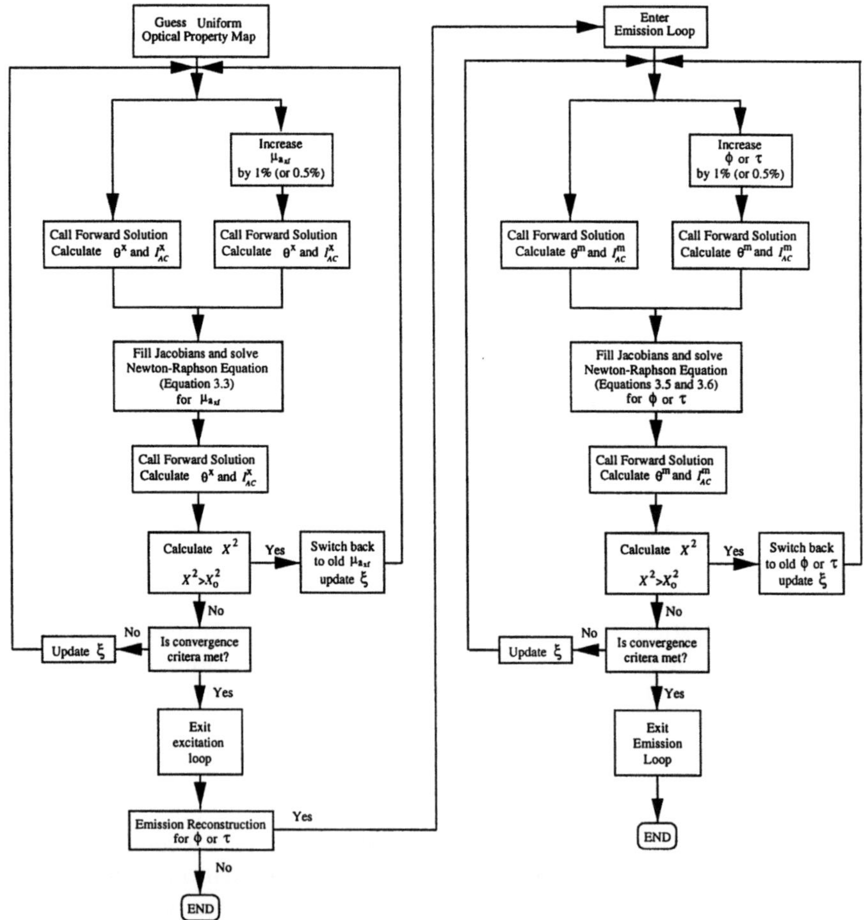

Fig. 3.2. Flow diagramm of the reconstruction algorithm. The excitation loop reconstructs $\mu_{a_{xf}}$ and the emission loop either reconstructs ϕ or τ

optical property map is given. Typical values are those found in normal tissues. From these values the forward solution is calculated to obtain θ^x and I^x_{AC} measurements at the detector positions for the excitation light. Next, two Jacobian matrices, $J(\theta^x, \mu_{a_{xf}})$ and $J(I^x_{AC}, \mu_{a_{xf}})$, are constructed where an entry of each matrix is defined as $j_{ij} = \partial\theta^x_i/\partial\mu_{a_{xf,j}}$ and $j_{ij} = \partial I^x_{AC_i}/\partial\mu_{a_{xf,j}}$, respectively. Each element of the matrix describes the response of the source-detector pair at position i to changes in $\mu_{a_{xf}}$ at each pixel j. The partial derivatives are numerically approximated by calling the forward solution (Cud2) a second time for a 1% increase of the original pixel value or $\mu_{a_{xf}}$ and subtracting the difference (i.e. $\partial\theta^x/\partial\mu_{a_{xf}} \approx \Delta\theta^x/\Delta\mu_{a_{xf}}$ and $\partial I^x_{AC}/\partial\mu_{a_{xf}} \approx \Delta I^x_{AC}/\Delta\mu_{a_{xf}}$).

To update $\mu_{a_{xf}}$, an iterative Newton-Raphson technique involving the Jacobians was used:

$$\left[\frac{J^T(\theta^x, \mu_{a_{xf}})\, J(\theta^x, \mu_{a_{xf}})}{\sigma_\theta^2} + \frac{J^T(I^x_{AC}, \mu_{a_{xf}})\, J(I^x_{AC}, \mu_{a_{xf}})}{\sigma_{I_{AC}}^2} + \xi I\right][\Delta\vec{\mu}_{a_{xf}}] =$$

$$= \left[\frac{J^T(\theta^x, \mu_{a_{xf}})}{\sigma_\theta^2}(\vec{\theta}^x_{obs} - \vec{\theta}^x_{cal}) + \frac{J^T(I^x_{AC}, \mu_{a_{xf}})}{\sigma_{I_{AC}}^2}(\vec{I}^x_{AC_{obs}} - \vec{I}^x_{AC_{cal}})\right] \tag{3.3}$$

where $\Delta\vec{\mu}_{a_{xf}}$ provides an increment change for updating $\mu_{a_{xf}}$, and ξ is a regularization parameter multiplied by the identity matrix I. Because the Jacobian matrices are ill conditioned due to the small sensitivity of $\mu_{a_{xf}}$ far away from the source and detector, the regularization parameter compensates by making the matrices more diagonally dominant. The parameter ξ is then adjusted by a Marquardt-Levenberg algorithm at every iteration [37]. In order to solve the linear algebraic equations (Eq. 3.3) for $\Delta\vec{\mu}_{a_{xf}}$, a LU decomposition back substitution method is used. The updated optical property map is then found by adding $\Delta\vec{\mu}_{a_{xf}}$ to the values from the previous iteration.

The forward solution is re-calculated using the new updated values in order to determine the reconstruction error:

$$\chi^2_x = \frac{1}{4nd}\sum_{k=1}^{4}\sum_{i=1}^{nd}\left[\left(\frac{\theta^x_{cal,i} - \theta^x_{obs,i}}{\sigma_\theta}\right)^2 + \left(\frac{I^x_{AC_{cal,i}} - I^x_{AC_{obs,i}}}{\sigma_{I_{AC}}}\right)^2\right] \tag{3.4}$$

where nd is the total number of detectors (56); θ^x_{cal} and $I^x_{AC_{cal}}$ are the predicted detector data; and θ^x_{obs} and $I^x_{AC_{obs}}$ are the simulated experimental detector data for the excitation light. Again, σ_θ and $\sigma_{I_{AC}}$ are the typical standard deviations of noise in the measurements taken to be 0.1° and 1%, respectively. For comparison, our multi-pixel measurements have errors in the order of 0.2° and 1.8%. The values of σ_θ and $\sigma_{I_{AC}}$ are scale factors that describe the confidence in the measurement. Every node on the grid will yield an equation and the only known variables are at the detector positions. Since there are many more nodes than detectors, the inversion scheme is underconstrained and not guaranteed to converge on the actual optical property map.

The entire procedure of iteratively adjusting $\mu_{a_{xf}}$ to minimize the χ^2_x error continues until a predetermined convergence criterion is met. The criterion is met if any of the following quantities: (i) the value of χ_x^2, (ii) the absolute change in χ_x^2, or (iii) the relative change in χ_x^2 becomes lower than 1.0×10^{-5} or a maximum of 50 iterations is reached. Typical times to complete the inverse solution are approximately 45 min on a Ultra 2 Sun workstation.

If only a map of $\mu_{a_{xf}}$ is desired, the program stops. Otherwise, the map of $\mu_{a_{xf}}$ is used in order to compute a second map of either ϕ or τ. Currenctly, the computation of the maps of ϕ and τ are conducted in two separate programs in order to evaluate their individual performances.

After the loop in which the map $\mu_{a_{xf}}$ is calculated, the forward solution is now calculated to obtain θ^m and I_{AC}^m measurements at the detector positions for the emission light using the homogeneous map described above except with the updated values of $\mu_{a_{xf}}$. Next, two Jacobian matrices are constructed: $J(\theta^m, \phi)$ and $J(I_{AC}^m, \phi)$ for the reconstruction of ϕ, or $J(\theta^m, \tau)$ and $J(I_{AC}^m, \tau)$ for the reconstruction of τ. Each element of these Jacobian matrices is defined as $j_{ij} = \partial \theta_i^m / \partial \phi_j$, $j_{ij} = \partial I_{AC_i}^m / \partial \phi_j$, $j_{ij} = \partial \theta_i^m / \partial \tau_j$, and $j_{ij} = \partial I_{AC_i}^m / \partial \tau_j$, respectively. Again, each element of the Jacobian matrix describes the response of the source-detector pair at position i to changes in ϕ or τ at each pixel j. The partial derivatives are approximated as described above where $\partial \theta^m / \partial \phi \approx \Delta \theta^m / \Delta \phi$, $\partial I_{AC}^m / \partial \phi \approx \Delta I_{AC}^m / \Delta \phi$, $\partial \theta^m / \partial \tau \approx \Delta \theta^m / \Delta \tau$, and $\partial I_{AC}^m / \partial \tau \approx \Delta I_{AC}^m / \Delta \tau$.

Equations 3.5 and 3.6 provide updates of ϕ and τ by adding $\Delta \vec{\phi}$ and $\Delta \vec{\tau}$, respectively, to the values of ϕ and τ from the previous iteration:

$$\left[\frac{J^T(\theta^m, \phi) \, J(\theta^m, \phi)}{\sigma_\theta^2} + \frac{J^T(I_{AC}^m, \phi) \, J(I_{AC}^m, \phi)}{\sigma_{I_{AC}}^2} + \xi I \right] [\Delta \vec{\phi}] =$$
$$= \left[\frac{J^T(\theta^m, \phi)}{\sigma_\theta^2} (\vec{\theta}_{obs}^m - \vec{\theta}_{cal}^m) + \frac{J^T(I_{AC}^m, \phi)}{\sigma_{I_{AC}}^2} (\vec{I}_{AC_{obs}}^m - \vec{I}_{AC_{cal}}^m) \right]$$

(3.5)

and

$$\left[\frac{J^T(\theta^m, \tau) \, J(\theta^m, \tau)}{\sigma_\theta^2} + \frac{J^T(I_{AC}^m, \tau) \, J(I_{AC}^m, \tau)}{\sigma_{I_{AC}}^2} + \xi I \right] [\Delta \vec{\tau}] =$$
$$= \left[\frac{J^T(\theta^m, \tau)}{\sigma_\theta^2} (\vec{\theta}_{obs}^m - \vec{\theta}_{cal}^m) + \frac{J^T(I_{AC}^m, \tau)}{\sigma_{I_{AC}}^2} (\vec{I}_{AC_{obs}}^m - \vec{I}_{AC_{cal}}^m) \right]$$

(3.6)

Again, the forward solution is re-calculated using the updated values in order to determine the reconstruction error:

$$\chi_m^2 = \frac{1}{4nd} \sum_{k=1}^{4} \sum_{i=1}^{nd} \left[\left(\frac{\theta_{cal,i}^m - \theta_{obs,i}^m}{\sigma_\theta} \right)^2 + \left(\frac{I_{AC_{cal,i}}^m - I_{AC_{obs,i}}^m}{\sigma_{I_{AC}}} \right)^2 \right]$$

(3.7)

where $\theta_{cal}^m, I_{AC_{cal}}^m$ are the predicted detector data and θ_{obs}^m and $I_{AC_{obs}}^m$ are the simulated experimental detector data for the emission light. The values for σ_θ and $\sigma_{I_{AC}}$ are the same as in the absorption reconstruction, $0.1°$ and 1.0%, respectively. The entire procedure of iteratively adjusting ϕ or τ to minimize the χ_m^2 error continues until the same convergence criterion as above is met or 50 iterations have passed.

4
Reconstruction of Absorption and Fluorescence Quantum Efficiency and Lifetime in a Scattering Medium

4.1
Results for the Reconstruction Algorithm

Three different types of simulations were separately performed. They included reconstructions on the basis of either absorption, $\mu_{a_{xf}}$, quantum efficiency, ϕ, or lifetime, τ. In all three classes of simulated experiments, an optical heterogeneity was 0.063 cm^2 which constituted 0.39% of the total area. During the reconstructions of both ϕ and τ unphysically high optical property values around the periphery and especially near the source locations were obtained. These values were replaced with the average background value with a constraint statement inside the program.

For all the reconstructions, values of $\mu_{a_{xf}}$ ranged from 0.010 to 0.200 cm^{-1}. These values of absorption correspond to an ICG concentration of approximately 0.076 to 1.53 µM [4]. These ICG concentrations are well below lethal levels and are approximately 5 to 60 times lower than the therapeutic concentrations currently administered with many photodynamic agents [42, 43].

4.2
Reconstruction of the Absorption Coefficient from Excitation Light

This section describes the simulations performed on the basis of absorption due to fluoropheres, $\mu_{a_{xf}}$, at the excitation wavelength. Figure 4.1 illustrates both simulated and reconstructed data as an optical tissue heterogeneity is moved diagonally towards the detectors at three different object locations. The heterogeneity contains 20-fold more absorbing dye than its surroundings (see Table 4.1). The experiment shows that the heterogeneity location is successfully reconstructed. However, since the program preserves the average pixel value, a smoothing of the reconstructed image is observed causing the magnitude of $\mu_{a_{xf}}$ for both the object and the surroundings to be incorrect as shown in Table 4.1. As the object moves closer to the detectors, less error in the reconstruction

Table 4.1. Optical Properties and Experimental Parameters Used as Inputs for the Forward Problem Along with Values of $\mu_{a_{xf}}$ Obtained from the Reconstructions for Three Different Object Locations

Case	Actual				Reconstructed		
	Background		Object		Background	Object	
	$\mu_{a_{xi}}, \mu_{a_{xf}}$ (cm^{-1})	μ_{s_x} (cm^{-1})	$\mu_{a_{xf}}$ (cm^{-1})	Position	$\mu_{a_{xf}}$ (cm^{-1})	$\mu_{a_{xf}}$ (cm^{-1})	Position
a	0.010	10.0	0.200	(11,7)	0.011	0.034	(11,7)
b	0.010	10.0	0.200	(13,5)	0.011	0.042	(13,5)
c	0.010	10.0	0.200	(15,3)	0.011	0.068	(15,3)

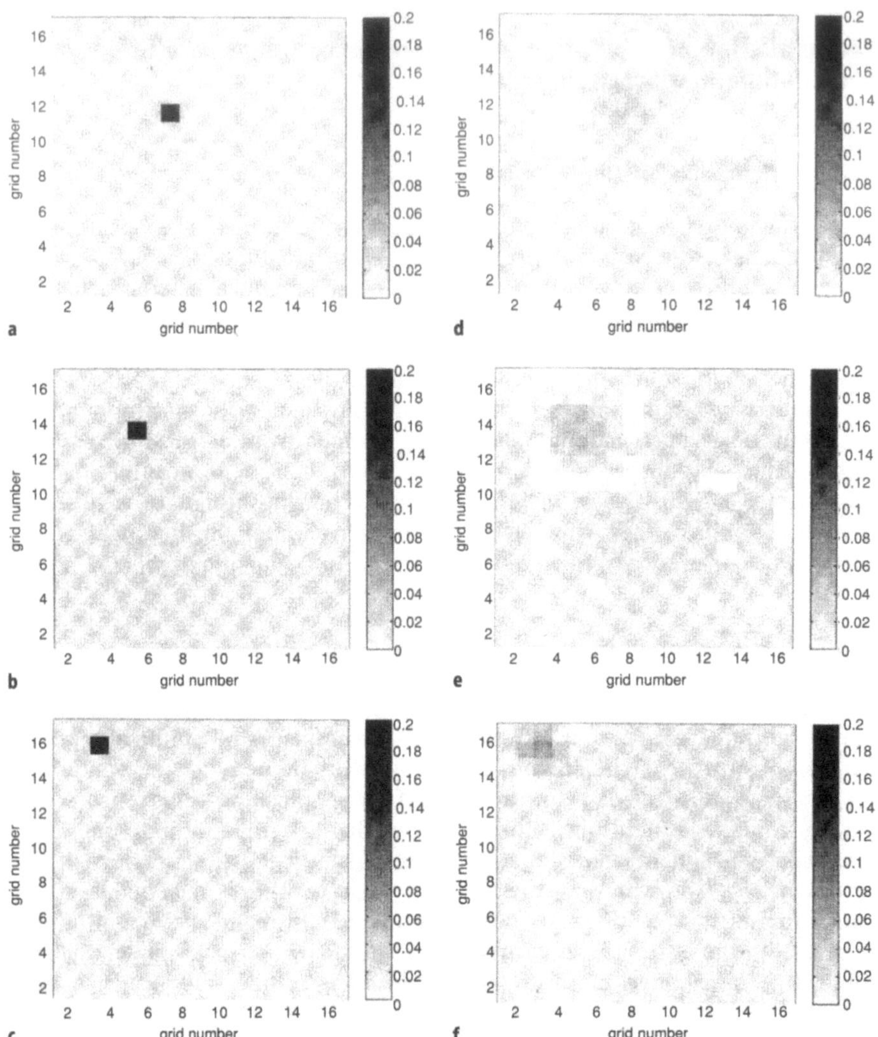

Fig. 4.1. Reconstructed spatial maps of an absorbing heterogeneity as a function of position for a 20:1 uptake ratio of dye inside the object. Figures **a**, **b** and **c** are actual spatial maps and Figures **d**, **e** and **f** are their corresponding reconstructions. The actual values of $\mu_{a_{xf}}$ in the object and the surroundings are 0.200 cm^{-1} and 0.010 cm^{-1}, respectively

is obtained since the value of $\mu_{a_{xf}}$ inside the object increases while the value in the background decreases. In fact, the value of $\mu_{a_{xf}}$ at the closest object position is 0.068 cm^{-1} (case c) is twice the value of the farthest position (0.034 cm^{-1}, case a). This reduction of error is expected since there is more signal yielding a larger disturbance to be measured at the detectors. In other words, a heterogeneity present close to the source and detector will have more influence on the inversion problem since the pixels closer to the source and detectors have more influence on the measured signal making the solution better defined. Successful reconstructions have also been performed for a more realistic uptake of 10 : 1 as well as higher resolutions (33 × 33 grid) [44].

4.3
Reconstruction of Quantum Efficiency from Emission Light

The performance of the inverse algorithm for reconstructing maps of quantum efficiency, ϕ, using detector data at the emission wavelength is discussed in this section. For all the inversions discussed here, the exact value of $\mu_{a_{xf}}$ inside the object was input into the initial homogeneous guess. This procedure causes the inversion algorithm to converge after one iteration in the excitation loop and the exact value of $\mu_{a_{xf}}$ to be used in the emission loop in order to only investigate the reconstruction on ϕ. Table 4.2 lists all the optical properties and experimental parameters used in the forward solutions to generate the simulated detector data and compares the values from the forward solution to those obtained by the inversion solution.

Figure 4.2 shows a the actual solution, b the reconstruction, and c the value of ϕ inside the object as a function of iteration number. The experiment ended

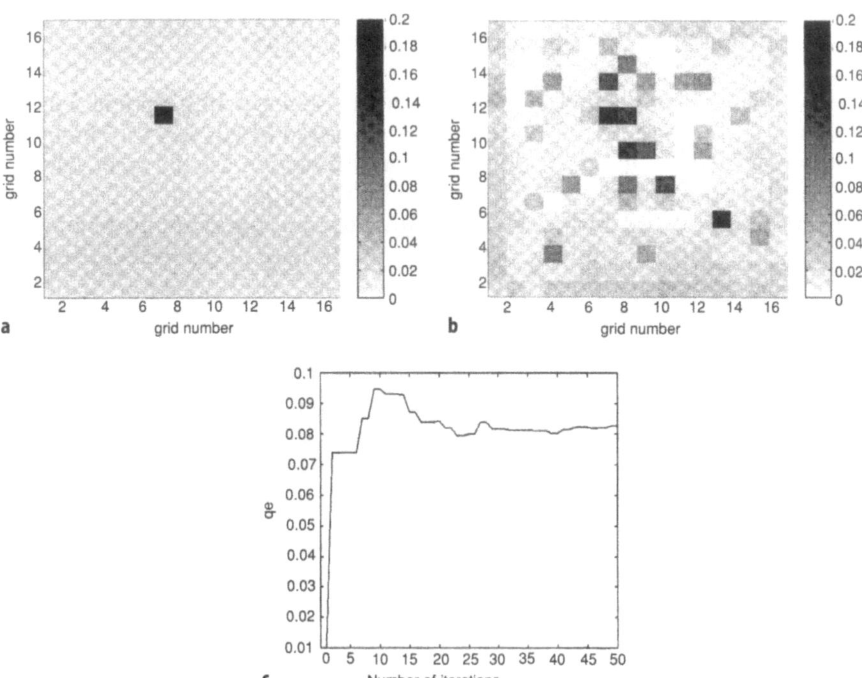

Fig. 4.2. Reconstructed spatial maps of fluorescent quantum efficiency, ϕ, for 10 : 1 uptake of dye inside the heterogeneity. Figure **a** represents the actual image, **b** shows the corresponding reconstruction and **c** shows the value of ϕ inside the object as a function of iteration number. The actual values of ϕ in the object and the surroundings were 0.100 cm^{-1} and 0.010 cm^{-1}, respectively, and the reconstructions yielded values of 0.082 cm^{-1} and 0.012 cm^{-1}, respectively

Table 4.2. Optical Properties and Experimental Parameters Used as Inputs for the Forward Problem Along with Values of ϕ Obtained from the Reconstructions

Actual								Reconstructed	
Background						Object		Background	Object
$\mu_{a_{xi}}, \mu_{a_{mi}},$ (cm^{-1})	$\mu_{a_{xf}}$ (cm^{-1})	μ_{s_x}, μ_{s_m} (cm^{-1})	$\mu_{a_{mi}},$ (cm^{-1})	τ (ns)	ϕ	$\mu_{a_{xf}}$ (cm^{-1})	ϕ	ϕ	ϕ
0.000	0.020	10.0	0.002	1.00	0.010	0.200	0.100	0.012	0.082

after 50 iterations without meeting any of the convergence criteria. The optical property map of the reconstruction shows that the location of the object is successfully located but that the value of ϕ inside the object only reaches 0.082 (18% smaller than the actual value of 0.100). Also, the background pixel values are noisy with an average value of 0.012 which is 20% larger than the actual value of 0.010.

4.4
Reconstruction of Lifetime from Emission Light

The performance of the inversion algorithm for reconstructing lifetime, τ, using detector data at the emission wavelength is discussed in this section. Similar to the reconstructions of ϕ, all the inversions in this section use the exact value of $\mu_{a_{xf}}$ inside both the object and background for the initial optical property map. This procedure causes the inversion algorithm to converge after one iteration in the excitation loop and the exact value of $\mu_{a_{xf}}$ to be used in the emission loop to investigate the reconstruction of τ only.

The lifetime reconstruction for a 10:1 uptake of dye, the current uptake ratio for available contrast agents, was investigated. Table 4.3 lists all the optical properties and experimental parameters used in the forward solution to generate the detector data and compares the values from the forward solution to those obtained by the inverse solution. This particular simulation had an arbitrary lifetime value of 1.00 ns inside the object and 10.0 ns everywhere else. Figure 4.3

Table 4.3. Optical Properties and Experimental Parameters Used as Inputs for the Forward Problem Along with Values of Lifetime Obtained from the Reconstructions

Actual								Reconstructed	
Background						Object		Background	Object
$\mu_{a_{xi}}, \mu_{a_{mi}},$ (cm^{-1})	$\mu_{a_{xf}}$ (cm^{-1})	μ_{s_x}, μ_{s_m} (cm^{-1})	$\mu_{a_{mi}},$ (cm^{-1})	ϕ	τ (ns)	$\mu_{a_{xf}}$ (cm^{-1}) τ (ns)		τ (ns)	τ (ns)
0.000	0.020	10.0	0.002	0.034	10.0	0.200	0.100	10.1	0.996

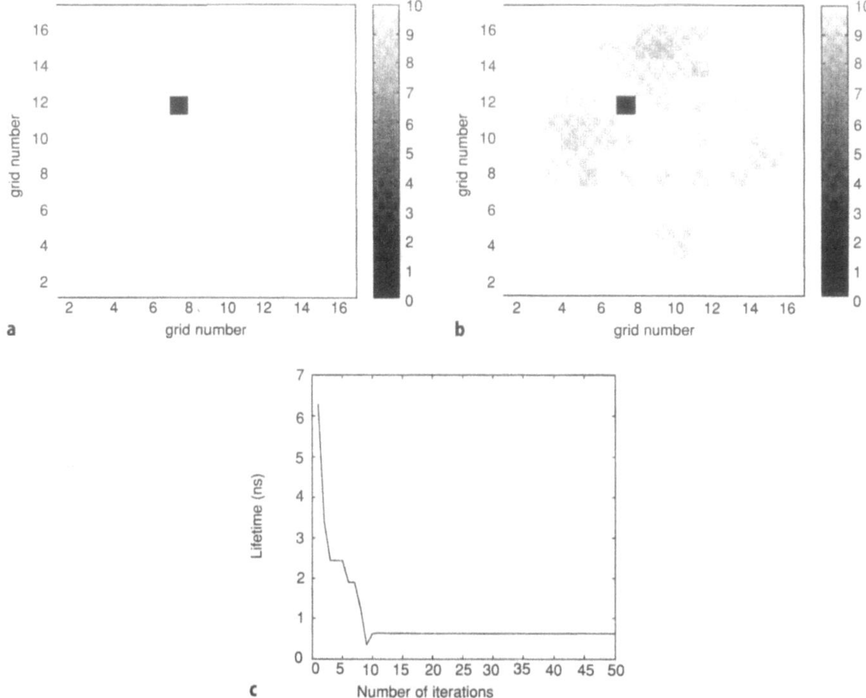

Fig. 4.3. Reconstructed spatial maps of fluorescent lifetime, τ, for 10:1 uptake of dye inside the heterogeneity. Figure **a** represents the actual image, **b** shows the corresponding reconstruction and **c** depicts the convergence of τ in the object as a function of iteration number. The actual lifetimes for the object and background are 1 and 10 ns, respectively. The reconstruction converges correctly to 1 ns in the object. The background converges to 10.067 ns which is 0.67 % larger than the actual value of 10 ns

shows a the actual spatial map, b the reconstructed spatial map, and c the average value of τ as a function of the iteration number inside the object. Because τ inside the object is lower than in the background, the object location is defined as the pixel with the lowest value of τ. Both the location and magnitude of τ inside the object are correctly found since the reconstructed value for pixel (11,7) is 0.996 ns, only 0.04 % lower than the actual value. The reconstruction for the background gives a good convergence of 10.1 ns even though the background contains a small amount of noise which is symmetric about the object. This noise is most likely an artifact from the large fluence that emerges from the four source locations.

4.5
Discussion

Numerical simulations were conducted in order to examine the resolution and accuracy of the reconstruction method. These results show the potential of photon migration for the optical detection of a heterogeneity inside a scattering medium. The simulated experiments based on $\mu_{a_{xf}}$ can locate an absorbing heterogeneity inside a tissue-simulating phantom; however, the magnitude of $\mu_{a_{xf}}$ inside the heterogeneity is smaller than the actual solution. The simulated experiments also show the ability to reconstruct the interior optical property maps for both quantum efficiency and lifetime using detector information at the emission wavelength. Reconstructing maps of lifetime not only provides object location but also may provide information regarding the environment (see Sect. 1.16 for discussion on the Stern-Volmer equation). This information about the local biochemistry can help to differentiate normal from diseased tissue.

To couple these inversion algorithms with experimental measurements, all the optical coefficients of the background need to be scaled by a factor of $(3/2)^{1/2}$ to account for the two-dimensional nature of the algorithm and the three-dimensional nature of the experiment [45]. The noise for the fluorescent measurements needs to be determined since it is unlikely to be the same as the excitation data ($0.1°$ for θ^x and 1% for I_{AC}^x). Also, both inversions assume that the noise is uniform over the entire region which may not be the actual situation, and updates to the inversion scheme, which take the noise as a function of position into account, need to be implemented.

To further improve the reconstructions, detector information at multiple modulation frequency needs to be incorporated into the algorithm. Multiple frequency information will provide more detector information helping with the underconstrained nature of the inversion problem.

4.6
Conclusion

Absorption and fluorescent methods of inducing contrast were examined using single pixel measurements. The results in this study show that fluorescence has improved contrast over absorption due to the additional mechanism from the kinetics of the fluorescence decay process. It has also been demonstrated that lifetime differences alter the propagation of the detected signal. Since targeted delivery of a contrast agent may not always be feasible, lifetime sensitive dyes which are specific to different environmental conditions found in normal and diseased tissue may enhance detection.

Acknowledgements. This work was supported in part by the National Institutes of Health (R01CA71413, R0167176-01 and K04CA687374-01).

References

1. Mordon S, Devoisselle JM, Maunoury V (1994) In vivo pH measurements and imaging of tumor tissue using pH sensitive fluorescent probe (5,6-carboxyfluorescein): instrumental and experimental studies. Photochem Photobiol 60:274–279
2. Russell DA, Pottier RH, Valenzeno DP (1994) Continuous noninvasive measurement of in vivo pH in conscious mice. Photochem Photobiol 59:309–313

3. Vinogradov SA, Lo LW, Jenkins WT, Evans SM, Koch C, Wilson DF (1996) Noninvasive imaging of the distribution in oxygen in tissue in vivo using near-infrared phosphors. Biophys 70:1609–1617
4. Sevick-Muraca E, Lopez G, Reynolds J, Troy T, Hutchinson C (1997) Fluorescence and absorption contrast mechanisms for biomedical optical imaging using frequency-domain techniques. J Photochem Photobiol 66:55–64
5. Cubeddu R, Canti G, Taroni P, Valentini G (1993) Time-gated fluorescence imaging for the diagnosis of tumors in murine model. Photochemistry and Photobiology 57:480–485
6. Cubeddu R, Canti G, Pifferi A, Taroni P, Valentini G (1997) Fluorescence lifetime imaging of experimental tumors in hematoporphyrin derivative sensitized mice. Photochemistry and Photobiology 66:229–236
7. Lim HW, Soter NA (1993) Clinical Photomedicine, Marcel Dekker, New York
8. Haskell RC, Svaasand LO, Tsay TT, Feng T, McAdams MS, Tromberg BJ (1994) Boundary conditions for the diffusion equation in radiative transfer. J Opt Soc Am A 11:2727–2741
9. Beauvoit B, Evans SM, Jenkins TW, Miller EE, Chance B (1995) Correlation between the light scattering and the mitochondrial content of normal tissues and transplantable rodent tumors. Analytical Biochemistry 226:167–174
10. Chance B, Liu H, Kitai T, Zhang Y (1995) Effects of solutions on optical properties of biological materials: models, cells and tissues. Analytical Biochemistry 227:351–362
11. Flock ST, Wilson BC, Patterson MS (1988) Hybrid Monte Carlo diffusion theory modeling of light distribution in tissue. Proc SPIE 908:20–28
12. Peters VG, Wyman DR, Patterson MS, Frank GL (1990) Optical properties of normal and diseased human breast tissues in the visible and near infrared. Phys Med Biol 35:1317–1334
13. Duderstadt JJ, Hamilton LJ (1976) Nuclear Reactor Analysis, Wiley, New York
14. Chandrasekhar S (1958) Radiative Transfer, Dover, New York
15. Ishimaru A (1973) Wave propagation and scattering in random media. Academic Press, New York, Vol. 1
16. Patterson MS, Chance B, Wilson B (1989) Time resolved reflectance and transmittance for the noninvasive measurement of tissue optical properties. Appl Opt 28:2331–2336
17. Reynolds J, Przadka A, Yeung S, Webb K (1996) Optical diffusion imaging: a comparative numerical and experimental study. Appl Opt 35:3671–3679
18. O'Leary M, Boas D, Chance B, Yodh A (1992) Refraction of diffuse photon density waves. Physical Review Letters 69:2658–2661
19. Lakowicz J (1983) Principles of Fluorescence Spectroscopy. Plenum Press, New York
20. Sevick EM, Burch CL (1994) Origin of phosphorescence signals reemitted from tissues. Opt Lett 19:1928–1930
21. Sevick-Muraca EM, Hutchinson C (1995) Probability description of fluorescent and phosphorescent signal generation in tissues and other random media. SPIE 2387:274–283 (1995)
22. Batchteler G, Drexhage KH, Arden-Jacob J, Han KT, Kollner M, Muller R, Sauer M, Seeger S, Wolfrum J (1994) Sensitive fluorescence using laser diodes and multiplex dyes. J Luminescence 60:511–514
23. Patterson MS, Pogue BW (1994) Mathematical model for time-resolved and frequency-domain fluorescence spectroscopy in biological tissues. Appl Opt 33:1963–1974
24. Hutchinson CL, Lakowicz JR, Sevick-Muraca EM (1995) Fluorescence life-time based sensing in tissues: A computational study. Biophys J 68:1574–1582
25. Wilson BC, Patterson MS (1986) The physics of photodynamic therapy. Phys Med Biol 31:327–360
26. Lee WW, Wilson D, Singerman L (1994) Correction of spatial distortion and registration in ophthalmic fluorescein angiography. IEEE 16:508–509
27. May SA (1994) Photonic appraoches to burn diagnostics. Biophotonics International, pp 44–50
28. van Staveren HJ, Moes CJM, van Marie J, Prahl SA, van Gemert MJC (1991) Light scattering in Intralipid-10% in the wavelength range of 400–1100 nm. Appl Opt 30:4507–4514

29. Yariv A (1985) Introduction to Optical Electronics, 3rd ed, Holt-Saunders International Editions, New York
30. Lopez G (1997) Absorption and fluorescent contrast mechanisms for the detection and diagnosis of breast cancer using single pixel frequency-domain photon migration techniques. Masters thesis, Purdue University, West Lafayette, IN
31. Reynolds JS, Troy T, Sevick-Muraca EM (1997) Multi-pixel techniques for frequency-domain photon migration imaging. Biotechnology Progress 13:669–680
32. O'Leary M, Boas D, Chance B, Yodh A (1994) Reradiation and imaging of diffuse photon density waves using fluorescent inhomogeneities. J Lumin 60 & 61:281–286
33. Boas DA, O'Leary MA, Chance B, Yodh AG (1994) Scattering of diffuse photon density waves by spherical inhomogeneities with turbid media: analytic solution and applications. Proc Natl Acad Sci USA 91:4887–91
34. O'Leary M, Boas D, Chance B, Yodh A (1995) Experimental images of heterogeneous turbid media by frequency-domain diffusion-photon tomography. Optics Letters 20:426–428
35. Yao Y, Wang Y, Pei Y, Zhu W, Barbour RL (1997) Frequency-domain optical imaging of absorption and scattering distributions by a Born iterative method. J Opt Soc Am A 14:325–342
36. Arridge SR, Schweiger M, Hiraoka M, Delpy DT (1993) Performance of an iterative reconstruction algorithm for near infrared absorption and scatter imaging. SPIE 1888:360–371
37. Pogue BW, Patterson MS, Jiang H, Paulsen KD (1995) Initial assessment of a simple system for frequency domain diffuse optical tomography. Phys Med Biol 40:1709–1729
38. Jiang H, Paulsen KD, Osterberg UL, Pogue BW, Patterson MS (1995) Simultaneous reconstruction of absorption and scattering maps in turbid media from near-infrared frequency-domain data. Opt Lett 20:2128–2130
39. Jiang H, Paulsen KD, Osterberg UL, Pogue BW, Patterson MS (1996) Optical image reconstruction using frequency domain data: simulations and experiments. J Opt Soc Am A 13:253–266
40. Paithankar DY, Chen AU, Pogue BW, Patterson MS, Sevick-Muraca EM (1997) Imaging of fluorescent yield and lifetime from multiply scattered light reemitted from random media. Appl Opt 36:2260–2272
41. Yorkey TJ, Webster JG, Tompkins WJ (1987) Comparing reconstruction algorithms for electrical impedance tomography. IEEE Trans Biomed Eng BME-34:843–852
42. Photodynamic Therapy: Basic Principles and Clinical Application, BW Henderson and TJ Dougherty, Eds, Marcel Dekker, Inc 1992
43. Marcus SL (1992) Photodynamic Therapy of Human Cancer. IEEE 80:869–889
44. Troy TL (1997) Biomedical Optical Imaging with Frequency Domain Photon Migration Measurements: Experiments and Numerical Image Reconstructions. PhD. Thesis, Purdue University, West Lafayette, IN, 1997
45. Burch CL (1993) Monte Carlo simulations of photon migration in highly scattering media. Masters Thesis, Vanderbilt University, Nashville, TN

Single-Molecule Detection in Biology with Multiplex Dyes and Pulsed Semiconductor Lasers

M. Sauer, J. Wolfrum

1
Single-Molecule Detection with Far-Field Fluorescence Microscopy Techniques

In recent times, researchers have made considerable efforts to achieve the detection of individual atoms and molecules in solids [1–6], on surfaces [7–10], and in liquids [11–20] using laser-induced fluorescence techniques. In particular, single-molecule detection in condensed phases has many important biological applications, including rapid DNA sequencing, medical diagnosis, and forensic analysis [21–23]. For the detection of single dye molecules it is essential to minimize the background from scattering and luminescent impurities in the solvent, i.e. the ability to detect fluorescence bursts emitted by individual dye molecules is not as much an issue of sensitive detection as it is of background reduction. The predominant source of background is Rayleigh scattering, as well as reflected light, which can be efficiently suppressed by suited optical filters. Since Raman scattering is directly proportional to the detection volume, this background signal can be greatly reduced if a confocal setup in combination with a detection volume of only a few femto liters or even less is employed [24–27]. The detection volume is usually defined by the intersection of the focused excitation laser and the image of a spatial filter. The reduction of the detection volume improves the signal-to-background ratio by many orders of magnitude without measurable photodestruction of the dye molecules under study. Poisson statistics predicts, for a concentration of less than 10^{-10} M, that the number of molecules fluctuates predominantly between 0 and 1 in an applied detection volume in the femto liter region. Hence, the probability of two or more molecules being present in the detection volume simultaneously is negligible at concentrations below 10^{-10} M. On the other hand, a small detection volume reduces the transition time, i.e. the measurement time. During the transition through the detection volume, the dye molecule is excited by the laser light from the ground state S_0 into high-lying vibrational levels of the first excited state S_1 which undergoes rapid nonradiative internal conversion to low-lying S_1 levels. The optical saturation limit is the maximum rate that a dye molecule can be cycled between S_0 and S_1 and is dependent on the fluorescence lifetime of the dye τ_f. Besides irreversible photodestruction several depopulation pathways such as intersystem crossing into the triplet state compete with the fluorescence emission thus reducing the number of emitted photons. Since no photon is obtainable during the triplet lifetime of the excited dye, the triplet quantum yield and lifetime is a very important photophysical constant for single-molecule experiments. In other words, only those fluorescent dyes which exhibit very small triplet yields and short triplet lifetimes are suited for single-molecule experiments. Therefore, most measurements on the single-molecule level in solution were performed with rhodamine dyes. Additionally, rhodamine dyes emit up to 10^5 photons before photodestruction, even in aqueous solutions [15–18].

Another technique for efficiently suppressing the background was pioneered by Keller and co-workers, who introduced time-gated detection techniques to eliminate Rayleigh and Raman scattering which appears simultaneously with the excitation pulse [15, 16, 28]. However, for an application of these very sensitive techniques in analytical chemistry, individual labeled analyte molecules have to be identified during their stay in the detection volume. In order to identify dye molecules in solution both spectral- and time-resolved detection methods can be used.

2
Time-Resolved Identification of Individual Dye Molecules

In practice, it is often necessary to identify different analytes in a given sample. Soper et al. [29] first reported the measurement of spectroscopic properties in solution at the single-molecule level. In these experiments, the differences in the spectral emission properties of individual molecules of Rhodamine 6G and Texas red in combination with two separate detectors were used to distinguish the two dyes. To distinguish different fluorescent dyes or to identify tagged analyte molecules, it is not necessary to collect as many photons as are necessary for an exact lifetime measurement with the time-correlated single-photon counting (TCSPC) technique but only a small fraction of it. The problem then consists of comparing raw data with certain well-defined patterns of the fluorescence decays measured with high precision. Hence, from a statistical point of view, one is confronted with a problem of classification. An optimal way of classification has already been developed in the framework of information theory and is described in detail elsewhere [30–34]. For the time-resolved identification of single dye molecules in solution an important question arises: How many photons are necessary to distinguish two or more dye labels with different lifetimes? A first approximation for the probability of misclassification can be obtained by Cramérs inequality [31]. By this relationship, an upper threshold for the probability of an erroneous classification of a dye molecule of type j as a dye molecule of type h is given (Eq. 1), where $p_i(j)$ are the probabilities of finding a photon in channel i, $(p_i(j) = \exp(-iT/\tau_j)/\sum \exp(-iT/\tau_j))$. N gives the total number of photons, k the total number of channels.

$$P(h/j) \leq \left[\min_{0<v<1} \sum_{i=1}^{k} p_i(h)^v p_i j^{1-v} \right]^N \tag{1}$$

For example, at a given number of 50 collected photons per single-molecule burst within a time-window of 10 ns (64 channels at 0.156 ns), an error probability of less than 17% is expected for two dyes with fluorescence lifetimes of $\tau_1 = 2.0$ ns and $\tau_2 = 4.0$ ns. Hence, in combination with an efficient maximum likelihood estimator (MLE) algorithm [35] (Eq. 2), the identification of different dye molecules on the single-molecule level should be possible due to their characteristic fluorescence lifetimes.

$$1 + (e^{T/\tau} - 1)^{-1} - m(e^{mT/\tau} - 1^{-1} = N^{-1} \sum_{i=1}^{m} iN_i \tag{2}$$

In Eq. 2, T is the width of each time channel, m the number of time channels, and N_i the number of photocounts in time channel i.

Zander et al. [36] first demonstrated the identification of single rhodamine dye molecules in water by their characteristic fluorescence lifetimes of 1.79 ± 0.33 ns (Rhodamine B) and 3.79 ± 0.38 ns (Rhodamine 6G) excited by a frequency doubled titanium/sapphire laser emitting at 514 nm. This was achieved by using confocal microscopy in combination with the technique of time-correlated single-photon counting. Using a modified efficient MLE-algorithm the classification probability was $> 90\%$ with 100 collected photons per burst.

These improvements have led to the observation and identification of various individual fluorescent dye and fluorescently labeled analyte molecules in solution using different excitation sources [18–20, 37–39]. Furthermore, the new techniques are ideally suited for biomedical multiparameter analysis. One of the most popular multiparameter applications is the rapid sequencing of large DNA fragments upon the detection and identification of single, fluorescently labeled nucleotides cleaved from a single DNA fragment [17, 18, 40, 41]. Nucleic acid fragment sizing [28], hybridization between primer and DNA targets [22], between peptides and receptors [21], as well as between antibodies and antigens [23], can, however, also be monitored on the single-molecule level. Unfortunately, with the implication of these sensitive techniques in biological analysis the problem arises that no matter how carefully the solvents for single-molecule experiments are purified, the background due to luminescent impurities sets detection limits [42].

3
Multiplex Dye Principle and Background Reduction

During the last few years, rhodamine dyes have been the fluorochromes of choice for single-molecule experiments. They are easily excitable by argon ion, frequency doubled Nd/YAG, and titanium/sapphire lasers in the green wavelength region. Owing to their structural rigidity, rhodamine dyes show high extinction coefficients, high fluorescence quantum yields and photostability, even in aqueous surroundings. Furthermore, the low triplet yield ($< 1\%$) and lifetime of approximately 4 µs are ideally suited for single-molecule experiments.

Although most of the background luminescence can be removed by the use of small detection volumes and appropriate optical filter systems, bursts from impurities which are excitable, especially in the green wavelength region, can prevent the definite detection and identification of single analyte molecules. Therefore, most of these ultrasensitive analytical techniques are restricted to purified solvent systems.

Due to the limited number of compounds which show intrinsic absorption and emission above 600 nm, the use of near-infrared (NIR) fluorescence detection is a desirable alternative. This fact, as well as the availability of low-cost diode lasers in this spectral region, has prompted current efforts to use NIR dyes for ultrasensitive detection methods. Using a titanium/sapphire laser as the excitation source, the observation of photon bursts from individual carbocyanine molecules (IR-132) in methanol has been reported [43]. Unfortunately, single-molecule detection of NIR dyes in pure water is hampered by the very low fluorescence quantum yield of most known NIR dyes in aqueous solution [44]. However, new dyes have been developed [45–47] which are easily excitable by diode lasers but which also exhibit absorption and emission wavelengths short enough to achieve high fluorescence quantum yields and lifetimes in the ns-range in aqueous surroundings. These dyes which show similar absorption and emission characteristics, but exhibit well distinct fluorescence lifetimes, are called multiplex dyes [48–50]. The multiplex dyes absorb in the wavelength range 630–670 nm and are most suitable for ultrasensitive detection techniques in combination with pulsed diode laser excitation. Recently [51], the first one-lane four-dye DNA sequencing in capillary gel electrophoresis (CGE) based on time-resolved fluorescence detection and identification of multiplex dyes with semiconductor technology has been successfully demonstrated. Time-resolved measurements of various new multiplex dyes in ethylene glycol using a pulsed laser diode and identification by a maximum likelihood (MLE) algorithm demonstrated that 40 collected photons per transit of a single dye molecule are sufficient to distinguish two rhodamine derivatives with an error probability of misclassification of less than 1% via their characteristic fluorescence lifetimes of 3.61 ± 0.45 ns (JA167) and 1.41 ± 0.30 ns (Cyanorhodamine B) [38].

4
Apparatus for Time-Resolved Identification of Individual Dye Molecules with Pulsed Diode Lasers

Figure 1 presents the confocal setup used for single-molecule detection (SMD) with semiconductor technology. In our system a pulsed diode laser (640 nm) served as the excitation source. The pulsing of the diode was performed by a self-matched tunable impulse generator. This system provided pulses of less than 400 ps (FWHM) duration with a repetition rate of 56 MHz. The beam was collimated, passed an excitation filter (639DF9; Omega Optics, Brattleboro, VT) and entered a conventional microscope (Axiovert 100TV; Zeiss, Germany) through the back port. The beam was coupled into an oil-immersion objective

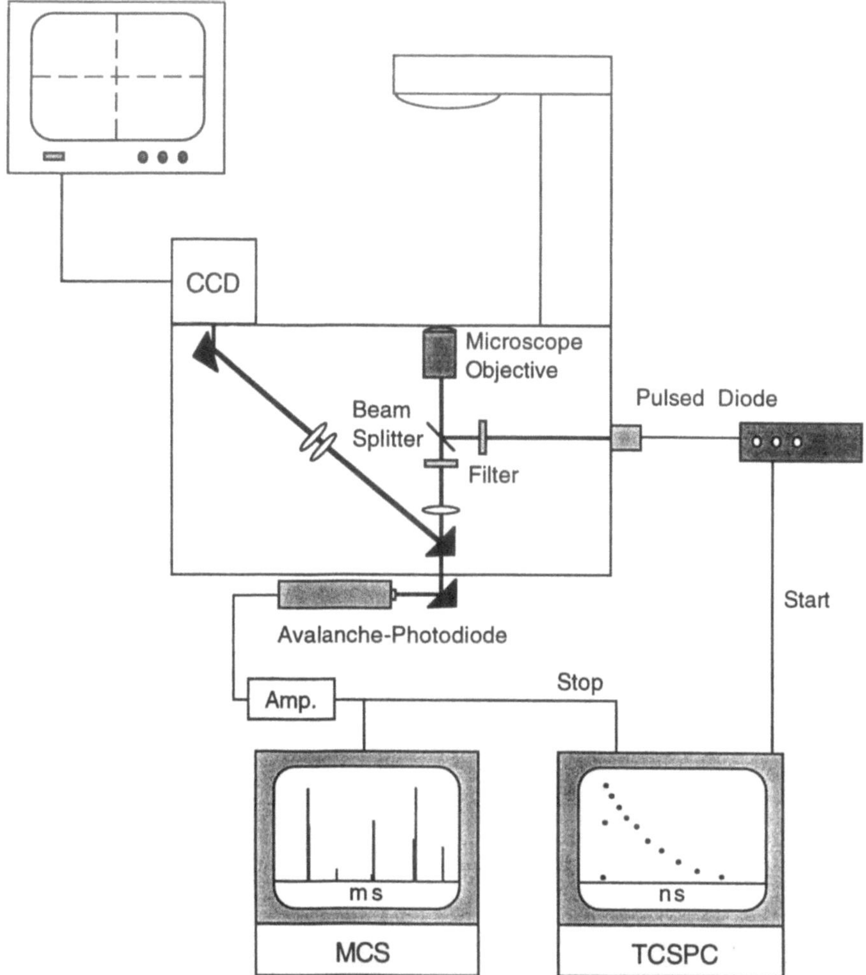

Fig. 1. Schematic diagram of optical and electronic apparatus employing confocal imaging. Two different data acquisition modes are used simultaneously. The MCS system produces a trace of the accumulated photons per bin (1 μs-20 ms). The SPC-330 card manages the time-correlated single-photon counting data with acquisition times of 625 μs for each fluorescence decay

$(100 \times NA = 1.4$; Olympus, Tokyo, Japan) by a dichroic beam splitter (645DRLP; Omega Optics, Brattleboro, VT). The maximum average power of the laser diode at the sample was 1.3 mW. The fluorescence signal was collected by the same objective, filtered by two band-pass filters (685HQ70; AF Analysentechnik, Tübingen, Germany) and imaged either through a CCD-camera for adjustment of microcapillaries or through the TV outlet on the bottom of the microscope onto a 100 μm or 150 μm pinhole directly in front of the avalanche-photodiode (SPAD; EG&G Optoelectronics, Canada).

The signal was amplified and split for a home-built PC-card-counter working as a multi-channel scaler (MCS) with a minimal integration time of 1 μs per bin, and a time-correlated single-photon (TCSPC) PC-interface card (SPC-330; Becker & Hickl, Berlin, Germany) to acquire time-resolved data. With the PC-card SPC-330 the signal was collected in up to 128 histograms with a minimal integration time of 625 μs each. Every event (burst) monitored by the MCS produced a decay trace collected by the PC-card SPC-330. The instrument response function (IRF) of the entire system was measured to be 420 ps (FWHM).

The raw data are presented without modification by any filtering algorithm. The fluorescence lifetimes of bulk solutions were also measured with this setup at concentrations of 10^{-8} M with subsequent deconvolution from the scatter and mono- or multiexponential fits. The solutions were prepared by diluting stock solutions with the appropriate amount of solvent down to the required concentration. The samples were transferred onto a microscope slide with a small depression and covered with a cover glass.

5
Time-Resolved Identification of Individual Labeled Mononucleotide Molecules in Water

Figure 2 shows the molecular structures of the three labeled mononucleotides Cy5-dCTP, MR121-dUTP, and Bodipy-dUTP, which are well suited for single-molecule experiments with pulsed diode laser excitation at 640 nm in water.

Due to their similar absorption and emission characteristics but unequivocal fluorescence lifetimes of 1.05 ns (Cy5-dCTP), 1.85 ns (MR121-dUTP) and 4.05 ns (Bodipy-dUTP) in aqueous solutions, they satisfactorily fulfill the re-

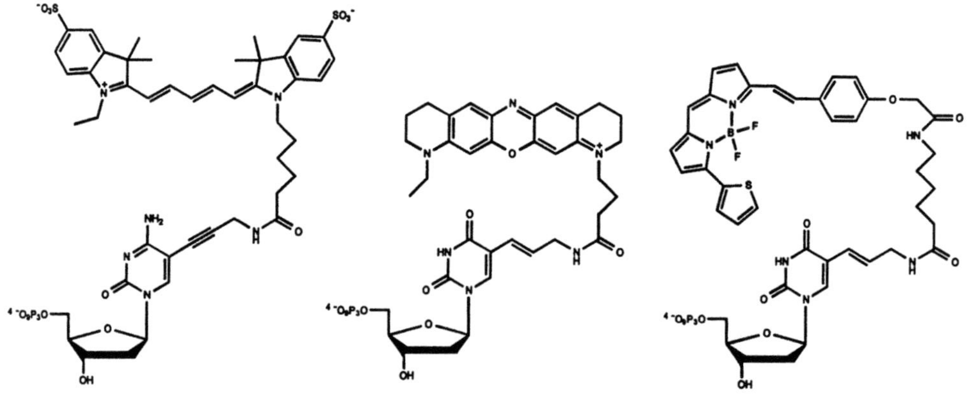

Cy5-dCTP   MR121-dUTP   Bodipy-dUTP

Fig. 2. Molecular structures of the labeled mononucleotides Cy5-dCTP, MR121-dUTP, and Bodipy-dUTP. The three labeled mononucleotides can be easily excited by a pulsed diode laser emitting at 640 nm and detected behind the same band-pass filter with comparable efficiency. Because the fluorescence lifetimes of 1.05 ns (Cy5-dCTP), 1.85 ns (MR121-dUTP), and 4.05 ns (Bodipy-dUTP) measured in bulk solutions are well distinct in water these nucleotides are ideally suited for testing the multipex dye principle

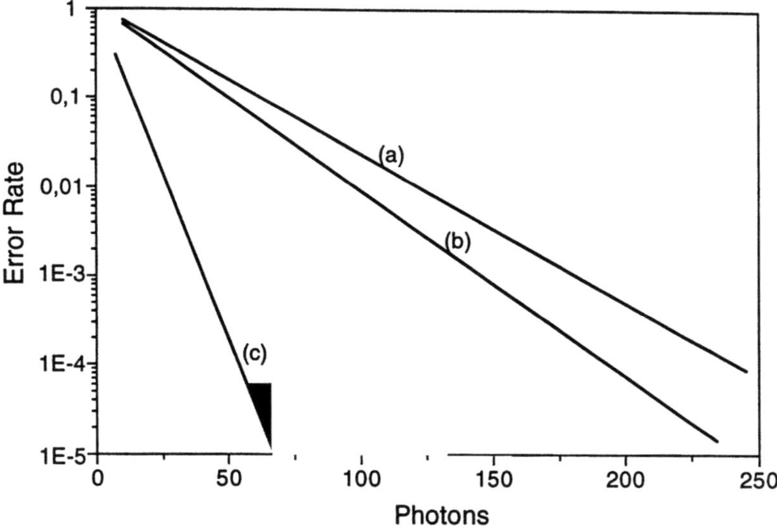

Fig. 3. Theoretical prediction of the error rate calculated by Eq. 1 for the identification of (a) Cy5-dCTP and MR121-dUTP molecules, (b) MR121-dUTP and Bodipy-dUTP molecules, and (c) Cy5-dCTP and Bodipy-dUTP molecules for increasing numbers of detected photons

quirements of multiplex dyes. Figure 3 shows the theoretically predicted error rate of misclassification between (a) Cy5-dCTP and MR121-dUTP, (b) MR121-dUTP and Bodipy-dUTP, and (c) Cy5-dCTP and Bodipy-dUTP for single-molecule identification for increasing number of detected photoelectrons per burst. On the basis of Fig. 3 an identification of individual mononucleotides with 50 collected photons per burst should be possible with an error rate of less than 20%. On the other hand, using two different mononucleotides dCTP and dUTP labeled with Cy5 and Bodipy, an unequivocal identification (error probability of 10^{-4}) should be possible, even with only 50 collected photons (Fig. 3c).

In Figure 4 the time-dependent fluorescence signals (1 ms/bin) observed from a 10^{-11} M aqueous solution of Cy5-dCTP with fluorescence burst rates of up to 150 photons in 1 ms (150 kHz) are shown. With our experimental setup an average background level of 1 kHz (upper part in Fig. 4) was obtained in pure water, arising mainly from Raman scattered photons passing the emission filter simultaneously with the excitation laser pulse. With this background we calculate a signal-to-background ratio of more than 100 for the most intense peaks of Cy5-dCTP in 1 ms.

Fluorescence bursts of 10^{-11} M aqueous solutions of MR121-dUTP and Bodipy-dUTP exhibit comparable signal-to-background ratios. As expected the background count rate of water is drastically reduced by excitation in the red region at 640 nm. Hence, we can use pure tap-water for our single-molecule experiments. From MCS-traces measured with smaller integration times we calculated characteristic diffusion times of approximately 300 μs for Cy5, MR121, and Bodipy labeled mononucleotides through the detection volume of

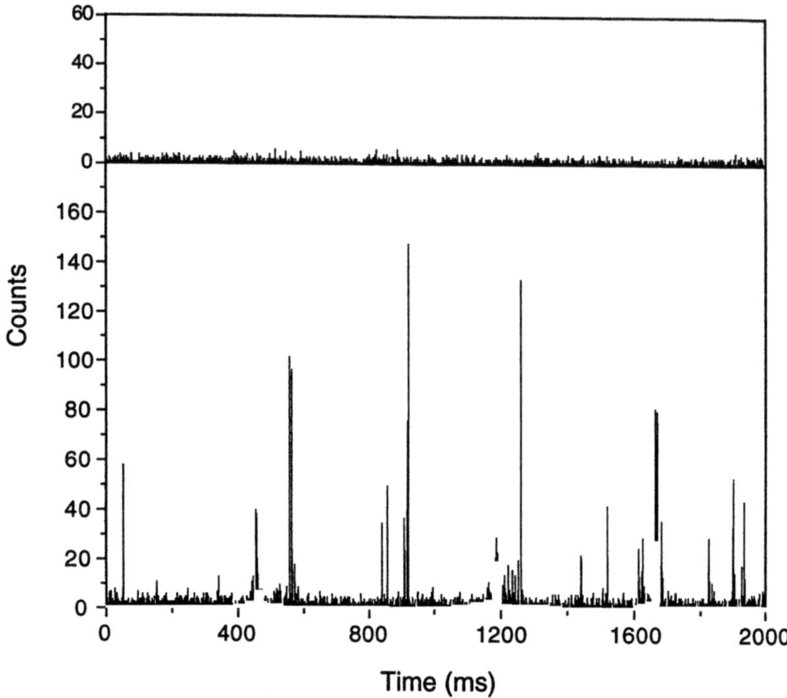

Fig. 4. Fluorescence signals observed from a 10^{-11} M aqueous solution of Cy5-dCTP at an average power of 630 µW at the sample. The most intense peaks reach 160 counts/ms. In the upper section the MCS trace of pure water with a background rate of 1 kHz is shown

a few femto liters. At this point it should be noted that the detection of single mononucleotide molecules is caused by their Brownian motion through the detection volume. Therefore, dependent of their trajectory, respectively, different burst sizes are detected. The highest bursts result from molecules with longer transition times through the center of the detection volume, where the laser intensity reaches its maximum value.

As can be seen in Fig. 5a if a single labeled mononucleotide molecule passes the detection volume, up to 200 counts are collected in the applied detection window of 10 ns of our TCSPC-system (64 channels at 0.156 ns). In most of the detected bursts we collected between 40 and 150 photoelectrons. However, there are also isolated bursts which contain even more than 200 photoelectrons. These bursts probably result from events where more than a single mononucleotide passes the detection volume simultaneously. Interestingly, some bursts were found which contained more than 100 photoelectrons but exhibited fluorescence lifetimes which could not be attributed to the used dye, respectively. Although the labeled conjugates were purified by HPLC, we assume that small modifications in the structure of some dye chromophores are responsible for the observed effect. We measured more than 50 time-resolved decays for individual Cy5-dCTP, MR121-dUTP, and Bodipy-dUTP molecules

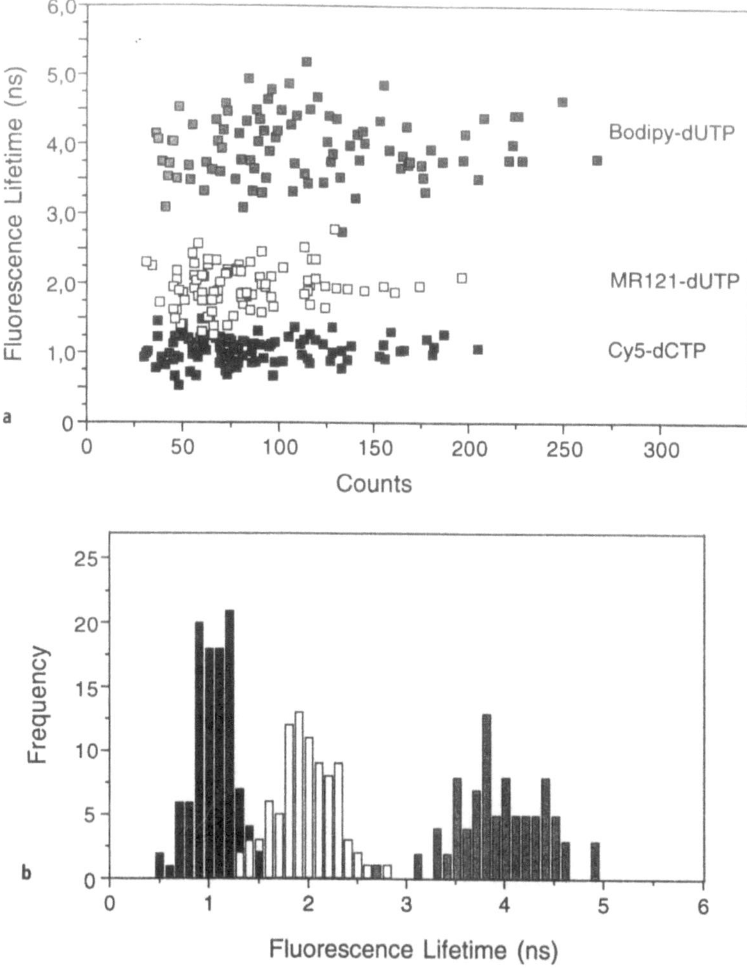

Fig. 5a, b. **a** Fluorescence lifetimes obtained by the MLE-algorithm for single Cy5-dCTP, MR121-dUTP, and Bodipy-dUTP molecules in water vs. the number of detected photons per burst. **b** Histogram of the determined lifetimes of single Cy5-dCTP, MR121-dUTP, and Bodipy-dUTP molecules in water. The fluorescence lifetimes measured on single-molecules of 1.05 ± 0.19 ns (Cy5-dCTP), 1.96 ± 0.30 ns (MR121-dUTP), and 3.93 ± 0.45 ns (Bodipy-dUTP) are in good agreement with the fluorescence lifetimes measured in bulk solutions of 1.05 ns, 1.85 ns, and 4.05 ns, respectively

and calculated the lifetimes using the MLE (Eq. 2). For these lifetimes the standard deviation σ_{exp} is derived from the distribution of lifetimes shown in Fig. 5b. Figure 5 demonstrates that three differently labeled mononucleotides can be identified in water at the single-molecule level due to their characteristic fluorescence lifetime of 1.05 ± 0.19 ns (Cy5-dCTP), 1.96 ± 0.30 ns (MR121-dUTP), and 3.93 ± 0.45 ns (Bodipy-dUTP). As predicted in Fig. 3, 50 collected photons per burst are sufficient to distinguish at least three differently labeled mononucleotides with an error probability of less than 20 % at the single-molecule level in water. By using an additional dye with a fluorescence lifetime of approximately 0.5 ns this technique can be applied successfully for time-resolved DNA sequencing strategies on the single DNA strand and other multi-parameter tests.

6
Potential for Single-Molecule DNA Sequencing in Microcapillaries

Keller's group in Los Alamos has proposed a very clever method for DNA sequencing on the single-molecule level [16–18, 40–42, 52]. The principle idea involves the incorporation of fluorescently labeled mononucleotides in a growing DNA strand, attachment of a single labeled DNA strand to a support (generally latex beads), movement of the supported DNA into a flowing sample stream and detection and identification of the individual fluorescently tagged DNA bases as they are cleaved from the DNA strand by an exonuclease enzyme. The DNA sequence is determined by the order in which differently labeled nucleotides are detected and identified. While the identification of four different fluorescence lifetimes on the single-molecule level will be possible in combination with the appropriate labels, the current bottleneck appears to be the enzymatically incorporation of four differently labeled nucleotides. However, the incorporation of two labeled nucleotides in a growing strand of M13 DNA seems to be possible [53]. No problems were observed with solubility and base-pairing. We assume that the linker length between the dye and the nucleotide, as well as the hydrophobicity of the chromophore, control the enzymatic acceptance and the degree of misincorporation during the replication. Unfortunately, most known suited dyes for single-molecule experiments exhibit a hydrophobic nature. Therefore they tend to form aggregates and complexes with DNA nucleotides and proteins, especially in aqueous surroundings [45–47, 54]. Hence, for a successful incorporation of labeled mononucleotides during enzymatic replication the used chromophores should be as hydrophilic as possible in combination with an optimized linker arm in order to avoid re-folding of the dye on the base.

On the other hand, the replication of the template with only two differently labeled nucleotides is sufficient to get the whole sequence information by changing the labeled nucleotides in a second experiment. Recently [47, 54, 55], it was suggested to label the four nucleotides with the same label and to distinguish between the bases by the fluorescence lifetimes of the attached dye due to different dye-nucleotide interactions. However, if the chromophore is influenced by the nucleotide, i.e. in close contact with the nucleotide, the enzymatic acceptance might deteriorate.

If a single labeled DNA strand attached to a support is available, in the next step all enzymatically cleaved nucleotides have to be detected and identified, i.e. to ensure that each molecule is delivered through the center of the detection volume for efficient excitation. Unfortunately, here a drawback is associated with the use of femto liter detection volumes because the diameter of the

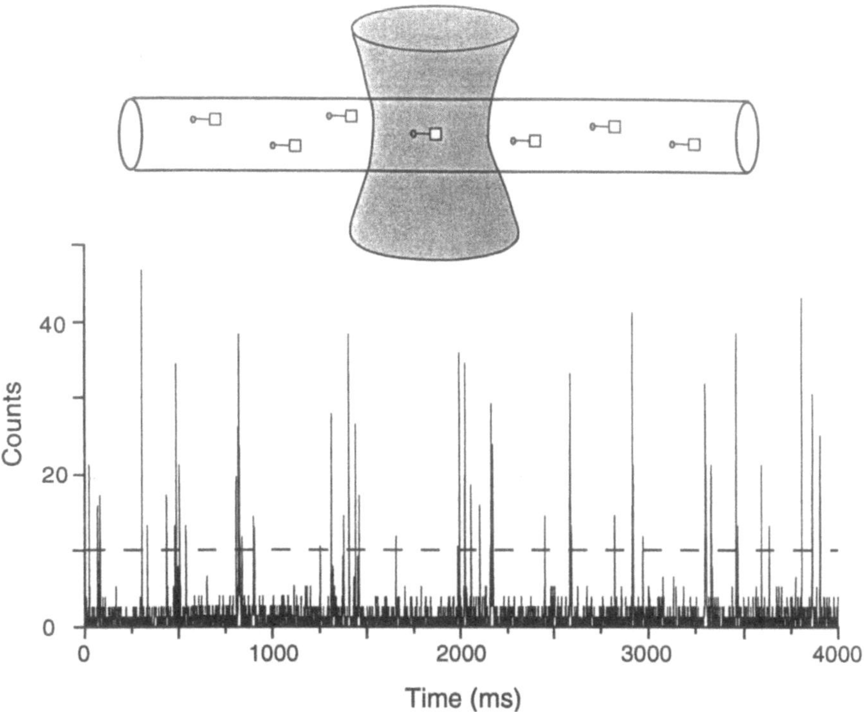

Fig. 6. Fluorescence signals observed from a Cy5-dCTP mononucleotide solution 10^{-11} M in 10 mM borate buffer in a microcapillary (i.d. approximately 500–600 nm) with 1 ms integration time per bin. Pure borate buffer exhibits a background rate of 800 Hz with small bursts of up to 7 counts/ms resulting from impurities in the buffer

detection area is much smaller than the diameter of the sample stream. Hence, microcapillaries or -channels with inner diameters smaller than the detection area ($<1\,\mu$m) have to be used [56–58] (Fig. 6). Additionally, the use of such channels makes it possible to manipulate the motion of single molecules by electrokinetic, electroosmotic or capillary forces.

On the other hand, the use of microchannels or -capillaries for confocal fluorescence microscopy of single-molecules is associated with other problems. While the refractive index differences on the outer walls can be matched by the use of the appropriate index-matching oil, the refractive index differences at the capillary inner wall and deviations of the beam profile generally result in higher background rates and smaller photon bursts.

The lower part in Fig. 6 shows the time-dependent fluorescence signals observed from a 10^{-11} M aqueous solution of Cy5-dCTP in a microcapillary with an inner diameter of approximately 500–600 nm released by electro-osmotic forces. In this experiment all molecules have to pass the center of the detection volume. Therefore the resulting burst size distribution (Fig. 7a) is peaked away from zero to approximately 30 counts per burst. As can be seen in Fig. 7a there are also some bursts which contain more than 55 photocounts.

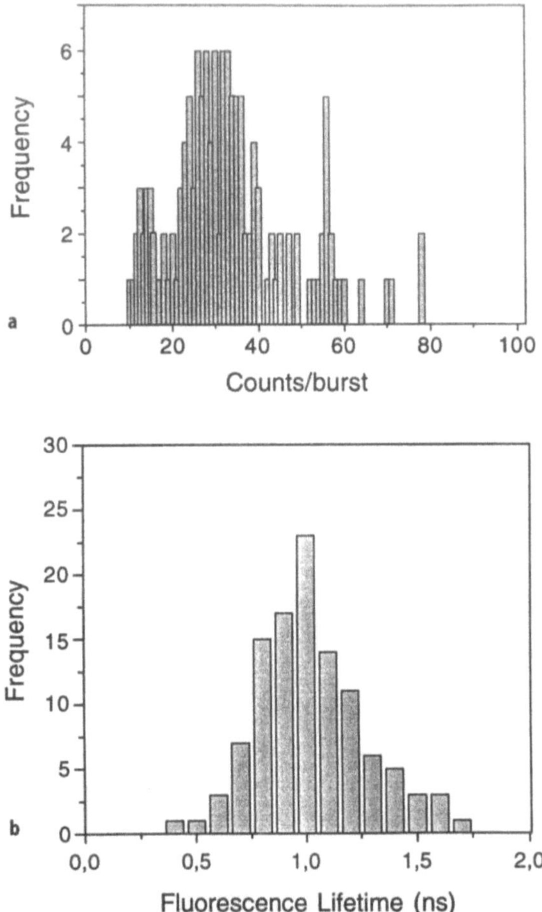

Fig. 7a, b. **a** Burst size distribution for Cy5-dCTP molecules observed in a microcapillary. As shown in Fig. 6 all events containing more than 10 photocounts per burst were used for preparation of the statistic. **b** Histogram of the determined lifetimes of individual Cy5-dCTP molecules in 10 mM borate buffer measured in the microcapillary

Probably these bursts result from events where two single mononucleotides pass the detection volume simultaneously during the integration time of 1 ms/bin. Nevertheless, the histogram in Fig. 7b demonstrates that an identification of individual mononucleotides due to the characteristic fluorescence lifetime of the attached dye of 0.98 ± 0.21 ns (Cy5-dCTP) is possible, even in microcapillaries with an inner diameter smaller than the detection area.

Although non-specific adsorption of mononucleotides on the capillary wall and the effect of electrostatic forces on single-molecule motion have to be carefully investigated and optimized, time-resolved identification of individual labeled analyte molecules in microcapillaries with inner diameters smaller than 1 μm can be useful for the DNA sequencing scheme on the single-molecule level.

7
Detection of Individual Tumor Marker Molecules in Neat Human Serum

As pointed out above, the background due to luminescent impurities sets detection limits for the application of single-molecule detection (SMD) techniques. In bioanalytical samples containing buffers, enzymes or biological extracts, in particular, luminescent impurities can decrease the sensitivity or even prevent the definite detection of individual labeled analyte molecules. Additionally, for applications in biological tests it would be desirable to detect and quantify analyte molecules on the single-molecule level in their natural environment, i.e. in a simplified clinical procedure. Furthermore, the technique should not require the separation of bound from free markers and should be insensitive to background sources naturally present in biological samples.

Using fluorescence correlation spectroscopy (FCS), the binding of fluorescently labeled antibodies or antigens has been successfully demonstrated with analyte concentrations in the nanomolar range [59]. The clinical standard antibody tests utilize enzyme-linked immunosorbent assays (ELISA) which achieve a sensitivity of a few nanogram antigen per milliliter corresponding to a concentration of 10^{-9} to 10^{-11} M. In all of these methods the analyte molecules have to be amplified after isolation from the serum by binding to specific materials, or different washing-steps are required. Measurements in human serum, e.g. direct detection and monitoring of individual labeled antibodies and their reactions with antigens, are complicated by the fact that besides inorganic ions and hydrophilic organic substances approximately 100 different proteins are present. Therefore human serum samples are strongly luminescent when irradiated in the blue or green region of the spectrum. Hence, detection of single chromophores, such as fluorescein or classical rhodamine dyes, is prevented in this environment.

In view of these limitations we tested diode laser based detection techniques in combination with the multiplex dye principle for the detection and identification of individual antigen molecules in human serum [23]. Applying a new rhodamine derivative JA169 (Fig. 8a) and the carbocyanine dye Cy5 the direct identification of individual antibody molecules (JA169-BM-7, Cy5-BM-7) and their specific binding to multiple epitopes of tumor associated glycoprotein molecules (MUC1) [60] could be obtained, even in undiluted human serum samples. For preparation of the labeled antibodies the dyes JA169 and Cy5 were converted into their N-hydroxysuccinimide esters. The activated dyes were added to monoclonal antibodies BM-7 in phosphate-buffered saline (PBS) pH 8.3 with different dye/protein ratios. Unbound dye was removed by gel filtration. The purification of the conjugates was performed by FPLC. The average number of dye molecules bound to an antibody was estimated spectroscopically. Purification and selection of the labeled antibody fractions are very important for time-resolved identification techniques due to the dependence of the measured fluorescence lifetime on the degree of labeling. While those antibody fractions which exhibit a dye/protein ratio greater than 2:1 exhibit higher single-molecule fluorescence bursts, the measured fluorescence lifetimes on the SMD-level show a broader distribution. Hence, the measurements were performed with antibody fractions which exhibit a dye/protein ratio of 1:1.

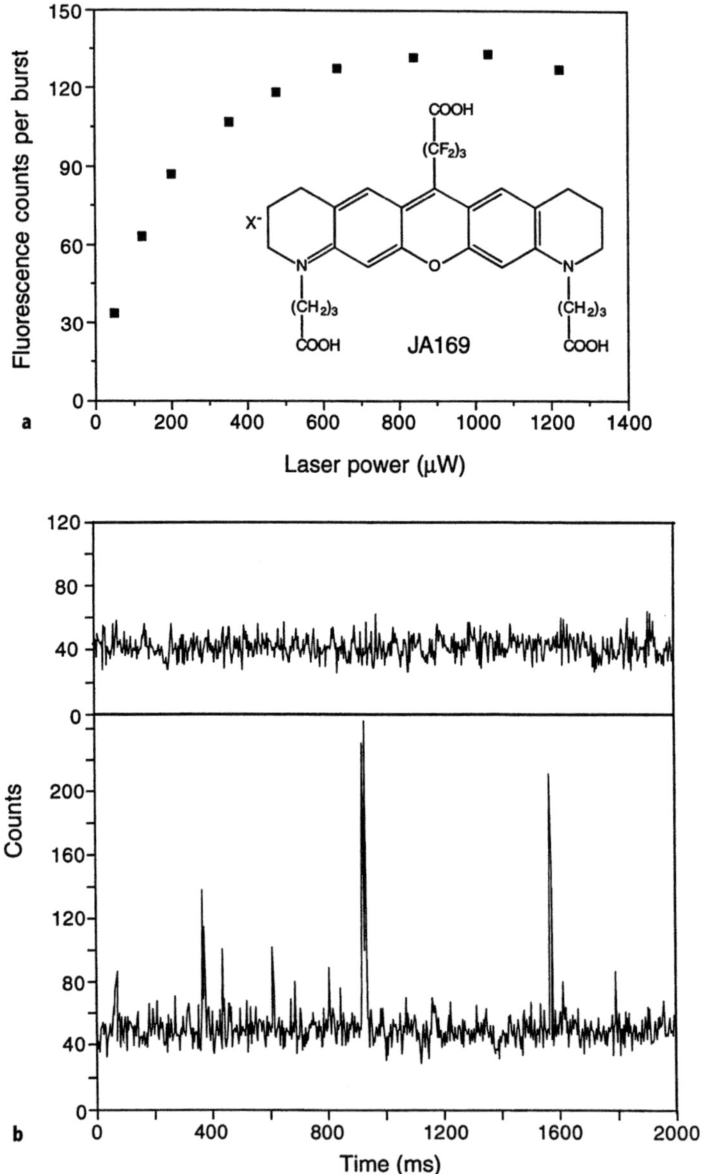

Fig. 8. a Fluorescence counts per single-molecule burst as a function of the average laser power at the sample for JA169 labeled BM-7 antibody molecules 5×10^{-11} M in PBS (pH 7.4, 1% bovine serum albumin, BSA). The measured fluorescence count rates are averaged values of 50 detection events. To obtain as many fluorescence photons as possible without photo-destruction of the individual labeled antibody molecules all measurements were carried out at 400 µW. **b** Fluorescence signals observed from a 10^{-11} M solution of JA169 labeled antibody molecules BM-7 in neat human serum at an average excitation power of 400 µW at the sample (2 ms integration time per bin) using pulsed diode laser excitation at 637 nm. In the upper section the MCS trace of neat human serum is shown

As indicated by the MCS-trace in the upper part of Fig 8b, human serum exhibits an almost burst-free background with count rates of approximately 20 kHz if the excitation is performed above 630 nm, i.e. there is a quasi continuum of photons which arises from relatively high concentrations of weakly luminescent or quenched compounds. In the lower part of Fig. 8b the time-dependent fluorescence signals of individual BM-7 antibody molecules labeled with JA169 in human serum are shown. The mean molecular transit time through the detection volume of approximately 20 femto liter was estimated to be 3–4 ms for both the Cy5 and the JA169 labeled antibody molecules. During this time we got up to 400 photocounts per individual antibody molecule with an average excitation energy of 400 µW at the sample. This result clearly demonstrates that even in neat human serum a real-time detection of individual antibody molecules is possible if the appropriate label in combination with a pulsed diode laser emitting above 635 nm is used.

In Fig. 9 a typical time-resolved photon burst of a 10^{-11} M solution of Cy5 labeled antibody molecules BM-7 in neat human serum is shown. Using an

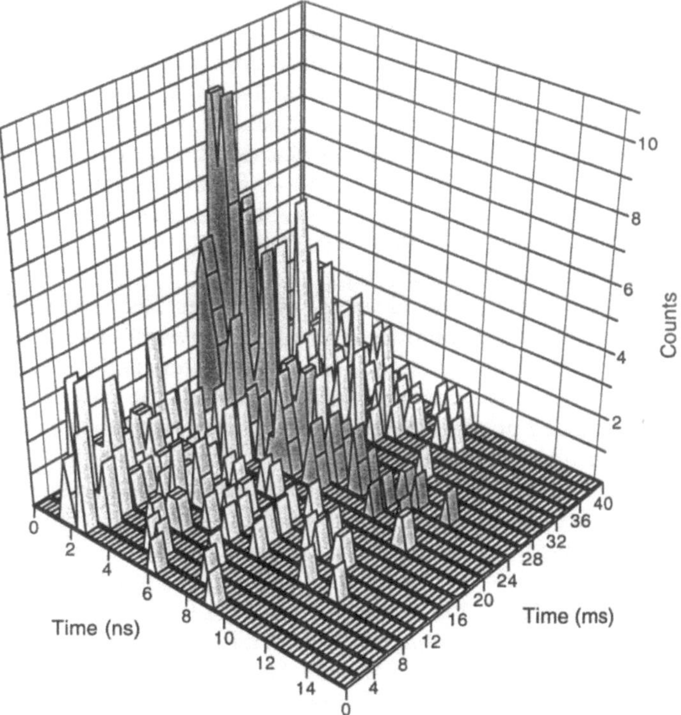

Fig. 9. Fluorescence decay curve of an individual Cy5 labeled BM-7 antibody molecule (10^{-11} M) in neat human serum with 2 ms integration time per decay using the PC-card SPC-330 (64 channels, 0.2 ns/channel). The 3D plot shows the time-resolved signals measured during the 2 ms integration time intervals. During the first 20 ms after the start of the experiment the decay curves of pure human serum are shown. The passage of a Cy5 labeled BM-7 antibody molecule between 20 and 26 ms is clearly indicated by the measured fluorescence decay

integration time of 2 ms per decay and an antibody concentration of 10^{-11} M the probability of finding more than one event in a measured decay is negligible. As can be judged by comparing the MCS-trace in Fig. 8b with the number of background events collected per 2 ms in Fig. 9, the signal-to-background ratio is further increased by an applied detection window of only 13 ns (64 channels at 0.2 ns). We achieved a signal-to-background ratio of over 10 in the time-resolved experiments.

In order to identify the individual labeled antibody molecules in neat human serum the luminescence decay of the pure serum has to be taken into account. Three different serum samples were measured which exhibit a multiexponential fluorescence decay which can be described by a biexponential model with a short lifetime τ_1 of 700 ps ($a_1 = 85-95\%$) and a longer lifetime τ_2 between 3.2 and 3.8 ns ($a_2 = 5-15\%$). The relative amplitudes of the components vary with different patient serums. Hence, we considered the luminescence decay of human serum by a biexponential fit with a constant short lifetime of $\tau_1 = 700$ ps and a variable lifetime τ_2 to describe the measured fluorescence decays. By applying this technique an identification of differently labeled antibody molecules is possible due to the different values of the lifetime τ_2 which describes the fluorescence lifetime of the attached label. The histogram in Fig. 10 was obtained from separate experiments in human serum containing only one class of labeled antibody molecules (Cy5-BM-7 or JA169-BM-7).

Taking only those decays for the construction of the histogram where $a_2\tau_2$ was greater than 0.7, the error of misclassification in this experiment was less than 10%. If we fit the data from Fig. 10 to a Gaussian, we obtain mean fluorescence lifetimes of 1.6 ns and 2.8 ns which is in good agreement with the data measured in bulk aqueous solutions of 1.5 ns and 2.7 ns for Cy5 and JA169 labeled antibody molecules, respectively.

These results demonstrate the advantage of time-resolved diode laser based single-molecule detection techniques in combination with suitable dye molecules. Due to the reduced background at 637 nm excitation and the fluorescence lifetimes of the dyes of a few nanoseconds, a discrimination between the background luminescence and the analyte fluorescence is possible, even in biological systems such as human serum.

However, to screen for the early-stage development of breast or ovarian tumors the specific binding of labeled BM-7 antibody molecules to MUC1 antigen molecules has to be monitored on the single-molecule level. The complete monoclonal antibody system for detection of the high molecular weight glycoprotein mucine (MUC1) is described elsewhere [60, 61]. It consists of the antibody BM-7 (IgG1) which is directed against multiple epitopes of the tumor-associated MUC1 molecules. MUC1, which shows increased concentrations in serum in the patients with breast or ovarian cancer, has a molecular weight between 360000 and 430000 Daltons and multiple epitopes for BM-7 antibodies. The defined normal value of a healthy person is in the range of 20 Units per ml, which is equal to a concentration of 3×10^{-10} M (1 Unit corresponds to 6 ng/ml). The conventional clinical method for tumor marker determination uses an ELISA test which exhibits a sensitivity limit of 7.5×10^{-11} M (5 Units/ ml). For an early-stage monitoring of tumor development it would be desirable to develop a

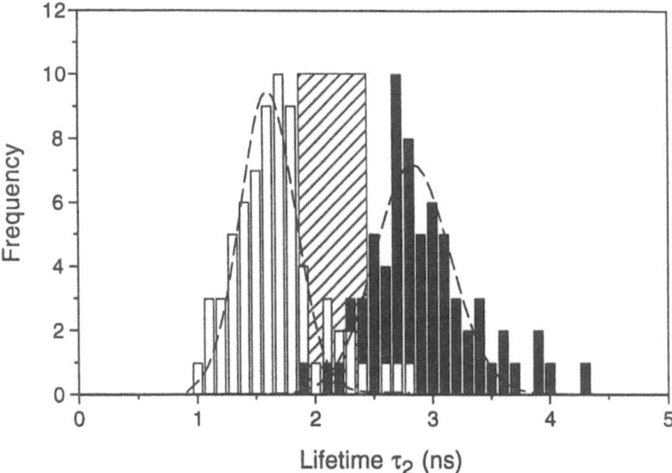

Fig. 10. Histogram of the lifetimes τ_2 (ns) of individual BM-7 antibody molecules labeled either with Cy5 or JA169 measured in human serum and corresponding Gaussian fits. The product of the fractional intensity a_2 and the lifetime τ_2 reaches values greater than 0.7 only if a labeled antibody molecule passes the detection volume. For pure human serum $a_2\tau_2$ was determined to be always smaller than 0.4 upon excitation at 637 nm. More than 60 data sets of both Cy5 and JA169 labeled BM-7 antibody molecules were analyzed. The average number of photons used for the calculation was in both cases approximately 120. As can be seen by the marked area, only about 20% of the measured single-molecule events exhibit a fluorescence lifetime τ_2 between 1.9 ns and 2.4 ns

fast method without several washing-steps which exceeds the sensitivity of the ELISA test. The use of FCS to monitor the specific binding of antibody molecules to the antigen in the concentration range below 10^{-11} M in neat serum samples is prevented by (a) the small changes in mass of the labeled antibody, (b) the high background signal of luminescent impurities, and (c) the low signal intensity, i.e. the low concentration range. Hence, we used the time-resolved fluorescence information by analyzing the fluorescence decay of individual analyte molecules. As can be seen in Fig. 10 single antibody molecules Cy5-BM-7 and JA169-BM-7 can be identified by the characteristic fluorescence lifetimes of the attached dyes. As can be seen from the marked area in Fig. 10 only about 20% of the measured single-molecule events exhibit a fluorescence lifetime τ_2 between 1.9 ns and 2.4 ns. On the other hand, a single mucin molecule can bind Cy5 and JA169 labeled antibody molecules BM-7 simultaneously. Hence, the time-resolved fluorescence detection of this individual antigen exhibits a fluorescence lifetime τ_2 which contains both lifetimes 1.6 ns and 2.8 ns. The biexponential fit of the decay results in a lifetime τ_2 of 1.9–2.4 ns (marked area of Fig. 10).

Figure 11 shows a titration of labeled antibody molecules with mucin monitored by time-resolved fluorescence detection of single-molecule events at different mucin concentrations in 1% BSA-PBS of pH 7.4. Each data point was calculated from approximately 100 single-molecule events collected during 2 min. With decreasing mucin concentration the portion of the single-molecule

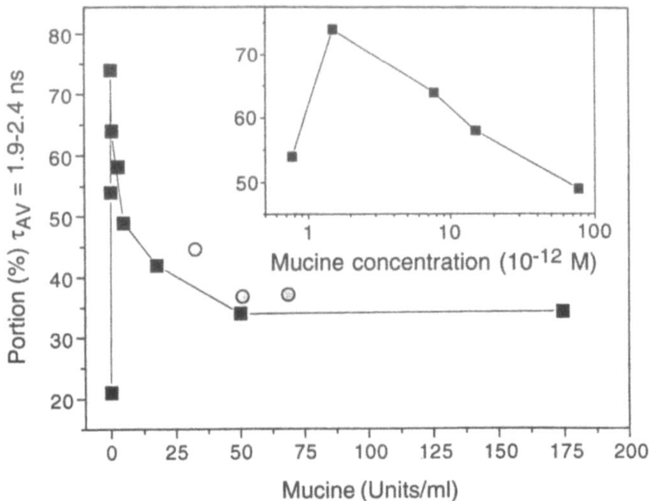

Fig. 11. The *rectangles* show the titration of a 5×10^{-11} M mixed solution of Cy5-BM-7 and JA169-BM-7 molecules in 1% BSA-PBS (pH 7.4) with mucin. The portion (%) of the measured fluorescence lifetime τ_2 between 1.9 ns and 2.4 ns vs. the mucin concentration shows a maximum at 1.5×10^{-12} M. The *circles* represent measured values in serum samples from different patients with mammary carcinomas. The amount of mucin in the serum samples was also measured with the ELISA technique. The inset shows an expanded view of the titration curve (0.05–5 Units/ml) and demonstrates the increased sensitivity of our technique compared to the detection limit of the conventional ELISA test of approximately 10^{-10} M MUC1 (corresponding to 5 Units/ml). For each rectangle or circle approximately 100 single-molecule events were analyzed

events which contain both fluorescence lifetimes 1.6 ns and 2.8 ns (fluorescence lifetime τ_2 between 1.9 ns and 2.4 ns) increases drastically if the mucin concentration is lower than the antibody concentration. The maximum of the titration curve in Fig. 11 is located at 0.1 Units of mucin per ml corresponding to a concentration of 1.5×10^{-12} M. This represents a sensitivity almost two orders of magnitude higher than the clinically used ELISA test.

At this mucin concentration each mucin molecule has bound approximately 33 antibody molecules. Further reduction of the antigen concentration results in a smaller portion of biexponential single-molecule events containing τ_2 between 1.9 ns and 2.4 ns because the epitopes of the tumor-associated MUC1 molecules are saturated by BM-7 antibody molecules. The circles in Fig. 11 represent values observed on serum samples from patients with a mammary carcinoma.

Unfortunately, the antibody BM-7 used presently in clinical ELISA tests reacts with both tumor-specific and non-specific epitopes of MUC1 molecules. Therefore, each serum sample exhibits mucin concentrations of at least 20 Units/ml serum and an early-stage diagnosis of tumor development is very difficult. On the other hand, until now a more sensitive antibody was not necessary because the pure sensitivity of the commonly used ELISA test prevented an early-stage diagnosis.

However, new antibodies are now under development which react only with tumor-specific epitopes of under-glycosilated MUC1 molecules which are not present in serum samples of a healthy person. Hence, the combination of the time-resolved technique with semiconductor lasers and microcapillaries presents a promising method for fast monitoring of early-stage development of breast and ovarian cancer in neat human serum. Just as antigen detection the described method allows the ultrasensitive detection and identification of specific DNA or RNA sequences by hybridization with two differently labeled primers.

Acknowledgements. The authors would like to thank the Bundesministerium für Bildung, Wissenschaft, Forschung und Technologie for financial support under grant 0310793A. Boehringer Mannheim GmbH is also gratefully acknowledged for the financial support and the generous disposal of the labeled mononucleotides MR121-dUTP and Bodipy-dUTP.

References

1. W.E. Moerner, L. Kador, Phys. Rev. Lett. 61, 2535, (1989)
2. L. Kador, D.E. Horne, W.E. Moerner, J. Phys. Chem. 94, 1237, (1990)
3. W.P. Ambrose, W.E. Moerner, Nature 349, 225, (1991)
4. W.E. Moerner, Th. Basché, Angew. Chem. 105, 537, (1993)
5. M. Orrit, J. Bernard, R. Personov, J. Phys. Chem. 97, 10256, (1993)
6. T. Plakhotnik, D. Walser, M. Pirotta, A. Renn, U. P. Wild, Science 265, 364, (1994)
7. E. Betzig, R.J. Chichester, Science 262, 1422, (1993)
8. W.P. Ambrose, P.M. Goodwin, J.C. Martin, R.A. Keller, Science 265, 364, (1994)
9. X. S. Xie, R.C. Dunn, Science 265, 361, (1994)
10. J.K. Trautman, J.J. Macklin, L.E. Brus, E. Betzig, Nature 369, 40, (1994)
11. T. Hirschfeld, Appl. Opt. 15, 2965, (1976)
12. D.C. Nguyen, R.A. Keller, J.H. Jett, J.C. Martin, Anal. Chem. 59, 2158, (1987)
13. K. Peck, L. Stryer, A.N. Glazer, R.A. Mathies, Proc. Natl. Acad. Sci. USA 86, 4087, (1989)
14. R.A. Mathies, K. Peck, L. Stryer, Anal. Chem. 62, 1786, (1990)
15. E.B. Shera, N.K. Seitzinger, L.M. Davis, R.A. Keller, S.A. Soper, Chem. Phys. Lett. 174, 553, (1990)
16. C.W. Wilkerson, P.M. Goodwin, W.P. Ambrose, J.C. Martin, R.A. Keller, Appl. Phys. Lett. 62, 2030, (1993)
17. P.M. Goodwin, R.L. Affleck, W.P. Ambrose, J.H. Jett, M.E. Johnson, J.C. Martin, J.T. Petty, J. A. Schecker, M. Wu, R.A. Keller, in Computer Assisted Analytical Spectroscopy; S.D. Brown, Ed.; Wiley and Sons: New York, Chapter 3, (1996)
18. P.M. Goodwin, W.P. Ambrose, R.A. Keller, Acc. Chem. Res. 29, 607, (1996)
19. R.D. Guenard, L.A. King, B.W. Smith, J. Winefordner, Anal. Chem. 69, 2462, (1997)
20. M.D. Barnes, C.-Y. Kung, W.B. Whitten, J.M. Ramsey, Anal. Chem. 69, 2115, (1997)
21. M. Eigen, R. Rigler, Proc. Natl. Acad. Sci. USA 91, 5740, (1994)
22. M. Kinjo, R. Rigler, Nucleic Acids Res. 23, 1795, (1995)
23. M. Sauer, C. Zander, R. Müller, B. Ullrich, K.H. Drexhage, S. Kaul, J. Wolfrum, Appl. Phys. B 65, 427, (1997)
24. R. Rigler, Ü. Mets, Proc. SPIE 1921, 239, (1992)
25. R. Rigler, Ü. Mets, J. Fluoresc. 4, 259, (1994)
26. S. Nie, D.T. Chu, R.N. Zare, Science 266, 1018, (1994)
27. S. Nie, D.T. Chu, R.N. Zare, Anal. Chem. 67, 2849, (1995)
28. A. Castro, E.B. Shera, Appl. Opt. 34, 3218, (1995)
29. S.A. Soper, L.M. Davis, E.B. Shera, J. Opt. Soc. Am. B 9, 1761, (1992)
30. S. Kullback, Information, Theory and Statistics, Wiley, New York, Chapter 6.4, (1959)

31. M. Köllner, Appl. Opt. 32, 501, (1993)
32. M. Köllner, J. Wolfrum, Chem. Phys. Lett. 200, 199, (1992)
33. M. Köllner, A. Fischer, J. Arden-Jacob, K.H. Drexhage, R. Müller, S. Seeger, J. Wolfrum, Chem. Phys, Lett. 250, 355, (1996)
34. J. Enderlein, P.M. Goodwin, A.V. Orden, W.P. Ambrose, R. Erdmann, R.A. Keller, Chem. Phys. Lett. 270, 464, (1997)
35. S.A. Soper, B.L. Legendre Jr., D.C. Williams, Anal. Chem. 67, 4358, (1995)
36. C. Zander, M. Sauer, K.H. Drexhage, D.-S. Ko, A. Schulz, J. Wolfrum, C. Eggeling, C.A.M. Seidel, Appl. Phys. B. 63, 517, (1996)
37. M. Sauer, C. Zander, K.H. Drexhage, J. Wolfrum, Chem. Phys. Lett. 254, 223, (1996)
38. R. Müller, C. Zander, M. Sauer, M. Deimel, D.-S. Ko, S. Siebert, J. Arden-Jacob, G. Deltau, N. J. Marx, K.H. Drexhage, J. Wolfrum, Chem. Phys. Lett. 262, 716, (1996)
39. M. Sauer, C. Zander, R. Müller, F. Göbel, A. Schulz, S. Siebert, K.H. Drexhage, J. Wolfrum, Proc. SPIE 2985, 61, (1997)
40. J.H. Jett, R.A. Keller, J.C. Martin, B.L. Marrone, R.K. Moyzis, R.L. Ratliff, N.K. Seitzinger, E.B. Shera, C.C. Stewart, J. Biomol. Struc. & Dynamics 7, 301, (1989)
41. L.M. Davis, F.R. Fairfield, M.L. Hammond, C.A. Hager, J.H. Jett, R.A. Keller, J.H. Halm, L.A. Krakowski, B. Marrone, J.C. Martin, H.L. Nutter, R.R. Ratliff, E.B. Shera, D. J. Simpson, S.A. Soper, C.W. Wilkerson, Los Alamos Sci. 20, 280, (1992)
42. R.L. Affleck, W.P. Ambrose, J.N. Demas, P.M. Goodwin, J.A. Schecker, J.A. Wu, R.A. Keller, Anal. Chem. 68, 2270, (1996)
43. S.A. Soper, Q.L. Mattingly, P. Vegunta, Anal. Chem. 65, 740, (1993)
44. S.A. Soper, Q.L. Mattingly, J. Am. Chem. Soc. 116, 3744, (1994)
45. M. Sauer, K.-T. Han, V. Ebert, R. Müller, A. Schulz, S. Seeger, J. Wolfrum, J. Arden-Jacob, G. Deltau, N.J. Marx, K.H. Drexhage, Proc. SPIE 2137, 762, (1994)
46. M. Sauer, K.-T. Han, R. Müller, S. Nord, A. Schulz, S. Seeger, J. Wolfrum, J. Arden-Jacob, G. Deltau, N.J. Marx, C. Zander, K.H. Drexhage, J. Fluoresc. 5, 247, (1995)
47. S. Nord, M. Sauer, J. Arden-Jacob, K.H. Drexhage, U. Lieberwirth, S. Seeger, J. Wolfrum, J. Fluoresc. 7, 15, (1997)
48. S. Seeger, G. Bachteler, K.H. Drexhage, G. Deltau, J. Arden-Jacob, K. Galla, K.-T. Han, R. Müller, M. Köllner, A. Rumphorst, M. Sauer, A. Schulz, J. Wolfrum, Ber. Bunsenges. Phys. Chem. 97, 1542, (1993)
49. M. Sauer, A. Schulz, S. Seeger, J. Wolfrum, J. Arden-Jacob, G. Deltau, K. H. Drexhage, Ber. Bunsenges. Phys. Chem. 97, 1734 (1993)
50. G. Bachteler, K.H. Drexhage, J. Arden-Jacob, K.-T. Han, M. Köllner, R. Müller, M. Sauer, S. Seeger, J. Wolfrum, J. Luminesc. 62, 101, (1994)
51. R. Müller, D.P. Herten, U. Lieberwirth, M. Neumann, M. Sauer, A. Schulz, S. Siebert, K.H. Drexhage, J. Wolfrum, accepted by Chem. Phys. Lett. (1997)
52. L.M. Davis, F.R. Fairfield, C.A. Harger, J.H. Jett, R.A. Keller, J.H. Hahn, L.A. Krakowski, B.L. Marrone, J.C. Martin, H.L. Nutter, R.l. Ratliff, E.B. Shera, D.J. Simpson, S.A. Soper, Genetic Anal. 8, 1, (1991)
53. P.M. Goodwin, H. Cai, J.H. Jett, S.L. Ishaug-Riley, N.P. Machara, D.J. Semin, A. Van Orden, R.A. Keller, Nucleosides & Nucleotides 16, 543, (1997)
54. M. Sauer, U. Lieberwirth, K.H. Drexhage, S. Nord, R. Müller, C. Zander, submitted to Chem. Phys. Lett. (1997)
55. C.A.M. Seidel, A. Schulz, M. Sauer, J. Phys. Chem. 100, 5541, (1996)
56. C. Zander, K.H. Drexhage, J. Fluoresc. 7, 37(S), (1997)
57. D.T. Chiu, A. Hsiao, A. Gaggar, R.A. Garza-López, O. Orwar, R.N. Zare, Anal. Chem. 69, 1801, (1997)
58. W.A. Lyon, S. Nie, Anal. Chem. 69, 3400, (1997)
59. J. Briggs, V.B. Elings, D.F. Nicoli, Science 212, 1267, (1981)
60. O.J. Finn, K.R. Jerome, R.A. Henderson, G. Pecher, N. Domenech, J. Magarian-Blander, S.M. Barrat-Boyes, Immun. Rev. 145, 61, (1995)
61. T.H. Brümmendorf, S. Kaul, J. Schumacher, R.P. Baum, R. Matys, G. Kliveny, S. Adams, B. Bastert, Cancer Res. 54, 4162, (1994)

CHAPTER 3

Time-Resolved Fluorescence of Conjugated Polymers

H. Bässler, M. Hopmeier, R.F. Mahrt

1
Introduction

Upon photo-excitation an organic molecule is promoted to a non-equilibrium state from which it relaxes back to the ground state. Time-resolved photo-luminescence spectroscopy is the method of choice for obtaining information on the relaxation pathway(s). If the primary excitation event were to create a vibrationally hot singlet (S_1) state or a higher lying S_n state, then relaxation can, in principle, involve vibrational cooling, internal conversion, intersystem cros-sing, readjustment of the molecular selection, and, finally, radiative or non-radiative decay to the ground state. In dense media solvation effects, as well as energy transfer and dissociation into a pair of charge carriers, can also occur. Which of these processes can be monitored by time-resolved spectroscopy depends on the time resolution of the experiment. Conventional methods with ns-time resolution yield the lifetime of the emitting states, the preceding history of that state remaining concealed, however. The advent of picosecond (ps) and femtosecond (fs) laser systems as excitation sources and of Streak camera and up-conversion techniques on the detection side allow the extension of the time resolution of fluorescence spectroscopy into the sub-picosecond regime. This has opened the possibility of studying ultra-fast photophysical relaxation processes such as the solvation of an excited chromophore in liquid solution, dephasing of optical excitations, and energy transfer in condensed systems. This article will focus on the latter aspect drawing heavily on the application of time and spectrally resolved fluorescence to elucidate the fate of optically excited states in conjugated polymers and their oligomeric counterpart structures.

After their successful synthesis, conjugated polymers were greeted with enthusiasm by solid state physicists who considered them to be almost ideal realizations of one-dimensional semiconductors. Since, for instance, polyacety-lene features a fairly sharp absorption threshold followed by a gradual decrease of the extinction coefficient at the high energy side of the spectrum it seemed straightforward to assign it to a van Hove singularity associated with a valence to conduction band transition [1, 2]. The onset of photoconductivity at the ab-sorption edge seemed to support this notion [3–5] as did the observation of a Stokes-shifted photoluminescence, considered to a signature of the radiative recombination of oppositely charged polarons, and the occurrence of transient sub-band gap absorption features that were attributed to these polarons [6, 7]. However, each of these phenomena is observed also with conventional glassy molecular solids, albeit in a different context, indicating that their interpretation is all but unambiguous. An example is the assignment of photoinduced IR absorption features to either transitions involving polaron/bipolaron levels and band states or transitions of radical ions [8].

This ambiguity will be illustrated in greater detail as far as absorption and fluorescence spectra are concerned. Stokes shift between absorption and emission is a signature of relaxation occurring between the events of absorption and emission. There are several possible reasons for this to happen. (i) If the equilibrium conformation of a chromophore in the excited state is significantly different from the ground state configuration, relaxation of the molecular skeleton occurs after excitation, the Stokes shift being a measure of that part of the excitation energy that goes into vibrational modes. In the physicists language the final excited state is referred to as an exciton-polaron. (ii) If, on the other hand, the primary excitations are charge carriers located at molecules with relaxed equilibrium geometry ("polarons") their collapse must give rise to photoluminescence that is off-set from absorption by the polaron binding energy of the radical cation and anion [6]. (iii) If the chromophores are chemically identical but physically different, e.g. because of statistical variations of their environment, they form an ensemble of absorbers with slightly different transition energies (because of different solvent shifts) manifested in inhomogeneously broadened absorption and emission bands. In a dense system energy transfer occurs among those chromophores. This will lead to spectral relaxation if the inhomogeneous band width is >kT [9, 10]. Case (i) implies an evolution of the Stokes shift on the time scale of a molecular vibration, i.e. in less than 100 fs. The same time scale should be involved in the formation of charged polarons [11, 12] (case ii). However, if luminescence is the result of charge pair recombination the kinetics should reflect the bimolecular character of charge encounter. Higher initial quantum energy might lead to slower recombination because excess energy might favor the formation of more distant charge pairs. The signature of case (iii) should be a red-shift of the emission as time progresses reflecting the *electronic* relaxation of the excitations within the manifold of states, henceforth called the density of states ("DOS"). Since spectral relaxation should slow down with time because fewer and fewer acceptor states become available one might expect to observe a concomitant variation of the fluorescence if probed at various spectral positions of the inhomogeneously broadened absorption/emission band. The time frame should be set by the jump time of an excitation among nearest neighbor chromophores and the lifetime of a singlet state ($\cong$ 1 ns). Spectrally resolved fluorescence spectroscopy with sub-ps time resolution is, therefore, the method of choice for distinguishing between various possible relaxation pathways. In the following, relevant experimental results will be discussed, test systems being conjugated polymers of the polyphenylenevinylene and polyparaphenylene type as well as an oligothiophene, present in the form of a single crystal and an evaporated film.

2
Experimental

While fluorescence recording employing Streak camera techniques has become a routine technique allowing spectra to be probed within a time interval extending from 5…20 ps (depending on the type of system used) to nanoseconds, the method of fluorescence up-conversion [13–15] has not yet fully reached this status. For this reason a brief description seems appropriate. The basic idea is to mix the fluorescence emitted from the sample as a response to an ultrashort laser pulse with an arbitrarily delayed part of the primary pulse in a non-linear crystal. The quantum energy of the resulting photon is the sum of excitation and emission quantum energies and its intensity reflects the temporal overlap of both signals. For this reason the time resolution is not determined by the response time of the detector as it is in the case of detection by a photomultiplier or a Streak camera but by the width of the exciting laser pulse and the group velocity dispersion within the non-linear crystal.

For the experiments to be described in the following sections a Kerr lens mode-locked titanium/sapphire laser was used delivering 150 fs pulses of $h\nu_0 = 1.56$ eV quantum energy at a repetition rate of 76 MHz. The primary beam was split into two parts, one of them passing a frequency doubler. To avoid sample degradation and non-linear annihilation effects inside the sample the output power at 3.12 eV was reduced to about 1 mw on a spot size of 100 μm in diameter. (In general, sample degradation in the course of such experiments may be a serious problem. Recall that the above intensity corresponds to a photon flux of 3×10^{19} quanta $(cm^2 s)^{-1}$ and 4×10^{11} quanta $(cm^2 pulse)^{-1}$, respectively. Considering the long data acquisition time required for achieving a good signal-to-noise ratio, photochemical degradation may occur depending on the photo-chemical stability of the system. Changing the excitation spot during the detection period can eliminate this problem). Luminescence emitted

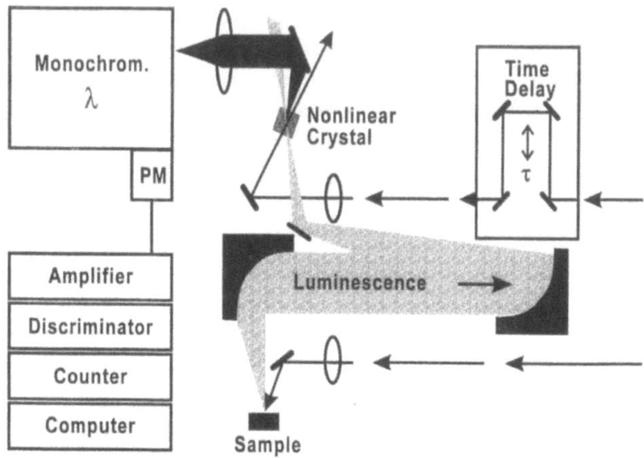

Fig. 1. Schematic view of an up-conversion setup

from the sample was collected by off-axis mirrors and focused dispersion free onto an optically non-linear crystal of β-barium borate (BBO). Photons with quantum energy $h\nu_s = h\nu_{fl} + h\nu_0$ where $h\nu_{fl}$ is the fluorescence quantum energy, are generated if fluorescence signal and reference pulse overlap temporally within the BBO crystal. Light at the sum frequency was dispersed by a monochromator and detected employing single photon counting techniques. The time frame that can be covered by this methodology, which is illustrated schematically in Fig. 1, is typically $\cong 250$ fs to 20 ps. In the time domain 20 ps to several ns, Streak camera techniques can be profitably used for signal detection although the most recent generation of Streak cameras has a time resolution extending to 0.5 ps.

3
Time-Resolved Fluorescence Experiments and Their Interpretation

3.1
Polyphenylenvinylenes

Polyphenylenevinylene (PPV) is a prototypical conjugated polymer. Its lowest excited singlet state is dipole allowed and is, therefore, amenable to fluorescence spectroscopy contrary to, for instance, that of polydiacetylenes [16].

PPV

Comparing the absorption spectra of distyrylbenzene, an oligomeric model compound with three phenylene rings, with that of standard PPV indicates that both are virtually identical except for a bathochromic shift of the latter. They can be understood in terms of a $S_1 \leftarrow S_0$ transition coupling to molecular vibrations, notably those of the phenylene ring. Plotting the maxima of the $S_1 \leftarrow S_0 0-0$ absorption bands measured on films of vapor-deposited oligophenylenevinylenes [17] vs. reciprocal chain length yields a linear relation extrapolating to $\Delta E(n \rightarrow \infty) = 2.27$ eV ($18\,300$ cm^{-1}) for a fully extended PPV chain. A transition energy of 2.50 eV ($20\,200$ cm^{-1}), measured for standard PPV, translates into $n \approx 8$. For PPV prepared in such a way as to favor intra-chain ordering, a transition energy of 2.42 eV ($19\,580$ cm^{-1}) was observed, equivalent to $n \cong 16$ [18]. This number is usually taken as a measure of the effective conjugation L_{eff}, i.e. the length of a perfectly ordered chain segment having the same transition energy. A polymer chain is thus considered as an array of ordered segments separated from each other by topological faults like kinks or twists that interrupt π-conjugation. Their statistical occurrence implies a statistical variation of L_{eff} and, by virtue of a qualitative particle-in-a-box-argument, of transition energy. This variation is reflected in the inhomogeneous broadening of the $S_1 \leftarrow S_0 0-0$ transition, a typical value being

$\cong 600 \div 800$ cm^{-1}. Chemical substitution gives rise to some shift of the transition energy as well as to an increase of the inhomogeneous width if it perturbs interchain alignment.

Conventionally recorded cw-fluorescence spectra of PPV bear out a Stokes shift between the maxima of the $S_1 \leftarrow S_0 0{-}0$ and $S_1 \rightarrow S_0 0{-}0$ bands of about 0.18 eV (1500 cm^{-1}). It increases as the temperature decreases and saturates below 100 K [19]. The inhomogeneous width of the emission band turns out to be about half that of the absorption band (Fig. 2). It is for this reason that fluorescence spectra reveal vibronic splitting even if absorption spectra are structureless in systems where inhomogeneous broadening is comparable to the dominant vibrational energy.

As mentioned in the introduction, the Stokes shift can, in principle, arise from structural relaxation of the molecular skeleton as a response to the changed π-electron distribution in the excited state or from spectral relaxation due to energy migration among energetically disperse chromophores [20]. In both cases the excited state is considered to be a coulombically coupled electron hole pair rather than a decoupled polaron pair as invoked by the proponents of the semiconductor model [2]. Spectroscopic techniques for distinguishing between these various possibilities are site-selective [18, 20] cw- and time-resolved fluorescence spectroscopy. The former technique was originally developed to allow the spectra of chromophore molecules matrix isolated in low temperature glasses to be observed subject only to homogeneous broadening [21]. It involves the use of a spectrally narrow laser, which makes it possible to excite selected chromophores from amongst a large ensemble contributing to an inhomogeneously broadened absorption. Only those whose transition energy is accidently resonant with the laser are excited and, provided that excitation is into the $S_1 \leftarrow S_0 0{-}0$ line, the resulting emission spectrum is a homogeneously broadened emission that provides information on the true molecular Stokes shift and on electron phonon coupling. An additional requirement for observing homogeneously broadened spectra is that any interchromophore interaction be minute. In a conjugated polymer this is not usually the case. One can, however, approach the isolated chromophore case if one ex-

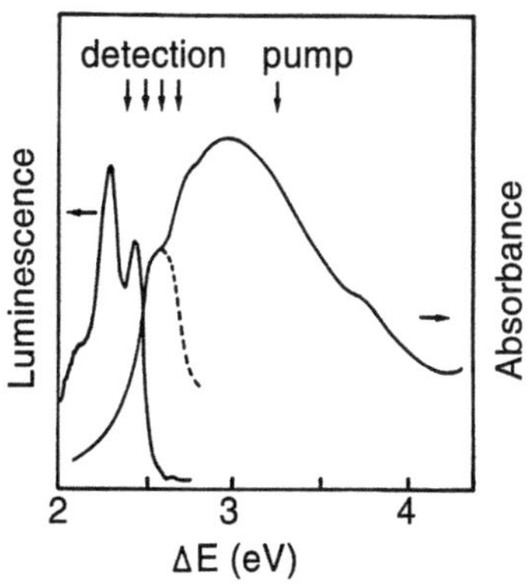

Fig. 2. Absorption and fluorescence spectra of PPV. *Arrows* indicate the spectral positions at which the temporal decay of the fluorescence, shown in Fig. 3a, was probed

cites into the very tail of the inhomogeneously broadened absorption where the segments with the largest effective conjugation length absorb and where the number of potential energy acceptors is too small to allow for energy transfer within an excited state lifetime. By applying this technique "true" fluorescence spectra of PPV, unaffected by spectral relaxation, have been measured and compared with the site-selective spectra of OPV embedded in a low temperature MTHF glass [18]. This comparison indicated that both are built on the same electronic origin and reveal the same vibronic structures, the only difference being some additional phonon broadening of the polymer spectra. This is a clear indication that the strength of excited state coupling to molecular vibration is the same for the oligomer and the polymer implying that the Stokes shift seen in the conventional cw-fluorescence spectra of the polymer is not caused by electron-phonon coupling but, rather, by electronic relaxation. The latter is an inherent property of any disordered organic bulk system, manifest also in its charge transporting properties [2].

To prove the latter hypothesis the fluorescence decay was monitored at various spectral positions overlapping with the lower energy portion of the absorption spectrum. When exciting the system with photons with excess energy a manifold of vibronic Franck–Condon S_1 states is populated that dissipate their excess vibrational energy into the local heat bath on a time scale ≤ 100 fs giving rise to a random population of a manifold of vibrationally cold S_1 states. At present little is known about the mechanism by which excited state cooling proceeds. One possibility is that energy transfer to an adjacent chromophore is involved that leaves the originally excited molecule in a vibrationally hot state that cools off on a ps-time scale. More work is required to further elucidate these processes. Förster-type energy transfer will commence thereafter. Probing the temporal profile of the emission at selected spectral positions will therefore provide a handle on the population as well as the depopulation dynamics within a certain energy slice of the density of excitonic states. What one expects to observe is a slowing down of the dynamics as the detection energy is scanned towards the absorption tail. (A cautionary note concerning the use of the term exciton is in order. Strictly speaking it denotes a mobile excitation in a crystal where translational symmetry exists. In the present context it is meant to describe an excitation that can migrate incoherently among localized sites.)

Figure 3a shows experimental results for a PPV bulk film fabricated via the precursor route and a 1% blend of soluble polyphenylphenylenevinylene (PPPV) in polycarbonate (PC) [23, 24]. The spectral position (ε_{det}) of fluorescence detection are indicated in Fig. 2. The spectral width of the detection window was 20 mn. It is obvious that there is emission overlapping with the absorption tail, and that the fluorescence decay is strongly dependent on ε_{det} and non-exponential. In the PPV case the initial decay becomes instrument limited if probed close to the center of the inhomogeneously broadened $S_1 \leftarrow S_0\,0\text{--}0$ band. Some rise phase is observed upon scanning ε_{det} towards the absorption tail, notably in the case of the PPPV-blend. The prolongation of the decay process continues if probed at energies beyond the absorption band. The experiments presented in Fig. 3c, carried out on undiluted PPPV using a Streak

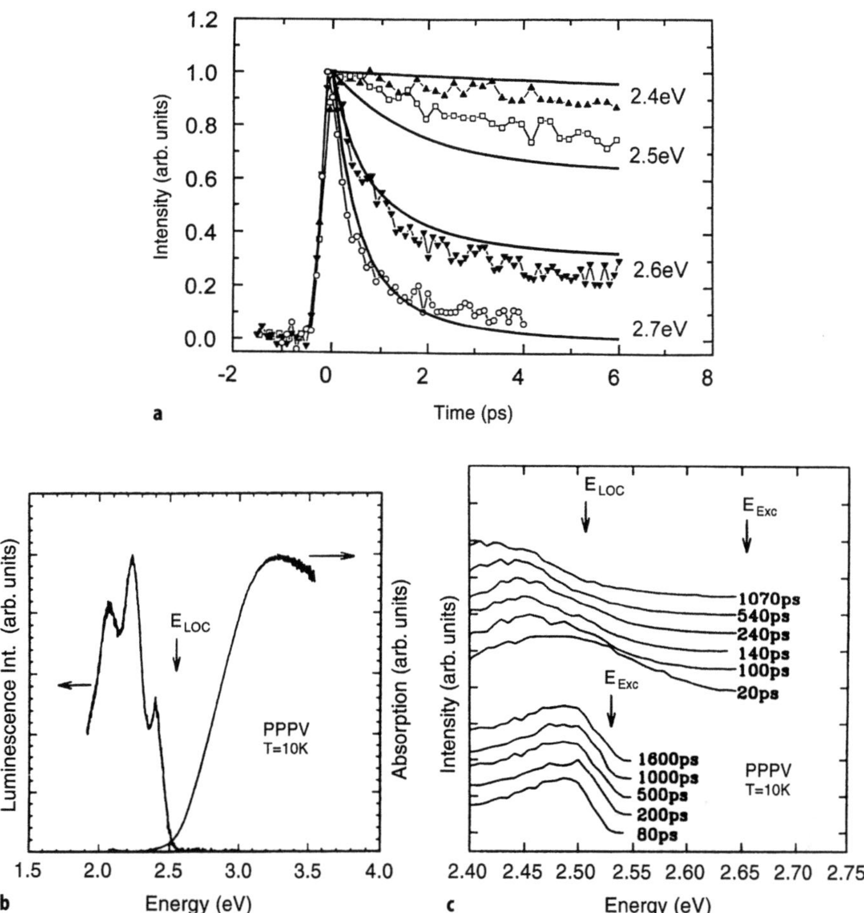

Fig. 3a–c. a) Temporal decay of the fluorescence of PPV excited by 150 fs pulses of 3.12 eV photons and detected at various spectral positions. *Solid lines* are the result of model considerations involving Monte Carlo simulations of the random walk of excitations within a three-dimensional random system (from [27]. b) Absorption and cw-fluorescence of a PPPV film. E_{LOC} is the demarcation energy below which quasi-resonant emission occurs. c) Decay of the photoluminescence of a 1% blend of PPPV in polycarbonate and a neat PPPV film, respectively, probed at various detection energies. Data in b) were obtained using the up-conversion technique, those shown in c) are the result of Streak-camera measurements (from [24])

camera, indicate that at $\varepsilon_{det} = 2.4$ eV the decay function approaches an exponential with a decay time of about 1 ns. This is close to the decay time of distyrylbenzene. It is also worth mentioning that emission spectra recorded within selected time windows after the exciting laser pulse bear out a monotonous red-shift with increasing decay time [25]. This will be illustrated in greater detail in Sect. 3.2. An appropriate conceptual framework for interpreting the above results is that of random walks of excitations within a distribution of chromophore states whose spectral envelope, reflected by the low energy tail of the $S_1 \leftarrow S_0 0{-}0$ band, is a Gaussian. Since the $S_1 \leftarrow S_0$ transition is dipole allowed it is straightforward to invoke dipole–dipole coupling as the dominant coupling mechanism for donor–donor interaction, expressed via Förster-type transition rates:

$$\gamma_{ij}^{DD} = \eta_0^{DD} \left[d^{DD}/(r_i - r_j)^6 \right] \vartheta(\varepsilon_{ip}, \varepsilon_j) \tag{1}$$

where $\vartheta(\varepsilon_i, \varepsilon_j)$ is a Boltzmann-type weighting factor:

$$\vartheta(\varepsilon_i, \varepsilon_j) = \begin{cases} \exp\left[-(\varepsilon_i - \varepsilon_j)/kT\right]; & \varepsilon_j > \varepsilon_i \\ 1 & ; \ \varepsilon_j < \varepsilon_i \end{cases}$$

that controls uphill processes. Downhill jumps are assumed to be temperature independent. This is appropriate for transitions with moderate coupling to phonons/vibrations. d^{DD} is the minimum distance among donor sites and η_0^{DD} is the transfer rate for $r_i - r_j = d^{DD}$. The problematic aspect of using Förster rates is that Eq. 1 has been developed for point dipoles. In conjugated polymers this condition is certainly not generally fulfilled since the nearest neighbor distance ($\sim 7\,\text{Å}$) is comparable to the spatial extension of the excited state [26].

Random walks on an energetically disordered system are dispersive if the width of the DOS is $\geqslant kT$. In this case energetic relaxation takes place that leads to a slowing down of the exciton motion. This renders transitions between sites asymmetric and precludes any straightforward analytic treatment. The alternatives are (i) effective medium concepts [9], (ii) Monte Carlo simulation [27] or (iii) a Monte Carlo master equation hybrid technique [24]. Application of methods (ii) and (iii) indicate that the experimental results can, in fact, be recovered provided that the existence of non-radiative traps is invoked. This is justified via independent spectroscopic evidence. Candidates are oxidation products that capture the electron from an on-chain excitation thus forming a geminate electron hole pair that can recombine non-radiatively or contribute to photoconductivity in the presence of a collecting electric field [28]. The probability that an exciton reaches a trap depends also on the initial starting energy of the excitations as donor–donor transfer does but prevents excitations from relaxing into the bottom states of DOS. This is one reason why almost no rise phase is seen upon fluorescence probing at lower detection energies, the other one being the overlap of the (fast) $S_1 \rightarrow S_0 0 \rightarrow 1$ emission component of excitations close to the center of the DOS with the more slowly developing $S_1 \rightarrow S_0 0{-}0$ emission from tail states, populated via energy transfer [29].

As far as a more detailed analysis of the experimental results is concerned the reader is referred to original work [24, 29]. It will suffice to address a few aspects only. Despite the inevitable simplifications introduced when modeling energy migration, several important features of the experiments are recovered in a quantitative way. They include the non-exponential fluorescence decay that does not follow a stretched exponential Kohlrausch–Williams–Watts (KWW) function either. The reason is that there is a discrete, fastest transfer rate set by downward energy transfer among nearest neighbor chains. This translates into an initial exponential portion of the decay pattern. It will be eventually merge into KWW-like behavior as expected for a random medium. In the long time limit the decay must again approach an exponential reflecting the intrinsic decay time of isolated chromophores that are electronically decoupled from both the ensemble of chromophores with higher transition energy and the reservoir of traps. Experimentally, this decay time is close to 1 ns. Note that the intrinsic fluorescence decay time of distyrylbenzene is 1.1 ns [30]. The initial $1/e$ decay time of the fluorescence signal increases by up to 3 orders of magnitude upon scanning the detection window from the center of the DOS towards the very tail states. The functional form of the entire decay pattern is in good agreement with model calculations. The Förster radius turns out to be about 8 Å. In a 3D bulk system the minimum dwell time of an excitation residing on a site having lower energy neighbors is ~ 0.3 ps. In a diluted system in which acceptor sites are likely to be segments of the same chain that time is ~ 1 ps. The observation of basically the same decay pattern in bulk PPV and in a diluted PPPV-blend also proves that inter- and intra-chain transfer processes are equivalent, their respective preponderance being a question of the system morphology.

The spectrally resolved fluorescence decay pattern is a reflection of the lifetime distribution of excitations at a given energy within the density of states. Since, however, the decay function cannot be expressed in a simple analytic way, one cannot retrieve that information by e. g. Laplace transformation. The only possibility is to construct a model system, determine both decay functions and lifetime spectra by the Monte Carlo master equation hybrid technique for a variety of system parameters and compare experiment and simulation [24, 29].

The model predictions also confirm that the Stokes shift of time-averaged fluorescence spectra in conjunction with the bathochromic shift of spectra recorded at different delay times after the laser pulse is a reflection of spectral relaxation and as such is an inherent property of any energy-dispersive hopping system. Since relaxation is the result of the interplay between exoenergetic and temperature activated jumps it is also intuitively plausible that the Stokes shift must increase with decreasing temperature and settle at a constant value when relaxation becomes frustrated because temperature-activated jumps require a time in excess of the excited state lifetime [31]. The smaller inhomogeneous width of emission as compared to absorption bands is a reflection of the same phenomenon [32].

Within the context of the experiments on PPV it seems appropriate to mention that the up-conversion technique can also be profitably used to obtain information on the effect of an external electric field on the dynamics of the fluorescent state. A strong electric field may affect the transition dipole moment and, hence, modify the absorption spectrum, or it can cause the excited state to ionize thereby forming an electron-hole pair (a radical anion-cation pair or, synonymously, a polaron pair) on adjacent or more distant chromophores. The first process occurs instantaneously, while the second one is a rate process that evolves on the time scale of an elementary charge transfer event. Since the latter destroys the primary excited state it must result in quenching of the prompt fluorescence. Monitoring the time evolution of fluorescence quenching is, therefore, a way to obtain information concerning the time scale on which charge transfer occurs and on how it depends on the applied electric field. The latter information is needed to determine how much energy it costs to dissociate an excited state. This question is of particular relevance for conjugated polymers. Would they behave like inorganic semiconductors in which, owing to their comparatively large dielectric constant, excitons are of the Wannier type and weakly bound they should be stable at lower temperatures only and an electric field of the order kV/cm should be sufficient for dissociation.

The fluorescence of a bulk PPPV as well as of PPPV-blends can, in fact, be quenched by electric fields exceeding 10^6 V/cm. The quenching efficiency rises quadratically with field [33]. At $F = 3 \times 10^6$ V/cm the fluorescence yield has dropped to $\cong 60\%$ of its original value. The effect diminishes at lower concentrations of the chromophore. Upon recording the fluorescence decay of a sample, sandwiched between two electrodes, via Streak camera techniques, i.e. on a time scale > 20 ps, no time evolution of the quenching effect is seen. This illustrates the necessity to employ the up-conversion technique in order to find out whether the effect is an instantaneous one, reflecting a change of the oscillator strength of the $S_1 \leftarrow S_0$ transition, or is a rate process. Performing the experiments with 200 fs time resolution showed, indeed, that the quenching effect evolves with time following a non-exponential time pattern with a maximum rate of ~ 1 ps^{-1} [30, 35]. By comparing the field dependence with the predictions of Monte Carlo simulations the exciton binding energy can be estimated to be 0.4 ± 0.1 eV. [33]. This is the energy needed to fully separate the electron hole pair comprising the S_1 state of PPPV into a radical cation/anion pair. It proves that the S_1 state is a tightly bound state requiring a large electric field for dissociation. Note that an energy of 0.2 eV is gained when an electron moves across an intermolecular distance of, say, 7 Å in a field of 2×10^6 V/cm.

3.2
Ladder-type Polyparaphenylene

The absorption and fluorescence spectra of oligophenylenes are usually fairly broad and bear out a significant Stokes shift. The reason is that the interplay between steric repulsion among adjacent rings and electronic coupling changes upon excitation. Therefore the ground state structure is twisted while the excited state structure is planar. This is equivalent to a strong coupling of the excited state to a torsional motion, expressed via a Huang–Rhys factor $S \gg 1$. It is reflected both in the band profile and the Stokes shift. A similar phenomenon has been observed with polyarylenevinylenes in which the arylene moiety is a biphenylene-unit [36]. The structural relaxation-pathway is closed if neighboring phenylene rings are covalently bridged as realized in the ladder-type polyparaphenylene (LPPP) [37] whose chemical structure has been included in Fig. 4. The absorption spectrum of LPPP, present in the form of either a bulk film or a 1:9 blend with polycarbonate, reveals a well-developed vibronic structure with a $S_1 \leftarrow S_0 0-0$ transition peak at 2.74 eV (blend, Fig. 4a) and 2.73 eV (bulk film). The upper portion of the absorption tail translates into a Gaussian width of 0.07 eV. The low energy tail is due to superposition of scattering – as borne out by photoluminescence excitation spectroscopy [38] – and absorption due to a defect. The cw emission spectrum carries a $S_1 \rightarrow S_0 0-0$

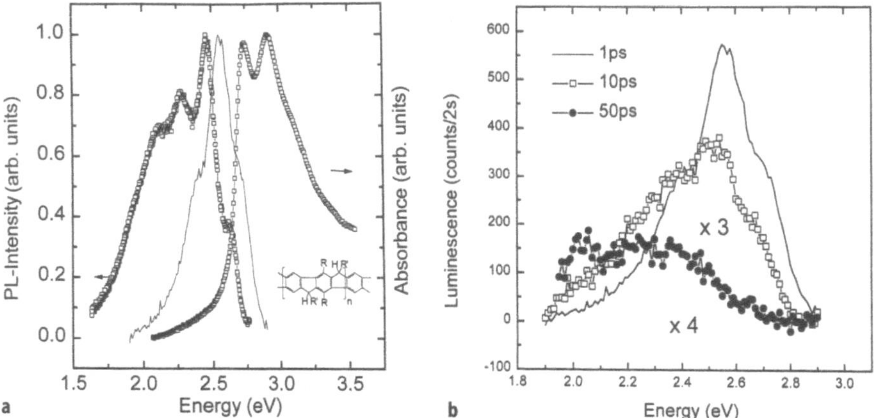

Fig. 4a, b. **a)** Absorption and cw-fluorescence spectra of a 1:9 LPPP/PC blend upon excitation at 3.1 eV. The *full curve* is the time-resolved spectrum of a neat film recorded 1 ps after the excitation pulse. The inset shows the chemical structure of LPPP: R and R' are n-hexyl and 1,4-decylphenyl substituents, respectively. **b)** Time-resolved fluorescence spectra of a neat LPPP film recorded at various delay times

feature at 2.66 eV (blend) and 2.65 eV (bulk film) followed by a vibronic progression of 0.18 eV and a superimposed shoulder whose fractional intensity depends on the concentration of LPPP and on the film preparation conditions indicative of it being due to a defect of a physical rather than a chemical nature. The existence of the fluorescent defect renders LPPP a particularly useful object for time-resolved fluorescence spectroscopy because it allows trap filling processes to be delineated via time-resolved emission spectroscopy.

LPPP bulk films were investigated employing the up-conversion technique in order to elucidate the excited state dynamics in much the same way as was done with PPV [37] (Laser pulses of 160 fs duration and a quantum energy of 3.1 eV were used for excitation). Figure 5 shows the fluorescence decay monitored at different spectral positions including positions above the center of the inhomogeneously broadened $S_1 \leftarrow S_0 0-0$ transition. Onset of the emission is instrument limited indicating that vibrational cooling of the initially excited Franck–Condon state is completed within 200 fs. The fastest 1/e decay time, measured at $\varepsilon_{det} = 2.83$ eV was 1.8 ps. It increases upon shifting the detection window towards lower energies. However, contrary to PPV, the increase of the 1/e decay is far less pronounced than with PPV. Time-dependent emission spectra offer an explanation for this observation. Figure 4b documents that the emission spectrum changes with time. Except for reabsorption effects that reduce the $S_1 \rightarrow S_0 0-0$ intensity, the spectrum recorded 1 ps after excitation is a mirror image of absorption. Adding the vibrational energy to the energy of the $S_1 \rightarrow S_0 0-1$ transition it becomes obvious that the center of the $S_1 \rightarrow S_0 0-0$ band, which gives rise to the high energy shoulder, is located near 2.73 eV, i.e. virtually coincident with the $S_1 \leftarrow S_0 0-0$ absorption maximum. Bearing in mind that the fastest 1/e decay time is 1.8 ps, it is obvious that emission events

Fig. 5. Fluorescence transients of a
LPPP neat film excited at 3.1 eV
and monitored at different spectral
positions

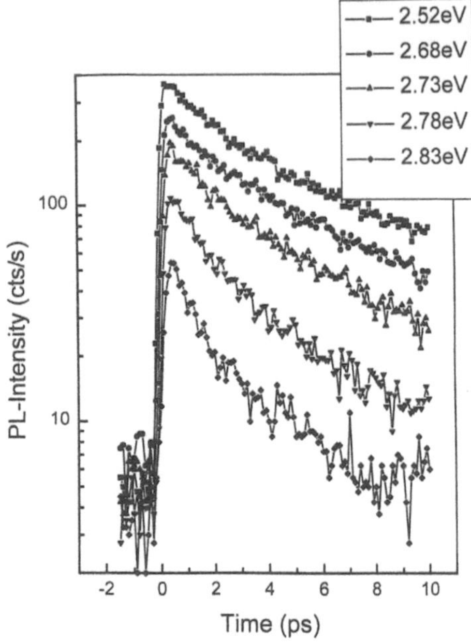

at 1 ps after excitation must occur from the manifold hopping states prior to any
spectral relaxation. This is clear evidence that the zero vibronic levels in absorption and emission are identical. It does not imply, of course, that there is no
relaxation of the molecular skeleton after excitation. It only means that the
Huang–Rhys factor for vibrational coupling is of the order unity which is a
typical value for rigid π-electron systems [16]. Note that the structural
relaxation energy inferred from a simulation of the fractional Huang–Rhys
factor turns out to be of the order 0.3 eV consistent with theoretical work on
oligomers of PPV and related long chain π-electron systems [39, 40]. Such
fractional Huang–Rhys factors are amenable to spectroscopy determination via
site-selective fluorescence spectroscopy on molecularly dispersed oligomers
[41]. It is worth noting that such work revealed spectra with sharp zero phonon
lines confirming the notion that the Huang–Rhys factors for coupling of the
excited state to both molecular vibrations and low energy phonons are small
enough to observe zero vibrational (i.e. 0–0) and zero phonon transitions.

The emission spectrum measured 10 ps after excitation is still similar to the
spectrum recorded after 1 ps except that the low energy wing is more
pronounced. After 50 ps only the broad low energy band centered at 2.1 eV. is
left. Since the temporal profile of the emission recorded at 1.95 eV bears out a
rising phase that matches the decay of the bulk emission (Fig. 6) it is straightforward to assign it to emission from a defect. Since it can be excited directly
below the absorption edge of the bulk material [38] the defect must already
exist in the ground state and cannot, therefore, be identified with a conventional
excimer. The broad band character of the emission, on the other hand, testifies
to the occurrence of some structural relaxation after excitation. This suggests

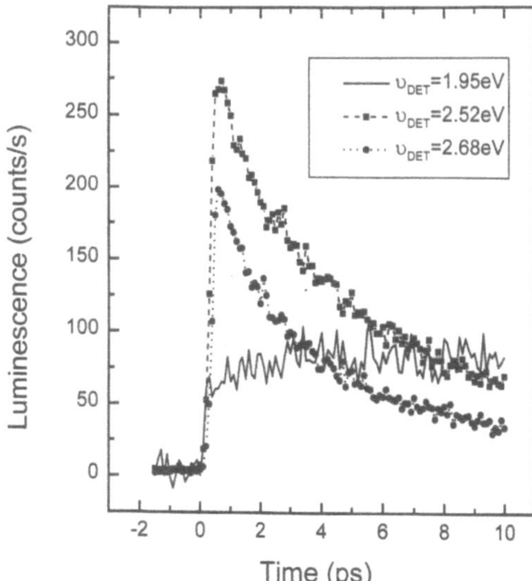

Fig. 6. Fluorescence transients of a LPPP neat film recorded at spectral positions within the bulk emission and the defect emission (1.95 eV), respectively

assignment to a pairwise arrangement of adjacent LPPP-segments in which the molecular planes are close to parallelity. This will both lower the excitation energy and allow for further structural relaxation after excitation.

Figure 7 shows the fluorescence decay curves monitored at different detection energies on a double logarithmic $\ln[I_0/I(t)]$ vs. time plot. Such a representation is appropriate to check whether or not the decay pattern follows a stretched exponential

$$I(t) = I_0 \left[- (t/t_0)^\beta\right] \; 0 < \beta < 1, \tag{2}$$

An exponential decay process would reproduce as a straight line with slope unity, its intersection with the horizontal $\ln I_0/I(t) = 1$ line yielding the decay

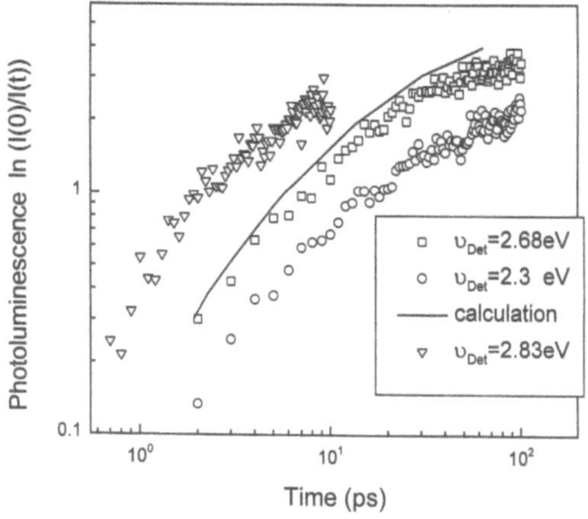

Fig. 7. Decay curves of the LPPP fluorescence at various detection energies plotted on a double logarithmic $\ln I_0/I$ vs. time scale. The *full line* is a result of the model calculation taken from [24]

time t_0. It is obvious that at early times the decay is exponential, the decay time increasing from 1.8 ps at $\varepsilon_{det} = 2.83$ eV to 7 ps at $\varepsilon_{det} = 2.68$ eV and 20 ps for $\varepsilon_{det} = 2.3$ eV. Before the emission intensity has decayed to 1/e of its peak value, however, gradual turn over to stretched exponential behavior is observed with β ranging from 0.4 to 0.25. It is remarkable that the functional shape of I(t) is in excellent agreement with the calculation originally intended to model the spectral relaxation in PPV [24] employing the Monte Carlo master equation hybrid technique (see dashed curve in Fig. 7), since those calculations were done on a statistical single chain with constant nearest neighbor distances generated via a self-avoiding random walk algorithm premised upon the existence of a Gaussian distribution of site energies This agreement indicates that the shape of a fluorescence decay function is quite insensitive to the actual sample topology. In particular, it shows that the three-dimensional nature of a bulk LPPP sample, as compared to a one-dimensional model chain has little if any effect on the relaxation pattern except for a shift of the time frame. Intra- and inter-chain jumps are thus proven to be equivalent.

3.3
Oligothiophene (6 T)

When measuring fluorescence spectra of a molecular solid, one is often confronted with the difficulty of separating intrinsic from extrinsic effects. It becomes all the more important the more ordered the sample is. Increasing degree of order implies an increasing mobility of the excited state in the form of a Frenkel-exciton and, concomitantly, an increasing likelihood that it encounters a trap before decaying radiatively on a host molecule. Recall that in an anthracene single crystal a relative concentration of 10^{-5} guest molecule is sufficient to lower the bulk emission by a factor of 2. Conventional fluorescence decay time measurements are unlikely to be useful for unravelling trapping kinetics because energy transfer may occur on a time scale that is shorter than the response time of the system. Picosecond or sub-ps time-resolved fluorescence spectroscopy will, however provide a handle on trapping phenomena as will be demonstrated choosing sexi-thiophene (6 T) as a model system.

α-T6

There are several reasons why oligothiophenes became fashionable as test objects for studying opto-electronic phenomena of conjugated oligomers and polymers. Apart from the advantage that an oligomer has a well-defined length, as compared to the statistically varying effective conjugation length of its polymeric counterpart, oligo-thiophene films of specified morphology can be prepared by various sublimation techniques including vapor phase epitaxy

[42]. Recently, even high quality single crystals have become available [43]. Evaporated 6 T films were used in the first organic field effect transistor [44, 45] and as emitters in light-emitting diodes [46]. On the fundamental side, the spectroscopic properties of the shorter oligomers have been extensively studied by Kohler and co-workers employing matrix isolation techniques to achieve high spectral resolution [47].

A peculiarity of solid films of 6 T, as well as of the single crystal, is that their absorption spectra differ in a characteristic way from those of their lowest membered analogues measured in a solid alkane matrix. While the latter reveal a strong $S_1 \leftarrow S_0 0-0$ transition followed by the usual vibronic progression, the origin of the film/crystal absorption is barely detectable [48, 49], and, most remarkably, coincident with the lowest two-photon allowed transition at $18335\ cm^{-1}$ [51]. Recent polarized absorption measurements [49, 52] in conjunction with a theoretical analysis of crystal field effects have led to the conclusion that this is a signature of Davydov splitting in a single crystal with four molecules in the unit cell, arranged in a herring-bone fashion, and belonging to the D_{2h} point group: The calculations indicate that the lowest Davydov component has gerade symmetry. Mixing with an energetically close transition of ungerade symmetry renders it weakly allowed also via one photon transition. The fundamental difference between single crystal and single molecule absorption spectra is thus identified as a crystal field effect.

A polarized cw-fluorescence spectrum of a 6 T single crystal taken at 4.2 K, is shown in Fig. 8. No similarity with the spectra of matrix-isolated single 3 T and 4 T molecules exists. There is a very weak $0-0$ feature at $18332\ cm^{-1}$, virtually

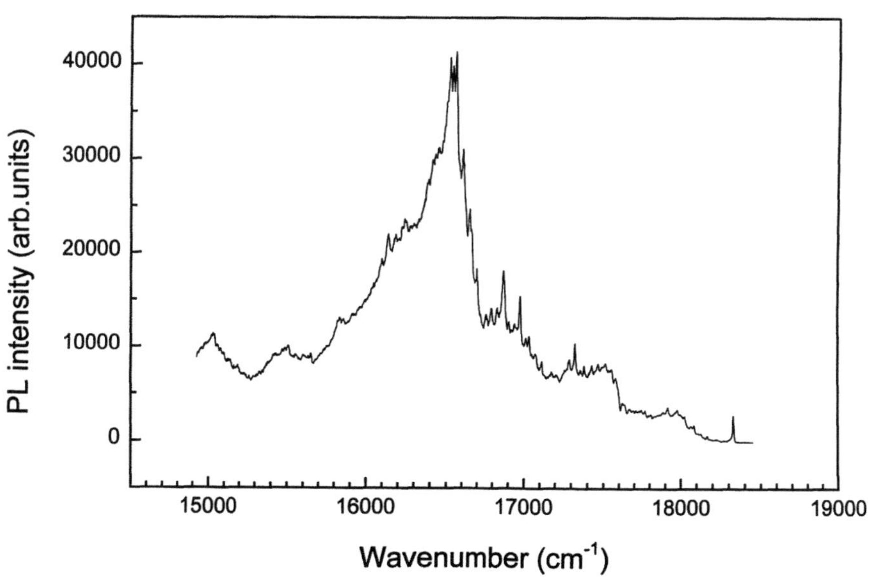

Fig. 8. $\vec{b}$-polarized fluorescence spectrum of a 6 T single crystal recorded at 4.2 K

coincident with the weak absorption origin, followed by a highly structural vibronic progression. The origin band is absent in spectra polarized along the $\vec{c}$ direction. The spectrum is a superposition of (weak) intrinsic emission and defect emission as is usually the case with low temperature fluorescence spectra of molecular crystals.

Time-resolved fluorescence spectra provide the clue for both disentangling intrinsic emission from defect emission and clarifying why the bulk emission spectrum of the 6 T crystal differs from that of the isolated molecule. This is illustrated in Fig. 9 on the basis of spectra of a well-ordered sublimed 6 T film in conjunction with the crystal spectrum taken at zero delay time. The spectra were recorded with a Streak camera at variable delay time and a gate width of 6 ps. It is particularly noteworthy that the t = 0 spectra of single crystal and film are virtually identical. This testifies to the high degree of molecular ordering in the film. The well-resolved vibronic structure in conjunction with a very weak 0–0 origin supports the notion that the emission arises from Herzberg–Teller coupling [53] of the dipole-forbidden lowest crystal transition to promoting modes that render that transition partially allowed. The main feature of the spectra is a $1492\,\mathrm{cm^{-1}}$ mode built on the $18332\,\mathrm{cm^{-1}}$ origin, which has

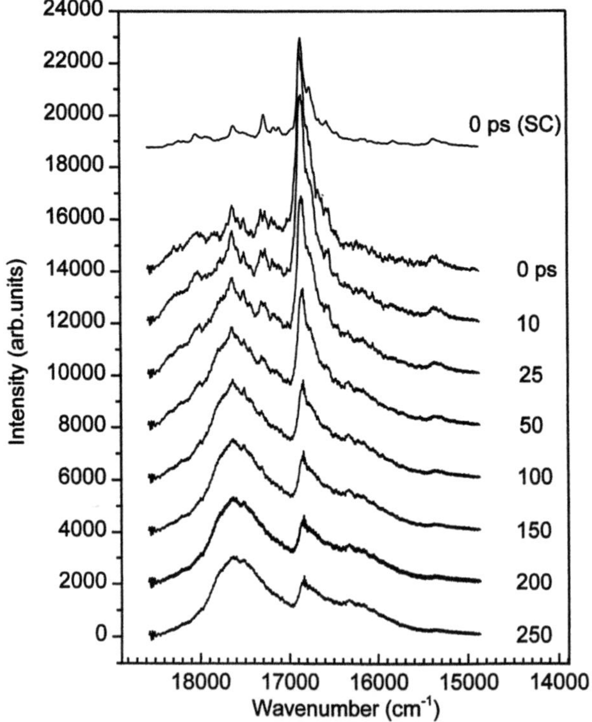

Fig. 9. Temporal evolution of the fluorescence spectrum of a sublimed 6 T film. Excitation occurred at $25000\,\mathrm{cm^{-1}}$ using a frequency doubled titanium/sapphire laser. The gate width was 6 ps. The crystal spectrum recorded at zero delay has been included for comparison

previously been identified as a ring stretching mode with ungerade symmetry [54]. This mode is also seen in the cw spectrum in Fig. 8. However, the remaining strong features of the cw-crystal spectrum, of which those at energies $\cong -1800 \text{ cm}^{-1}$ and 3200 cm^{-1} below the origin are the most prominent ones missing in the transient spectrum taken at $t = 0$ indicating that these have to be identified as emission from trap(s) of most likely structural origin ("X-traps" [55]). Since traps are localized states, split off for the exciton band, they are not subject to the same selection rules. Therefore the lowest excited state of a 6 T-defect does not carry the gerade symmetry of the lowest Davydov component but is a molecular 1^1B_u state that does not require coupling a promoting ungerade molecular vibration to become (partially) allowed. Consequently, the trap emission resembles that of an isolated molecule as far as the intensity distribution within the vibronic manifold is concerned.

The gradual transition from the Herzberg–Teller-like bulk fluorescence spectrum to the dipole-allowed trap emission is demonstrated in Fig. 9 on the basis of the spectra of the 6 T-film. The dominant intrinsic vibronic feature near 16800 cm^{-1}, decays with a 1/e decay time of about 25 ps and broad features centered at 17800 cm^{-1} and 16500 cm^{-1} evolve with the same time constant (Fig. 10). After some spectral relaxation occurring on a time scale > 250 ps the

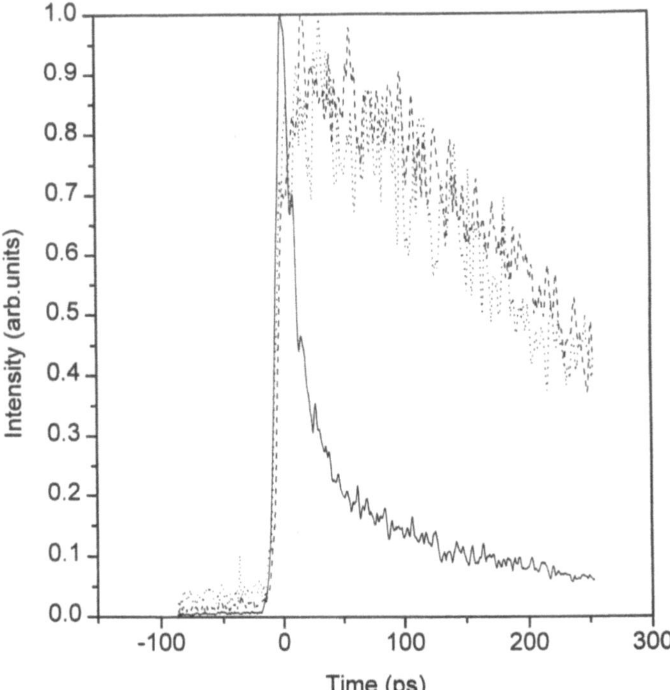

Fig. 10. Time dependence of the fluorescence emission from a 6 T film recorded at 16920 cm^{-1} (*solid line*), 17700 cm^{-1} (dashed line), and 16260 cm^{-1} (*dotted line*)

cw-emission spectrum evolves. Apart from the spectral shift, its shape is (i) characteristic of an inhomogeneously broadened molecular transition of shorter oligothiophenes [47] and (ii) agrees with the fluorescence spectra of polythiophene films [56]. The different and, at first glance, incompatible structures of the fluorescence spectra of crystalline 6 T, on the one hand, and of disordered films and a related polymer are thus identified as a signature of the suspension of selection rules occurring when a Frenkel exciton in a crystalline structure is captured by a trap.

4
Concluding Remarks

The emphasis of this article has been on the incoherent time-resolved fluorescence spectroscopy covering the time frame extending from 0.3 ps to several ns employing up-conversion and Streak-camera techniques, the objective being the unraveling of the dynamics of excited singlet states in random organic solids. A general feature of systems like conjugated polymers, molecular organic glasses, or crystalline systems containing structural faults is their built in disorder, manifest in inhomogeneously broadened absorption spectra. The energetic randomness in conjunction with the dense packing of the chromophores gives rise to energy transfer processes that lead to spectral relaxation because an excitation has the tendency to reach the lowest possible metastable state before decaying radiatively or non-radiatively. It inevitably leads to non-exponential relaxation dynamics, at least at low temperatures, because the number of potential intrinsic chromophores that can act as acceptors in donor–donor transfer processes decreases continuously as relaxation proceeds. To cover the full spectrum of transition rates the spectroscopic detection method has, therefore, to extend from the time scale of typical jump times of an exciton in a molecular crystal (~ 1 ps) [57] to times set by the intrinsic excited state lifetime. Probing the temporal rise and decay pattern of fluorescence at variable spectral positions provides a handle on the population and depopulation dynamic of a state at a specified energy within the inhomogeneously broadened density of states. Energy transfer processes can thus be monitored without requiring the incorporation of an external sensor molecule which might perturb the local environment and, concomitantly, affect the transfer process one wants to probe. The disadvantage of the method is that there is no analytic treatment that leads to a closed form of mathematical description permitting an experimentalist to extract information such as the distribution of excited state lifetimes. One has, instead, to rely on statistical techniques such as Monte Carlo simulation in order to analyze the experiment in terms of model predictions.

Acknowledgements. The authors gratefully acknowledge collaboration with Profs E. O. Göbel, H. Kurz, K. Müllen, W. W. Rühle and C. Taliani. This work was supported by the Deutsche Forschungsgemeinschaft (SFB 383), the European Community (SELMAT), Stiftung Volkswagenwerk and the Fonds der Chemischen Industrie.

References

1. A.J. Heeger, S. Kivelson, R.J. Schrieffer, W.P. Su, Rev.Mod.Phys. 60, 782 (1988)
2. A.J. Heeger in: "Primary Photoexcitations in Conjugated Polymers: Molecular Excitons versus Semiconductor Band Model, edited by N.S. Saricifti, World Scientific Singapore 1997
3. J. Opfermann, H.H. Hörhold, Z.Phys.Chem. (Leipzig) 259, 1089 (1978)
4. S. Tokito, T. Tsutsui, R. Tanaka, S. Saito, Jpn. J. Appl. Phys. 25, L680 (1980)
5. C.H. Lee, G. Yu, A.J. Heeger, Phys. Rev. B. 47, 15543 (1993)
6. K. Fesser, A.R. Bishop, D.K. Campbell, Phys. Rev. B. 27, 4804 (1983)
7. R.H. Friend, D.D.C. Bradley, P.O. Townsent, J. Phys. D 20, 1367 (1987)
8. M. Deussen, H. Bässler, Chem. Phys. 164, 247 (1992)
9. B. Movaghar, M. Grunewald, B. Ries, H. Bässler, D. Würtz, Phys. Rev. B 33, 5543 (1986)
10. H. Bässler in Primary Photoexcitations in Conjugated Polymers: Molecular Excitons versus Semiconductor Band Model, edited by N.S. Saricifti, World Scientific Singapore 1997
11. U. Lemmer, E.O. Göbel, in: Primary Photoexcitations in Conjugated Polymers: Molecular Excitons versus Semiconductor Band Model, edited by N.S. Saricifti, World Scientific Singapore 1997
12. X.Q. Zhou, G.C. Cho, U. Lemmer, W. Kütt, K. Wolter, K. Kurz, Solid State Electron. 32, 1591 (1989)
13. J. Shah, IEEE J. Quantum Electron, 24, 276 (1988)
14. H. Mahr, M.D. Hirsch, Opt. Commun. 13, 96 (1975)
15. M.A. Kahlow, W. Jarzeba, T.A. DuBruih, P.F. Barbara, Rev. Sci. Instrum. 59, 1098 (1988)
16. H. Bässler in Electronic Materials: The Oligomer Approach, edited by K. Müllen, G. Wegner, VCH, Heidelberg, 1997
17. A. Schmidt, M.C. Anderson, D. Dunphy, T. Wehrmeister, K. Müllen, N.R. Armstrong, Adv. Mat. 7, 722 (1995)
18. S. Heun, R.F. Mahrt, A. Greiner, U. Lemmer, H. Bässler, D.A. Halliday, D.D.C. Bradley, P.L. Burn, A.B. Holmes, J. Phys. Condens. Matter 5, 247 (1993)
19. K. Pichler, D.A. Halliday, D.D.C. Bradley, P.L. Burn, R.H. Friend, A.B. Holmes, J. Phys. Condens. Matter 5, 7155 (1993)
20. U. Rauscher, H. Bässler, D.D.C. Bradley, M. Hennecke, Phys. Rev. B 42, 9830 (1990)
21. R.I. Personov in Spectroscopy and Excitation Dynamics of Condensed Molecular Systems, edited by V.M. Agranovich, R.M. Hochstrasser, North Holland, Amsterdam, 1983, p 555
22. H. Bässler, Phys. Stat. Sol (B), 175, 15 (1993)
23. R. Kersting, U. Lemmer, R.F. Mahrt, K. Leo, H. Kurz, H. Bässler, E.O. Göbel, Phys. Rev. Lett. 70, 3820 (1993)
24. B. Mollay, U. Lemmer, R. Kersting, R.F. Mahrt, H. Kurz, H.F. Kauffmann, H. Bässler, Phys. Rev. B 50, 10769 (1994-I)
25. U. Lemmer, R.F. Mahrt, Y. Wada, A. Greiner, H. Bässler, E.O. Göbel, Chem. Phys. Lett. 209, 243 (1993)
26. A. Horvat, G. Weiser, H. Bässler, Phys. Stat. Sol (B) 173, 755 (1992)
27. M. Scheidler, U. Lemmer, R. Kersting, S. Karg, W. Riess, B. Cleve, R.F. Mahrt, H. Kurz, H. Bässler, E. Göbel, P. Thomas, Phys. Rev. B 54, 5536 (1996)
28. H. Antoniadis, L.I. Rothberg, F. Papadimitrakopoulos, M. Yan, M.E. Galvin, M.A. Abkowitz, Phys. Rev. B 50, 14911 (1994)
29. R. Kersting, B. Mollay, M. Rusch, J. Wenisch, G. Leising, H.F. Kauffmann, J. Chem. Phys. 106, 2850 (1997)
30. I.B. Berlman, Handbook of Fluorescence Spectra of Aromatic Molecules, 2nd Ed., Academic Press, New York, 1971
31. R. Richert, H. Bässler, B. Ries, B. Movaghar, D. Würtz, Phil. Mag. Lett. 59, 95 (1989)
32. A.I. Rudenko, H. Bässler, Chem.Phys.Lett. 182, 1269 (1991)
33. M. Deussen, M. Scheidler, H. Bässler, Synth. Met. 73, 123 (1995)

34. R. Kersting, U. Lemmer, M. Deussen, H.J. Bakker, R.F. Mahrt, E.O. Göbel, Phys. Rev. Lett. 73, 1440 (1994)

35. M. Deussen, P. Haring Bolivar, G. Wegmann, H. Kurz, H. Bässler, Chem. Phys. 207, 147 (1996)

36. R.F. Mahrt, H. Bässler, Synth. Met. 45, 107 (1991)

37. R.F. Mahrt, T. Pauck, U. Lemmer, U. Siegner, M. Hopmeier, R. Hennig, H. Bässler, E.O. Göbel, P. Haring, Bolivar, G. Wegmann, H. Kurz, U. Scherf, K. Müllen, Phys. Rev. B. 54, 1759 (1996-I)

38. U. Lemmer, S. Heun, R.F. Mahrt, U. Scherf, M. Hopmeier, U. Siegner, E.O. Göbel, K. Müllen, H. Bässler, Chem. Phys. Lett. 240, 373 (1995)

39. D. Beljonne, Z-Shuai, R.F. Friend, J.L. Brédas, J. Chem. Phys. 102, 2042 (1995)

40. Z. Shuai, J.L. Brédas, W.P. Su, Chem. Phys. Lett. 228, 301 (1994)

41. T. Pauck, H. Bässler, J. Grimme, U. Scherf, K. Müllen, Chem. Phys. Lett. 210, 219 (1996)

42. W. Gebauer, M. Sokolowski, E. Umbach, Chem. Phys. in press

43. G. Horowitz, B. Backet, A. Yassar, P. Lany, F. Demanze. J.L. Fave, F. Garnier, Chem. Mater. 7, 1335 (1995)

44. D. Fichou, G. Horowitz, B. Xu, F. Garnier, Synth. Met.4 8, 167 (1992)

45. F. Geiger, M. Stoldt, H. Schweizer, P. Bäuerle, E. Umbach, Adv. Mater. 5, 1684 (1993)

46. R. N. Marks, F. Biscarini, R. Zamboni, C. Taliani, Europhys. Lett. 32, 523 (1995)

47. D. Birnbaum, D. Fichou, B.E. Kohler, J. Chem. Phys. 96, 165 (1992)

48. R.N. Marks, R.H. Michel, W. Gebauer, R. Zamboni, C. Taliani, R.F. Mahrt, M. Hopemeier, J. Phys. Chem. in press

49. M. Muccini, E. Lunedei, D. Beljonne, J. Cornil, J.L. Bredas, C. Taliani, J. Chem. Phys. in press

50. N. Periasamy, R. Danieli, G. Ruani, R. Zamboni, C. Taliani, Phys. Rev. Lett. 68, 919 (1992)

51. M. Muccini, E. Lunedei, C. Taliani, F. Garnier, H. Bässler, Synth. Met. 84, 863 (1997)

52. R.N. Marks, R. Michel, M. Muccini, M. Murgia, R. Zamboni, C. Taliani, G. Horowitz, F. Garnier, M. Hopmeier, M. Oestreich, R.F. Mahrt, Chem. Phys. in press

53. G. Herzberg, E. Teller, Z. Phys. Chem. B 21, 410 (1933)

54. A Degli Espositi, O. Moze, C. Taliani, J.T. Tomkinson, R. Zamboni, F. Zerbetto, J. Chem. Phys. 104, 9704 (1996)

55. A. Pröpstl, H.C. Wolf, Z. Naturforsch. 18a, 822 (1963)

56. Z. Vardeny, E. Ehrenfreund, J. Shinar, F. Wudl, Phys. Rev. B. 35, 2498 (1987-I)

57. H.C. Wolf, H. Port, J. Lum. 12/13, 33 (1976)

CHAPTER 4

Low-Temperature Photophysics of Permethylated n-Heptasilane:

The Borderline Between Excitation Localization and Delocalization in a Conjugated Chain

M. K. Raymond, Th. F. Magnera, I. Zharov, R. West, B. Dreczewski, A. J. Nozik, J. R. Sprague, R. J. Ellingson, J. Michl

1
Introduction

The delocalization of electronic excitation in fully saturated molecules plays an important role in determining their photophysical and photochemical properties. Probably the best known example of saturated structures with extensive σ delocalization are the polysilanes, $(RR'Si)_n$, saturated polymers with a silicon backbone, whose valence $\sigma\sigma^*$ absorption lies in the near UV region [1]. For examples of computational studies, see ref [2]. They show a variety of interesting optical properties, such as thermochromism and solvatochromism, which result from the sensitivity of their electronic structure to conformational changes. They have high third-order hyperpolarizability, are photosensitive, and show promise as photolithographic materials. They have high hole conductivity, which is useful in xerography, etc.

Electronic excitation of the random coils present in room temperature solutions of polysilanes is believed to be localized in backbone segments of lengths apparently dictated by irregularities in the chain conformation in a manner that is presently not understood in detail [3]. Energy transfer from the shorter to the longer segments appears to play an important role in polysilane photophysics. The longest segments, responsible for much of the polymer fluorescence, are believed to contain two or three dozen silicon atoms, while polymer absorption seems to be dominated by somewhat shorter chain segments. In order to obtain a better understanding of the optical properties of polysilanes, it appears necessary to first study the spectroscopic, photophysical, and photochemical properties of peralkylated oligosilane chain segments of well-defined length, and as far as possible, of defined conformation. We have chosen to work with permethylated oligosilanes, which are the simplest conformationally [4, 5] and also the most accessible synthetically.

The intense first absorption peak of the Si_nMe_{2n+2} chain is generally believed to be due to a HOMO-LUMO excitation of the $\sigma\sigma^*$ type [1]. Although, at least in the shorter chains, excitations of the $\sigma\pi^*$ type are of comparable energy and mix with $\sigma\sigma^*$ excitations in ways that are sensitive to chain conformation [4, 6, 7], most of the observed absorption intensity is of $\sigma\sigma^*$ origin. The strong first absorption peak has a Franck-Condon allowed shape and as n grows, it acquires intensity and moves regularly to lower energies [1, 8, 9]. This provides convincing evidence for sigma bond delocalization at the equilibrium ground state geometry, which optimizes σ conjugation by providing very similar SiSi bond lengths and SiSiSi valence angles at all interior silicon atoms.

The narrow, Franck-Condon allowed fluorescence peaks of the peralkylated polymers [10], and of the long oligosilanes, $Si_{16}Me_{34}$ [11] and $Si_{10}Me_{22}$ [12], mirror their absorption peaks with a minimal Stokes shift, suggesting that the relaxed excited state keeps a geometry very close to that of the ground state, and enjoys essentially complete excitation delocalization. This fluorescence has high quantum yields and short radiative lifetimes.

In contrast, the fluorescence spectra of short permethylated oligosilanes, Si_6Me_{14} [13, 14], Si_5Me_{12} [14], Si_4Me_{10} [14], and an open $(SiMe_2)_6$ chain that has been constrained to an all-anti conformation [15], while all similar to each other, are dramatically different from those of the long oligosilanes. The Franck-Condon forbidden fluorescence band of these shorter chains is very much broader, is strongly Stokes shifted, and its energy and shape are nearly independent of chain length. The fluorescence has lower quantum yields and a longer radiative lifetime. No emission has been observed from the even shorter permethylated chains, Si_3Me_8 and Si_2Me_6 [14], even at very low temperatures.

The initial suspicion that short peralkylated oligosilane segments have a different lowest excited state with an anomalously long radiative fluorescence lifetime actually arose upon examination of the fluorescence of peralkylated high-molecular-weight polysilanes [10], and the first reported observation and assignment of an anomalous emission in a short oligosilane were made on a nine-silicon-atom chain carrying a complex pattern of both methyl and phenyl substituents [16].

There is hardly any doubt that the emitting relaxed excited singlet state of the shorter oligosilanes has a highly localized excitation and an equilibrium geometry substantially different from that of the ground state [13], and a qualitative rationalization of such excitation trapping has been offered [8, 14]. Based on fluorescence polarization measurements on Si_6Me_{14} [13] and on calculations [8, 14], it has been proposed that the geometrical distortion that is responsible for excitation localization is the stretching of one of the SiSi bonds. More recently a computational search [17] has indeed located a single minimum in the potential energy surface of the lowest singlet excited state of Si_4H_{10}, with the central SiSi bond stretched by 0.2 Å and the other two not changed much, supporting the original proposal. Since in principle any one of several inequivalent internal SiSi bonds could be stretched in a longer chain, it would not be surprising to find excited state isomers differing in the location of the excitation ("excitation localization isomerism"). Since both the SiSi bonding HOMO and the SiSi antibonding LUMO have larger amplitudes towards the ends of a conjugated linear chain, it is likely that the initial downhill slope of the HOMO-LUMO excited S_0 potential energy surface will correspond to the stretching of a terminal SiSi bond, and that the excitation will localize there first. If this is not the most stable localized excited state, isomerization into a species with excitation localized in one of the interior bonds by a suitable combination of bond stretching and compression motions can follow.

In fact, clearly distinct yet fairly similar dual fluorescent emissions have been observed [15] both from Si_6Me_{14} and from an open $(SiMe_2)_6$ chain structurally constrained to an all-anti conformation. As the introduction of a conformational constraint had very little effect, the alternative possibility that the two emissions originate in two distinct conformers was eliminated. At very low temperatures, the excitation appears to be localized in one of the SiSi bonds, while at somewhat higher temperatures, it travels adiabatically to another SiSi bond, where its energy is lower. At sufficiently high temperatures, this happens faster than radiative or radiationless deactivation can occur, and fluorescence from the initially formed isomer is then no longer observable. Very recently, an essentially identical dual fluorescence behavior has been observed for Si_5Me_{12} and also for an $(SiMe_2)_5$ chain constrained to an all-anti geometry [18]. However, no evidence is available so far on the location of the stretched bonds in the various localized excited states in any of these chains.

It has been noted that the low-temperature fluorescence of Si_7Me_{16} is similar to that of the long oligosilanes, while its room-temperature fluorescence resembles that of the short oligosilanes, suggesting that seven silicon atoms represent the critical borderline length for relaxed excitation delocalization in a permethylated oligosilane [8].

There is considerable analogy between the phenomena described here and the behavior of excitons in extended systems of low dimensionality [19]. For instance, theory suggests that in a cyclic one-dimensional system exciton self-trapping will occur when the length of the perimeter crosses a critical value [20].

Although the nature of the localized excited state of the shorter oligosilanes is thus gradually becoming elucidated, many questions remain. Those that we address presently are: (i) Is it really at the seven-silicon chain that the relaxed excited state of a permethylated oligosilane switches from being localized on one or another SiSi bond, and Franck-Condon forbidden, to being delocalized throughout the chain and Franck-Condon allowed? (ii) How does this transition occur; e.g., does the seven-silicon chain have two relaxed minima in its excited state potential energy surface, one with a localized and one with a delocalized geometry ("bond-length isomerism"), interconverting adiabatically? In order to answer these questions, we have synthesized a series of permethylated oligosilanes and examined their photophysics, with particular emphasis on Si_7Me_{16}, Si_8Me_{18}, and $Si_{10}Me_{22}$. Since quantitative measurements had to be done at a series of quite low temperatures, the techniques that had to be developed are somewhat uncommon. They may be useful to other investigators, and are therefore described in considerable detail.

2
Experimental

2.1
Sample Origin and Purification

Literature procedures were used for the synthesis of Si_3Me_8 [21], Si_4Me_{10} [21], Si_5Me_{12} [21], Si_6Me_{14} [21], and $Si_{10}Me_{22}$ [22]. The syntheses of Si_7Me_{16} [23] and Si_8Me_{18} [23] deviated from the literature and are described below.

All samples were purified by preparative gas chromatography (3′ × 1/4″ SE-52 or 6′ × 1/8″ OV-1, both on 80–100 mesh Chromosorb W-HP) to better than 99.9% purity, as checked by gas chromatography on a capillary column (0.2 mm × 50 m, cross-linked silica HP-5) and by ^{1}H NMR.

2.2
Synthesis of n-Hexadecamethylheptasilane

Si_7Me_{16} was prepared by the reaction of two equivalents of pentamethyldisilanyllithium, obtained by a modified literature procedure [24], with one equivalent of 1,3-dichlorohexamethyltrisilane. Pentamethyldisilane [25] (66 g, 0.5 mol) was heated with di-*tert*-butylmercury [26] (31.4 g, 0.1 mol) at 120 °C under inert atmosphere in a flask connected to a gas burette. After isobutane stopped evolving (9 h, 87% of the theoretical amount), excess pentamethyldisilane was removed under reduced pressure and the residue was distilled under vacuum, providing bis(pentamethyldisilanyl)mercury as an air- and light-sensitive yellow oil, b.p. 131–132 °C/0.1 mm Hg. This material (12 g, 26 mmol) was dissolved in dry degassed hexane (100 ml) and finely cut Li metal (2 g, 26 mmol) was added. After stirring at room temperature for 12 h the reactant was consumed and the resulting hexane solution of pentamethyldisilanyllithium was filtered to remove lithium and mercury metals. To 50 mL of the solution of pentamethyldisilanyllithium (3.6 g, 26 mmol), 1,3-dichlorohexamethyltrisilane [25] (2.45 g, 10 mmol) in dry degassed hexane (10 ml) was slowly added at room temperature. After 12 h of stirring, GC-MS analysis showed 65% of Si_7Me_{16} along with 32% of 1-chloroundecamethylpentasilane. Additional solution of pentamethyldisilanyllithium in hexane (15 ml) was added to the reaction mixture and stirring was continued for an additional 12 h, resulting in 94% overall conversion to Si_7Me_{16}, which was isolated by column chromatography on silica gel with hexane as eluent (3.4 g, 78%).

2.3
Synthesis of n-Octadecamethyloctasilane

1,6-Dichlorododecamethylhexasilane was first prepared by stirring and heating dodecamethylcyclohexasilane (10.0 g, mmol) with excess tin tetrachloride (4.8 ml) to 180 °C for 80 min under inert atmosphere. After cooling, solids were removed by filtration, washed with hexanes, solvent was removed from the combined washings, and the residue was distilled on a Kugelrohr apparatus to yield 3.4 g (29% yield) of 1,6-dichlorododecamethylhexasilane. This was refluxed with 3.6 equivalents of trimethylsilyl chloride and excess sodium in benzene. After 50 h the yield of Si_8Me_{18} was 28%, as determined by GC. Ethanol was added to quench the sodium, and the organic layer was extracted with water. The benzene was evaporated and the product was purified by Kugelrohr distillation followed by column chromatography on silica with hexane. This provided a sample of about 90% purity, suitable for further purification by preparative GC.

Table 1. Volume Contraction of Cyclopentane/Isopentane (3/7, v/v) with Temperature

T/K	Volume	T/K	Volume
292	100.0	155	84.7
275	97.3	146	83.9
248	94.0	132	82.5
221	90.9	115	81.1
202	88.9	107	80.5
171	86.3	77	77.2

2.4
Solvent

All measurements were performed in a mixture (3:7 by volume) of 99% cyclo-pentane (Fisher) purified by stirring with sulfuric acid overnight, neutralizing, washing with water, drying, and distilling with sodium, and 99+% isopentane (Fisher) [27]. This solvent mixture remains fluid down to 90–95 K and forms a clear glass at lower temperatures, stable at least to 30 K. Room-temperature absorbance of the solvent in a 1 cm cell did not exceed 0.4 at 200 nm. Concentrations were corrected for solvent contraction (Table 1). Before emission measurements, all solutions were degassed by at least three freeze-pump-thaw cycles on a 10^{-6} Torr vacuum line.

2.5
Cells and Temperature Control

Sample solutions were contained in Suprasil cells held in a quartz dewar vessel with Suprasil windows, either filled with liquid nitrogen, or purged with a stream of boil-off nitrogen gas warmed to the desired temperature. Sample temperatures were read with a thermocouple held against the cell with Teflon tape.

Measurements below 77 K were performed in sealed cells held with Cry-con grease in a copper block mounted in a closed-cycle cryostat (Air Products Displex 202). The thermocouple was held against the cell with Cry-con grease.

The accuracy of the temperature determinations is estimated at ±2 K.

2.6
Spectral Measurements

Absorption spectra were measured with a Cary 17 spectrometer modified by Olis, Inc. Emission and excitation spectra were measured at front-surface geo-metry using a 15° angle between the exciting and observed beam, with a 300-W xenon arc and a 0.5 m SPEX 1702 double monochromator for excitation, and a ISA Yvon-Spex 0.75 m monochromator and a Centronics 04283S-25 photomul-tiplier for detection. Excitation intensity corrections were based on comparison of absorption and uncorrected fluorescence excitation spectra of dilute solutions of anthracene, 9,10-diphenylanthracene, and naphthalene, known [28] to have identical absorption and true excitation spectra. Emission intensity corrections were based on an Oriel 63358 standard irradiance lamp. Dilute solutions were used in all measurements ($< 10^{-4}$ M for fluorescence, optical density below 0.35 at excitation wavelengths for excitation spectra). The spectra were concentration independent in the range used.

2.7
Fluorescence Quantum Yields

These were measured with excitation at the peak wavelength in the excitation spectrum, whose optical density was below 0.35, using 10^{-5}–10^{-6} M solutions of the oligosilane containing about 10^{-7} M 9,10-diphenylanthracene as an internal standard (quantum yield of unity [29]), and were double-checked by relative intensity measurements without the standard at a series of temperatures, using optical density above 1.8 to avoid the need for corrections for temperature-induced concentration and absorption shape changes. Measurements were performed with the weakest possible excitation intensities and were repeated after various periods of irradiation to guard against possible effects of photodecomposition.

2.8
Fluorescence Decay Curves

These were measured with a time-correlated single-photon counting spectrometer. Quadrupled light from a Spectra-Physics 3800 Nd:YAG laser mode-locked at 82 MHz was used for excitation of Si_6Me_{14} and Si_7Me_{16} at 266 nm (sample optical density 0.05–0.1). The maximum pulse energy was 0.13 nJ, and with 80 ps pulse width, the maximum peak power was 1.7 W. For the excitation of Si_8Me_{18}, 4 MHz output from a cavity-dumped synchronously pumped Spectra-Physics 3500 dye laser was doubled to give 290–310 nm light. A minimum average power of 80 mW created a pulse energy of 0.2 nJ. The maximum peak power was 13–20 W and the pulse width was 10–15 ps.

Neutral density filters were used to attenuate the power by 0–90%. The spot size at the sample had a diameter of about 0.2 mm. The emitted light was collected at a 90° angle and sent through a monochromator to the detector. The detection system used an R3809U-01 standard microchannel plate photomultiplier tube sensitive from 160 to 910 nm in an E3059–02 socket, a C2773 cooler (all from Hamamatsu), a Tennelec TC 454 constant fraction discriminator and a Tennelec TC 864 time to amplitude converter. The number of counts reaching the detector was kept below 0.1% of the repetition rate by appropriate use of neutral density filters. The instrument response function was measured with the oligosilane solution as a scattering solution and the monochromator set to the excitation wavelength or to the wavelength of the strongest Raman band of the solvent mixture in order to eliminate contributions to scattered light that did not originate from the sample. Standard Laplace transform convolution techniques were used. The reproducibility of the subnanosecond decay rate constants was better than 8%. The shortest decay times that could be measured reliably were less than 100 ps. The literature value of the fluorescence lifetime of 9,10-diphenylanthracene (8.35 ns [28]) was reproduced exactly.

2.9
Low-Temperature Flow Cell

In order to avoid sample photodecomposition in the fluorescence decay measurements, a flow cell (Fig. 1) was built from about 3 cm long and 1 cm thick rectangular Suprasil quartz tubing of 3 × 5 mm internal dimensions, connected to Teflon tubing at both ends. The tube was placed inside the temperature-controlled optical dewar vessel described above. The solution was degassed and kept under nitrogen. It was stored in a precooled container located in another dewar vessel, from which it was pushed at a rate of 0.5 ml/min by pressure of nitrogen gas. It passed through a thermally insulated transfer line into 0.3 m of thin Teflon tubing coiled within the optical dewar to assure that its temperature was equilibrated before it reached the cell. It was not recirculated. This flow cell worked well at temperatures down to 90–95 K, but could not be used at lower temperatures since the solution was too viscous to flow.

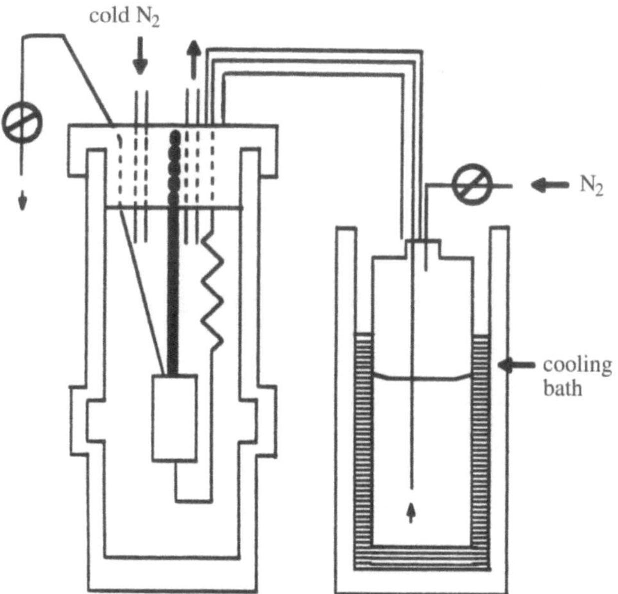

Fig. 1. Low-temperature flow cell

3
Results

3.1
Room- and Low-Temperature Spectra of Permethylated Linear Oligosilanes

The absorption and corrected fluorescence spectra of permethylated linear oligosilanes in a hydrocarbon solvent are shown in Fig. 2 (77 K) and Fig. 3 (room temperature). At low temperature, corrected fluorescence excitation spectra agree so well with the absorption spectra that they are difficult to discern in the plot, but the room-temperature excitation spectra are clearly distinct from the absorption spectra. The fluorescence shape was independent of concentration and excitation wavelengths, except for $Si_{10}Me_{22}$ at room temperature, where the peak shifted by up to 300 cm^{-1} as the wavelength of excitation changed. The excitation spectra were independent of concentration and the monitored wavelength, again with the exception of $Si_{10}Me_{22}$, where very minor shape differences were observed as a function of the latter. The fluorescence quantum yields drop with decreasing chain length and increasing temperature, and this is reflected in the noise level and in the absence of some of the curves in Figs. 2 and 3. Peak positions in the absorption, fluorescence, and excitation spectra are collected in Table 2. The fluorescence results shown for Si_5Me_{12} and Si_6Me_{14} reflect the low-temperature limit (at somewhat higher temperatures another moderately red-shifted fluorescence appears [15]). No phosphorescence was observed for any of the samples.

3.2
Temperature Dependence of Si_7Me_{16} Fluorescence

A comparison of Figs. 2 and 3 makes it clear that Si_7Me_{16} is indeed the borderline member of the series. After a search at a series of temperatures, we con-

Table 2. Peak Positions (cm^{-1}) in the Absorption, Excitation, and Fluorescence Spectra of Si_nMe_{2n+2} in Cyclopentane/Isopentane (3/7, v/v) at Low (LT)[a] and Room (RT) Temperature

n	$\tilde{\nu}_{ABS}(LT)$[b]	$\tilde{\nu}_{ABS}(RT)$	$\tilde{\nu}_{EXC}(RT)$	$\tilde{\nu}_{FL}(LT)$[c]	$\tilde{\nu}_{FL}(RT)$
2	53100[d]				
3	47100	46200			
4	43100	42600	26800		
5	40100	40000	27600		
6	38100	38500	37900	29600	27000
7	36600	37200	36200	35660	34000
8	35500	36700	35500	34790	33400
10	34100	35700	34000	33600	32200
16[e]	32100	34100	31750	30500	

[a] $\tilde{\nu}_{EXC}(LT) = \tilde{\nu}_{ABS}(LT)$. [b] At 77 K. [c] At 30–35 K. [d] Ref. [8]. [e] In 3-methylpentane [11].

Fig. 2. Absorption, emission, and excitation (dashed) spectra of permethylated oligosilanes in cyclopentane/isopentane (3/7, v/v). At 77 K, except for Si_4Me_{10} (17 K, neat) and Si_5Me_{12} and Si_6Me_{14} (30 K). Peak intensity normalized to unity

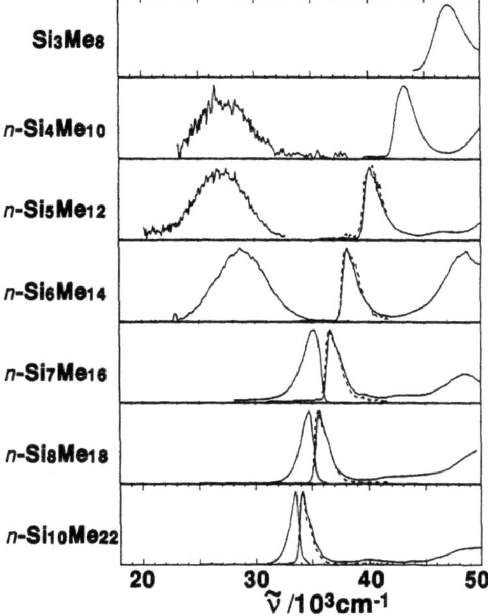

Fig. 3. Absorption, emission, and excitation (dashed) spectra of permethylated oligosilanes in cyclopentane/isopentane (3/7, v/v) at room temperature. Peak intensity normalized to unity

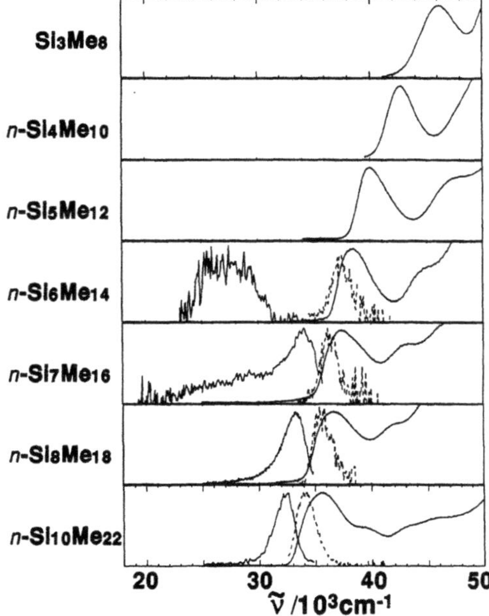

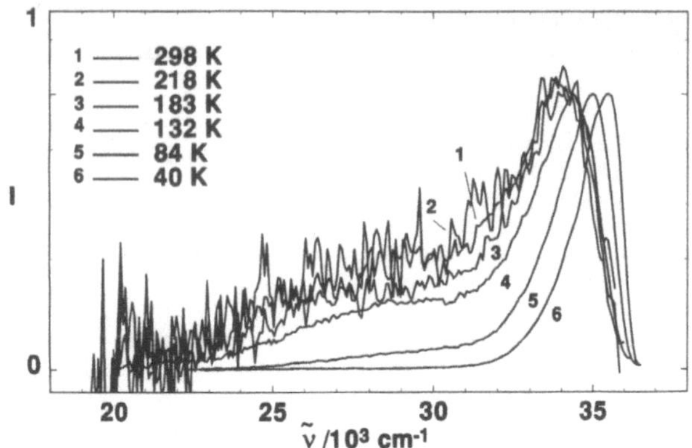

Fig. 4. Fluorescence of Si_7Me_{16} in cyclopentane/isopentane (3/7, v/v) at a series of temperatures. Peak intensity normalized to unity

cluded that Si_8Me_{18} only exhibits emission from a delocalized state and Si_6Me_{14} only shows emission from two or more localized states. In contrast, the fluorescence of Si_7Me_{16} (Fig. 4) is strictly from a delocalized state at 40 K, while emissions from both types of states are observed at room temperature.

Further investigation has therefore concentrated on Si_7Me_{16}. Its fluorescence has been recorded at a series of temperatures (Fig. 4) and decomposed into contributions of the "delocalized" and "localized" types. The procedure is approximate since both the normal and the anomalous emission broaden and shift somewhat as temperature increases, as illustrated in Fig. 5 for 105 K and 132 K. The decomposition was done by Gaussian broadening, shifting, and scaling the 40 K emission spectrum and then subtracting it from the 132 K fluorescence

Fig. 5. Decomposition of the dual fluorescence of Si_7Me_{16} in cyclopentane/isopentane (3/7, v/v) at two temperatures into emissions A (normal) and B (anomalous)

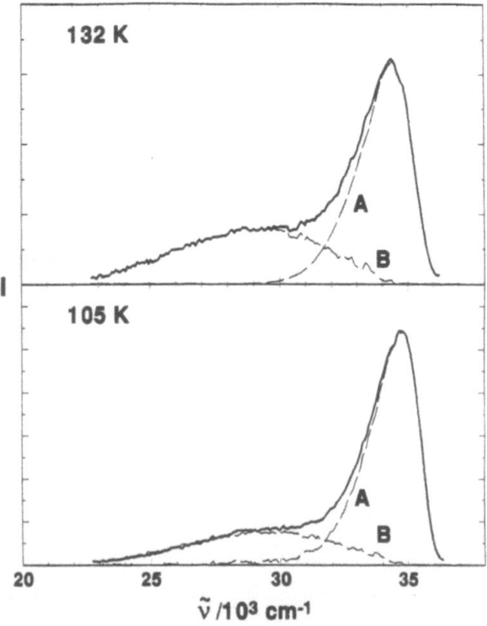

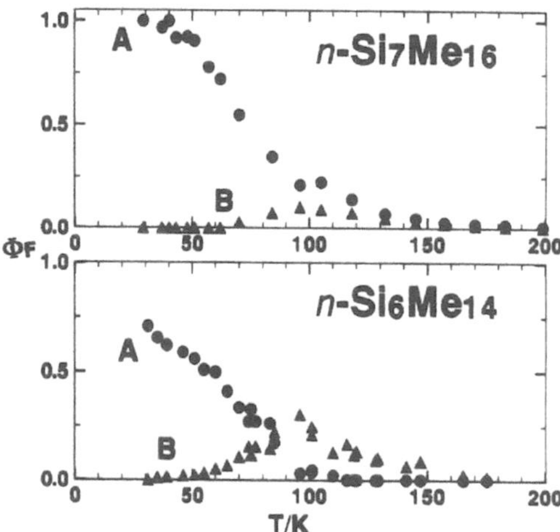

Fig. 6. Quantum yields of fluorescence of Si_6Me_{14} (by permission from ref. [15]) and Si_7Me_{16} in cyclopentane/isopentane (3/7, v/v). Circles, emission A, triangles, emission B

spectrum to yield the broad curve of the anomalous emission B (Fig. 5), with a width at half-height of about 7600 cm⁻¹ and a Stokes shift of about 10000 cm⁻¹. This anomalous emission curve was then scaled and subtracted from the dual emission curves observed at other temperatures to obtain the normal emission A. Figure 6 gives the total fluorescence quantum yield as well as the separate quantum yields of the two individual Si_7Me_{16} emissions A and B in the range of temperatures in which they could be measured reliably by integration of the two areas that resulted from the decomposition.

For comparison, the quantum yields of the two emissions A and B from Si_6Me_{14} [15] are also shown. In this case, both A and B arise from localized excited states and dominate at different temperatures. Their spectral shapes have already been published [15] and are not repeated here (the spectrum shown in Fig. 2 is emission A).

3.3
Fluorescence Decay Measurements

In most instances, the fluorescence decay curves could only be measured reliably at temperatures above 90–95 K, since photochemical decomposition of samples that were not flowing was too rapid. Decays of Si_6Me_{14} [13] and Si_8Me_{18}, which do not show multiple emissions at these temperatures, were fitted well by single exponentials (Table 3). The decay of the two emissions from Si_7Me_{16} was more complicated. In the initial period, the normal emission from the delocalized excited state decayed faster than the anomalous emission from the localized state, but its decay was clearly non-exponential, slowed down in time, and in the long-time limit its rate approached that of the anomalous emission, which decayed nearly exponentially all along. At higher temperatures, the limiting single exponential decay rate was reached faster and it was higher (Table 3), as illustrated in Fig. 7. No induction period was observed for either the normal or the anomalous emission.

Table 3. Single-Exponential Fluorescence Lifetimes of Si_nMe_{2n+2} in Cyclopentane/Isopentane (3/7, v/v)

Si_6Me_{14}[a]		Si_7Me_{16}[b]		Si_8Me_{18}[c]	
T/K	τ_{FL}/ps	T/K	τ_{FL}/ps	T/K	τ_{FL}/ps
77	1190[d]	98	890	77	460[d]
100	870	106	610	85	580[c]
108	700	123	270	106	530
117	410	136	240	113	445
129	245	139	130	125	350
146	110	153	~ 100	133	320
162	~ 90	140	230		
163	130				
188	~ 60				

[a] Anomalous fluorescence.

[b] Anomalous fluorescence (nearly single exponential). The normal fluorescence shows a strongly non-exponential decay (Fig. 7).

[c] Normal fluorescence.

[d] A less reliable value (sample not flowing).

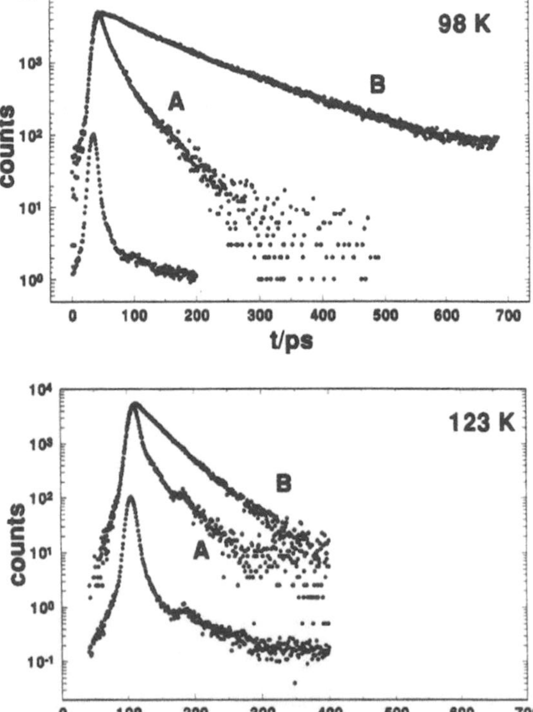

Fig. 7. Fluorescence decay curves of Si_7Me_{16} in cyclopentane/isopentane (3/7, v/v) at different temperature. A: normal, B: anomalous emission. Raw data, no deconvolution. The experimental response to the excitation pulse is shown at the bottom on a different scale

4
Discussion

4.1
Identification of the Borderline Case

Figures 2 and 3 provide a good overview of the chain-length dependence of the spectral properties of permethylated oligosilanes, and yield an answer to the first question posed in the Introduction: a permethylated oligosilane chromophore supports relaxed delocalized excitation down to a chain of seven silicon atoms, and it supports relaxed localized excitation up to a chain of seven silicon atoms. In spite of a fairly diligent search at a series of temperatures, we have failed to find any indication that a chain with fewer than seven silicon atoms has an equilibrium geometry in the excited state in which the excitation is delocalized, and no indication that a chain with more than seven silicon atoms has one in which the excitation is localized.

4.2
Nature of the Borderline Case

The answer to the second question is less obvious. The dual emission observed from Si_7Me_{16} makes it clear that the potential energy surface for the lowest singlet of linear-chain Si_7Me_{16} contains at least two minima from which fluorescence is possible, one (A) at a geometry nearly identical with the equilibrium geometry of the ground state (delocalized excitation), the other (B) at a geometry that is quite different and presumably differs by the stretching of one of the SiSi bonds (localized excitation). One would not expect more than a single fluorescence from a localized excited state to have much intensity, even though Si_7Me_{16} contains several inequivalent SiSi bonds, any one of which could presumably stretch, since its anomalous emission only appears at temperatures above about 70 K. Already at temperatures above 90 K the conversion from the less stable to the most stable localized excited isomer must be very fast in Si_5Me_{12} [18] and Si_6Me_{14} [15], since no emission from the former is observed, and it is thus likely to be very fast in Si_7Me_{16} as well. Therefore, only in a narrow temperature window, 70–90 K, is there any likelihood that multiple emissions from isomeric localized states could be observed. Although in Fig. 4 the spectrum recorded at 84 K appears somewhat blue shifted relative to those recorded at higher temperatures, the differences are too small to permit any claims at this time. It is also possible that several excitation localization isomers of Si_7Me_{16} are isoenergetic and are in rapid equilibrium at higher temperatures, and that their emissions overlap.

If we view two conformers as two distinct ground-state molecules G and G', we can ask whether (i) both fluorescent excited potential surface minima A and B occur within one molecule (conformer) or whether (ii) they each occur in a distinct molecule (conformer). The former situation (i) has just been encountered with Si_5Me_{12} [18] and Si_6Me_{14} [15] (at the low temperatures chosen for Fig. 2 only emission A is seen in both cases) and their conformationally constrained analogues (excitation localization isomerism, bond stretch isomerism) but appears unlikely in Si_7Me_{16}, as there would have to be a distinct barrier in the potential energy surface between a geometry in which all SiSi bonds have equal lengths and a geometry in which one of them has been stretched by about 0.2 Å. Even the excited state barriers that separate the geometries of Si_5Me_{12} and Si_6Me_{14} in which one or another bond is stretched are exceedingly small, and thermally induced travel over them competes successfully with fluorescence at all temperatures above 100 K.

The latter situation (ii) is quite interesting. None of the chains shorter than Si_7Me_{16} exhibit significant fluorescence from more than one conformer, but $Si_{16}Me_{34}$ clearly does [11], and the slight variations in the fluorescence spectrum of $Si_{10}Me_{22}$ as a function of exciting wavelength, and in its excitation spectrum as a function of monitoring wavelength, suggest that more than one conformer of this chain fluoresces, too. The changeover from a single fluorescent conformer to two or more could well take place at Si_7Me_{16}. The existence of two fluorescent conformers in a saturated molecule would offer a unique opportunity to study the dynamics of conformer interconversion in real time. Before we address this issue, we shall briefly describe conformational isomerism in the ground state of permethylated oligosilanes as it is understood today.

4.3
Oligosilane Conformers

There is much computational and considerable experimental evidence that the oligosilanes occur as mixtures of conformers in the ground state, at least at room temperature. It is known from matrix isolation work [14] that room-temperature vapors of Si_4Me_{10}, Si_5Me_{12}, and Si_6Me_{14} contain several conformers, and that even at low temperatures, only the one assigned as all-anti fluoresces with easily detectable intensity. In the cases of Si_4Me_{10} [4] and Si_5Me_{12} [5], the conformers have been studied in considerable detail. Calculations suggest that conformational isomerism is due to the existence of three stable arrangements around a SiSi bond. In the order of increasing energy, they are the anti (a_+, a_-), gauche (g_+, g_-), and ortho (o_+, o_-) mirror image (enantiomeric) pairs of structures, characterized by SiSiSiSi dihedral angles near ± 165, ± 55, and $\pm 90°$, respectively. Their availability causes the number of stable conformers to increase rapidly as the chain length increases, and eight enantiomeric conformer pairs have been identified computationally in Si_5Me_{12} [5]. All conformers of these short oligosilanes are racemic and consist of equal amounts of left-handed and right-handed helices, but certain conformers of longer oligosilanes are expected to be of the meso type, with a plane of symmetry relating two helices of opposite handedness within the same molecule. For simplicity, we sometimes also refer to the enantiomeric pairs of structures as single conformers in the following discussion.

The perfect agreement between the absorption and excitation spectra in the series of oligosilanes measured at low temperature (Fig. 2) is most simply understood by postulating that at the lowest temperatures the equilibrium composition of the conformer mixture is totally dominated by a single fluorescent conformer, characterized by a strong low-energy absorption peak and by intense low-temperature fluorescence. Its structure has been established as anti (a_+, a_-) by detailed spectral analysis of matrix-isolation results in the case of Si_4Me_{10} [4, 6] and as all-anti by comparison of the spectra of a free with a conformationally constrained chain in the case of Si_5Me_{12} (a_+a_+, a_-a_-) [18] and Si_6Me_{14} $(a_+a_+a_+, a_-a_-a_-)$ [15]. According to calculations [5], the a_+a_- structure of Si_5Me_{12} and the $a_+a_0a_-$ structure of Si_6Me_{14} are also stationary points on the S_0 potential energy surface, but they are merely transition states for conformational interconversion. As a result, each of these three oligosilanes has only a single all-anti pair of conformers.

No thorough conformational searches have been performed for longer permethylated oligosilanes, but it is reasonable to assume that the racemic mixture of $a_+a_+a_+a_+$ and $a_-a_-a_-a_-$ is the dominant low-temperature conformer in Si_7Me_{16}, and that the situation is similar in somewhat longer chains. In these, however, we anticipate the existence of stable all-anti forms other than those of a single handedness, such as $a_+a_+a_-a_-$ in Si_7Me_{16} and $a_+a_+a_-a_-a_-$, $a_+a_+a_+a_-a_-$ in Si_8Me_{18}.

The disparity between the room-temperature absorption and excitation spectra of the permethylated oligosilanes (Fig. 3) is caused by changes in the absorption spectra upon warming, while the excitation spectra remain unchanged, except for a slight shift and broadening. The difference can thus be readily attributed to the presence of non-fluorescent higher energy conformers at elevated temperatures, which contribute to the absorption but not to the excitation spectra. These non-fluorescent conformers presumably contain ortho and/or gauche links.

4.4
Temperature Dependence of Quantum Yields

Although at first sight the dependence of the dual fluorescence quantum yields of Si_6Me_{14} and Si_7Me_{16} on temperature (Fig. 6) is quite similar, there are significant differences between the two that suggest that the two cases are described by different kinetic schemes. In both cases, as temperature approaches 0 K, the fluorescence quantum yield of the high-temperature form goes to zero and that of the low-temperature form seems to extrapolate to unity, or close to unity, but in very different ways (the results for conformationally constrained Si_6Me_{14} [15] are nearly identical with those for free Si_6Me_{14} and are not shown here). Note that the quantum yield of unity in the low-temperature limit would imply that intersystem crossing in the all-anti conformer is not competitive at low temperatures, in spite of the presence of a fair number of relatively heavy silicon atoms in the molecules. It is unlikely that intersystem crossing would become more competitive at higher temperatures, when all the thermally activated radiationless decay processes are much faster. However, no direct information on the nature of non-radiative processes in twisted conformers is available, and fast intersystem crossing is one of the possible reasons for their essentially exclusively dark decay.

For Si_6Me_{14}, a qualitative kinetic scheme has already been outlined (Scheme 1) [15]. The two fluorescing species, A and B, responsible for the emis-

sions A and B, respectively, both have localized excitation, presumably one in
one SiSi bond (say, a terminal one), and the other in another (say, an internal
one). The limiting value of unity quantum yield of the low-temperature form A
is approached slowly, and the value continues to rise well after there is no observable emission from the high-temperature form B. This behavior has been
attributed to a radiationless decay process from A, with a very small activation
energy. The abrupt increase in the slope of the quantum yield of the low-temperature form A at about 60 K, which corresponds with an onset of the emission
of the high-temperature form B, has been attributed to the opening of a new
channel, the thermally activated conversion of the low-temperature form A into
the high-temperature form B, which is a better emitter. As the temperature increases further to about 100 K, the emission yield of the low-temperature form
A goes to zero, while that of the high-temperature form B peaks. At even higher
temperatures, the emission yield of the high-temperature form B slowly drops,
and becomes hard for us to measure accurately above 180 K or so.

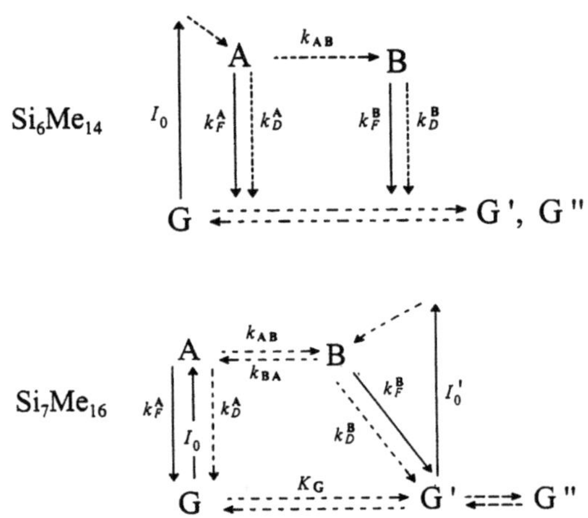

This behavior is compatible with Scheme 1 in which the initial vertically
excited state of Si_6Me_{14} "immediately" relaxes into the low-temperature form A,
which then either fluoresces, or undergoes a thermally activated radiationless
decay (presumably by climbing up the lowest excited singlet potential energy
surface to a nearby conical intersection with the ground state), or a thermally
activated irreversible conversion to the more stable high-temperature form B,
which in turn can fluoresce or undergo a thermally activated radiationless decay, presumably by climbing to another conical intersection with the ground
state. After return to the ground state through either "funnel", the molecule
may relax to its initial equilibrium geometry, or to the equilibrium geometry of
another conformer, or to that of the photochemical products, dimethylsilylene
($SiMe_2$) and an abridged oligosilane, Si_5Me_{12}. The likely geometries at which
the funnels are located are suggested by the results of calculations for 2-
methyltrisilane [30] and are quite close to the anticipated geometries of the
stretched-bond minima. A qualitative schematic sketch of a potential energy

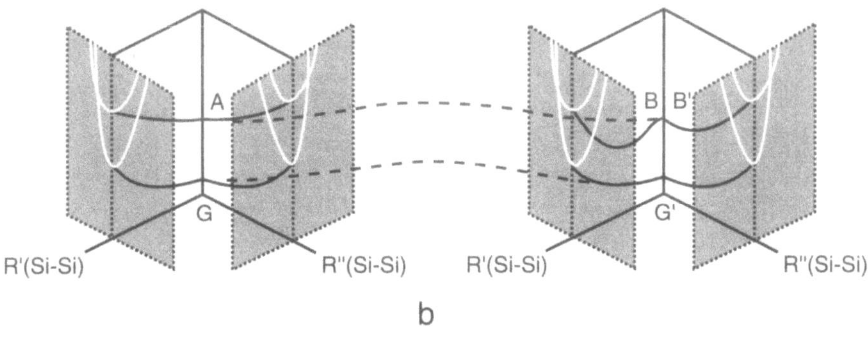

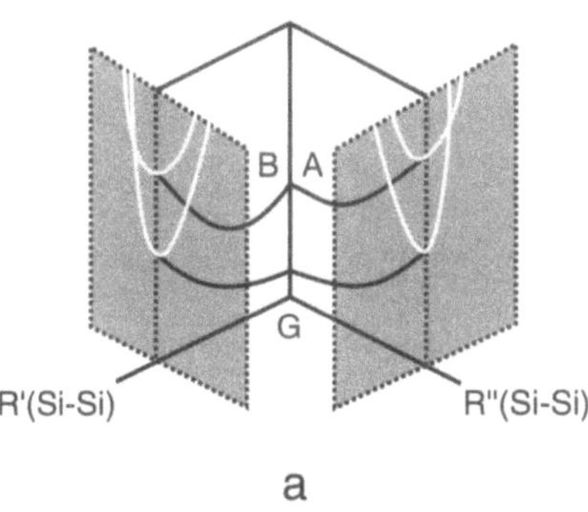

Fig. 8. A schematic sketch of cuts through multidimensional S_0 and S_1 potential energy surfaces proposed for Si_6Me_{14} (**a**) and Si_7Me_{16} (**b**). The two principal coordinates shown, R' and R", correspond to the stretching of one and another SiSi bond. The additional ones (shaded planes) lead to a S_0-S_1 touching and correspond to less well defined motion involving the separation of an $SiMe_2$ unit from the chain. The dashed lines in Part **b** represent cuts through the surfaces in yet another dimension, corresponding to G-G' conformer interconversion

diagram that accommodates the tentatively proposed kinetic scheme is shown in Fig. 8a.

A quantitative analysis is hampered by two factors. First, as the temperature increases, a growing but unknown fraction of the ground-state molecules is undoubtedly present not in the form of the fluorescent all-anti conformer G $(a_+a_+a_+, a_-a_-a_-)$, but in the form of twisted non-fluorescent conformers G', G", etc. (not shown in Fig. 8a), whose presence most likely decreases the fluorescence quantum yield, as they probably absorb at the excitation wavelength with some unknown extinction coefficients. Second, since the sample does not flow at temperatures below 90–95 K, we have not yet been able to obtain reliable fluorescence decay curves at low temperatures, where they would contain information about the excited species A and about the kinetics of its conversion into B. At temperatures at which fluorescence decay measurements have been possible, the expected rise times for the fluorescence of B are well below our instrumental limit of 50–100 ps. In spite of the uncomfortable number of free kinetic variables involved at temperatures for which decay curves are available, a rough analysis was performed and left no doubt that the observations can be fitted by the kinetic model of Scheme 1, in which only one fluorescent species G absorbs light, and the low-temperature form of the excited state A converts irreversibly into the high-temperature form B by shifting excitation from one SiSi bond to another, i.e., by stretching one SiSi bond and shortening another.

For Si_7Me_{16}, the two emitting species A and B differ more profoundly. The high-temperature form B has localized excitation, and the form A that is present both at low and high temperatures (the "low-temperature form") has delocalized excitation. The limiting value of unity for the emission quantum yield of the low-temperature form A is reached at about 40 K, a little below the temperature at which the emission of the high-temperature form B disappears. The fluorescence quantum yield of the low-temperature form A drops as the temperature increases, but never becomes lower than that of the high-temperature form B, and above 130 K or so, the two occur at a constant ratio. It thus appears that the two excited species A and B interconvert adiabatically at temperatures higher than about 130 K, and are in rapid equilibrium. The radiationless decay of the low-temperature form A appears to have a higher activation energy than was the case in Si_6Me_{14}, permitting the fluorescence quantum yield to remain very close to unity at higher temperatures than was the case for Si_6Me_{14}.

In this case, the number of kinetic parameters to be determined is even higher. In order to make progress, we must also consider the fluorescence decay curves. Fortunately, we can now observe both emissions at temperatures above 90–95 K, where the sample still flows and reliable measurements are easier. However, above 170 K or so, all lifetimes decrease rapidly and we reach the limit of our instrumentation.

4.5
Temperature Dependence of Fluorescence Decay Curves

The combination of the fluorescence quantum yield (Fig. 6) and decay (Fig. 7) measurements for Si_7Me_{16} suggest the model outlined in Scheme 1 and the qualitative potential energy diagram of Fig. 8b as the simplest interpretation of the results (more complicated models are easily thought of): At the lowest temperatures, only the most stable all-anti ground state conformer G is present. Its excitation produces a delocalized excited state (the low-temperature emitting species A). This can fluoresce, producing the normal emission, it can undergo radiationless decay with thermal activation, and it can reversibly convert to the high-temperature form B adiabatically, with thermal activation. In Fig. 8b, this is shown by a dashed potential energy curve, which runs in yet another dimension than those shown individually for conformers G and G'. It represents the degree of freedom that interconverts conformers G and G' by changing a SiSiSiSi dihedral angle or angles. Form B, responsible for the anomalous emission, is the excited state of a less stable ground state conformer G'. This is absent at the lowest temperatures, but present in significant amounts at temperatures above about 60 K. The structure of G' is not known, but a proposal is made below. The excited state B, reached either adiabatically from A, or by the absorption of light from G', can fluoresce, undergo radiationless decay to the ground state, or adiabatically convert back to A without loss of electronic excitation. It is possible that light absorption by G' initially leads to a different localized excited isomer (B' in Fig. 8b), as was the case in Si_5Me_{12} and Si_6Me_{14}, but at the relatively high temperatures required to populate G' sufficiently for observation, B' relaxes to B too fast to be observed. Although B' is shown in Fig. 8b, at present we have no experimental evidence for its existence.

At higher temperatures, other conformers, G", etc., not shown in Fig. 8b, are present and absorb light, but do not fluoresce. Their presence contributes to the general decrease of the total fluorescence quantum yield with temperature. At sufficiently high temperatures, the interconversion of the excited conformers A and B is rapid and a stationary state is reached within the excited state lifetime. Soon after pulsed excitation, an excess of form A is present since the ground state conformer G is more stable and perhaps also has a higher absorption coefficient. This excess is lost rapidly, perhaps since A has a higher radiative rate constant, but certainly because it also converts to B. The rapid loss of the excess population of A is responsible for the fast initial decay of its fluorescence (Fig. 7). Once the stationary state for the interconversion of A and B is reached, they decay with the same lifetime. Figure 7, as well as results obtained at other temperatures, suggest that in the long-time limit emissions A and B indeed decay with similar slopes, and that the limit is reached faster at higher temperatures, as expected. Unfortunately, we again run into the instrumental limit at short decay times, and at present the data are not as conclusive as would be desirable.

A detailed kinetic analysis encounters the difficulties already noted in the case of Si_6Me_{14}, and the number of free kinetic parameters is excessive. We have attempted a quantitative analysis nevertheless and found that all the observations can be easily accommodated with the kinetic model of Scheme 1. Further progress will require a reduction of the number of unknowns, by procedures such as making fluorescence decay measurements at lower temperatures and constraining the chain to a single conformation.

At this point, it seems appropriate to suggest a likely structure of the fluorescent G and G' conformers of Si_7Me_{16}. First, we have already established that in chains up to and including Si_6Me_{14}, the only all-anti conformer possible according to our calculations [5] (actually, an enantiomeric conformer pair) is also the only fluorescent conformer among the several present at higher temperatures. We hypothesize that a single gauche or ortho link in an oligosilane of a moderate length, less than the length of an exciton in an infinite chain (perhaps two or three dozen Si atoms [1]), suffices to suppress fluorescent emission by providing a dark decay channel. Thus, we assume that in such oligosilanes only the all-anti conformers fluoresce efficiently. Second, counting a mirror image pair as a single conformer, we note that Si_7Me_{16} is the shortest permethylated oligosilane for which two distinct all-anti conformers are expected (the chiral $a_+a_+a_+a_+$, $a_-a_-a_-a_-$ pair and the meso $a_+a_+a_-a_-$ structure), and is also the first one for which two conformers are emissive, G and G'. It therefore appears logical to postulate that the racemic $a_+a_+a_+a_+$, $a_-a_-a_-a_-$ pair is the more stable G conformer and the meso form $a_+a_+a_-a_-$ is the somewhat less stable G' conformer. If this is correct, Si_8Me_{18} also has two emissive conformers (two pairs of enantiomers), Si_9Me_{20} has four (three pairs of enantiomers and one meso form), etc., and for these longer chains the differences in the free energies of the various all-anti conformers are likely to be smaller. Our initial observations indeed indicate that Si_8Me_{18} as well as $Si_{10}Me_{22}$ have multiple emissive conformers even at liquid nitrogen temperature, where Si_7Me_{16} has essentially only one. Unfortunately, this suggests that a detailed analysis will be difficult unless conformationally fixed chains are synthesized.

5
Conclusions

An all-anti conformer of Si_7Me_{16}, presumably the racemic $a_+a_+a_+a_+$, $a_-a_-a_-a_-$ pair, is the shortest permethylated oligosilane that can support delocalized excitation in its relaxed lowest singlet excited state, and another conformer of Si_7Me_{16}, presumably the meso form $a_+a_+a_-a_-$, is the longest permethylated oligosilane that can support localized excitation in its relaxed lowest singlet excited state. Thus, the heptasilane Si_7Me_{16} is the shortest permethylated oligosilane that has more than one strongly fluorescent conformer, and its two emissive excited conformers rapidly interconvert adiabatically at temperatures above about 130 K. In all shorter members of the series, the only all-anti conformer possible (the all-a_+, all-a_- enantiomeric pair) is the only species that fluoresces significantly. In longer members of the series, a larger number of fluorescent conformers is expected.

Tentative, fairly complex, and significantly different kinetic schemes have been suggested for the photophysics of the emissive conformers of Si_6Me_{14} and Si_7Me_{16}, and illustrative potential energy diagrams consistent with the observations have been proposed.

Acknowledgements. This project was supported by a grant from USARO administered jointly with NSF/DMR (DAAH04–94-G-0018), and by the Japan High Polymer Center within the framework of the Industrial Science and Technology Frontier Program funded by the New Energy and Industrial Technology Development Organization. AJN, JRS, and RJE acknowledge support from the U.S. Department of Energy, Office of Energy Research, Chemical Sciences Division.

References

1. Miller RD, Michl J (1989) Chem Rev 89:1359
2. Cui CX, Karpfen A, Kertesz M (1990) Macromolecules 3:3302
   Jalali-Heravi M, McManus SP, Zutaut SE, McDonald JK (1991) Chem Mater 3:1024
   Welsh WJ (1993) Adv Polym Technol 12:379
3. Klingensmith KA, Downing JW, Miller RD, Michl JJ (1986) Am Chem Soc 108:7438
   Teramae H, Matsumoto N (1996) Solid State Commun 99:917
4. Albinsson B, Teramae H, Downing JW, Michl (1996) J Chem Eur J 2:529
   Neumann F, Teramae H, Downing JW, Michl JJ Am Chem Soc, in press

5. Albinsson B, Antic D, Neumann F, Ottosson CH, Michl J, unpublished results
6. Imhof R, Teramae H, Michl J (1997) Chem Phys Lett 270:500
7. Teramae H, Antic D, Crespo R, Michl J Chem Phys, in press
8. Plitt HS, Downing JW, Raymond MK, Balaji V, Michl JJ (1994) Chem Soc Faraday Trans 90:1653
9. Kishida H, Tachibana H, Tokura Y (1996) Phys Rev B 54:14254
10. Sun Y-P, Miller RD, Sooriyakumaran R, Michl JJ (1991) Inorg Organomet Polym 1:3
11. Sun Y-P, Hamada Y, Huang LM, Maxka J, Hsiao J-S, West R, Michl JJ (1992) Am Chem Soc 114:6301
12. Sun Y-P, Wallraff GM, Miller RD, Michl JJ (1992) Photochem Photobiol 62:333
13. Sun Y-P, Michl JJ (1992) Am Chem Soc 114:8186
14. Plitt H, Balaji V, Michl J (1993) Chem Phys Lett 213:158
15. Mazières S, Raymond MK, Raabe G, Prodi A, Michl JJ (1997) Am Chem Soc 119:6682
16. Thorne JRG, Williams SA, Hochstrasser RM, Fagan PJ (1991) Chem Phys 57:401
17. Teramae H, Michl J (1997) Chem Phys Lett 76:127
18. Raymond MK (1997) PhD Dissertation, Uni. of Colorado, Boulder, CO
    Mazières S, Raymond MK, Raabe G, Prodi A, Fogarty H, Michl J, unpublished results
19. Rashba EI (1982) in Excitons, Rashba EI, Sturge MD, Eds, North-Holland Publishing Co: Amsterdam, Holland, Chapter 13
    Ueta M, Kanzaki H, Kobayashi K, Toyozawa Y, Hanamura E (1986) Excitonic Processes in Solids, Springer-Verlag, Berlin
    Song KS, Williams RT (1993) Self-Trapped Excitons, Springer-Verlag, Berlin
    Higai S, Sumi H (1994) J Phys Soc Japan 63:4489
20. Rashba EI (1994) Synth Met 64:255
21. Kumada M, Ishikawa MJ (1963) Organomet Chem 1:153
22. Maxka J, Huang L-M, West R (1991) Organometallics 10:656
23. Kumada M, Tamao K (1968) Adv Organomet Chem 6:19
24. Sekiguchi A, Nanjo M, Kabuto C, Sakurai H (1995) Organometallics 14:2630
24a. Bravo-Zhivotovskii DA, Pigarev SD, Vyazankina OA, Vyazankin NS (1984) Izv Akad Nauk SSSR, Ser Khim 2414
25. Kumada M, Ishikawa M, Maeda SJ (1964) Organomet Chem 2:478
26. Blaukat U, Neumann WPJ (1973) Organometal Chem 49:23
27. Nickel B, Karbach H-J (1990) Chem Phys 148:155
28. Berlman I (1971) Handbook of Fluorescence Spectra of Aromatic Molecules, Academic Press, New York
29. Williams ATR, Winfield SA, Miller JN (1983) Analyst 108:1067. Other authors suggest a somewhat lower value of 0.9 for the room temperature quantum yield: Eaton D., in Handbook of Organic Photochemistry, Scaiano JC, Ed, CRC Press, Inc: Boca Raton FL (1989) Vol 1, p 235. At the temperatures of interest to us, 1.0 seems to be the most likely value, and this is supported by the convergence of the fluorescence quantum yields of both Si_7Me_{16} and Si_8Me_{18} to unity at low temperatures when this value is adopted for the standard
30. Venturini A, Vreven T, Bernardi F, Olivucci M, Robb MA (1995) Organometallics 14:4953

CHAPTER 5

Characterization of Membrane Mimetic Systems with Fluorescence

M. Hink, A. J. W. G. Visser

1
Introduction

One of our major goals in signal transduction research is to understand how a living cell processes information in response to external stimuli leading ultimately to proliferation, differentiation or apoptosis. The plasma membrane has a crucial role in this process which is reflected by the fact that membrane interactions involving proteins and lipid cofactors are continuously modulated (switched off, turned on) enabling transmission of signals at the right moment and along the correct pathway. To increase our understanding of the complexity of the signal transduction network in living cells, experimental data on spatial-temporal organization of the signalling partners are required. Because of their sensitivity, microspectroscopic techniques are the methods of choice since they provide quantitative information on molecular interactions and dynamic events involving signalling molecules carrying fluorescent probes.

Fluorescence correlation spectroscopy (FCS) is such a technique developed for studying dynamic processes of fluorescently marked molecules under equilibrium conditions. FCS measures fluctuations of fluorescence intensity in an open volume element created by a focused laser beam. Any dynamic process which is accompanied by a fluorescence change can be measured. The fluorescence intensity fluctuations can be caused either by diffusion of fluorescently labeled biomolecules in and out of the light cavity, by chemical reaction kinetics such as arising from association/dissociation of a biomolecular complex, by conformational transitions in macromolecules or by flow of fluorescent molecules. Fast correlation of the intensity fluctuations is then used to evaluate the particle number (concentration), the reaction dynamics and the diffusion rates of biomolecules.

Prior to applying FCS to relatively complex cellular systems, more simple model membranes have to be first investigated with this technique. We will therefore focus on micelles and vesicles as membrane mimetic systems. Micelles are clusters of detergent molecules which are formed in water when the detergent concentration exceeds a certain threshold value, called the critical micelle concentration (CMC). Micellization is a consequence of the amphiphilic properties of the detergent molecules, resulting in the exposure of the hydrophilic parts to the aqueous environment, while the hydrophobic parts are in contact with each other in the micellar core. Although micelles do not mimic the exact biological bilayer, they have the advantage that they form a simple experimental system. Vesicles or liposomes, on the other hand, mimic the properties of a biological bilayer more closely than micelles. In this chapter it is demonstrated that FCS is an extremely sensitive technique for determining the overall size or distribution of sizes of these membrane mimetic systems.

2
Principles of FCS

The concepts for FCS date as far back as the 1970s [1–5]. It is, however, due to recent developments in confocal microscopy, such as diffraction-limited laser beams, confocal excitation and detection optics and rapid avalanche photodiode detectors, that the measurements can now be performed both routinely and rapidly [6–11].

Briefly, the principle of FCS in the case of translational diffusion is as follows. A tiny open volume element is created by a focused laser beam. The volume can be as small as 0.2 fL when the focal spot has a diffraction-limited radius of about 200 nm. When a fluorescent molecule enters the confocal volume, the molecule will become excited and a burst of fluorescent photons will be created. When the molecule diffuses away out of the excitation cavity, the process can be repeated, either originating from the same molecule which diffuses back into the volume, or from another entering one. In FCS the fluctuations in fluorescence intensity of single fluorophores are analyzed. These intensity fluctuations can be rapidly measured with a fast photon-detector like an avalanche photodiode. The autocorrelation function, which can be instantaneously determined from the fluorescence intensity fluctuations using a fast correlator, provides information on diffusion properties of the fluorescent molecules. One can easily visualize that small-sized molecules move more rapidly in the confocal volume element than large macromolecules. In order to give an idea about the characteristic diffusion times in a 0.2 fL volume element (with a pinhole of 40 μm diameter) the following data are supplied: the dye rhodamine 6G (molecular mass 550 Da) has a diffusion time of 40 μs, while that of a fluorescent multienzyme complex (molecular mass 550 kDa) has a slower diffusion time of ca. 400 μs. This is the basis of the advantage of the technique: one can measure biomolecular interactions at the unsurpassed sensitivity of single molecules. If a small, fluorescent ligand interacts with a macromolecule it can be immediately determined by a retarded diffusion and the interaction can be fully quantitated. Investigated examples are antigen/antibody, DNA-primer/DNA-target, hormone/receptor, inhibitor/enzyme, protein/protein, lipid/protein and protein/cell interactions. Using this technique, not only transport properties but also chemical reactions like association-dissociation kinetics [12, 13] and photochemical reactions can be investigated [14].

In Fig. 1A two possible causes of fluorescence intensity fluctuations are presented. The first one is connected with transport of single molecules in and out of the laser cavity. This diffusion phenomenon will depend on the volume of the light cavity V_c, which in turn can be controlled by the choice of magnifying objective lenses. The larger the volume the longer the fluorescent molecule resides therein resulting in longer diffusion times. On the other hand, FCS events related to chemical relaxation such as the association/dissociation rates of a protein-bound ligand or conformational transitions in a single protein molecule are not dependent on the size of the cavity volume, since these phenomena take place in the still smaller molecular territory. Chemical relaxation can be measured only when the fluorescence quantum efficiencies of the molecules involved are different. In principle the off-rate constant (k_{off}) of a binding equilibrium reaction can be determined during the residence time in the volume element (example 1 in Fig. 1A). Alternatively, the interconversion rate between two macromolecular conformations can be determined (example 2 in Fig. 1A). The cross sections depicted in Fig. 1B account for the relative volumes of the light cavity V_c and of the particle territory V_T which determine whether we have large average fluorescence intensities ($V_c \gg V_T$) or large fluctuations ($V_c \ll V_T$).

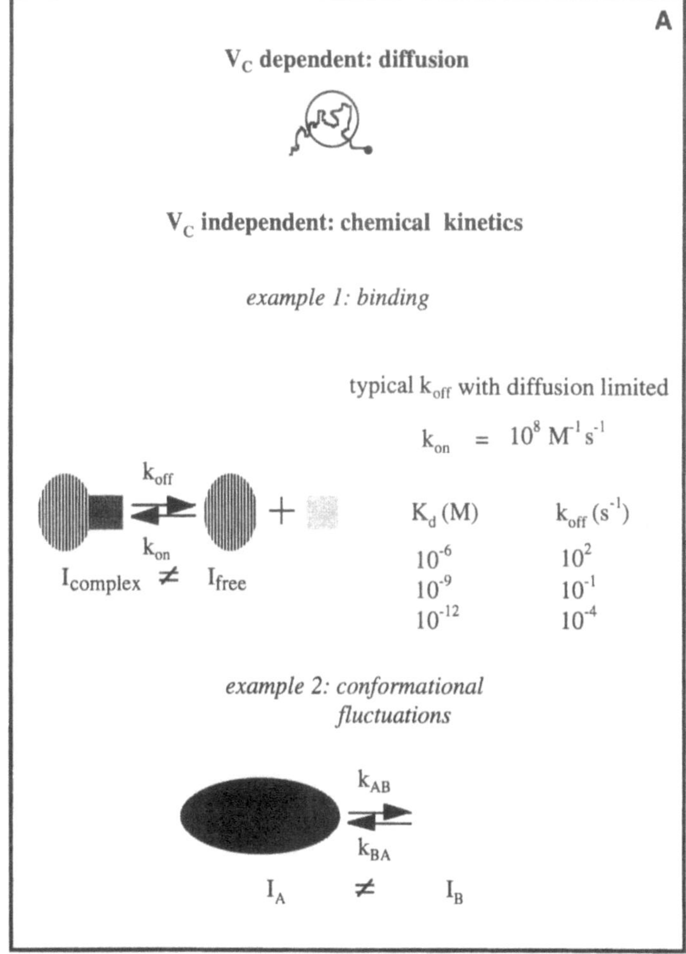

Fig. 1A. Physical concepts of fluorescence correlation measurements

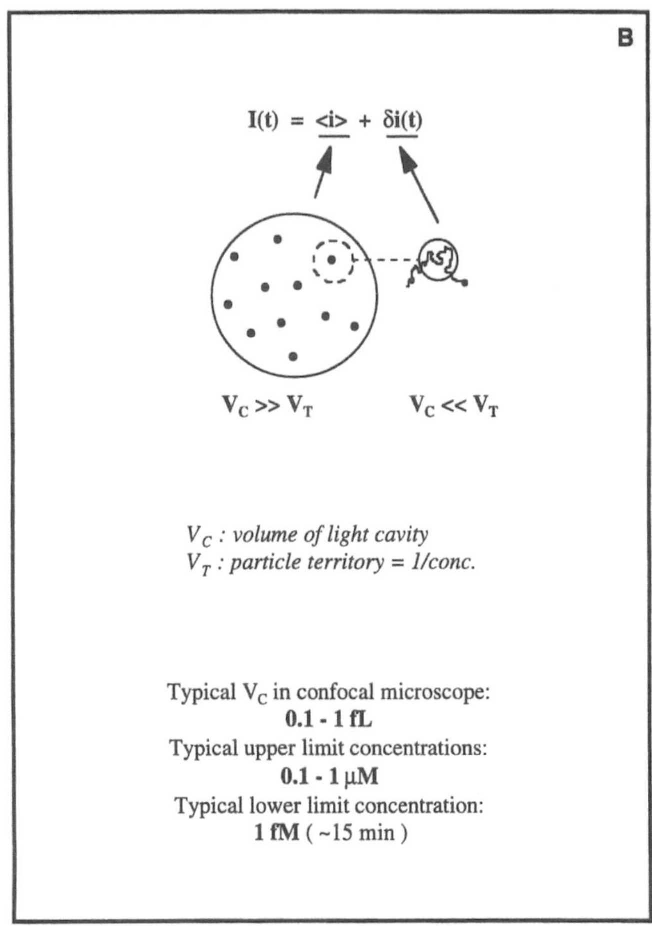

Fig. 1 B. Physical concepts of fluorescence correlation measurements

The particle territory is the reciprocal of the concentration. Figure 1 C gives a kinetic picture. Kinetically there are two characteristic times to consider: i) the reciprocal of the encounter frequency τ_r^{-1} of the particle to enter the light cavity of radius R and ii) the residence time of the particle in the cavity τ_{diff}. D is the translation diffusion coefficient and n is the particle density (number of particles per cm^3). The ratio τ_{diff}/τ_r is simply the ratio of volumes of light cavity and particle territory and reflects the probability of finding a particle inside the cavity (details can be found in [9]).

The diffusion constant for translational movement can be determined from the autocorrelation function $G(\tau)$, which is related to the fluctuations of the fluorescence intensity. The collected fluorescence intensity $I(t)$ is given by:

$$I(t) = q \cdot \sigma_{abs} \cdot Q_f \int p(r) \cdot C(r, t) \, dV \tag{1}$$

C

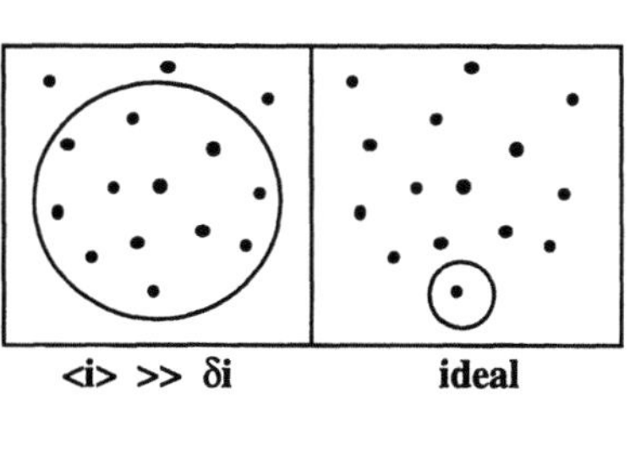

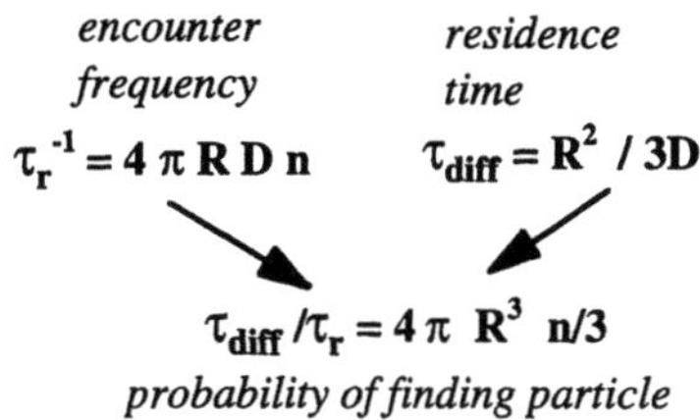

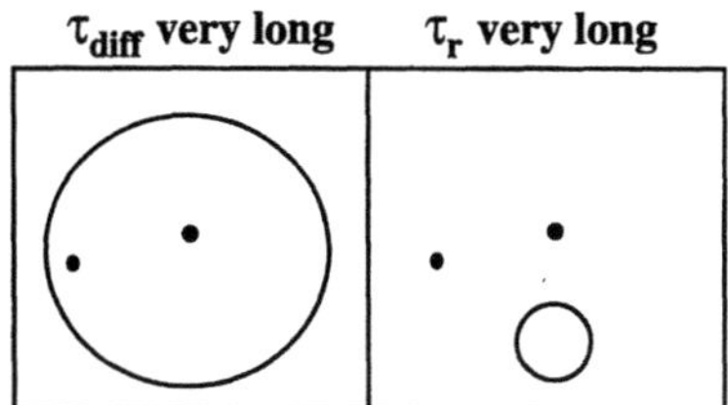

Fig. 1C. Physical concepts of fluorescence correlation measurements

where $C(r, t)$ is the concentration of fluorescent molecules at position r and time t. Parameter q signifies the detection quantum efficiency of the detector and the attenuation of the fluorescence in the passage from the sample to the detector area. Equation 1 also includes the excitation cross section σ_{abs}, the fluorescence quantum yield Q_f and a function $p(r)$ representing the product of the intensity distribution of the exciting light, of a function defining the volume from which the fluorescence is measured, and of a function describing the distribution of the sample. The normalized autocorrelation function $G(\tau)$ relates the fluorescence intensity at a time t to that τ seconds later:

$$G(\tau) = \frac{\langle I(t) \cdot I(t+\tau)\rangle}{\langle I\rangle^2} = \frac{\langle I\rangle + \langle \delta I(t) \cdot \delta I(t+\tau)\rangle}{\langle I\rangle^2} \tag{2}$$

where δI denotes the fluctuation of the fluorescence intensity around the mean value $\langle I \rangle$. As can be seen from Eq. 2, the time-dependent part of $G(\tau)$ is related to the relative fluctuations of the system. Substitution of Eq. 1 into Eq. 2 followed by integration in Fourier space yields Eq. 3 [15]:

$$G(\tau) = 1 + \frac{1}{N_{\mathrm{m}}\left(1 + \dfrac{4D}{\omega_1^2 \cdot \tau}\right)\sqrt{1 + \dfrac{4D}{\omega_1^2 \cdot \tau}}} \tag{3}$$

where D denotes the translational diffusion constant ($\mathrm{m^2\,s^{-1}}$) and both ω_1 as ω_2 are constants corresponding to the radial and axial radii ($\mathrm{e^{-2}}$ point of the Gaussian beam) of the sample volume element (Fig. 2). It should also be noted that Eq. 3 is only obtained when the excitation intensity profile is Gaussian in all three dimensions.

Because the relative fluorescence fluctuations increase as the number of fluorescent particles (N_{m}) decreases, it is important to perform experiments with a minimum amount of fluorescent probe. However, high background intensities, dominated by Raman scattering of water molecules and Rayleigh-scattered laser light, had made it impossible to reduce the fluorophore concentration without the background dominating the fluorescence signal. In recent years an increase in the sensitivity of optical and electronical instrumentation has resulted in detection volumes below 1 femtoliter leading to a diminished contribution of background, relatively larger fluorescence fluctuations and a

Fig. 2. Schematic view of the sample volume element

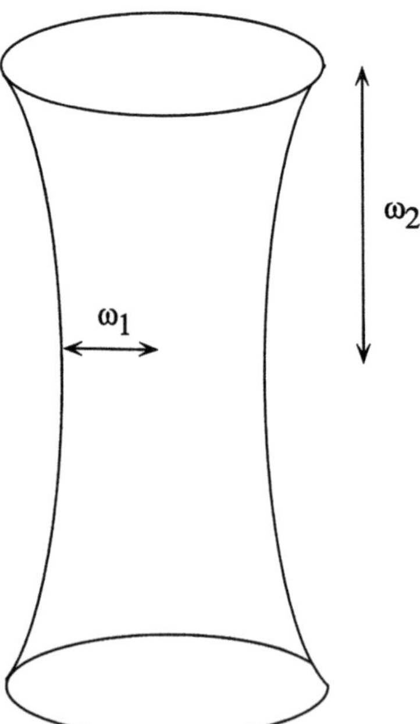

reduced photophysical destruction of the probe. Nowadays fluorophore concentrations between 10^{-7} M and 10^{-12} M can easily be measured with FCS and even lower concentrations have been used [16].

The translational diffusion constant D is related to the diffusion time τ_{diff} (s) as follows:

$$D = \frac{\omega_1^2}{4 \cdot \tau_{\text{diff}}} \tag{4}$$

where τ_{diff} is the time needed to diffuse over a distance ω_1. The square of the laser beam waist ω_1 (m^2) can be obtained by calibration with a Rhodamine 6G solution having a known diffusion constant D of $2.8 \cdot 10^{-10}$ m^2 s^{-1}. The value for ω_1 together with the experimental value for the structural parameter (SP, usually between 6 and 10, defines the size of the detection volume element; see Fig. 2) can be used to calculate the confocal detection volume $V(L)$:

$$V = \omega_1^2 \cdot 2 \cdot \pi \cdot \omega_2 . \tag{5}$$

The hydrodynamic radius of the fluorescent particle r_h (m) can be calculated by Eq. 6:

$$r_h = \frac{k \cdot T}{6\tau \cdot d \cdot \eta} \tag{6}$$

where k is the Boltzmann constant (J $\cdot$ K^{-1}), T the temperature (K) and η the viscosity (Pa $\cdot$ s). Assuming a spherical particle, the number of detergent molecules S in the micelle can be calculated according to:

$$S = \frac{4\pi \cdot r_h^3 \cdot N_a}{6m} \tag{7}$$

where ϱ is the mean density of the micelle (g $\cdot$ cm^{-3}), N_a is Avogadro's number and m the molecular mass of a surfactant molecule (g $\cdot$ mol^{-1}).

3
Experimental Setup and Data Analysis

FCS measurements were performed with a Zeiss-EVOTEC ConfoCor inverted confocal microscope (Fig. 3). An air-cooled argon ion laser supplied two excitation wavelengths (488 and 514 nm) which were filtered by a 485 ± 15 nm and 515 ± 15 nm interference filter, respectively (Omega) and reflected by proper Zeiss dichroic filters. The excitation light was focused by a water immersible objective (C-Apochromat 40×, 1.2 NA) into the sample of investigation. The laser power was set with neutral density filters to less than 1 mW. The fluorescent light was collected by the same objective, passed through the dichroic mirror and the appropriate band-pass filter (Omega). The light passed through a motor-controlled pinhole (typically 40 µm diameter) in the image plane (to reject out-of-focus light) and finally hit an avalanche photodiode in single-photon counting mode whose signal was processed by a correlator (ALV-5000). The sampling time of the correlator was between 200 ns and 3400 s using 288 (logarithmic spaced) time channels. The temperature of the experiments was always 20 °C.

Data analysis was performed with the FCS ACCESS software package developed by EVOTEC/Zeiss, Inc. (version 1.0.10), using either one- or two-component fit models. These fit models determine the average number of fluorescent molecules in the detection volume as well as their characteristic diffusion times. The models also determine values for the triplet fraction and triplet lifetime. For analysis the curves were fitted with a fixed value for the structural parameter SP, obtained from the Rhodamine 6G calibration.

4
Probes and Membranes

Fluorescent probes which have the ability to be incorporated in membrane systems are selected. Therefore these probes should contain a large hydrophobic (or lipophilic) part. The following probes were used as fluorescent markers: BP-HPC [C5-HPC(2-(4,4-difluoro-5,7-diphenyl)-4-bora-3a,4a-diaza-s-indacene-3-pentanoyl)-1-hexadecanoyl-*sn*-glycero-3-phosphocholine], in which the fluorescent BODIPY group was coupled to one of the fatty acyl chains of phosphati-

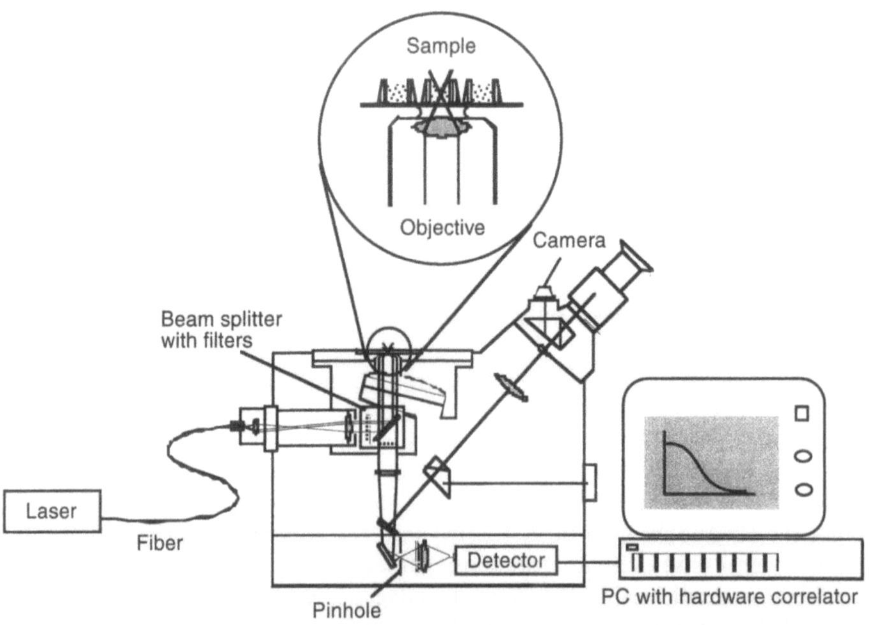

Fig. 3. Inverted microscope set up for fluorescence correlation measurements

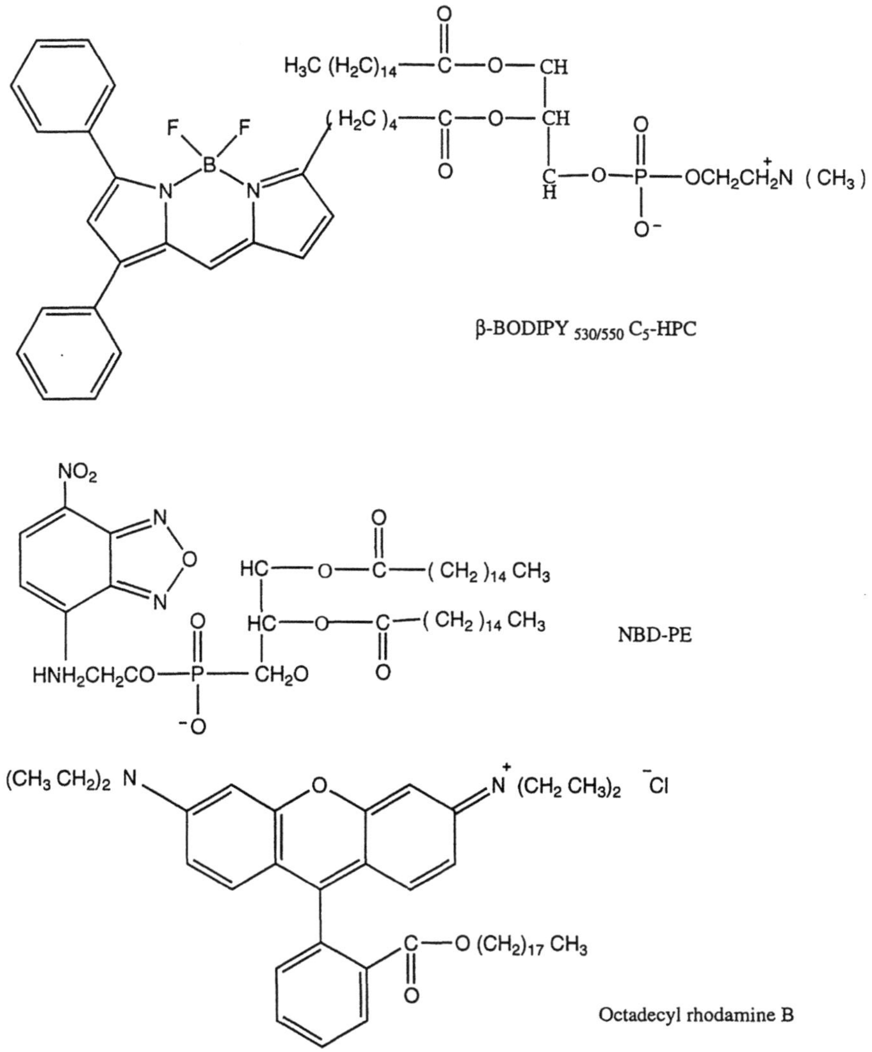

β-BODIPY $_{530/550}$ C$_5$-HPC

NBD-PE

Octadecyl rhodamine B

Fig. 4. Molecular structures of the fluorescent probes used in FCS experiments

dylcholine], NBD-PE [N-(7-nitrobenz-2-oxa-1,3-diazol-4-yl)-1,2-dihexadeca-noyl-*sn*-glycero-3-phosphoethanolamine, triethylammonium salt, which is a headgroup-labeled phosphatidylethanolamine] and ODRB (octadecyl rhoda-mine B chloride, which consists of a fluorophore coupled to a long hydrophobic alkyl chain). The fluorescent probes were purchased from Molecular Probes, Inc. and the structural formulas are presented in Fig. 4. All probes were dis-solved in stock in ethanol and stored at − 10 °C. BP-HPC and ODRB were excit-ed at 514 nm, while NBD-PE was excited at 488 nm.

CTAB (cetyltrimethylammonium bromide), digitonine, deoxycholate, SDS (sodium dodecyl sulfate), Triton X-100, Thesit (polyoxyethylene-9-lauryl ether) and Tween 80 (polyoxyethylene sorbitan mono-oleate) were obtained from different commercial sources. The structural formulas are presented in Fig. 5. The detergents were dissolved in PBS (8.8 g L^{-1} NaCl, 0.2 g L^{-1} KCl, 1.45 g L^{-1} $Na_2HPO_4 \cdot 2H_2O$ and 0.5 g L^{-1} KH_2PO_4 in nanopure water). The detergent con-

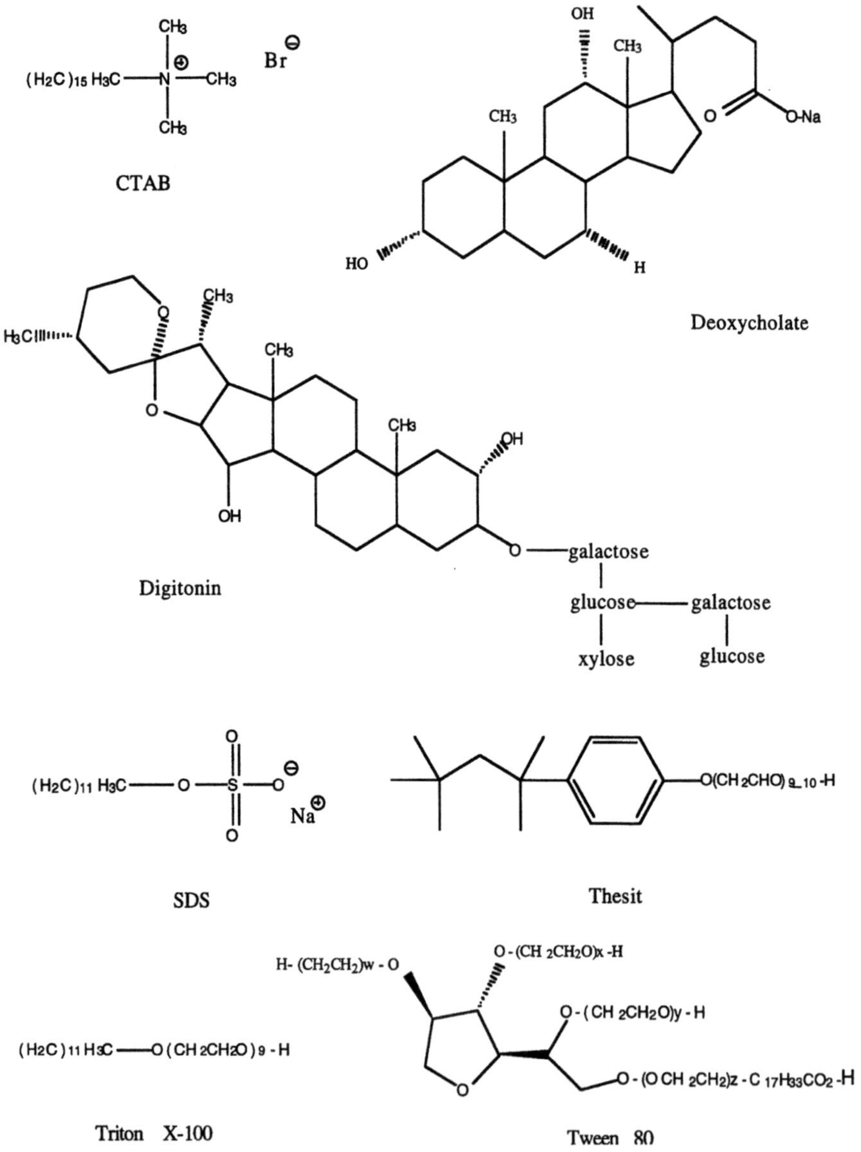

Fig. 5. Molecular structures of the surfactants used in this work

centrations varied between two and ten times (1–15 mM) the CMC value. All solutions were set at the appropriate pH using 1 M HCl, except for digitonin which was always dissolved at pH = 8.3. Subsequently, the fluorescent probe was added (maximal 3 vol%) and the sample was vortexed for 30 s or sonified for 1 min to optimize the probe distribution and then incubated for at least 10 min in the dark at room temperature before measurement.

DOPC vesicles (1,2-dioleoyl-*sn*-glycero-3-phosphocholine) [Avanti Polar Lipids, Inc.] were prepared from stocks in chloroform by two different methods. Using the first method, the chloroform was evaporated with N_2 gas to minimize oxidation. The lipids were dried in a vacuum desiccator for 20 min, 20 µL of the fluorescent probe (BP-HPC) and 980 µL of PBS buffer were added and the mixture was sonified for 1 min. The mixture was subsequently frozen with liquid N_2 and thawed with hot water (60 °C). This cycle was repeated five times. The mixture was then transferred to a syringe and connected to a membrane filtration apparatus. The mixture was pushed 35 times through a polycarbonate membrane with pores of 100 nm in diameter. In the second method the dried lipids (including BP-HPC) were dissolved in ethanol and rapidly injected into a PBS solution using a Hamilton syringe. This procedure should result in a very homogenous population of vesicles [17].

5
Translational Diffusion of Micelles

From the diffusion time of a fluorescent particle through the detection volume, the translational diffusion constant and the apparent size of the particle can be determined (cf. Eq. 7). The aggregation number can then be calculated. Figure 6 illustrates an example of normalized autocorrelation curves of various ODMR-loaded micelles in PBS buffer. The differences in translational diffusion times between the various aggregates are clearly visible reflecting the size differences among them. The diffusion time τ_{diff} is listed for each micelle in the legend under Fig. 6. Figure 6 also contains the effect of different probe concentrations (at constant surfactant concentration) on the autocorrelation

Table 1. Translational Diffusion Constants (10^{-11} m^2 s^{-1}), Aggregation Numbers and Hydrodynamic Radii of Several Detergents at pH = 7.0[a]. Loaded with 10 nM of Fluorescent Probe

	NBD-PE	ODRB	Aggregation number	Hydrodynamic radius (nm)
Buffer and probe	25 ± 1	19 ± 2	–	–
CTAB	6.4 ± 0.3	6.3 ± 0.3	319 ± 41	3.7 ± 0.2
Deoxycholate	9.9 ± 0.5	11 ± 1	50 ± 13	2.2 ± 0.2
Digitonin	5.2 ± 0.2	4.5 ± 0.3	170 ± 32	4.8 ± 0.3
SDS	4.9 ± 0.3	5.9 ± 0.3	357 ± 51	3.7 ± 0.2
Thesit	6.3 ± 0.2	5.3 ± 0.3	136 ± 22	3.1 ± 0.2
Triton X-100	5.8 ± 0.2	5.5 ± 0.4	92 ± 19	3.5 ± 0.3
Tween 80	5.1 ± 0.4	4.2 ± 0.4	250 ± 68	5.6 ± 0.5

[a] Except the Digitonin solution which had pH = 8.3.

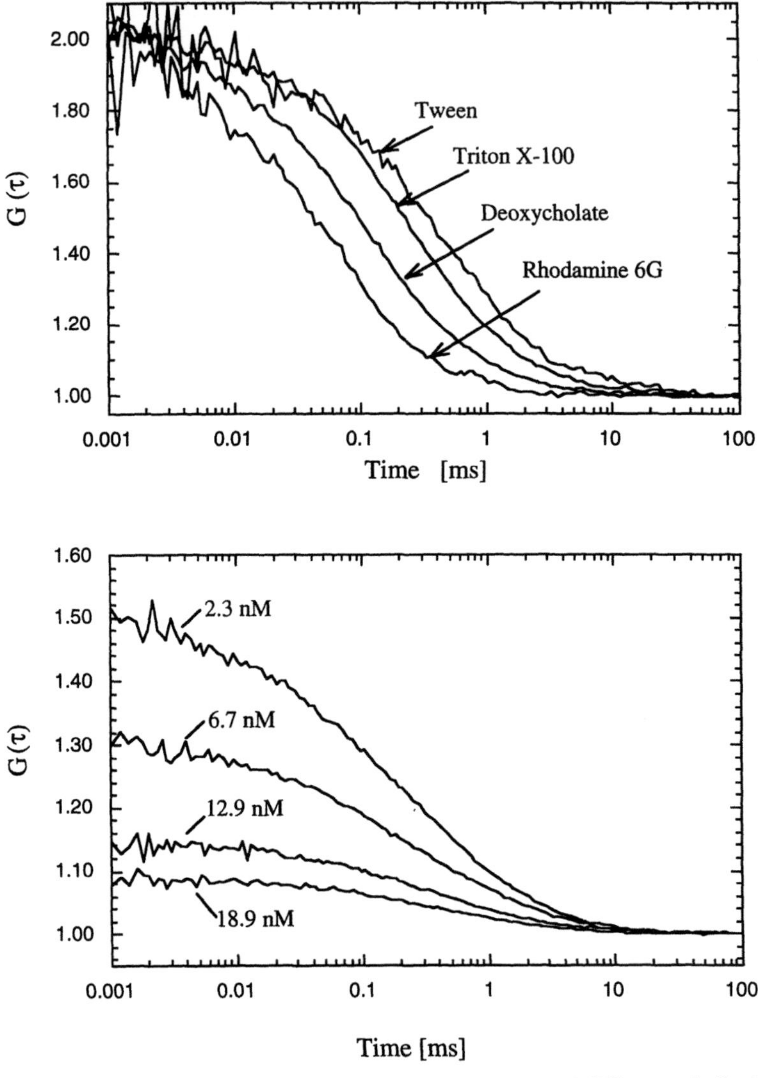

Fig. 6. *Upper panel*: autocorrelation curves for Rhodamine 6G and different micelles loaded with ODRB. The following diffusion times are obtained: τ_{diff} (Rhodamine 6G) = 65 ± 2 µs, τ_{diff} (deoxycholate) = 131 ± 4 µs, τ_{diff} (Triton-X100) = 273 ± 4 µs, τ_{diff} (Tween 80) = 466 ± 22 µs. The effect of increasing amplitude of the autocorrelation curve with lower dye concentration is shown for ODRB in Triton-X100 in the *lower panel*

curve in Triton X100 micelles accounting for the fact that the amplitude of the autocorrelation curve increases with lower relative probe concentration. Table 1 presents the translational diffusion constants for ODRB and NBD-PE dispersed in several detergents under various experimental conditions. Loading the micelles with either ODRB or NBD-PE resulted in slightly different diffusion constants. Table 1 also contains the parameters which characterize the micellar system: aggregation number and hydrodynamic radius. With the exception of the nonionics, Triton X100 and Thesit, the aggregation numbers are higher than the ones obtained from other experiments such as light-scattering or sedimentation equilibrium [18]. Apparently, the incorporation of the fluorescent probe is perturbing the micellar system in some cases requiring extra detergent molecules to fully "solubilize" the fluorescent moiety. On the other hand, FCS measurements are very good in a comparative sense, because the relative size differences between the micelles studied are in good agreement with literature values.

6
Comparison of Fluorescence Anisotropy Yields with Intramicellar Diffusion Coefficients

Time-resolved fluorescence anisotropy measurements can provide complementary information on the same micellar system that was used in FCS. In fluorescence anisotropy decay experiments, the time-correlation function of the emission transition moment is determined (for details on experiment and data analysis see [19]). In contrast to FCS (where one is dealing with measurements on single micelles), the experiments have to be performed on a whole ensemble of micelles in a normal 3 mL quartz cuvette. Assuming that the micelle rotates as a spherical particle, the overall rotational correlation time ϕ_{mic} obtained from fluorescence anisotropy decay would lead to the same hydrodynamic radius as obtained from FCS (cf. Eq. 6) via the Stokes-Einstein relationship (see Eq. 8):

$$\phi_{\mathrm{mic}} = \frac{4\,\pi\,r_{\mathrm{h}}^3 \cdot \eta}{4k \cdot T} \tag{8}$$

In Fig. 7 the experimental fluorescence anisotropy decay of BP-HPC in Thesit micelles is presented together with a simulated anisotropy decay curve. The excitation wavelength was 480 nm and approximately the same filter set was chosen as used for the FCS experiment. The simulated curve is characteristic for a spherical micelle having a single, isotropic rotational correlation time in which the hydrodynamic radius has been determined from FCS. From the experimental fluorescence anisotropy decay it can be clearly observed that additional depolarization processes are present during the 40 ns duration of the fluorescence anisotropy decay experiment. This was noted earlier [19] and the fluorescence anisotropy decay $r(t)$ can, in fact, be described by a Soleillet product of three correlation functions, provided that the time scales of these processes are different:

$$r(t) = r_0 \cdot C_{\mathrm{rot}}(t) \cdot C_{\mathrm{int}}(t) \cdot C_{\mathrm{lat}}(t) \tag{9}$$

where r_0 is the fundamental anisotropy, C_{rot} the correlation function describing the overall micellar rotation, C_{int} the correlation function accounting for internal motion and C_{lat} that describing intramicellar lateral diffusion. The overall micellar rotation is characterized by a rotational correlation time ϕ_{mic} as given

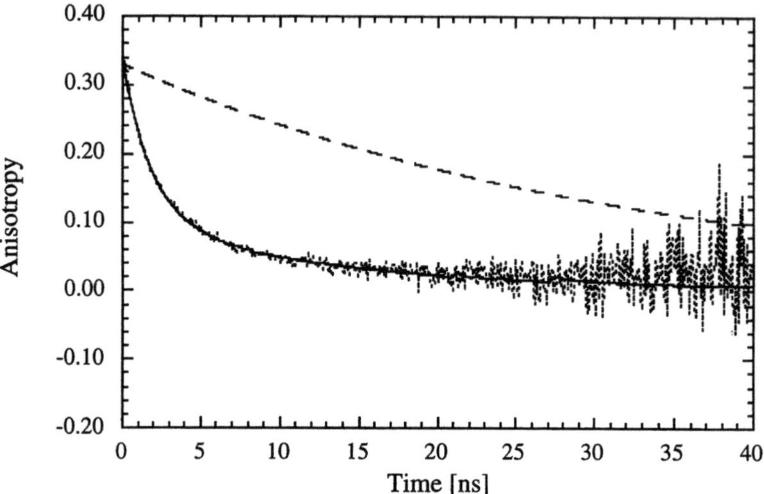

Fig. 7. Experimental (…) and fitted (—) fluorescence anisotropy decay curves of BP-HPC in thesit micelles at 21 °C. The recovered parameters of a fit according to the model of Eq. 9 are listed in Table 2. The initial anisotropy $r_0 = 0.335$. For comparison a simulated fluorescence anisotropy decay is presented (---) in which the sole depolarization mechanism is overall rotation of the micelle with $\phi_{\mathrm{mic}} = 31$ ns

in Eq. 8. Details of the internal motion in terms of a Gaussian diffusion model has been presented previously [19]. It is assumed that the probe orientation is Gaussian distributed (over a given angular width θ_g) around the micellar radius and the dynamics is reflected by an internal diffusion rate constant $D_\perp$. The intramicellar correlation function C_{lat} can be characterized by a single correlation time ϕ_L which describes lateral diffusion over a distance corresponding to the micellar radius r_h and is inversely related to the lateral diffusion coefficient D_L (cf. Eq. 4):

$$D_L = \frac{r_h^2}{4 \cdot \phi_L} \tag{10}$$

Analysis of the fluorescence anisotropy decay in terms of a model given by Eq. 9 requires five adjustable parameters (ϕ_{mic}, ϕ_L, r_0, $D_\perp$ and θ_g). It need almost not be said that this will be a hopeless task, given the high correlation between the parameters. Two important assumptions can, however, be made. The overall micellar rotational correlation time ϕ_{mic} can be obtained from FCS and fixed during the analysis. The fundamental anisotropy r_0 can be obtained from a global analysis of several decay curves and also be fixed. This will lead to the determination of only three parameters which can be easily accomplished. The fitted decay curve is presented in Fig. 7 and the numerical results are given in Table 2. If we compare the recovered parameters with the ones from DPH-PC (DPH = diphenylhexatriene) in the same thesit micelles [19], then the following differences are apparent (for clarity the parameters for DPH-PC are also listed

Table 2. Optimized Parameters from Analysis of Anisotropy Decay of Fluorescent Lipid Probes in Thesit Micelles

Probe	ϕ_{mic} (ns)	ϕ_L (ns)	D_L (m^2 s^{-1})	θ_g (rad)	$\langle P_2 \rangle$	$\langle P_4 \rangle$	$D_\perp$ (μs)$^{-1}$
DPH-PC[a]	37	88	$6.5 \cdot 10^{-11}$	0.46	0.73	0.35	29
BP-HPC	31	22	$1.1 \cdot 10^{-10}$	0.65	0.54	0.13	57

[a] The data of DPH-PC were as obtained in [19]. ϕ_{mic} was indirectly obtained from NBD-PE in thesit micelles at a different temperature [10]. ϕ_{mic} was calculated from Eqs. 6 and 8, independently determined from FCS and fixed in the analysis.

in Table 2). The angular range (θ_g) over which the BP probe is moving is larger than for DPH, and also the local order of the BP moiety is less than the DPH moiety which is reflected by smaller parameters $\langle P_2 \rangle$ and $\langle P_4 \rangle$. The dynamics of the restricted BP reorientation is faster than the dynamics of restricted DPH motion. The explanation is that the BP moiety is bulkier than the DPH moiety and therefore requires more space in the micellar interior. In contrast to DPH, BP is not a perfect cylinder and some surfactant-packing constraints may also exist allowing more motional freedom for the BP moiety. The intramicellar lateral diffusion (D_L) is much faster than the lateral diffusion of fluorescent lipids observed in planar lipid bilayers such as black lipid membranes (BLMs) or large vesicles [20, 21]. In BLMs the diffusion coefficients were found to be in the order of 10^{-11} m^2 s^{-1}, while large vesicles (above the phase transition temperature) has a $D \approx 10^{-12}$ m^2 s^{-1}. Clearly, the interior of a micelle is more fluid than that of a lipid bilayer.

7
Size Distributions of Liposomes from FCS

The method of preparation strongly affects the size and the size heterogeneity of DOPC liposomes (or vesicles) loaded with BP-HPC. The average diffusion time τ_{diff} for the liposomes (10 determinations), prepared by the injection method, was typically 2.1 ± 0.7 ms ($D \approx 7.2 \cdot 10^{-12}$ m^2 s^{-1}). The liposomes prepared by membrane filtration had shorter diffusion times and were more homogenous in size, since the standard deviation (10 determinations) of τ_{diff} was reduced ($\tau_{diff} \approx 0.8 \pm 0.2$ ms, $D \approx 1.4 \cdot 10^{-11}$ m^2 s^{-1}). Assuming spherically shaped liposomes with a mean density of 1.05 g cm^{-3} in an aqueous solution (with viscosity of $1.0 \cdot 10^{-3}$ Pa $\cdot$ s), liposomes with a diameter of 38 nm and a aggregation number of 21 000 were calculated. These vesicles, formed by membrane filtration (100 nm pores), were much smaller than those prepared by injection, which had an average diameter of 70 nm.

Given the relatively large error in the diffusion times in the non-filtrated vesicles, we have performed an analysis using a distribution of τ_{diff} values in the range 0.1–10 ms. The CONTIN program was used for this [22]. The results of one analysis are presented in Fig. 8. It can be clearly observed that the filtrated liposomes have only a single distributed diffusion time at 600 µs. The non-

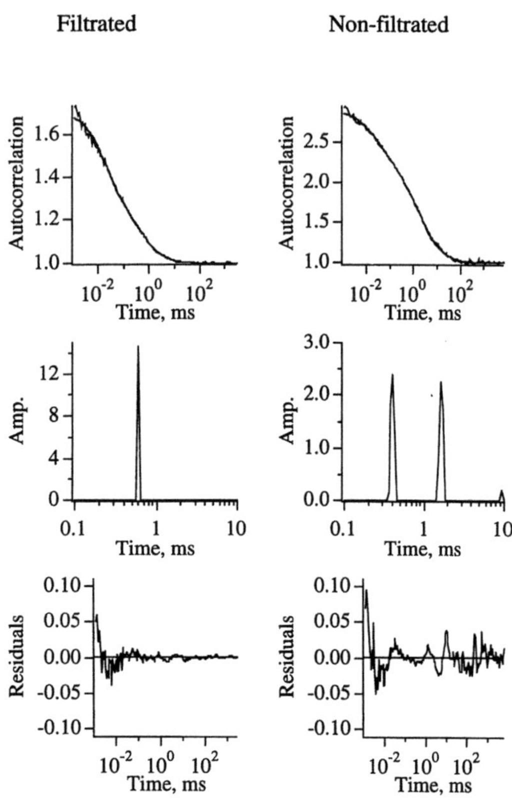

Fig. 8. Analysis of fluorescence autocorrelation traces of 0.1 nM BP-HPC in 1.3 µM DOPC vesicles into a distribution of diffusion times using the CONTIN program. Top: experimental and fitted autocorrelation traces. Middle: distribution of diffusion times. The time between 0.1 and 10 ms was divided into 100 points logarithmically spaced. Bottom: residuals between experimental and fitted autocorrelation traces. *Left*: filtrated vesicles. *Right*: non-filtrated vesicles

filtrated liposomes are not homogeneous in size, since one fraction has a $\tau_{diff} =$ 400 µs and another fraction (of similar amplitude) has a $\tau_{diff} = 2.2$ ms, while even a third tiny fraction of large aggregates ($\tau_{diff} = 9.4$ ms) can be distinguished. FCS in combination with an analysis in a distribution of diffusion times is therefore capable of checking the homogeneity of the membrane preparations.

8
Conclusions and Perspectives

We have demonstrated that FCS is very suitable for characterizing membrane systems. It is a rapid method giving information on the overall size or the size distribution of single micelles or vesicles. The combination of FCS with time-resolved fluorescence anisotropy leads to insight into intramicellar lateral diffusion. The next challenging step is to apply the technique to the plasma membrane of a living cell [11, 23]. We shall call the technique fluorescence correlation microscopy (FCM) when it is applied to cellular systems. Preliminary FCM experiments carried out by us on living cells indicate that some technical problems have to be solved first before the methodology can be routinely applied. Cellular autofluorescence and photobleaching of the fluorescent probe are important perturbing factors which should be eliminated before FCM is feasible. Furthermore, in order to study cellular systems in real time, rapid scanning through the cell in three dimensions combined with efficient data collection needs systematic investigation [23–26]. Dual color fluorescence cross-correlation spectroscopy [27] is a new technique for probing molecular interactions that carry different fluorescent labels. The occurrence of fluorescence resonance energy transfer (FRET) between two different, juxtaposed fluorescent signaling molecules can be perfectly measured using this concept. This approach allows the observation of interacting signaling elements and provides us with nm resolution in confocal microscopy.

Acknowledgements. We thank Arie van Hoek and Eward Pap for help in the fluorescence anisotropy decay experiment and analysis, Frank Vergeldt for help with the CONTIN program, Wim Wolkers for assistance with the membrane filtration and Nina Visser for preparing Fig. 1. Mr. Peter Bremer (Carl-Zeiss Jena, Inc.) is gratefully acknowledged for making available the Zeiss-EVOTEC ConfoCor system in the initial stage of this work. This research was supported by an investment grant from The Netherlands Organization for Scientific Research (NWO).

References

1. Elson EL, Magde D (1974) Fluorescence correlation spectroscopy. I. Conceptual basis and theory. Biopolymers 13:1–27
2. Magde D, Elson EL, Webb WW (1974) Fluorescence correlation spectroscopy. II. An experimental realization. Biopolymers 13:29–61
3. Ehrenberg M, Rigler R (1974) Rotational Brownian motion and fluorescence intensity fluctuations. Chem Phys 4:390–401
4. Aragon SR, Pecora R (1975) Fluorescence correlation spectroscopy and Brownian rotational diffusion. Biopolymers 14:119–138
5. Magde D, WebbWW, Elson E (1978) Fluorescence correlation spectroscopy. III. Uniform translation and laminar flow. Biopolymers 17:361–376
6. Thompson NL (1991) Fluorescence correlation spectroscopy. In: Topics in Fluorescence Spectroscopy, Lakowicz JR (ed), Vol 1, pp 337–378, Plenum Press, New York
7. Rigler R, Widengren J, Mets Ü (1992) Interaction and kinetics of single molecules as observed by fluorescence correlation spectroscopy. In: Fluorescence Spectroscopy Wolfbeis OS (ed), pp 13–24, Springer Verlag, Berlin
8. Rigler R, Mets Ü, Widengren J, Kask P (1993) Fluorescence correlation spectroscopy with high count rates and low background: analysis of translational diffusion. Eur Biophys J 22:169–175
9. Eigen M, Rigler R (1994) Sorting single molecules: applications to diagnostics and evolutionary biology. Proc Natl Acad USA 91:5740–5747
10. Bastiaens PIH, Pap EHW, Widengren J, Rigler R, Visser AJWG (1994) Fluorescence methods to study lipid-protein association. The interaction of protein kinase C with lipid-loaded mixed micelles. J Fluorescence 4:377–384
11. Berland KM, So PTC, GrattonE (1995) Two-photon fluorescence correlation spectroscopy: method and application to the intracellular environment. Biophys J 68:694–701
12. Kinjo M, Rigler R (1995) Ultrasensitive hybridization analysis using fluorescence correlation spectroscopy. Nucleic Acids Res 23:1795–1799

13. Rauer B, Neumann E, Widengren J, Rigler R (1996) Fluorescence correlation spectrometry of the interaction kinetics of tetramethylrhodamin α-bungarotoxin with *Torpedo californica* acetylcholine receptor. Biophys Chem 58:3–12
14. Widengren J, Mets Ü, Rigler R (1995) Fluorescence correlation spectroscopy of triplet states in solution: a theoretical and experimental study. J Phys Chem 99:13368–13379
15. Aragon SR, Pecora R (1976) Fluorescence correlation spectroscopy as a probe of molecular dynamics. J Chem Phys 64:1791–1803
16. Rigler R (1995) Fluorescence correlation, single molecule detection and large number screening. Applications in biotechnology. J Biotechnology 4:259–264
17. Batzri S, Korn ED (1973) Single bilayer liposomes prepared without sonication. Biochim Biophys Acta 298:1015–1019
18. Neugebauer J (1994) A guide to the properties and uses of detergents in biology and biochemistry. Calbiochem-Novabiochem International.
19. Pap EHW, Ketelaars M, Borst JW, van Hoek A, Visser AJWG (1996) Reorientational properties of fluorescent analogues of the protein kinase C cofactors diacylglycerol and phorbol ester. Biophys Chem 58:255–266
20. Fahey PF, Koppel DE, Barak LS, Wolf DE, Elson EL, Webb WW (1977) Lateral diffusion in planar lipid bilayers. Science 195:305–306
21. Fahey PF, Webb WW (1978) Lateral diffusion in phospholipid bilayer membranes and multilamellar liquid crystals. Biochemistry 17:3046–3053
22. Provencher SW (1982) CONTIN: a general purpose constrained regularization program for inverting noisy linear algebraic and integral equations. Comp Phys Commun. 27:229–242
23. Petersen NO, Hoddelius PL, Wiseman PW, Seger O, Magnusson K-E (1993) Quantitation of membrane receptor distributions by image correlation spectroscopy: concept and application. Biophys J 65:1135–1146
24. Koppel DE, Morgan F, Cowan AE, Carson JH (1994) Scanning concentration correlation spectroscopy using the confocal laser microscope. Biophys J 66:502–507
25. Berland KM, So PTC, Chen Y, Mantulin WW, Gratton E (1996) Scanning two-photon fluctuation correlation spectroscopy: particle counting measurements for detection of molecular aggregation. Biophys J 71:410–420
26. Huang Z, Thompson, NL (1996) Imaging fluorescence correlation spectroscopy: nonuniform IgE distributions on planar membranes. Biophys J 70:2001–2007
27. Schwille P, Meyer-Almes F-J, Rigler R (1997) Dual-color fluorescence cross-correlation spectroscopy for multicomponent diffusional analysis in solution. Biophys J 72: 1878–1886

CHAPTER 6

Excited State Probing of Supramolecular Systems on a Submicron Scale

P. Vanoppen, J. Hofkens, L. Latterini, K. Jeuris, H. Faes, F. C. De Schryver,
J. Kerimo, P. F. Barbara, A. E. Rowan, R. J. M. Nolte

1
Introduction

Imaging techniques, such as Confocal Fluorescence Microscopy (CFM) or Near-field Scanning Optical Microscopy (NSOM) [1, 2] are essential techniques to study complex heterogeneous organic thin films by mapping their optical properties. They are complementary techniques having different advantages and disadvantages [3]. CFM is relatively easy to apply and combines a lateral resolution approaching $\lambda/2$ with the possibility of layered imaging in the z-direction. In contrast, NSOM has significantly better spatial resolution and offers simultaneous optical and topographic images. Although confocal microscopy has been mostly used for biological applications, the technique has proven to be useful, for example, in the study of colloids [4, 5], polymer blends [6] and liquid crystals [7, 8].

The development of near-field microscopy was to a large extent focused on the ability to detect and spectroscopically analyze fluorescence of single molecules[9]. Less is known about the application of NSOM for the fluorescence mapping of heterogeneous organic films. Only a limited number of examples have been reported [10–12].

This contribution reviews some recently published results on composite polymer films [13, 14] and on the optical and morphological investigation of porphyrin wheels [15].

In the field of CFM, experiments are reported in which confocal images, together with excited state lifetimes, are obtained by the combination of a CFM with a single photon counting (SPC) setup for objects not requiring a resolution beyond the diffraction limit. Additionally, CFM and SPC were combined with a single beam optical trapping technique. The latter technique was first demonstrated by Askin [16]. More recently, optical trapping has been used for three-dimensional manipulation of microparticles [17] and has also been applied to biological samples [18]. CFM combined with optical trapping can be used for fixation of individual microparticles in solution in a non-destructive way and for studying their properties using a spectroscopic technique. We have applied this technique to acquire spectroscopic data on micrometer size, dye-labeled latex particles in solution.

NSOM was used to map the fluorescence of nanometer objects, namely 14 nm dye-labeled latex particles distributed in a polymer film. These experiments show the capabilities of the NSOM instrument.

Schenning et al. [19] recently described a facile technique for the construction of ring-shaped assemblies of porphyrins on a graphite surface. We report the investigation of the morphology and optical properties of these solvent assisted self-assembled porphyrin wheels from the compound, bis (21H, 23H-5(4-pyridyl)-10,15,20-tri(4-hexadecyloxyphenyl)porphyrin) platinum dichloride (PtP) on a glass surface. It will be shown that NSOM is indispensable in determining the local molecular orientation and photophysics of the porphyrin molecules in the ring-shaped assemblies.

2
Experimental Methods

2.1
Scanning Confocal Microscopy (SCM) and Laser Trapping Coupled to a Time-Correlated Single Photon Counting (TSPC) Setup

The scanning confocal microscope consists of a Nikon Diaphot 300 inverted microscope connected to a Biorad MRC 600 scanning unit. The optical image is built up by scanning the excitation laser beam over a fixed sample.

As an excitation source a high repetition rate, mode-locked, solid state titanium/sapphire laser (Tsunami, SP model 3960) pumped by a continuous argon ion laser (488 nm) (model 2080) is used which permits a fast measurement of fluorescent decay curves with high time resolution. The light output of the Tsunami apparatus is frequency doubled in a second harmonic generator (SP model 3980) and the pulse width at half maximum of this frequency doubled light is less than 2 picoseconds, implying that the time resolution is predominantly determined by the high speed photodetector and the electronic circuits used. Part of the excitation light is focused (f 20 mm) on a Hamamatsu silicon avalanche photodiode and a variable neutral density filter is adjusted until an output pulse of 180 mV is measured. This output signal is delayed and connected to a constant fraction discriminator (CFD) (EG&G Ortec model 934) whose output is connected to the TAC (time-to-amplitude converter) STOP due to reverse TAC operation. Before entering the Biorad scanning unit the size of the beam waist is reduced by a spatial filter (f 150 mm/0.2 µm/f 80 mm).

Light emitted by the sample passes through the objective, several mirrors and a circular variable aperture before reaching the photodetector. Fluorescent images are acquired using a long-pass filter in front of the detector. Comos software is used for the image acquisition and to control the scanning electronics settings. For time-resolved data collection a cooled (Hamamatsu C2773 cooler) simplex MCP (multi channel plate, Hamamatsu R3809U-50) is used as the detection system and is placed close to the entrance/exit aperture of the Biorad scanning unit. The fluorescence emanating from the sample (microscope) is directed by a 100 % reflective mirror (placed in the first filter block) towards the MCP. In front of the detector two adjustable pinholes are placed. Wavelength selection takes place by narrow band-pass filters (FWHM 10 nm, center wavelength 500 nm or 550 nm, CVI). The MCP output is first attenuated and then amplified by a Hewlett-Packard amplifier (model 8447D). The output pulse is connected to a CFD (Tenelec TC 454) and connected with the TAC START. The TAC output is sent to a 100 MHz Wilkinson type 8192 channel analogue-to-digital converter. The fluorescence decay curves are collected in 1/2 K data points of the multichannel analyzer. The instrument response function IRF (t, λ_{ex}, λ_{ex}) is measured with a metal-coated mirror positioned above a 10x/0.45 NA objective and has a FWHM of 66.5 picoseconds. Confocal images and time-resolved data for the coumarin derivative on latex particles in a polyvinyl alcohol film were obtained using a 60×/1.4 NA oil immersion objective.

A single beam optical trapping system is coupled to the other side port of the Nikon microscope by a dichroic mirror that reflects the infrared light (700 mW) of a diode pumped CW Yag laser (1064 nm) but transmits visible light below 650 nm. For the trapping experiments a 100×/1.3 NA oil immersion lens is used. To properly fill the aperture of the oil immersion objective, and in order to make the beam convergent, two planconvex lenses (f_1 150 mm and f_2 100 mm) are used with their flat surfaces facing one another. The second lens f_2 is placed on an x, y, z translation stage and its position is changed to optimize the focal spot size and depth (Fig. 1A).

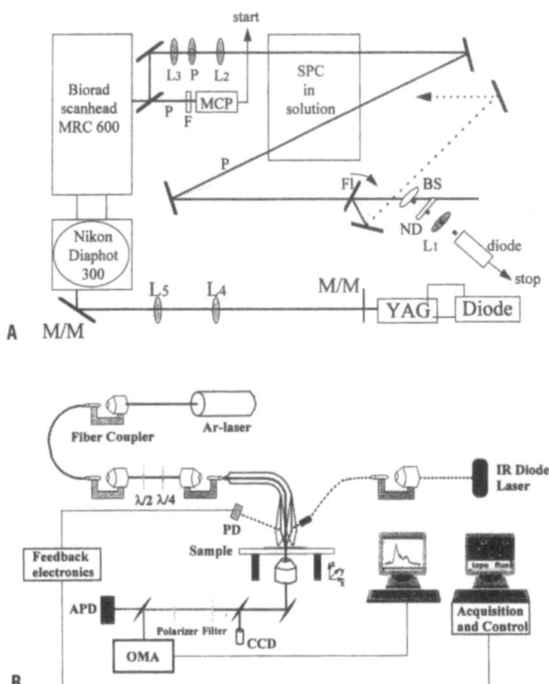

Fig. 1A, B. A Schematic presentation of the system used for measuring decay times of single trapped labeled particles; abbreviations: L1 to L3 biconvex lenses (with respectively f 20 mm, f 150 mm and f 80 mm); L4 and L5 planconvex lenses (f 100 mm and f 150 mm); BS beam splitter, ND neutral density filter, P pinhole, F band-pass filter, F1 metal-coated flipper mirror, M/M metal-coated mirror. **B** Schematic drawing of the setup for the NSOM

Fluorescence decays were analyzed by a non-linear least-squares iterative reconvolution method, based on the Marquardt algorithm [20]. Petri dishes containing solutions of single exponential reference compounds were placed on the microscope stage above a $10\times/0.45$ NA objective. The single and global curve analysis methods used in the analysis of the decays have been discussed extensively in earlier papers [21, 22].

2.2
Near-Field Scanning Optical Microscopy (NSOM)

The NSOM is a commercially available instrument from TopoMetrix (Aurora) modified for fluorescence imaging purposes (Fig. 1B). As an excitation source, a multiple line argon ion laser (Spectra Physics, model 2025) is used. The laser light is coupled into a single-mode optical fiber (3 M, type FS-SN 3224). Before coupling the light into the NSOM probe, it is first coupled out of the single-mode fiber to allow for the insertion of wave plates and filters. The filters are used to inhibit the coupling of fluorescence coming from the fibercoating into the NSOM probe. The wave plates are used to create linearly polarized light to excite the sample and are positioned between the single-mode fiber and the NSOM probe in order to keep the length of the fiber in which the polarization has to be controlled as short as possible. The experiments are performed either with home-made or with commercially available (TopoMetrix) NSOM probes. The collection of the fluorescence light is performed with and $60\times$, 0.7 NA, dry or an $100\times$, 1.4 NA oil immersion objective (Nikon) and detected by means of an Avalanche PhotoDiode module (APD) (EG&G, model SPCM-200-CD1718). The excitation light was blocked by using a holographic laser notch filter (Kaiser Optical Systems). The NSOM fluorescence spectra were acquired with a Chromex 250IS imaging spectrograph coupled to a cooled ICCD (Princeton Instruments, model ICCD-576-S/RB-EM).

The laser diode used for the shear-force feedback in the commercial instrument was replaced by an external laser diode with a wavelength of 980 nm (EG&G, model C86147 E). The light emitted by the diode is coupled into a single-mode optical fiber (Newport, model F-SBA). The other end of this fiber is positioned close to the NSOM probe in order to monitor its dithering, necessary for establishing shear-force feedback. The IR light in the detection path is blocked by means of two 900 nm short-pass filters (CVI, model SPF-900)

3
Applications

3.1
Fluorescence Imaging and Spectroscopy of Latex Particle Polymer Composite Films Fixed on a Surface

Polymer films consisting of dye-labeled latex particles with sizes of 500, 220 ± 10 nm and 14 ± 3.7 nm randomly dispersed in a poly (vinyl alcohol) film were investigated.

Figure 2A–C shows the photobleaching of single fluorescent latex particles (excitation of the coumarin derivative used for labeling at 415 nm and emission > 475 nm). The two particles indicated as bead 1 and bead 2 in this figure were selectively excited and imaged over 15 s. This process was repeated until complete photobleaching occurred. The bleaching constant derived from this experiment is 3×10^{-3} $^{-1}$.

When photons are accumulated in time-resolved fluorescence experiments this bleaching effect has to be taken into account. Figures 2D–I present the confocal fluorescence images of latex beads with a size of 220 nm with their respective fluorescence decays. The labeled latex particles are excited at 420 nm (repetition frequency 800 kHz). First, a fluorescence image is obtained with the

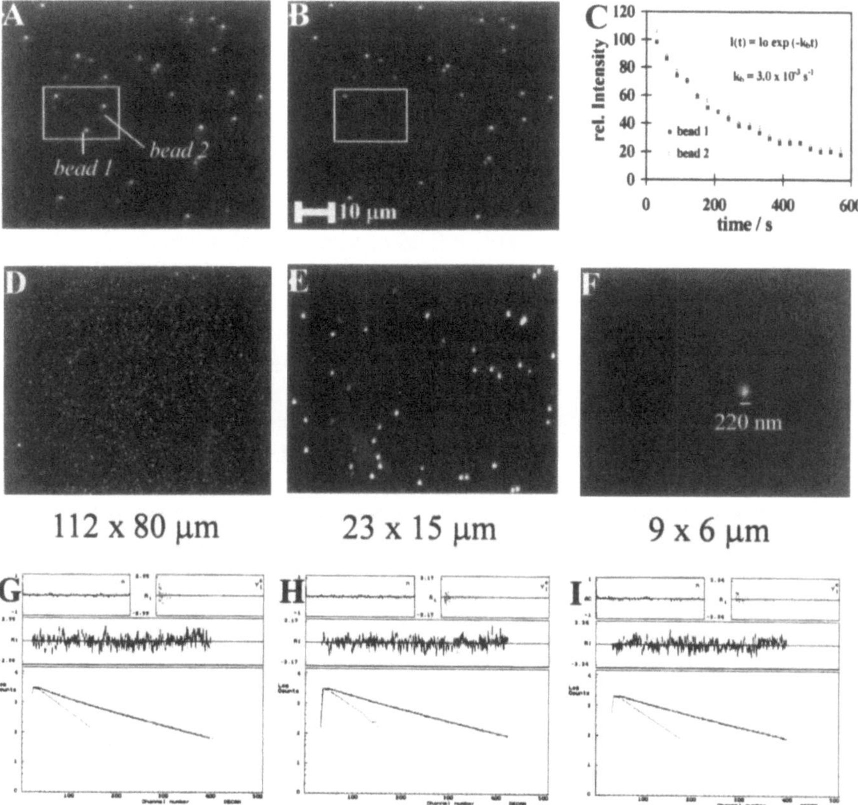

Fig. 2A–I. (A) and (B) are CFM images of a polymer composite film consisting of dye-labeled latex particles randomly dispersed in a PVA film. Two particles at the center of (A) are selectively excited at 415 nm and imaged over 15 s. The process is repeated until complete bleaching occurs. (C) The relative intensity of the coumarin derivative is observed to decrease exponentially with time with a bleaching constant of 3×10^{-3} s^{-1}. (D–I) Confocal fluorescence images and corresponding time-resolved fluorescence decay curves of 220 nm labeled latex particles in a PVA film. Complete field of view (D, G), a few particles (E, H) and single particle (F, I)

Table 1

	τ_s (ns)	τ_r (ns)	χ^2	$Z(\chi^2)$	σ
(i) all beads (no zoom)	2.7	1.2	1.0	0.4	1.1
(ii) a few beads	2.8	1.2	0.9	−0.2	1.0
(iii) one single bead	2.9	1.2	1.3	3.2	1.0

Table 2

	τ_s (ns)	τ_r (ns)	χ^2	$Z(\chi^2)$	σ
(i) all beads (no zoom)	2.7	1.2	1.1	1.1	1.0
(ii) a few beads	2.9	1.2	0.9	−0.9	1.0
(iii) one single bead	3.0	1.2	1.3	3.3	1.1

slowest scan speed (F1-slow: 3 s per frame) and zoom factor 1 (no zoom). The laser beam is scanned over the entire field of view and photons are directed to the PMT inside the Biorad MRC 600. Subsequently, the 460 DCLP dichroic is replaced by a 100 % reflective mirror that directs the emitted light to a simplex MCP. In this way a decay curve is obtained after scanning (F4-fast 2: 1/4 s per frame) of an ensemble of latex particles. A higher magnification is reached by a zoom facility without loss of resolution. However, at higher zoom factors a smaller area of the sample is illuminated with the same laser intensity and bleaching is higher. Finally, one single latex particle is imaged and the corresponding fluorescence decay recorded (Fig. 2F, I). The time-resolved decay curves of ensembles of labeled latex particles are single curve and globally analyzed. The results of single curve and global analysis are summarized in Tables 1 and 2.

Figure 3 shows the NSOM topography and simultaneously acquired fluorescence image of dye-labeled 14 ± 3.7 nm large latex particles dispersed in an approx. 25 nm thick PVA film. This sample is appropriate to show the capabilities of imaging heterogeneous organic films with high sensitivity and a resolution beyond the diffraction limit. Spots of high light intensity which appear in the fluorescence image can be attributed to fluorescence of these sites. Some of these spots clearly correlate with interface structures exhibited in the topography image as indicated, for example, by the two arrows. The protrusions in the topography image at the site of high intensities in the fluorescence image can be explained by latex particles which are located at the polymer-air interface and which emerge from the embedding PVA matrix. Remarkably, the latex particles accounting for most of the fluorescence spots in Fig. 3 can hardly be seen in the topography image. This observation shows the unique capability of NSOM, compared to other scanning probe techniques, namely the ability to

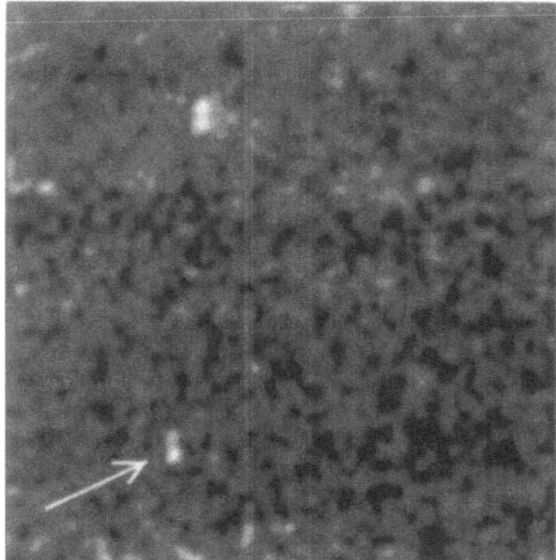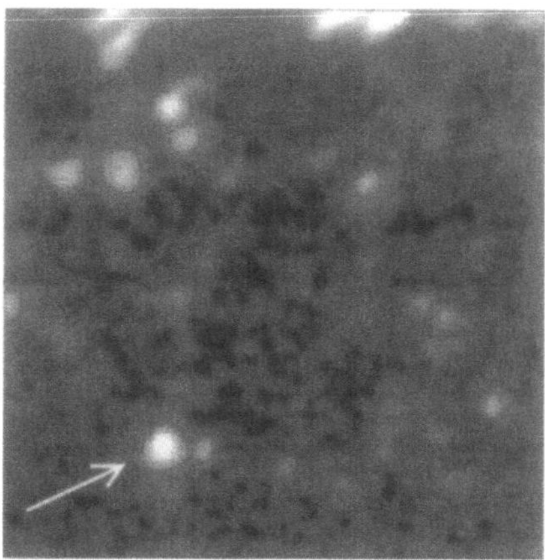

Fig. 3. NSOM topography (*left*) and simultaneously acquired fluorescence image (*right*) of a section of an ~ 25 nm thin polymer film containing 14 nm small dye-labeled latex beads in a PVA matrix. The excitation wavelength is 458 nm. Image size: $5 \times 5 \ \mu m^2$. The black-white contrast in the topography image corresponds to 21 nm and to 2070 cps for the fluorescence image

detect sub-interface particles. NSOM is capable of detecting properties of the sample under investigation in spatial domains which are located a few tenths of a nanometer beneath a sample-air interface.

3.2
Fluorescence Imaging and Spectroscopy of Latex Particles in a Laser Trap

By combining confocal laser scanning microscopy and TCSP with a single beam optical trapping technique it is possible, for example, to obtain time-resolved fluorescence data of dyes on the surface of a particle trapped in an aqueous solution. By making use of the advanced collection schemes possible in the scanning unit, series of images can be collected separated by a timestep (every 10 s an image was taken) or a Z-step (2 μm variations in the Z-direction) while keeping the trapped object in focus. Figure 4B represents a time series of 10 consecutive frame scans (slow scan mode) collected with a time interval of 10 s for a solution containing 1 μm labeled latex particles. In this way the Brownian motion of the surrounding particles can be visualized. The confocal image in Fig. 4A is a snapshot of the fluorescent latex particles in solution. The particle marked with the circle is the one in the laser trap. The latex particles in image B, with the exception of the trapped one, look blurry due to random motion in the x, ẏ as well as in the x, z direction. Only the trapped particle has sharp contours since it remains in focus. In Fig. 4C a Z-series (step-variation 2 μm) consisting of five merged images is presented. The trapped 4.5 μm latex particle remains in focus while the surrounding particles come in and out of focus. The latex particles (4.5 and 1 μm) are trapped with highly focused IR light (700 mW, 1064 nm, spot size 1 μm) and simultaneously excited at 420 nm (repetition rate

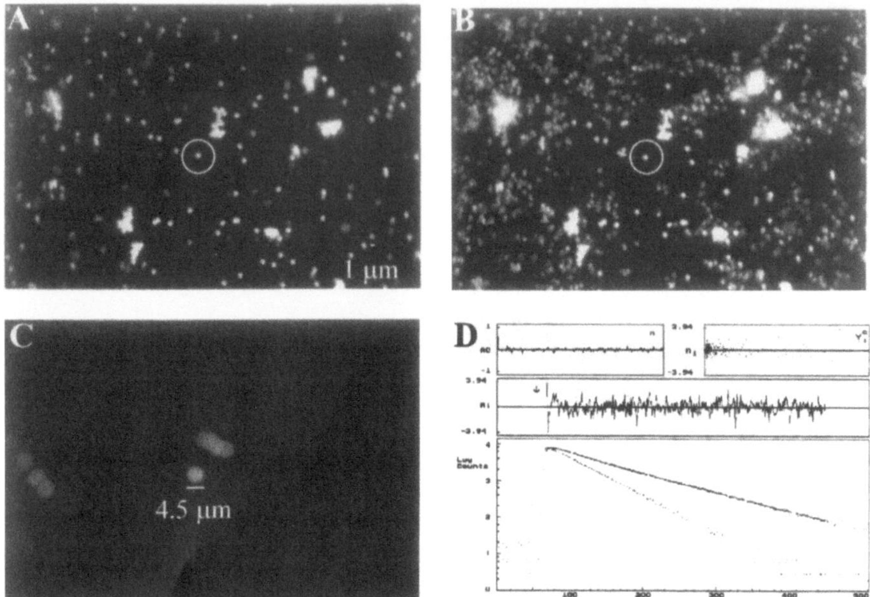

Fig. 4A–D. (A) Snapshot of a solution containing 1 µm labeled latex particles; the particle in the circle is trapped. (B) Time series of 10 consecutive scans with a time interval of 10 s, projected in a two-dimensional image for a solution containing 1 µm labeled latex particles (scan speed: 3 frames/s, image size 115 × 77 µm). (C) A Z-series of 5 consecutive frame scans projected in a two-dimensional image for a solution containing 4.5 µm latex particles with a step variation of 2 µm). Scan speed: 3 frames/s. (D) Time-resolved fluorescence decay curve of a 4.5 µm coumarin labeled latex particle trapped in solution with coumarin 485 in methanol as the reference. Excitation wavelength is 420 nm, and wavelength selection is done with a 550 ± 5 nm narrow band-pass filter. t_s = 2.30 ns, t_r = 1.3 ns and $Z(\chi^2)$ = 2.4

800 kHz) and their fluorescence detected by a cooled simplex MCP. Wavelength selection is performed with narrow band-pass filters. Figure 4D corresponds to the fluorescence decay of a single 1 µm trapped latex particle together with the residuals and the autocorrelation function. In these experiments, Coumarin 485 in methanol is used as the reference.

3.3
Microscopic Investigation of a Supramolecular System: Evaporated Porphyrin Thin Films

When PtP (Fig. 5) is dissolved in chloroform and evaporated on a glass substrate, wheel-shaped structures are formed. The morphology and optical properties of these porphyrin wheels are investigated using AFM, NSOM and CFM.

Confocal Fluorescence Microscopy of Porphyrin Wheels. CFM was used to study a broad range of porphyrin thin films prepared under various conditions.

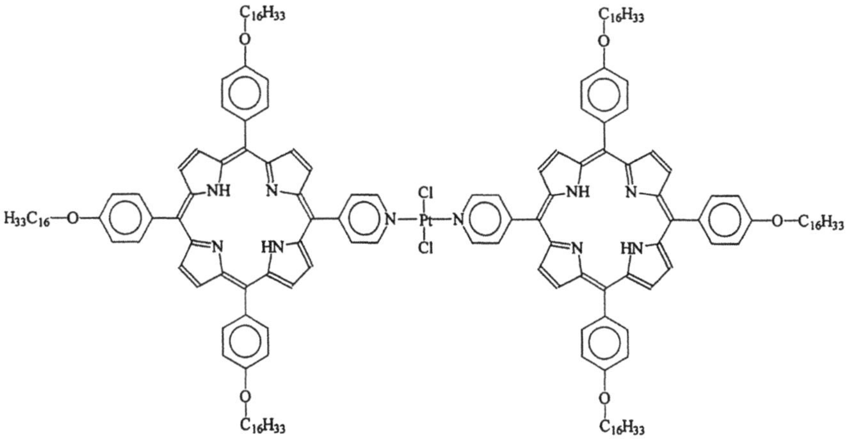

Fig. 5. Chemical structure of PtP

Time-resolved fluorescence spectroscopy was combined with confocal microscopy to monitor the time-resolved emission dynamics of porphyrin rings. Confocal images in transmission- and fluorescence-mode showed that the morphology of the structures, formed upon evaporation of the solvent, strongly depend on the concentration of the starting solution. In the low concentration regime highly interrupted (like beads on a chain) rings could be observed while in the high concentration regime continuous depositions of porphyrin material were found. However, at a concentration of 2×10^{-6} M of PtP in chloroform, nice and well formed rings were found on the surface (Fig. 6). The images show that isolated rings exist together with complex structures built up by several rings fused together leading occasionally to very complex structures like the ring of rings depicted in Fig. 6C. It is likely that these various structures result from an organization that occurs in the growth mechanism. Some of the rings have a large individual deposition on one side which may function as a nucleation site for its growth. Such a large structure is clearly seen in Fig. 6D on the upper right-hand side of the large ring in the center of the image. In all of the experiments performed with CFM no fluorescence could be detected from inside the rings.

Time-resolved fluorescence decays were taken from different regions of the sample and compared. The fluorescence decays obtained from individual rings, of an area of a sample including many fused rings, and of larger structures occasionally found on the edge of the rings, are indistinguishable within the limits of experimental error. The similarity of these data reveals that the porphyrin material which comprises the rings, the large composite aggregates, and the material between the rings have a very similar structure at the spatial resolution of CFM.

In Fig. 6E, F polarized fluorescence confocal images of porphyrin rings are presented. The horizontal and vertical polarized images are identical within the limits of experimental error. The same result was obtained using different excitation polarizations. The absence of anisotropy in the CFM images indi-

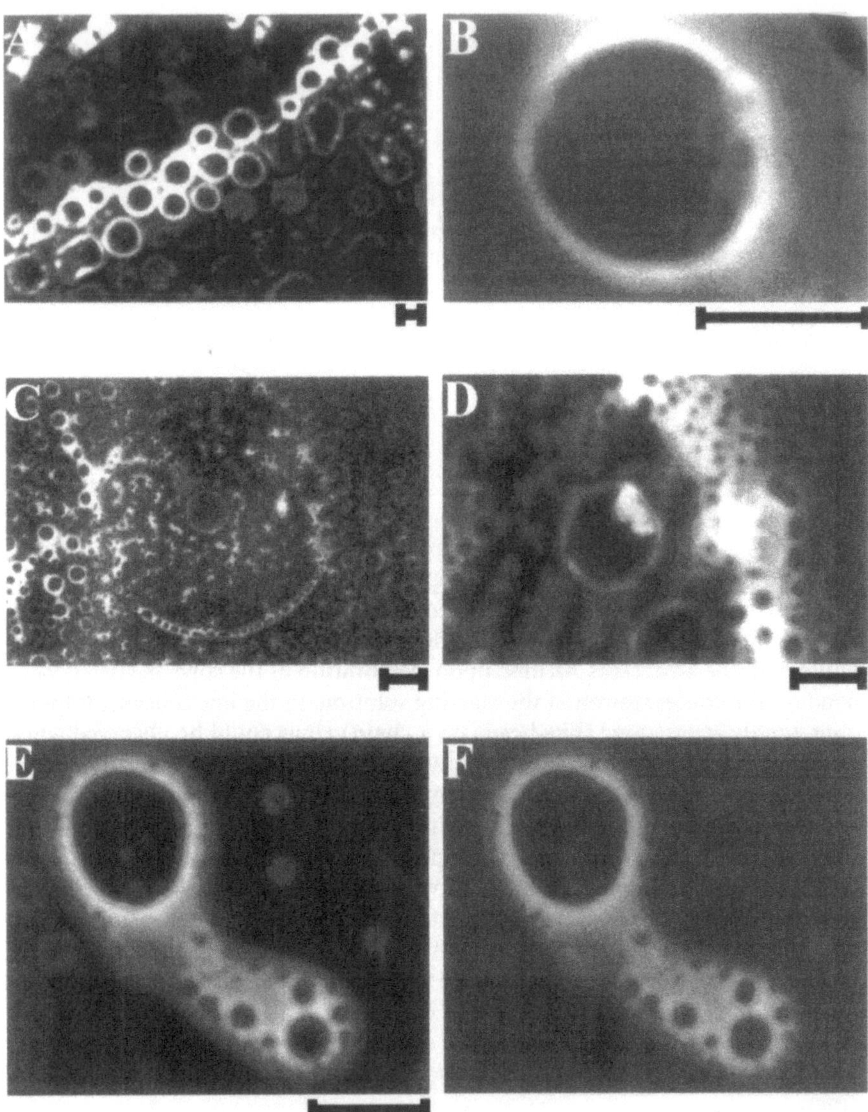

Fig. 6A–F. Confocal fluorescence images of evaporated films of 2×10^{-6} M solutions of PtP/CHCl$_3$ on glass representing the different types of structures observed. The scale bar below each image represents 10 µm. (**E, F**) Confocal fluorescence polarization images of evaporated films of PtP in a mixture of CHCl$_3$ and MeOH on glass. The black scale bar corresponds to 10 µm

cates that no ordered structures could be observed within the spatial resolution (ca. 400 nm) and the instrumental sensitivity.

Two mechanisms for the ring formation can be proposed which both explain the variety of structures observed in evaporated films of PtP in chloroform. A first possible model, the "bubble"-model, is based on ring formation induced by gas bubbles formed during the evaporation of the solution used to prepare the film. At the circumference of the gas bubbles, nucleation and growth of an amorphous porphyrin phase should occur, which eventually results in ring formation after evaporation of the solvent is complete. Bubble-induced aggregation is a general and well-documented phenomenon [23]. In a second model, the "hole"-model, the ring formation can be induced by dynamic processes during the de-wetting of the substrate by the chloroform solutions. During evaporation of the solvent, dry patches nucleate and grow at the substrate/solvent interface. The hole formed in this way expands (due to capillary forces) and introduces the displacement and accumulation of porphyrin material. This ends in ring structures when the evaporation is complete. The self assembly of material induced by dry holes, and nucleation during wetting and de-wetting processes, has recently been reported [24]. The processes, which will be discussed in detail elsewhere, are complex phenomena, which require an organizing structure (bubble/hole) and involve hydrodynamic and surface effects.

Near-Field Scanning Optical Microscopy and Spectroscopy of Porphyrin Wheels. With NSOM more information could be obtained concerning the morphologies and optical properties of evaporated films of PtP on a glass substrate. The use of high-resolution fluorescence imaging and fluorescence polarization imaging combined with local spectroscopy leads to more insight into the local molecular aggregation behavior of the porphyrin compound in the films.

Figure 7 displays three sets of images consisting of topography and fluorescence maps corresponding to the three typical structures observed when varying the concentration of the solution used to prepare these samples.

As observed by CFM, solvent cast films of the 2×10^{-6} M solutions of PtP in chloroform consistently formed ring assemblies. These assemblies could be imaged straightforwardly by fluorescence NSOM and, simultaneously, shear-force microscopy. For example, Figs. 7D, E portray simultaneous NSOM and shear-force topographic images of several ring-shaped assemblies that range in height from 50 to 300 nm and have fluorescence intensities as high as 7,000 counts per s. Figure 7F is a zoom of the fluorescence image in Fig. 7E. Fluorescence intensity and topography are highly correlated indicating that the porphyrin material is primarily restricted to the ring assemblies. It should be emphasized that the height of the material inside the rings and between the rings is equal to that of the glass substrate within experimental uncertainty (approximately 5 nm) for most of the films prepared from the 2×10^{-6} M solution. This was established by studying the edge of the film where it was possible to unambiguously determine the height of the glass substrate. Correspondingly, the NSOM fluorescence intensity outside the rings and inside the rings is indistinguishable from the background level in the NSOM ex-

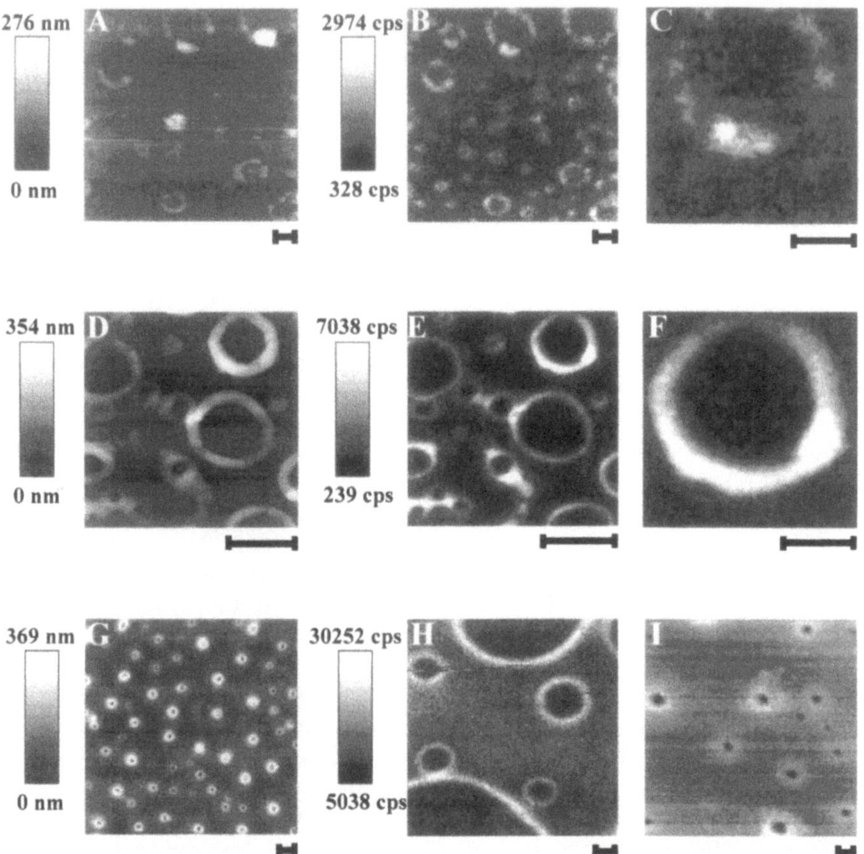

Fig. 7A–I. Near-field fluorescence and topographic images of evaporated films of PtP/CHCl₃ solutions on glass. The black scale bar below each image represents 1 μm. (A, B) Topographic and simultaneously acquired near-field image taken on a sample prepared by solvent casting of a 2×10^{-7} M solution. (C) Enlarged view of the top middle ring displayed in B. The color scale at the left-hand side of B is also appropriate for C. (D, E) Topographic and corresponding near-field fluorescence image of a sample prepared by solvent casting of a 2×10^{-6} M PtP/CHCl₃ solution on glass. (F) Enlarged view of the top right ring displayed in E. The color scale at the left side of E is also appropriate for F and H. (G–I) Topographic and near-field images of a sample prepared by solvent casting a 2×10^{-4} M PtP/CHCl₃ solution. G Topography image taken with an uncoated fiber probe. (G, H) Near-field fluorescence images taken, respectively, at the center and at the edge of the film. The color scale at the left-hand side of H is only appropriate for I

periment. The resolution in the NSOM images is substantially better than the simultaneous topographic images which are significantly broadened due to a tip effect from the flat and smooth end of the NSOM probe. Well-resolved topographic images were obtained using uncoated, sharp fibertip probes and with commercially available Si₃N₄ AFM probes. The well-resolved topographic images exhibited ring diameters varying between 100 nm to 10 μm. The height of the rings was approximately linearly correlated with the ring diameters and varied in the range of 10 nm for the smallest rings to over 200 nm for the largest rings. The width of the rings also increased with the ring height. The apparent ring widths may still be erroneous due to convolution of the tip shape.

When the solution used for preparing the sample contained a lower concentration (2×10^{-7} M) of PtP, a different film structure was observed. Figures 7 A, B correspond to the topographic and simultaneously acquired fluorescence image of such a sample. The similarity of the fluorescence and the topographic images also holds for these experiments. Some of the rings observed in this experiment are built up from individual isolated particles like beads on a necklace. Interestingly, these incomplete rings seem to have exactly one larger particle on their edges. In general the diameter of the rings and the height of the particles comprising the ring structure are somewhat smaller than the measured values for the well-formed rings. Similar topographic images were obtained using non-contact AFM as is expected for artifact free imaging. Figure 7 C is a zoom in of the fluorescence image in Fig. 7 B. The structure observed in this case can easily be interpreted in terms of a bubble- or hole-induced ring growth mechanism. Each ring corresponds to a hole formed during the de-wetting process or to the bubble formed during evaporation. The presence of a larger particle at the edge of a ring can be explained by the nucleation site for the bubble or hole formation. The incomplete ring structures with their relatively small heights may be the result of the limited amount of material available at this particular concentration during the ring growth, in contrast with the situation at 2×10^{-6} M where sufficient material is available to form complete rings.

Figure 7 E shows that some of the rings have some excess of fluorescent material built up in flat regions outside the rings. This excess material is especially prevalent in the regions between nearby rings and where several rings are fused together. Smaller rings are often found to be fused to larger rings. Additionally, the height in the topography images and the fluorescence intensity at the intersections of the two rings is substantially higher than in either of the two rings away from the intersection. The excess material between the rings is even more prominent in the films produced from more concentrated PtP/CHCl$_3$ solutions, e.g. 2×10^{-4} M as shown in Fig. 7 H. Here the excess material forms a continuous film between the rings. The excess material is observed in both the topographic and fluorescence NSOM images. The regions inside the rings still lack fluorescence and are at a lower elevation (presumably at the height of the substrate) than the film. The continuous excess material is especially prevalent at the thick edges of the evaporated film as shown in Fig. 7 I. In this region "volcano"-shaped assemblies form with a pinhole "crater" at each ring. Topographic images such as in Fig. 7 G taken with relatively sharp uncoated probes show analogous volcano-like assemblies on top of a continuous film. The diameters of the rings observed at the edges of the droplet are substantially smaller than the ones found at the center of the droplet and at lower concentrations. A possible explanation for this observation can be related to the solution density, that should increase with the PtP concentration. This should affect the solution vapor pressure and ultimately its evaporation rate and also the dynamics of the liquid film on the glass.

Further insight into the local molecular organization of the porphyrin material in these films was obtained from fluorescence spectroscopy. The fluorescence spectrum of a bulk film deposited by evaporation of a 10^{-6} M chloroform solution of PtP on glass is compared to the corresponding data for the solution itself in Fig. 8A. The emission spectrum has the expected Q band with an emission maximum at 654 nm with a vibronic band at 725 nm. The emission spectrum of the film is noticeably red shifted relative to the solution consistent with an aggregated structure involving a head-to-tail organization [25]. Almost identical emission spectra were observed for films deposited with more concentrated solutions up to 10^{-4} M.

Emission spectra at localized regions of the film have been recorded with excitation from an uncoated NSOM probe (excitation spot size approximately 500 nm). Figure 8B shows a spectrum recorded during a 60 s irradiation of a porphyrin ring at the position marked by "X" on the image displayed as an inset. The maximum of the localized emission spectrum of the ring (Fig. 8B) is, within the limits of experimental error, identical to the emission spectrum of the bulk films indicating that there is little spectral variation with the position

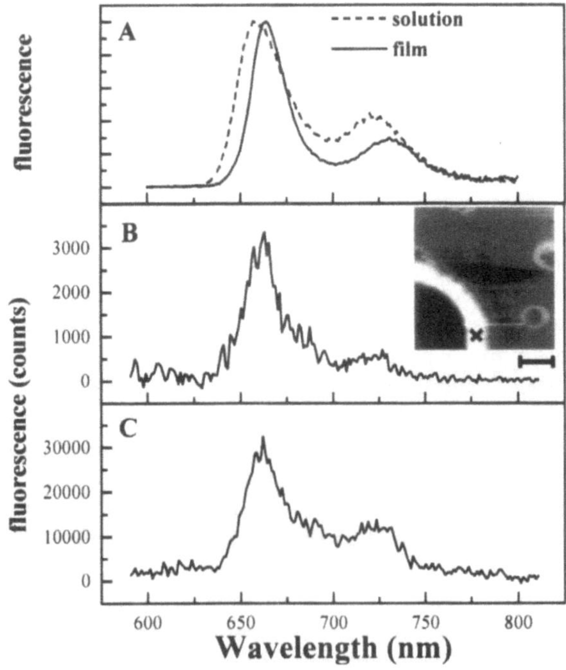

Fig. 8A–C. A Normalized emission spectra of a 10^{-6} M PtP/CHCl$_3$ solution deposited on a glass slide. B Spectrum taken on the position of the black cross marked in the image displayed as an inset. The scale bar below this image corresponds to 2 µm. Acquisition time 60 s. C Spectrum taken on an area of the sample similar to that shown in Fig. 7I. Spectra shown in B, C were taken on a sample prepared by evaporating a 2×10^{-4} M solution in CHCl$_3$ and are corrected for background fluorescence

in the film. This suggests that the aggregation properties of the porphyrin material is similar throughout the films. This conclusion is further supported by the similarity of the bulk spectra for the films deposited with different concentrations of the porphyrin and correspondingly very different morphologies. This is shown in Fig. 8C which portrays a spectrum that was recorded for especially thick regions of the films as shown in Fig. 7I. This spectrum was recorded with continuous irradiation and simultaneous scanning to ensure a broad sampling of the scanned region.

A different measure of the local molecular orientation of the porphyrin films is available in the analysis of the anisotropy of the fluorescence. For example, Figs. 9A, B show fluorescence NSOM images that were recorded with a polarizer in the collection optics oriented respectively in the horizontal or vertical direction of the image. The nearly identical patterns of the two images, disregarding the absolute intensities, indicate that the emission of the porphyrin

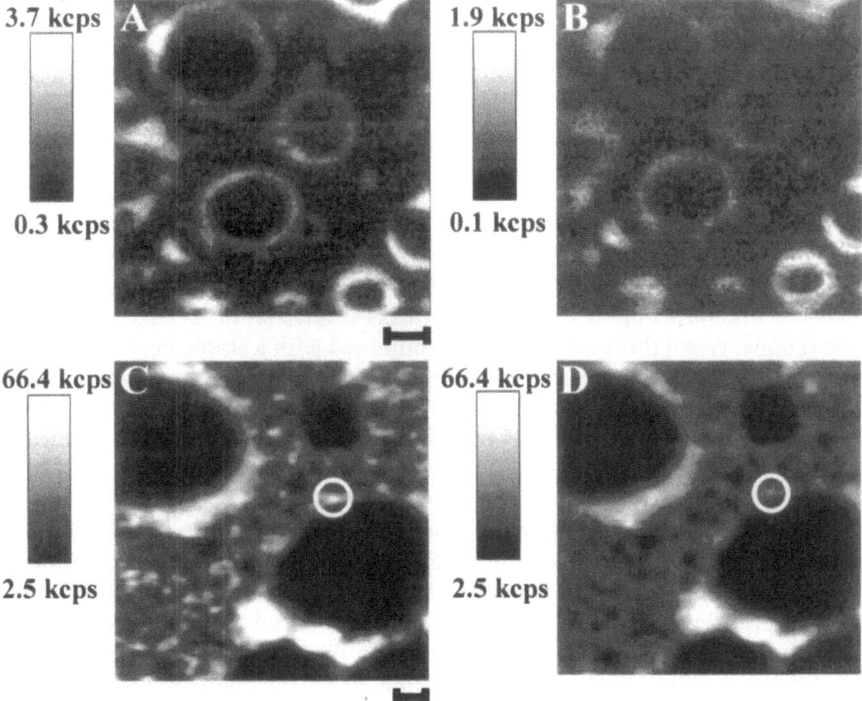

Fig. 9A–D. NSOM fluorescence polarized emission images acquired on a sample prepared by evaporating a 2×10^{-6} M solution in $CHCl_3$. The black scale bars represent 1 μm. Images **A** and **B** are acquired using randomly polarized excitation light and with a polarizer in the detection optics. For image **B** the polarizer is turned 90° with respect to its position for image **A**. The purpose of the images is to compare the intensity distributions in both images despite the difference in contrast. **C, D** Images acquired using linearly polarized excitation light. Image **C** represents the polarized emission image where detection is parallel to the excitation, while in **D** the detection is perpendicular to the excitation. The blue circle indicates an area showing polarized emission

wheels is not polarized on the ~100 nm distance scale. Separate experiments using confocal microscopy, as described above, verified that the porphyrin ring fluorescence is unpolarized on the <1 µm distance scale, even though the presence of aggregation in solution was observed [15]. Since fluorescence polarization measurements in solution demonstrate that individual porphyrin aggregates have a substantially anisotropic fluorescence, the absence of fluorescence anisotropy in the thin film ring experiments strongly suggests that the individual aggregates that comprise the ring structures are randomly oriented and must have a domain size much less than 1 µm.

Preliminary observations using linearly polarized light to excite the sample, however, reveal the presence of small domains giving rise to polarized emission upon photoselection. Figures 9 C, D represent, respectively, the parallel and perpendicular polarization images with respect to the excitation polarization. From a comparison of the intensity distribution in the two images it can be stated that it was possible to photoselect aggregates randomly distributed in the film with measured sizes ranging from about 250 nm down to 50 nm which is the resolution limit of the microscope. One of the aggregates is indicated with a white circle in the images.

4
Conclusions

In this contribution an overview of the advanced microscopic instrumentation, together with a few applications of these techniques, are given. It is shown that confocal microscopy combined with a SPC setup allows the imaging of heterogeneous organic samples and the spectroscopic analysis of localized regions of the sample. When this system is again combined with a single beam laser trapping setup, one is capable of extending the possibilities in terms of studying single microparticles in solution. The capabilities of the instrument are tested using dye-labeled latex particles distributed in a polymer film or dissolved in solution.

Additionally, a description of the NSOM instrument is presented and its performance is tested using 14 nm dye-labeled latex particles distributed in a polymer film. Furthermore a microscopic and spectroscopic investigation of films containing porphyrin wheels reveals the molecular organization and the optical properties of these films. Several features lead to new insights in the growth mechanism of the porphyrin rings. Spectroscopy and polarization experiments using NSOM give information about the aggregation of the porphyrin compound in the film and in the ring. It is shown that although ordered structures are present on the nanometer scale no molecular order can be observed on the length scale of the rings. The data are also discussed in terms of two possible mechanisms of ring formation. One model is based on bubble-induced aggregation while the other model is based on hole-assisted assembly during de-wetting phenomena. The system studied is, furthermore, one of the first optical visualizations in such detail of such models related to surface wetting of organic materials.

Acknowledgements. This work was supported by FWO and DWTC through IUAP-4-11 and by the Dutch Foundation for Chemical Research (SON) with financial aid from the Dutch Foundation for Chemical Research (NWO). PFB was supported by a fellowship from KU Leuven. J.H. and L.L. thank FWO for post-doctoral fellowships. J.K. received partial support from the U.S. office of Naval Research, 3M corporation and the U.S. National Science Foundation.

References

1. For an introduction see, e. g., a) D.W. Pohl In Adv. Opt. Electron. Microscopy 12 (1991) 243 (Academic Press New York). b) D. Courjon, C. Bainier Rep. Prog. Phys. 57 (1994) 989. c) Near-field Optics; D.W. Pohl, D. Courjon Eds.; NATO ASI Series, Series E 242; Kluwer Dordrecht, The Netherlands, 1993
2. E. Betzig, J.K. Trautman, Science 257 (1992) 189
3. J.K. Trautmann and J.J. Macklin, Chem. Phys. 205 (1996) 221
4. M.H. Chestnut, Current Opinion in Colloid and Interface Science 2(2) (1997) 158–161
5. N. Garbow, J. Müller, K. Schätzel, T. Palberg, Physica A 235 (1997) 291
6. L. Li, L. Chen, P. Bruin, M.A. Winnik, T.C. Jao, Acta Polymerica 47(9) (1996) 407–410
7. S.A. Carter, J.D. LeGrange, W. White, J. Boo, P. Wiltzius, J. Appl. Phys. 81(9) (1997) 5992
8. K. Amundson, A. van Blaaderen, P. Wiltzius, Physical Review E 55 (2) (1997) 1646
9. a) E. Betzig, R.J. Chichester Science 262, (1993) 1422. b) W.P. Ambrose, P.M. Goodwin, J.C. Martin, R.A. Keller Phys. Rev. Lett. 72 (1994) 60. c) J.K. Trautman, J.J. Macklin, L.E. Brus, E. Betzig Nature 369 (1994) 40. c) S.X. Xie, R.C. Dunn Science 265, (1994), 361. d) W.P. Ambrose, P.M. Goodwin, J.C. Martin, R.A. Keller Science 265 (1994) 364
10. a) M.H.P. Moers, H.E. Gaub, N.F. Van Hulst Langmuir 10 (1994) 221. b) M.H.P. Moers, N.F. Van Hulst, A.G.T. Ruiter, B. Bölger Ultramicroscopy 57 (1995) 298
11. a) D.A Vanden Bout, J. Kerimo, D.A. Higgins, P.F. Barbara, Acc. Chem. Res. 30 (1997) 204. b) D.A. Higgins, J. Kerimo, D.A. Vanden Bout, P.F. Barbara, J. Am. Chem. Soc. 118 (1996) 4049
12. M. Rücker, P. Vanoppen, F.C. De Schryver, J.J. Ter Horst, J. Hotta, H. Masuhara, Macromolecules 28 (1995) 7530
13. H. Faes, J. Hofkens, F.C. De Schryver, submitted
14. M. Rücker, F.C. De Schryver, P. Vanoppen, K. Jeuris, S. De Feyter, J. Hotta, H. Masuhara, NIM B, 131 (1997) 30
15. J. Hofkens, L. Latterini, P. Vanoppen, K. Jeuris, H. Faes, S. De Feyter, J. Kerimo, A.E. Rowan, R.J.M. Nolte, P.F. Barbara, F.C. De Schryver, J. Phys. Chem., in press
16. a) A. Ashkin, Phys. Rev. Lett. 24 (1970) 156–159. b) A. Askin and J.M. Dziedzic, Appl. Phys. Lett. 19 (1971) 283. c) A. Ashkin, Phys. Rev. Lett. 40 (1978) 729. d) A. Ashkin, J.M. Dziedzic, J.E. Bjorkholm, S. Chu, Opt. Lett. 11 (1986) 288
17. a) N. Kitamura, K. Sasaki, H. Misawa, H. Masuhara, Microchemistry: Spectroscopy and Chemistry in Small Domains, H. Masuhara, Ed. (Elsevier Science B.V.) (1994) pp 23–34. b) K. Sasaki, M. Koshioka, H. Misawa, N. Kitamura, H. Masuhara, Opt. Lett. 16 (1991) 1463–1465. c) K. Sasaki, M. Koshioka, H. Misawa, N. Kitamura, H. Masuhara, Jpn. J. Appl. Phys.30 (1991) 907. d) H. Misawa, K. Sasaki, M. Koshioka, N. Kitamura, H. Masuhara, Appl. Phys. Lett. 60 (1992) 310
18. a) S.M. Block, Nature 360 (1992) 493–495. b) K. Svodoba, S.M.Block, Ann. Rev. Biophys. Biomol. Struct. 23 (1994) 247–285. c) K. Svodoba, S.M. Block, Cell 77 (1994) 773–784
19. A.P.H.J. Schenning, F.B.G. Benneker, H.P.M. Geurts, X.Y. Liu, R.J.M. Nolte, J. Am. Chem. Soc. 118 (1996) 8549
20. D.W. Marquardt, J. Soc. Indust. Appl. Math. 11 (1963) 431
21. J.R. Knutson, J.M. Beechem, L. Brand, Chem. Phys. Lett. 102 (1983) 501

22. K.A. Zachariasse, G. Duveneck, W. Kühnle, P. Reynders, G. Striker, Chem. Phys. Lett. 133 (1987) 390
23. a) R.L. Judd, C.H. Lavdas, J. Heat Transfer 102 (1980) 461. b) Vinogradova, O.I., Colloids Surfaces A: Physicochem. Eng. Aspects 82 (1994) 247
24. a) P.C. Ohara, J.R. Heath, W.M. Gelbart, Angew. Chem. 109 (1997) 1120; Angew. Chem. Int. Ed. Engl. 36 (1997) 1078. b) C. Sykes, C. Andrieu, V. Détappe, S. Deniau, J. Phys. III France 4 (1994) 775. c) F. Brochard-Wyart, J. Daillant, Can. J. Phys. 68 (1990) 1084. d) G.P. de Gennes, Rev. Mod. Phys. 57 (1985) 827
25. a) M. Kasha, Radiat. Res. 20 (1963) 55. b) S. Eriksson, B. Källebrin, S. Larsson, J. Matensson, O. Wennerström, Chem. Phys. 146 (1990) 165

CHAPTER 7

Three-Photon Excitation of Fluorescence

J. R. Lakowicz, I. Gryczynski

1
Introduction

Most experiments in optical spectroscopy rely on the interaction of a single photon with the absorbing molecule. However, since 1931 it has been known that molecules can simultaneously interact with two photons. Two-photon absorption was predicted theoretically by Goppert-Mayer in 1931 [1]. This phenomenon only occurs at high light intensities, and was not studied experimentally until after the appearance of lasers in the 1960s and 1970s. Since then, two-photon spectroscopy has been used to study the excited state symmetry of organic chromophores [2–4]. Two-photon excitation has been applied to study such systems as hydrocarbons [5, 6], porphyrins [7], polyenes [8, 9], protein bound chromophores [10] and indole derivatives [11, 12]. Two-photon excitation has also become a tool in analytical chemistry [13, 14]. Until recently, the main emphasis of two-photon spectroscopy has been to study the symmetry properties of electronic states. This emphasis was due to selection rules for electronic transitions, which are different for one-photon and two-photon processes.

Studies of electronic symmetry, whether by one-photon (1PE) or two-photon (2PE) processes, often rely on excitation with linearly or circularly polarized light. Excitation with linearly polarized light is also used in fluorescence spectroscopy. However, the experiments and interpretation are distinct for studies of electronic symmetry or fluorescence. In two-photon spectroscopy, information about the symmetry of excited states is obtained from the differential absorption of linearly and circularly polarized light in fluid solution [3, 15]. In contrast, in time-resolved fluorescence one measures the anisotropy of the emission with excitation provided by linearly polarized light, which provides information about the rotational dynamics of the fluorophore. Hence, in time-resolved fluorescence the zero-time anisotropy, resulting from multi-photon excitation, is a parameter of high interest. In the past few years, two-photon excitation has been applied to study time-dependent intensities and anisotropies for a wide range of fluorophores including membranes labeled with DPH [16], labeled DNA [17, 18], proteins [19], fluorescence standards [20–22], dioxane [23] and alkanes [24–27].

The interest in multi-photon excitation has increased dramatically over the past few years as the result of its application to fluorescence microscopy. This is a result of the increased availability of ps and fs lasers. Additionally, multi-photon excitation has become an important tool in fluorescence microscopy and cellular imaging. In two-photon excitation the intensity of the fluorescence depends on the square of the local optical power. Consequently, the excitation is localized to the focal point. The conditions equivalent to confocal microscopy can be achieved without the complexity associated with confocal systems [28–31].

In recent experiments, we have shown that higher order excitation is possible. Using the femtosecond (fs) pulses from a titanium/sapphire laser, we have observed simultaneously three-photon excitation of a variety of biochemical fluorophores. In this paper, we present our latest results on three-photon excitation of biofluorophores including tyrosine and tryptophan, stained DNA- or DPH-labeled membranes. Of course, an improvement in the localization of excitation for the three-photon processes is expected compared to two-photon excitation.

2
Experimental Methods

Multi-photon excitation was accomplished as described in detail previously [16–27]. The basic idea is to use long wavelengths, approximately twice as long as those used for single-photon excitation. While this approach works in practice, it is important to note that the selection rules are different for different modes of excitation (1PE, 2PE or 3PE). Hence, there is no reason to expect that the two-photon absorption spectrum is at twice the one-photon wavelengths. Multi-photon excitation is accomplished using the focused output of a cavity-dumped dye laser with a pulse width near 5 ps, or a titanium/sapphire laser with a pulse width near 100 fs.

Time-resolved fluorescence intensity and anisotropy decays were measured in the frequency domain, as described previously [32–37]. In frequency-domain measurements the sample is excited with intensity-modulated light, where the modulation frequency is comparable to the inverse of the mean decay time (Fig. 1). Because of the finite lifetime at the excited state, the emission is delayed relative to the excitation. This time delay is measured as a phase shift. At higher modulation frequencies, the phase angle becomes larger. In frequency-domain measurements, one also measures the modulation of the emission, relative to the modulated excitation. As the light modulation frequency is increased, the modulation of the emission decreases.

The phase angle and modulation of the emission, measured over a range of modulation frequencies, is the frequency response of the sample. These data are analyzed by non-linear least squares to recover the intensity decay law. The intensity decay is analyzed in terms of the multi-exponential model, as in Eq. 1:

$$I(t) = \sum_i \alpha_i \exp\left(- t/\tau_i\right)_0 \tag{1}$$

where τ_i is the decay time, α_i is the pre-exponential factor, $\sum \alpha_i = 1.0$

Anisotropy decays were also measured using the frequency-domain method [36]. In this case, excitation is with vertically polarized light. One measures the

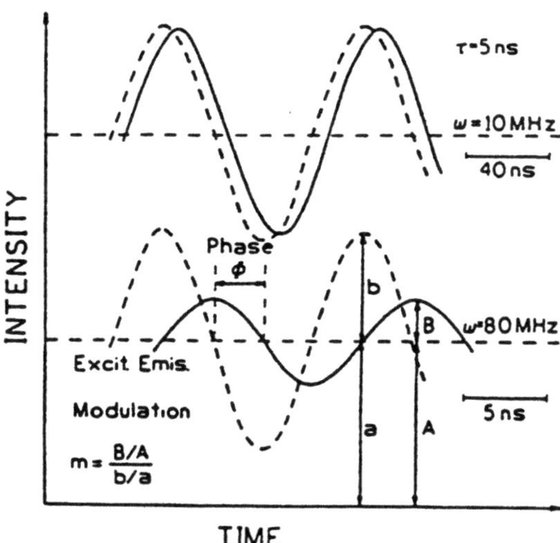

Fig. 1. Principles of frequency-domain measurements of an intensity decay

phase angle difference between the vertically and horizontally polarized components, and the amplitudes of the polarized and modulated emission. These data, measured over a range of modulation frequencies, are used to recover the anisotropy decay.

$$r(t) = \sum r_{0j} \exp(-t/\Theta_j) \tag{2}$$

where r_{0j} is the amplitude associated with each correlation time (θ_j).

The fundamental anisotropy ($r_0 = \sum r_{0j}$) reveals the relative orientation of the electronic transitions for absorption and emission. As shown elsewhere [38], the value of r_0 can be related to the angle (β) between these transitions. The value of r_0 can be measured in frozen solution where there are no rotational differences. Alternatively, r_0 can be found from the time-resolved anisotropy decay using Eq. 2.

3
Results

3.1
Photoselection for Multi-Photon Excitation

The fundamental or time-zero anisotropy (r_0) is a measure of the displacement of the emission transition moment from the direction of the polarized excitation. The anisotropy theory for two-photon excitation is complex [39–41], and there is no theory for three-photon excitation. For many fluorophores the anisotropy behavior can be understood in terms of polarized photoselection with collinear transitions for the two- and three-photon excitation. In these cases the fundamental anisotropy is given by Eqs. 3, 4 and 5 for single-, two- and three-photon excitation, respectively.

$$r_{01}(\beta) = \frac{2}{5}\left(\frac{3}{2}\cos^2\beta - \frac{1}{2}\right) \tag{3}$$

$$r_{02}(\beta) = \frac{4}{7}\left(\frac{3}{2}\cos^2\beta - \frac{1}{2}\right) \tag{4}$$

$$r_{02}(\beta) = \frac{2}{3}\left(\frac{3}{2}\cos^2\beta - \frac{1}{2}\right) \tag{5}$$

The angular displacement (β) between the absorption and emission transitions, and need not be identical for each mode of excitation. The angular distributions of the excited state populations are shown in Fig. 2. For single-photon excitation, with $\beta = 0$, the excited state population is preferentially along the z-axis, which is parallel to the polarized excitation. However, this orientation is not complete, as $\cos^2$ photoselection provides a maximum of 3-fold more intensity along the z-axis than along the equivalent x- or y-axes, where is the angle from the z-axis. Two- and three-photon excitation provide a more strongly oriented population, as can be seen from the narrower angular distribution (Figure 2). For single-, two- and three-photon excitation, the maximal anisotropies for $\beta = 0$ are 0.40, 0.571 and 0.667, respectively. Hence, observation of a

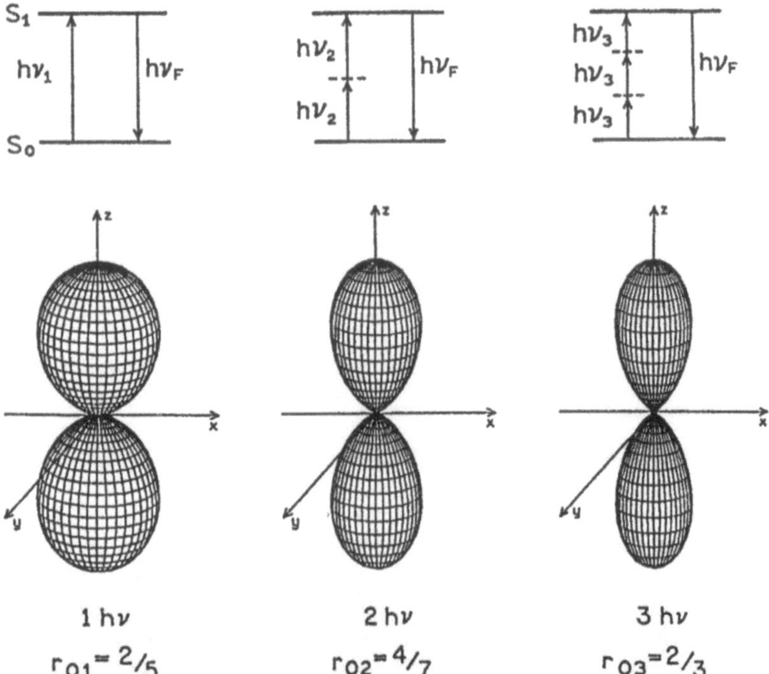

Fig. 2. Excited state orientation distribution from single-, two-, and three-photon excitation. We have assumed collinear electronic absorption and emission transitions

larger anisotropy than found for two-photon excitation suggests a higher mode of excitation. It should be noted that most fluorophores do not display the theoretically maximal anisotropies, but usually somewhat lower values due to a non-zero value of β. One should also note the importance of anisotropy measurements for determination of the mode of excitation. With the high focused powers needed for multi-photon excitation it is often difficult to obtain clear evidence for a quadratic or cubic dependence on intensity. However, the anisotropy measurements are independent of the total intensity.

3.2
Three-Photon Excitation of DPH

Diphenylhexatriene (DPH) is widely used as an extrinsic membrane probe. It was of interest to determine whether one could observe three-photon excitation of DPH, and whether the required illumination intensities would perturb We examined the emission spectrum of DPH in triacetin resulting from focused illumination of 860 nm provided by a titanium/sapphire femtosecond laser [42]. The emission spectrum (not shown) was essentially identical to that observed for DPH with one-photon excitation at 287 nm or 360 nm.

DPH does not display one-photon absorption above 420 nm. Hence it seemed unlikely that DPH would display two-photon excitation at wavelengths above 840 nm. To determine the mode of excitation we examined the dependence of the DPH emission intensity on the laser power. At 860 nm the emission intensity was dependent on the cube of the laser power (Fig. 3), which indicated

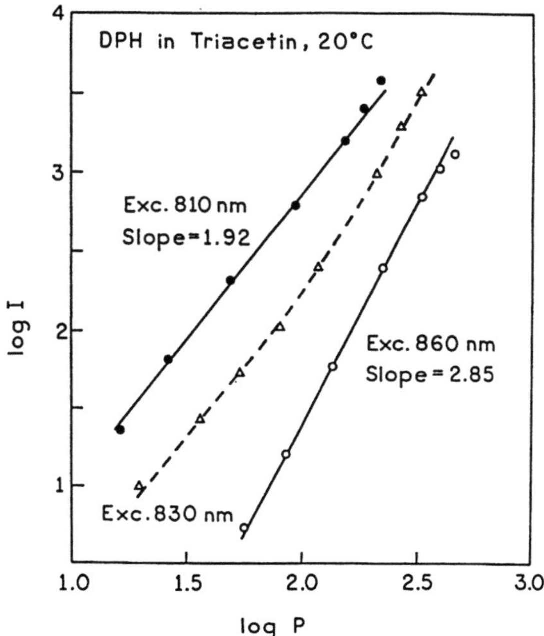

Fig. 3. Dependence of the emission intensity of DPH on laser power for excitation at 810, 830 and 860 nm

that DPH displayed three-photon excitation. The fact that the exponent is slightly less than 3 (2.85) does not invalidate our conclusion of three-photon excitation. Various photochemical and physical processes can account for this minor discrepancy. For instance, the DPH molecules may be sensitive to the intense focused illumination, so that there is a power-dependent decrease in the emission intensity. The most important point is that the dependence of the DPH emission intensity on incident power is greater than a square dependency

To date, three-photon excitation has been observed for a number of fluorophores, always at wavelengths longer than twice the single-photon absorption. At an intermediate wavelength of 830 nm the mode of excitation is between two- and three-photon excitation, and the slope depends on laser power. This suggests that under our experimental conditions of 830 nm excitation, some of the DPH molecules display simultaneous absorption of two photons, and other DPH molecules display simultaneous three-photon absorption. Under these conditions, the fundamental anisotropy is no longer a molecular parameter, but depends on laser power (not shown).

It is well known that the intensity decay of fluorophores is characteristic of the excited state and sensitive to the fluorophore's environment. Hence, we used intensity decay measurements to determine if DPH was emitting from the same excited states for 1PE, 2PE and 3PE, and whether DPH was being perturbed by the intensely focused excitation. The intensity decay of DPH in triacetin at 40 °C displayed the same single exponential with a decay time of 7.2 ns for one-, two- and three-photon excitation [42]. Frequency-domain anisotropy decays of DPH are shown in Fig. 4 for two- (810 nm) and three-photon (860 nm) excitation. The recovered time-zero anisotropies 0.532 and 0.628 for two- and three-photon excitation, respectively, are in agreement with those measured in frozen solution. The same rotational correlation times were observed for both wavelengths. Local heating of the sample would be expected to decrease the correlation time. The similar lifetimes and correlation time of DPH with 1PE, 2PE, and 3PE suggest the absence of heating or other adverse effects with three-photon excitation.

Anisotropy decays are sensitive to the local viscosity surrounding a fluorophore. Local heating during multi-photon excitation would be expected to decrease the rotational correlation time. The same correlation time near 0.57 ns was obtained for 2PE or 3PE of DPH, indicating the absence of local heating. The differential polarized phase angles shown in Fig. 4 provide additional evidence for three-photon excitation. As shown elsewhere [43], the maximum possible differential phase angle (Δ) for any light modulation frequency is given by Eq. 6:

$$\tan \Delta_i = \frac{3 r_{0i}}{2 \left[(1 + 2r_{0i}) (1 - r_{0i}) \right]^{1/2}}$$

where i indicates the number of simultaneously absorbed photons. The maximum value for two-photon excitation with $r_{02} = 0.571$ is 41.8°. For the measured r_{02} value of 0.528 the maximum two-photon value of $\tan \Delta_2$ is 38.8°. The highest observed value of DPH is 42.5°, indicating that excitation at 860 nm is due to more than two- photons.

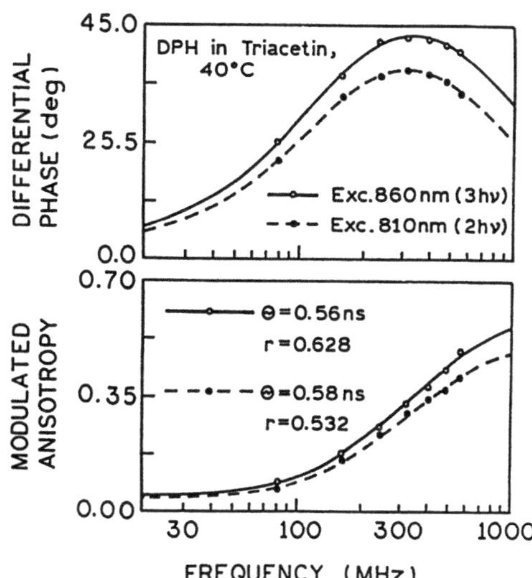

Fig. 4. Frequency-domain anisotropy decay of DPH in triacetin at 40 °C for two-photon (810 nm) and three-photon (860 nm) excitation

We next examined the steady state anisotropies of DPH-labeled membranes, to determine whether three-photon excitation could still be observed. Additionally, we reasoned that adverse effects of the focused illumination on the bilayers would be revealed by a change in the temperature-dependent anisotropies and/or transition temperatures. Anisotropies of DPH are shown in Fig. 5 for bilayers of DMPG or DMPG/chol. Also shown are the anisotropies observed with one-photon excitation. As expected, the anisotropies are higher for

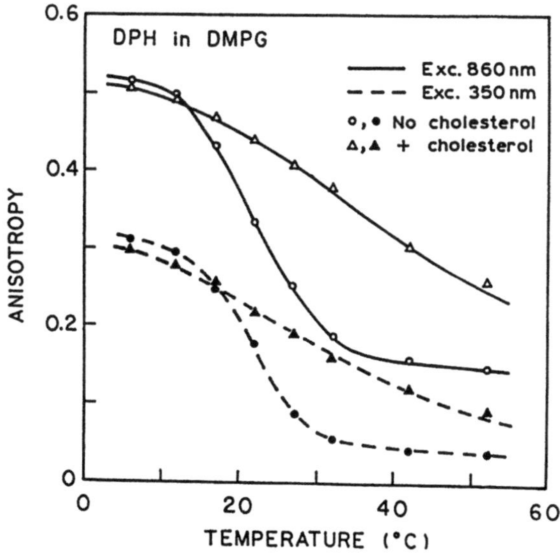

Fig. 5. Temperature-dependent anisotropies of DPH in DMPG and DMPG/chol bilayers for three-photon excitation at 860 nm (——) and one-photon excitation at 350 nm (– – –)

three-photon as compared with one-photon excitation. The temperature-dependent profiles and transition temperatures are the same for one- and three-photon excitation. This result indicates that the bilayers are not significantly heated with three-photon excitation. Hence it appears that three-photon excitation could be observed in DPH-labeled cells in a fluorescence microscope without significant damage.

3.3
Three-Photon Excitation of Proteins

When illuminated with high intensity pulses near 600 nm, one can expect two-photon excitation of proteins, as has been observed [19, 44]. More remarkable is the observation of three-photon excitation of proteins using wavelengths near 900 nm. Conveniently, the optimal output of a titanium/sapphire laser is from 800–900 nm, making these fs lasers ideal for three-photon excitation of proteins.

The emission spectra of N-acetyl-L-tyrosine amide (NATyr) and N-acetyl- L-tryptophanamide (NATA) for excitation at 280 nm (one-photon) and 840 (three-photon) are shown in Fig. 6. These emission spectra are essentially identical for both modes of excitation, indicating that emission occurs from the same state independent of the excitation wavelength. The fluorescence intensi-

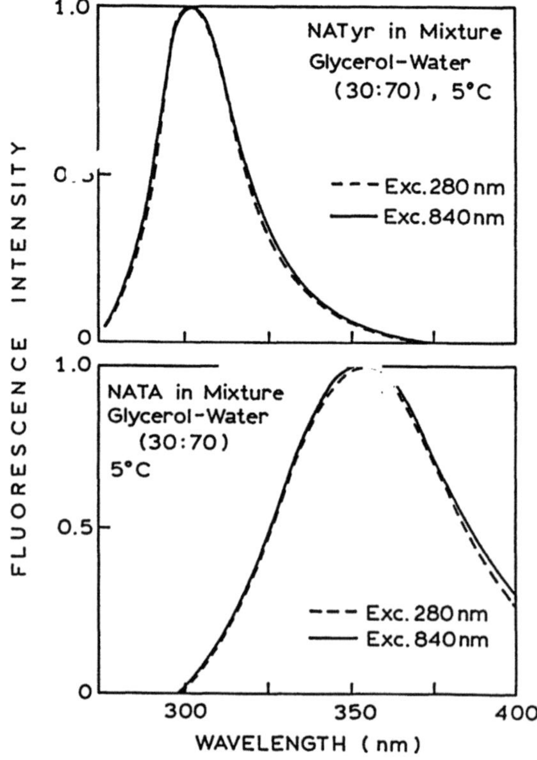

Fig. 6. Emission spectra of NATyr (top) and NATA (bottom) with one- and three-photon excitation

ties for NATyr and NATA were found to depend on the cube of the laser power at 840 nm (Fig. 7). Hence, both tyrosine and tryptophan can be excited by near-infrared light pulses from a titanium/sapphire laser.

One may question why we examined NATyr and NATA in a mixture of glycerol and water. This was done for the following reasons. Firstly, the more viscous environment with glycerol is one step closer to the environment of tyrosine or tryptophan in proteins. Secondly, the more viscous solution allowed measurements of the time-zero anisotropy.

The time-resolved intensity and anisotropy decays of NATA are presented in Fig. 8. Essentially, the same lifetimes and rotational correlation times were found for both one-photon and three-photon excitation. However, the time-zero anisotropy is positive for 280 nm excitation ($r_{01} = 0.131$) and negative for 840 nm excitation ($r_{03} = -0.063$), which indicates a different direction for the electronic transition(s) for 1PE and 3PE excitation. A similar result was found for two-photon excitation, where the excitation anisotropy spectrum was rather different for 2PE versus 1PE excitation [45]. The negative value of initial anisotropy has been confirmed by time-domain measurements [46].

To evaluate the feasibility of three-photon excitation of proteins we examined a mutant of troponin C (TnC) which contained a single tryptophan residue in place of phenylalanine 22 (F22 W). Figure 9 shows the emission spectrum of TnC F22 W excited at 855 nm [47]. The single tryptophan residue of TnC F22 W displays an unstructured emission spectrum with an emission maximum near

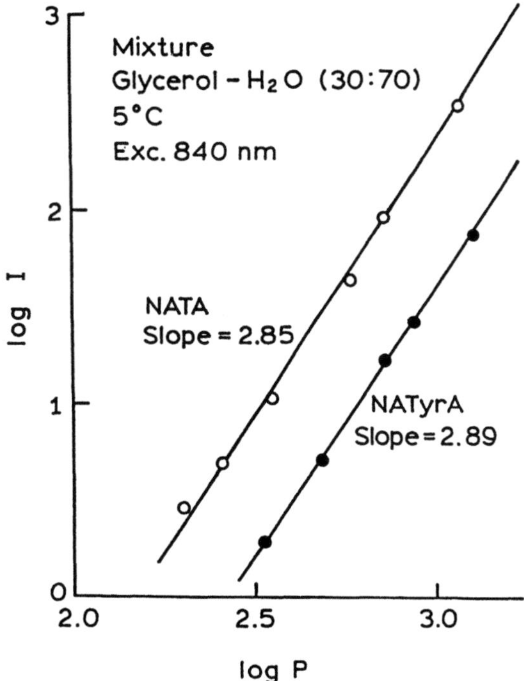

Fig. 7. Effect of laser power at 840 nm on the emission intensity of NATyr and NATA

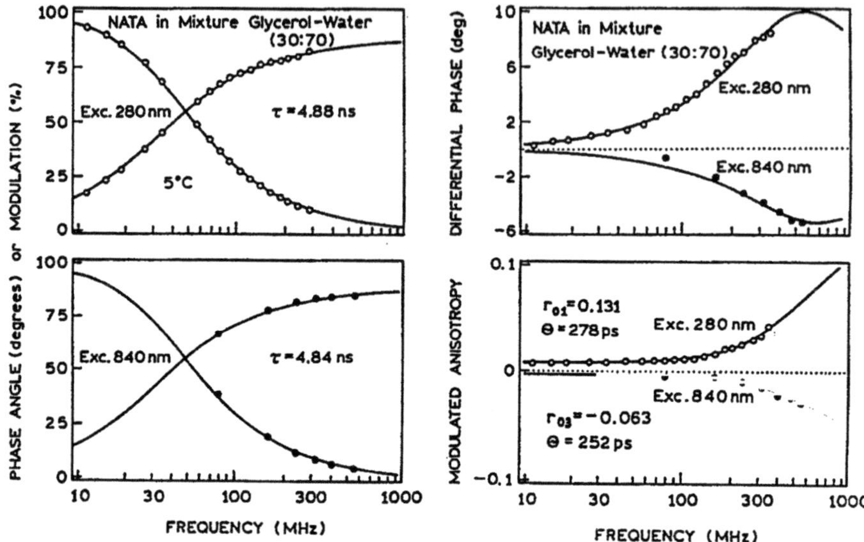

Fig. 8. Frequency-domain intensity (left) and anisotropy (right) decays of NATA for excitation at 280 and 890 nm. For the intensity decays (left), the increasing values are phase angles and the decreasing values are the modulation

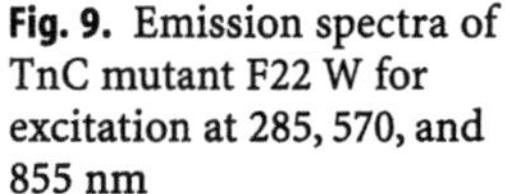

Fig. 9. Emission spectra of TnC mutant F22 W for excitation at 285, 570, and 855 nm

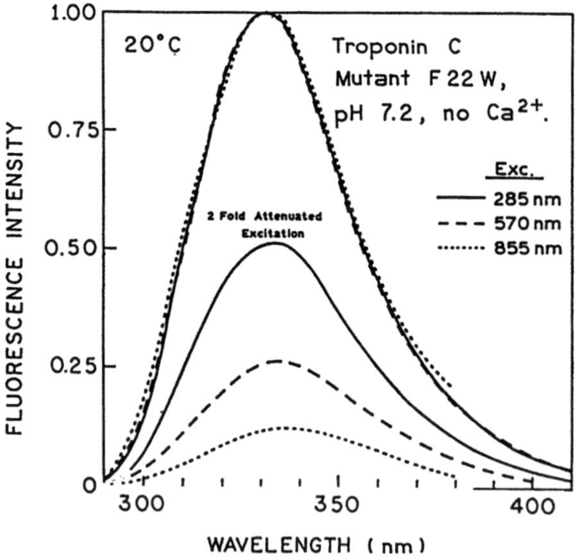

330 nm. We were initially surprised that we were able to observe tryptophan emission at this long excitation wavelength. Two-photon excitation of TnC F22 W is not expected for excitation at 855 nm because the one-photon absorption does not extend beyond 305 nm, so that two-photon excitation is not expected above 610 nm. The emission spectrum observed with 855 nm excitation is essentially identical to that found with one-photon excitation at 285 nm (Fig. 9). When the laser beam at 885 nm was attenuated two-fold, the TnC intensity decreased eight-fold, suggesting three-photon excitation. TnC F22 W displayed the same intensity decay with one-, two-, or three-photon excitation (Fig. 10), which indicates that the protein is not being heated significantly by the intense light needed for three-photon excitation.

We also examined the frequency-domain anisotropy decay (Fig. 11) and steady state anisotropy (Fig. 12) of TnC F22 W for one-, two- and three-photon excitation. For three-photon excitation the differential phase angles and modulated anisotropies were negative (Fig. 11). This suggests that the fundamental anisotropy is negative for three-photon excitation, a fact confirmed by the steady state data (Fig. 12). These results suggest that three-photon absorption

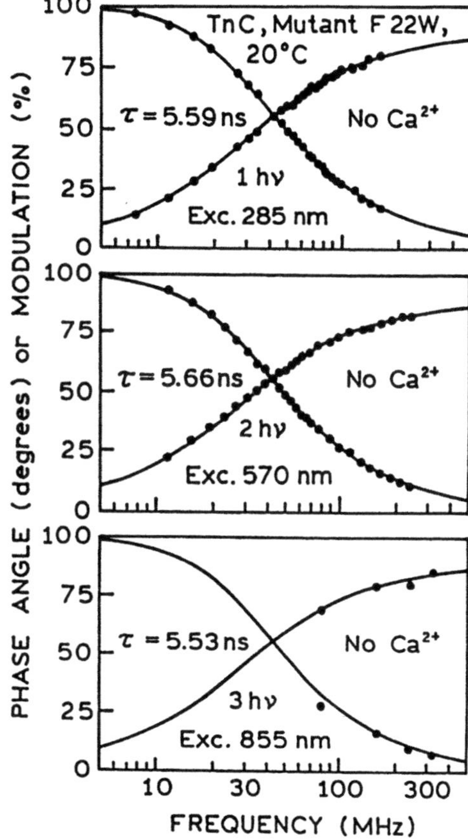

Fig. 10. Frequency-domain intensity decays of TnC F22 W without Ca²⁺ for one-, two-, and three-photon excitation

Fig. 11. Frequency-domain anisotropy decays of TnC F22 W for one-, two-, and three-photon excitation

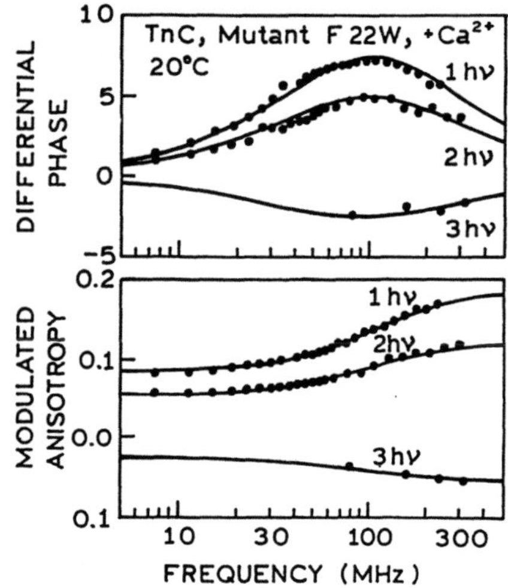

Fig. 12. Excitation anisotropy spectrum of TnC mutant F22 W for one-, two-, and three-photon excitation

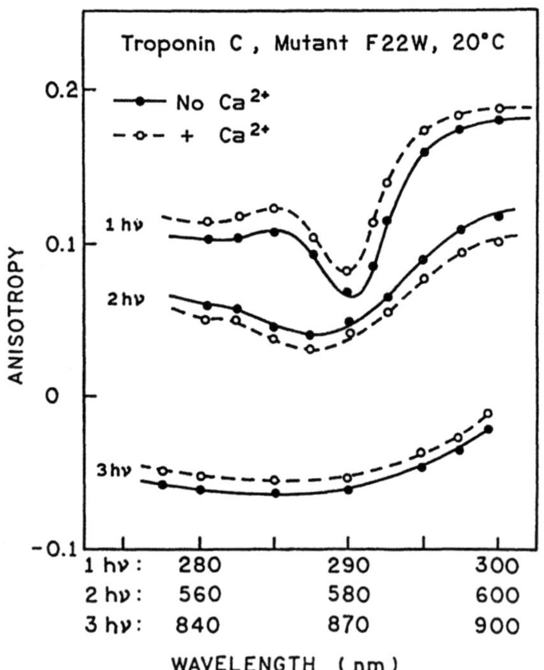

occurs to the ^1L_b state, followed by emission from the ^1L_a state of indole, which is oriented 90° from the ^1L_b absorption. The observation of different anisotropies for one-, two- and three-photon excitation indicate that distinct spectroscopic information is available with each mode of excitation.

The observation of protein fluorescence with the fundamental output of a titanium/sapphire laser has significant implications for fluorescence microscopy. It will now be possible to observe protein fluorescence in a microscope, without the need for quartz optics. In fact, the first observations of protein fluorescence [48] in a microscope appeared shortly after our reports on three-photon excitation of tryptophan and proteins [46, 47].

3.4
Three-Photon Excitation of DNA Stained DAPI

The intrinsic fluorescence of DNA is weak, and not useful for biophysical studies of DNA. Hence, fluorophores which label DNA are in widespread use. Now we describe the fluorescence spectral properties of the DNA stain (4′,6-diamidino-2-phenylindole, hydrochloride) when excited with the fundamental output of a titanium/sapphire laser [49]. Emission spectra are shown in Fig. 13 for DAPI-DNA for excitation at 360, 830, and 885 nm. The emission spectra are essentially equivalent, irrespective of the excitation wavelength. We examined the nature of the long wavelength excitation by attenuating the peak laser power using neutral density filters. A twofold attenuation of the incident light for 360 nm excitation results in a twofold attenuation of the emission (Fig. 13). For

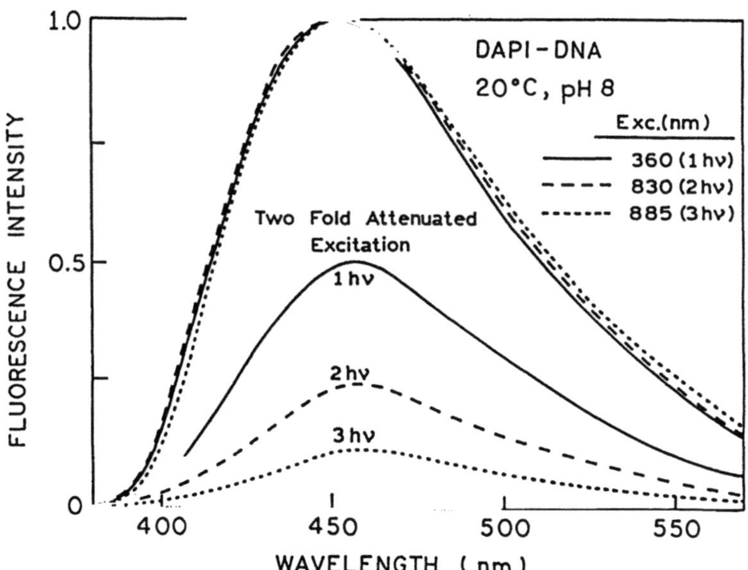

Fig. 13. Normalized emission spectra of DAPI-DNA for excitation at 360, 830, and 885 nm. Also shown are the emission spectra with a twofold attenuation of the excitation

830 nm excitation, a twofold attenuation results in a fourfold decrease in the DAPI emission. At 885 nm excitation, a twofold decrease in the incident intensity results in an eightfold decrease in the signal from DAPI. This dependence on the square and cube of the laser power suggests that the emission is due to two- and three-photon excitation, at 830 and 855 nm, respectively.

The intensity and especially the anisotropy decays are known to be sensitive indicators of the conformation and dynamics of biological molecules. In principle, photobleaching is not expected to result in changes in decay time, because the fluorophores are no longer fluorescent. However, we have noticed that other fluorophores, in particular a calcium probe [50], display lifetime changes with intense illumination, which we interpret as phototransformation of the probes. Hence the intensity decay can be an indicator of undesired photo effects on macromolecules.

We examined these time-resolved decays with long wavelength excitation. We reasoned that detrimental effects of the increase in long wavelength illumination would be revealed by changes in the decay times or correlation times. These data could thus determine whether imaging of macromolecules would be possible without unacceptable photon effects. Frequency-domain intensity and anisotropy decays of DAPI-DNA are shown in Fig. 14 for 2PE and 3PE excitation [49]. The intensity decay parameters (α_i and τ_i values) are essentially the same for 2PE and 3PE excitation [49] and are similar to those observed previously for 1PE at 360 nm.

Anisotropy decays obtained for 830 nm (two-photon) and 885 nm (three-photon) excitations reveal the same correlation times and the data can be fitted simultaneously. However, the zero-time anisotropies are distinct for two- and three-photon excitation. These results suggest that the DNA molecules do not

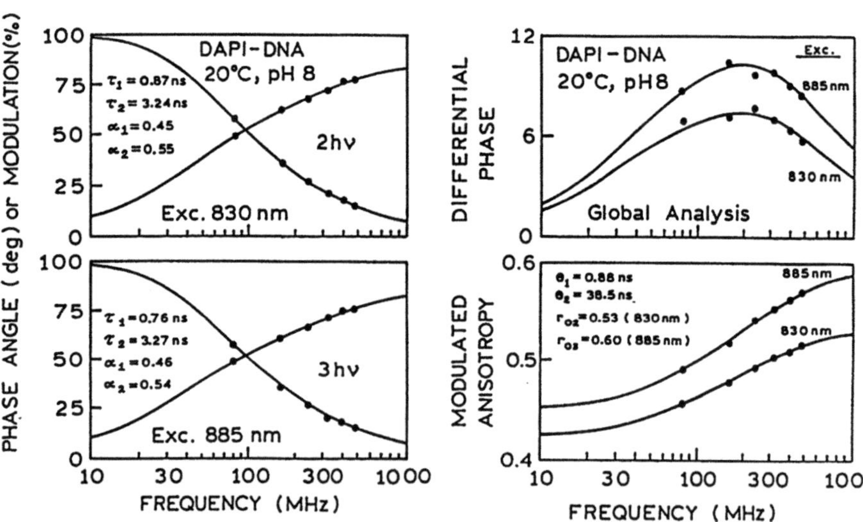

Fig. 14. Frequency-domain intensity (left) and anisotropy (right) decays of DAPI-DNA with two- and three-photon excitation

suffer adverse effects on significant heating when used with multi-photon excitation at commonly available titanium/sapphire wavelengths, thus enabling multi-photon imaging. In fact, the three-photon images of DAPI-stained DNA have been reported [49].

Since DAPI is often used for cellular imaging it is of interest to determine the effect of excitation wavelength on the mode of excitation. This was accomplished by examining the effects of a twofold attenuation of the incident power on the observed intensity. At the shorter wavelengths available from our titanium/sapphire laser (810 and 820 nm), a 2.25-fold decrease in laser power resulted in a fivefold decrease in intensity (Fig. 15, top). At longer wavelengths above 880 nm a 2.25-fold attenuation of the incident light resulted in an 11-fold decrease in emission intensity, consistent with three-photon excitation. At intermediate wavelengths the intensity ratio was variable. These data display the characteristics of a phase transition with a constant ratio above and below the transition wavelength. DAPI displayed transition midpoints near 855 nm.

Fig. 15. Effect of a 2.25-fold excitation attenuation on the emission intensity of DAPI-DNA. DNA samples were in 10 mM tris, pH = 8. I_0 and I are the emission intensities observed with full and attenuated excitation, respectively. Also shown: wavelength-dependent anisotropy of DAPI-DNA

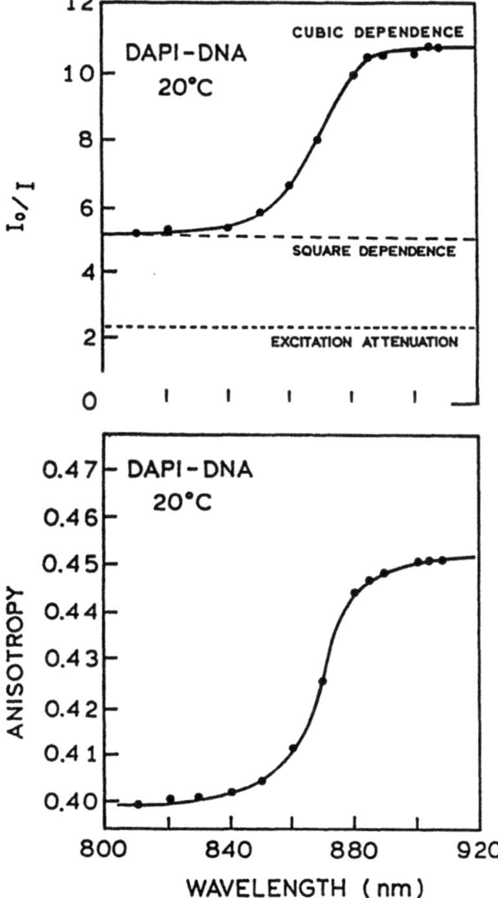

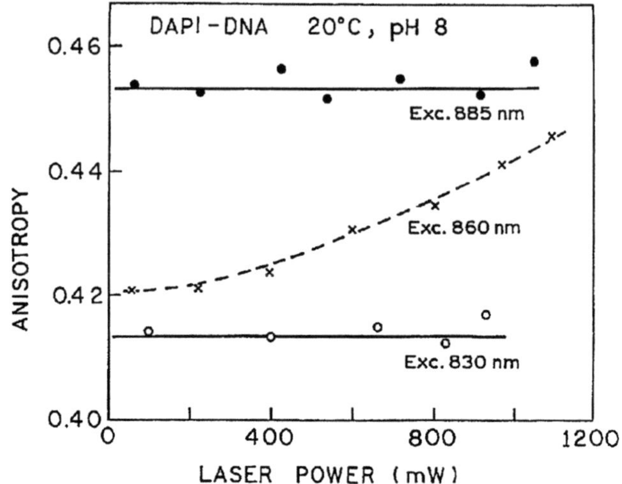

Fig. 16. Dependence of the steady-state anisotropy of DAPI-DNA on incident power at 830, 860, and 885 nm

However, as described below, these transition wavelengths are apparent quantities which depend on the experimental conditions.

The transitions from two- to three-photon excitation can also be observed from the fluorescence anisotropy. As described above, for collinear electronic transitions one can expect the anisotropy to be larger for three-photon excitation than for two-photon excitation. Wavelength dependent steady state anisotropies of DAPI are shown at the bottom of Fig. 15. The anisotropies increased with increasing excitation wavelength. Once again the wavelength-dependent values display a transition from two- to three-photon excitation, and constant values above and below the transition regions.

We also examined the dependence of the anisotropy on incident power. For DAPI (Fig. 16) the anisotropy is different for excitation at 830 and 885 nm, but each value is independent of laser power over a wide range of power. This suggests that at these wavelengths one can be reasonably certain of the mode of excitation. At an intermediate wavelength of 860 nm the anisotropy of DAPI-DNA depends on laser power. This suggests that the mode of excitation changes from 2PE to 3PE as the power is increased. Because the anisotropy is independent of fluorescence intensity, anisotropy measurements are recommended for clarifying the mode of excitation in multi-photon microscopy. Furthermore, the anisotropy of a known compound can be used to estimate the illumination intensity at the focal point of the microscopy.

3.5
Spatial Localization of the Three-Photon Excitation

It is known that two-photon excitation results in strongly localized excitation at the focal point of the laser beam due to the quadratic dependence on excitation intensity. Hence, a three-photon process should provide greater spatial localization than two-photon excitation due to the cubic dependence on intensity. We

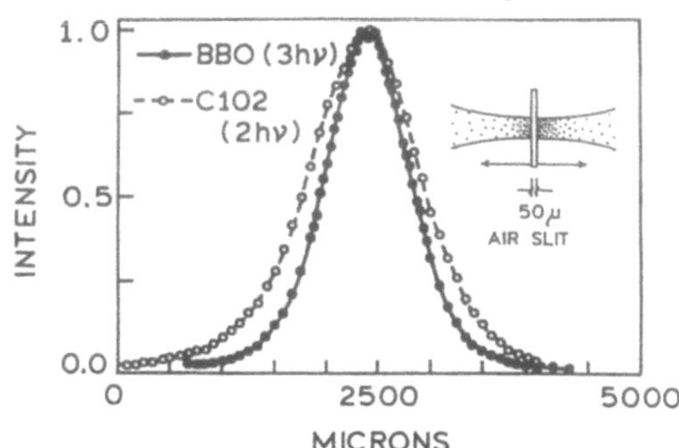

Fig. 17. Spatial distribution of the emission of BBO (•) and coumarin 102 (○) along the laser beam for excitation at 885 nm

examined the spatial distribution of 2,5-bis(4-biphenyl)oxazole (BBO) emission when excited at 885 nm. At this wavelength BBO displays pure three-photon properties as proven by power-dependent emission intensity and anisotropy ($r_0 = 0.65$) [51].

For comparison we also examined the spatial distribution of the emission from coumarin 102 which displays two-photon excitation at this same wavelength of 885 nm. Both BBO and coumarin 102 were dissolved in toluene. The excited volume was substantially smaller for BBO than for coumarin 102, both along the direction of the incident beam (Fig. 17) and across the focused beam (Fig. 18). Since these spatial profiles were obtained at the same excitation wavelength, the difference in size can only be due to the nature of the excitation process. Also, the smaller excited volume with three-photon excitation of BBO could be observed visually by looking at the emission of BBO or coumarin 102.

4
Conclusions

Presented above are examples which show that the fundamental output of a titanium/sapphire laser is a valuable light source for three-photon excitation of fluorescence. Several intrinsic and extrinsic fluorophores which absorb in the UV region can be excited and studied with near-infrared pulses from this laser.

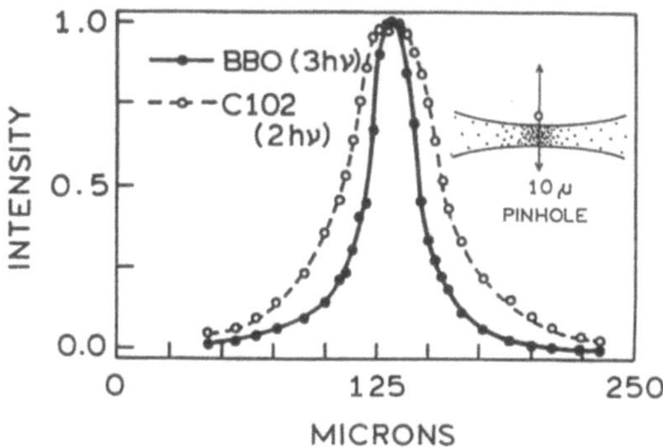

Fig. 18. Spatial distribution of the emission of BBO (•) and coumarin 102 (○) across the focused laser beam for excitation at 885 nm. The emission image was enlarged threefold, so that the effective pinhole size is about 3.5 μm

The high pulse repetition rate (~ 80 MHz) is almost ideal for time-resolved spectroscopic measurements and image collection in fluorescence microscopy. One may question the efficiency of fluorescence signal with three-photon excitation. Of course, the intensity of three-photon-induced fluorescence is a few orders of magnitude lower than fluorescence signals obtained with two-photon excitation. However, we have already demonstrated that micromolar concentrations of such fluorophores as indo-1 can be easily detected in spectroscopic conditions (2 cm focusing lens and not expanded laser beam) with three-photon excitation [53, 54].

Comparison of fluorescence localization with two- and three-photon excitation (Figs. 17 and 18) clearly shows the advantage of a higher excitation mode in potential microscopy applications. In fact, our group and several others have reported fascinating images of biological specimens [48, 49, 55, 56] obtained with three-photon excitation. Hence, three-photon excitation has already found uses in fluorescence spectroscopy and microscopy.

What are the perspectives of further development of higher multi-photon excitation? Currently, the regenerative amplifiers together with optical parametric amplifiers seem to be the most promising light sources for three- or more-photon excitation. However, these systems require titanium/sapphire lasers to provide the seed pulses and are expensive. In addition, the lower repetition rate of the output pulses may limit time-resolved and imaging applications. Importantly, simpler and lower cost titanium/sapphire lasers are under development, and may soon be widely available for multi-photon spectroscopy and imaging.

Acknowledgements. This work was supported by a grant from the National Center for Research Resources, RR-08119.

References

1. M. Goppert-Mayer, Über Elementarakte mit zwei Quantensprungern, *Ann. Phys.* 9, 73–294, (1931)
2. D.M. Friedrich, W.M. McClain, Two-photon molecular electronic spectroscopy, *Annu. Rev. Phys. Chem.* 31, 559–577 (1980)
3. M.J. Wirth, A. Koskelo, M. J. Sanders, Molecular symmetry and two-photon spectroscopy, *Appl. Spectros.* 35, 14–21 (1981)
4. W.L. Peticolas, Multiphoton spectroscopy, *Annu. Rev. Phys. Chem.* 18, 233–261 (1967)
5. R.D. Jones, P.R. Callis, Two-photon spectra of inductively perturbed naphthalene, *Chem. Phys. Letts.* 144, 158–164 (1989)
6. M. J. Wirth, A.C. Koskelo, C.E. Mohler, B. L. Lentz, Identification of methyl derivatives of naphthalene by two-photon symmetry parameters, *Anal. Chem.* 53, 2045–2048 (1981)
7. M.B. Masthay, L.A. Findsne, B.M. Pierce, D.F. Bocian, J.S. Lindsey, R.R. Birge, A theoretical investigation of the one- and two-photon properties of porphyrins, *J. Chem. Phys.* 84, 3901–3915 (1986)
8. H.L.-B. Fang, R.J. Thrash, G.E. Leroi, Observation of the low-energy ^1A_g state of diphenylhexatriene by two-photon excitation spectroscopy, *Chem. Phys. Letts.* 57, 59–63 (1978)
9. B. Hudson, Linear polyene electronic structure and spectroscopy, *Ann. Rev. Phys. Chem.* 25, 437–460 (1974)

10. R.R. Birge, Two-photon spectroscopy of protein-bound chromophores, *Acc. Chem. Res.* 19, 138–146 (1986)

11. B.E. Anderson, R.D. Jones, A.A. Rehms, P. Ilich, P.R. Callis, Polarized two-photon fluorescence excitation spectra of indole and benzimidazole, *Chem. Phys. Letts.* 125, 106–112 (1986)

12. A. Rehms, P.R. Callis, Resolution of L_a and L_b bands in methyl indoles by two-photon spectroscopy, *Chem. Phys. Letts.* 140, 83–89 (1987)

13. W.D. Pfeffer, E.S. Yeung, Laser two-photon excited fluorescence detector for microbore liquid chromatography, *Anal. Chem.* 58:2103–2105 (1986)

14. M.J. Wirth and H.O. Fatunmbi, Very high detectability in two-photon spectroscopy, *Anal. Chem.* 62:973–976 (1990)

15. W.M. McClain, Excited state symmetry assignment through polarized two-photon absorption studies of fluids, *J. Chem. Phys.* 55, 2789–2796 (1971)

16. J.R. Lakowicz, I. Gryczynski, J. Kusba, E. Danielsen, Two-photon induced fluorescence intensity and anisotropy decays of diphenylhexatriene in solvents and lipid bilayers, *J. Fluoresc.* 2, 247–258 (1992)

17. J.R. Lakowicz, I. Gryczynski, Fluorescence intensity and anisotropy decay of the 4',6'-diamidino-2-phenylindole-DNA complex resulting from one-photon and two-photon excitation, *J. Fluoresc.* 2, 117–122 (1992)

18. I. Gryczynski, J.R. Lakowicz, Fluorescence intensity and anisotropy decays of the DNA stain HOECHST 33342 resulting from one-photon and two-photon excitation, *J. Fluoresc.* 4, 331–336 (1994)

19. J.R. Lakowicz, I. Gryczynski, Tryptophan fluorescence intensity and anisotropy decays of human serum albumin resulting from one-photon and two-photon excitation, *Biophys. Chem.* 45, 1–6 (1993)

20. J.R. Lakowicz, I. Gryczynski, Characterization of p-bis(o-methylstyryl) benzene as a lifetime and anisotropy decay standard for two-photon induced fluorescence, *Biophys. Chem.* 47, 1–7 (1993)

21. A. Kawski, I. Gryczynski, Z. Gryczynski, Fluorescence anisotropies of 4-dimethylamino--diphenylphosphinyl-*trans*-styrene in isotropic media in the case of one-and two-photon excitation, *Z. Naturforsch.* 481, 551–556 (1993)

22. J.R. Lakowicz, I. Gryczynski, Z. Gryczynski, E. Danielsen, M.J. Wirth, Time-resolved fluorescence intensity and anisotropy decays of 2,5-diphenyloxazole by two-photon excitation and frequency-domain fluorometry, *J. Phys. Chem.* 98, 3000–3006 (1992)

23. J.R. Lakowicz, I. Gryczynski, K. Nowaczyk, Two-photon excitation of dioxane; time-resolved measurements of excited state complex formation with water, *Spectrochimica Acta.* In press, 1997

24. J.R. Lakowicz, I. Gryczynski, Fluorescence intensity decays of cyclohexane and methylcyclohexane with two-photon excitation from a high repetition rate frequency-doubled dye laser, *Biospectroscopy 1*, 3–8 (1995)

25. I. Gryczynski, J.R. Lakowicz, Quenching of methylcyclohexane fluorescence by methanol, *Photochem. Photobiol.* 62, 426–432 (1995)

26. I. Gryczynski, A. Razynska, J.R. Lakowicz, Two-photon induced fluorescence of linear alkanes; a possible intrinsic lipid probe, *Biophys. Chem.* 57, 291–295 (1996)

27. R.A. Agbaria, I. Gryczynski, H. Malak, J.R. Lakowicz, Two-photon induced fluorescence of cholestane, *Biospectroscopy 2*, 213–224 (1996)

28. W. Denk, J.H. Strickler, W.W. Webb, Two-photon laser scanning fluorescence microscopy, *Science* 248, 73–76 (1990)

29. D.W. Piston, D.R. Sandison, W.W. Webb, Time-resolved fluorescence imaging and background rejection by two-photon excitation in laser scanning microscopy, *Proc. SPIE* 1640, 379–389 (1992)

30. S.W. Hell, S. Lindek, E.H.K. Stelzer, Enhancing the axial resolution in far-field microscopy: two-photon 4Pi confocal fluorescence microscopy, *J. Mod. Opt.* 41:675–681 (1994)

31. S.W. Hell, M. Schrader, P.E. Hanninen, E. Soini, Resolving fluorescence beads at 100–200 nm axial distance with a two-photon 4Pi-microscope operating in the near infrared, *Opt. Commun.* 120:129–133 (1995)

32. J.R. Lakowicz, B.P. Maliwal, Construction and Performance of a Variable-Frequency Phase-Modulation Fluorometer. *Biophys. Chem.* 21:61–78 (1985)
33. G. Laczko, J.R., Lakowicz, I. Gryczynski, Z. Gryczynski, H. Malak, A 10 GHz Frequency-Domain Fluorometer, *Rev. Sci. Instrum.* 61:2331–2337 (1990)
34. J.R. Lakowicz, E. Gratton, G. Lackzko, H. Cherek, M. Limkeman, Analysis of Fluorescence Decay Kinetics from Variable-Frequency Phase Shift and Modulation Data. *Biophys. J.* 46:463–477 (1984)
35. E. Gratton, J.R. Lakowicz, B. Maliwal, H. Cherek, G. Laczko, M. Limkeman, Resolution of Mixtures of Fluorophores Using Variable-Frequency-Domain Phase-Modulation Data. *Biophys. J.* 46:479–486 (1984)
36. J.R. Lakowicz, H. Cherek, J. Kusba, I. Gryczynski, M.L. Johnson, Review of Fluorescence Anisotropy Decay Analysis by Frequency-Domain Fluroescence Spectroscopy, *J. Fluoresc.* 3:103–116 (1993)
37. J.R. Lakowicz, B.P. Maliwal, E. Gratton, Recent Developments in Frequency-Domain-Fluorometry, *Anal. Instrum.* 14:193–223 (1985)
38. Principles of Fluorescence Spectroscopy, J.R. Lakowicz, Plenum Press, New York. Chapter 5, pp. 111–143
39. C.Wan, C.K. Johnson, Time-resolved anisotropic two-photon spectroscopy, *Chem. Phys.* 179:513–31 (1994)
40. P.R. Callis, On the theory of two-photon induced fluorescence anisotropy with application to indole, *J. Chem. Phys.* 99:27–37 (1993)
41. S-Y Chen, B.W. Van Der Meer, Theory of two-photon induced fluorescence anisotropy decay in membranes, *Biophys. J.* 64:1567–75 (1993)
42. H. Malak, I. Gryczynski, J.D. Dattelbaum, J.R. Lakowicz, Three-photon induced fluorescence of diphenylhexatriene in solvents and lipid bilayers, *J. Fluoresc.* 77(2):99–106 (1977)
43. J.R. Lakowicz, I. Gryczynski, E. Danielsen, Anomalous differential polarized phase angles for two-photon excitation with isotropic depolarizing rotations, *Chem. Phys. Letts.* 191(1,2):47–53 (1992)
44. B. Kierdaszuk, I. Gryczynski, A. Modrak-Wojcik, A. Bzowska, D. Shugar, J.R. Lakowicz, Fluorescence of tyrosine and tryptophan in proteins using one- and two-photon excitation, *Photochemistry and Photobiology* 61(4), 319–324 (1995)
45. J.R. Lakowicz, I. Gryczynski, E. Danielsen, J.K. Frisoli, Anisotropy spectra of indole and N-acetyl-L-tryptophanamide observed for two-photon excitation of fluorescence, *Chem. Phys. Lett.* 194, 282–287 (1992)
46. I. Gryczynski, H. Malak, J.R. Lakowicz, Three-photon excitation of a tryptophan derivative using a fs-Ti:sapphire laser, *Biospectroscopy* 2:9–15 (1996)
47. I. Gryczynski, H. Malak, J.R. Lakowicz, H.C. Cheung, J. Robinson, P.K. Umeda, Fluorescence spectral properties of Troponin C mutant F22 W with one-, two-, and three-photon excitation, *Biophysical J.* 71:3448–3453 (1996)
48. S. Maiti, J. B. Sher, R. M. Williams, W. R. Zipfel, W. W. Webb, Measuring Serotonin distribution in live cells with three-photon excitation, *Science* 275:530–532
49. J.R. Lakowicz, I. Gryczynski, H. Malak, M. Schrader, P. Engelhardt, H. Kano, S.W. Hell, Time-resolved fluorescence spectroscopy and imaging of DNA labeled with DAPI and Hoechst 33342 using three-photon excitation, *Biophysical Journal* 72:567–578 (1997)
50. J.R. Lakowicz, H. Szmacinski, K. Nowaczyk, W. J. Lederer, M.L. Johnson, Fluorescence lifetime imaging of intracellular calcium in COS cells using Quin-2, *Cell Calcium* 15:7–27 (1994)
51. I. Gryczynski, H. Malak, S. W. Hell, J.R. Lakowicz, Three-photon excitation of 2,5-bis (4-biphenyl) oxazole: steady state and time-resolved intensities and anisotropies, *Journal of Biomedical Optics* 1(4):473–480 (1996)
52. S.W. Hell, S. K. Bahlmann, M. Schrader, A. Soni, H. Malak, I. Gryczynski, J.R. Lakowicz, Three-photon excitation in fluorescence microscopy, *J. Biomed. Optics* 1:71–74 (1996)
53. I. Gryczynski, H. Szmacinski, J.R. Lakowicz, On the possibility of calcium imaging using Indo-1 with three-photon excitation, *Photochem. Photobiol.*, 62:804–808 (1995)

54. H. Szmacinski, I. Gryczynski, J.R. Lakowicz, Three-photon induced fluorescence of the calcium probe Indo-1, *Biophys. J.*, 70:547–555 (1996)
55. D.L. Wokosin, V. E. Centonze, S. Crittenden, J. White, Three-photon excitation fluorescence imaging of biological specimens using an all-solid state laser, *Bioimaging* 4:208–214
56. C. Xu, W. Zipfel, W. W. Webb, Three-photon excited fluorescence and applications in nonlinear microscopy, *Biophys. J.*, 70:A429 (1996)

**Part 2
Analytical Fluorescence Probes,
and Environmental Research**

Fluorescence Properties of Crown-Containing Molecules

M. V. Alfimov, S. P. Gromov

1
Introduction

Crown compounds may be considered as the most interesting and promising components of photosensitive systems [1–6]. Crown compounds contain several heteroatoms with lone pairs that can participate in formation of coordination bonds with metal cations. The complexing ability of crown compounds depends strongly on their molecular and spatial structures.

For example, photochemical behavior in the absence and presence of metal ions is one of the most intriguing properties of crown-containing dyes [5]. If a crown ether and a dye are connected by a covalent bond in such a way that one or two heteroatoms of the macrocycle are in conjugation with the dye chromophore, the additivity in the manifestation of specific properties of such a crowned dye may be violated. Photostructural transformations in the dye moiety can change the efficiency of metal cation binding by the crown-ether moiety, whereas the complexation, in turn, may lead to a change in the spectral and photochemical characteristics of the dye moiety.

Interdependent reactions of this type can occur in supramolecular structures formed by self-assembling crown-containing dyes and metal ions. This phenomenon can be used for the development of new types of photophysical and photochemical molecular devices [7].

Owing to its high sensitivity and selectivity, fluorescent spectroscopy is widely used in various areas of physics, chemistry, biology and medicine. Development of fluorescent techniques both for determination of trace amounts of metals and for identification of metal ions in various complex systems such as, for example, as biological objects is of great interest [8, 9]. Introduction of groups into organic luminophores capable of forming chelate complexes with metal ions was a first step in this direction [9–11]. The purpose was to obtain derivatives whose fluorescence intensity and/or fluorescence wavelength can undergo changes upon formation of metal complexes. The use of fluoroionophores containing an organic luminophore and a crown ether, which exhibit high selectivity and high stability constants, provides new powerful tools for research [12–19].

These studies have shown that the values of the fluorescence band shifts on complexation are determined by the following factors [6, 12]:

1. The surface charge density of a metal cation. In a series of cations of identical ionic radius, the spectral shift increases with increasing charge.
2. Geometrical complementarity between a cation and a crown ether cavity. For cations of an identical charge, the better this conformity, the greater the value of the spectral shift.
3. The nature of a solvent. All fluoroionophores under consideration exhibit an appreciable solvatochromism. As a result, the value of the ionochromic effect may in a certain degree depend on the solvent polarity.

In subsequent works, as a basis for the construction of crown-containing fluorophores, systems were chosen in which an effective intramolecular fluorescence quenching can occur by one of the basic mechanisms: via intermediate formation of an excimer, exciplex or by photoinduced electron transfer. An excimer is known to be formed by the interaction of an excited molecule with an unexcited molecule of the same kind, whereas an exciplex is formed by interaction of the excited molecule with an unexcited molecule of a different kind. All of these processes are summarized in Fig. 1.

A crown-ether moiety can be linked to a fluorophore in such a way that the formation of a complex with a metal ion would strongly affect the quenching efficiency. Systems in which complexation retards the fluorescence quenching and leads to an enhancement of the fluorescence intensity are of specific interest.

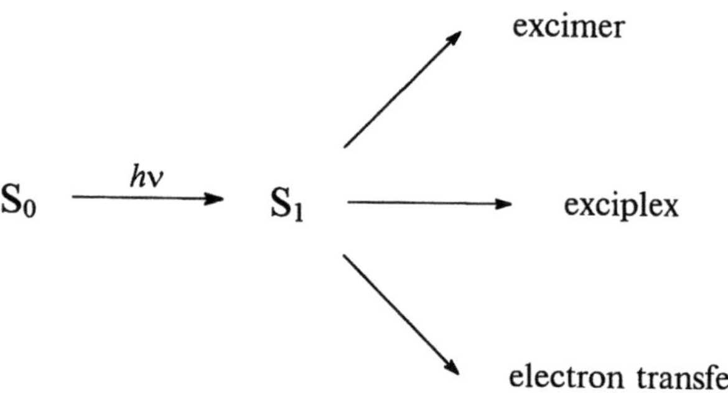

Fig. 1. Different pathways of deactivation of electronic excited states (fluorescence quenching) for crown-containing fluoro-ionophores

2
Crown-Containing Arenes

Examples are provided by derivatives of anthracene, 1 and 2, which exhibit very weak fluorescence in methanolic solutions (for example, $\varphi_f = 0.003$ for 1). In this event, the anthracene fluorescence is efficiently quenched by intramolecular electron transfer from the nitrogen lone pair of the crown-ether moiety to the electronically excited anthracene moiety [20, 21]. Addition of alkali-metal salts to solutions leads to a significant enhancement of the fluorescence quantum yields (by a factor of 47 in the case of potassium ion), while other spectral parameters remain unaffected. The effect of restoring the anthracene fluorescence on incorporating alkali-metal cations into the monoaza-crown ether is

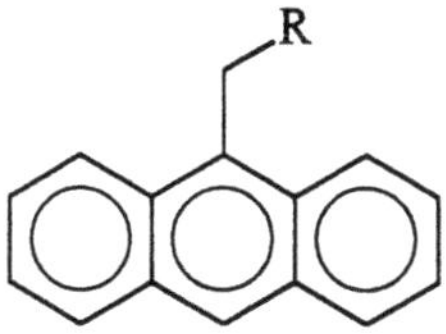

$$R = -N$$

1 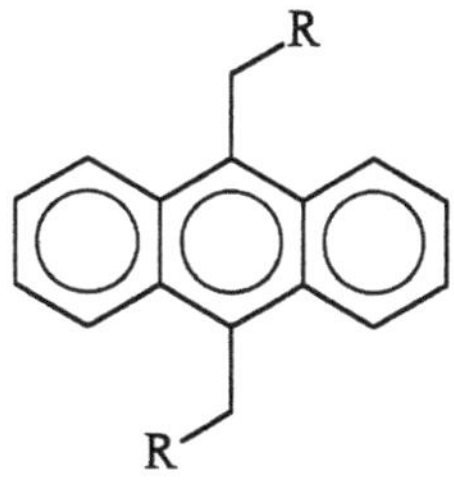   2

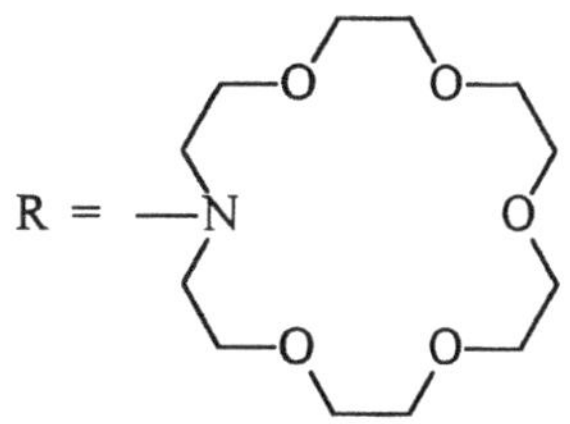

assumed to be connected with stopping the above-mentioned electron transfer by the metal cations.

It is obvious that in the case of compound 2, the fluorescence enhancement can only occur on blocking the lone pairs of the nitrogen atoms of both crown-ether moieties. This was achieved by using dications of polymethylene diammonium, which are capable of forming pseudo-cyclic complexes with 2. As an example, the fluorescence quantum yield of 2 was found to increase by two orders of magnitude in the presence of dications with four methylene units [21].

On replacement of the azacrown-ether fragment in compound 1 by a benzo-15-crown ether moiety, the process of photoinduced electron transfer becomes inefficient [22]. In this case, the complexation with metal cations does not practically influence the fluorescence quantum yield. However, on introduction of an acceptor group CN in the anthracene nucleus, the intramolecular quenching becomes significant and formation of a complex with a metal cation results in an enhancement of the fluorescence quantum yield.

A number of anthracene-containing cryptands were synthesized [23, 24], in which the anthracene nucleus is connected at positions 9 and 10 with the nitrogen atoms of diaza-18-crown-6 ether through identical polymethylene chains. For these compounds, besides weak fluorescence connected with the anthracene nucleus, a greater long-wavelength fluorescence is observed that is caused by formation of an exciplex between the electronically excited anthracene and the nitrogen atoms of the diazacrown-ether moiety. Upon complexation the nitrogen lone pairs are blocked by metal cations and the formation of the exciplex is suppressed. As a consequence, the fluorescence band of the exciplex disappears and the anthracene fluorescence enhances strongly, the overall quantum yield also being sharply increased.

The fluorescence properties of a cylindrical receptor 3 with two sites of binding [25–27] are of interest. Similar to anthracenocryptands, compound 3 displays a dual fluorescence. However, in this case, the authors ascribed the long-wavelength component of the emission spectrum to the intramolecular eximer formed as a result of interaction between two anthracene rings arranged in an appropriate pattern to each other (see Table 1). The exciplex formation between the anthracene ring and the nitrogen atom of the crown-ether moiety appears to be complicated by steric hindrance owing to the rigidity of the molecule. Upon formation of a complex with two Rb$^+$ ions, the fluorescence caused by excimer formation enhances strongly. This cylindrical receptor is able to recog-

Table 1. Photophysical Characteristics of **3** in the Absence and Presence of $MClO_4$

	3	Na^+	K^+	Rb^+	Cs^+
Φ_{ft}	0.06	0.22	0.21	0.24	0.21
Φ_{fm}	0.03	0.22	0.13	0.02	0.02
Φ_{fe}	0.03	–	0.08	0.22	0.19
λ_f	530	–	570	560	530

Φ_{ft}, Φ_{fm} and Φ_{fe} – quantum yields of overall, monomer, and excimer fluorescence; λ_f – excimer fluorescence maximum.

nize linear organic dications. As an example, heptamethylenediammonium dication is located on complexation between two anthracene rings thus hindering the eximer formation. In this compound the intensity of the excimer fluorescence falls sharply on complexation, while the intensity of anthracene fluorescence (monomer fluorescence) grows.

3   4

Bisanthraceno-crown ether **4** was also found to display a dual fluorescence consisting of a "monomer" and excimer emission [28]. The compound was found to form a complex of 1:2 stoichiometry with Na^+, and a positive cooperative effect (when the first complexation facilitates complexation of the second cation) was discovered for the complexation process in methanol and acetonitrile. In acetonitrile, the ability of **4** to complex Na^+ is appreciably higher in the exited state than in the ground state. The effect observed is assumed to be due mainly to a shaping of the cavity that is triggered by intramolecular excimer formation.

A cylindrical receptor **5** with two binding sites was synthesized by using dimethylnaphthalenes [29]. In CH_2Cl_2 solution at room temperature, the disappearance of dimethylnaphthalene fluorescence and the appearance of a broad and weak fluorescence band at 438 nm is characteristic of 5. This appears to be connected with the intramolecular charge transfer between the nonbonding orbital of the amine moieties and the orbitals of the naphthalene rings. Addition of an acid causes the protonation of the four amine units. As a consequence, the CT fluorescence disappears, whereas the dimethylnaphthalene fluorescence peak at 342 nm reappears and its intensity increases by a factor of about 800. Hence, compound 5 may be considered as a sensitive fluorescence sensor highly responsive to protons.

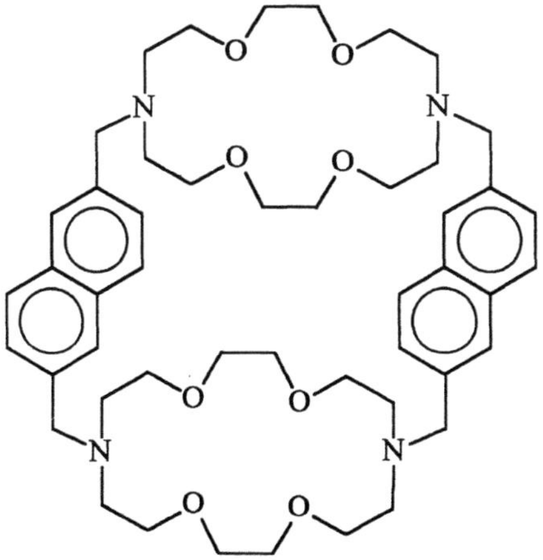

5   6

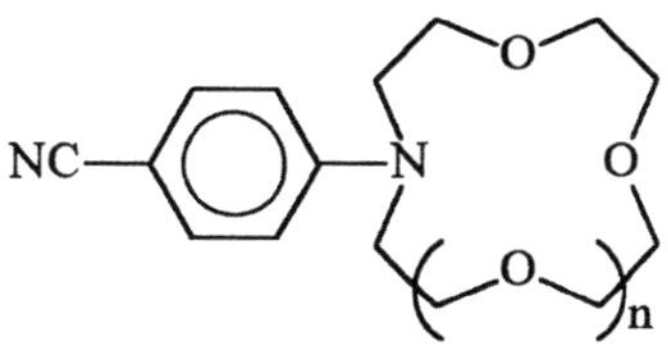

Compounds in which a twisted intramolecular charge transfer (TICT) state is formed on electronic excitation may be considered as perspective fluoroionophores. Simple examples of such systems are crown-containing benzonitriles 6 [30].

These compounds display a dual fluorescence. A short-wavelength component of the emission spectrum of 6 corresponds to the fluorescence from the initial exited state, whereas the long-wavelength component is connected with fluorescence from the TICT state. In the presence of metal cations, the intensity of the long-wavelength fluorescence decreases because a metal cation bound to the crown-ether moiety destabilizes the TICT state. The intensity of the short-wavelength fluorescence increases strongly because one of the main channels of its quenching becomes inefficient upon complexation.

3
Crown-Containing Coumarins

In reported works [31, 32], the properties of fluoroionophores 7 and 8 based on coumarin 153 are considered. On complexation of compounds 7 with alkali, alkaline-earth and transition metals, the shifts in long-wavelength of the fluorescence band of 7 and an appreciable increase in the fluorescence quantum yields were observed. These effects seem to be connected with increasing the degree of the intramolecular charge transfer in the molecule (increase in conjugation in the chromophore) caused by direct interaction of the metal cation, which is within the cavity of the crown-ether moiety, with the carbonyl group of the coumarin.

n = 1,2

7

8

A significant effect of fluorescence self-quenching is characteristic of compound 8. The mechanism of the self-quenching was not investigated; however, it was assumed that the quenching is connected with intramolecular interaction of two coumarin moieties in the ground state. Such a mechanism is usually used for the description of the concentration quenching of the fluorescence of some laser dyes. The excimer mechanism of self-quenching of 8, by analogy with the self-quenching of anthracene derivatives, may be an alternative mechanism.

On complexation of 8 with metal cations of small ionic radius, the fluorescence quantum yield changed insignificantly. However, on binding of large cations, such as Ba^{2+} and K^+, which appear to hinder the interaction of two coumarin rings, the fluorescence intensity is enhanced.

4
Crown-Containing Diarylethylenes

Crown ether 9 with a fluorescent stilbene-cap was the first of the synthesized fluoroionophores of this type [33]. As opposed to the ions of large ionic radius, such as K^+, Rb^+ and Cs^+, complexation of *trans*-9 with Li^+ and Na^+ ions in acetonitrile was found to result in fluorescence quenching. Upon UV irradiation, *cis-trans* isomerization takes place leading to a considerable change in the ability of the compound to extract the alkali metals from water to benzene. It should be noted that UV irradiation which produces photoisomerized *cis*-9 leads to an enhancement of extractability of Li^+ and Na^+ (i.e., to a conformational change in the crown-ether moiety).

R = H, CN, Ph

9   10

Recently, examples of crown-containing derivatives of stilbene **10**, which exhibit the twisted intramolecular charge transfer states have been published [34, 35]. An analysis of time evolution of the transient absorption spectra of **10** (R = Ph) in the sub-picosecond range after excitation pulse has shown that a biexponential decay is observed thus indicating the existence of two fluorescing exited states – the usual Franck–Condon state and a TICT state. Complexation of compounds **10** with Ca^{2+} results in small blue shifts of the fluorescence bands, whereas the fluorescence quantum yields remain unaltered. The effect of metal ions on *trans-cis* photoisomerization of compound **10** was not studied in these works.

11

5
Crown-Containing Merocyanine Dyes

In a series of works [36–39], the fluorescence of azacrown-containing merocyanine **11** , which is an analog of the well-known laser dye DCM, containing the aza-15-crown-ether fragment instead of the dimethylamino group, and its complexes with alkali and alkaline-earth metal cations was studied. The fluo-

Table 2. Photophysical Characteristics of Dye **11** and Its Complexes with Metal Cations

Compound	λ_f (nm)	Φ_f	τ_f (ns)
11	621	0.73	2.20
11 · Li⁺	606	0.57	1.88
11 · Na⁺	611	0.65	2.06
11 · Mg²⁺	610	< 0.05	2.10
11 · Ca²⁺	608	0.27	1.94
11 · Ba²⁺	611	0.40	1.99

rescence spectra of the free and complexed compound are alike in shape and only a small hypsochromic shift of 10–20 nm is observed (see Table 2). Complexation was found to induce fluorescence quenching to an extent dependent on the nature of the metal cation. In contrast to the quantum yields, only a slight decrease in the excited-states lifetimes was observed.

It should be noted that compounds having the hydrogen atom instead of the azacrown-ether moiety exhibit a very weak emission band ($\varphi_f = 3 \cdot 10^{-4}$) located at a much shorter wavelength ($\lambda_f = 490$ nm) than that of free or complexed compound **11**. Hence, the fluorescence emission of **11** may be assumed to result almost solely from an intramolecular charge transfer from the donor moiety (amino group) to the acceptor moiety (dicyanomethylene group). Excitation seems to lead first to a nonemissive "locally" excited (LE) and then to the ICT state having a high fluorescence quantum yield (see Fig. 2).

The absence of marked differences between the lifetimes of **11** and its complexes with metals indicates that the radiative and nonradiative rate constants and the quantum yield of the ICT state are almost unaffected by complexation. However, the overall quantum yield of a given complex depends on the competition between the nonradiative decay from the LE state to the ground state and the intramolecular charge transfer. The competition strongly depends on the ability of the cation to monopolize the lone pair of the crown nitrogen atom, i.e., to decrease the potentiality of charge transfer.

Slight changes in position and shape of the fluorescent spectra of **11** on cation binding as well as in its fluorescence lifetime (whereas the quantum yield is decreased) may be explained by the photoejection of the cation. Direct evidence of photoejection of Li⁺ and Ca²⁺ ions from a crown-ether-linked merocyanine was obtained by picosecond spectroscopy. The appearance of a delayed absorption band (at 530 nm) after subpicosecond excitation appears to be compelling evidence for cation ejection from the excited complexes. It is assum-

Fig. 2. Kinetic scheme showing emission properties of dye **11**

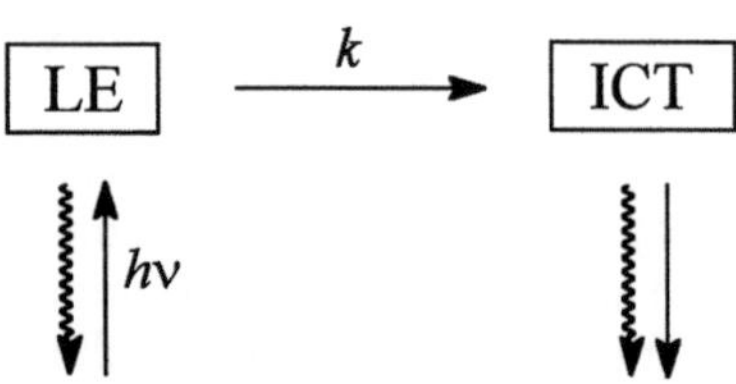

Table 3. Photophysical Characteristics of Dye 12 and Its Complexes with Metal Cations

Compound	λ_f (nm)	Φ_f	$k_{nr} \cdot (10^{-8} \, s^{-1})$
12	642	0.33	3.2
$12 \cdot Li^+$	612	0.44	2.0
$12 \cdot Na^+$	611	0.38	2.5
$12 \cdot Mg^{2+}$	573	0.48	1.6
$12 \cdot Ca^{2+}$	574	0.64	1.1
$12 \cdot Ba^{2+}$	576	0.59	1.3

ed that owing to photoinduced charge transfer from the nitrogen atom of the crown to the dicyanomethylene group, the nitrogen atom of the crown becomes positively polarized, and the resulting repulsion with the complexed cation causes its photoejection. The authors noted that in fact the photophysical data indicate that the coordination bond between the cation and the nitrogen atom of the crown is broken, but some of these cations may still interact with the oxygen atoms of the crown. The fact is that upon photoexcitation an equilibrium is reached on a time scale of a few picosecond, which is much shorter than the time required for cation separation by diffusion.

Unusual properties are exhibited by crown-containing benzoxazinone 12 and its complexes with metal ions [40–43]. Alkaline-earth cations markedly shift the emission wavelength (from 642 nm for the free molecule to 578 nm in the presence of calcium perchlorate) and enhance the fluorescence intensity of 12 (see Table 3). Since the crown can act as a second electron donor (in addition to the dimethylamino group), the observed hypsochromic shift is assumed to be connected with a reduction in the donor character of the crown nitrogen atom, which is involved in the conjugation of the molecule, upon complexation with a metal ion.

12

It is the opinion of the authors that the differences in the fluorescence intensity of 12 and its derivatives, in which the crown-ether moiety is replaced by a hydrogen atom or an NMe_2 group, are connected with differences in their nonradiative decay rates k_{nr}. A large value of k_{nr} for a dye with the NMe_2 group ($k_{nr} = 4.8 \cdot 10^8 \, s^{-1}$) seems to indicate the relatively free rotation of the group in respect to a crown-ether moiety. The properties of the complexes must be close to the properties of the unsubstituted compound because the nitrogen lone pair takes part in the formation of the bond with a metal ion. In this case, k_{nr} for

complexes must have a value intermediate between the values characteristic of
12 and its derivative with the hydrogen atom ($k_{nr} = 1.1 \cdot 10^8$ s^{-1}). This conclusion
has been confirmed experimentally.

At the same time some other properties, for example, strong dependence of
the fluorescence spectrum on the solvent polarity, may be explained in the
framework of the model of the TICT state.

Crown-containing benzodiazinones 13 and 14 have a similar structure
[44–46]. Upon addition of magnesium and calcium perchlorate in acetonitrile
solution, a small hypsochromic shift (up to 34 nm) is observed. In the case of
dye 14, the fluorescence quantum yield on complexation appreciably increases,
while for 13, on the contrary, the yield falls. However, the lifetimes of the exited
state condition of 13 and 14 and their complexes with Ca^{2+} are about identical.

13

14

On going from 13 to 14, a hypsochromic shift, a large value of the Stokes shift,
and a greater nonradiative constant are obtained. This indicates the existence of
a TICT state for dye 14. For a correct description of physical properties of these
dyes it is necessary to also take into account the rotation isomerism and photoisomerization around the C=C bond.

6
Crown-Containing Styryl and Butadienyl Dyes

The intense absorption bands in the visible region of the spectrum of the *trans-*
isomers, the relative high values of the quantum yields of the fluorescence (φ_f)
and the *trans-cis* isomerization ($\varphi_{t\text{-}c}$) allows one to use such compounds for the
luminescent information recording [47]. However, thermal instability of non-
fluorescent *cis*-isomers is an essential disadvantage: *cis*-isomers cannot be isolated and kept for a long time.

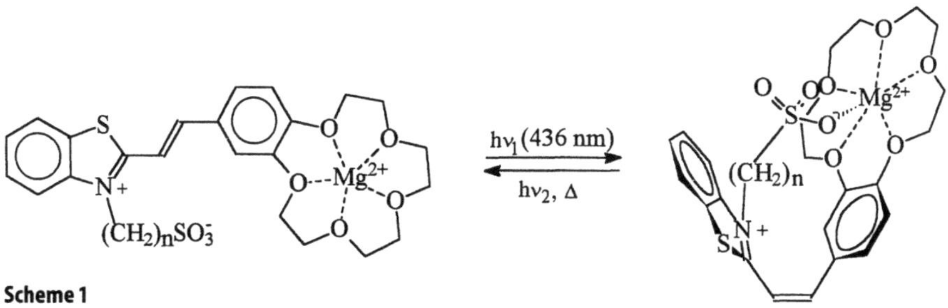

Scheme 1

Earlier, we investigated the interconnected reactions of photoisomerization and complexation of crown-containing styryl dyes (CSD) [48]. Formation of a so-called anion-"capped" complex was observed, in which an additional coordination bond is formed between the SO_3^- group of the N-sulfoalkyl substituent of the heterocyclic ring of the *cis*-form of a styryl dye and a metal cation located in the crown-ether cavity (Scheme 1) [49–51].

This leads to a strong hypsochromic shift of the absorption long-wavelength band and, simultaneously, to a sharp increase in the thermal stability of the *cis*-form of the dye.

Photochemical studies [52] have also shown that such molecules in solutions in the presence of metal cations are capable of self-organizing in the pairs of the "head-to-tail" type, so that the double bonds are located one above the other. It was found that upon irradiation of solutions of such dimeric complexes, a photochemical regio- and stereospecific [2+2]-cycloaddition of the "head-to-tail" type takes place. As a result, a nonfluorescent cyclobutane isomer is formed that is only one of the 11 theoretically possible isomers of cyclobutane (see Fig. 3) [53–55].

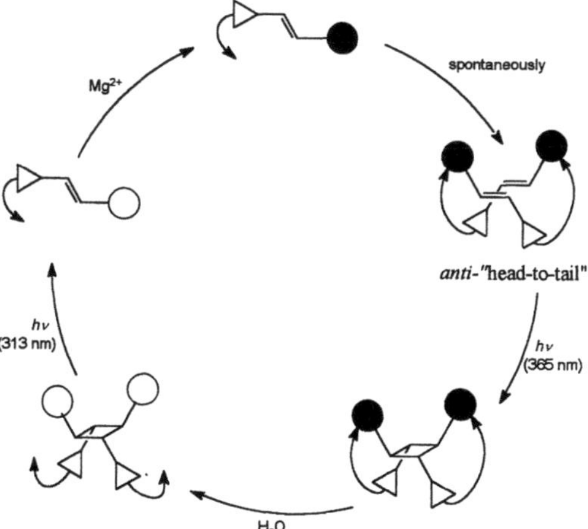

Fig. 3. Transformations of CSD upon complexation and illumination
○ – benzocrown fragment; ● – benzocrown fragment with M^{n+}; ▷ – heterocyclic residue;
⌒ – $(CH_2)_nSO_3^-$, n = 3 – 4

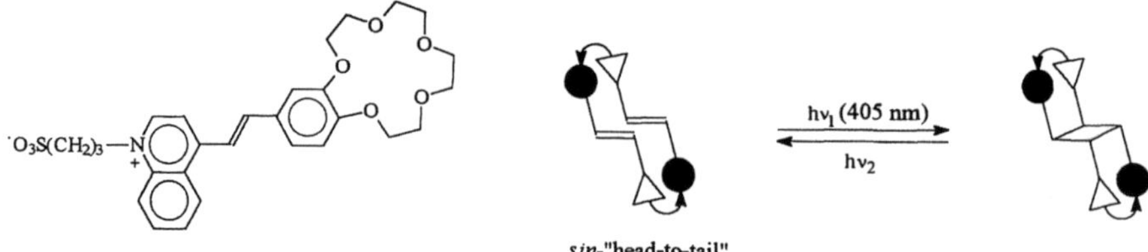

Fig. 4. Dimeric complex and cyclobutane formed by using CSD with a residue of 1,4-disubstituted quinolinium

While dimerization of the *anti*-"head-to-tail" type plays a main role in the case of crown-containing styryl dyes with a benzthiazole residue, this reaction is impossible for 1,4-disubstituted quinoline owing to steric hindrance. Though in the previous example dimerization of the type *syn*-"head-to-tail" is of secondary importance for the same reasons, in the latter case it becomes basic. On irradiation of solutions of such dimeric complexes, a photochemical stereospecific [2+2]-cycloaddition ($\Phi = 0.15$) occurs to also form a nonfluorescent isomer of cyclobutane (see Fig. 4) [56].

The effects observed permit us to consider the CSD as promising elements on development of fluorescent photoswitching molecular devices.

The *trans*-isomers of CSD are weak fluorescent in MeCN and MeOH at 293 K (fluorescence quantum yield $\varphi_f = 0.002-0.1$) [57, 58], whereas *cis*-isomers of CSD exhibit no fluorescence. The fluorescence intensity of *trans*-isomers of CSD increases significantly on cooling (up to $\varphi_f = 0.6$ at 173 K) in viscous solvents or in a polymeric matrix [57, 59]. An increase in the electron-donating ability of a heterocyclic residue and a crown-ether moiety of CSD results in a red shift of the fluorescence bands and in an increase in φ_f. The enhancement of the fluorescence lifetime from 150 ps in MeOH up to 2.3 ps in glycerol [60–62] is characteristic of CSD. We attribute the observed changes in the fluorescence lifetimes to a barrier *trans-cis* isomerization occurring in the exited singlet state.

As a rule, the addition of salts of alkali and alkaline-earth metals leads to a hypsochromic shift of the fluorescence maxima and to quenching of the fluorescence. The decrease in φ_f cannot be explained by changing the photoisomerization quantum yield, because complexation of CSD is not accompanied by an appreciable change in φ_{t-c}. It is known that the introduction of a methoxy group into the conjugation chain of a chromophore of an organic luminophore usually results in an increase in φ_f [63]. Hence, it could be expected that complexation with a metal ion (when the lone pairs of the oxygen atoms participate in the formation of the coordination bonds and thus are removed from conjugation with the chromophore) will cause an opposite effect, i.e., a decrease in φ_f. We actually observed this in our experiments [57].

In some cases, an unusual behavior of CSD was observed: the position of λ_f changes only slightly after the addition of metal salts [57, 64]. A small shift or even the absence of the shift of the fluorescence spectrum on complexation in comparison with a significant shift of the long-wavelength absorption band, as was found, for example, for *trans*-15 and Na$^+$ (see Fig. 5), may be explained by

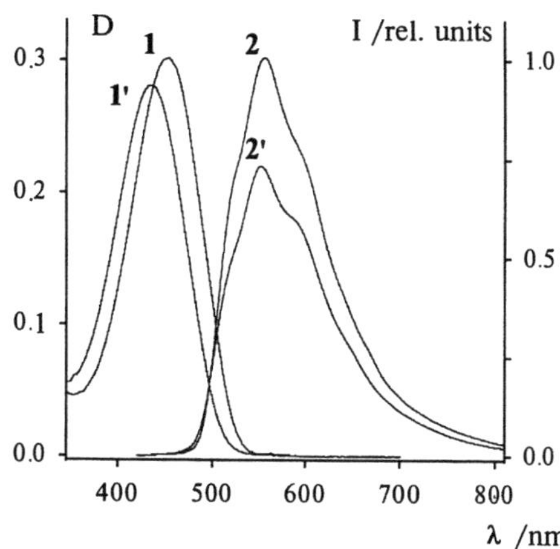

Fig. 5. Absorption spectra (*1, 1'*) at $T = 293$ K and corrected fluorescence spectra (*2, 2'*) at 173 K of *trans*-15 in EtOH. Concentrations of $NaClO_4$ are 0 (*1, 2*) and 0.01 mol L^{-1} (*1', 2'*), respectively

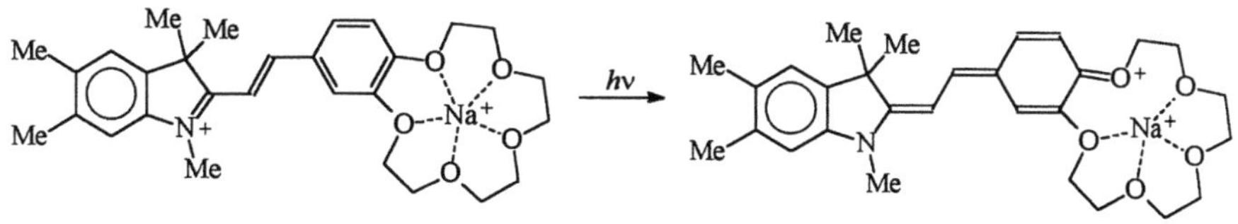

(*trans*-15) · Na^+

Scheme 2

changing the character of coordination of a metal ion and its position in the macroheterocycle. On transition in the first exited state of *trans*-15, the electronic density on the oxygen atom incorporated in the chromophore and located at the *para*-position with respect to the C=C bond decreases; as a consequence, the coordination bond of the metal ion with this oxygen atom becomes weak (Scheme 2).

This leads to such a change in the equilibrium position of the metal ion in the macroheterocycle that it is bound predominantly to four of the five oxygen atoms of the 15-crown-5 ether. The phenomenon observed can be defined as a photoinduced recoordination of the metal cation.

We observed an interesting example of recoordination in the ground and excited states of *trans*-16 (see Scheme 3) [65]. Analysis of the absorption spectrum shape (see Fig. 6, *a*) has shown that the long-wavelength shoulder of the band with a maximum at 410 nm cannot be attributed to absorption of the free ligand molecules. This conclusion is confirmed by the fact that the calculated spectrum practically coincides with the experimental one recorded after the addition to a solution of *trans*-16 of a large amount of $Ca(ClO_4)_2$ ($C_M = 0.5$ mol L^{-1}).

The origin of the low-intensity band can be explained by the existence of a thermodynamic equilibrium between two forms of the complex. In the one form, which corresponds to the long-wavelength band that peaked at 410 nm, the Ca^{2+} ion interacts effectively with the nitrogen atom of the crown-ether moiety. In the other form, the coordination bond with the nitrogen atom is

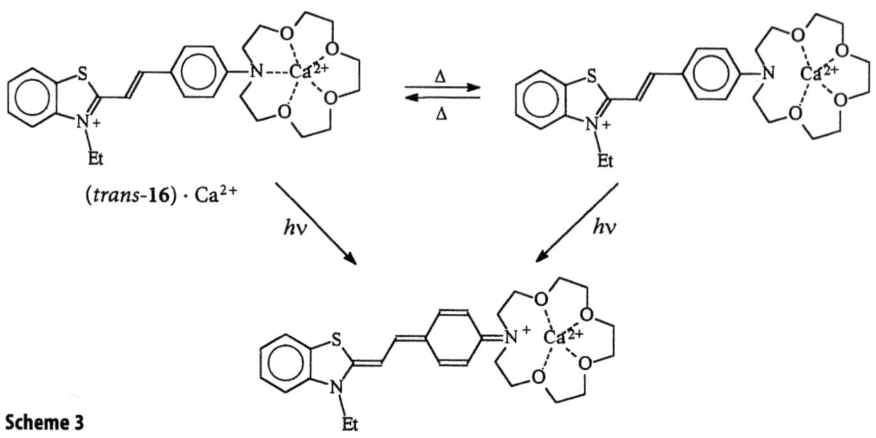

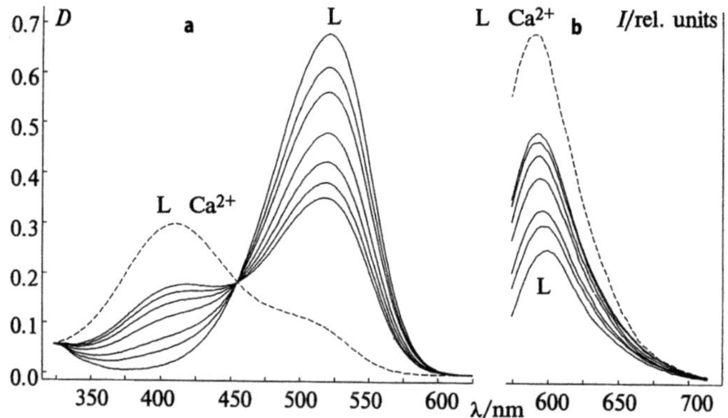

Scheme 3

Fig. 6. Dependencies of the absorption **a** and fluorescence **b** spectra of *trans*-16 (C_L = $1.0 \cdot 10^{-5}$ mol L^{-1}) in MeCN on the concentration of Ca^{2+} ions (C_M/C_L = 50, 100, 200, 300, 400, 500) at a constant concentration of the perchlorate ions C_A = 0.01 mol L^{-1}. The dashed curve shows the calculated spectra of the complex

weakened, but the interaction with the oxygen atoms is retained. In the latter case, the effect of the cation on the electronic structure of the chromophore is much weaker. In spite of the fact that there are common features in the structure of compounds **10**, **11**, **16** and even **15** (first of all the presence of the C=C double bond and the crown-ether moiety) and the possibility of proceeding the photoinduced recoordination reaction in their complexes with metal ions, the above example of thermal recoordination reaction is unique.

Figure 6,*b* shows the fluorescence spectrum of *trans*-16 as a function of the Ca^{2+} concentration. Since photoexcitation of *trans*-16 leads to an increase in the electron density on the nitrogen atom, it may be assumed that in the excited state the equilibrium is shifted toward the complex in which the corresponding coordination bond is absent. This form of the complex seems to provide a main

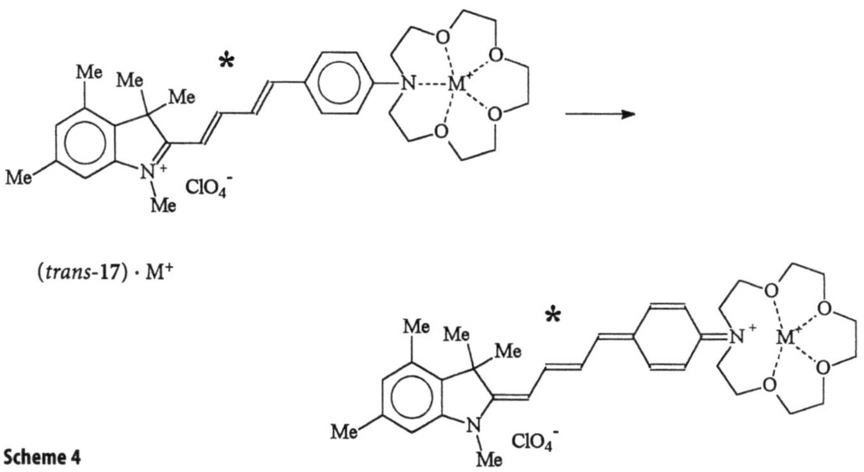

$(trans\text{-}17) \cdot M^+$

Scheme 4

contribution to the fluorescence spectrum and therefore the position of the fluorescence band of *trans*-16 changes only slightly upon complexation.

It is known that butadienyl dyes are capable of generating laser radiation with a high efficiency [66]. Therefore, studies of the fluorescence characteristics, complexing ability and generating properties of dyes containing two double bonds and a crown-ether moiety are of great interest.

We investigated laser radiation of synthesized butadienyl dyes on excitation by the second harmonic (347.2 nm) of a ruby laser. It was found that the dyes of the type **17** generate laser radiation in the 687–730 nm spectral region. Substitution of the NMe_2 group by a crown-ether moiety results in an 8-fold increase in the efficiency of laser radiation. The fluorescence spectra of complexes of butadienyl dyes with Na^+, Mg^{2+}, Ca^{2+}, Sr^{2+}, and Ba^{2+} ions are similar to the spectra of free dyes; however, complexation leads to an increase in the fluorescence intensity. It is supposed that in this case recoordination also takes place, i.e., the breaking of the coordination bond between the nitrogen atom and the metal cation in the exited state of the complex (Scheme 4) [67].

The efficiency of laser radiation generation increases by a factor of 1.3–2.3 on complexation of butadienyl dyes with metal ions.

The data obtained by us reveal characteristics of the effects of the nature of a heterocyclic residue and crown-ether moiety, temperature and medium on the fluorescence of ionophores of the styryl and butadienyl series. This has provided the basis for a desired influence with the help of synthesis on their fluorescent parameters and cation-complexing ability.

7
Conclusions

The results of studies of crown-containing fluoroionophores discussed above show the potentialities of these compounds and variability in principles of their design. Certain of the compounds discussed are presently of little practical interest. Other compounds are designed for practical use as selective fluoroiono-phores, including determination of cations of great ionic radius and small organic cations, ecological monitoring of environment, applications in medicine and biology, development of new systems for recording and reconstruction of information, new types of molecular photophysical and photochemical devices, and novel laser dyes with improved generation characteristics, etc. It may be safely suggested that photophysics and photochemistry of crown-containing fluoroionophores will be further developed and that these compounds will be widely used in the future.

References

1. Yatsimirskii, A. K. Itogi nauki I tekhniki. Bioorganicheskaya khimiya [Achivements in Science and Technology. Bioorganic Chemistry], VINITI, Moscow, 1990, 17, 148 pp. (in Russian)
2. Alfimov, M. V.; Gromov S. P. in Supramolecular Chemistry, Eds. V. Balzani and L. De Cola, NATO ASI Ser., Dordrecht, 1992, 371, 343
3. Shinkai, S. in Comprehensive Supramolecular Chemistry, Ed. J.-M. Lehn, Pergamon Press, New York, 1996,p 1
4. Supramolecular Reactivity and Transport: Bioorganic Systems, in Comprehensive Supramolecular Chemistry, Ed. J.-M. Lehn, Pergamon Press, New York, 1996,pp 4, 5
5. Gromov, S. P.; Alfimov, M. V. Izv. Akad. Nauk, Ser. Khim. 1997, 641 [Russ. Chem. Bull., 1997, 46, 611 (Engl. Trans.)]
6. Inoue, Y.; Hakushi, T. in Cation Binding by Macrocycles, Eds. Inoue, Y.; Gokel, G. W. Marcel Dekker, New York, 1990
7. Balzani, F. Scandola, in Comprehensive Supramolecular Chemistry, Ed. J.-M. Lehn, Pergamon Press, New York, 1996,p 10, Ch. 27
8. Hoelzl Wallach, D. F., Steck, T. L. Anal. Chem. 1963, 35, 1035
9. Rink, T. J. Pure Appl. Chem. 1983, 55, 1977
10. Tsien, R. Y.; Pozzan, T.; Rink, T. J. Trends Biochem. Sci. 1984, 263
11. Grynkiewicz, G.; Poenie, M.; Tsien, R. Y. J. Biol. Chem. 1985, 260, 3440
12. Löhr, H. G.; Vögtle, F. Acc. Chem. Res. 1985, 18, 65
13. Sousa, L. R.; Larson, J. M. J. Am. Chem. Soc. 1977, 99, 307
14. Larson, J. M.; Sousa, L. R. J. Am. Chem. Soc. 1978, 100, 1943
15. Nishida, H.; Katayama, Y.; Katsuki, H.; Nakamura, H.; Takagi, M.; Ueno, K. Chem. Lett. 1982, 11, 1853
16. Thanabal, V.; Krishnan, V.; J. Am. Chem. Soc. 1982, 104, 3643
17. Nakashima, K.; Nakatsuyi, S.; Akiyama, S.; Tanigawa, I.; Kaneda, T.; Misumi, S. Talanta, 1984, 31, 749
18. Nakashima, K.; Nagaoka, Y.; Nakatsuji, S.; Kaneda, T.; Tanigawa, I.; Hirose, K.; Misumi, S.; Akiyama, S. Bull. Chem. Soc. Jpn., 1987, 60, 3219
19. Forrest, H.; Pacey G. E. Talanta, 1989, 36, 335
20. De Silva, A. P.; De Silva, S. A. J. Chem. Soc., Chem. Comm. 1986, 1709
21. De Silva, A. P.; Sandanayake, K. R. A. S. Angew. Chem., Int. Ed. Engl. 1990, 29, 1173
22. De Silva, A. P.; Sandanayake, K. R. A. S. Tetrahedron Lett. 1991, 32, 421
23. Konopelski, J. P.; Kotzyba-Hibert, F.; Lehn, J.-M.; Desvergne, J.-P.; Fages, F.; Castellan, A. J. Chem. Soc., Chem. Comm. 1985, 433
24. Fages, F.; Desvergne, J.-P.; Bouas-Laurent, H.; Marsau, P.; Lehn, J.-M.; Kotzyba-Hibert, F.; Albrecht-Gary, A.-M.; Al-Joubbeh, M. J. Am. Chem. Soc. 1989, 111, 8672
25. Fages, F.; Desvergne, J.-P.; Bouas-Laurent, H.; Lehn, J.-M.; Konopelski, J. P.; Marsau, P.; Barrans, Y. J. Chem. Soc., Chem. Comm. 1990, 655

26. Fages, F.; Desvergne, J.-P.; Kampke, K.; Bouas-Laurent, H.; Lehn, J.-M.; Meyer, M.; Albrecht-Gary, A.-M.; J. Am. Chem. Soc. 1993, 115, 3658

27. Fages, F.; Desvergne, J.-P.; Bouas-Laurent, H.; Lehn, J.-M.; Barrans, Y.; Marsau, P.; Meyer, M.; Albrecht-Gary, A.-M. J. Org. Chem. 1994, 59, 5264

28. Marquis, D.; Desvergne, J.-P. Chem. Phys. Lett. 1994, 230, 131

29. Ballardini, R.; Balzani, V.; Credi, A.; Gandolfi, M. T.; Kotzyba-Hibert, F.; Lehn, J.-M.; Prodi, L. J. Am. Chem. Soc. 1994, 116, 5741

30. Létard, J.-F.; Delmond, S.; Lapouyade, R.; Braun, D.; Rettig, W.; Kreissler, M. Rec. Trav. Chim. Pays-Bas 1995, 114, 517

31. Bourson, J.; Pouget, J.; Valeur, B. J. Phys. Chem. 1993, 97, 4552

32. Bourson, J.; Badaoui, F.; Valeur, B. J. Fluorescence 1994, 4, 275

33. Shinkai, S.; Miyazaki, K.; Nakashima, M.; Manabe, O. Bull. Chem. Soc. Jpn. 1985, 58, 1059

34. Dumon, P.; Jonusauskas, G.; Dupuy, F.; Pée, Ph.; Rullière, C.; Létard, J.-F.; Lapouyade, R. J. Phys. Chem. 1994, 98, 10391

35. Mathevet, R.; Jonusauskas, G.; Rullière, C.; Létard, J.-F.; Lapouyade, R. J. Phys. Chem. 1995, 99, 15709

36. Bourson, J.; Valeur, B. J. Phys. Chem. 1989, 93, 3871

37. Martin, M. M.; Plaza, P.; Dai Hung, N.; Meyer, Y. H.; Bourson, J.; Valeur, B. Chem. Phys. Lett. 1993, 202, 425

38. Martin, M. M.; Plaza, P.; Meyer, Y. H.; Bégin, L.; Bourson, J.; Valeur, B. J. Fluorescence 1994, 4, 271

39. Martin, M. M.; Plaza, P.; Meyer, Y. H.; Badaoui, F.; Bourson, J.; Lefèvre, J.-P.; Valeur, B. J. Phys. Chem. 1996, 100, 6879

40. Fery-Forgues, S.; Le Bris, M.-T.; Guetté, J.-P.; Valeur, B. J. Chem. Soc., Chem. Commun. 1988, 384

41. Fery-Forgues, S.; Le Bris, M.-T.; Guetté, J.-P.; Valeur, B. J. Phys. Chem. 1988, 92, 6233

42. Fery-Forgues, S.; Bourson, J.; Dallery, L.; Valeur, B. New J. Chem. 1990, 14, 617

43. Fery-Forgues, S.; Le Bris, M. T.; Mialocq, J.-C.; Pouget, J.; Rettig, W.; Valeur, B. J. Phys. Chem. 1992, 96, 701

44. Cazaux, L.; Faher, M.; Picard, C.; Tisnés, P. Can. J. Chem. 1993, 71, 1236

45. Cazaux, L.; Faher, M.; Picard, C.; Tisnés, P. Can. J. Chem. 1993, 71, 2007

46. Cazaux, L.; Faher, M.; Lopez, A.; Picard, C.; Tisnés, P. J. Photochem. Photobiol. A: Chem. 1994, 77, 217

47. Sovremennye sistemy registratsii informatsii (Modern Systems for Information Recording). Ed. A. V. El'tsov, „Sintez" Co., St. Petersburg, 1992, 328 (in Russian)

48. Gromov, S. P.; Fomina, M. V.; Ushakov, E. N.; Lednev, I. K.; Alfimov, M. V. Dokl. Akad. Nauk SSSR, 1990, 314, 1135. [Dokl. Chem. 1990, 314 (Engl. Transl.)]

49. Gromov, S. P.; Fedorova, O. A.; Ushakov, E. N.; Stanislavskii, O. B.; Alfimov, M. V. Dokl. Akad. Nauk SSSR, 1991, 321, 104. [Dokl. Chem. 1991, 321 (Engl. Transl.)]

50. Alfimov, M. V.; Gromov, S. P.; Lednev, I. K. Chem. Phys. Lett., 1991, 185, 455

51. Barzykin, A. V.; Fox, M. A.; Ushakov, E. N.; Stanislavsky, O. B.; Gromov, S. P.; Fedorova, O. A.; and Alfimov, M. V., J. Am. Chem. Soc., 1992, 114, 6381

52. Ushakov, E. N.; Stanislavskii, O. B.; Gromov, S. P.; Fedorova, O. A.; Alfimov, M. V. Dokl. Akad. Nauk SSSR, 1992, 323, 702. [Dokl. Chem. 1992, 323 (Engl. Transl.)]

53. Alfimov, M. V., Gromov, S. P.; Stanislavskii, O. B.; Ushakov, E. N.; Fedorova, O. A. Izv. Akad. Nauk, Ser. Khim. 1993, 1449 [Russ. Chem. Bull. 1993, 42, 1385 (Engl. Transl.)]

54. Gromov, S. P.; Fedorova, O. A., Alfimov, M. V., Mol. Cryst. Liq. Cryst., 1994, 246, 183

55. Gromov, S. P.; Ushakov, E. N.; Fedorova, O. A.; Buevich, A. V.; Alfimov, M. V. Izv. Akad. Nauk, Ser. Khim. 1996, 693 [Russ. Chem. Bull. 1996, 45, 654 (Engl. Transl.)]

56. Gromov, S. P.; Fedorova, O. A., Ushakov, E. N.; Buevich, A. V.; Alfimov, M. V. Izv. Akad. Nauk, Ser. Khim. 1995, 2225 [Russ. Chem. Bull. 1995, 44, 2131 (Engl. Transl.)]

57. Gromov, S. P.; Fedorova, O. A., Alfimov, M. V.; Druzhinin, S. I.; Rusalov, M. V.; Uzhinov, B. M. Izv. Akad. Nauk, Ser. Khim. 1995, 2003 [Russ. Chem. Bull. 1995, 44, 1992 (Engl. Transl.)]

58. Druzhinin, S. I.; Rusalov, M. V.; Uzhinov, B. M.; Alfimov, M. V.; Gromov, S. P.; Fedorova, O. A. Proc. Indian Acad. Sci. (Chem. Sci.), 1995, 107, 721

59. Druzhinin, S. I.; Rusalov, M. V.; Uzhinov, B. M.; Alfimov, M. V.; Gromov, S. P.; Fedorova, O. A. J. Appl. Spectroscopy, 1995, 62, 69 (in Russian)
60. Alfimov, M. V.; Buevich, O. E.; Gromov, S. P.; Kamalov, V. F.; Lifanov, A. P.; Fedorova, O. A. Dokl. Akad. Nauk SSSR, 1991, 319, 1149. [Dokl. Chem. 1991, 319 (Engl. Transl.)]
61. Alfimov, M. V.; Kamalov, V. F.; Struganova, I. A.; Yoshihara, K. Chem. Phys. Lett., 1992, 195, 262
62. Alfimov, M. V.; Kamalov, V. F.; Struganova, I. A.; Yoshihara, K. J. Fluorescence 1994, 4, 61
63. Bolotin, B. M.; Krasovitskii, Organicheskie lyuminofory [Organic Luminophores], Khimiya, Moscow, 1984, 334 pp. (in Russian)
64. Gromov, S. P.; Golosov, A. A.; Fedorova, O. A., Levin, D. E.; Alfimov, M. V. Izv. Akad. Nauk, Ser. Khim. 1995, 129 [Russ. Chem. Bull. 1995, 44, 124 (Engl. Transl.)]
65. Ushakov, E. N.; Gromov, S. P.; Fedorova, O. A.; Alfimov, M. V. Izv. Akad. Nauk, Ser. Khim. 1997, 484 [Russ. Chem. Bull. 19 97, 46, 463 (Engl. Transl.)]
66. Kato, K. IEEE J. Quantum Electron., 1980, QE16, 1017
67. Gromov, S. P.; Sergeev, S. A.; Druzhinin, S. I.; Rusalov, M. V.; Uzhinov, B. M.; Kuz'mina, L. G.; Churakov, A. V.; Howard, J. A. K.; Alfimov, M. V. Izv. Akad. Nauk, Ser. Khim. 1998, in press [Russ. Chem. Bull. 1998, 47, in press (Engl. Transl.)]

CHAPTER 9

Recent Developments in Luminescent PET (Photoinduced Electron Transfer) Sensors and Switches

A. Prasanna de Silva, A. J. M. Huxley

1
Introduction

Photoinduced electron transfer [1–3] has proved to be a durable principle for adaptation into luminescent systems capable of molecular recognition [4–7]. This leads to sensory and switching molecular devices [8–17] for applications in information gathering and processing [18]. Scheme 1 outlines the relationship between sensors and switches in the context of information handling. Sensors enable us to receive chemical information by means of a physical message whereas switches allow us to manipulate a physical signal by means of chemical stimuli. In this short review, we set our own work in the context of research in luminescent PET systems from around the world during the past year. Comprehensive reviews of earlier research in this specific field are available for those wishing to explore it in depth [4–12, 14–16].

Scheme 2 presents the principal concepts and terms employed in this field. It must be noted that the direction of PET is immaterial to the efficiency of luminescence switching in the context of this status report. A related point is that luminescence can be successfully switched "off" by any one of several PET pathways as illustrated for the "on-off" system in Scheme 2. The available values of key parameters arising from the examples reviewed here are collected in Table 1 for readers requiring quantitative information.

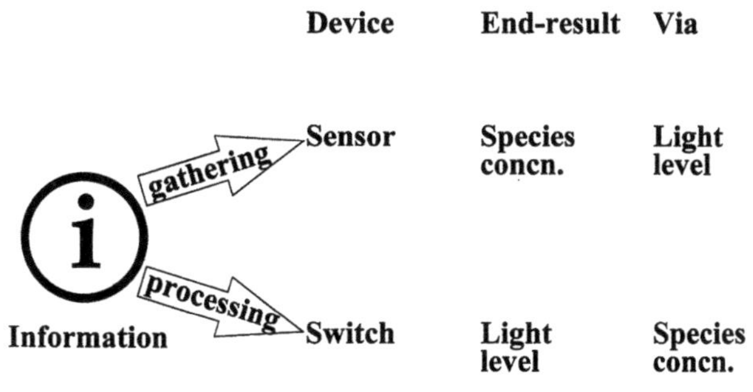

Scheme 1. Information handling with sensors and switches. A chemical species at a sufficient concentration serves to trigger a light signal

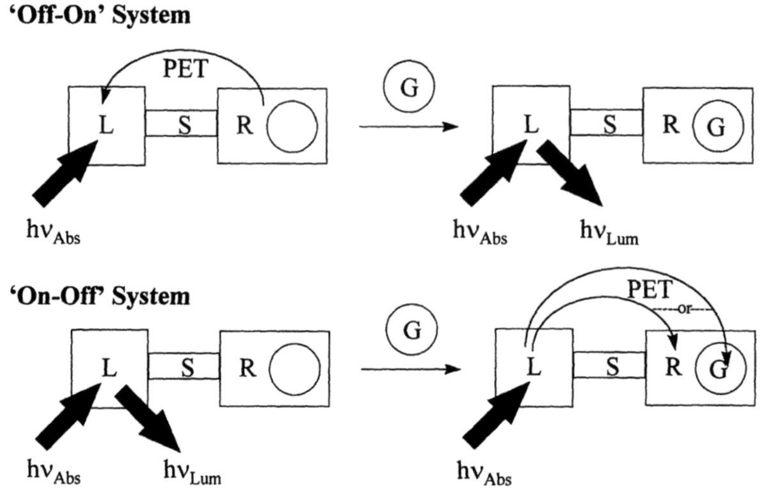

'Off-On' System

'On-Off' System

Scheme 2. Two common luminescent PET sensor/switch systems and their mechanism of action. Both cases involve sharp switching between a luminescent ("on") and a quenched state ("off"). L = Lumophore, S = Spacer, R = Receptor, G = Guest species, PET = Photoinduced electron transfer

2
Mono/Homomultireceptor Systems

Simple "lumophore-spacer-receptor" formats continue to attract attention, sometimes with surprising consequences. Briefly, the design principle is the guest-induced recovery of fluorescence or another form of luminescence which has previously been quenched by PET involving the receptor. The opposite idea of guest-induced luminescence quenching can be equally useful, if not visually as spectacular.

Fluorescence "off-on" switching by Na^+ is the goal of **1**, a recent contribution from our laboratory [19]. The receptor unit within **1** takes inspiration from Tsien's sensor for physiological Na^+ which operates by Na^+-induced wavelength shifts of the ICT (internal charge transfer) fluorophore [20]. Cation-responsive fluorescent systems continue to receive attention [21, 22], especially with regard to red emitters [23, 24, 24a]. The switching efficiency of **1** can be ascribed to the strong suppression of PET upon Na^+ binding. This, in turn, is due to the Na^+-induced decoupling of the diaza-15-crown-5 nitrogens from the methoxybenzene groups which increases the oxidation potential of the receptor. It so happens that the 2-pyrazoline fluorophore in **1** possesses an ICT excited state. Nevertheless, the design is according to PET principles. Hence, no ion-induced wavelength shifts are seen with **1**.

Order-of-magnitude enhancement of fluorescence, though at rather short wavelengths, is seen when protons or Zn^{2+} bind completely to **2** [25]. The fact that no ion-induced wavelength shifts are seen is in line with fluorescent PET

Table 1. Key Parameters for PET Sensors and Switches[a]

Parameter	1	2	3	4	5	6	7	8	10
	Target = Na^+ Other = K^+, H^+ Solvent = MeOH:H_2O (1:1)	Target = Zn^{2+} Other = H^+, Cu^{2+} Solvent = H_2O	Target = Hg^{2+} Other = Cu^{2+} Solvent = H_2O	Target = Histidine Other = Acetate Solvent = H_2O	Target = Cu^{2+} or Ni^{2+} or Zn^{2+} Solvent = MeCN	Target = Fe^{3+} Other = Zn^{2+} Solvent = MeCN	Target = Ca^{2+} Other = Ba^{2+} Solvent = MeCN	Target = Ca^{2+} Other = Ba^{2+} Solvent = MeCN	Target = Hydroquinone Solvent = PhCN
λ_{Abs} (log ε)	389 (4.40)	261 (2.86)	368		360		382 (3.68), 362 (3.75)	384 (3.67), 364 (3.72)	650 (3.66), 590 (3.75)
λ_{Exc}	389	210	368		360	400	350	350	514
λ_{Lum}	480	290			~ 405, 430	~ 510	457	~ 392, 412	652, 720
ϕ_{Max} 0.22	0.31				0.33		0.26	0.10	0.25
LE_{Target}	16	50,[b] 8[c]	0.06	~ 0.6	230–330	34	0.6	2.5	> 25
LE_{Other}	4.3, 23	~ 8, 0.12,[d] 0.29[e]	0.25	~ 1.0		55	0.52	1.9	
log β_{Target}	3.1	0.46 (18)	> 6.0	2.9	0.014 (> 14)				
log β_{Other}	< 1.8. 4.2	~ 6.6 &3.2	4.25	≪ 2.9	0.33				

Table 1 (continued)

Parameter	12	14 & 15	16	17	18	19	20.Tb^{3+}	21.Eu^{3+}	22.Eu^{3+}
	Target = H$^+$& Na$^+$Other = H$^+$ or Na$^+$ Solvent = MeOH	Target = Bu$_3$N Solvent = CH$_2$Cl$_2$:MeCN (9:1)	Target = Glucosamine Other = Glucose, H$^+$ Solvent = EtOH:H$_2$O (1:2)	Target = GABA Other = Glutamate, H$^+$ Solvent = MeOH:H$_2$O (3:2)	Target = H$^+$ Solvent = MeOH:H$_2$O (1:4)	Target = H$^+$ Solvent = MeOH:H$_2$O (1:1)	Target = H$^+$ Solvent = H$_2$O	Target = K$^+$ Other = Na$^+$ Solvent = MeOH	Target = Fructose Solvent = MeOH:MeCN (500:1)
λ_{Abs} (log ε)		~ 430 (3.70) 340 (3.90)		393 (3.93), 372 (3.93)	384 (3.98), 364 (4.00)			345 (4.44), 355 (4.44)	345 (4.77), 333 (4.80)	
λ_{Exc} 274	377	277	372	372		350	345	340	274
λ_{Lum}	406, 428	343, (432)	425	402, 424, 449	398, 420, 443	370–550	493, 544	~ 590, 616	614
ϕ_{Max}	0.22			0.024	0.58		0.49	0.47	
LE$_{Target}$	91	> 100	1.5	2.2	0.22,[g] 34[h]	~ 0.06,[g] 10[h]	16	18	9
LE$_{Other}$	3.7 or 2.1		1.0	1.1, 70				9	
log β_{Target}	2.7[f]		3.3	1.6	5.0, 7.9	4.4, 7.5	7.2	4.3	
log β_{Other}			–,[i] 7.3	–,[i] 7.4				2.9	

[a] The wavelengths of maximum absorption λ_{Abs} and luminescence λ_{Lum}, as well as the excitation wavelength λ_{Exc} are given in nm. Extinction coefficients ε are in units of M^{-1} cm^{-1}. The maximum quantum yield of luminescence ϕ_{Max} and the luminescence enhancement factors with the target species LE$_{Target}$ or others LE$_{Other}$ are unitless. The binding constants for the target β_{Target} and other species β_{Other} are in units of M^{-1}. [b] Dizinc complex. [c] Monozinc complex. [d] Dicopper complex. [e] Monocopper complex. [f] For Na$^+$ in the presence of H$^+$. [g] For the "on-off" arm in the low pH region. [h] For the "off-on" arm in the high pH region. [i] Immeasurably small.

sensor behavior from the tertiary amine receptors to the benzene fluorophores. The rigidity of the benzene fluorophores are not affected by ion binding. "Loose bolt" effects are rarely large in magnitude [26]. Although ion-induced rigidification has been previously implicated in large fluorescence enhancements [27], subsequent analyses have reassigned the responsibility to PET processes [28]. Interestingly, no excimer emissions are seen from this homo-bifluorophoric system under any of the reported conditions. The fluorescence intensity-pH profile of **2** shows some similarity with that of Garcia–Espana and Pina's polyazaparacyclophane [29] which also employs the benzene fluorophore. The presence of multiple amine receptors is responsible for the response over a wide pH range. The dizinc complex of **2** is devoid of free amine receptors and hence shows pH-independent fluorescence. Open shell metal ions such as Cu^{2+} and Ni^{2+} cause quenching of the fluorescence of **2** most probably by a combination of PET and EET (electronic energy transfer) [30].

Fluorescence quenching with Cu^{2+} is also the observation with Czarnik's **3** [31]. However, **3** is distinguished by the fact that Cu^{2+} and Hg^{2+} are the only ions from a large palette to cause this effect. The selectivity of **3** towards Hg^{2+} vs. Cu^{2+} is >56 as judged by the two binding constants. The combined forces of rational design [32] and serendipity are at work here since systems with less or more rigid receptors degrade the selectivity. The idea of using the fluorophore in an additional role of rigidifying the receptor is a useful one. There are several examples of fluorophores serving as rigid backbones in bireceptor systems capable of linear recognition. Fabbrizzi's **4** is of this type where the pair of tetracoordinated Zn^{2+} centers serve to hold a deprotonated imidazole guest between them [33]. The signaling arises via fluorescence quenching because of PET to the anthracene fluorophore from the imidazole guest close by. Histidine can be specifically sensed in the presence of any other natural amino acid at pH 9.6. In the absence of imidazole functionalities, the fluorescence intensity-pH profile of **4** with or without Zn^{2+} can be understood according to PET principles as in the case of **2** [25] and related polyamines [29]. Older examples of PET active receptor pairs connected to the opposite ends of a fluorophore cause selective switching "on" of fluorescence upon encountering 1,4-butanediammonium ions [34] or glucose [35] rather than structurally close relatives. An earlier monoreceptor version of **4** is also available for the signaling of redox active benzoates via fluorescence quenching [36].

At a general level, **2** and **3** are the latest in a long line of cases where redox-active metal ion guests produce fluorescence quenching. Bharadwaj's **5** [37] and Samanta's **6** [38] are recent examples which report the opposite effect with remarkably large fluorescence enhancement factors. The cryptand receptor in **5** appears to stifle the redox activity of guests (perhaps by resisting ligand reorganization) as evidenced by cyclic voltammetric silence over a wide potential range. It therefore appears that the metal ion guest's only role is to suppress PET from the receptor to any of the anthracene fluorophores. No such topological stifling is possible with **6**. Ramachandram and Samanta's design relies on the relatively electron-deficient fluorophore to refrain from PET to the metal ion, e.g., Fe^{3+}. The presence of aliphatic amine receptors within **5** and **6** raises the specter that fluorescence recovery is an artifact due to protons adventitiously pro-

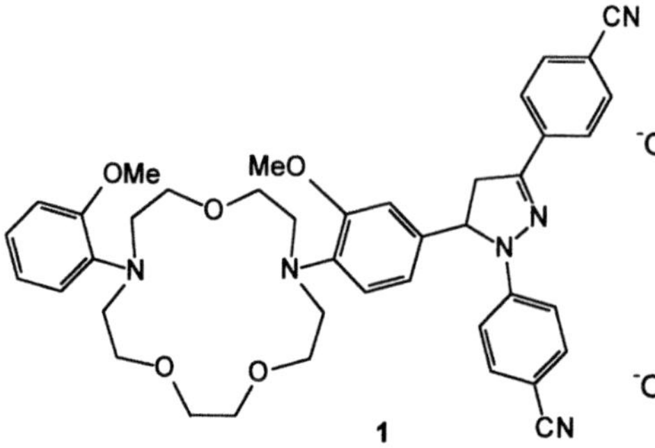
CN
OMe
MeO
O
N
N
N
N
O
O
CN

1

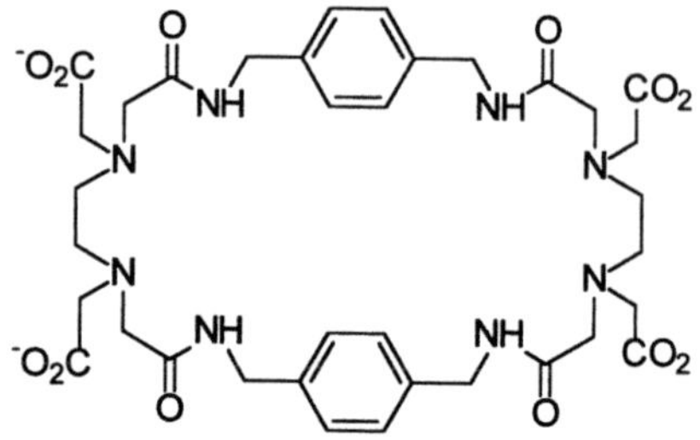
O
O
$^{-}O_2C$
NH
NH
CO_2
N
N
N
N
NH
NH
$^{-}O_2C$
CO_2
O
O

2

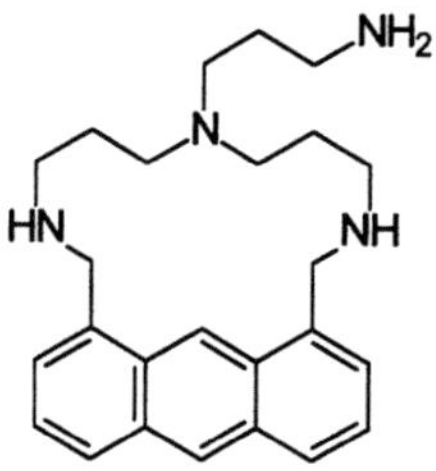
NH_2
N
HN
NH

3

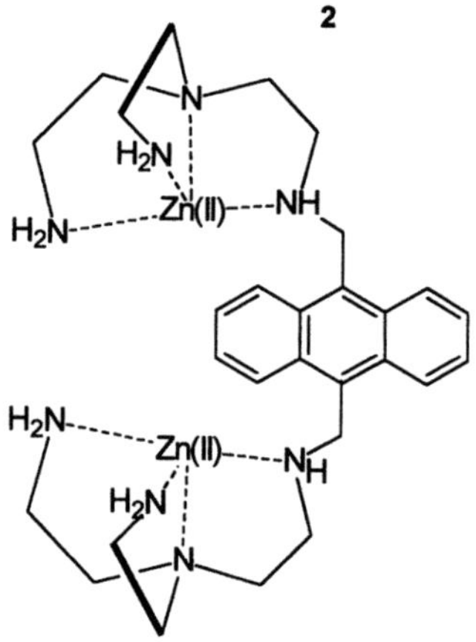
N
H_2N
H_2N
Zn(II)
NH
H_2N
Zn(II)
NH
H_2N
N

4

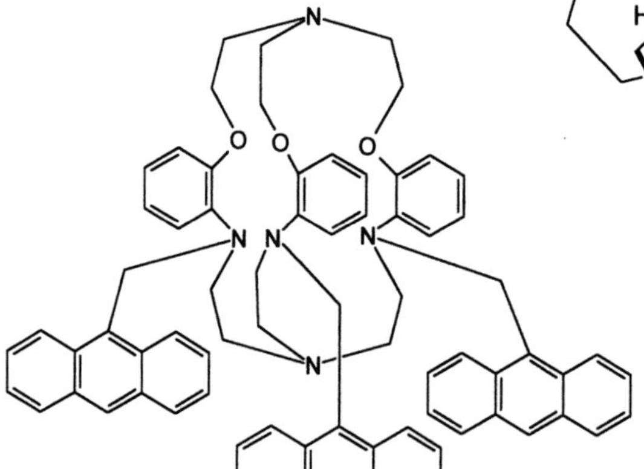
N
O
O
O
N
N
N
N

5

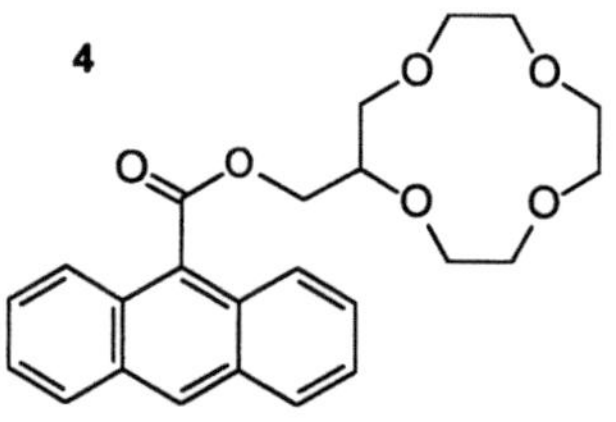
O
O
O
O
O
O

7

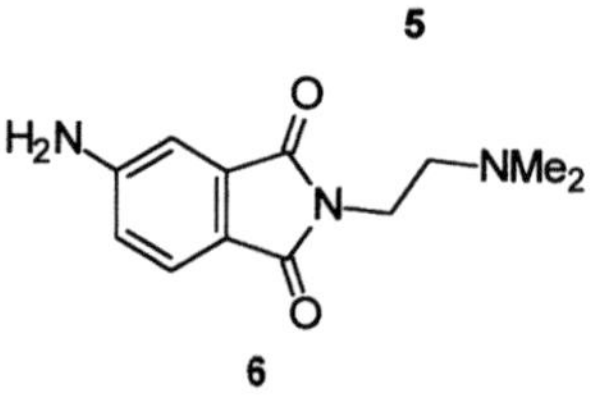
O
H_2N
N
NMe_2
O

6

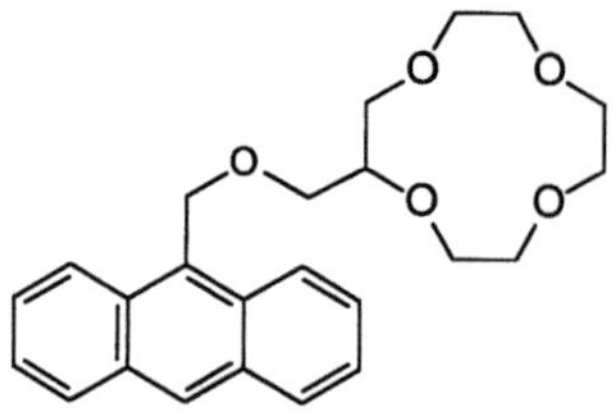
O
O
O
O
O

8

duced during the use of metal salts. This issue appears to have been addressed in the case of **5**, but not yet for **6**.

A pair of "fluorophore-spacer-receptor" systems **7** and **8** studied by Tanaka and co-workers [39] display moderate, but opposite, ion-induced fluorescence intensity effects. Weak PET-type effects from the benzylic oxygen [40] appear to underlie the fluorescence enhancement of **8** caused by Ca^{2+}. The process of rationalizing the behavior of **7** may begin by considering older reports concerning the unusual emission behavior of 9-anthroic acid and 9-anthroate esters [41, 42]. This appears to be caused by rotation of a carboxylate acid/ester moiety in the excited state towards planarity with the anthracene tricycle. The opposite phenomenon, i.e. rotation of groups towards orthogonality, is the hallmark of TICT-type (twisted internal charge transfer) behavior [43, 44]. In contrast, the 9-anthroate anion displays typical anthracene-like fluorescence [41, 42]. Ca^{2+} binds not only to the 12-crown-4 receptor but also to one or both of the ester oxygens as in Gokel's lariat ethers [45] This can influence the excited state planarization process. On a separate point, the fluorophores in **7** and **8** are not as electron deficient in their excited states as in Tanaka's previous case **9** [46]. Hence, PET from counter ions such as SCN^- are not dominant.

Strong fluorescence effects due to non-ionic species are uncommon with the exception of saccharide [47, 48] and steroidal guests [49]. Given this context it is refreshing that the fluorescence of D'Souza's **10** is enhanced by a factor of > 25 upon binding hydroquinone [50]. Charge transfer and hydrogen bonding contribute to the recognition process. The action of **10** can be considered to be more attractive (both theoretically and visually) than the benzoquinone-induced quenching of fluorescence seen in **11** [51], also from the D'Souza stable. Once complexed to hydroquinone, the PET process in **10** is calculated to be endergonic, whereas it is exergonic before complexation. Interestingly, the fluorescence quantum yield of **10** once switched "on" is nearly double that of the model compound *meso*-tetraphenylporphyrin. This can be attributed to the charge transfer interaction between the quinone unit of **10** and the hydroquinone guest. The even higher fluorescence quantum yield of one-electron reduced **10** is offered as supporting evidence. However, this is surprising in view of the marked tendency of radical centers to act as fluorescence quenchers via electron exchange [52, 53].

We close this section by drawing attention to interesting receptors for amino alcohols [54] and group 14 tetramethyls [55] which may have potential for PET sensor development in the footsteps of the aminomethyl boronic acid receptors for glucose [47, 48]. On the other hand, acridone's photostability [56] suggests that it can be accommodated as a fluorophore in robust PET systems.

3
Heteromultireceptor Systems

The previous section focused on fluorophores connected via spacers to one or more similar receptor units. On some of these occasions, these units combined to reversibly capture non-protonic guests. Now we shift our attention to systems equipped with a pair of dissimilar receptors. Such systems can be useful not

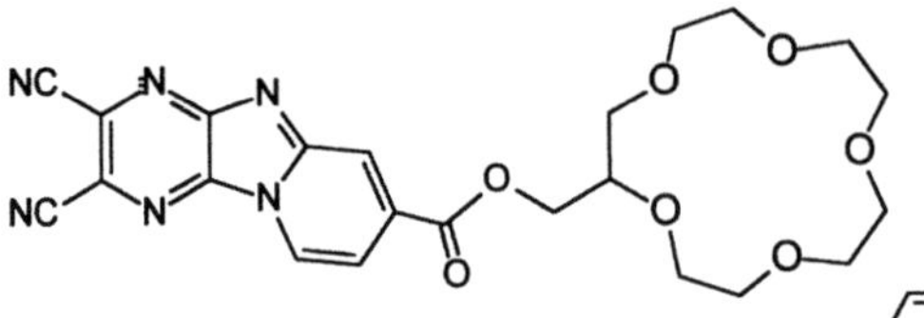

9

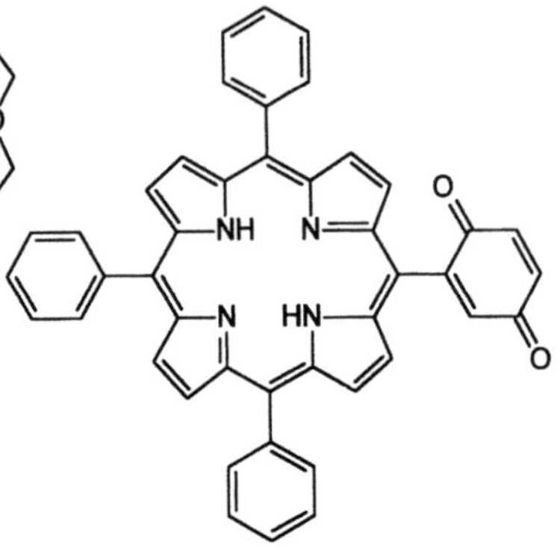

10

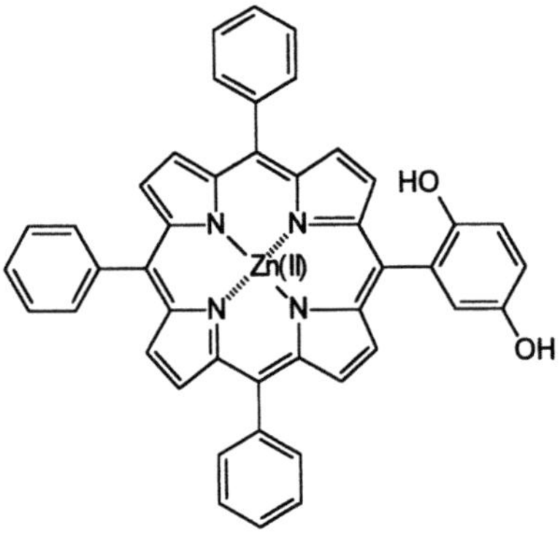

11

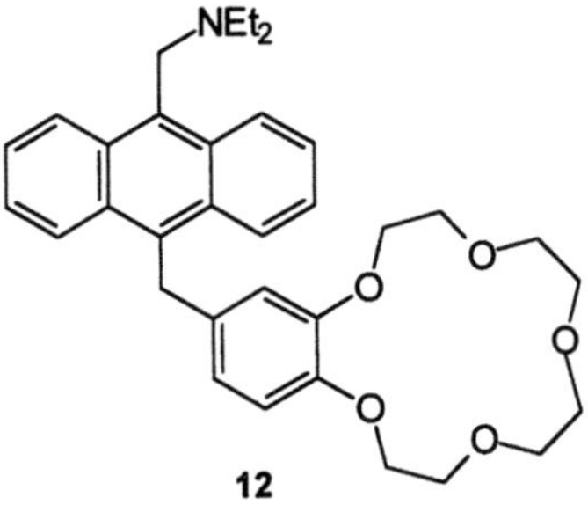

12

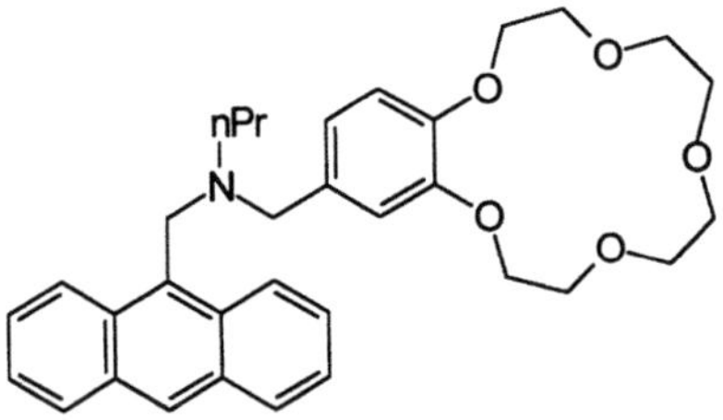

13

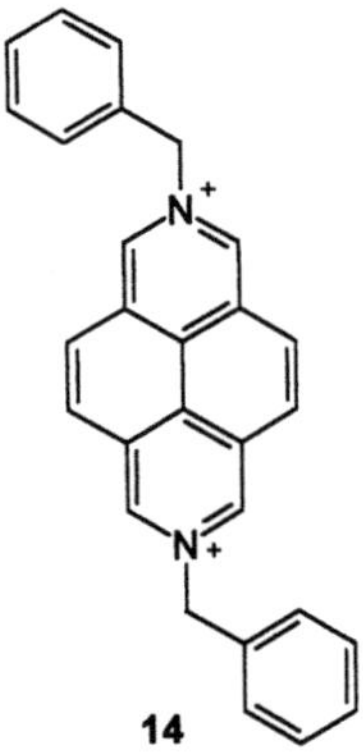

14

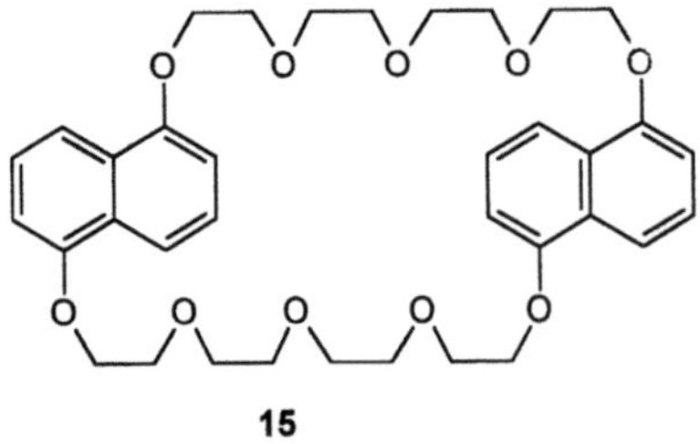

15

only as selective sensors but also as sophisticated switches. Logic gates come into this category and these have the potential for use in molecular computation. For instance, AND logic gates will produce an output signal only if two (or more) input signals are supplied simultaneously. Our **12** behaves as an "off-on" digital AND gate [57] by producing a fluorescence output in response to the simultaneous presence of H^+ and Na^+ inputs at closely separated spatial addresses. It will also act as a sensor for coincident H^+ and Na^+ in a micro-environment. Cross-talk between the two input channels is essentially eliminated by positioning the receptors with mutually exclusive selectivity characteristics on the opposite sides of the fluorophore. Molecules with "all or none" responses [58] are ideal for the development of digital technology. Molecular AND logic gates with a textbook truth table such as **12** are likely to participate in molecular-scale arithmetic operations of the future [59]. AND gate **13**, the forerunner of **12**, introduced the subject of molecular photonic logic [60] but cross-talk prevented attainment of the ideal truth table. Iwata and Tanaka note that **9** [46], mentioned in Sect. 2, also shows a logic response since fluorescence quenching is effected only when Ca^{2+} and SCN^- are present as inputs.

One approach to the execution of molecular arithmetic operations will require XOR logic gates to partner AND gates. An important step has been taken by Balzani and Stoddart towards this objective [61]. The pseudorotaxane formed by the association of the components **14** and **15** is non-fluorescent due to PET-type charge transfer interactions. It is dissociated by complexing **14** with tributylamine or by protonating **15**. If the two inputs are added in stoichiometric amounts, they neutralize each other thus allowing the non-fluorescent pseudorotaxane to reassemble. We look forward to higher generation XOR gates where the inputs can remain independent instead of undergoing mutual annihilation. In the arithmetic scenario, inputs cannot afford to neutralize each other since they must feed an AND gate operating in parallel.

Fluorescent AND logic behavior is also achieved with ammonium and glucose functionalities as inputs by Cooper and James [62]. Their emphasis is to selectively sense the simultaneous occurrence of these two functionalities in the form of protonated glucosamine at pH 7.2. Fortunately, the protonation of the aza-18-crown-6 ether moiety within **16** is moderate enough at this pH value. The excellent design of **16** can be appreciated by the fact that two previously successful fluorescent PET sensing motifs [34, 63, 64] have been integrated within **16** to produce a unique behavior. Very recently, Shinkai and co-workers have also described a fluorescent sensor for sugar uronic acids which employs related ideas [64a]. Though constructed for catalytic applications, Canary's modular heterobireceptor system [65] has the potential to be elaborated into similar fluorescent AND logic gates.

Cases like **12** [57] and **16** [62] are particularly effective for their intended objectives because both uncomplexed receptors in each case are PET active towards the fluorophore. On the other hand, our sensor **17** for γ-aminobutyric acid (GABA) zwitterion [64] is only the first step towards this vital target because the guanidinium receptor for the carboxylate terminal of GABA is devoid of PET capability in these circumstances. The substantial fluorescent signaling action of **17** can be identified with the presence of the aza-18-crown-6 ether unit close to the anthracene fluorophore. This motif has featured in the literature before [64] and after [62] the report on **17**, though with different purposes.

So far this section has dealt with systems containing qualitatively different receptor units. This difference can be relaxed considerably while still preserving relatively complex fluorescence switching functions. Our "off-on-off" sensor **18** [67] illustrates this approach. Its genealogy can be traced to our first fluorescent pH sensors based on 9-dialkylaminomethylanthracene [68]. The new outgrowth involves a pyridyl receptor which develops PET activity upon protonation [69]. The fluorescence of **18** is switched "off" by PET to the pyridinium unit at low pH or from the amino group at high pH. Thus it is clear that **18** and relatives can directly indicate microenvironments whose analyte status is within a chosen concentration window. A similar line of reasoning has independently led S. A. de Silva and his collaborators to an even more efficient example **19** [70]. The shorter spacers within **19** vis-a-vis **18** lead to a faster PET rate to the pyridinium unit and hence to a lower fluorescence signal in the "off" state at low pH. As a bonus, the fluorescence of **19** is switched "on" by Zn^{2+} by virtue of its bis(2-pyridylmethyl)amino N_3 atom array.

Amine and pyridine units also feature within **20**, though it is the terpyridine dicarboxylate entity which will selectively capture Tb^{3+} (except at very low pH) whereas the amino groups will choose H^+. The complexation of Tb^{3+} converts the multireceptor array **20** into a luminescent "off-on" pH sensor [71]. The relatively long emission lifetime of 0.57 ms permits the sensor signal to be disentangled from the fluorescence noise ubiquitous in biomatrices. The sensitivity advantage of time-gated observation of delayed luminescence has been well appreciated by bioanalytical chemists [72]. The terpyridine dicarboxylate moiety within **20** serves in several roles besides that of ligand for the lanthanide ion. It is the photon antenna which compensates for the weak light absorption characteristic of Tb^{3+} and relatives. It is also the shield which prevents excessive hydration of the Tb^{3+} lumophore and subsequent nonradiative deexcitation. It also appears to receive the electron from one of the unprotonated amino groups when driven by its triplet state energy. The efficiency of this PET process involving a triplet state [73] is increased by a lifetime enhancement due to a degree of thermally aided back energy transfer from the $^5D_4$ excited state of Tb^{3+} which lies only 5.0 kcal mol^{-1} below. The Eu^{3+} complex of **20** shows very poor proton-induced luminescence enhancement factors which stands in marked contrast to the corresponding Tb^{3+} complex. The rather low energy of the $^5D_0$ excited state of Eu^{3+} precludes the triplet lifetime enhancement which enabled PET in the case with Tb^{3+}.

We find that the sensory profile of $21.Ln^{3+}$ is inverted relative to its structural cousin $20.Ln^{3+}$. Now the luminescence of the Eu^{3+} complex switches "on" with K^+ whereas the Tb^{3+} complex totally fails [74]. Such a dramatic turnaround appears to be triggered by a relatively small reduction of the triplet state energy of the terpyridyl ester compared to the terpyridyl dicarboxylate by 0.9 kcal mol^{-1}. This increases thermally aided back energy transfer from the $^5D_4$ excited state of the Tb^{3+} to such an extent that the ligand triplet state suffers facile quenching by molecular oxygen. As noted under the discussion of **20**, Eu^{3+} complexes are energetically incapable of participating in such a mechanism in these cases. The PET process in the Eu^{3+} complex of **21** appears to occur from

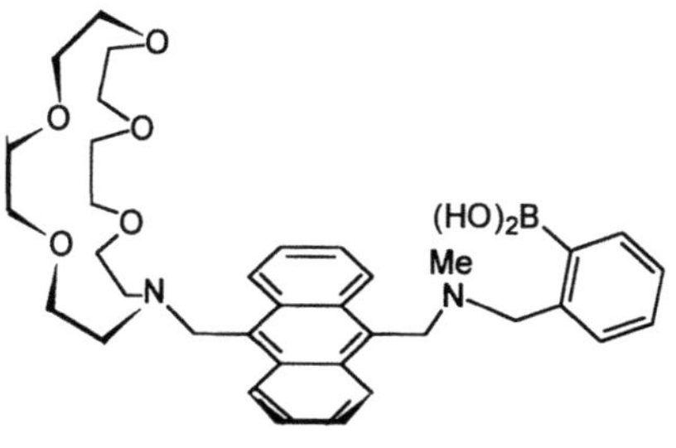

16

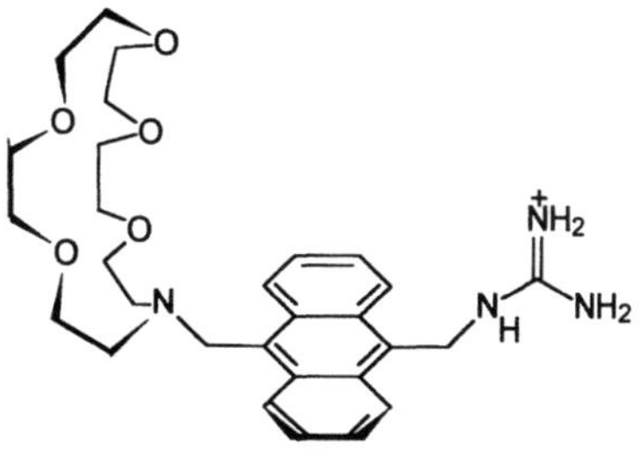

17

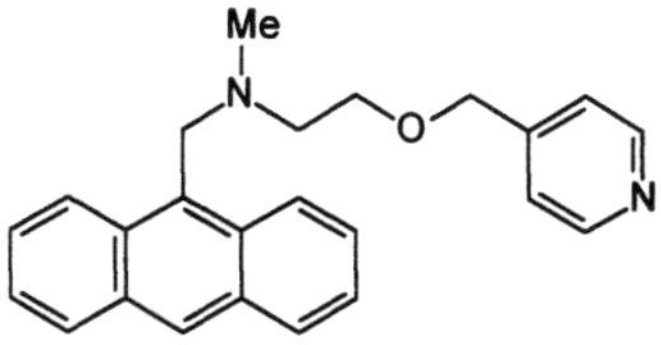

18

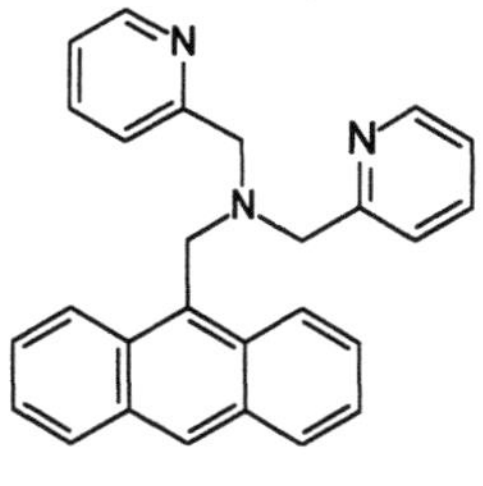

19

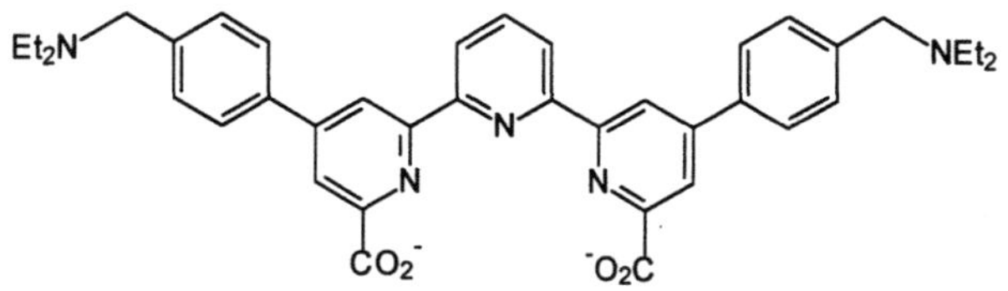

20

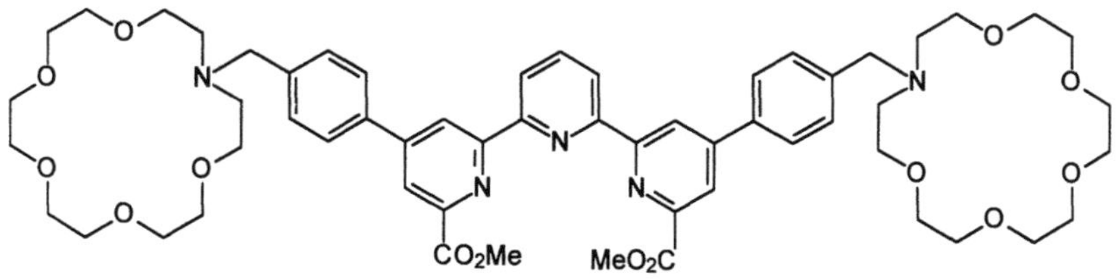

21

one of the aza-18-crown-6 nitrogens to the $^5D_0$ excited state of Eu^{3+}. PET processes from electron-rich aryl moieties to Eu^{3+} have been known for some time [75]. The intervening π-system can serve as a super-exchange medium. The sensing success of $21 \cdot Eu^{3+}$ and the failure of $20 \cdot Eu^{3+}$ are thought to rest on the easier reducibility of the Eu^{3+} center when ligated to neutral esters rather than to carboxylate anions.

22

A rather similar outcome, though assigned to a different reason, is available in Shinkai's **22** [76]. Lanthanide complexes of *22* extend saccharide sensing from fluorescence [47, 48] to delayed luminescence. The Eu^{3+} complex of **22** shows nearly an order-of-magnitude enhancement of luminescence in the presence of fructose whereas the luminescence of the Tb^{3+} version barely doubles. Shinkai and colleagues suggest that different EET routes (from phenacyl or phenoxy groups) are responsible for this divergence. Our conclusions on the results with **21** indicate that the possibility of direct PET to the Eu^{3+} center mediated by the intervening aromatic unit also deserves serious consideration in the case of **22**.

4
Conclusion

Luminescent PET sensors and switches have continued to increase in numbers and scope during the past year. Several previously elusive analytes have been apprehended by purposely designed systems. The selectivity of guest capture and signal transduction has also seen improvement. Fluorescent PET systems have even crossed into lanthanide chemistry with major consequences for interference-free monitoring of various analytes via delayed luminescence. They have consolidated their foothold in the important research field of molecular-scale logic gates and related devices. We see no reason why the trend of growth will not continue.

Acknowledgements. We are grateful to the following institutions for their support of our research efforts: Queen's University of Belfast (Northern Ireland), SERC/EPSRC (UK), Department of Education (Northern Ireland), European Social Fund, IAESTE, NATO (CRG921408 held jointly with J.-P. Soumillion at Universite Catholique de Louvain, Belgium) and Zeneca Colours.

References

1. Photoinduced Electron Transfer. Parts A-D; Fox, M.A.; Chanon, M., Eds.; Elsevier: Amsterdam, 1988
2. Photoinduced Electron Transfer. Parts I-V; Mattay, J., Ed.; Top. Curr. Chem. 1990, 156, 158; 1991, 159; 1992, 163; 1993, 168
3. Kavarnos, G.J. Fundamentals of Photoinduced Electron Transfer; VCH; Weinheim: New York, 1993
4. de Silva, A.P.; Gunaratne, H.Q.N.; Gunnlaugsson, T.; Huxley, A.J.M.; McCoy, C.P.; Rademacher, J.T.; Rice, T.R. Chem. Rev. 1997, 97, 1515
5. Valeur, B. in Topics in Fluorescence Spectroscopy. Vol. 4. Probe Design and Chemical Sensing; Lakowicz, J.R.; Ed.; Plenum, New York, 1994, p. 21
6. Bissell, R.A.; de Silva, A.P.; Gunaratne, H.Q.N.; Lynch, P.L.M.; Maguire, G.E.M.; McCoy, C.P.; Sandanayake, K.R.A.S. Top. Curr. Chem. 1993, 168, 223
7. Czarnik, A.W. Acc. Chem. Res. 1994, 27, 302
8. Bissell, R.A.; de Silva, A.P.; Gunaratne, H.Q.N.; Lynch, P.L.M.; Maguire, G.E.M.; Sandanayake, K.R.A.S. Chem. Soc. Rev. 1992, 21, 187
9. Fabbrizzi, L.; Poggi, A. Chem. Soc. Rev. 1995, 24, 197
10. de Silva, A.P.; Gunaratne, H.Q.N.; Gunnlaugsson, T.; McCoy, C.P.; Maxwell. P.R.S.; Rademacher, J.T.; Rice, T.E. Pure Appl. Chem. 1996, 68, 1443
11. de Silva, A.P.; Gunnlaugsson, T.; Rice, T.E. Analyst 1996, 121, 1759
12. de Silva, A.P.; Gunaratne, H.Q.N.; Gunnlaugsson, T.; Huxley, A.J.M.; McCoy, C.P.; Rademacher, J.T.; Rice, T.E. Adv. Supramol. Chem. 1997, 4, 1
13. Balzani, V.; Scandola, F. Supramolecular Photochemistry; Ellis-Horwood: Chichester, 1991
14. Fluorescent Chemosensors of Ion and Molecule Recognition, ACS Symp. Ser. 538; Czarnik, A.W., Ed.; American Chemical Society: Washington DC, 1993
15. Chemosensors of Ion and Molecule Recognition, NATO ASI-C Ser. Desvergne, J.-P.; Czarnik, A.W., Eds.; Kluwer: Dordrecht, 1997
16. de Silva, A.P.; Gunnlaugsson, T.; McCoy, C.P. J. Chem. Educ. 1997, 74, 53
17. Shen, Y.; Sullivan, B.P. J. Chem. Educ. 1997, 74, 685
18. de Silva, A.P.; McCoy, C.P. Chem. Ind. 1994, 992
19. de Silva, A.P.; Gunaratne, H.Q.N.; Gunnlaugsson, T. Chem. Commun. 1996, 1967
20. Minta, A.; Tsien, R.Y. J. Biol. Chem. 1989, 264, 19449
21. Gocmen, A.; Erk, C. J. Incl. Phenom. Mol. Recognit. Chem. 1996, 26, 67
22. Karsli, N.; Erk, C. Dyes Pigm. 1996, 32, 85
23. Oguz, U.; Akkaya, E.U. Tetrahedron Lett. 1997, 38, 4509
24. Akkaya, E.U.; Turkyilmaz, S. Tetrahedron Lett. 1997, 38, 4513
    Isgor, Y.G.; Akkaya, E.U. Tetrahedron Lett. 1997, 38, 7417
25. Inoue, M.B.; Medrano, F.; Inoue, M.; Raitsimring, A.; Fernando, Q. Inorg. Chem. 1997, 36, 2335
26. Langhals, H.; Demmig, S.; Huber, H. Spectrochim. Acta 1988, 44A, 1189
27. Masilamani, D.; Lucas, M.; Morgan, K. (quoted by Stinson, S.C.) Chem. Eng. News 1987, 65(45), 26
28. Masilamani, D.; Lucas, M.E. in [14], p 162
29. Bernardo, M.A.; Parola, A.J.; Pina, F.; Garcia-Espana, E.; Marcelino, V.; Luis, S.V.; Miravet, J.F. J. Chem. Soc., Dalton Trans. 1995, 993
30. Fabbrizzi, L.; Licchelli, M.; Pallavicini, P.; Perotti, A.; Taglietti, A.; Sacchi, D. Chem. Eur. J. 1996, 2, 75
31. Yoon, J.Y.; Ohler, N.E.; Vance, D.H.; Aumiller, W.D.; Czarnik, A.W. Tetrahedron Lett. 1997, 38, 3845
32. Hancock, R.D.; Martell, A.E. Chem. Rev. 1989, 89, 1875
33. Fabbrizzi, L.; Francese, G.; Licchelli, M.; Perotti, A.; Taglietti, A.; Sacchi, D. Chem. Commun. 1997, 581
34. de Silva, A.P.; Sandanayake, K.R.A.S. Angew. Chem., Int. Ed. Engl. 1990, 29, 1173

35. James, T.D.; Sandanayake, K.R.A.S.; Shinkai, S. Angew. Chem., Int. Ed. Engl. 1994, 33, 2207
36. De Santis, G.; Fabbrizzi, L.; Licchelli, M.; Poggi, A.; Taglietti, A. Angew. Chem., Int. Ed. Engl. 1996, 35, 202
37. Ghosh, P.; Bharadwaj, P.K.; Mandal, S.; Ghosh, S. J. Am. Chem. Soc. 1996, 118, 1553
38. Ramachandram, B.; Samanta, A. Chem. Commun. 1997, 1037
39. Iwata, S.; Matsuoka, H.; Tanaka, K. J. Chem. Soc., Perkin Trans. 1 1997, 1357
40. Castellan, A.; Lacoste, J.-M.; Bouas-Laurent, H. J. Chem. Soc., Perkin Trans. 2 1979, 411
41. Werner, T.C.; Hercules, D.M. J. Phys. Chem. 1969, 73, 2005
42. Werner, T.C. in Modern Fluorescence Spectroscopy. Vol. 2; Wehry, E.L., Ed.; Plenum: New York, 1977, p 277
43. Rettig, W. Angew. Chem., Int. Ed. Engl. 1986, 25, 971
44. Rettig, W. Top. Curr. Chem. 1994, 169, 253
45. Gokel, G.W. Crown Ethers and Cryptands; Royal Society of Chemistry: Cambridge, 1991
46. Iwata, S.; Tanaka, K. J. Chem. Soc., Chem. Commun. 1995, 1491
47. Sandanayake, K.R.A.S.; James, T.D.; Shinkai, S. Pure Appl. Chem. 1996, 68, 1207
48. James, T.D.; Sandanayake, K.R.A.S.; Shinkai, S. Angew. Chem., Int. Ed. Engl. 1996, 35, 1910
49. Hamada, F.; Minato, S.; Osa, T.; Ueno, A. Bull. Chem. Soc. Jpn. 1997, 70, 1339
50. D'Souza, F.; Deviprasad, G.R.; Hsieh, Y.Y. Chem. Commun. 1997, 533
51. D'Souza, F. J. Am. Chem. Soc. 1996, 118, 923
52. Blough, N.V.; Simpson, D.J. J. Am. Chem. Soc. 1988, 110, 1915
53. Bystryak, I.M.; Likhtenshtein, G.L.; Kotelnikov, A.I.; Hankovskii, H.O.; Hideg, K. Zh. Fiz. Khim. 1986, 60, 2796
54. Schrader, T. Angew. Chem., Int. Ed. Engl. 1996, 35, 2469
55. Garel, L.; Dutasta, J.-P.; Collet, A. New J. Chem. 1996, 20, 1265
56. Bahr, N.; Tierney, E.; Reymond, J.-L. Tetrahedron Lett. 1997, 38, 1489
57. de Silva, A.P.; Gunaratne, H.Q.N.; McCoy, C.P. J. Am. Chem. Soc. 1997, 119, 7891
58. Kimura, K.; Mizutani, R.; Yokoyama, M.; Arakawa, R.; Matsubayashi, G.; Okamoto, M.; Doe, H. J. Am. Chem. Soc. 1997, 119, 2062
59. Ball, P. New Scientist 1997, 2 August, 32
60. de Silva, A.P.; Gunaratne, H.Q.N.; McCoy, C.P. Nature 1993, 364, 42
61. Credi, A.; Balzani, V.; Langford, S.J.; Stoddart, J.F. J. Am. Chem. Soc. 1997, 119, 2679
62. Cooper, C.R.; James, T.D. Chem. Commun. 1997, 1419
63. James, T.D.; Sandanayake, K.R.A.S.; Shinkai, S. J. Chem. Soc., Chem. Commun. 1994, 477
64. de Silva, A.P.; de Silva, S.A. J. Chem. Soc., Chem. Commun. 1986, 1709
    Takeuchi, M.; Yamamoto, M.; Shinkai, S. Chem. Commun. 1997, 1731
65. dos Santos, O.; Lajmi, A.R.; Canary, J.W. Tetrahedron Lett. 1997, 38, 4383
66. de Silva, A.P.; Gunaratne, H.Q.N.; McVeigh, C.; Maguire, G.E.M.; Maxwell, P.R.S.; O'Hanlon, E. Chem. Commun. 1996, 2191
67. de Silva, A.P.; Gunaratne, H.Q.N.; McCoy, C.P. Chem. Commun. 1996, 2399
68. de Silva, A.P.; Rupasinghe, R.A.D.D. J. Chem. Soc., Chem. Commun. 1985, 1669
69. de Silva, A.P.; Gunaratne, H.Q.N.; Lynch, P.L.M. J. Chem. Soc., Perkin Trans. 2 1995, 685
70. de Silva, S.A.; Zavaleta, A.; Baron, D.E.; Allam, O.; Isidor, E.V.; Kashimura, N.; Percarpio, J.M. Tetrahedron Lett. 1997, 38, 2237
71. de Silva, A.P.; Gunaratne, H.Q.N.; Rice, T.E. Angew. Chem., Int. Ed. Engl. 1996, 35, 2116
72. Hemmila, I. Applications of Fluorescence in Immunoassays; Wiley: New York, 1991
73. Bissell, R.A.; de Silva, A.P. J. Chem. Soc., Chem. Commun. 1991, 1148
74. de Silva, A.P.; Gunaratne, H.Q.N.; Rice, T.E.; Stewart, S. Chem. Commun. 1997, 1891
75. Abu-Saleh, A.; Meares, C.F. Photochem. Photobiol. 1984, 39, 763
76. Matsumoto, H.; Ori, A.; Inokuchi, F.; Shinkai, S. Chem. Lett., 1996, 301

CHAPTER 10

Photostability of Fluorescent Dyes for Single-Molecule Spectroscopy:

Mechanisms and Experimental Methods for Estimating Photobleaching in Aqueous Solution

C. Eggeling, J. Widengren, R. Rigler, C. A. M. Seidel

1
Introduction

Laser-induced fluorescence detection is used for various ultrasensitive analytical techniques in chemistry, biology, and medicine by probing reagents that are either autofluorescing or tagged with a fluorescent dye molecule [1–29]. Applying different microscopic techniques with tight spatial and spectral filtering, various groups have directly visualized a variety of single fluorescent dye molecules (rhodamines [5–7, 15, 16, 30–32] and coumarins [17, 18, 33]) dissolved in liquids by using coherent one- and two-photon excitation. Due to the high sensitivity and specificity, fluorescence is probably the most important optical readout mode in biological, scanning confocal microscopy. These unique features of fluorescence are critically dependent on the availability of appropriate fluorophores [34–40]. Photophysical parameters in general, and photobleaching or the photostability in particular, play a very important role not only for the accuracy of single-molecule detection (SMD) by laser-induced fluorescence [17, 18, 41–44] and in dye laser chemistry [45, 46], but in practically all applications of fluorescence spectroscopy, where a high sensitivity or a high signal rate is crucial [47]. Key properties for the suitability of a fluorescent dye are its absorption coefficient, fluorescence-, triplet-, and photobleaching quantum yield [17, 18, 41–44, 47–53]. Photobleaching is a dynamic, mostly irreversible process in which fluorescent molecules undergo photoinduced chemical destruction upon absorption of light, thus losing their ability to fluoresce. Hence, the statistical accuracy of the detection is limited, because for every absorption process there is a certain fixed probability Φ_b of the molecule to be bleached. The probability P to survive n absorption cycles and become bleached in the $(n + 1)^{th}$ cycle is given by:

$$P\,(\text{survival of } n \text{ absorptions}) = (1 - \Phi_b)^n\,\Phi_b \qquad (1)$$

This distribution is called geometric distribution, which is the discrete counterpart of the exponential distribution. Due to this exponential nature of the photodestruction process with the standard deviation $(1 - \Phi_b)/\Phi_b$ the relative fluctuation of the number of detected photons due to a single molecule transit can be as high as 100% [54, 55]. The mean number of survived absorption cycles μ is equal to the standard deviation shown in Eq. 2.

$$\mu = \frac{(1 - \Phi_b)}{\Phi_b} \approx \frac{1}{\Phi_b} \tag{2}$$

Despite the central importance of good fluorophores, information on the photophysical parameters is sparse, and little is known about how to rationally design good dyes. One reason is that some of the relevant parameters that influence the photostability are not known or have not yet been determined in enough detail. The quantum yield of triplet formation, the triplet lifetime, and absorption properties of excited electronic states belong to this category. Experimental techniques, which give information on populated transient species, are transient absorption [56–60] and fluorescence correlation spectroscopy [49, 50, 61, 62]. It is a problem that the above parameters tend to be sensitive to the environment, making it in some cases difficult to compare different investigations and to predict the properties for a certain environment. In particular, data in water, the medium of central importance in SMD and confocal microscopy, are difficult to obtain by conventional techniques due to the limited solubility of many organic fluorescent dyes.

The objective of this paper is the formal and quantitative description of photobleaching based on clear definitions. We will start with a brief overview of the molecular and kinetic aspects of photobleaching reactions to define the area to be covered here (Sects. 2 and 3). Subsequently a simple formalism will be developed to calculate the fluorescence signal of a dye excited by a focused laser beam and to point out the importance of appropriate photophysical parameters for an efficient SMD (Sect. 3). Section 4 will discuss the experimental techniques and will give the theory for data analysis of photobleaching. In Sects. 5 and 6 we attempt to tabulate quantitative information for future work on all aspects of photobleaching relevant to single molecule spectroscopy in aqueous solution. Finally, we present some experimental results on strategies to minimize photobleaching. We hope that this presentation will serve as a guide to experimental work in new directions, since photobleaching is the ultimate limit of fluorescence-based single-molecule spectroscopy.

2
Photobleaching: An Overview

2.1
Definitions

In order to quantify the photobleaching of a dye solution, we introduce two parameters – the quantum yield of photobleaching Φ_b and the photobleaching probability p_b. Usually, photophysical and photochemical reactions are characterized by a quantum yield. While the chemical reaction yield depends on the total equivalents of reactant, the quantum yield depends on the number of photons absorbed by this reactant. By definition, the quantum yield of the photobleaching reaction is the number of molecules which have been photobleached divided by the total number of photons absorbed during the same time interval (Eq. 3) [63].

$$\Phi_b = \frac{\text{number of photobleached molecules}}{\text{total number of absorbed photons}} . \tag{3}$$

It is important to note that the total number of absorbed photons also includes those photons absorbed in a second step leading to higher excited electronic states, which open up new bleaching channels [17, 44, 63–68]. The definition of the quantum yield Φ_b (Eq. 3) does not make sense for SMD experiments, since the total number of absorbed photons cannot be measured for a single molecule or generally for a low dye concentration. However, since the fluorescence emission is usually related to the first excited singlet state S_1, it is crucial for the accuracy of experiments like single-molecule fluorescence detection to know the probability of photobleaching p_b at a certain applied excitation irradiance [44, 52]. This is equal to the number of photobleached molecules divided by the mean number of molecules in the S_1 state for a given time interval (Eq. 4).

$$p_b = \frac{\text{number of photobleached molecules}}{\text{mean number of molecules in the } S_1} . \tag{4}$$

Solely in the case of a single-quantum photoreaction the two parameters are equivalent, whereas there are marked differences for two- or multi-quantum reactions as shown below.

Unfortunately, the total number of absorbed photons, as well as the mean number of molecules in the S_1 state, cannot be measured directly and precisely under the experimental conditions of SMD, making it impossible to determine Φ_b and p_b directly. However, the above definitions can be expressed in kinetic terms, that is, as a function of rate constants for the different deactivation processes of the electronically excited state from which the photochemical reaction occurs. If a dye solution is illuminated, it is possible to measure the number of irreversibly photobleached molecules as a decrease in the dye concentration $c(t)$ with time t. Since under appropriate conditions the rate of the decrease of $c(t)$ at time t is proportional to $c(t)$, the photobleaching reaction can be treated as a quasi-unimolecular reaction, characterized by an effective pseudo first-order bleaching rate constant k_z [18, 41, 44, 48].

$$dc(t)/dt = - k_z c(t) \tag{5a}$$

This bleaching reaction results in an exponential decrease in the dye concentration $c(t)$ with the initial concentration $c(0)$ at time $t = 0$.

$$c(t) = c(0) \exp(- k_z t) \tag{5b}$$

In the following section, expressions are successively derived to describe the dependence of the experimentally accessible parameter k_z on the cw irradiance. In this way the photobleaching parameters Φ_b and p_b for an evaluated (assumed) certain number of absorbed photons can be calculated.

2.2
Photobleaching Reactions

Fluorescent molecules emit only a limited, dye-specific number of photons before they are photobleached. There are multiple chemical reaction pathways for the alteration of the chromophore. Some of them only change the fluorescence properties (e.g. dealkylation reactions), while others produce nonfluorescent species. In this paper we focus on irreversible photobleaching reactions. An overall review of photochemical reactions is available [69–72].

Photobleaching reactions can be discussed in three categories where the fluorescent molecule M may either be in the first excited electronic singlet $^1M_1$ or in the triplet state $^3M_1$. Thereby the triplet state, $^3M_1$, is of major importance for the photostability of a fluorescent molecule, which will be discussed at the end of this chapter and in Sect. 6.1.

Unimolecular, First-Order Reactions. Dependending on the individual molecular potential energy surfaces of an electronically excited molecule, these reactions are important for vibronically excited (competition with relaxation) as well as for relaxed states (thermally activated reactions). Therefore, two reaction classes must be considered which depend solely on the properties of the fluorophore and the solvent environment.

(1) Rearrangement and isomerization reactions [70];

$$^{(1,3)}M_1 \rightarrow X$$

Typical examples are *cis-trans* interconversions in double-bonded systems like stilbenes [73–80], polymethine dyes [81–86] and Vitamin A.

(2) Part of the second class of reactions are photochemical fragmentation reactions such as homo- and heterolytic dissociation reactions producing radicals $(X^{\bullet}, Y^{\bullet})$ or ions [ion pair (M^+e^-)], respectively, [69] as well as elimination reactions [87, 88].

$$^{(1,3)}M_1 \rightarrow X^{\bullet} + Y^{\bullet} \text{ or } M^+ e^-$$

These single-quantum dissociation reactions are important decomposition pathways for the fluorescent aromatic amino acids tyrosine and tryptophan [89–91].

Multi-Photon Photolysis. In SMD and confocal microscopy high irradiances must be applied which result in an increasing probability of multi-photon absorption with a high probability of a photobleaching reaction from the excited electronic state. Thereby, subsequent heterolytic dissociation forming a radical ion pair (M^+e^-) is especially favored in polar solvents like water. Two-quantum photoexcitation can be achieved in two ways, which must be carefully distinguished.

(1) Two-step excitation as a result of successive absorption of two photons via a real intermediate state (e.g. S_1 or T_1) [92].

$$^1M_0 \xrightarrow{h\nu} {}^{(1,3)}M_1 \xrightarrow{h\nu} {}^{(1,3)}M_n \longrightarrow (M^+e^-)$$

(2) Two-photon absorption as a result of a simultaneous absorption of two photons via a virtual state [93], whereby the photon energy determines the energy of the state reached. Two-photon excitation (TPE) in the NIR (700–1100 nm) with fs lasers is widely used for fluorescence imaging [94, 95].

However, two-photon excitation, especially in the UV, can produce highly excited electronic molecular states close to or even above the ionization threshold of 7 to 8 eV [65].

$$^1M_0 \xrightarrow{\ 2\,h\nu\ } {}^1M_n \longrightarrow M^+e^-$$

More details on multi-photon photolysis are given in Sect. 6. General follow-up reactions of radical cations have been reviewed in [96].

Bimolecular Reactions. This section covers the most relevant photoinduced reactions, proceeding via the formation of an encounter complex. In variable distances all types of chemical reactions can subsequently take place [70, 71], such as energy transfer with follow-up reactions (sensitized reactions), electron transfer, proton transfer, coupled electron and proton transfer, short-range electron exchange energy transfer, and addition reactions.

$$^{(1,3)}M_1 + R \ \rightarrow \ X + Y \quad \text{or} \quad {}^{(1,3)}M_1 + R \ \rightarrow \ Z$$

The reaction partner R may be a solvent molecule, an added reactant (e.g. a quencher, a proton donor etc.), an impurity, another dye molecule, or dissolved atmospheric oxygen. The latter is probably being the most important reactant. Upon photon absorption, bimolecular photobleaching reactions and also two-step excitation compete with other molecular deactivation processes. Thus, most of these reactions of electronically excited molecules take place in long-lived (vibronically relaxed) states, i.e. in the first excited singlet state S_1 or in the first excited triplet state T_1. Compared to singlet states, which have short lifetimes in the order of 1 to 20 ns for most of the organic fluorophores, the "triplet trap" is especially "dangerous" for the molecule due to its long lifetime of several µs in an air-saturated aqueous solution. Thus, the triplet state is very important in photobleaching reactions (see also Sect. 6.1).

The kinetic treatment of bimolecular photodegradation processes can depend strongly on the local and global concentrations of the reactants involved. Several different cases can be distinguished that deserve attention:

(1) A pseudo first-order reaction can be expected for the case of a bimolecular reaction if there is an excess concentration of the reaction partner.
(2) In second-order reactions, a time-dependent concentration of both reactants is found. Several mechanisms are important in photochemistry. The excited dye molecule may react with another dye molecule being either in an electronically excited state or in its ground state, e.g. self quenching, triplet-triplet annihilation, and radical ion pair formation [70, 71, 97–99], or react with other partners such as oxygen.
(3) The photobleaching can be governed by combined (complex) kinetic laws such as: (a) simultaneous second- and first-order reaction [100]; (b) equilibrium between a local depletion of the reactant and the recovery by diffu-

sion (e.g. local consumption of oxygen) [101–103]; (c) heterogeneous chemical microenvironments resulting in different fluorophore populations with specific properties.

The reactions (3a–c) play a very important role in microscopy, where non-single-exponential photobleaching decays are widely found [98, 99, 104–106]. Sometimes the bleaching characteristics are even different from one sample to another [98]. This more complex kinetic behavior can be rationalized in view of the high density and the heterogeneous microenvironment of the fluorophore when bound to targets of interest (such as DNA, RNA, protein, or other cellular components). Furthermore, such photoreactions are frequently observed in solution experiments with higher dye concentrations (in the order of at least 10 µM), such as in preparative organic photochemistry [107] or, unintentionally, in mechanistic studies using transient absorption spectroscopy [97, 108–110].

2.3
Solvent Dependence

In many cases the solvent has an important influence on the photophysical and photochemical properties of fluorescent dyes [111–114]. One effect is related to the influence of the solvent on the molecular energy levels of the dye. Besides non-specific solute-solvent interactions due to dielectric interactions, static specific interactions such as hydrogen bonding play an important role. Since the properties of the electronic states are determined by individual electronic charge distributions, the energies of the states may be shifted differently if the solvent is changed. Thus, all radiative, internal conversion, and intersystem crossing processes between two states may be altered; new reaction channels may be opened or others may become impossible. This means that the fluorescence quantum and triplet quantum yield of many dyes is critically solvent dependent. Water in particular is a very polar solvent with hydrogen-donating and -accepting properties and thus favors polar states due to high solvation energies (see also Sect. 5).

Energetic high frequency vibrational modes of exchangeable hydrogen atoms at heteroatoms also cause, to some extent, non-radiative depopulation of the S_1 state, which is partly responsible for the drop in the fluorescence quantum yield of dyes emitting in the red or NIR spectral region (besides the energy gap law). The use of deuterated solvents increases the fluorescence quantum yield by approx. 5%. For further details see [115, 116].

The solvent viscosity has an important influence on isomerization reactions and internal conversion processes coupled with large amplitude motions. Furthermore, the viscosity of the solvent determines the diffusion limited rate constant of bimolecular reactions, e.g. quenching of the triplet state by oxygen. Finally, the oxygen concentration in different solvents can vary by at least one order of magnitude due to the solvent-specific solubility of atmospheric oxygen [58].

In view of the various solvent effects, it is not surprising that the photostabilities in most cases are very solvent dependent. In general, the photostability of many organic dyes in organic solvents is higher than in water. Widengren et al. [43] reported an approximately 1.5 times higher photostability of Rhodamine 6G in ethanol than in water for very high excitation irradiances ($I > 300$ kW/cm^2). Soper et al. [117] determined the quantum yield of photobleaching Φ_b for some rhodamines in ethanol and in water. Thereby, they found an increase of Φ_b of 1 to 2 orders of magnitude in water. For dyes excited in the UV region, such as coumarin, pyrene derivatives, or aromatic amino acids, this increase is even greater [118].

2.4
Covered Topics

Since we are interested in single-molecule spectroscopy, where ultra-low concentrations of the fluorophore in water (< 10 pM) are used, we confine ourselves to an overview of photoreactions occurring under these conditions. For a more general case, further investigations (e.g. variation of the dye concentration, addition of selective triplet quenchers, solvent variation) are necessary to give more insights into the mechanisms. In the following, we will look more closely at first-order photoreactions and at the dependence of the fluorescence on the laser power to develop strategies to increase the photostability by varying the oxygen concentration, adding stabilizer, or changing the excitation conditions (wavelength, pulse length, one-photon or two-photon excitation) for very dilute aqueous dye solutions.

3
Fluorescence Dependence on Photophysical and Photochemical Parameters

In single-molecule spectroscopy, two photodynamic properties of the fluorophores – photostability and fluorescence saturation – impose limitations on the achievable fluorescence flow and the resulting signal-to-background ratio.

Figure 1 A shows the electronic energy diagram of a dye molecule with five electronic levels, ground singlet state, S_0, first excited singlet state, S_1, lowest excited triplet state, T_1, and higher excited singlet and triplet states, S_n and T_n. Photobleaching reactions are assumed to be possible from all excited states with the microscopic photobleaching rate constants k_{bS}, k_{bT}, k_{bSn}, and k_{bTn}, respectively. Thereby, it is assumed that the molecules, once they have been photobleached, can no longer participate in the excitation-emission cycles (photochemically irreversible bleaching).

The rate constants for excitation from a state i to a state f are proportional to the irradiance I [W/cm^2] and to the absorption cross section $\sigma_{if}(\lambda)$ [cm^2] at a wavelength λ (Eq. 6),

$$k_{(T)if}(\lambda) = I\,\sigma_{(T)if}(\lambda)\,\gamma \tag{6}$$

where $\gamma = \lambda/(h\,c_1)$ (h is the Planck constant and c_1 the velocity of light).

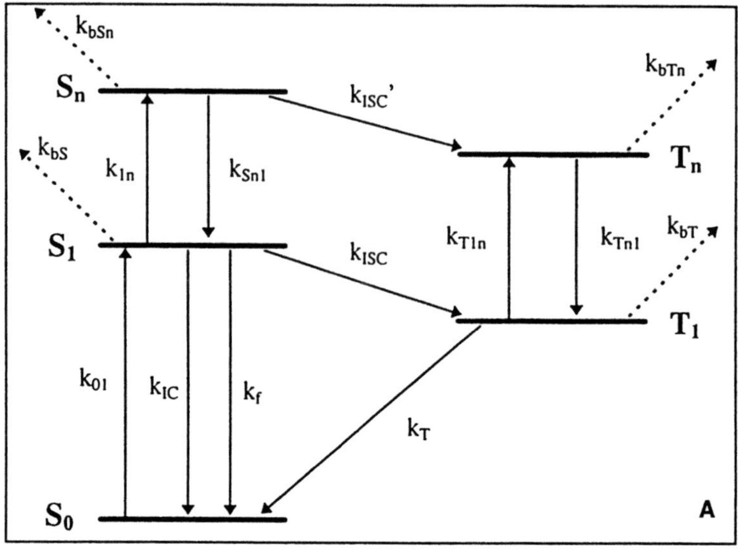

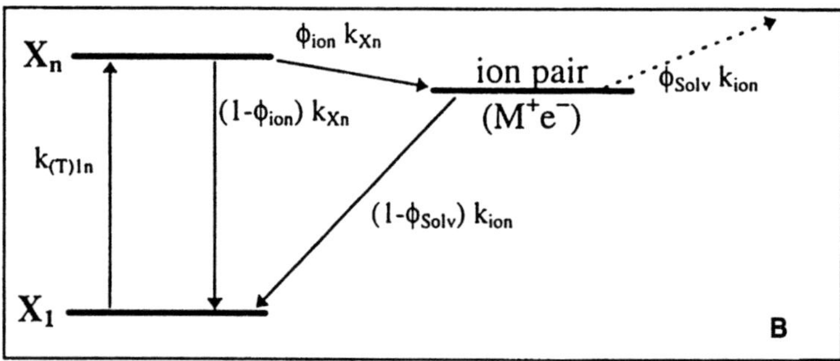

Fig. 1 A, B. (A) Electronic energy diagram of a dye molecule with five electronic levels, regarding photobleaching from every electronic level (details see text). (B) Basic model of the photobleaching reaction from a higher electronic excited state X_n (X: S or T) involving the formation of an ion pair (M^+e^-) and following excitation from the first electronic excited state X_1

k_f, k_{IC}, and k_{ISC} are the rate constants for depopulation of S_1 by fluorescence emission, internal conversion, and intersystem crossing to T_1, respectively, resulting in a lifetime τ_0 of the S_1 state (fluorescence lifetime) (Eq. 7).

$$\tau_0 = 1/k_0 = 1/(k_f + k_{IC} + k_{ISC}) \tag{7}$$

The depopulations of T_1 to S_0, S_n to S_1, and T_n to T_1', are described by the rate constants k_T, k_{Sn1} and k_{Tn1}, respectively. The rate constant k_{ISC}' for intersystem crossing from S_n to T_n is ignored here due to the short lifetimes of these states.

The mean population of the S_1 state for a 2-level system (S_0, S_1), a 3-level system (S_0, S_1, T_1) and a 5-level system (S_0, S_1, S_n, T_1, T_n) upon cw excitation will be derived in order to study the influence of the selected system on the theoretically achievable fluorescence count rate. Depending on the triplet parameters and the applied irradiance, different level schemes should be used. While for a low irradiance a 3-level system is sufficient to describe the fluorescence signal, a 5-level system is necessary for higher irradiances, because the population of S_n and T_n due to two-step excitation becomes important. This opens up new bleaching pathways, because in polar solvents like water the higher electronic states, S_n and T_n, of organic molecules couple efficiently with ionic states due to the high solvation energy [63–66, 119], as illustrated in the inset of Fig. 1B, where X represents either the singlet state S or the triplet state T.

S_1 **Steady State Concentrations.** The kinetic behavior of the electronic level system of a fluorescent dye upon excitation can be expressed by first-order differential equations for each respective level model. Under cw excitation the system can be described by steady state population probabilities S_{1eq} of a dye molecule to be in S_1 [44, 50]. The obtained steady state population probabilities S_{1eq}^2, S_{1eq}^3, and S_{1eq}^5, for a 2-level, 3-level, and a 5-level system, respectively, are given in Eqs. 8–10:

$$S_{1eq}^2 = \frac{k_{01}}{k_{01} + k_0} \tag{8}$$

$$S_{1eq}^3 = \frac{k_{01}\, k_T}{k_{01}\,(k_{ISC} + k_T) + k_0\, k_T} \tag{9}$$

$$S_{1eq}^5 = k_{Tn1}\, k_{Sn1}\, k_{Sn1}\, k_T\, k_{01}/X \tag{10}$$

with $X = k_{Tn1}\,[k_T\,(k_{Sn1}\,(k_0 + k_{01}) + k_{01}\, k_{1n})] + (k_{T1n} + k_{Tn1})\,(k_{ISC}\, k_{Sn1}\, k_{01})$.

Fluorescence saturation follows from the fact that a molecule cannot be in an electronically excited state and in the ground state at the same time; i.e. a single molecule can emit only a limited number of fluorescence photons in a certain time interval. Thus, the saturation characteristics of the fluorescence flow are determined by ground state depletion due to the finite excited state lifetimes of the S_1 and T_1 state, as well as by the triplet quantum yield (see below). However, methods like Raman or resonance-Raman spectroscopy populating ultra short-lived excited (virtual) states can, in principle, achieve even higher signal rates than fluorescence, if the excitation cross section for this pump process is high enough. The detection of single Rhodamine 6G molecules adsorbed on selected nano-particles using surface-enhanced resonance-Raman scattering (SERS) is possible due to enhancement factors of 10^{14} to 10^{15}, which raise the absorption cross sections of approx. 10^{-30} cm^2 known from conventional Raman spectroscopy to values comparable with those of allowed electronic S_0–S_1 transitions (10^{-16} cm^2) [120, 121].

To study the important effect of the triplet state on the achievable signal of a fluorophore, the population probabilities S_{1eq} of a 2-level (without the triplet state) and 3-level system must be compared [56, 57]. Taking the rate constants of the dye Rhodamine 6G as an example (except k_{ISC}), Fig. 2A shows the calculated relative fluorescence flow $F_{eq}^3/F_{eq}^2 = (1 - \Phi_{isc})\, S_{eq}^3/S_{eq}^2$ of the 3-level system nor-

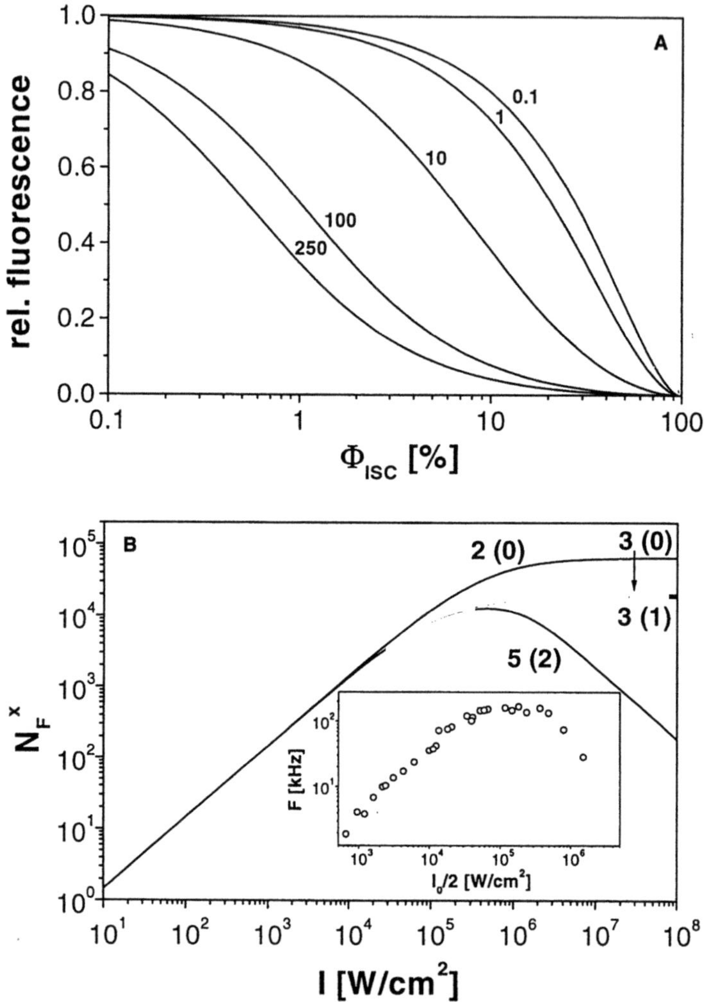

Fig. 2 A, B. (A) Calculated relative fluorescence flow F^3_{eq}/F^2_{eq} of a dye molecule considering triplet state population (electronic 3-level system) normalized to the case without triplet state population (electronic 2-level system) as a function of the triplet quantum yield $\Phi_{ISC} = k_{ISC}/k_0$ for five different excitation irradiances given in kW/cm² (fixed parameters: $k_0 = 2.5 \times 10^8$ s⁻¹ + k_{ISC}, σ_{01} (514.5 nm) = 22.2×10^{-17} cm², $k_T = 5 \times 10^5$ s⁻¹). (B) Calculated mean number of emitted fluorescence photons N^x_F from a single dye molecule (Rhodamine 6G) following cw excitation. Three different electronic level systems are considered: a 2-level system without photobleaching (2 (0)), a 3-level system with (3 (1)) and without photobleaching (3 (0)), and a 5-level system with photobleaching (5 (2)). The values are calculated from Eq. 19 using the following parameters: $\Phi_f = 0.98$, σ_{01} (514.5 nm) = 22.2×10^{-17} cm², $k_0 = 2.56 \times 10^8$ s⁻¹, $k_{ISC} = 1.1 \times 10^6$ s⁻¹, $k_T = 4.9 \times 10^5$ s⁻¹, $k_b = 310$ s⁻¹, σ_{1n} (514.5 nm) = 0.77×10^{-17} cm², σ_{S1n} (514.5 nm) = 0.77×10^{-17} cm², σ_{T1n} (514.5 nm) = 3.85×10^{-17} cm², $1/k_{Sn1} = 1/k_{Tn1} = 200$ fs, $p_{bXn} = 5.5 \times 10^{-5}$ and $t_t = 0.26$ ms. (Inset B) Average fluorescence flow of Rhodamine 6G (ca. 10^{-9} M) in water measured at a confocal epi-illuminated microscope for excitation by a cw argon ion laser at 528 nm as a function of the applied irradiance $I_0/2$. The focal area was determined by FCS at low irradiance ($\omega_0 = 0.51$ µm, $t_t = 0{,}26$ ms)

malized by the flow of the 2-level system as a function of the triplet quantum yield $\Phi_{ISC} = k_{ISC}/k_0$ for five different excitation irradiances given in kW/cm².

Triplet quantum yields of less than 10% result in no significant drop of the relative fluorescence flow for low irradiances. However, the importance of the triplet parameters becomes apparent at high irradiances in the order of 100 to 250 kW/cm² typical for SMD and confocal microscopy, where even a very low triplet quantum yield in the order of 1% causes a significant fluorescence decrease of approx. 50%.

Photobleaching Models. Since the rate constants of photobleaching are usually much smaller than the other relevant rate constants, it is most convenient to use the steady-state probabilities S_{0eq}, S_{1eq} and T_{1eq} for the analysis of photobleaching [43, 44, 48]. Using this kinetic model, the whole electronic level system acts as a 1-level system with respect to photobleaching from any state, allowing no selective depopulation of a special electronic state. Depending on the applied irradiance, two kinetic cases for photobleaching should be distinguished.

A 3-level system is sufficient for low irradiances. Because the probability of a molecule to be in a first excited electronic state, S_1 and T_1, is very low, the absorption of a second photon by S_1 and T_1 ($k_{(T)1n}$) can be neglected. If bleaching occurs only from S_1 and T_1, the effective bleaching rate constant k_z^3 is proportional to the steady-state concentration S_{1eq}^3 (Eq. 9) and depends on the microscopic rate constants k_{bS} and k_{bT}, as well as the triplet parameters. This dependence can be expressed by the composite microscopic rate constant k_b of photobleaching from S_1 and T_1 (Eq. 11).

$$k_z^3 = k_b S_{1eq}^3 \quad \text{with } k_b = (k_{bS} + k_{bT}\, k_{ISC}/k_T) \tag{11}$$

Using Eqs. 3 and 4, equivalent expressions of the photobleaching probability $p_b^3(I)$ and the photobleaching quantum yield $\Phi_b^3(I)$ for a 3-level system are obtained, which are independent of the excitation irradiance I (Eq. 12).

$$p_b^3(I) = \frac{k_z^3}{k_0 S_{1eq}^3} = \frac{k_b}{k_0} = \Phi_b^3(I) = \text{constant} \tag{12}$$

However, a 5-level system is necessary for high irradiances, because the population of S_n and T_n becomes important due to two-step excitation. As shown in Fig. 1 B, S_n and T_n open up new channels for photobleaching. Thus, the effective bleaching rate constant k_z^5 is now a sum of the composite microscopic bleaching rate constants k_b of the first (Eq. 11) and k_{bn} of the higher excited electronic states, S_n and T_n (Eq. 13).

$$k_z^5 = (k_b + k_{bn} I)\, S_{1eq}^5 \tag{13}$$

with $k_{bn} = (k_{bSn}/k_{Sn1})\, \sigma_{1n}\, \gamma + (k_{ISC}/k_T)(k_{bTn}/k_{Tn1})\, \sigma_{T1n}\, \gamma$

According to the scheme of multi-photon photolysis (Fig. 1 B), the composite bleaching constant k_{bn} of the higher excited electronic states (Eq. 13) can be expressed by the microscopic bleaching rate constants k_{Xbn} of the states X_n (k_{Sbn} for S_n and k_{Tbn} for T_n) with a total lifetime $\tau_{Xn} = 1/k_{Xn}$. A microscopic description assumes the following picture: Once excited with the rate constant $k_{(T)1n}$ to the state X_n, a molecule M can either directly relax to X_1 with a quantum yield $(1 - \Phi_{ion})$ or form a radical cation electron pair (M^+e^-) with a quantum yield Φ_{ion}. The formed radical ion pair has two depopulation reactions described by the rate constant k_{ion}: (1) fast geminal charge recombination to the first excited state X_1 [65]; (2) an escape reaction characterized by the quantum yield of solvation Φ_{solv} to form a solvated electron and a free radical cation with a subsequent, presumably irreversible decomposition [63, 64, 66, 119]. Therefore, it is important to realize that Φ_{ion} and Φ_{solv} increase with increasing electronic excitation. Thus, the rate constants for depopulation of X_n to X_1, k_{Xn1}, and for photolysis via X_n, k_{bXn}, can be realized by this model.

$$k_{Xn1} = (1 - \Phi_{ion})\, k_{Xn} + \Phi_{ion}\, (1 - \Phi_{solv})\, k_{ion} \tag{14}$$

$$k_{bXn} = \Phi_{ion}\, \Phi_{solv}\, k_{ion} \tag{15}$$

The resulting probability p_{bXn} of photolysis via X_n is given by Eq 16.

$$p_{bXn} = \frac{\Phi_{ion}\, \Phi_{solv}\, k_{ion}}{(1 - \Phi_{ion})\, k_{Xn} + \Phi_{ion}\, k_{ion}} = \frac{k_{bXn}}{k_{Xn1} + k_{bXn}} \tag{16}$$

The final expression of the overall photobleaching probability $p_p^5(I)$ for a 5-level system is obtained using the definition (Eq. 4) and the irradiance dependent expression for k_z^5 (Eq. 13).

$$p_p^5(I) = \frac{k_z^5}{k_0 S_{1eq}^5} = \frac{k_{bn}I + k_b}{k_0} = \frac{k_{bn}}{k_0}\, I + p_b^3 = \frac{k_{bn}}{k_0}\, I + \Phi_b^3 \tag{17}$$

To conclude, the photobleaching probability $p_p^5(I)$ allows a more general description of the photobleaching reaction. For a low irradiance it reduces to the classical expression of the quantum yield Φ_b^3, while the importance of the influence of two-step photolysis on photobleaching at high irradiances is also described.

Fluorescence Signal. To theoretically describe the fluorescence signal of a dye solution excited by a focused laser beam as in SMD and confocal microscopy, one has to consider the shape of the excitation profile, which in the case of a focused laser in the TEM-00 mode [122] is Gaussian in the radial ($1/e^2$-radius ω_0) and Lorentzian in the axial direction [122–124]. However, to get to a straightforward analytical solution, one has to approximate this profile. As done before in fluorescence correlation spectroscopy (FCS), where triplet and photobleaching rate constants of Rhodamine 6G were determined [43, 44, 49], we made the simple assumption of a rectangular excitation profile with a radial radius ω_0 and the overall irradiance $I_0/2$. Thereby, $I_0 = P/(0.5\,\pi\omega_0^2)$ is the irradiance in the focal plane if an excitation power P is applied. In this way, average excited state populations $S_{1eq}(I_0/2)$ and $T_{1eq}(I_0/2)$ and an average apparent photobleaching rate constant $k_{z,av}(I_0/2)$ can be defined over the whole excitation volume. The above formalism can now be used in a convenient way to optimize the experimental conditions for SMD (laser irradiance and beam diameter) with respect to two aspects: the number of collected fluorescence photons and the signal-to-background ratio for a given mean transit time $t_t = 4/3$ $\omega_0^2/4D$ through the assumed rectangular excitation profile with the diffusion constant D [1]. The fluorescence flow F and mean number of fluorescence photons N_F^x emitted by a single dye molecule depend $p^x(t)$ on the absolute probability that the molecule is still intact after being excited by the irradiance I for the time t (fluorescence quantum yield Φ_F) [44, 48].

$$N_F^x = \Phi_F k_0 S_{1eq}^x \int_0^{t_t} p^x(t)\, dt \tag{18}$$

If the absolute probability of still being intact after time t is described by the expression $p^x(t) = \exp(-k_z^x(I)\, t_t)$, N_F^x may easily be calculated assuming a x-level system ($x = 2, 3, 5$). This stepwise approach takes into account saturation with singlet and triplet levels as well as one- and two-photon photobleaching (Eq. 19).

$$N_F^x(I) = \Phi_F k_0 S_{1eq}^x\, 1/k_z^x(I)\, (1 - \exp(-k_z^x(I)\, t_t)) \tag{19}$$

Considering four different level models, the dependence of the mean number of emitted fluorescence photons N_F^x on the irradiance I is given in Fig. 2B, taking, for example, rate constants of Rhodamine 6G.

The irradiance dependence of $N_F(I)$ has three characteristic phases:

(1) An identical linear dependence of N_F for low irradiances is observed for all models.
(2) The saturation of N_F begins to occur at irradiances of approximately 10^5 W/cm². The irradiance at which saturation starts to appear and the height of the saturation levels decrease stepwise with the number of included reactions in the level model: 2 (0) simple 2-level model; 3 (0) 3-level model without photobleaching; 3 (1) 3-level model with photobleaching via S_1 and T_1; 5 (2) 5-level model with photobleaching via all excited states. Thus, it is evident that this decrease becomes more severe by an increased possibility of triplet state population (i.e. a higher intersystem crossing rate constant k_{ISC} and a larger lifetime $1/k_T$ of the triplet state) and of photobleaching (i.e. higher microscopic bleaching rate constants k_{bX} and k_{bXn}). Since Eq. 19 calculates the fluorescence signal of a single molecule (i.e. the number of photons in a burst), triplet state population and all photochemically reversible and irreversible one-photon reactions result solely in a decrease of the maximum calculated fluorescence signal.
(3) If the irradiance exceeds 5×10^5 W/cm², the single-molecule theory of a 5-level system even predicts a decrease of fluorescence due to two-step photolysis.

However, in practice, the dependence of the fluorescence signal on the irradiance is often measured in fluorophore solutions of typically 10^{-8} M taking the average of many molecules passing through the detection volume (e.g. inset Fig. 2B, Figs. 9 and 10A). In this case, in addition to the photobleaching seen during the passage of a single fluorescent molecule (Eq. 19), an accumulated depletion of the dye concentration in the absolute surrounding of the focused laser beam is to be expected. This will lead to a drop in the mean number of fluorescent molecules in the detection volume, which consequently results in a decrease in the detected fluorescence flow F already within milliseconds after onset of excitation illumination (see discussion in Sect. 6 and inset in Fig. 9). One has to keep in mind that this depletion will also contribute to the experimentally observed drop in F at very high irradiances as shown in the inset of Fig. 2B.

Nevertheless, the predicted irradiance dependence of N_F is in good agreement with experimental results observed for the irradiance dependence of the average fluorescence flow F of an aqueous Rhodamine 6G solution (10^{-9} M) (inset Fig. 2B). Even keeping the depletion effect in mind, this result still gives a first hint that the probability of photobleaching may indeed be dependent on the irradiance.

It is important to note that the excitation profile can also be approximated by a rectangular excitation profile with the radius $\omega_0/\sqrt{2}$ and the overall irradiance I_0 ("top hat") [122, 125], giving the same power throughput and defining average values $S_{1eq}(I_0)$ $T_{1eq}(I_0)$, and $k_{z,av}(I_0)$ over the whole excitation volume. However, here we chose the previously described rectangular model (radius ω_0 and overall irradiance $I_0/2$) for all subsequent analyses, since it had a closer agreement with numerical simulations of the total fluorescence flow of an assumed aqueous Rhodamine 6G solution considering a 3-dimensional Gaussian excitation volume as considered in FCS-theory. Nevertheless, both cases only represent an approximation of reality, whereby the deviation gets highest at very high irradiances, where uneven saturation of the dye fluorescence over the whole excitation profile takes place.

4
Experimental Methods To Measure Photobleaching

Using absorption or fluorescence detection, several experimental setups have been reported to monitor photobleaching under continuous or pulsed illumination by a lamp or laser at low and high irradiances: (1) cell-bleaching: stirred dye solution in a cell [18, 41, 44, 126–129]; (2) monitoring of the fluorescence signal in a microscope as a function of the applied irradiance [17, 18] or of the illumination time [98, 99, 104–106, 130, 131]; (3) digital imaging in a microscope: sequential fluorescence images of a substrate [104–106, 131]; (4) fluorescence signal in a flow cell as a function of the linear flow velocity [42, 117, 132]; (5) fluorescence correlation spectroscopy (FCS) in a confocal microscope [17, 18, 43, 44].

In view of the comments made in Sect. 2.2, several aspects of reaction mechanisms should be considered to ensure the appropriate and intended investigation of photobleaching with the above setups: (1) continuous data acquisition to allow for precise kinetic data analysis; (2) kinetic and product analysis should take into account the possibility that the photobleaching products are fluorescent or absorb at the analyzed wavelength; (3) establishment of pseudo first-order (monomolecular) reaction conditions: low concentrations of the fluorescent dye ($\leq 10^{-6}$ M) to avoid bimolecular reactions of the dye molecules with themselves and to minimize oxygen consumption during illumination; and (4) solution of the diffusion problem, because the measured fluorescence signal is determined by fluorescent dye molecules diffusing into the illumination volume in continuous exchange of the bleached molecules.

In view of these requirements, we prefer a combination of complementary methods with fluorescence detection to fulfill all requirements. Thus, in the next sections two methods for low and high irradiances, cell-bleaching and FCS, respectively, will be discussed in detail. They can provide the most detailed information regarding kinetic analysis of the photobleaching process and the influence of the excitation irradiance, and of additive compounds on these kinetics. Although product studies of photobleaching reactions would, in principle, be possible using the cell-bleaching experiment, the results of such product studies will not be discussed here. Nevertheless, product studies of photobleaching reactions have been carried out for, e.g., Rhodamine 6G [133] or coumarins [134–137] implying high dye concentrations ($>10^{-3}$ M). However, these necessary high dye concentrations will cause deterioration of the monitored kinetics and interfere with the requirement of pseudo first-order reactions (see point 3 above).

Cell-bleaching. The cell-bleaching method is a convenient and reliable method for low irradiances avoiding non-linear effects [18, 41, 44]. In this process, the excitation light, which is also monitored by a photodiode to correct for possible fluctuations, is focused into a quartz cell containing the constantly stirred dye solution. Since mixing is usually much faster than photobleaching with the rate constant k_z, the cell-bleaching experiment measures selectively the time-dependent decrease of the spectrally filtered fluorescence flow $F(t)$ of the whole solution, detected perpendicular to the excitation light by a photodiode or a photomultiplier. The continuous data acquisition by a PC facilitates a precise analysis. A sample volume of approx. 1 ml allows the product analysis to be made by additional methods. Two typical bleaching curves of different rhodamine dyes in water excited with $P = 2$ W at 514 nm are given in Fig. 3.

The registered fluorescence flow $F(t)$ is proportional to several factors: the concentration $S_{1eq} c(t)$ of dye molecules in S_1 (Eq. 9), the fluorescence quantum yield $\Phi_f = k_f/k_0$ of the dye, the fluorescence detection efficiency g_d of the setup, the illumination volume V_{ill} and the inverse of the fluorescence lifetime τ_0 (Eq. 20) [18, 44].

$$F(t) = F(0) \exp(-k_z t) = g_d \, \Phi_f S_{1eq} \, c(0) \, V_{ill} \, 1/\tau_0 \exp(-k_z t) \tag{20}$$

The final dependence of the quantum yield of photobleaching, Φ_b^3, on the experimental parameters of the cell-bleaching method (Eq. 21) assumes the 3-level system and almost all molecules to be in the ground state S_0 ($S_{0eq} = 1$) due to the low applied excitation irradiance. It also takes into account that only a fraction R_{ill} of the dye solution with the total volume V_{sol} is illuminated at a time in the cell, where $R_{ill} = V_{ill}/V_{sol} = A_{ill} \, b_c/V_{sol}$ (A_{ill} is the focal area of the excitation light beam and b_c the optical path length in the cell).

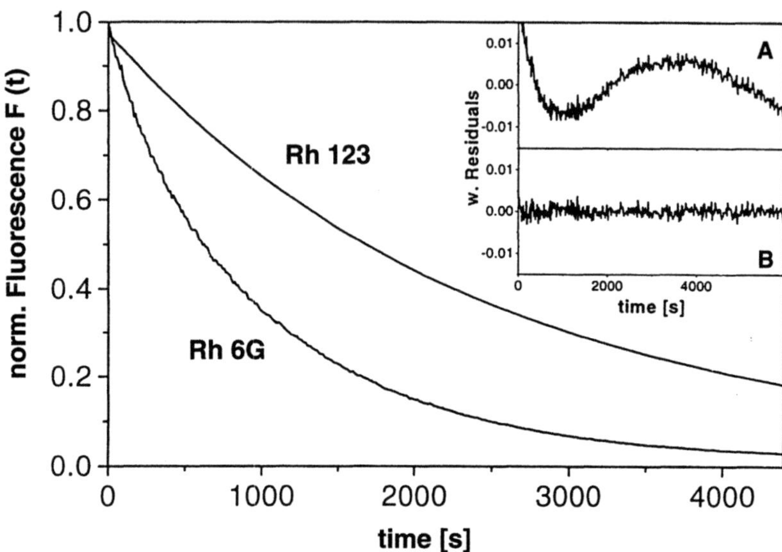

Fig. 3. Time-dependent exponential decrease of the fluorescence signal $F(t)$ of a stirred Rhodamine 6G (Rh 6G) (1 ml, 10^{-6} M in water) and Rhodamine 123 (Rh 123) solution (1 ml, 5×10^{-8} M) in a cell excited by a cw argon ion laser with $P = 2$ W at 514.5 nm. The insets (A) and (B) show the weighted residuals of a single (A) and double (B) exponential fit to the Rh 123 data: single: $F(t) = 0.02 + 0.96 \exp(-t/2390 \text{ s})$; double: $F(t) = 0.27 \exp(-t/630 \text{ s}) + 0.91 \exp(-t/2885 \text{ s})$

$$\Phi_b^3 = \frac{V_{\text{sol}}}{\sigma_{01}\gamma b_e} \frac{k_z^3}{P} \tag{21}$$

where P [W] is the excitation power ($P = I A_{\text{ill}}$), which can be measured by a power meter.

A detailed analysis of the bleaching curves in Fig. 3 by a fit with one and two exponential decay constants clearly indicates that only a double exponential fit is adequate as shown by the residuals in the insets **A** and **B**. Such non-single exponential decays were also found for the bleaching curves of the other dyes investigated (see below Table 1). Several observations explain this unexpected behavior. The fluorescence flow of the aqueous dye solutions drops even without illumination reaching a steady-state level, whereby this initial signal drop increases with decreasing dye concentration and correlates with the hydrophobic properties of the dyes. In agreement with findings of Dörre [12] and Enderlein [8] this suggests that this drop is caused by non-specific adsorption of the dye to the cell surface and not by precipitation due to the limited solubility of the organic dyes in water [18, 44]. In the presence of detergents such as Tween 20, which suppress non-specific adsorption, single exponential bleaching decays are obtained, which supports the above interpretation.

In the data analysis of photobleaching in pure water, we treated the decrease of the signal flow by two processes: (1) a reversible adsorption-desorption equilibrium to the cell surface with the rate constants k_{ads} and k_{des} leading to a dye concentration c_{solv} in solution and c_{ads} on the surface; (2) irreversible photobleaching of the free dye with the rate constant k_z^3 producing a nonfluorescent, bleached dye molecule concentration c_b.

$$c_{ad} \underset{k_{ads}}{\overset{k_{des}}{\rightleftarrows}} c_{solv} \xrightarrow{k_z^3} c_b$$

A similar kinetic scheme is well known from the monomer excimer equilibrium kinetics [138]. Applying the initial conditions ($c_{solv}(0) = 1$, $c_{ad}(0) = 0$, and $c_b(0) = 0$) for $t = 0$, the transient behavior of the fluorescent dye $c_{solv}(t)$ during illumination can be described by two exponentials ($c_{solv}(t) = a_1 \exp(-\lambda_1 t) + a_2 \exp(-\lambda_2 t)$). The rate constants k_{des}, k_{ads}, and k_z^3 can be extracted from the amplitudes a_x and rate contants λ_x via the following relationships [125]: $a_{1,2} = (G \pm E \mp 2k_{des})/(2G)$ and $\lambda_{1,2} = (E \pm G)/2$ with the defined constants $G = ((k_{ads} + k_{des} - k_z^3)^2 + 4 k_{des} k_z^3)^{1/2}$ and $E = (k_{ads} + k_{des} + k_z^3)$.

Fluorescence correlation spectroscopy (FCS). Fluorescence correlation spectroscopy (FCS) uses the fluorescence fluctuations arising from fluorescent molecules excited by a focused laser beam to obtain information about dynamic processes at the molecular level [139–142]. The time dependent fluorescence signal $F = \delta F + \langle F \rangle$ described by fluorescence fluctuations $\delta F(t)$ about an average value $\langle F \rangle$ is thereby analyzed in the form of a normalized autocorrelation function $G(t_c)$, where t_c denotes the correlation time.

$$G(t_c) = \frac{\langle F(t) F(t + t_c) \rangle}{\langle F(t) \rangle^2} = 1 + \frac{\langle \delta F(t) \, \delta F(t + t_c) \rangle}{\langle F(t) \rangle^2} \tag{22}$$

If one strives for the ultimate in sensitivity for SMD, a high excitation irradiance must be applied, which is achieved using laser excitation in a confocal epi-illuminated fluorescence microscope. In this way, the observation of even single fluorescent molecules in a small open volume element in the order of a few femtolitres is possible because of the high spatial and spectral discrimination. These experimental conditions, with their possibility of SMD has drastically increased the capabilities of FCS [1, 26, 124, 143] and has during the last years and thus 25 years after its introduction [139–141] given the technique a wide range of applications. Therefore, FCS can be used as a tool to provide precise kinetic information on all processes influencing the fluorescence flow of the molecule in the detection volume, e.g. translational (i.e. spot size of laser focus) and rotational diffusion, photophysical (fluorescence and triplet) and photochemical (chemical equilibrium reactions, photostability) parameters. In principle, dynamic events from the nanosecond to well over the millisecond range can be investigated, as long as they are faster than the translational diffusion and manifest themselves by changes in fluorescence intensity. Furthermore, it is evident from Eq. 22, that the amplitude $G(t_c = 0)$ of the autocorrelation function is equivalent to the normalized variance of the fluorescence signal (second order central moment of light intensity). Therefore, FCS can also be used to obtain a statistical average value of the molecule number N in the detection volume without any external calibration standard. To summarize, FCS can give two kinds of information for the analysis of a photobleaching reaction: (1) decrease of fluorescent molecules N in the detection volume due to photobleaching; and (2) effective average rate constant of photobleaching $k_{z,av}$.

The theoretical basis and first experimental realizations of FCS were presented 25 years ago [139–141], and have been reviewed by Thompson [142]. In our experimental setup of a confocal epi-illuminated fluorescence microscope [43, 44], the excitation light of a cw argon ion laser at 514.5 nm is used, which is focused by a lens ($f = 600$ mm) in front of the microscope and coupled into the objective (Zeiss Plan-Neofluar 63 × NA 1.2, water immersion) via a dichroic mirror. The fluorescence light of the sample droplet hanging beneath the objective is collected, spectrally filtered, imaged onto a pinhole (diameter 250 µm) by the same objective, divided afterwards by a beam splitter, and detected by two avalanche photodiodes, whose pulses are computed by a PC-adapted correlator card.

To study the photobleaching behavior of rhodamines by FCS, the radius of the excitation volume was deliberately expanded by a factor of three compared to SMD experiments [6, 15, 17, 31]. This increases the probability of observing photobleaching of the dye molecules due to longer transit times through the excitation volume. Figure 4A shows typical autocorrelation curves $G(t_c)$ of tetramethylrhodamine (TMR) (10^{-9} M in water) for three cw excitation irradiances, that differ by one order of magnitude.

For the low irradiance (full points), photobleaching is negligible and $G(t_c)$ shows only one decay revealing the characteristic diffusion time through the detection volume. A higher irradiance has two effects on $G(t_c)$ (full and dashed line): (1) an additional exponential decay in the µs-time scale is observed due to transitions between the singlet state and the, in practice, "dark" triplet state which is more populated with increasing excitation [49, 50]; (2) photobleaching becomes evident by an apparently reduced diffusion time [43, 44]; i.e. there is another limit for the duration of the fluorescence emission of a molecule than its dwell time in the detection volume.

The theoretical description of autocorrelation curves follows the above experiment. If for low irradiances translational diffusion is the only noticeable process causing fluorescence fluctuations, the time-dependent part $G_D(t_c)$ of the normalized correlation function $G(t_c)$ (Eq. 22) is given by:

$$G_D(t_c) = \frac{1}{N}\left(\frac{1}{1 + t_c/\tau_D}\right)\left(\frac{1}{1 + (\omega_0/z_0)^2\, t_c/\tau_D}\right)^{1/2} \tag{23}$$

whereby a spatial 3-dimensional Gaussian distribution of the detected fluorescence $(W(x,y,z) = \exp(-2(x^2 + y^2)/\omega_0^2)\,\exp(-2(z^2)/z_0^2))$ is assumed [26, 124, 144]. The characteristic time for diffusion τ_D of the fluorescent molecules through the detection volume is related to the radial $1/e^2$ radius ω_0 via their translational diffusion coefficient D: $\tau_D = \omega_0^2/4D$.

Since for high irradiances, photochemical reactions, which are restricted to the excitation volume, generate additional signal fluctuations with characteristic time constants, FCS theory must take into account the following processes: (1) triplet formation in a reversible photochemical equilibrium [49, 50]; (2) ir-

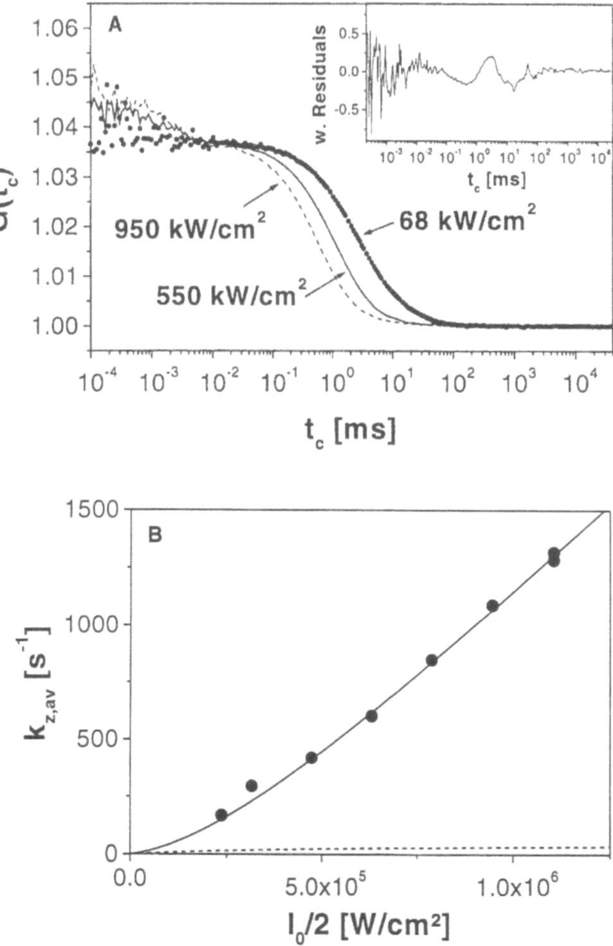

Fig. 4A, B. (A) Fluorescence autocorrelation curves $G(t_c)$ of TMR (10^{-9} M in water) for three different cw excitation irradiances $I_0/2$ at a wavelength of 514.5 nm. For illustration the different curves are normalized on each other. The inset shows the weighted residuals of a fit of Eq. 25 to the data at 550 kW/cm²: $N = 19.9$, $\omega_0 = 1.63$ µm (fixed), $\omega_0/z_0 = 4$ (fixed), $T_{1eq} = 0.16$, $t_T = 2.6$ µs, $A = 0.71$, $k_{z,av} = 604$ s^{-1}. (B) Irradiance dependence of the phenomenological bleaching rate constant $k_{z,av}(I_0/2)$ of TMR (10^{-9} M in water) (solid points) determined from the measured FCS curves using Eq. 25. The data are well described by the 5-level model (Eq. 13) (full curve) and do not fit to the 3-level model (Eq. 11) (dashed curve). Fixed fit parameters: $k_0 = 4.35 \times 10^8$ s^{-1}, $\sigma_{01} = 1.33 \times 10^{-16}$ cm², $k_{ISC} = 5.4 \times 10^5$ s^{-1}, $k_T = 4.0 \times 10^5$ s^{-1}; 3-level model: $k_b = 140$ s^{-1}. Variable fit parameters: $k_{bXn} = 1.2 \times 10^7$ s^{-1}, $1/k_{Xn1} = 220$ fs, $k_b = 220$ s^{-1}.

reversible first-order photobleaching reaction to a nonfluorescent product with the effective bleaching rate constant k_z (Eq. 5b) [43, 44, 125]. Under our experimental conditions, photobleaching can still be treated as a chemical pseudo-equilibrium reaction. Because the total sample volume is much larger than the excitation volume, the depletion of fluorophores within the excitation volume will be balanced by a net inflow from outside due to the concentration gradient formed. Yet, it is important to note that k_z is dependent on the irradiance (Eq. 12) and thus on space. To obtain an analytical solution for such an auto-correlation function, we made the simple assumption of a rectangular excitation profile with the radius ω_0 and the height $2z_0\sqrt{\pi/8}$ as mentioned previously (Sect. 3) [44]. This defines an average effective photobleaching rate constant $k_{z,\,av}(I_0/2)$ over the whole excitation volume with the focal irradiance $I_0 = I$ ($z = 0$) (Eq. 24).

$$k_z \approx \begin{cases} k_{z,\,av}(I_0/2); & x^2 + y^2 \le \omega_0^2; \ |z| \le z_0\sqrt{\pi/8} \\ 0; & \text{elsewhere} \end{cases} \tag{24}$$

In this way the full autocorrelation function $G(t_c)$ is given by [44, 125]:

$$G_D(t_c) = 1 + \frac{G_D(t_c)}{(1 - T_{leq})} \times$$

$$[1 - A(1 - T_{leq}) + A(1 - T_{leq})\exp(-k_{z,\,av}(I_0/2)\,t_c) - T_{leq} + T_{leq}\exp(-t_c/t_T)]$$

$$\text{with } \frac{1}{t_T} = \left(k_T + \frac{k_{01}k_{ISC}}{k_{01} + k_0}\right) \text{ and } T_{leq} = \frac{k_{ISC}}{k_T}S_{leq} \tag{25}$$

T_{1eq} is the average equilibrium fraction of molecules in T_1 with an average triplet correlation time t_T [49, 50]. On the basis of the assumptions we made on the average effective photobleaching rate constant $k_{z,\,av}$, the homogenous distribution of $k_z(I)$ on all molecules over the whole excitation volume is changed into an overall bleaching reaction with $k_{z,av}(I_0/2)$ of only a fraction A of all excited molecules.

The residuals of a nonlinear fit of the experimental data to Eq. 25 (inset of Fig. 4) and the obtained value of A demonstrate the satisfactory agreement between theoretical (A = 0.8) and experimental (A $\approx$ 0.75) values. In this way, a number of $k_{z,\,av}$ values were determined for several irradiances using FCS.

The dependence of $k_{z,\,av}$ on the irradiance in Fig. 4B was analyzed using the theory for a 3-level (Eq. 11, dotted line) or 5-level system (Eq. 13, full line) by varying the one- and two-photon bleaching constants k_b and k_{bn}. These results clearly demonstrate that in the case of TMR only a 5-level system is adequate to describe photobleaching, which is also consistent with the data shown in Fig. 2B. Considering multi-photon photolysis (Fig. 1B), recovery times $t_{Xn1} = 1/k_{Xn1}$ (Eq. 14) in the order of a few hundred femtoseconds are obtained.

To summarize, the cell-bleaching method and FCS are complementary methods, which allow for detailed studies of many aspects of photobleaching without interference by diffusion. Concerning confocal microscopy and SMD, the advantage of FCS is its performance under identical experimental conditions, giving simultaneous information on the photobleaching rate as well as the formation and decay rates of the triplet state (Eq. 25) [44, 50]. This contributes to a better understanding of photobleaching given the central importance of the triplet states in photobleaching processes (see Sect. 6).

5
Photostabilities of Organic Fluorescent Dyes in Aqueous Media

In this section we shall quantitatively discuss the photobleaching properties of organic fluorophores in aqueous media. Table 1 lists the photobleaching quantum yields, Φ_b, the corresponding average number of survived excitation cycles, μ (Eq. 2), if specified the S_0–S_1 absorption cross sections at the excitation wavelength λ, $\sigma_{01}(\lambda)$, and the wavelengths of maximum absorption, $\lambda_{e,max}$, and maximum fluorescence emission, $\lambda_{f,max}$, for a number of fluorescent organic dyes in alphabetical order measured by us or reported in literature. Furthermore, the molecular structures of the mentioned dyes are shown in Fig. 5.

Being aware that Φ_b might have a non-linear irradiance dependence, the applied irradiance has been added in parentheses to the listed values allowing for a detailed examination of this problem. At a first glance, it is apparent that some rhodamine derivatives, such as Rhodamine 123 and tetramethylrhodamine (TMR), are the most photostable fluorescent dyes, having numbers of survived cycles μ of the order of 10^6. The dyes with the lowest photostability in Table 1 are coumarin derivatives [including AMCA (Coumarin 120)] being approximately 2 to 3 orders of magnitude less stable. Surprisingly, in view of their wide use in microscopy, the dyes Fluorescein and Texas Red (Sulforhodamine 101, being an exception in the rhodamine group) have also a quite low value of $\mu = 20000$, which constitutes a major problem. B-Phycoerithrin (B-PE) is a highly fluorescent compound belonging to a class of proteins (phycobiliproteins) found in the light-harvesting structures of red algae and cyanobacteria. B-PE is comprised of three polypeptide subunits containing 34 bilin chromophores, which results in a 5 to 10 times higher absorption cross section than fluorophores with a single chromophore. In SMD, Wu et al. [147] observed abrupt photobleaching of B-PE in a single step. They were also able to demonstrate by photon pair correlation measurements, that the molecule behaves as a single-quantum system and not as a collection of 34 independent chromophores. Keeping in mind, that rapid energy transfer occurs in these proteins, a single photochemically generated trap in the molecule will act as an efficient sink for subsequent excitations. This result rationalizes the rather moderate photostability ($\mu \approx 20000$) of B-PE, which is determined by the photostability of a single bilin chromophore. The photostability of the widely used dye Cyanin 5 (Cy 5), which is important for fluorescence detection in the red spectral region, is fairly high ($\mu \approx 200000$) and comes close to those of the rhodamines.

The extraordinary photostability of rhodamines can be explained by several arguments: (1) the triplet quantum yield of rhodamines is extremely low (below 1%) [50, 57] leading to a decreased photobleaching probability, since the triplet state is proposed to be a major bleaching channel [97, 99, 149].

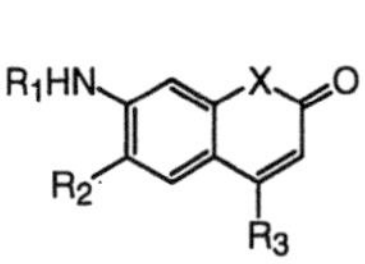

Carbostyryl-124:
$R_1 = H$, $R_2 = H$, $R_3 = CH_3$, $X = NH$
Coumarin-120:
$R_1 = H$, $R_2 = H$, $R_3 = CH_3$, $X = O$
Coumarin-307:
$R_1 = Et$, $R_2 = CH_3$, $R_3 = CF_3$, $X = O$

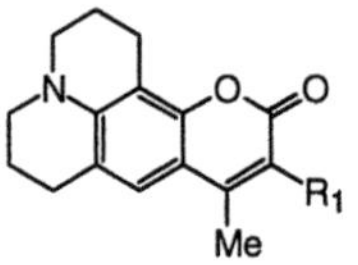

Coumarin-102:
$R_1 = H$
Coumarin-39:
$R_1 = CH_3$

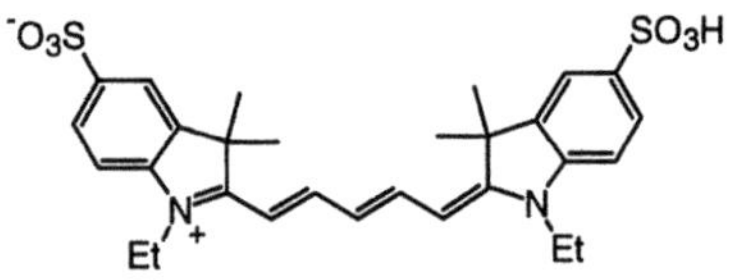

Cyanin 5

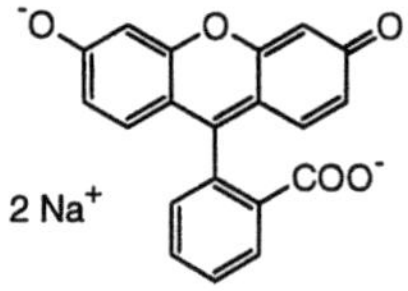

Fluorescein

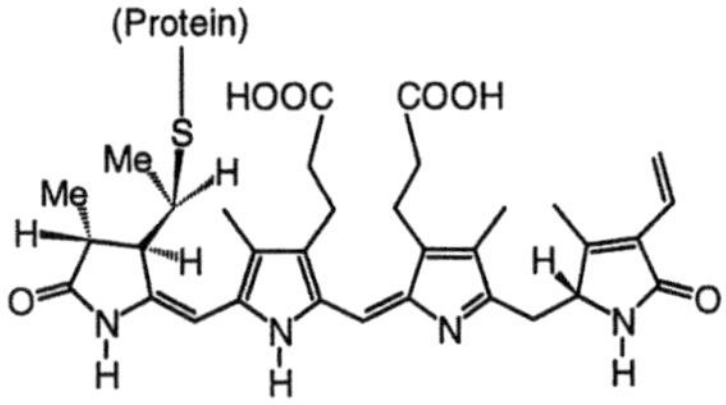

B-phycoerithrin

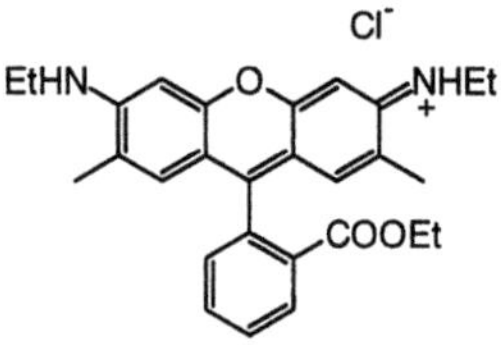

Rhodamine 6G

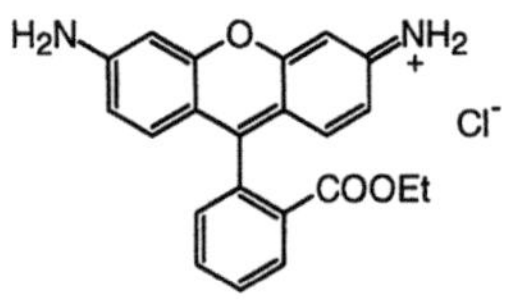

Rhodamine 123

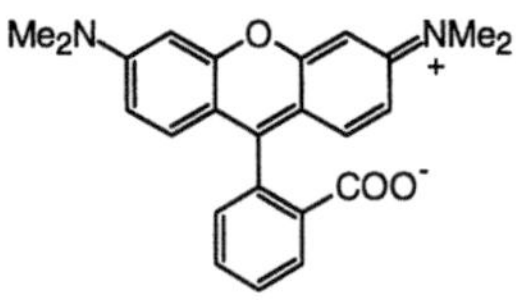

Tetramethylrhodamine

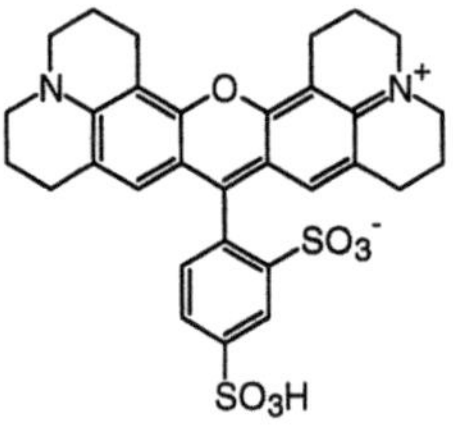

Texas red

Fig. 5. Molecular structures of the fluorescent dyes mentioned in Tables 1–3

Nevertheless, the population of the triplet state is a dye-specific property which cannot be discussed in detail here. Several factors are important: (1) singlet-triplet energy gap, spin-orbit coupling in electronic systems, heavy atom effect [71, 72]. (2) There is a correlation of Φ_b with the singlet and triplet energies E_S and E_T of the dyes, which are significantly different for rhodamines ($E_S = 2$ eV [118] and $E_T = 1.86$ eV [58]) and coumarins ($E_S = 3$ eV [118] and $E_T = 2.5$ eV [150, 151]). A state that contains a higher energy may be more chemically reactive. (3) Coumarin and carbostyryl chromophores, for example, are chemically quite reactive [136], whereas the xanthene chromophore in rhodamines is chemically stable and very rigid, which minimizes internal conversion processes.

The photochemical properties of rhodamines are mainly governed by two effects: (1) those rhodamines (e.g. Rhodamine B [152, 153]), which have no esterified carboxyl group in the phenyl ring, exhibit a pH- and solvent-dependent equilibrium between a colorless lacton form, and fluorescent zwitterionic and cationic forms [152–154]. This effect is absent for rhodamines with an esterified carboxyl group (Rhodamine 6G). (2) The fluorescence quantum yield Φ_f of rhodamines and its solvent dependence are also determined by the alkylation of the two amino groups. If both amino groups have n-alkyl groups, the population of a non-emissive state with charge-transfer character is possible from the S_1-state. This transition is governed by a subtle interplay of steric and electronic factors; e.g. Φ_f of the TMR-zwitterion drops from approx. 95 % in non-polar solvents to approx. 60 % in basic water, whereas Φ_f of Rhodamine 6G displays no significant decrease [155]. The controversy over the molecular aspects of the nature of this state is far from being settled. Even for prominent model systems such as 4-N,N-dimethylaminobenzonitrile (DMABN), at least three explanations for the reduced fluorescence quantum yield of the locally excited state in polar solvents have been discussed in the scientific community: (1) twisted intramolecular charge transfer state (TICT) with a rotated amino group (for a review see [156]); (2) vibronical coupling by the pyramidalization mode of the nitrogen (pseudo Jahn Teller model) enables the transition [157–159]; (3) formation of a solute-solvent exciplex [160]. Furthermore, ab initio calculations propose additional reaction coordinates such as a non-linear distortion of the CN-group or a breathing mode of the benzene ring [161, 162]. The main conclusion one can draw is, that it is very important for a high fluorescence quantum yield to have a rigid fluorophore with few functional groups interacting with the molecular environment.

Even though the mean number of survived absorption cycles of Coumarin-120 is only 3000, the photostability was found to be sufficiently high to allow SMD by coherent two-photon excitation at 700 nm (see Sect. 6) [17, 18, 33]. However, due to the exponential nature of the photodestruction process, the possibility of a single dye molecule to emit a certain minimal number of photons necessary for an accurate detection is much smaller for coumarins than for rhodamines. Therefore, current projects for single-molecule DNA sequencing based on fluorescence detection are using rhodamine dyes as labels [3, 12].

If one looks more closely at the values of Φ_b of Rhodamine 6G (Rh 6G) determined by various groups [43, 44, 117, 126–128, 148], agreement between the

Table 1. Photostabilities of Various Dyes in Aqueous Solution

Dye	Φ_b (Irradiance [W/cm^2])[a]	μ[b]	$\sigma_{01}(\lambda)$ [10^{-17} cm^2][c]	$\lambda_{e,max}$[d]	$\lambda_{f,max}$[e]
Carbostyryl-124	1.4×10^{-3} (low)[f] [18]	700	6.3 (335 nm)[g]	340[h]	400[h]
Coumarin-39	1.2×10^{-3} (low)[f] [18]	800	4.7 (397 nm)[g]	380[h]	443[h]
Coumarin-102	4.3×10^{-4} (low)[f] [18]	2300	7.0 (397 nm)[g]	393[h]	465[h]
Coumarin-120	4.3×10^{-4} (low)[f] [18]	2300	7.0 (397 nm)[g]	344[h]	445[h]
Coumarin-307	1.5×10^{-4} (low)[f] [18]	6500	5.4 (397 nm)[g]	395[h]	510[h]
Cyanin 5	5.0×10^{-6} (low)[f]	200000	96.3 (647 nm)[g]	650[h]	670[h]
	5.0×10^{-6} (low)[f]	200000	1.5 (515 nm)[g]		
Fluorescein	–	–	–	500[i]	521[i]
Lit. values	2.7×10^{-5} (1.8×10^5)[j] [42]	37000	30.0 (488 nm) [42]		
	3.6×10^{-5} (1.2×10^4)[k] [41]	27800	–		
	8.0×10^{-5} (4.0×10^3)[l] [43]	12500	–		
	1.0×10^{-4} (3.0×10^4)[l] [43]	10000	–		
	–	–	26.0 (488 nm) [50]		
B-phycoerithrin	–	–	–	532 [147]	575 [147]
Lit. values[m]	1.1×10^{-5} (1.5×10^3)[j] [42]	90900	540 (515 nm) [42]		
	6.6×10^{-6} (8.8)[j] [132]	152000	578 (515 nm) [132]		
Rhodamine 6G	1.2×10^{-6} (low)[f] [44]	833000	22.2 (515 nm)[g]	524[h]	550[h]
Lit. values	9.0×10^{-6} (low)[n] [127]	111000	–		
	7.0×10^{-6} (low)[n] [126]	143000	–		
	1.9×10^{-5} (3.0×10^5)[j] [117]	52600	22.0 (515 nm) [117]		
	6.0×10^{-6} (4.0×10^4)[l] [43]	167000	–		
	1.5×10^{-5} (1.0×10^6)[l] [43]	66700	–		
	5.8×10^{-7}, 3.9×10^{-6} or		–		
	2.0×10^{-5} (low)[o] [148]		–		
	2.5×10^{-6} (low)[k] [128]	400000	–		
	–	–	17.0 (515 nm) [50]		

Table 1 (continued)

Rhodamine 123	6.4×10^{-7} (low)[f] [44]	1570000	12.3 (515 nm)[g]	500 [h]	524[h]
TMR	3.3×10^{-7} (low)[f] [44]	3070000	13.3 (515 mn)[g]	553[h]	576[h]
Lit. values	5.6×10^{-6} (1.5×10^5)[j] [117]	179000	16.0 (515 mn) [117]		
	2.0×10^{-6} (6.0×10^4)[p] [145]	500000	–		
Texas red	–	–	–	595 [40]	615 [40]
Lit. values	5.5×10^{-5} (1.5×10^5)[j] [117]	18200	37.0 (585 nm) [117]		

[a] Φ_b: quantum yield of photobleaching (Eq. 3) at a certain excitation irradiance I (low: I $< 10^3$ W/cm^2).

[b] The number of survived excitation cycles μ was calculated from Φ_b: $\mu = 1/\Phi_b$.

[c] $\sigma_{01}(\lambda)$: absorption cross section at a certain excitation wavelength λ.

[d] $\lambda_{e,max}$: wavelength [nm] of maximum light absorption.

[e] $\lambda_{f,max}$: wavelength [nm] of maximum fluorescence emission.

[f] The quantum yield of photobleaching, Φ_b, was calculated from the fluorescence decreases measured at our cell bleaching set-up at low irradiances ($<10^2$ W/cm^2). Their standard deviation is in the order 20% obtained by repeated measurements.

[g] The absorption cross sections of the dyes, σ_{01}, were determined from absorption measurements of exactly weighted in dye solutions.

[h] Determined from own measurements in water.

[i] Measured in ethanol [146].

[j] Determined from measurements in a flow cell.

[k] Obtained from the fluorescence decrease in cell-bleaching experiments.

[l] Determined from fluorescence correlation measurements in a confocal microscope.

[m] The photostabilities of R-phycoerithrin ($\Phi_b = 1.1 \times 10^{-5}$), allophycocyanin ($\Phi_b = 1.1 \times 10^{-5}$), and C-phycocyanin ($\Phi_b = 1.1 \times 10^{-5}$) were also measured at low irradiance ($I < 10^3$ W/cm^2) and 515 nm [132].

[n] Obtained from cell-bleaching experiments using absorption detection.

[o] Measured on a fused silica surface. The different values result from different adsorption geometries of the dye on the surface.

[p] Determined from diffusion measurements in membranes.

reported values is not satisfactory. Being aware that different experimental approaches, dye concentrations, excitation irradiances, and definitions of Φ_b have been applied, the wide range of values of Φ_b between 0.1×10^{-5} and 2.0×10^{-5} can be rationalized to some extent. As shown above for TMR (Fig. 4), FCS measurements indicate that photobleaching of Rh 6G also depends on the applied irradiance [43, 44, 125]. Thus, it is advantageous to use the probability of photobleaching p_b (Eqs. 4 and 17) to describe the non-linear dependence as illustrated for both dyes in Fig. 6.

The plotted photobleaching probability of Rh 6G (full line), which is approximately 10^{-5} at high irradiances (400 kW/cm²), is in good agreement with previous reports [43, 117] (full symbols in Fig. 6) also listed in Table 1. However, there is evidence from cell-bleaching measurements at very low dye concentrations [open circles (our cell-bleaching experiments) [44] and stars [128], from Table 1] and by extrapolated FCS data (full line), that at low irradiances, p_b drops by a factor of 5 ($p_b \approx 2 \times 10^{-6}$). The slight difference between results obtained by cell-bleaching and FCS indicates that the extrapolation of the FCS measurements to low enough irradiances introduces some uncertainty. It is evident that the results obtained by previous cell-bleaching measurements [126, 127] at higher concentrations ($> 10^{-5}$ M) (crosses in Fig. 6, from Table 1) are not appropriate to describe photobleaching under the conditions of SMD, because these results were influenced by dye-dye photoreactions and other effects (see Sect. 2.2) [101–103].

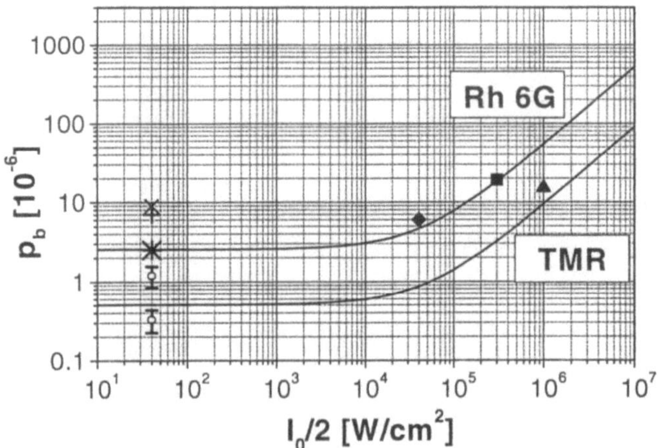

Fig. 6. Dependence of the probability of photobleaching p_b on the cw excitation irradiance $I_0/2$ calculated (Eq. 17) for both dyes (full curves) using the following parameters partially determined from FCS measurements: Rhodamine 6G (Rh 6G): $k_0 = 2.56 \times 10^8$ s⁻¹, $\sigma_{01} = 2.22 \times 10^{-16}$ cm², $k_{ISC} = 1.1 \times 10^6$ s⁻¹, $k_T = 4.9 \times 10^5$ s⁻¹, $k_b = 650$ s⁻¹, $p_{bXn} = 5.5 \times 10^{-5}$, $1/k_{Xn1} = 200$ fs; TMR: $k_0 = 4.35 \times 10^8$ s⁻¹, $\sigma_{01} = 1.33 \times 10^{-16}$ cm², $k_{ISC} = 5.4 \times 10^5$ s⁻¹, $k_T = 4.0 \times 10^5$ s⁻¹, $k_b = 220$ s⁻¹, $p_{bXn} = 2.5 \times 10^{-5}$, $1/k_{Xn1} = 220$ fs; for both dyes: $\sigma_{S1n} = 0.77 \times 10^{-17}$ cm², $\sigma_{T1n} = 3.85 \times 10^{-17}$ cm². For comparison the probability of photobleaching p_b obtained by own cell bleaching measurements (open points with error bars) (e.g. $P = 2$ W at a focal diameter of 2.5 mm) or from previous reports (various symbols) are given (see Table 1)

Comparing the different experiments, the results of photobleaching measurements are easily susceptible to various influences induced by only slight changes in the experimental conditions, which is for example due to the usually low probability of the bleaching process. These slight changes can sometimes not even be detected by other methods. Thus, it is crucial to carefully control the purity of the solvent, added compounds, and the oxygen concentration, as well as the concentration of the dye itself. Another important parameter is the applied irradiance. Due to the high irradiance necessary for an efficient SMD in a confocal microscope, additional photobleaching by two-step excitation occurs even at cw-excitation in the visible spectral range. The non-linear effects are especially severe for pulsed excitation. Hence, special attention should be paid to the pulse duration and the repetition frequency of the laser. Furthermore, one should take a closer look at absorption spectra of the lower excited states, S_1 and T_1, to reduce two-step excitation by a more proper choice of the excitation wavelength where the absorption cross sections of S_1 and T_1, σ_{S1n} and σ_{T1n}, are as low as possible. Nevertheless, in principal, it is best to choose the wavelength of maximum absorption of the S_0–S_1 transition, since it requires the least excitation irradiance to obtain a certain probability of being in the S_1-state and thus to obtain a certain fluorescence signal. This would also minimize the probability of two-step excitation and optimize the signal-to-background ratio.

6
Strategies To Prevent Photobleaching of Fluorescent Dyes

6.1
Variation of the Oxygen Concentration

Theory. Under our experimental conditions, quenching of excited states of organic dye molecules by dissolved oxygen is a very important deactivation pathway, which may also subsequently lead to photobleaching as outlined below. Oxygen has a triplet electronic ground state $^3O_2$ with unpaired electrons according to Hund's rule. In contrast, most stable organic compounds show a singlet spin multiplicity in the electronic ground state. This is doubtless one of the major reasons why the reactions of these compounds with oxygen are generally slow under physiologic conditions without light. Photochemical reactions can produce three different oxygen species in the first excited singlet state having paired electrons. The state $O_2(^1\Delta_g)$, denoted as "singlet oxygen", has the lowest energy and the longest lifetime, which makes this state important in chemical reactions. The electronic transitions between the triplet ground state and the excited singlet states are spin-forbidden transitions, which is the main reason for the long lifetime τ of the singlet oxygen. It varies with the nature of the solvent; e.g. $\tau(O_2(^1\Delta_g), \text{water}) = 3.3 - 7.4\ \mu s$ [163]. In oxygenation processes the two different species of molecular oxygen have quite different chemical reactivity. The triplet (ground state) oxygen reacts preferably with photo-produced radicals and radical ions, whereas the singlet oxygen $O_2(^1\Delta_g)$ may be treated as an ethene analog undergoing electron-pair reactions; i.e. singlet oxygen efficiently reacts with singlet species.

The photochemical quenching reactions have been extensively examined [71, 164–166]. Oxygen can affect excited singlet states as well as triplet states. The principal interaction mechanism of the excited molecule $^{(3,1)}M_1$ with oxygen $^3O_2$ is based on a short-range electron-exchange energy transfer (Dexter mechanism).

Considering quenching of an excited singlet state, $^1M_1$, the formed encounter complex in Eq. 26 may, via the equilibrium in Eq. 27, lead (a) to the formation of singlet oxygen $O_2(^1\Delta_g)$, (b) to oxygen-enhanced intersystem crossing, isc, or (c) to radiationless deactivation.

$$^1M_1 + {}^3O_2 \rightleftharpoons {}^3[M - O_2]
\begin{array}{ll}
\longrightarrow {}^3M_1 + O_2(^1\Delta_g) & (26\,a) \\
\longrightarrow {}^3M_1 + {}^3O_2 & (26\,b) \\
\longrightarrow M + {}^3O_2 & (26\,c)
\end{array}$$

$$^5[M - O_2] \xrightarrow{\text{isc}} {}^3[M - O_2] \xrightarrow{\text{isc}} {}^1[M - O_2] \qquad (27)$$

Spin statistics (probability 3/3) allows in principle diffusion-controlled quenching of singlet-excited molecules (Eq. 26) with a rate constant of approx. 1×10^{10} M^{-1} s^{-1} in water [167, 168]. Since the oxygen concentration dissolved in water under 1 bar air pressure is only 0.29 mM at 20 °C [58], the rate is approx. 3×10^6 s^{-1}. However, free-energy calculations for reaction 26a introduce the limitation that the difference between the singlet and triplet energy of the dye, $E_{S\text{-}T}$, should be larger than the energy of singlet oxygen, $E(O_2(^1\Delta_g)) = 0.98$ eV [164, 165]. In practice, for many fluorescent dyes, discussed in Table 1, this condition is not fulfilled, i.e. the reaction is energetically impossible. Remarkable counterexamples are polycyclic aromatics such as anthracene with $E_{S\text{-}T} = 1.45$ eV [169].

Quenching of an excited triplet state $^3M_1$ may be possible via the following reaction pathways (Eq. 28), which obey the Wigner-rule for spin-allowed reactions:

$$^3M_1 + {}^3O_2 \xrightleftharpoons{1/9} {}^1[M - O_2] \longrightarrow M + O_2(^1\Delta_g) \qquad (28\,a)$$

$$^3M_1 + {}^3O_2 \xrightleftharpoons{3/9} {}^3[M - O_2] \longrightarrow M + {}^3O_2 \qquad (28\,b)$$

$$^3M_1 + {}^3O_2 \xrightleftharpoons{5/9} {}^5[M - O_2] \qquad (28\,c)$$

In contrast to singlet states, spin statistics predict that quenching of triplet states (Eq. 28) should be, in principle, limited to 4/9 (1/9 + 3/9) of the diffusion-controlled rate ($< 1.3 \times 10^6$ s^{-1}), since the encounter of two triplets leads with a probability of 5/9 to a quintet encounter complex. The latter cannot proceed to products but can only redissociate (Eq. 28c). Free-energy considerations for reaction 28a require that the triplet energy of the dye, E_T, is larger than $E(O_2(^1\Delta_g))$, which is always the case for the dyes excited in the UV-VIS spectral range; i.e. oxygen is a very effective triplet quencher. Reaction 28b corresponds to an exergonic relaxation process, where the rate constant is determined by the energy gap law. For large free energies ΔG this results in a decrease of the rate constant with an increase of ΔG (Marcus inverted region [169]). Although most experimental data seem to support the spin-statistical and energetic arguments for oxygen quenching of singlet and triplet states, exceptions are also known, which are explained by intersystem crossing (Eq. 27) and charge-transfer interactions in the encounter complex [170].

To conclude, the excited state triplet quenching by oxygen (Eq. 28a) is the main energetic- and spin-allowed pathway for the formation of singlet oxygen, which can readily chemically react with other molecules in the singlet state. This photodestruction reaction can take place in two ways: (1) as a follow-up reaction in the reaction complex with the dye molecule just quenched; or (2) due to the long lifetime of 3–7 µs of singlet oxygen, reaction of a free diffusing singlet oxygen with another singlet dye molecule. Since the triplet lifetime of dye molecules in air-saturated water (typical lifetime of milliseconds to seconds) is determined by oxygen quenching, every transition to the triplet state is coupled with a certain probability of photobleaching by singlet oxygen produced by quenching. Furthermore, in case of some dye classes, e.g. polycyclic aromatics, singlet oxygen production via reaction 26a should also be considered with the quenching efficiency depending on the length of the fluorescence lifetime.

Another important class of photobleaching channels are reactions of oxygen in its triplet ground state (if present in a high concentration) with dye radicals or radical ions. Such radicals may either be produced by an electron- or by a hydrogen-transfer reaction in the dye triplet state (important for reactions of alcohols [163]) or by multi-photon photolysis as outlined in Sect. 2.2.

Hence, there are several reasons to select dyes for SMD that have a low triplet quantum yield, because the probability of photobleaching reactions by singlet and triplet oxygen may be minimized and a maximal fluorescence flow can be achieved (see Fig. 2A). However, reported measurements determining triplet parameters (e.g. Rhodamine 6G [44, 50, 57, 97]) or monitoring triplet state properties [171, 172] are very rare and values of these parameters are solvent dependent.

Experiments. The role of oxygen in the photobleaching process of fluorescent dyes in aqueous solution was investigated by measurements done under oxygenated and deoxygenated conditions at low and high irradiances. To change the dissolved oxygen concentration, $c(O_2)$, the sample solution was bubbled with nitrogen or oxygen for cell-bleaching experiments. Thus, three different conditions were achieved: (1) air saturation: $c(O_2) \approx 2.9 \times 10^{-4}$ M at 20°C, assuming 0.21 atm O_2 [58]; (2) oxygen saturation: $c(O_2) \approx 1.4 \times 10^{-3}$ M at 20°C, assuming 1 atm O_2 [58]; (3) oxygen free: $c(O_2) \approx 0$. For two coumarin dyes and three rhodamines, Rhodamine 6G (Rh 6G), tetramethylrhodamine (TMR), and Rhodamine 123 (Rh 123) (see Fig. 5 for molecular structures), the oxygen effects on the quantum yields of photobleaching Φ_b at low irradiances are listed in Table 2.

Considering the coumarins, the oxygen effects are very dye-specific. The stability of Coumarin-120 (7-amino-4 methylcoumarin, C-120) is highest in solution purged of oxygen by nitrogen bubbling. In an air- and in an oxygen-

Table 2. Photostabilities of Coumarin and Rhodamine Dyes in Aqueous Solution at Different Oxygen Concentrations[a]

Dye	Φ_b (air saturated)[b]	Φ_b (oxgen saturated)[c]	Φ_b (oxygen free)[d]
Coumarin-120	3.4×10^{-4}	44.0×10^{-4}	1.3×10^{-4}
Coumarin-307	$1.5 \times 10^{-4\,e}$	5.2×10^{-4}	63.0×10^{-4e}
Rhodamine 6G	1.2×10^{-6}	12.5×10^{-6}	6.0×10^{-6e}
TMR	3.3×10^{-7}	16.0×10^{-7}	5.4×10^{-7}
Rhodamine 123	6.4×10^{-7}	65.0×10^{-7}	3.0×10^{-7}

[a] The quantum yield of photobleaching, Φ_b, was calculated from the fluorescence decreases measured at our cell-bleaching set-up at low irradiances ($< 10^2$ W/cm^2). Their standard deviation is in the order 20% obtained by repeated measurements.

[b] The concentration of oxygen in an air-saturated aqueous solution is approximately $c(O_2) = 2.9 \times 10^{-4}$ M at 20°C, assuming 0.21 atm O$_2$ [58].

[c] The dye solutions were oxygen saturated by purging them with oxygen for 30 min. The concentration of oxygen in an oxygen-saturated aqueous solution is approximately $c(O_2) = 1.4 \times 10^{-3}$ M at 20°C, assuming 1 atm O$_2$ [58].

[d] Oxygen was removed from the dye solutions by purging them with nitrogen for 30 min.

[e] Decay to another fluorescing species.

bubbled solution, Φ_b is increased by a factor 3 and 30, respectively. This proves the enormous photodestructive efficiency of oxygen. Previous photodegradation studies at higher dye concentrations [107, 134–137] showed a stepwise oxidation of the chromophore at the methyl group and at the C_3–C_4 double bond. Compared to C-120, the exchange of the methyl group at C_4 for a trifluoromethyl group in Coumarin-307 (7-ethylamino-6-methyl-4-trifluoromethycoumarin, C-307) stabilizes the dyes in air-saturated solution by a factor of 4 and in oxygen-saturated solution by a factor of 9. However, C-307 is extremely unstable in a nitrogen atmosphere, where a new product with a red-shifted fluorescence is formed. In the nitrogen atmosphere, the effective quencher oxygen is absent (for the theory of this, see Sect. 6.1), which results in a tremendous build up of the triplet-state population (see Fig. 7 for the case of TMR) [50]. Thus, the obtained results for coumarins identify the reactive triplet state as a major source for oxidative and reductive photodegradation and suggest the stabilization of the dyes by the exchange of oxygen for a non-oxidizing triplet quencher.

The three rhodamine dyes differ less in terms of their oxygen effects on Φ_b in aqueous solution. If the stability in an air-saturated solution is compared with that in an oxygen-saturated solution, Φ_b increases by a factor of 5 to 10. This identifies the reaction of the triplet state with oxygen (reaction 28a) as an important reaction channel for photobleaching. In contrast to Rh 6G, the removal of oxygen has no destabilizing effect for the most stable rhodamines, TMR and Rh 123.

Comparable oxygen effects were also found for the rhodamines at high irradiances using FCS [43]. In FCS experiments the oxygen concentration in the sample droplet was changed by equilibrating the solution in pure argon/oxygen gas for at least 15 min. The influence of the cw irradiance $I_0/2$ and the oxygen content is shown for typical autocorrelation cures of TMR (Fig. 7).

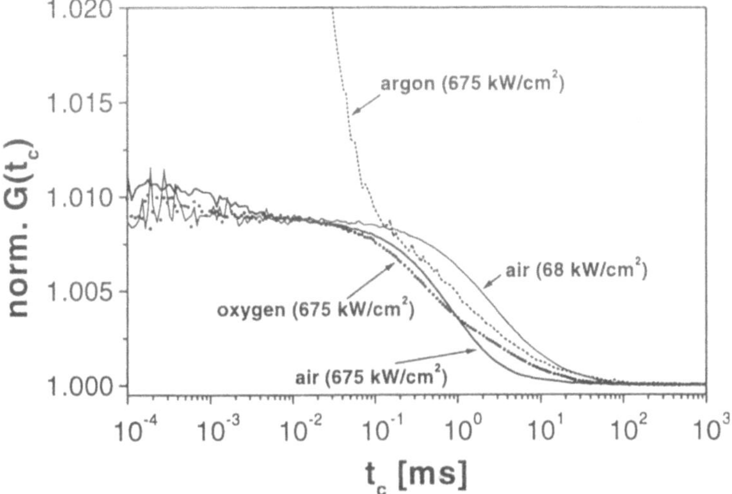

Fig. 7. Influence of the oxygen concentration and cw excitation irradiance $I_0/2$ on the fluorescence autocorrelation curves $G(t_c)$ of TMR (ca. 10^{-9} M in water). Three different oxygen concentrations (pure oxygen, air and argon atmosphere) and excitation irradiances of 675 kW/cm² or 68 kW/cm² at a wavelength of 514.5 nm were applied. For comparison, the different curves are normalized on each other. The amplitude of the autocorrelation curve of TMR in an argon atmosphere is 1.062

If the autocorrelation curve in air at a moderate irradiance ($I_0/2 = 68$ kW/cm²) is used as a standard, photobleaching of TMR is evident by the apparently reduced diffusion time for all oxygen concentrations at high irradiances ($I_0/2 = 675$ kW/cm²). Nevertheless, there are characteristic differences between the three curves obtained at high irradiances. The lower oxygen concentration in the argon atmosphere results on one hand in a reduced rate of photobleaching and is paralleled on the other hand by a strong build-up of the triplet state population. This causes a decrease of the mean excited state population, S_{1eq}, to the same or even higher extent. Therefore, the probability of photobleaching for TMR can remain unchanged or may even be increased, which was also observed for Rh 6G by Widengren and Rigler [43]. Unfortunately, it is difficult to determine the rate constant of triplet deactivation k_T under these conditions, which prevents an exact calculation of p_b. However, due to the small value of k_T below 0.01×10^6 s^{-1}, it is likely that saturation and multi-photon excitation have already occurred at very moderate irradiances.

For oxygenated TMR solutions, $k_{z,av}$ can be measured precisely as a function of $I_0/2$ (Eq. 25). Using Eqs. 13, 16, and 17, $k_{z,av}$ can be calculated from the data as well as the probability of photobleaching p_b^5, which is defined by the probability for photolysis p_{bXn} and by the quantum yield of photobleaching Φ_b^3 (compare Fig. 4B). In agreement with the cell-bleaching measurements in Table 2, Φ_b^3 is increased in the oxygenated solution by a factor of more than three, whereas p_{bXn} remains unchanged. This result implies that oxygen has no major influence on the efficiency of multi-photon photolysis. Due to the ultra-short lifetimes of the highly excited states, our model (Fig. 1B) suggests that p_{bXn} is mainly determined by the intramolecular processes of ionization with Φ_{ion} and of radical solvolysis with Φ_{solv} and not by external effects, which is consistent with the above FCS results.

To conclude, cell-bleaching and FCS measurements reveal distinct effects of the oxygen concentration on the achieved fluorescence flow and on photobleaching. Reduction of the oxygen concentration as a strategy to prevent photobleaching should be practiced with caution, in case no other non-reactive triplet quencher is added. These results show that the oxygen concentration dissolved under air pressure is in most cases a good compromise between photobleaching and prevention of fluorescence saturation due to accumulation of the dyes in their triplet states. The investigations also provide further evidence that additional photobleaching mechanisms exist which are independent of oxygen.

6.2
Effect of "Stabilizers"

A lot of effort has been made to develop protective agents which can be added to a dye solution to increase its photostability [47]. This is thought to be accomplished by blocking some of the reaction pathways for photobleaching: (1) quenching of the reactive molecules such as singlet oxygen or the lowest excited triplet state, T_1, of the dye; or (2) scavenging some of the photo-products such as radicals before irreversible decomposition.

However, one has to keep in mind that a lot of these additive compounds only reach their stabilization effect when they are added in high concentrations (>1 mM). This can lead to simultaneous quenching of the fluorescence of the dye. Thus, no increase in the mean number of fluorescence photons N_F emitted by a single molecule is obtained, a fact first pointed out by Hirschfeld [41]. For long illumination times t and low excitation irradiances, this number N_F can be expressed by the fluorescence and photobleaching quantum yield, Φ_f and Φ_b^3, respectively, using Eq. 12 and 19. Thus, if an illumination time t is regarded, which is long enough for complete bleaching of the dye, the total number of fluorescence photons N_F emitted by a single dye is independent of Φ_f.

$$N_F(t \rightarrow \infty) = \frac{\Phi_f}{\Phi_b^3} = \frac{k_f/k_0}{k_b/k_0} = \frac{k_f}{k_b} \tag{29}$$

In other words, when an additive compound quenches the fluorescence of a dye and thus reduces the fluorescence quantum yield Φ_f without affecting the bleaching rate constant k_b, the photobleaching quantum yield Φ_b^3 decreases in the same proportion. This is because both quantum yields are normalized by the same rate constant k_0, which is the sum of all depopulation processes of the excited singlet state S_1. Such fluorescence quenching agents prolong the bleaching process at the cost of a reduced fluorescence signal. Hence, no additional fluorescence photons are gained and N_F is kept constant. However, in a lot of applications of SMD with a limited observation time of the molecule or with interest in monitoring fast dynamic processes influencing the fluorescence signal, a high fluorescence flow is desired. In such cases, protective agents which quench the fluorescence are not appropriate.

In view of this restriction, we investigated several additives at low as well as at high excitation irradiance in our own cell-bleaching and FCS measurements. Furthermore, there are various reports on agents which have been tested for their suitability as stabilizers in confocal microscopy. The summary of our results and of the effects reported in the literature are listed in Table 3 for a number of agents and commercial products in alphabetical order.

Regarding Eq. 29, the stabilizing effect of the agents should be judged by several effects as listed in Table 3: (1) proposed mechanism for stabilization; (2) use for specific fluorescent dyes and applications; (3) concentration of agent; (4, 5) stabilization as well as fluorescence quenching effect quantified by the factors in parentheses; (6) the most important row giving the net stabilization effect with regards to N_F (see Eq. 29), and further comments.

Surprisingly, only a few additives really lead to a decreased quantum yield of photobleaching Φ_b and an increase in N_F without quenching the fluorescence of the dye, such as Cystamine (MEA) for Coumarin-120 and Fluorescein, Cysteine and DABCO (1,4-diazabicyclo[2.2.2]octane) for some coumarins, and the commercial stabilizers Mowiol and ProLong with unknown composition. On the other hand the radical scavengers reacting primarily as antioxidants, such as ascorbic acid, Cysteamine, and n-propyl-gallate (n-PG), and proposed triplet quenchers, such as cyclo-octa-tetraene (COT), show no effect at low irradiances but have strong stabilization properties at high irradiances [178]. It is important to state that the latter stabilizers affect the probability of photobleaching p_b^5; mainly in such a way, that Φ_b^3 remains approximately constant, but an increase in photobleaching due to non-linear effects is prevented. Taking Rhodamine 6G (Rh 6G) as an example (see Fig. 6), under standard FCS measuring conditions, p_b^5; may drop by a factor of 5 by eliminating the non-linear effects.

The stabilizing effects of n-PG were found at much lower concentrations than previously reported, which reflects that the effects exerted by anti-fading compounds depend very much on the specific environment, where they are acting. In contrast to low concentrations, n-PG in millimolar concentrations exhibits a disadvantageous effect on the fluorescence generation due to charge-transfer-induced dynamic and static quenching. COT and surprisingly Cysteamine were found to quench the triplet state of Rh 6G in ethanol and water, respectively. This led to a pronounced increase in the fluorescence emission rate per molecule. However, for Cysteamine the beneficial effect was neutralized above millimolar concentrations by an increased reduction of neutral Rh 6G molecules. These effects, found for Rh 6G in FCS measurements [178], prove the involvement of radical and triplet states in the photobleaching pathways. The influence of added ascorbic acid on the FCS curves of an aqueous Rh 6G solution is illustrated in Fig. 8.

The comparison of the normalized autocorrelation curves obtained at moderate and at high irradiances indicates that the addition of this compound prevents the appearance of an apparently reduced diffusion time (see also Fig. 4A)

Table 3. Photostabilization Properties of Several Additive Compounds to Dye Solutions

Compound (postulated mechanism)/ investigated dye [a,b] (concentration of dye)	Concentration of compound	Stabilization effect (factor) at reported I	Fluorescence quenching effect (factor)	Net effects[c]/comments
Ascorbic acid ($C_6H_8O_6$): (radical scavenger [173, 174, ow] and antioxidant [174, ow]				
Cou-120 ($< 10^{-6}$ M) [ow]	10 mM (pH 5)	low I: (2)	(2)	no net effect (low I)
Cou-120 ($< 10^{-6}$ M, O_2-sat.[b]) [ow]	10 mM (pH 5)	low I: (3.6)[d]	(2)	still more unstable than air saturated solution
Cou-307 ($< 10^{-6}$ M, O_2-sat.[b]) [ow]	10 mM (pH 5)	low I: (1.7)[d]	no	still more unstable than air saturated solution
Cou-307 ($< 10^{-6}$ M, O_2-free[b]) [ow]	10 mM (pH 5)	low I: (7)[e]	no	still more unstable than air saturated solution
Fluo-labeled tubulin [174]	4 mM	yes, I#	#	prevents photodissolution of microtubules
Rh6G ($< 10^{-7}$ M) [ow]	4 mM (pH 5 or 8.7)	low I: no high I: strong	no	strong net effect only at high I (see Fig. 7)
Cartenoids: ($^1O_2$ quencher[b] [175,176], crocetin [175], and crocin [176] in contrast to most derivatives both water soluble)				
COT [b]: (triplet quencher of Rh6G [178], hardly water soluble)				
Rh6G ($< 10^{-7}$ M, ethanol) [178]	<3 mM	high I: strong	no	fluorescence quenching at higher compound concentrations
Cysteamine (MEA[b], C_2H_7NS): (radical scavenger [99], $^1O_2$ quencher [b] [99,177] and triplet quencher of Fluo [99] and Rh6G [178])				
Cou-120 ($<10^{-6}$ M) [ow]	10 mM	low I: (2)	no	strong net effect (low I)
Fluo (10^{-8} M) [99]	100 mM	high I: strong	no	strong net effect (high I)
Rh6G ($<10^{-7}$ M) [178, ow]	$\leq$ 10 mM	low I: no high I: strong	no	strong net effect only at high I
Cysteine ($C_3H_7NO_2S$): (dye laser additive [179])				
Cou-1, Cou-311 (5 x 10^{-3} M) [179]	10 mM	low I: strong	#	lasing stabilization [179]

Table 3 (continued)

DABCO[b]: ($^1O_2$ quencher[b] [99,136,149,180,ow] and triplet quencher of Cou [136,149])

Cou-120, Cou-311, Cou-1 (5×10^{-3} M or 10^{-5} M) [136,149]	10 mM	low I: strong	no	strong net effect at high dye concentrations
Cou-120, Cou-307, Rh6G ($<10^{-6}$ M) [ow]	20 mM	low I: no	no	no net effect at low dye concentrations
Cou-120 ($<10^{-6}$ M, O_2-sat.[b]) [ow]	20 mM	low I: (2)[d]	no	still more unstable than air saturated solution
Fluo (10^{-8} M), FITC-m [99,104–106]	100–200 mM	high I: strong	(2)	hardly any net effect
POPOP[b] (#) [180]	50 mM	low I: (3)	#	

Glutathione (reduced)[b]: (molecular oxygen scavenger [177])

n-Propyl-Gallate (n-PG [b]): (antioxidant [131] and reduction of oxidized dye [178])

Cou-120, Cou-307 ($<10^{-6}$ M) [ow]	10 mM	low I: (2)	(2)	no net effect (low I)
Rh6G ($<10^{-6}$ M) [ow]	10 mM	low I: (5)	(5)	no net effect (low I)
Rh6G ($<10^{-9}$ M) [178]	< 10 mM	high I: strong	concentration dependent	strong net effect at low compound concentration and high I; quenching of fluorescence at higher compound concentration (> 1 mM)
FITC-m [104,106,131]	200 mM	high I: strong	strong	hardly any net effect
Phycobiliproteins (10^{-6} M) [132]	24 mM	low I: yes	yes	slight net effect

4-POBN[b]: (radical scavenger [99])

Fluo (10^{-8} M) [99]	100 mM	high I: strong	strong	hardly any net effect

PPD[b]:

FITC-m [104–106,181]	10 mM	high I: strong	yes	positive net effect

Urate and Vitamin E analog, Trolox ($C_{15}H_{20}O_4$): (antioxidant, protection against peroxyl radicals [173])

Table 3 (continued)

Compound (postulated mechanism)/ investigated dye [a,b] (concentration of dye)	Concentration of compound	Stabilization effect (factor) at reported I	Fluorescence quenching effect (factor)	Net effects[b]/comments
Mowiol: (commercial product; Hoechst, Germany):				
FITC-m [104]	#	high I: yes	no	strong net effect
Slowfade: (commercial product; Molecular Probes, Eugene, Or)				
FITC-m [40,104]	#	high I: strong	strong	hardly any net effect
ProLong: (commercial product; Molecular Probes, Eugene, Or)				
Fluo, Texas Red [40]	#	strong, I#	no	strong net effect

[a] Unless otherwise stated, all reported dyes were dissolved in air-saturated water.

[b] COT: Cyclooctatetrane (C_8H_8). Cou: Coumarin. DABCO: 1,4-diazabicyclo[2.2.2]octane (Triethylenediamine, $C_6H_{12}N_2$). Fluo: fluorescein. FITC-m: Fluorescein isothiocyanate stained cells mounted in buffered glycerol. I: excitation irradiance (high: $I \geq 10^4$ W/cm^2 used in fluorescence microscopy, low: $I < 10^3$ W/cm^2 used e.g. in cell-bleaching experiments). Glutathione (reduced): γ-L-Glutamyl-L-cysteinylglycine ($C_{10}H_{17}N_3O_6S$). MEA: mercaptoethylamine. $^1O_2$: singlet oxygen. O_2-sat.: oxygen-saturated solution. O_2-free: oxygen-free solution. 4-POBN: α-(4-Pyridyl 1-oxide)-N-tert-butyl nitrone. POPOP: 2,2'-(1,4-phenylene)-bis(4-phenyloxazole). PPD: p-phenylenediamine (1,4-Diamino-benzol, $C_6H_8N_2$). Rh: Rhodamine

[c] Estimated net effect with regards to the total number of fluorescence photons N_F emitted by a single dye molecule (Eq. 29).

[d] Compared to oxygen-saturated dye solution without additive.

[e] Compared to oxygen-free dye solution without additive.

ow: own work.

#: not mentioned.

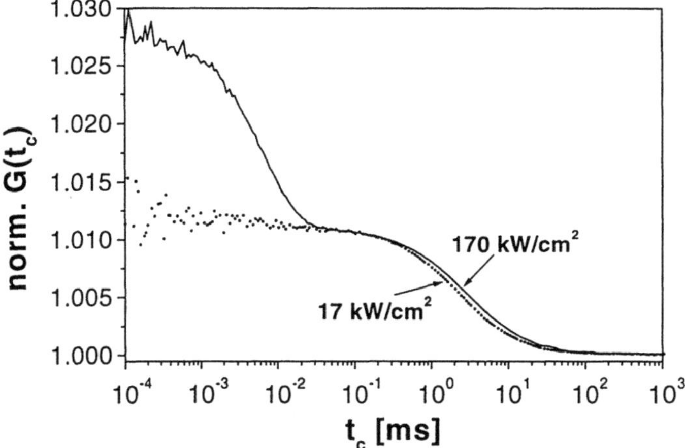

Fig. 8. Fluorescence autocorrelation curves $G(t_c)$ of Rhodamine 6G (10^{-9} M in water) with ascorbic acid (4 mM, pH 8.7) for two different cw excitation irradiances $I_0/2$ at a wavelength of 514.5 nm. For illustration the different curves are normalized on each other

and thus of a photobleaching reaction (see Sect. 4). Ascorbic acid acts as an antioxidant, but may also reduce the concentration of dissolved oxygen. The latter fact leads to a strongly increased triplet population. The uneven saturation of the dye fluorescence leads to a broadening of the fluorescence profile over the detection volume which becomes evident by the apparently increased diffusion time at the high irradiance.

These results show that FCS and the cell-bleaching method can be used as complementary methods to map photophysical and photochemical properties of "stabilizers" in order to better understand how they act and when they may be useful. Since many triplet and radical quenchers quite often unintentionally quench also the singlet state of the dye, it is important to use small concentrations (<1 mM) to minimize these additional effects.

6.3
Effects of the Excitation Mode on the Efficiency of SMD

So far, dyes have been used for single-molecule detection and identification [5–7, 15, 25, 32, 182–184] by one-photon excitation (OPE), where the S_0–S_1 absorption maxima are in a wavelength region ranging from green to NIR, i.e. fluoresceins [25, 182], boradipyrromethane dyes [25, 182], rhodamines [4–6, 30], oxazines [7], and carbocyanines [183]. However, there are problems in detecting single dye molecules with a S_0–S_1 absorption maximum in the UV region in solution by OPE. In the following, we discuss the results of our approach for SMD of the widely used dye Coumarin-120 (C-120) as a typical example.

One-photon excitation (OPE) in the UV. A comparison of OPE for C-120 and rhodamines shows that, in contrast to rhodamines, C-120 has a large Stokes shift of 6750 cm^{-1}, which allows a good spectral separation of the fluorescence from Raman scattering. This property makes C-120 in principal a good candidate for SMD, due to a low background level. A serious problem encountered with most coumarins is their low photochemical stability as shown in Table 1. The quantum yield of photobleaching on low irradiance OPE, Φ_b^3, is in the order of 10^{-3}, which is three orders of magnitude larger than that of rhodamines. An additional problem with OPE in the UV region may be two-step photolysis at high excitation irradiances [44, 118, 125].

According to Fig. 2 B, such effects become easily apparent by studying the dependence of the fluorescence flow on the irradiance. In the absence of saturation of the S_0–S_1 transition (i.e. $S_{1eq} \approx 0$) and of photobleaching, Eq. 30 gives the linear dependence of the collected time-averaged fluorescence photon flow F_P of a dilute dye solution on the focal irradiance I_0.

$$F_p = \alpha\, \sigma_{01} c N_A \times \Phi_f \times (\gamma I_0)^\beta \int_V W(x, y, z)\ dx\ dy\ dz \tag{30}$$

α is a unitless constant, which is proportional to the system efficiency for the detection of emitted fluorescence photons, σ_{01} is the one-photon absorption cross section, c is the molar concentration of the fluorescent dye, N_A is Avogadro's constant, Φ_f is the fluorescence quantum yield, $\gamma = \lambda/(hc_1)$ (Eq. 6) and β is a unitless constant for the relationship between F$_p$ and I$_0$. The spatial distribution of the detected fluorescence photons for OPE can be described by a three-dimensional Gaussian distribution $W(x, y, z)$ (Eq. 23). In the case of OPE, β is expected to be equal to 1. However, saturation and bleaching effects will cause significant deviations from this linearity at higher irradiances.

We used pulsed excitation by a frequency doubled titanium: sapphire laser at 350 nm with a repetition rate $f = 76$ MHz to measure the dependence of the fluorescence photon flow, F_P, of a 5×10^{-8} M aqueous C-120 solution on the focal quasi-cw (qcw) irradiance $I_{0,qcw}$ using time-correlated single-photon counting. Varying the excitation irradiance by five orders of magnitude, the time-resolved signal decays were measured for a fixed integration time. Each integrated signal curve was fitted by a single exponential fluorescence decay to calculate the fluorescence photon flow F_P (Fig. 9).

Up to a qcw-irradiance of approx. 5×10^3 W/cm^2, the expected linear dependence ($\beta = 1.04$) of the fluorescence flow (open circles) on $I_{0,qcw}$ is observed. However, at a higher qcw-irradiance the fluorescence signal saturates, whereas the background signal (full circles), measured on pure water, increases linearly over the entire investigated irradiance range. At a qcw-irradiance of approx. 3×10^4 W/cm^2, a rather sharp maximum of the fluorescence signal is reached and a further increase of the excitation energy results in a dramatic decrease of the fluorescence.

The calculated value of the steady state probability S_{leg}^3; of 2% at the maximum of the fluorescence signal proved that we were far from saturation. Thus, two-photon photolysis (Sect. 2.2) must limit the achievable signal level. This interpretation is supported by several additional findings: (1) after sudden illu-

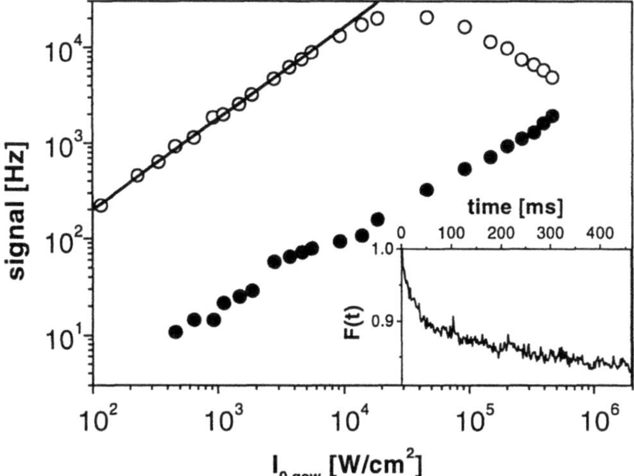

Fig. 9. Average fluorescence flow of Coumarin-120 (5×10^{-8} M) in water (open circles) and background signal of pure water (full circles) for OPE by a pulsed titanium: sapphire laser at 350 nm (pulse length 250 fs, repetition rate 76 MHz) as a function of the applied focal quasi-cw irradiance $I_{0,qcw}$. A linear fit to the fluorescence signal at low $I_{0,qcw}$ (black line) results in a slope of 1.04. The focal area and detection volume were characterized by FCS at low irradiance ($\omega_0 = 0.54$ μm, $z_0/\omega_0 = 5$). (Inset) The time-dependent fluorescence decrease of an aqueous Coumarin-120 solution (5×10^{-9} M) in a confocal epi-illuminated microscope following pulses excited with 51.4 kW/cm^2 at 350 nm

mination at high irradiances by opening a closed shutter (shutter experiment), the fluorescence flow drops significantly within several hundred milliseconds (inset Fig. 9), which is a typical depletion effect (see Fig. 2B with discussions in Sect. 3); (2) a decrease in the laser frequency shifts the saturation irradiance to higher values [18]. The observed dependence on the temporal distance of the laser pulses shows the critical role of the long-lived T_1-state in the pathway of two-step photolysis, which is very efficient for the high laser frequency.

Coherent two-photon excitation (TPE) in the NIR. Due to the limitations observed in OPE, we investigated the applicability of coherent two-photon excitation (TPE) at 700 nm for SMD of C-120. TPE has, for example, been used successfully for SMD of Rhodamine B in water [16], SMD of diphenyloctatetraene trapped in a n-tetradecane matrix [185], FCS in cells [186], as well as in fluorescence microscopy [94]. A particular feature of TPE is its dependence on the square of the applied irradiance. This permits selective excitation in the focal plane. This inherent sectioning effect avoids excessive photodamage of the fluorophore as well as excitation of impurities anywhere else in the sample.

The main task of SMD in liquids is to maximize the number of detected fluorescence photons and at the same time to minimize the background signal, because the statistical accuracy of dye characterization via fluorescence spectra or lifetimes depends on the number of collected fluorescence photons [54, 187].

Therefore, it is a prerequisite for successful SMD with TPE, that the TPE cross section of the fluorophore is sufficiently high. It should therefore be possible to reach the saturation of the dye excitation before other limiting nonlinear effects (see below) or heating of the solvent occur. From this viewpoint, coumarins are ideal candidates for SMD by TPE, because they have high TPE cross sections of approximately 20×10^{-50} cm^4 s in the wavelength region around 700 nm [188–190]. Furthermore, very high excitation irradiances should be applied to achieve a sufficient photon flow. This can be achieved by using pulsed excitation with a short pulse width and a high repetition rate, supplying a very high irradiance per pulse and on the other hand a high quasi cw-irradiance.

Analogous to Eq. 30, which neglects saturation and photobleaching, Eq. 31 describes the quadratic dependence of the collected time-averaged fluorescence photon flow F_P on the focal quasi cw-irradiance $I_{0,\mathrm{qcw}}$ [17, 93, 188–190].

$$F_p = \alpha \, \delta_{01} c N_A \frac{\Phi_f}{2} \frac{0{,}588}{\tau_p f} (\gamma I_{0,\mathrm{qcw}})^\beta \int_V W^2(x, y, z) \, dx \, dy \, dz \tag{31}$$

α, c, N_A, γ, Φ_f are used analogously to Eq. 30, δ is the two-photon absorption cross section, and β is a unitless constant for the relationship between F_P and $I_{0,\mathrm{qcw}}$ (for TPE β should be equal to 2). The factor 1/2 takes into account that two photons are needed for excitation. The periodic train of excitation pulses is characterized by the repetition rate f and the excitation pulse width τ_p. The spatial distribution of the detected fluorescence photons for TPE is described by a squared three-dimensional Gaussian distribution $W^2(x, y, z)$.

We used the confocal fluorescence microscope with excitation light from the pulsed titanium: sapphire laser at 700 nm ($\tau_p \approx 300$ fs, $f = 76$ MHz) for TPE of C-120 in water (5×10^{-8} M). The qcw-irradiance dependence of the photon flow, which was determined in the same way as described for OPE, is shown in Fig. 10A.

For an irradiance below $I_{0,\mathrm{qcw}} = 7 \times 10^6$ W/cm^2, the fluorescence flow (open circles in Fig. 9A) shows the expected quadratic dependence on $I_{0,\mathrm{qcw}}$ [17, 18, 33, 188–190] and the fluorescence decay curves (decay inset 1 of Fig. 10A) can be described by a single-exponential decay with the fluorescence lifetime of 5.0 ns typical for C-120 in water [191]. Above this irradiance $I_{0,\mathrm{qcw}}(\mathrm{sat}) = 7 \times 10^6$ W/cm^2, the fluorescence starts to saturate. If $I_{0,\mathrm{qcw}}$ exceeds 7×10^7 W/cm^2, an additional component with a short decay time becomes apparent (decay inset 2 of Fig. 10 A) and the fluorescence flow decreases. Furthermore, the background signal of pure water (solid circles, Fig. 10 A) increases dramatically above this irradiance. The additional component occurring only above a certain threshold irradiance is assigned to continuum generation ("white light") [17, 18, 192, 193], which is caused by the solvent and not by the dissolved dye. This results in a strong increase in the background signal and in a decrease in fluorescence limiting the applicable irradiance.

To investigate the photobleaching behavior of C-120 for TPE, fluorescence correlation curves were measured at a slightly expanded radius of the excitation volume ($\omega_0 = 0.98$ μm, $z_0/\omega_0 = 10.7$) to determine the average number of molecules N_{DV} in the observed detection volume (DV) as a function of the applied

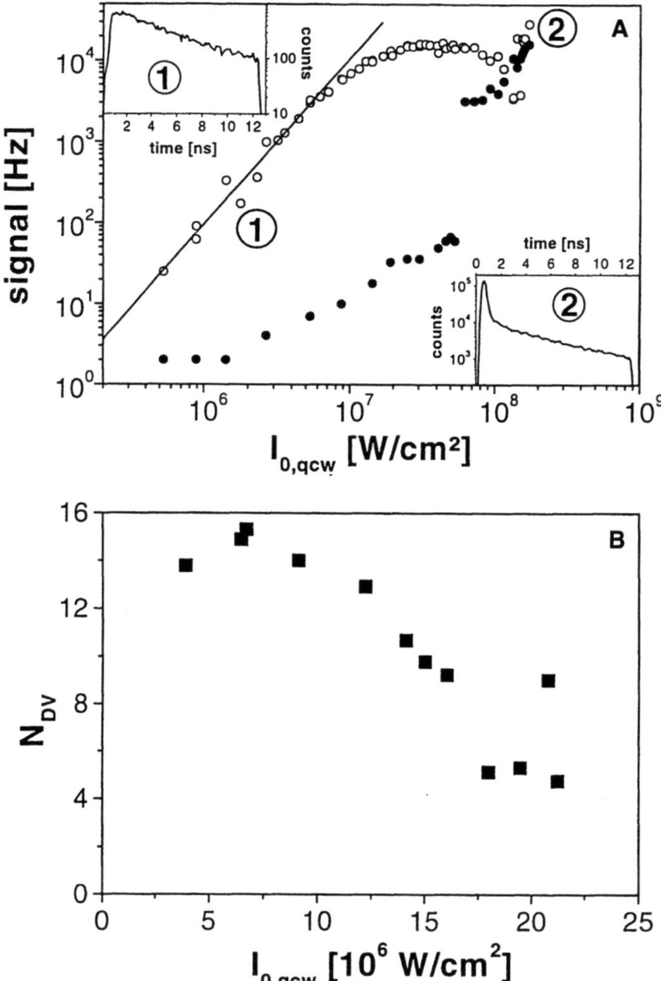

Fig. 10 A, B. (A) Average fluorescence flow of Coumarin-120 (5×10^{-8} M) in water (open circles) and background signal of pure water (solid circles) for TPE by a pulsed titanium: sapphire laser at 700 nm (pulse length 250 fs, repetition rate 76 MHz) as a function of the applied focal quasi-cw irradiance $I_{0,qcw}$. A linear fit to the logarithmic fluorescence flow at low irradiances $I_{0,qcw}$ results in a slope of 1.99. The focal area and detection volume were characterized by FCS at an excitation irradiance $I_{0,qcw} = 6 \times 10^7$ W/cm² ($\omega_0 = 0.53$ µm, $z_0/\omega_0 = 8.5$) (Inset A) Fluorescence decay curves of the same measurement for two different focal quasi-cw irradiances $I_{0,qcw}$ ((1) 2.0×10^6 kW/cm², (2) 1.5×10^8 kW/cm²). The curves were recorded for a corresponding time length of 45 s. (B) Dependence of the average number N_{DV} of Coumarin-120 molecules in the detection volume on the applied focal quasi-cw irradiance $I_{0,qcw}$ for TPE determined by FCS. The different fluorescence correlation curves were recorded under the same optical conditions and with the same dye solution (10^{-9} M in water). The detection volume was characterized from FCS at low irradiance ($\omega_0 = 0.98$ µm, $z_0/\omega_0 = 10.7$)

qcw-laser irradiance $I_{0,\text{qcw}}$. Thereby, N_{DV} was calculated by fitting the measured autocorrelation curves to Eq. 23 and using the diffusion term with $\tau_{\text{D}} = \omega_0^2/8D$ because of TPE [17, 18]. To correct the decrease in the amplitude $G(t_c = 0)$ caused by background signal, the ratio of the background flow I_{B} to the total signal flow I_{S} ($I_{\text{S}} = F + I_{\text{B}}$) is taken into account by using the correction term $(1 - I_{\text{B}}/I_{\text{S}})^2/N$ instead of $1/N$ [194]. The obtained absolute number of molecules N_{DV} (Fig. 10B) as well as the radial radius ω_0 of detection volume decrease slightly. This might indicate slight photobleaching.

However, we note that the normalized number of molecules $N_{\text{N}}(I_{0,\text{qcw}}) = (N_{\text{DV}}(I_{0,\text{qcw}}) \times \omega_{0,\text{low irradiance}})/\omega_0(I_{0,\text{qcw}})$ remains approximately constant (≈ 15) over the whole range of applied laser irradiances. This is important because we cannot fully exclude self-focusing effects of the laser beam at high irradiances, which is known to occur for continuum generation. Fluorescence bleaching curves obtained by instantaneous illumination in shutter experiments support the arguments for slight bleaching effects of C-120. The effect of enhanced photobleaching using TPE was reported by Sánchez et al. [195] for the imaging of single Rhodamine B molecules dispersed on a glass substrate even at low irradiances. This brings us to the still open question as to whether the photobleaching efficiency of OPE and TPE should be equal at low irradiances or whether the different initially populated excited states may have a marked difference in their photophysical and photochemical parameters.

Finally, we could show that both techniques, OPE and TPE, were suitable for SMD and identification of a fluorescent dye, which has its one-photon $S_0 - S_1$ absorption maximum in the near UV [17]. The signal-to-background ratio of 1300 for TPE was more than three times higher than that of OPE, due to the more efficient rejection of background fluorescence. On the other hand the efficiency of TPE is limited by the competing non-linear processes of the solvent and by dye-specific multi-photon photolysis. Experiments are underway to study the dependence of TPE efficiency for SMD in liquids on the laser pulse width. The applicable laser irradiance and, thus, the accuracy of SMD of Coumarin-120 by OPE is limited by very efficient two-step photolysis.

7
Concluding Remarks

On a theoretical and experimental basis we have surveyed various effects of the photophysical and photochemical parameters on the fluorescence signal, achievable in single-molecule spectroscopy and scanning confocal microscopy. We have discussed two complementary experimental methods, cell-bleaching and FCS, which allow for the study of kinetic and other aspects of photobleaching simultaneously. The proved non-linear dependence of the photobleaching rate on the irradiance, which corresponds to a non-constant, increased yield of photobleaching at or close to saturation irradiances, should be borne in mind, when trying to reach optimal signal-to-background conditions for FCS and SMD experiments. The definition of the photobleaching probability $p_{\text{P}}^5(I)$ (Eq. 4) based on a molecular model with five electronic levels allows for a general description of the observed photobleaching effects at cw-excitation. For a low irradiance it reduces to the classical expression of the quantum yield Φ_b, while the influence of two-step photolysis on photobleaching, being of importance at high irradiances, is also described within the theoretical framework. The most photostable fluorescent dyes in aqueous solution found so far are Tetramethyl-rhodamine and Rhodamine 123 with an average of more than 10^6 mean survived absorption cycles.

Two extremes in SMD should be considered with respect to a maximum fluorescence signal. In solution the number of detected fluorescence photons N_F (Eq. 19) is limited by the dwell time t_t of the free diffusing molecule/particle, which depends on its diffusion constant and on the size of the detection volume. For short observations times ($t_t \approx 100$ µs – 10 ms), rhodamines are sufficiently stable even at high irradiances. If, however, immobile molecules/particles are investigated, photobleaching determines the absolute number of detectable photons within the illumination time. Thus, strategies to maximize the fluorescence signal should be examined and the addition of stabilizers should be taken into account. At very low background levels, excitation irradiances below saturation can be recommended to avoid multi-photon photolysis via excited singlet and triplet states. Considering the effects of oxygen on the fluorescence signal, reduction of the oxygen concentration as a strategy to prevent photobleaching should be practiced with caution. On one hand, oxygen is a very effective triplet quencher, on the other hand, the produced singlet oxygen is chemically very reactive, which opens effective photodestruction pathways. For certain dyes a reduced triplet quenching may even result in an increase in the photobleaching quantum yield, due to the high triplet state population. Thus, added stabilizing agents should perform several functions: selective triplet and no fluorescence quenching, as well as scavenging of dye radicals. However, many stabilizers presently known are not appropriate or cannot be applied in water.

Acknowledgements. We are grateful to J. Troe, J. Wolfrum, and K. H. Drexhage for generous support of this work. We thank C. Zander and M. Sauer for gradual collaboration. We also thank M. Köllner, A. Schönle, and S. Hell for fruitful discussions. We thank J. Mertz for his comments on photobleaching using TPE. We are grateful to K.-O. Greulich and S. Monajembashi for loaning us a Zeiss Ultrafluar 100x microscope objective. We would like to thank the Bundesministerium für Bildung und Forschung for financial support under grants 0310806.

References

1. Eigen, M.; Rigler, R. Proc.Natl.Acad.Sci.USA 1994, 91, 5740 – 5747
   Rigler, R. J.Biotech. 1995, 41, 177–186
3. Keller, R.A.; Ambrose, W.P.; Goodwin, P.M.; Jett, J.H.; Martin, J.C.; Wu, M. Appl.Spec. 1996, 50, 12A–32A
4. Jett, J.H.; Keller, R.A.; Martin, J.C.; Marrone, B.L.; Moyzis, R.K.; Ratliff, R.L.; Seitzinger, N.K.; Shera, E.B.; Stewart, C.C.J. J.Biomol.Struct.Dynam. 1989, 7, 301–309
5. Soper, S.A.; Davis, L.M.; Brooks Shera, E. J.Opt.Soc.Am.B 1992, 9, 1761–1769

6. Zander, C.; Sauer, M.; Drexhage, K.H.; Ko, D.S.; Schulz, A.; Wolfrum, J.; Brand, L.; Eggeling, C.; Seidel, C.A.M. Appl.Phys.B 1996, 63, 517–523

7. Müller, R.; Zander, C.; Sauer, M.; Deimel, M.; Ko, D.-S.; Siebert, S.; Arden-Jacob, J.; Deltau, G.; Marx, N.J.; Drexhage, K.H.; Wolfrum, J. Chem.Phys.Lett. 1996, 262, 716–722

8. Enderlein, J.; Goodwin, P.M.; van Orden, A.; Ambrose, W.P.; Erdmann, R.; Keller, R.A. Chem.Phys.Lett. 1997, 270, 464–470

9. Sauer, M.; Arden-Jacob, J.; Drexhage, K.-H.; Göbel, F.; Lieberwirth, U.; Mühlegger, K.; Müller, R.; Wolfrum, J.; Zander, C. Bioimaging 1998, 6(1), 14–24

10. Chen, D.; Dovichi, N.J. Anal.Chem. 1996, 68, 690–696

11. Ambrose, W.P.; Goodwin, P.M.; Jett, J.H.; Johnson, M.E.; Martin, J.C.; Marrone, B.L.; Schecker, J.A.; Wilkerson, C.W.; Keller, R.A.; Haces, A.; Shih, P.-J.; Harding, J.D. Ber.Bunsenges.Phys.Chem. 1993, 97, 1535–1542

12. Dörre, K.; Brakmann, S.; Brinkmeier, M.; Han, K.-T.; Riebeseel, K.; Schwille, P.; Stephan, J.; Wetzel, T.; Lapczyna, M.; Stuke, M.; Bader, R.; Hinz, M.; Seliger, H.; Holm, J.; Eigen, M.; Rigler, R. Bioimaging 1997, 5, 139–152

13. Eggeling, C.; Fries, J.R.; Brand, L.; Günther, R.; Seidel, C.A.M. Proc.Natl.Acad.Sci.USA 1998, 95, 1556–1561

14. Affleck, R.L.; Ambrose, W.P.; Demas, J.N.; Goodwin, P.M.; Schecker, J.A.; Wu, M.; Keller, R.A. Anal.Chem. 1996, 68, 2270–2276

15. Mets, Ü.; Rigler, R. J.Fluores. 1994, 4, 259–264

16. Mertz, J.; Xu, C.; Webb, W.W. Opt.Lett. 1995, 20, 2532–2534

17. Brand, L.; Eggeling, C.; Zander, C.; Drexhage, K.H.; Seidel, C.A.M. J.Phys.Chem. 1997, 101, 4313–4321

18. Eggeling, C.; Brand, L.; Seidel, C.A.M. Bioimaging 1997, 5, 105–115

19. Sauer, M.; Zander, C.; Müller, R.; Ullrich, B.; Drexhage, K.H.; Kaul, S.; Wolfrum, J. Appl.Phys.B 1997, 65, 427–431

20. Lyon, W.A.; Nie, S. Anal.Chem. 1997, 69, 3400–3405

21. Zander, C.; Drexhage, K.H. J.Fluores. 1996, 7, 37S–39S

22. Goodwin, P.M.; Ambrose, W.P.; Keller, R.A. Acc.Chem.Res. 1996, 29, 607–613 and further articles of this issue

23. Li, L.-Q.; Davis, L.M. Appl.Opt. 1995, 34, 3208–3217

24. Lermer, N.; Barnes, M.D.; Kung, C.-Y.; Whitten, W.B.; Ramsey, J.M. Anal.Chem. 1997, 69, 2115–2121

25. Nie, S.; Chiu, D.T.; Zare, R.N. Anal.Chem. 1995, 67, 2849–2857

26. Rigler, R.; Widengren, J. In BioScience; Klinge, B., Owman, C., Eds.; Lund University Press: Lund, Sweden, 1990; 180–183

27. Dickson, R.M.; Cubitt, A.B.; Tsien, R.Y.; Moerner, W.E. Nature 1997, 388, 355–358

28. Dickson, R.M.; Norris, D.J.; Tzeng, Y.L.; Moerner, W.E. Science 1996, 274, 966–969

29. Xu, X.-H.; Yeung, E.S. Science 1997, 275, 1106–1109

30. Shera, E.B.; Seitzinger, N.K.; Davis, L.M.; Keller, R.A.; Soper, S.A. Chem.Phys.Lett. 1990, 174, 553–557

31. Rigler, R.; Mets, Ü. SPIE 1992, 1921, 239–248

32. Sauer, M.; Drexhage, K.H.; Zander, C.; Wolfrum, J. Chem.Phys.Lett 1996, 254, 223–228

33. Brand, L.; Eggeling, C.; Seidel, C.A.M. Nucleosides & Nucleotides 1997, 16, 551–556

34. Raue, R.; Harnisch, H. Heterocycles 1984, 21, 167–190

35. Patonay, G.A., Antoine, M.D. Anal.Chem. 1991, 63, 321A–327A

36. Fabian, J.; Zahradnik, R. Angew.Chem. 1989, 6, 693–848

37. Griffiths, J. Colour and Constitution of Organic Molecules; Academic Press: London, New York, San Franciso, 1976

38. Fabian, J.; Hartmann, H. Light Absorption of Organic Colorants. Theoretical Treatment and Empirical Rules; Springer-Verlag: Berlin, Heidelberg, New York, 1980

39. Prasanna de Silva, A.; Nimal Gunaratne, H.Q.; Gunnlaugsson, Th.; Huxley, A.J.M.; McCoy, C.P.; Rademacher, J.T.; Rice, T.E. Chem.Rev. 1997, 97, 1515–1566

40. Haugland, R.P. Handbook of Fluorescent Probes and Research Chemicals; Molecular Probes, Inc.: Eugene, OR, 1996

41. Hirschfeld, T. Appl.Opt. 1976, 15, 3135–3139
42. Mathies, R.A.; Stryer, L. In Applications of Fluorescence in the Biomedical Sciences; Alan R. Liss: New York: 1986; 129–140
43. Widengren, J.; Rigler, R. Bioimaging 1996, 4, 149–157
44. Eggeling, C.; Widengren, J.; Rigler, R.; Seidel, C.A.M. Anal.Chem. 1998, 70, 2651–2659
45. Antonov, V.S.; Hohla, K.L. Appl.Phys.B 1983, 30, 109–116
46. Antonov, V.S.; Hohla, K.L. Appl.Phys.B 1983, 32, 9–14
47. Tsien, R.Y.; Waggoner, A. In Handbook of Biological Confocal Microscopy; Pawley, J.B., Ed.; Plenum Press: New York, 1995; 267–279
48. Mathies, R.A.; Peck, K. Anal.Chem. 1997, 62, 1786–1791
49. Widengren, J.; Rigler, R.; Mets, Ü. J.Fluores. 1994, 4, 255–258
50. Widengren, J.; Mets, Ü.; Rigler, R. J.Phys.Chem. 1995, 99, 13368–13379
51. Enderlein, J. Appl.Opt. 1995, 34, 514–526
52. Enderlein, J.; Robbins, D.L.; Ambrose, W.P.; Goodwin, P.M.; Keller, R.A. J.Phys.Chem.B 1997, 101, 3626–3632
53. Enderlein, J. Phys.Lett.A 1996, 221, 427–733
54. Köllner, M. Appl.Opt. 1993, 32, 806–820
55. Köllner, M. PhD-thesis Ruprecht-Karls-Universität Heidelberg, Germany, 1993
56. Thiel, E.; Drexhage, K.H. Chem.Phys.Lett 1992, 199, 329–334
57. Thiel, E. Habilitation-thesis Universität-GH Siegen, Germany, 1995
Heupel, M.; Thiel, E.J. Fluoresc. 1997, 7, 371–375
58. Murov, S.L.; Carmichael, I.; Hug, G.L. Handbook of Photochemistry; Marcel Dekker Inc.: New York, 1993
59. Carmichael, I.; Hug, G.L. J.Phys.Chem.Ref.Data 1986, 15, 1–250
60. Jasny, J.; Sepit, J.; Karpiuk, J.; Gilewski, J. Rev.Sci.Instrument. 1994, 65, 3646–3652
61. Kask, P.; Piksarv, P.; Mets, Ü. Eur.Biophys.J. 1985, 12, 163–166
62. Widengren, J.; Dapprich, J.; Rigler, R. Chem.Phys. 1997, 216, 417–426
63. Nikogosyan, D.N. Int.J.Radiat.Biol. 1990, 57, 233–299
64. Khoroshilova, E.V.; Nikogosyan, D.N. J.Photochem.Photobiol.B:Biology 1990, 413–427
65. Reuther, A.; Nikogosyan, D.N.; Laubereau, A. J.Phys.Chem. 1996, 100, 5570–5577
66. Hart, E.J.; Anbar, M. The Hydrated Electron; Wiley Interscience: New York, 1970
67. Rylkov, V.V.; Cheshev, E.A. Opt.Spectrosc.(USSR) 1987, 63, 608–611
68. Rylkov, V.V.; Cheshev, E.A. Opt.Spectrosc.(USSR) 1987, 63, 462–465
69. Becker, H.G.O. Einführung in die Photochemie; Georg Thieme Verlag: Stuttgart, New York, 1983
70. von Bünau, G.; Wolff, T. Photochemie; VCH: Weinheim, Germany, 1987
71. Gilbert, A.; Baggott, J. Essentials of Molecular Photochemistry; Blackwell Scientific Publications: Oxford, 1991
72. Turro, N.J. Modern Molecular Photochemistry; University Science Books: Mill Valley, CA, 1991
73. Schroeder, J.; Troe, J. In Reaction Dynamics in Cluster and Condensed Phases; Kluver Academic: 1994; 361
74. Schroeder, J.; Troe, J.; Vöhringer, P. Z.Phys.Chem. 1995, 188, 287
75. Mohrschladt, R.; Schroeder, J.; Schwarzer, D.; Troe, J.; Vöhringer, P. J.Chem.Phys. 1994, 101, 7566
76. Waldeck, H. Chem.Rev. 1991, 91, 415
77. Arai, T.; Tokumaru, K. In Advances in Photochemistry; Neckers, D.C., Volman, D.H., von Bünau, G., Eds.; John Wiley & Sons, Inc.: New York, 1995; 1–58
78. Görner, H.; Kuhn, H.J. Adv.Photochem. 1995, 19, 1
79. Voth, G.A.; Hochstrasser, R.M. J.Phys.Chem. 1996, 100, 13034
80. Saltiel, J.; Sun, Y.-P. In Photochromism, Molecules, and Systems; Dürr, H., Bouas-Laurent, H., Eds.; Elsevier: Amsterdam, 1990; 64
81. Ernst, L.A.; Gupta, R.K.; Mujumdar, R.B.; Waggoner, A.S. Cytometry 1989, 10, 3–10
82. Mujumdar, R.B.; Ernst, L.A.; Mujumdar, S.R.; Waggoner, A.S. Cytometry 1989, 10, 11–19
83. Krieg, M.; Redmond, R.W. Photochem.Photobiol. 1993, 57, 472–479

84. Korobov, V.E.; Chibisov, A.K. Russ.Chem.Rev. 1983, 52, 27–42

85. Noukakis, D.; Vanderauweraer, M.; Toppet, S.; DeSchryver, F.C. J.Phys.Chem. 1995, 99, 11860–11866

86. Dipaolo, R.E.; Scaffardi, L.B.; Duchowicz, R.; Bilmes, G.M. J.Phys.Chem. 1995, 99, 13796–13799

87. Nau, W.M.; Adam, W.; Scaiano, J.C. J.Am Chem.Soc. 1996, 118, 2742–2743

88. Nau, W.M.; Adam, W.; Scaiano, J.C. Chem.Phys.Lett 1996, 253, 92–96

89. Shimizu, O. Photobiochem.Photobiophys. 1982, 4, 347–357

90. Grabner, G.; Köhler, G.; Marconi, G.; Monti, S.; Venuti, E. J.Phys.Chem. 1990, 94, 3609

91. Baugher, J.F.; Grossweiner, L.I. J.Phys.Chem. 1977, 81, 1349–1354

92. Rahn, R.; Schroeder, J.; Troe, J.; Grellmann, K.H. J.Phys.Chem. 1989, 93, 7841–7846

93. Göppert-Mayer, M. Ann.Phys. 1931, 9, 273–295

94. Denk, W.; Strickler, J.H.; Webb, W.W. Science 1990, 24, 73–76

95. Denk, W.; Piston, D.W.; Webb, W.W. In Handbook of Biological Confocal Microscopy; Pawley, J.B., Ed.; Plenum Press: New York, 1995

96. Schmittel, M.; Burghart, A. Angew.Chem.(Int.Ed.Engl.) 1997, 36, 2550–2589

97. Korobov, V.E.; Chibisov, A.K. J.Photochem. 1978, 9, 411–424

98. Song, L.; Hennink, E.J.; Young, I.T.; Tanke, H.J. Biophys.J. 1995, 68, 2588–2600

99. Song, L.; Varma, C.A.G.O.; Verhoeven, J.W.; Tanke, H.J. Biophys.J. 1996, 70, 2959–2968

100. Grellmann, K.H.; Scholz, H.-G. Chem.Phys.Lett. 1979, 62, 64–71

101. Gauglitz, G.; Lorch, A.; Oelkrug, D. Z.Naturforsch.A 1982, 37a, 219–223

102. Polster, J. Reaktionskinetische Auswertung spektroskopischer Messdaten; Vieweg: Braunschweig, Germany, 1995

103. Platsch PhD-thesis Universität Würzburg, Germany, 1986

104. Longin, A.; Souchier, C.; Ffrench, M.; Bryon, P.-A. J.Histochem.Cytochem. 1993, 41, 1833–1840

105. Johnson, G.D.; Davidson, R.S.; McNamee, K.C.; Russell, G.; Goodwin, D.; Holborow, E.J. J.Immunol.Methods 1982, 55, 231–242

106. Krenik, K.D.; Kephart, G.M.; Offord, K.P.; Dunnette, S.L.; Gleich, G.J. J.Immunol.Methods 1989, 117, 91–97

107. Kunjappu, J.T. J.Photochem.Photobiol. 1991, 56, 365–368

108. Dunne, A.; Quinn, M.F. J.Chem.Soc.Faraday 1959, I, 1104–1110

109. Dempster, D.N.; Morrow, T.; Quinn, M.F. J.Photochem. 1973, 2, 343–359

110. Stevens, B.; Sharpe, R.R.; Bingham, W.S.W. Photochem.Photobiol. 1967, 6, 83–89

111. Suppan, P. J.Photochem.Photobiol.A 1990, 50, 293–330

112. Suppan, P.; Ghoneim, N. Solvatochromism; Royal Society of Chemistry: Cambridge, England, 1997

113. Reichardt, Ch. Chem.Rev. 1994, 94, 2319–2358

114. Reichardt, Ch. Solvents and solvent effects in organic chemistry; VCH: Weinheim, Germany, 1988

115. Drexhage, K.H. In Topics in Applied Physics, Vol.1; Schäfer, F.P., Ed.; Springer Verlag: Berlin, 1973; 144–192

116. Zander, C.; Drexhage, K.H. In Advances in Photochemistry, Vol.20; Neckers, D.C., Volman, D.H., von Bünau, G., Eds.; John Wiley & Sons, Inc.: New York, 1995; 59–78

117. Soper, S.A.; Nutter, H.L.; Keller, R.A.; Davis, L.M.; Brooks Shera, E. Photochem.Photobiol. 1993, 57, 972–977

118. Seidel, C.A.M. PhD-thesis Ruprecht-Karls-Universität Heidelberg, Germany, 1992

119. Anbar, M.; Hart, E.J. J.Am.Chem.Soc. 1964, 86, 5633–5637

120. Nie, S.; Emory, S.R. Science 1997, 275, 1102–1106

121. Kneipp, K.; Wang, Y.; Kneipp, H.; Perelman, L.T.; Itzkan, I.; Dasari, R.R.; Feld, M.S. Phys.Rev.Lett. 1997, 78, 1667–1670

122. Siegmann, A.E. Lasers; University Science Books: Mill Valley, Ca., 1986

123. Qian, H.; Elson, E.L. Appl.Opt. 1991, 30, 1185–1195

124. Rigler, R.; Mets, Ü.; Widengren, J.; Kask, P. Eur.Biophys.J. 1993, 22, 169–175

125. Eggeling, C. Diploma-thesis Georg-August-Universität Göttingen, Germany, 1996

126. Beer, D.; Weber, J. Opt.Commun. 1972, 5, 307–309
127. Rosenthal, I. Opt.Commun. 1978, 24, 164–166
128. Mathis, H. Personal communication, 1997
129. Chem.Abstr. 1984, 100, 77083
130. Enderlein, J.; Klose, E.; Krahl, R.; Strehmel, B.; Weissflog, R. Exp.Tech.Phys. 1991, 39, 559–565
131. Giloh, H.; Sedat, J.W. Science 1982, 217, 1252–1255
132. White, J.C.; Stryer, L. Anal.Biochem. 1987, 161, 442–452
133. Kuznetsova, N.A.; Kaliya, O.L.; Shevtsov, V.K.; Luk'yanets, E.A. Zhurnal Strukturnoi Khimii 1986, 27, 97–101
134. Jones II, G.; Bergmark, W.R.; Jackson, W.R. Opt.Commun. 1984, 50, 320–324
135. Jones II, G.; Bergmark, W.R. J.Photochem. 1984, 26, 179–184
136. Kunjappu, J.T.; Rao, K.N. J.Photochem. 1987, 39, 135–143
137. Winters, B.H.; Mandelberg, H.I.; Mohr, W.B. Appl.Phys.Lett. 1974, 25, 723–725
138. Birks, J.B. Photophysics of Aromatic Molecules; J.Wiley & Sons Ltd.: London, 1970
139. Magde, D.; Elson, E.L.; Webb, W.W. Phys.Rev.Lett. 1972, 29, 705–708
140. Elson, E.L.; Magde, D. Biopolymers 1974, 13, 1–27
141. Ehrenberg, M.; Rigler, R. Chem.Phys. 1974, 4, 390–401
142. Thompson, N.L. In Topics in Fluorescence Spectroscopy, Volume 1: Techniques; Lakowicz, J.R., Ed.; Plenum Press: New York, 1991; 337–378
143. Rigler, R.; Widengren, J.; Mets, Ü. In Fluorescence Spectroscopy. New Methods and Applications; Wolfbeis, O.S., Ed.; Springer-Verlag: Berlin, 1993; 13–24
144. Aragón, S.R.; Pecora, R. J.Chem.Phys. 1976, 64, 1791–1803
145. Schmidt, Th.; Schütz, G.J.; Baumgartner, W.; Gruber, H.J.; Schindler, H. Proc.Natl. Acad.Sci.USA 1996, 93, 2926–2929
146. Brackmann, U. Lambdachrome Laser Dyes; Lambda Pysik GmbH: Göttingen, Germany, 1994
147. Wu, M.; Goodwin, P.M.; Ambrose, W.P.; Keller, R.A. J.Phys.Chem. 1996, 100, 17406–17409
148. Huston, A.L.; Reimann, C.T. Chem.Phys. 1991, 149, 401–407
149. von Trebra, R.; Koch, T.H. Chem.Phys.Lett. 1982, 93, 315–317
150. Priyadarsini, K.I.; Naik, D.B.; Moorthy, P.N. Chem.Phys.Lett 1988, 148, 572–576
151. Priyadarsini, K.I.; Naik, D.B.; Moorthy, P.N. J.Photochem.Photobiol.A:Chemistry 1990, 46, 251–261
152. Hinckley, D.A.; Seybold, P.G.; Borris, D.P. Spectroch.Acta 1988, 44A, 1053–1059
153. Hinckley, D.A.; Seybold, P.G.; Borris, D.P. Spectroch.Acta 1986, 42A, 747–754
154. Karpiuk, J.; Grabowski, Z.R.; Schryver, F.C. J.Phys.Chem. 1994, 98, 3247–3256
155. Vogel, M.; Rettig, W.; Sens, R.; Drexhage, K.H. Chem.Phys.Lett. 1988, 147, 452–460
156. Rettig, W. In Topics in Current Chemistry; Springer-Verlag: Berlin, Heidelberg, Germany, 1994; 254–299
157. Zachariasse, K.A.; Grobys, M.; Tauer, E. Chem.Phys.Lett 1997, 274, 372–382
158. Zachariasse, K.A.; Grobys, M.; von der Haar, Th.; Hebecker, A.; Il'ichev, Yu.V.; Jian, Y.-B.; Morawski, O.; Kühnle, W. J.Photochem.Photobiol.A:Chemistry 1996, 102, 59–70
159. Zachariasse, K.A.; Grobys, M.; von der Haar, Th.; Hebecker, A.; Il'ichev, Yu.V.; Morawski, O.; Rückert, I.; Kühnle, W. J.Photochem.Photobiol.A:Chemistry 1997, 105, 373–383
160. de Lange, M.C.C.; Thorn Leeson, D.; van Kuijk, A.H.; Huizer, A.H.; Varma, C.A.G.O. Chem.Phys. 1993, 177, 243
161. Sobolewski, A.L.; Domcke, W. Chem.Phys.Lett. 1996, 250, 428
162. Sobolewski, A.L.; Domcke, W. Chem.Phys.Lett. 1996, 259, 119
163. Braun, A.M.; Maurette, M.-T.; Oliveros, E. Photochemical Technology; John Wiley & Sons Ltd: Chichester, England, 1991
164. Wilkinson, F.; Helman, W.P.; Ross, A.B. J.Phys.Chem.Ref.Data 1993, 22, 113–219
165. Wilkinson, F.; Helman, W.P.; Ross, A.B. J.Phys.Chem.Ref.Data 1995, 24, 663–1021
166. Nau, W.M.; Scaiano, J.C. J.Phys.Chem. 1996, 100, 11360–11367
167. Vaughan, W.M.; Weber, G. Biochem. 1970, 9, 464–473
168. Lakowicz, J.R. Principles of fluorescence spectroscopy; Plenum Press: 1983

169. Kavarnos, G.J.; Turro, N.J. Chem.Rev. 1986, 86, 401–449
170. Kikuchi, K.; Sato, C.; Watabe, M.; Ikeda, H.; Takahashi, Y.; Miyashi, T. J.Am.Chem.Soc. 1993, 115, 5180
171. Pavlopoulos, T.G. Appl.Opt. 1992, 31, 1261–1266
172. Pavlopoulos, T.G.; Golich, D.J. J.Appl.Phys. 1988, 64, 521–527
173. Glazer, A.N. FASEB J. 1988, 2, 2487–2491
174. Vigers, G.P.A.; Coue, M.; McIntosh, J.R. J.Cell Biol. 1988, 107, 1011–1024
175. Matheson, I.B.C.; Rodgers, M.A.J. Photochem.Photobiol. 1982, 36, 1–4
176. Manitto, P.; Speranza, G.; Monti, D.; Gramatica, P. Tetrahedron Lett. 1987, 28, 4221–4224
177. Sheetz, M.P.; Koppel, D.E. Proc.Natl.Acad.Sci.USA 1979, 76, 3314–3317
178. Widengren, J.; Eggeling, C.; Seidel C.A.M. in preparation 1998
179. von Trebra, R.; Koch, T.H. Appl.Phys.Lett. 1982, 42, 129–131
180. Liphardt, B.; Lüttke, W. Opt.Commun. 1983, 48, 129–133
181. Johnson, G.D.; de C.Nougueira Araujo, G.M. J.Immunol.Methods 1981, 43, 349–350
182. Nie, S.; Chiu, D.T.; Zare, R.N. Science 1994, 266, 1018–1021
183. Soper, S.A.; Mattingly, Q.L.; Vegunta, P. Anal.Chem. 1993, 65, 740–747
184. Lee, Y.-H.; Maus, R.G.; Smith, B.W.; Wineforder, J.D. Anal.Chem. 1994, 66, 4142–4149
185. Plakhotnik, T.; Walser, D.; Pirotta, M.; Renn, A.; Wild, U.P. Science 1996, 271, 1703–1707
186. Berland, K.M.; So, P.T.C.; Gratton, E. Biophys.J. 1995, 68, 694–701
187. Köllner, M.; Fischer, A.; Arden-Jacob, J.; Drexhage, K.H.; Müller, R.; Seeger, S.; Wolfrum, J. Chem.Phys.Lett. 1996, 250, 355–360
188. Fischer, A.; Cremer, C.; Stelzer, E.H.K. Appl.Optics 1995, 34, 1989–2002
189. Xu, C.; Webb, W.W. J.Opt.Soc.Am.B 1996, 13, 481–491
190. Xu, C.; Williams, R.M.; Zipfel, W.; Webb, W.W. Bioimaging 1997, 4, 198–208
191. Seidel, C.A.M.; Schulz, A.; Sauer, M.H.M. J.Phys.Chem. 1996, 100, 5541–5553
192. Alfano, R.R. The Supercontinuum Laser Source; Springer: New York, 1989
193. Friedrich, D.M. J.Chem.Education 1982, 59, 472–481
194. Koppel, D.E. Phys.Rev.A 1974, 10, 1938–1945
195. Sanchez, E.J.; Novotny, L.; Holtom, G.R.; Xie, X.S. J.Phys.Chem.A 1997, 101, 7019–7023

Analysis of Chemical Dynamics and Technical Combustion by Time-Resolved Laser-Induced Fluorescence

H.-R. Volpp, C. Schulz, J. Wolfrum

1
Introduction

Combustion as one of the oldest and most successful technologies of mankind is still the most important technology providing the energy supply for our modern society. However it leads to the release of unwanted pollutants such as unburned hydrocarbons and nitric oxides which affect our environment [1]. Significant improvements of combustion processes are necessary to lower pollutant formation while, at the same time, maintaining a high efficiency for the conversion of chemical to thermal, mechanical and electrical energy. Until now, the construction and optimization process of technical combustion devices in general is an essentially empirical process calling, to a large degree, on the experience of engineers and technicians employing methods based on cut and trial processes and global performance measurements. With the growing number of performance and environmental protection requirements that must be met, this kind of approach is reaching its limits. Combustion processes consist of a complex interaction of chemistry and transport processes [2]. As a consequence, the development of numerical tools that can accurately predict combustion and its characteristics for practical applications has to be based on a detailed understanding of the underlying chemical reaction system and physical processes.

In recent years, the development of laser-based diagnostic techniques has contributed significantly to the understanding of the kinetics and dynamics of elementary chemical reactions as well as to the understanding of the large variety of phenomena that occur in complex gas-phase chemical reaction systems [3–5]. In the following we will start by highlighting the experimental possibilities that laser-based fluorescence methods have created for studies of microscopic details of bimolecular reactions. After this we will present selected examples in which laser diagnostic methods are applied to the investigation of technical combustion processes [2].

2
Laser-Induced Fluorescence Studies of Chemical Dynamics

2.1
The Concept of Elementary Reactions

Thermodynamic laws can be used to determine the equilibrium state of a chemical reaction system like the H_2/O_2 combustion, which is described by the overall reaction equation:

$$2\,H_2 + O_2 = 2\,H_2O \,. \tag{1}$$

If chemical reactions are fast compared to all other physical processes like molecular diffusion, heat conduction and flow, thermodynamics alone allows the local description of even complex systems [2, 6]. However, in most cases, chemistry occurs on time scales which are comparable with those of flow and molecular transport. As a consequence, chemical kinetics information is required, e.g. rate coefficients $k(T)$ of the individual elementary reactions which occur on the molecular level have to be known. For example, the reaction mechanism of the overall reaction (1) consists of 38 elementary reactions involving a variety of reactive intermediates like H, O atoms and OH radicals [2]. The fundamental concept of using a set of elementary reactions to describe a macroscopic chemical net reaction can be traced back to the landmark work of Max Bodenstein at Heidelberg University who in the summer of 1894 studied the photo-induced decomposition of hydrogen iodine using the sun as a natural photolysis-light-source [7]. Since then, the advent of pulsed lasers in particular with their high temporal and spectral resolution has dramatically expanded the experimental possibilities for direct thermal reaction kinetics investigations. In addition, laser "pump-and-probe" techniques, which combine pulsed laser photolysis for reactive species generation with time-resolved laser-induced fluorescence detection of reaction products, have paved the way for detailed studies of the molecular dynamics of elementary reactions [3].

2.2
Chemical Kinetics and Molecular Dynamics: The Thermal Rate Coefficient $k(T)$ and the Reaction Cross Section $\sigma(E_{c.m.})$

Since the first systematic investigations of the temperature dependence of chemical reactions, it is well known that the rate of reaction can depend strongly, and in a nonlinear way, on temperature [8]. As suggested by Arrhenius in 1889, the temperature dependence of the reaction rate coefficient can be expressed in terms of the following equation (Arrhenius law) [9, 10]:

$$k(T) = A \exp(-E_a/k_B\,T) \tag{2}$$

where E_a, the Arrhenius activation energy, and A, the pre-exponent or frequency factor, are implicitly independent of temperature (k_B is the Boltzmann constant). However, with improving quality of kinetic data, significant deviations from the "Arrhenius-behavior" have been observed like, for example, for reactions of free radicals with saturated molecules which have been extensively studied using Laser Induced Fluorescence (LIF) [11]. For this class of reactions, the plots of ln k(T) vs. (1/RT) were found to exhibit an upward curvature [11]. Such an upward curvature in the "Arrhenius plot" means that the Arrhenius activation energy defined by:

$$E_a = -d \ln k(T) \, d \, (1/k_B \, T) \tag{3}$$

increases with T. Kinetic data of reactions with curved Arrhenius plots can be represented by a modified Arrhenius equation of the empirical form:

$$k(T) = A' \, (T/298)^n \, \exp(-\theta/T) \tag{4}$$

The activation energy now depends on temperature according to: $E_a = k_B (\theta + nT)$. For bimolecular elementary reactions, the modified Arrhenius equation Eq. 4 can be given theoretical justification either by using transition state theory (TST) [11, 12] or molecular collisional models [13]. In the latter case, the rate coefficient k(T) can be obtained by a thermal average over the energy-dependent reactive cross section $\sigma(E_{c.m.})$. Therefore, in terms of the translational energy $E_{c.m.}$ of the relative motion of the reactants, k(T) is given by the expression [10, 14]:

$$k(T) = \left(\frac{8}{\pi\mu \, [k_B T]^3} \right)^{1/2} \int_0^\infty E_{c.m.} \, \sigma(E_{c.m.}) \, \exp(-E_{c.m.}/k_B T) \, dE_{c.m.} \tag{5}$$

where μ is the reduced mass of the two reactants. A simple collision model like, for example, the line-of-centers (LOC) model, in which $\sigma(E_{c.m.})$ is proportional to $(E_{c.m.} - E_0)/E_{c.m.}$, leads via. Eq. 5 to a rate coefficient of the modified Arrhenius form from Eq. 4 with n = 0.5. E_0 denotes the reaction threshold, which is the minimum reagent energy for which reaction is possible. The relationship between the LOC-excitation function and the corresponding thermal rate coefficient k(T) is shown schematically in Fig. 1a.

In reactions which involve H atoms, a curvature of the Arrhenius plot takes place at low temperatures, as has been observed in kinetics studies of the $H + H_2$ reaction [15]. This can originate from quantum mechanical tunneling effects (see below), which allows reaction to also take place at energies below E_0.

2.3
The Chemical Dynamics of the Three-Atom Benchmark Reaction: $H + D_2 \rightarrow HD + D$

The gas-phase reaction of H atoms with deuterium (D_2):

$$H + D_2 \rightarrow HD + D \tag{6}$$

is one of the isotopic variations of the simplest chemical reaction $H + H_2$.

This reaction has been of fundamental interest ever since 1928 when London interpreted for the first time the exchange reaction between a H atom and a H_2 molecule in terms of a H_3-potential energy surface (PES) calculated from quantum mechanics [16]. Since then the $H + H_2$ reaction has become "the prototype test system" towards the development of rigorous quantum mechanical atom–diatom scattering methods [17]. In the case of the $H + H_2$ reaction, it is the high accuracy of the available global PES representations, i.e. the Liu–Siegbahn-Truhlar–Horowitz (LSTH)-PES [18] and the DMBE version [19] (based on a double many-body expansion) that ensures that the comparison of the experimental dynamic results indeed tests the theoretical method rather than the PES. The PES for this three-electron case are known with higher accuracy than any other reactive system. A potential energy profile along the minimum energy path for the $H + D_2 \rightarrow HD + D$ reaction is depicted in Fig. 1b.

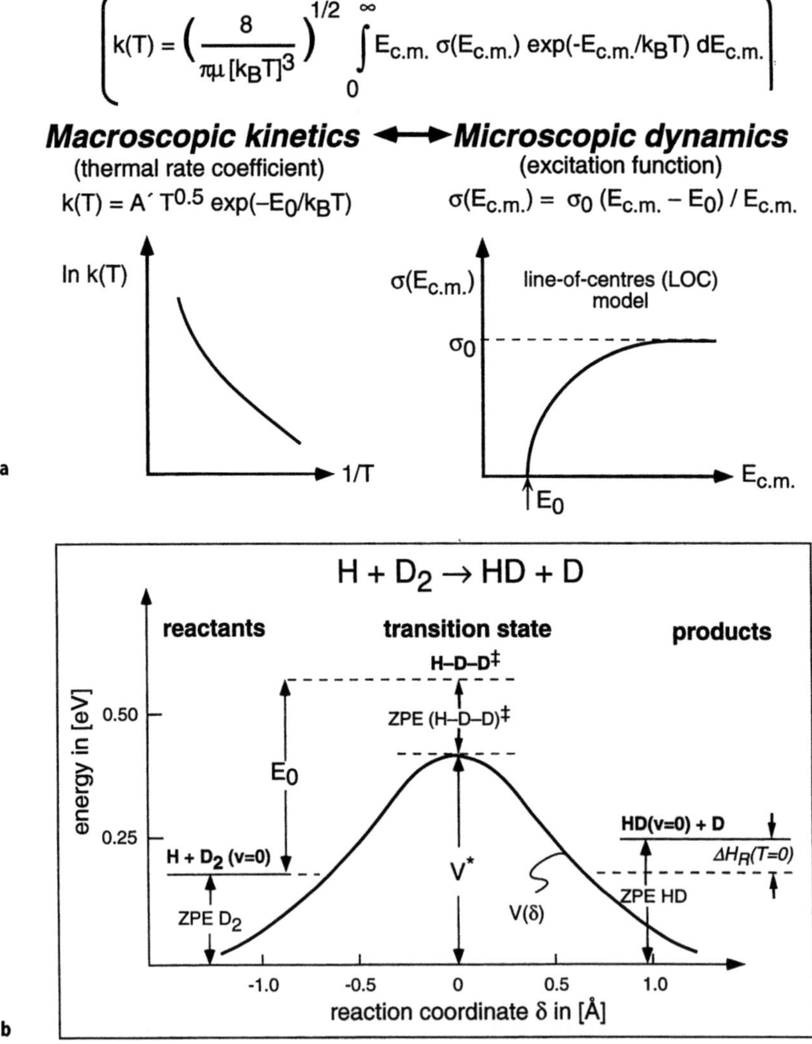

Fig. 1 a, b. **a** Relationship between the bimolecular thermal rate coefficient $k(T)$ and the reaction cross section σ as a function of the reagents' translational energy $E_{c.m.}$ (schematically depicted for a LOC-excitation function). **b** Potential energy profile along the minimum energy path for the $H + D_2 \rightarrow HD+D$ reaction on the LSTH-(ab initio)-PES

About half a century after the first kinetics experiment carried out by Farkas [20], experimental and theoretical results for the H + H$_2$ reaction were reviewed by Levine [21]. More recent reviews can be found [22–24]. Advances in the study of the reaction dynamics of the H+H$_2$ system include measurements of quantum state-specific differential cross sections [25], which allow for detailed comparison with quantum mechanical scattering (QMS) calculations [26]. Only recently was a full three-dimensional QMS study of the H + D$_2$ → HD + D reaction carried out in which absolute reaction cross sections were calculated for different collision energies. The results of this study allow for direct comparison with the results of the molecular dynamics studies to be presented in the following.

2.4
The Laser Photolysis/Laser-Induced Fluorescence "Pump-and-Probe" Technique

In early experimental studies of chemical reactions with high threshold energies, translationally energetic "hot" reactants were generated by nuclear reactions [10, 27] or flash-lamp photolysis of appropriate precursor compounds [28]. The photodissociation of hydrides can be used as a source of translationally hot H atoms [29]. First experiments in which the laser photolysis/laser-induced fluorescence (LP/LIF) method was used to study the reactions H + O$_2$, H + CO$_2$ and + H$_2$O were reported [30]. In these experiments, hot H atoms were generated by laser photolysis of HBr/O$_2$, HBr/CO$_2$ and HBr/H$_2$O mixtures and OH product formation was monitored by LIF. Later on this technique was adapted to measure absolute single collision reactive cross sections for reactions of translationally energetic O(^{1}D) atoms and OH radicals with H$_2$/D$_2$/HD [31] and CO [32]. In the present work we used a combination of a single collision and a moderated hot H atom pulsed laser "pump/probe" method in which energetic H atoms with known translational energies were produced by laser photolysis of HCl and H$_2$S at wavelengths of 193 and 248 nm. Experiments were carried out using the setup depicted in Fig. 2.

Experimental Conditions. Measurements were carried out in a stainless steel flow cell, through which HX/D$_2$ as H atom precursor/reactant mixture together with a moderator gas were continuously pumped. The flow rate was high enough to ensure renewal of the gas mixture in the cell between two successive photolysis ("pump") laser shots. HCl and H$_2$S were used as H atom precursors. HCl, H$_2$S and D$_2$ flow rates were regulated by a calibrated mass flow controller, The cell pressure was measured by a MKS Baratron. In the measurements of single collision absolute reaction cross sections, no moderator gas was used. The HX/D$_2$ ratios were typically between 1:5 and 1:20. These experiments were carried out at a low total pressure of p_{tot} = 60 – 120 mTorr and at short pump-probe delay times of Δt = 80 – 130 ns so as to avoid translational relaxation of the hot H atoms. A schematic description of the LP/LIF "pump/probe" method for the single collision studies is shown in Fig. 3a. The moderated hot H atom experiments were carried out in excess of Ar and N$_2$ moderator gas. In these experiments the total pressure was typically p_{tot} = 1^{-5} Torr with a H$_2$S/D$_2$ partial pressure between 50 and 350 mTorr.

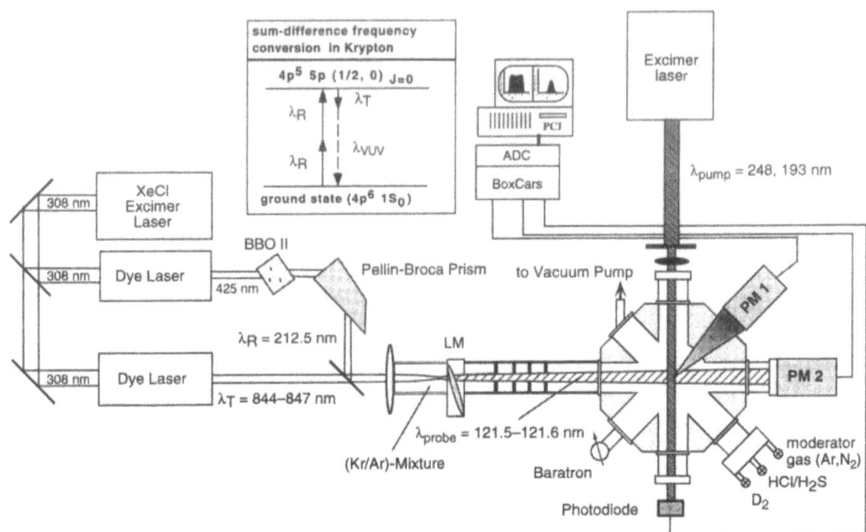

Fig. 2. Experimental apparatus used for the H + D$_2$ → HD + D reaction studies. The Kr four-wave mixing scheme to generate tunable vacuum-UV laser radiation for H and D atom LIF detection is shown as an inset

Generation of Translationally Energetic H Atoms. Hot H atoms with a non-equilibrium velocity distribution were generated by pulsed laser photolysis (duration of the laser pulse was about 15 ns) of HCl/H$_2$S at two different excimer laser "pump" wavelengths (193 and 248 nm). An aperture was used to skim off a homogeneous part of the excimer laser profile to provide a photolysis beam of about 3–7 mJ/pulse which was slightly focused and directed through the flow cell. At the laser intensities used in the present study a linear dependence of the photolytically produced H atom concentrations on the intensity of the photolysis laser was observed.

H Atom and D Atom Vacuum-Ultraviolet Fluorescence Detection. H and D atoms were detected with sub-Doppler resolution via ($2p^2 \leftarrow 1s^2S$) laser-induced vacuum-ultraviolet (VUV) fluorescence. Narrow-band ($\Delta\nu_{VUV} = 0.4$ cm^{-1}) VUV laser "probe" radiation, tunable around the H (121.567 nm) and D (121.534 nm) atom Lyman-α transitions, was generated by the Wallenstein method for resonant third-order sum-difference frequency conversion of pulsed dye laser radiation (pulse duration ~15 ns) in a phase-matched Kr-Ar mixture [33]. In the Kr mixing scheme (shown as an inset in Fig. 2) used to generate the VUV radiation ($\omega_{VUV} = 2\omega_R - \omega_T$), the laser radiation of ω_R ($\lambda_R = 212.55$ nm) is resonant with the Kr 4p–5p (1/2, 0) two-photon transition and held fixed during the experiments, while ω_T is tuned from 844 nm to 848 nm to generate the VUV laser radiation. The generated VUV radiation was carefully separated from the unconverted laser radiation by a lens monochromator (denoted as LM in Fig. 2) followed by a light baffle system. The VUV probe beam was aligned to overlap the photolysis beam at right angles in the viewing region of a LIF detector. The delay time between the photolysis and probe pulse was controlled by a pulse generator. The H and D LIF signals were measured through a band pass filter by a solar blind photomultiplier (PM1 in Fig. 2) positioned at right angles to both photolysis and probe laser. The VUV-probe beam intensity was monitored after passing through the reaction cell with an additional solar blind photomultiplier (PM2 in Fig. 2). The LIF signal,the VUV-probe beam intensity and the photolysis laser intensity were recorded with a boxcar system and transferred to a microcomputer where the LIF signal was normalized to both photolysis and probe laser intensities. In order to obtain a satisfactory S/N ratio, each point of the measured H and D atom Doppler profiles shown in Fig. 3b, c was averaged over 30 laser shots. All measurements were carried out at a repetition rate of 6 Hz.

2.5
Determination of the Excitation Function and the Reaction Threshold of the H + D$_2$ → HD + D Reaction

Absolute Reaction Cross Sections. Absolute reaction cross sections at single collision energies were obtained using the method of Bersohn and co-workers [34]. As described in detail [34], the absolute reactive cross section σ for the H + D$_2$ → HD + D reaction can be determined using the expression:

$$\sigma(E_{c.m.}) = S_D/(S_H v_{rel} [D_2] \Delta t) \tag{7}$$

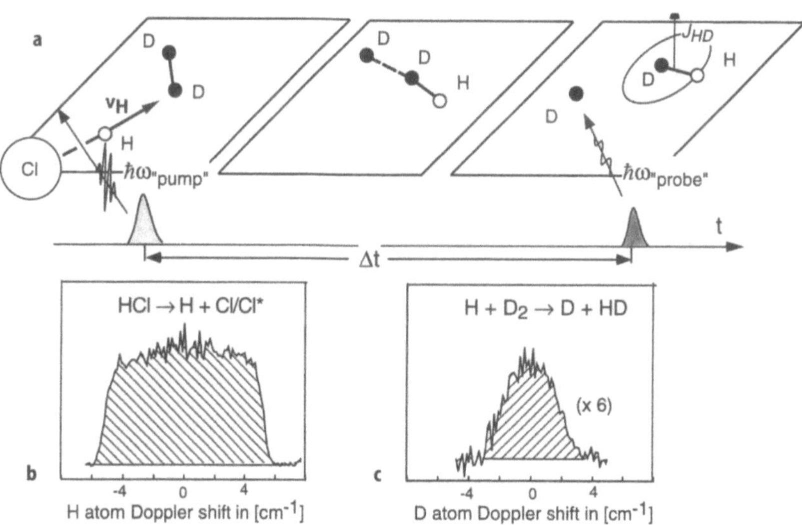

Fig. 3a–c. a A schematic description of the LP/LIF "pump/probe" method for the single collision studies of translationally energetic H atoms with D$_2$. **b** Fluorescence excitation spectra of H atoms with E$_{c.m.}$ = 1.6 eV generated by pulsed HCl laser photolysis at 193 nm in the presence of 55 mTorr D$_2$. **c** Fluorescence excitation spectra of D atoms produced in the subsequent H + D$_2$ reactive collisions. D atoms were detected via VUV-LIF 130 ns after pulsed H atom photolysis

where v_{rel} is the relative velocity, $E_{c.m.} = 1/2\ \mu v_{rel}^2$ stands for the corresponding average center-of-mass collision energy of the reactants and μ is the reduced mass of the $H-D_2$ collision pair. $E_{c.m.}$, and hence v_{rel}, can be calculated from the photolysis laser wavelength, the $H-X$ bond dissociation energy of the hot H atom precursor molecule and the internal state distribution of the X fragment. Δt is the time delay between pump and probe laser pulses. $[D_2]$ denotes the concentration of the D_2 reagent present in excess, which is therefore essentially constant. S_D and S_H are the integrated areas of the corresponding Doppler profiles (see Fig. 3 b, c), which are a measure of the relative concentrations of the photolytically produced H atom reagent $[H]_{\Delta t=0}$ and the D atom reaction products $[D]_{\Delta t}$. Figures 3b and 3c show H atom reagent and D atom product Doppler profiles observed when a room-temperature mixture of HCl and D_2 was irradiated by laser light with a wavelength of 193 nm. For the three HX/photolysis wavelength combinations ($H_2S/248$ nm; $HCl/193$ nm; $H_2S/193$ nm) the following absolute reaction cross sections were measured:

$$\sigma(0.9\ eV) = 0.97 \pm 0.11\ \text{Å}^2$$

$$\sigma(1.6\ eV) = 1.23 \pm 0.32\ \text{Å}^2$$

$$\sigma(2.0\ eV) = 1.28 \pm 0.28\ \text{Å}^2$$

Moderated Hot H Atom Experiments. When translationally hot H atoms are generated by pulsed laser photolysis of H_2S precursor molecules in a large excess of a moderator gas, the initially present nascent non-equilibrium H atom velocity distribution evolves towards the thermal equilibrium distribution determined by the temperature of the moderator gas. In this case the velocity distribution of the H atoms has to be described by a time-dependent distribution function for which the time evolution is given by a linearized Boltzmann equation. When, in addition, a reagent is present, reactive collisions occur in competition with the translational relaxation [35]. The non-equilibrium kinetics of the formation of D atoms in the moderated hot H atom reaction $H + D_2 \rightarrow HD + D$ are then described by the following rate equation:

$$\frac{d\,[D]_t}{dt} = \left\{ \int_0^\infty \sigma(v_{rel})\,v_{rel} f(v_{rel}, t)\,dv_{rel} \right\} [H]_t\,[D_2] \tag{8}$$

Here it has been assumed that D_2 is present in excess over H atoms so that the D_2 concentration remains constant in time. The term in braces represents the reaction rate constant which, under the translational non-equilibrium conditions of the present experiments, is time-dependent. After replacing $[H]_t$ by $[H]_{t=0} - [D]_t$ and introducing the new variable $\chi_D(t) = [D]_t/[H]_{t=0}$ – representing the fractional yield of D atoms produced – the following expression:

$$\chi_D\,(\Delta t) = [D_2] \times \int_0^{\Delta t} \left\{ \int_0^\infty \sigma(v_{rel})\,v_{rel} f(v_{rel}, t)\,dv_{rel} \right\} (1 - \chi_D(t))\,dt \tag{9}$$

can be obtained as a solution of Eq. 8 for the initial condition $\chi_D(t = 0) = 0$. The delay time Δt between the pump laser pulse, which generates the H atom reagents, and the probe laser pulse, which detects the D atom products, corresponds to the reaction time. In order to perform the integration in Eq. 9, one has to know the excitation function, i.e. the reaction cross section σ_R as a function of the relative velocity or collision energy, as well as the time dependence of the relative velocity distribution function $f(v_{rel}, t)$. On the other hand, when $\chi_D(\Delta t)$ and $f(v_{rel}, t)$ can be measured, information about the actual form of the excitation function can be obtained.

The experimental method used in the present studies allowed the direct determination of both the D atom product yield $\chi_D(\Delta t)$ and the time-dependent relative velocity distribution function $f(v_{rel}, t)$. $\chi_D(\Delta t)$ was obtained by calculating the ratio of the integrated areas of the D atom product measured at reaction time Δt and the nascent H atom reagent Doppler profile measured at a reaction time close to zero (before significant reaction could occur). Each H atom Doppler profile measured at a given time t (see Fig. 4a) reflects, via the linear Doppler shift $v - v_0 = v_z v_0/c$, the laboratory frame distribution $f_H(v_z, t)$ of the velocity component v_z. Therefore, evaluation of H atom Doppler line shapes measured at different pump/probe delay times allows the time evolution of the v_z velocity component $f_H(v_z, t)$ to be derived. A symmetric double sigmoidal function was used as a fitting function to evaluate the measured H atom Doppler line shapes (see Fig. 4a). The time-dependent relative velocity distribution function $f(v_{rel}, t)$ as shown in Fig. 4b was obtained by differentiation of $f_H(v_z, t)$ followed by a laboratory to center-of-mass transformation [36].

Analyses of the data were first tried using a simple model function like the LOC excitation function as shown in Fig. 1a. However, it proved impossible to

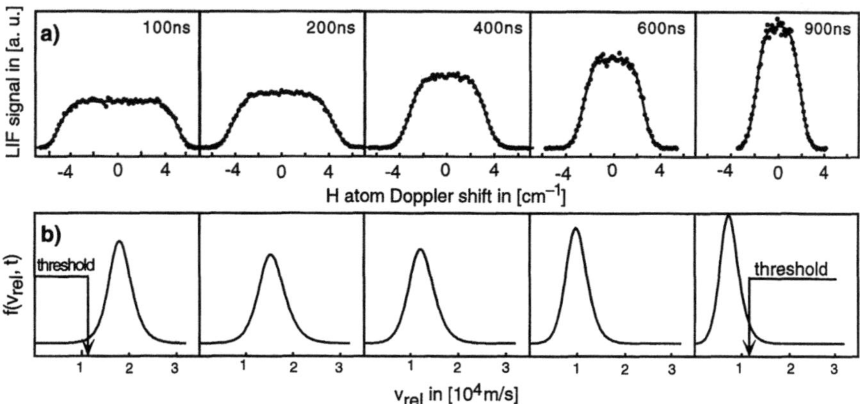

Fig. 4a, b. H atom translational moderation studies. **a** H atom Doppler profiles measured at pump/probe delay times between 100 and 900 ns after 193 nm photolysis of H_2S in the presence of 1.4 Torr Ar moderator gas. Solid lines are results of a fit to the time evolution of H atom laboratory velocity distribution $f_H(v_z, t)$. **b** Evolution of the corresponding time-dependent distribution function $f(v_{rel}, t)$ of the $H - D_2$ reagents' relative velocity. Arrows mark the $H + D_2 \rightarrow HD + D$ reaction threshold

find a set of parameters in the framework of that model which provided a good description of both the non-equilibrium kinetics and the single-collision data sets. As the excitation function at high collision energies ($1.6\,\text{eV} \leq E_{c.m} \leq 2.7\,\text{eV}$) is already well-characterized by the single collision reaction cross section measurements of the present work and [34], the excitation function given by Eq. 10 was tested.

$$\sigma(E_{c.m.}) = \begin{cases} \dfrac{1.52}{1 + \exp\left\{ -\dfrac{(E_{c.m.} - \varepsilon)}{\eta} \right\}} - 0.25 & \text{for } E_{c.m.} \geq E_0 \\ \\ 0 & \text{for } E_{c.m.} < E_0 \end{cases} \tag{10}$$

This form of the excitation function is arbitrary but ensures that the cross section reaches asymptotically a constant value of $\langle \sigma \rangle = 1.26\,\text{Å}^2$ for $E_{c.m.} \geq 1.6$ eV. $\langle \sigma \rangle$ represents the average value of the single collision cross sections measured in the present work and in [34] at collision energies of $E_{c.m.} = 1.6, 2.0, 2.7$ eV. The model parameters ε and η determine the position of the reaction threshold energy ($E_0 = \varepsilon - 1.625\eta$) and the steepness of the rise of the function below 1.6 eV. The aim of this approach was to obtain information about the reaction cross section in the less well-characterized region below $E_{c.m.} = 1.6$ eV.

Using the time-dependent parameterization of the H atom speed distribution $f(v_{rel}, t)$ determined for each set of experimental conditions, D atom fractional yields predicted by the model excitation function (Eq. 10) could be calculated (dashed line in Fig. 5a), via Eq. 9, for the conditions corresponding to each measured data point $\chi_D(\Delta t)$. The integration over t in Eq. 9 was performed

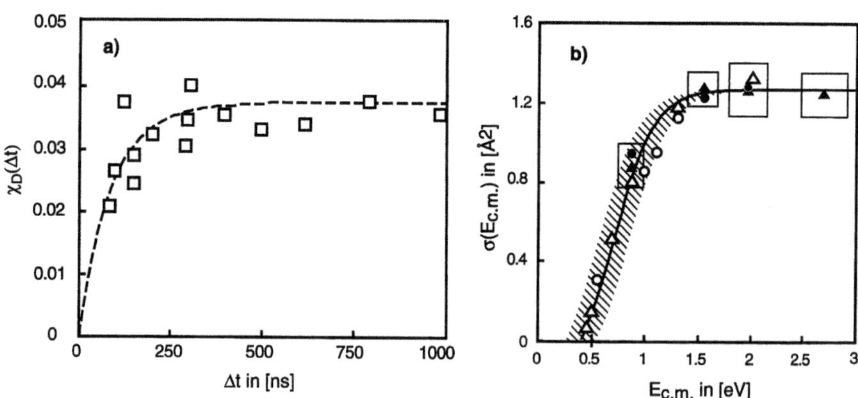

Fig. 5a, b. a Plot of D atom product yields χ_D vs. reaction time observed in the hot H atom moderation study of the $H + D_2 \rightarrow HD + D$ reaction. (□) represent experimental results. The *dashed line* is the result of a non-equilibrium kinetics simulation of the experimental moderation/reaction conditions with the excitation function $\sigma(E_{c.m.})$ shown in **b**. The *solid line* in **b** represents the $H + D_2 \rightarrow HD + D$ excitation function $\sigma(E_{c.m.})$ determined in the H atom moderation/reaction experiments. (■) and (▲) are experimental single collision reaction cross sections from [34] and from the present study, respectively. (○) and (△) are reaction cross sections from QCT [39] and recent exact 3D-QMS [40] calculations on the LSTH–PES

numerically, using a fourth-order Runge–Kutta algorithm incorporating error and adaptive step-size control [37]. A measure of the quality of a given set of excitation function parameters was obtained by computing the global mean squared deviation of both the non-equilibrium kinetics data sets and the single collision cross sections from the corresponding values calculated from the trial excitation function. Optimization of the parameters ε and η was performed by a non-linear least-squares fit to minimize the sum of these two mean squared deviations. This procedure yielded the excitation function shown as the solid line in Fig. 5b, which corresponds to a reaction threshold energy of $E_0 = 0.40 \pm 0.14$ eV (one standard deviation). The statistical uncertainty in each of the model parameters ε and η allowed the confidence region (one standard deviation) of the global excitation function to be estimated [37], which is shown as the shaded area in Fig. 5b.

Comparison with Chemical Dynamics Calculations. The single collision reaction cross sections $\sigma(1.6$ eV$)$ and $\sigma(2.0$ eV$)$ measured in the present study are in excellent agreement with the values in [34] and confirm the results of the early quasi-classical trajectory (QCT) calculations on the LSTH-PES, in which it was found that the excitation function shows a broad maximum of ~ 1.3 Å^2 around $E_{c.m.} \approx 2$ eV [38].

In Fig. 5b, comparison is made between the excitation function obtained in the present study and QCT [39] and accurate 3D-QMS [40] reaction cross sections calculated using the LSTH-PES. The theoretical reaction cross sections represent rotationally averaged quantities; in order to account explicitly for the experimental conditions, the reactive cross sections were averaged over a room-temperature Boltzmann distribution of the D_2 rotational states. So both the quasi-classical and the quantum mechanical cross sections can be directly compared with the experimental results. Therefore – because of the very high accuracy of the LSTH-PES – any difference regarding the agreement with experimental could in principle be attributed directly to the type of theoretical concept used in the dynamic calculations.

The fact that the QMS results show in general a somewhat better agreement with experimental results clearly favors the quantum mechanical approach over the classical one. QMS calculations account, per se, for tunneling as well as for zero point energy constrains. This is not the case for QCT calculations. The absence of tunneling in the QCT calculations leads – for collision energies close to the reaction threshold – to reaction cross sections which are too small and hence to reaction rate coefficients which are too low. On the other hand QCT calculations (without further modifications) do not account for vibrational adiabatic zero point constrains, leading in some cases to trajectories associated with products which have less than the required zero point energy. The latter effect tends to make QCT reaction cross sections in general too high (but again the influence of such "zero point energy-violating" trajectories on the reactive cross section is most pronounced at energies close to the reaction threshold). As a consequence, good agreement between results from QCT and QMS calculations can in some cases simply be the result of the fortunate cancellation of these two opposing errors in the classical treatment.

3
Laser Diagnostics of Combustion Processes

The experimental possibilities for studying processes in technical combustion devices have expanded quite dramatically in recent years as a result of the development of various pulsed high-power laser sources which provide high temporal, spectral and spatial resolution. Laser spectroscopic methods are especially important for non-intrusive measurements in technical systems in which complex chemical kinetics are coupled with transport processes. New detection techniques with high sensitivity allow the measurement of multidimensional temperature and species distributions which can be used for comparison with detailed mathematical modeling of laminar and turbulent reactive flows including heat and species transport. Validated mathematical models can later be used to find optimal conditions for the various combustion parameters which lead to decreased pollutant formation and fuel consumption. In addition, in situ laser diagnostics offer new ways for an active control of industrial combustion systems in order to maintain optimum process conditions [41].

3.1
Laser-Induced Ignition Processes

One of the key factors for improving the performance of many technical combustion devices is an optimal control of the ignition process. Optimized reproducible ignition ensures an efficient and safe operation. Methanol has a considerable potential as an alternative fuel which can be produced from biomass or natural gas. Experimental studies on CO_2 laser-induced thermal ignition of CH_3OH/O_2-mixtures were performed [43] to supply quantitative data for comparison with results from direct numerical simulations of unsteady CH_3OH oxidation under laminar flow conditions.

The experimental setup is depicted in Fig. 6a. In a quartz cell equipped with SrF_2 windows, CH_3OH/O_2-mixtures were ignited using a cw CO_2-laser (Edinburgh Instruments) in the pulsed mode. The coincidence of the 9P(12) CO_2-laser line in the (001)–(020) band with the R(12) CO stretch fundamental band of the methanol molecule at 9.6 µm allows controlled heating and ignition of the mixture. OH radicals formed during flame propagation were excited in the $(v' = 3, v'' = 0)$ vibrational band of the $OH(A^2\Sigma^+-X^2\Pi)$-transition around 248 nm using two tunable KrF excimer lasers (laser wavelength tunable in the range 247.9–248.9 nm with a bandwidth of typically 0.5 cm^{-1}). Each excimer laser beam was formed into a light sheet 30 mm in height and 400 µm thick using quartz lenses. The light sheets were spatially overlapped with a time delay of 100 ns to separate the signals induced by the different laser pulses. The laser-induced fluorescence was collected using achromatic UV lens systems ($f = 100$ mm, $f_\# = 2$). Reflection filters built of four narrow-band dielectric mirrors (at 297 ± 6 nm, transmission $> 90\%$, blocking 5×10^4) were used to spectrally isolate the $(v' = 3, v'' = 2)$ fluorescence band of the OH radical for detection. Fluorescence was detected by gated image-intensified CCD cameras. Excitation of two different optical transitions starting from the $N'' = 8$ and $N'' = 11$ rota-

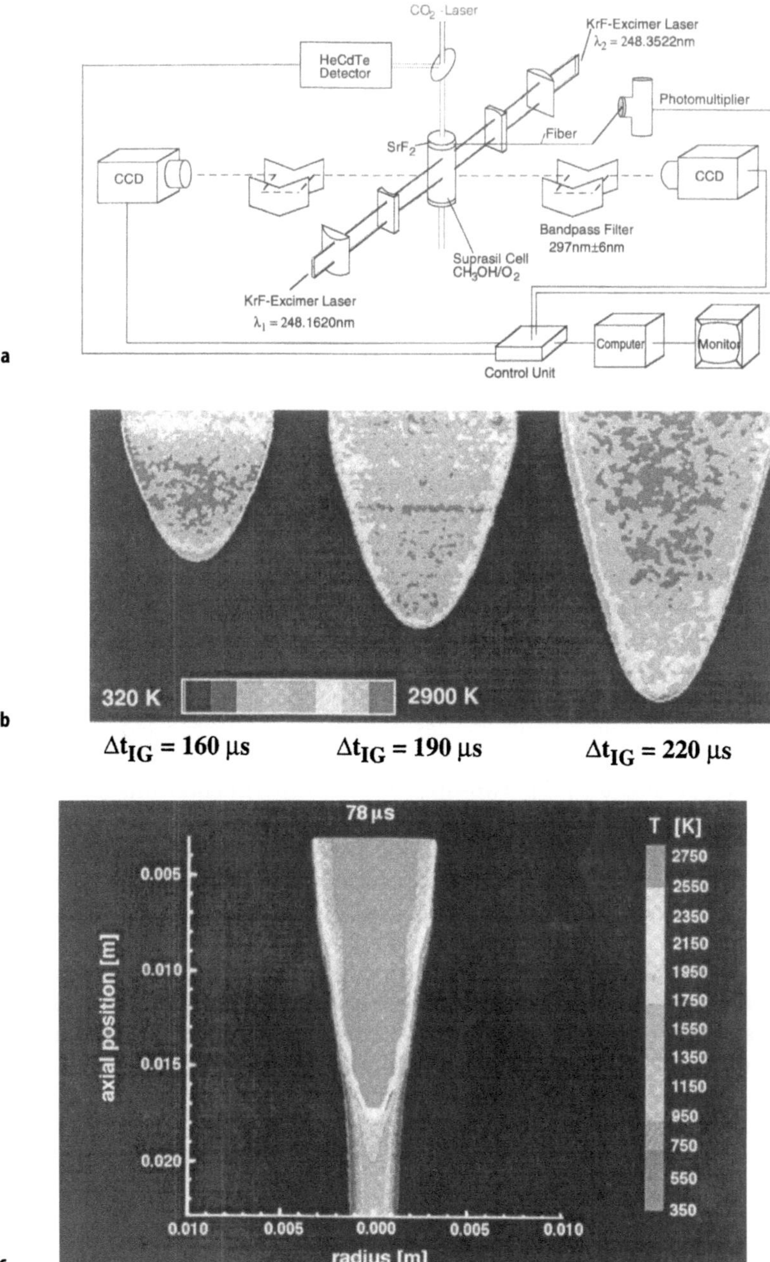

Fig. 6a–c. Thermal ignition of methanol/oxygen mixtures. **a** Experimental setup. Mixtures were ignited by a CO_2 laser beam. Flame propagation is visualized by 2D-LIF of OH radicals. For 2D-temperature measurements different OH rotational levels are probed simultaneously using two tunable KrF-excimer lasers. **b** Development of the 2D-temperature field during the ignition of a methanol/oxygen mixture ($\phi = 0.9$, p = 300 mbar). **c** Numerical simulation of the temperature field 78 μs after CO_2 laser ignition (data shown with kind permission from U. Maas and I. Gran)

tional levels of the OH($X^2\Pi$, $v'' = 0$)-vibrational state, allowed the measurement of spatially corresponding LIF image pairs. Assuming a Boltzmann distribution for the population of the OH ($X^2\Pi$, $v'' = 0$)-rotational states, the ratio of the two OH fluorescence images can be converted into a OH temperature field (see Fig. 6b). Calculating temperatures from a two-line LIF measurement requires careful consideration of a number of effects. The influence of fluorescence quenching, usually the dominating deactivation process for electronically excited OH radicals under atmospheric pressure conditions, is reduced in the experiments by the fast predissociation of the excited state [44] to less than 3%. The tunable excimer laser emits polarized light, which could induce an anisotropic LIF signal, depending on the optical transition excited [45]. Collisions can redistribute the spatial alignment; this means that the ratio of the LIF images could be dependent on the nature of the collider gas composition. This effect is accounted for with a calibration procedure, where temperatures measured by the two-line LIF-method were compared to point-wise coherent anti-Stokes Raman spectroscopy (CARS) temperature measurements. A good agreement between the results of these two techniques was observed [46]. This also shows that a possible variation of OH rotational energy transfer rates does not have a significant impact on the measured temperature distribution under the present experimental conditions.

For the direct numerical simulation of this two-dimensional reactive flow the system of coupled ordinary differential and algebraic conservation equations was solved numerically by spatial discretization using finite differences [43, 47]. A reaction mechanism containing 22 chemical species and 173 elementary chemical reactions [48] and a multi-species transport model was used. A calculated temperature distribution during the ignition process after the onset of ignition is shown in Fig. 6c [49]. Two-dimensional experimental imaging of the temperature as well as the numerical simulation show a conical flame front. This can be explained by the presence of a preheated channel formed by the CO_2-laser beam. As a consequence, flame propagation is faster in the axial than in the radial direction. Fast axial flame propagation is caused mainly by successive ignition along the cell axis, due to different induction times which follow the axial temperature gradient. Typical propagation speeds of the ignited CH_3OH/O_2-mixtures investigated are 30 m/s in the radial direction and 130 m/s in the axial direction. This observation is an important step towards the understanding of knock phenomena in automobile engines, where unwanted self-ignition occurs in hot spots during the adiabatic compression phase. This local ignition forms shock waves which produce temperature jumps, that can accelerate the flame propagation and sometimes lead to a transition to detonations.

3.2
Laser Diagnostics of Turbulent Premixed Natural Gas Flames

To increase the rate of chemical conversion in natural gas combustion mostly turbulent flow conditions are applied. Similar to the laminar case, turbulent reactive flows can be described by solving the conservation equations for total and species mass, momentum and enthalpy (Navier–Stokes equations) [2]. However, solving the Navier–Stokes equations by direct numerical simulation (DNS) is even in the days of modern parallel computers with a capacity of about 10^{12} computing steps/s a very demanding task. For a realistic system of liquid hydrocarbon oxidation in an internal combustion engine one would need up to 10^{20} computing steps. Therefore at present and for the near future DNS of three-dimensional turbulent reactive flows in technical combustion systems will not be possible. Instead, turbulence has to be modelled using averaged Navier–Stokes equations, while mean reaction rates are evaluated by means of probability density functions (PDFs) [50]. A procedure for the mathematically correct simplification of chemical kinetics in complex reactive flow systems has recently been developed and tested [51]. For the experimental validation of such models, multidimensional non-intrusive measurements in turbulent flames are necessary.

In order to understand the complex interaction of turbulence and chemistry, detailed measurements of the flame front structure have been performed. Such measurements are also important from a theoretical point of view. Up to a certain degree of turbulence, it is possible to calculate the structure of turbulent flames, using the flamelet approach [52]. The flamelet model assumes that a turbulent flame front can be described as an ensemble of laminar flamelets. The limit of this assumption has to be verified experimentally, because there is no general agreement in the literature. Analyzing a turbulent flame front structure to find the limits of the flamelet model requires systematic determination of the flame front thickness as a function of the turbulence parameters.

In premixed flames with Karlovitz numbers Ka below one – the Karlowitz number is given by $Ka = (\delta_n/l_k)^2$ where δ_n is the laminar flame front thickness and l_k is the size of the smallest turbulent eddies (Kolmogorov scale) [2] – it was found that the local flame front structure (temperature and hydroxyl radical concentration profiles normal to the flame front) is very similar to the one calculated for laminar premixed flames. This supports the idea of Damköhler [53], who suggested that the three-dimensional problem of turbulent combustion can be reduced to a one-dimensional "flamelet" model [54]. For Ka >1, however, recent experiments [55] show that in contrast to the expectations of Damköhler (thicker flame fronts for Ka >1, Klimov–Williams criterion) the mean thermal flame front thickness is significantly reduced compared to laminar unstrained calculations of flames with the same stoichiometry. Recent calculations support this observation of thinning and nearly unchanged OH profiles with increasing strain rate [56]. DNS studies showed that due to the increase of viscosity with temperature, an enhanced dissipation of the small-scale turbulent eddies happens after entering the flame front and their energy content is too low to influence the flame front [57]. Thus, only the preheat zone structure is somewhat modified but not the hot part unless local extinction of the flame front occurs.

In order to investigate these dependencies, flames in a large turbulent Bunsen burner with a maximum power of 150 kW, a diameter of 80 mm and grid-generated turbulence were studied (see Fig. 7). Defined turbulence parameters ($Re_{grid} \leq 100\ 000$, $Re_t \leq 15\ 300$, geometrical and turbulent Reynolds number, respectively), which are homogeneous over a large area, could be selected using different turbulence grids. A small H_2/air flame around the burner head stabilizes the flame. Temperature alone is not an unambiguous parameter to describe the situation in a reactive flow at high turbulence. For this reason

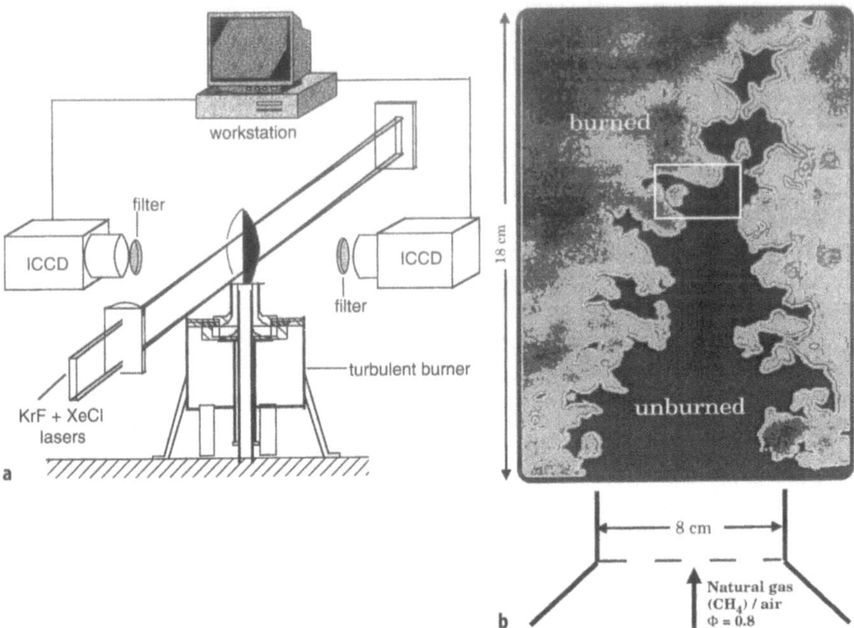

Fig. 7a,b. Investigation of turbulent premixed natural gas/air flames. **a** Experimental setup for simultaneous measurements of 2D-temperature fields via Rayleigh scattering and OH distribution via LIF. **b** OH concentration field measured in the large-scale turbulent burner

simultaneous measurements of the local temperature and the concentration distribution of OH radicals formed in the flame front were performed.

Temperature fields were measured using the 2D-Rayleigh scattering technique employing broadband KrF excimer laser radiation at about 248.6 nm. The Rayleigh signal is proportional to the number density of all molecules times their Rayleigh cross sections [58]. For premixed natural gas (CH$_4$ mole fraction 97.5%)/air flames the variation of the effective Rayleigh cross section for the reactants, products and the intermediates is constant within 2%. Thus, neglecting the very small pressure changes in an atmospheric pressure flame, the Rayleigh signal is inversely proportional to the gas temperature (see also Sect. 3.3). For the calibration, a Raleigh density image $R_{cal}(x, y)$ of the ambient air at room temperature ($T_{cal} = 300$ K) was used. Taking a correction factor C to allow for the different Rayleigh cross sections of ambient air and the natural gas/air mixture, the temperature field can be calculated via:

$$T_{fl}(x, y) = C\, T_{cal}\, \{R_{cal}(x, y)/R_{fl}(x,y)\} \tag{11}$$

where $R_{fl}(x, y)$ and $T_{fl}(x, y)$ denote the single-shot Rayleigh intensity image and the temperature distribution in the flame, respectively. Additional corrections had to be applied to subtract interfering Mie scattering and parasitic reflection of laser radiation. OH- and hydrocarbon-fluorescence, however, was negligible compared to the Rayleigh scattering signal. The accuracy of the measurements were estimated to be about 10%.

For OH concentration measurements a tunable narrow-band XeCl excimer laser (wavelength range 307.7–308.5 nm, bandwidth 0.5 cm^{-1}) was used. In order to measure OH concentration fields quantitatively saturated LIF was performed [59]. The use of the saturated LIF technique reduces the influence of variations in quenching rate and laser intensity by a factor of 10. Self absorption effects were checked by comparing OH images, which were simultaneously recorded for the $(A^2\Sigma^+; v' = 0 - X^2\Pi; v'' = 0)$- and $(A^2\Sigma^+; v' = 0 - X^2\Pi; v'' = 1)$-fluorescence transition at 308 nm and 343 nm, respectively, with two intensified CCD cameras [60].

The beams of the two excimer lasers were overlapped and focussed to light sheets of about 200 µm thickness (see Fig. 7a). Light emitted perpendicular to the laser beams was collected by UV lens systems and imaged onto two gated intensified CCD cameras. The short gate times of the synchronized cameras, which detected the OH-LIF (10 ns) and the Rayleigh signal (50 ns) completely suppressed detection of flame luminosity. Interference between LIF and Rayleigh images was avoided by using a delay of 300 ns between the two laser pulses. This time delay is small compared to the chemical reaction time scales and negligible compared to the turbulence time scale. A single-shot cross section through a large area of a turbulent flame is shown in Fig. 7b. For this image the laser sheet was widened in height and OH fluorescence was used as a qualitative marker of the flame front structure. The flame front is strongly corrugated. From 128 individual measurements length scales were determined using the method of crossing interval length [61].

A typical profile normal to the flame front measured with a high spatial resolution (see Fig. 8) shows the gradients for the temperature and OH concentration. The temperature rises significantly before the OH concentration increases. Comparison with simulations for a laminar unstrained flame shows similarities between the measured local turbulent flame front structure and the modeling results. The measured flame front thickness, defined as the steepest gradient extrapolation of the temperature rise, is around 700 µm for this flame. The temperature fluctuations in the burned gas region represent noise from the measurement. The absolute OH concentrations are overpredicted by the model because no strain has been implemented in the calculations.

Flames in the region in which broadened or broken flame fronts are expected still have nearly laminar flame fronts in most parts of the flame. A global broadening, which was expected to come from homogeneously distributed eddies with Kolmogorov scales smaller than the laminar flame front thickness [62], could not be observed. At single locations, deviations from a laminar flame front can be seen. One example is shown in Fig. 9, where the field of view is 6×6 mm. A tongue of hot gas with T between 800–1200 K is wrapped around a vortex of approx. 2 mm diameter. On this tongue no OH is seen. The temperature at which reaction starts (e.g. where the OH concentration has reached 20%) is about 1400–1500 K at the base of the flame tongue in contrast to about 1200 K on parts of the flame without strong wrinkling. The frequency of more or less pronounced temperature protrusions or slightly broadened flame fronts grows with the mean stretch factor. The results fit well together with the idea of coherent turbulent structures, where vortex tubes with high strain rates are expected to be sparsely distributed in a turbulent field [63].

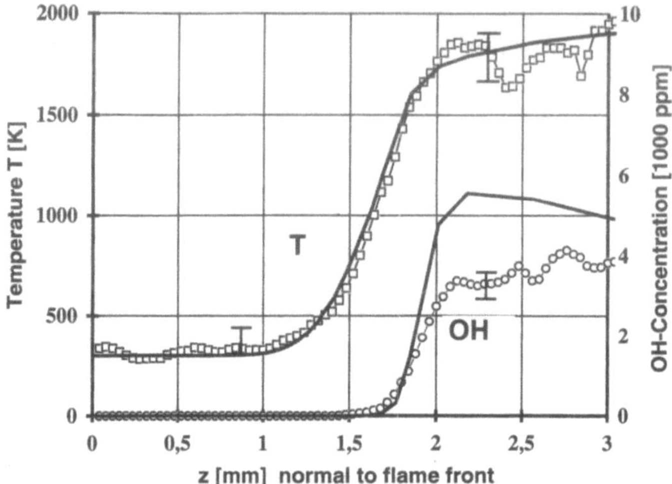

Profile normal to flame front ($\phi = 0.8$, Ret = 760, Ka = 0.8)

line : laminar unstrained calculation (Warnatz/Muris)

dots : experimental example

Fig. 8. Temperature and OH concentration profiles normal to the flame front in a large scale premixed turbulent natural gas/air flame. Comparison of experimental results with results from model calculations for a laminar unstrained flame

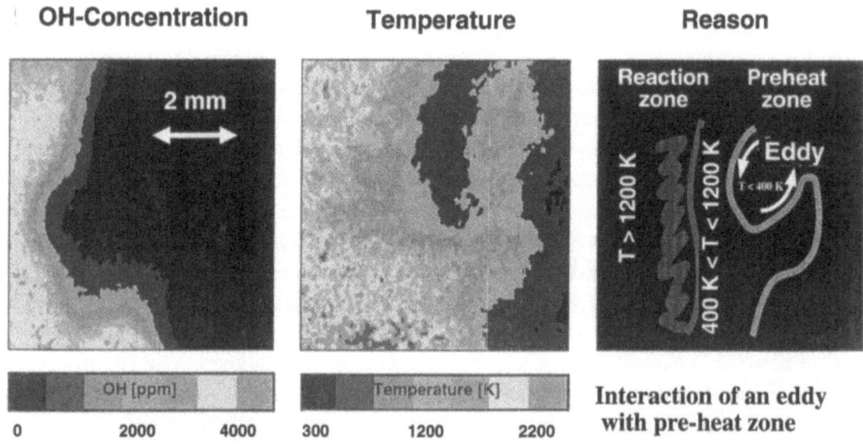

Fig. 9. Deviations from the assumptions of the flamelet approach visualized by simultaneously recorded temperature and OH concentration fields in a highly turbulent premixed natural gas/air flame (Re$_t$ = 1650, K$_a$ = 3.1). A turbulence-induced vortex ('eddy') is interacting with hot gases in the pre-heat zone resulting in high local temperatures in an area where no OH is present

3.3
Imaging of NO Concentrations and Temperatures in Otto Engines

The reduction of the emission of nitric oxide (NO), as one of the major pollutants in combustion, is of particular interest in improving the environmental acceptance of internal combustion engines. A substantial amount of NO released by combustion processes is formed in spark-ignition (SI) and compression-ignition (CI) engines used in cars and trucks for which further drastic legislative regulations of the NO release are expected in the next years. However, use of a nowadays standard three-way catalyst is not possible with future direct-injected engines operated under lean-burn conditions since the air/fuel ratio deviates significantly from unity. It is therefore necessary to investigate possibilities which would already reduce the formation of nitric oxide during the combustion process itself before it is released from the engine into catalytic converters or plasma-based exhaust gas treatments. For a detailed understanding of NO formation processes in engines, quantitative NO concentration distributions have to be measured with high spatial and temporal resolution directly in the engine.

High Arrhenius activation energies of the NO forming elementary reactions (like, e.g., $O(^3P) + N_2 \rightarrow NO + N$; $E_a = 319$ kJ/mol) lead to a very pronounced temperature dependence with a highly non-linear increase of the rate of NO formation with increasing temperature [2]. Experimental investigations must therefore also include in situ measurements of the local temperature. Due to the strong temperature dependence of NO formation simultaneously recorded temperature and NO concentration data provide a valuable possibility for testing capabilities of the combustion models.

As described in the previous section, the LIF method allows the measurement of OH radical concentrations. But, in addition, many combustion-related species, like CH, CN, NH, O_2 and NO have absorption spectra in the ultraviolet spectral region, where they can be excited with commercially available laser sources. Detection limits in the ppm range are easily obtained. However, the quantitative determination of concentrations is often complicated, by complex pressure, temperature and interference effects. Quantitative interpretation of the results of LIF measurements in technical devices therefore requires detailed knowledge of collisional quenching and spectral line-broadening effects.

Various strategies using different laser excitation wavelengths, laser sources and fluorescence detection bands were investigated for in-cylinder engine NO-diagnostics. NO concentration measurements in spark-ignition (SI) and diesel engines using tunable ArF excimer laser radiation for exciting the $(D^2\Sigma^+; v' = 0 - X^2\Pi; v'' = 1)$-transition of the NO molecule at 193 nm (ε-bands) were reported by several groups [64]. However, the strong (D-X)-transition cannot be used for engine measurements during the combustion process since the necessary short-wavelength laser radiation is strongly absorbed by other intermediate combustion products. In [65], NO was detected via LIF by exciting the $(A^2\Sigma^+; v' = 0 - X^2\Pi; v'' = 0)$-transition around 225 nm (γ-bands) using the Raman-shifted output of a tunable KrF-excimer laser. In this case, use of the longer excitation wavelength allowed quantitative NO LIF measurements. However, for engines operated with liquid fuels, attenuation of the 225 nm laser radiation during the actual combustion phase is still. Recently, Schulz et al. [66] have reported a new LIF detection scheme for NO. In this scheme, a tunable KrF-excimer laser is used to excite the $NO(A^2\Sigma^+; v' = 0 - X^2\Pi; v'' = 2)$-transition around 248 nm. For this wavelength the absorption due to other combustion products is further reduced and the strategy allowed quantitative concentration measurements of the in-cylinder NO formation in an iso-octane driven SI engine [67].

The detection scheme based on excitation with a tunable KrF-excimer laser at 248 nm was characterized by measurements in stabilized high-pressure flames [68]. Fluorescence signals were recorded with an image-intensified CCD camera after dispersion with a spectrometer. This yielded the entire emission spectrum for a given laser wavelength. Tuning the laser and composing the emission spectra into an excitation-emission map (Fig. 10) reveals all spectroscopic information necessary to select the best excitation-emission wavelength combination. In a combustion environment, especially at high pressures when the lines are broadened, a laser tuned to a particular NO absorption line can also excite other species. A carefully selected combination of the excitation and detection wavelength is necessary for a specific detection of the desired species. In the case of NO, the excitation laser is tuned to the O_{12}-bandhead of the $(A^2\Sigma^+;$ $v' = 0 - X^2\Pi; v'' = 2)$-transition. This allows efficient excitation because several rovibrational transitions are excited simultaneously. Furthermore, the fluorescence excitation spectrum of molecular oxygen has a local minimum at that wavelength. NO fluorescence emitted in the $(A^2\Sigma^+; v' = 0 - X^2\Pi; v'' = 0)$- and $(A^2\Sigma^+;$ $v' = 0 - X^2\Pi; v'' = 1)$-bands at shorter wavelengths is detected. This scheme, represented by the rectangles in Fig. 10, provides a possibility to avoid interference with fluorescence from O_2, OH, hydrocarbons, Rayleigh scattering and Stokes-shifted Raman signals.

The use of a tunable KrF-excimer laser for LIF detection of NO allows the simultaneous measurement of Rayleigh signals. As described in Sect. 3.2, these can be used to determine spatially-resolved temperature distributions. Rayleigh signal intensities, $I_{Rayleigh}$, can be converted into temperature fields using the expression [69]:

$$I_{Rayleigh} = N\,V\,\sigma_{eff}\,E_{Laser}\,\Omega\,\eta_{Rayleigh} \tag{12}$$

where N is the number density of light-scattering molecules and related to the gas temperature T via the perfect gas equation of state ($N \propto 1/T$). V describes the laser illuminated scattering volume, E_{Laser} is the energy of the laser pulse and σ_{eff} is the effective Rayleigh scattering cross section. Ω denotes the acceptance solid angle of signal detection. $\eta_{Rayleigh}$ is the overall quantum efficiency of the detection system and accounts, e.g., for losses of Rayleigh-scattered photons due to the collection optics (incl. filters) and also includes the photoelectron efficiency of the light detection device. To convert Rayleigh signal intensities into temperature information via the ideal gas law the effective scattering cross section σ_{eff} has to be known. σ_{eff} represents a summation over the mole fraction-weighted Rayleigh cross sections of all species present in the gas mixture. For combustion conditions in engines, however, σ_{eff} is markedly different for the unburned and burned gases. Therefore, Rayleigh images must be analyzed with

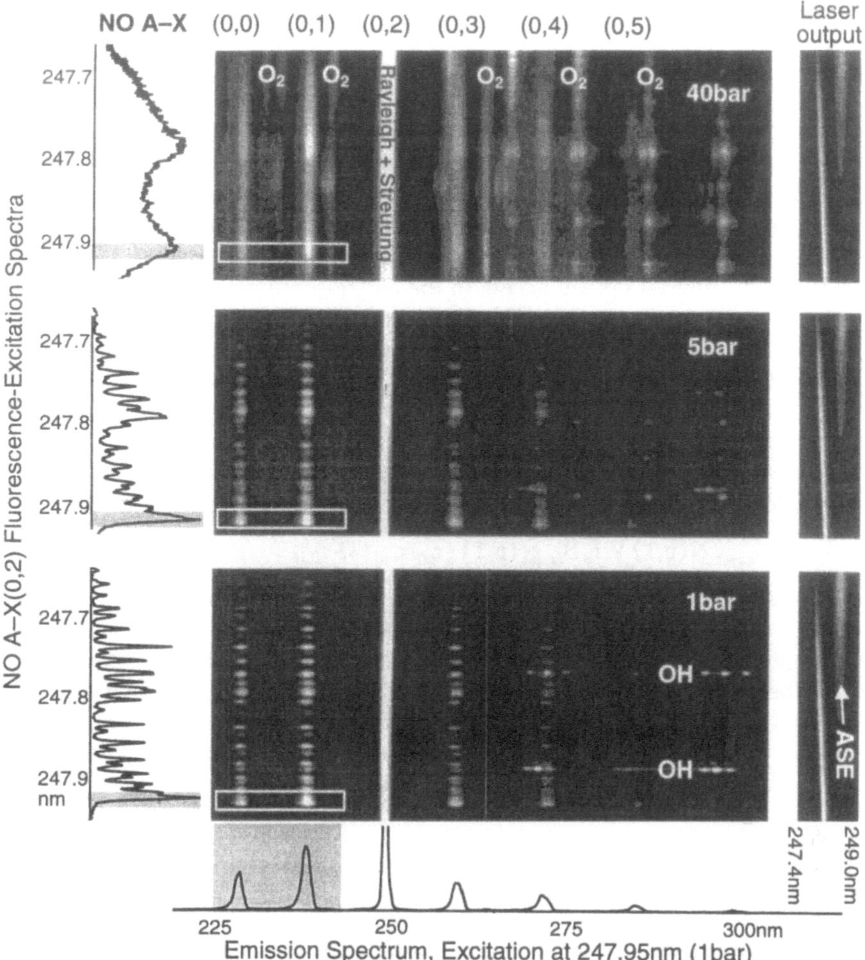

Fig. 10. Excitation-emission spectra obtained in a lean methane/air flame at 1, 5 and 40 bar. The spectra on the left show profiles along the NO A-X (0, 1) emission lines for excitation wavelengths between 247.65 and 248.0 nm at 1, 5 and 40 bar, respectively. The emission spectrum at the bottom was obtained at 1 bar after excitation at 247.95 nm. The NO A-X (0, v″) progression is clearly resolved with the (0, 2) band overlapped with the Rayleigh peak (off-scale). The small frames on the right show enlarged excitation-emission spectra of the laser output obtained via Rayleigh scattering. The intensity of the narrow-band part (diagonal line) drops significantly at wavelengths shorter than 247.8 nm

the appropriate scattering cross sections which can be determined for the major gas components. An direct absolute determination of $\eta_{Rayleigh}$ is rather difficult and usually a reference image recorded at ambient pressure and room temperature is used to eliminate $\eta_{Rayleigh}$ from the data reduction. Because of the elastic character of Rayleigh scattering ($\lambda_{Rayleigh} \approx \lambda_{Laser}$), reflections of the laser beam at surfaces must be carefully avoided. To account for remaining amounts of reflected light, background images for subtraction from the actual data, can be recorded, e.g., with the measurement area flooded with helium. Helium has a very small Rayleigh cross section and thus the measured signal can be attributed to the undesired reflections.

The simultaneous application of Rayleigh scattering and LIF for measurements of temperatures and NO concentrations using the NO $(A^2\Sigma^+; v' = 0 - X^2\Pi;$ $v'' = 2)$-absorption band at 247.95 nm in an internal combustion engine is shown in Fig. 11. Tunable narrow-band laser pulses of this wavelength with pulse energies of typically 60 mJ were generated with a KrF-excimer laser and expanded to a thin light sheet for NO excitation inside the engine. The engine under study allowed optical access to the entire combustion chamber through cylinder walls made out of synthetic quartz [67]. Perpendicular to the light sheet an image-intensified CCD camera was mounted to detect the fluorescence signals. On the opposite side of the engine, a second CCD camera was used to detect the Rayleigh scattering signal, generated by the same laser pulse as the LIF signals.

In case of linear LIF, the measured laser-induced fluorescence intensity, I_{LIF}, can be described by [68]:

$$I_{\mathrm{LIF}} = E_{\mathrm{Laser}}\, N(p,T)\, V\, f_B\, B_{ik}\, g_\lambda(p,T)\, (\tau_{\mathrm{eff}}/\tau_{\mathrm{rad}})\, \Omega\, \eta_{\mathrm{LIF}} \tag{13}$$

where $N(p,T)$ is the total number density of NO molecules in the electronic ground state and V is the laser illuminated excitation volume. f_B denotes the Boltzmann fraction of the rovibrational level from where the transition starts, B_{ik} is the corresponding Einstein coefficient for absorption, $g_\lambda(p,T)$ is the spectral overlap integral of the molecular absorption line with the excitation laser emission spectral profile. The ratio of the effective fluorescence lifetime τ_{eff} to

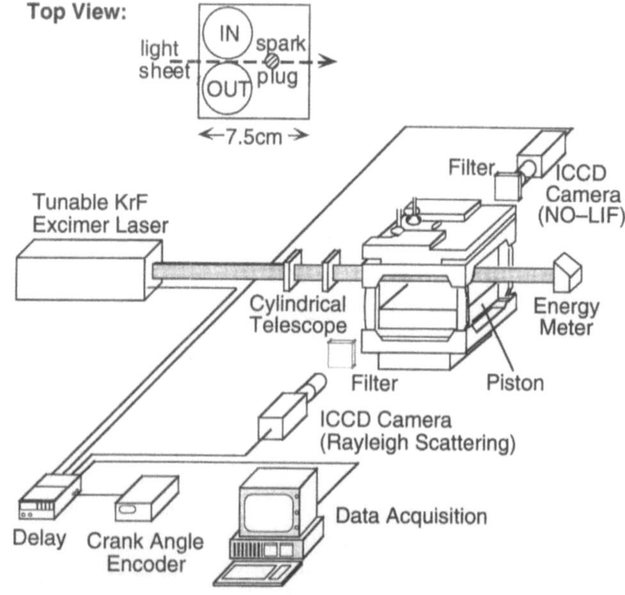

Fig. 11. Experimental setup for the simultaneous single-shot measurement of NO concentration and temperature fields in a transparent SI engine. The insert shows a top view to illustrate the position of the laser light sheet

the radiative lifetime τ_{rad} accounts for fluorescence losses due to radiationless deactivation of excited molecules by collisions. η_{LIF} is the overall quantum efficiency of the LIF detection system, E_{Laser} and Ω have the same meaning as in Eq. 12.

For the analysis of NO-LIF signals measured in a running engine, in particular, the variation of $g_\lambda(p,T)$ with pressure must be carefully studied. NO spectral line-widths vary from 0.5 cm^{-1} at atmospheric pressure to more than 10 cm^{-1} under engine conditions. In the latter case all rotational lines blend (see spectra inserted in the left part of Fig. 10) and the resolved rotational structure found at atmospheric pressure is lost. Only a detailed modeling of the absorption spectrum allows the evaluation of the overlap integral. For each rovibronic transition a Voigt profile with a pressure- and temperature-dependent linewidth and frequency-shift was calculated. This effect has to be considered when quantitative measurements are carried out in an object under varying pressure conditions [68].

Collision quenching leads to a decrease of τ_{eff} by radiationless deactivation of excited molecules and thus to a decrease of the fluorescence intensity. The fluorescence quenching rate scales linearly with pressure, and is a function of temperature and gas composition. It turns out, however, that behind the flame front, i. e. the region where NO is predominantly formed in engines, the quenching rates are nearly independent of temperature and gas composition. Only the variation with pressure must therefore be included in the data analysis. To achieve a high detection sensitivity, the optical system used must cover a large acceptance solid angle Ω. For the NO measurements, specially designed achromatic lens systems with $f_\# = 2$ were used. Additionally, the optical system must have a high transmission in order to ensure a high overall quantum efficiency η_{LIF}. Therefore, anti-reflective coatings for the lenses and high efficiency broadband reflection filters were used. To be able to fully calculate species concentrations from the LIF signals, or vice versa, all factors in Eq. 13 must be known very precisely. In practical applications it is usually easier to calibrate the overall detection efficiency by adding known amounts of NO to the engine in order to determine the necessary calibration factors for the measurement of absolute NO concentrations.

Image pairs with corresponding NO concentrations and temperature fields, measured simultaneously in the transparent engine, are presented in Fig. 12 which show that the formation of NO occurs in high temperature areas. While the overall spatial distribution of NO and temperature is strongly correlated, the profiles in Fig. 12 indicate that the temperature distribution is more uniform than the NO concentration distribution. This is an important result for the comparison with mathematical models which are developed for engine design. Due to the strong non-linear temperature dependence of the NO production rate, careful control and homogenization of the combustion conditions should allow significant reduction in primary NO formation.

Figure 13 shows a series of averaged (over 25 – 30 single frames) NO concentration images at different times after ignition, taken around top dead center. The engine is fuelled with stoichiometric iso-octane/air mixtures at 1000 rpm. The area where NO is formed expands rapidly into the combustion chamber. In

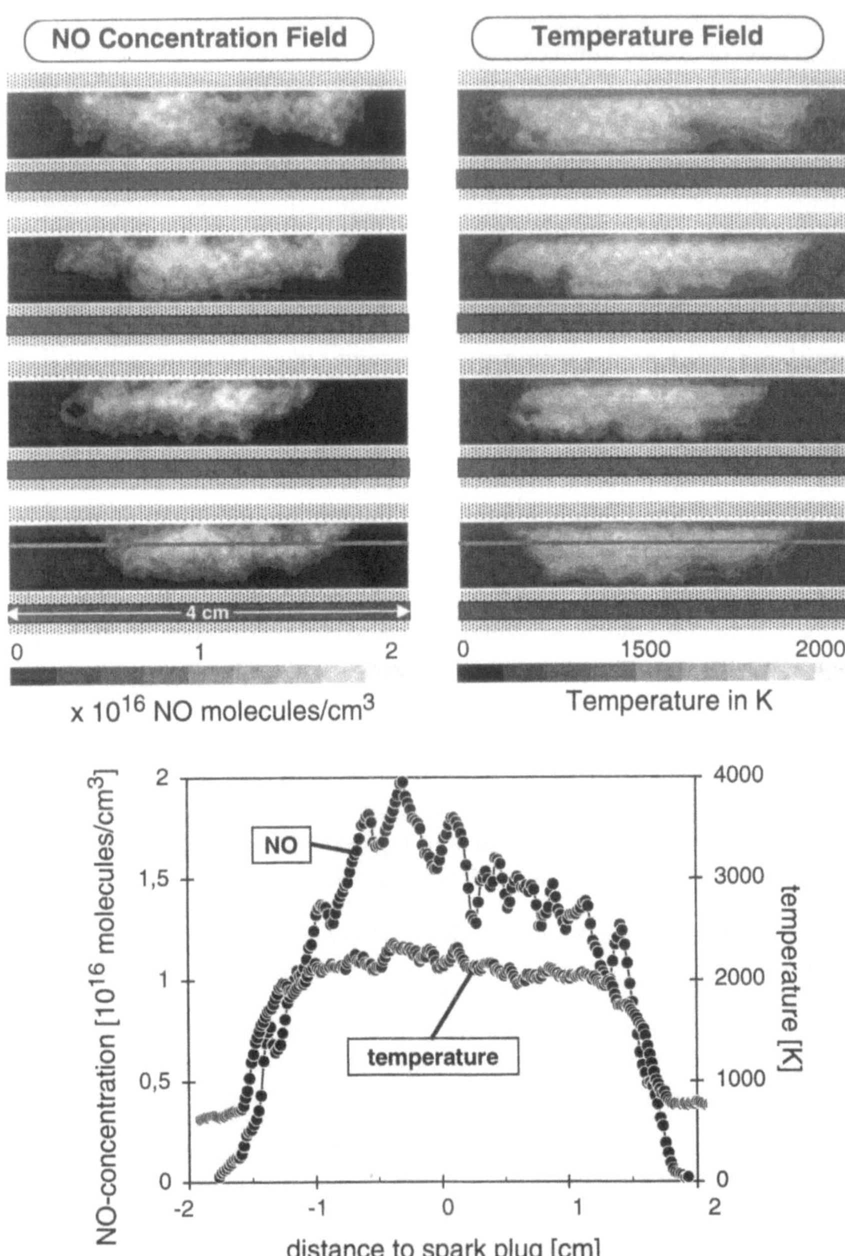

Fig. 12. Simultaneous single-shot measurements of absolute NO concentration and temperature fields in the transparent SI engine fueled with propane/air at $\lambda = 1.0$. Ignition at 340°ca, detection at 352° ca, with the engine running at 1000 rpm. The horizontal line in the lowest picture indicates the position of the profiles shown below

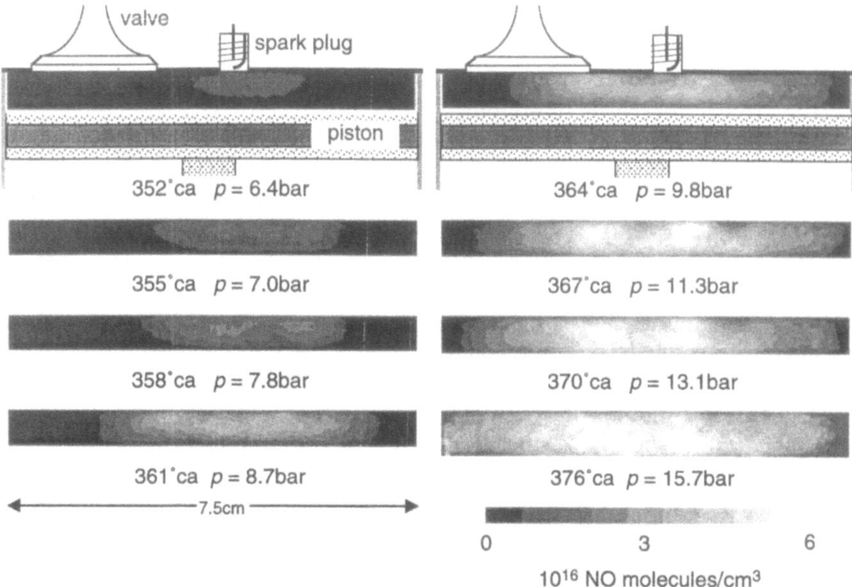

Fig. 13. Absolute NO concentration fields in the SI transparent engine fueled with iso-octane/air at $\lambda = 1.0$ as a function of the crank angle. Ignition at 340°ca, engine speed 1000 rpm. The figure shows averages over 25–30 frames

addition to the spatial growth, the absolute concentration of NO rises very rapidly. The total amount of NO within the cylinder increases with time due to the increasing volume of the post-flame region and the slow NO formation in the post-flame gases. The highly temperature-dependent NO formation rate of the Zeldovich mechanism [2] causes a strong dependence of the NO concentration on the residence time of the hot gases. Thus the highest NO concentrations are found in the middle of the combustion chamber where the hot gases necessary for NO formation have been present for the longest time.

3.4
Imaging of Fuel and OH Distributions in Engines

Lean-burn engines are designed to decrease the formation and release of pollutants from automobiles. To ensure reliable ignition of the fuel/air mixture in a lean-burn engine, the goal is to stratify the fuel inside the cylinder to ensure an ignitable mixture at the spark plug. Mixture formation is also related to the release of unburned hydrocarbons and soot from diesel engines. In addition, the degree of mixing determines how the fuel is ignited and how the engines operate under extreme conditions like cold start. In all cases it is desirable to obtain information about the fuel distribution and the ignition process simultaneously. LIF imaging of fuel can be used to visualize the mixing process and the fuel distribution. Furthermore, once the mixture has ignited, it is necessary to follow the development of the flame front and to measure its characteristic scales. As described in Sects. 3.1 and 3.2, time-resolved LIF measurements of OH radical distributions are suitable for studies of such ignition processes and for the localization of the developing flame fronts.

Commercial gasoline, which can be excited in the UV spectral region, subsequently emits a very strong fluorescence. However, due to the composition of many, up to several hundreds of components, the emitted fluorescence is not specific and can vary for different fuels. The influence of temperature, quenching, pressure and enrichment of different components during combustion is hard to quantify. Therefore, in order to obtain quantitative data, fuels containing only one fluorescing compound should be used. For this compound the spectral properties have to be characterized in separate experiment.. In the quantitative fuel distribution measurements to be described in the following, high purity, non-fluorescing iso-octane was used as a fuel and doped with 5% of 3-pentanone, as a fluorescence tracer. 3-Pentanone was chosen as a tracer because its boiling point (375 K) closely matches the boiling point of the iso-octane fuel (372 K). In addition, the influence of pressure and temperature on the fluorescence quantum yield of 3-pentanone after $S_0 \rightarrow S_1$ excitation has been studied in detail [70], which allows quantification of the fluorescence intensities measured under different engine operating conditions.

After laser-excitation at 248 nm the fluorescence quantum yield of 3-pentanone shows an increase of about 50% in the pressure range 1–5 bar, when N_2 is used as a quenching gas. At higher pressure the quantum yield becomes pressure-independent. The influence of the initial rise of the quantum yield decreases when longer excitation wavelengths are used. The pressure-dependence almost disappears for an excitation wavelength of 308 nm. On the other hand, when O_2 was used as a quenching gas, a slow decrease of the fluorescence intensity with pressure was observed. The initial increase in the signal intensity with increasing N_2 pressure might be attributed to vibrational relaxation processes in the optically excited S_1 state of 3-pentanone to levels with a higher fluorescence quantum yield. In the case of O_2, quenching of the S_1 state could compete with vibrational relaxation leading to the observed decrease in the fluorescence intensity. Efficient quenching of electronically excited organic molecules by $O_2(X^3\Sigma_g^-)$ is well known [70, 71].

A strong variation in the fluorescence signal with temperature was observed after excitation of 3-pentanone at 248 nm. To investigate this effect, UV absorption spectra of 3-pentanone were recorded at different temperatures in separate experiments (Fig. 14a). It was found that the maximum of the absorption band, which at T = 373 K is located at a wavelength of 275 nm, is shifted towards longer wavelengths with increasing temperature while the width of the band remains almost constant (with a FWHM of about 30 nm). As a consequence, the absorption strength at a wavelength of 248 nm (which is located on the left-hand side of the absorption maximum shown in Fig. 14a) will decrease with increasing temperature. Therefore, the observed decrease in the fluorescence signal after excitation at 248 nm is to be expected. Accordingly, if 3-pentanone is excited on the long-wavelength side of the absorption maximum, a positive temperature dependence should be observed, as was found for a excitation

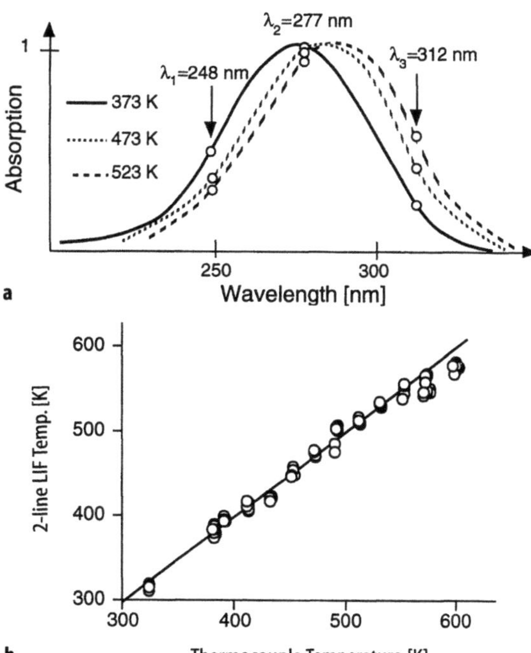

Fig. 14a, b. Temperature dependence of 3-pentanone fluorescence. **a** Absorption spectrum of 3-pentanone at different temperatures at atmospheric pressure. **b** Temperature measurement using 2-line-fluorescence of 3-pentanone. The temperature calculated from these 2-line measurements is shown as a function of the corresponding thermocouple readings

wavelength of 308 nm [72]. For an excitation wavelength of 277 nm, which is close to the absorption maximum, a minimal temperature dependence is expected.

The observed temperature dependence of the absorption strength of 3-pentanone and the corresponding fluorescence intensity offers the possibility for a new type of temperature measurement. To demonstrate the feasibility of the new method, the ratio of the fluorescence intensity following 248 nm and 312 nm excitation was measured. These measurements were carried out at atmospheric pressure in synthetic air for the temperature range 350–640 K. Fluorescence intensities $I_{312\,nm}(T)$ and $I_{248\,nm}(T)$ were measured as a function of temperature. After normalization of the data with respect to the signal strengths observed at T=383 K for each pair of excitation pulses, the ratio $I_{312\,nm}(T)/I_{248\,nm}(T)$ was calculated. In separate experiments the relative fluorescence intensities were measured as a function of temperature for both excitation wavelengths. This calibration data set was used to obtain temperatures from the fluorescence intensity ratio data. Fig. 14b shows the comparison of the temperature obtained from the fluorescence intensity ratio and the thermocouple readings. Each measurement point represents the average of 15 successive laser shots. Over the whole temperature range investigated a good agreement between the results of the fluorescence intensity ratio temperature measurement method and the thermocouple measurement was found.

Fuel distribution and flame propagation were investigated in a four-cylinder engine with a four-valve cylinder head with only minor modifications compared to the series engine [73]. Fuel was injected into the intake ports of every cylinder. Injection timing, pressure traces and crank angle were monitored cycle by cycle. Optical access was possible via quartz windows (33x4 mm) mounted in the cylinder head on both sides of each cylinder. The laser beams illuminated a plane 10 mm below the spark plug and the signals generated in the laser light sheet plane were monitored through a window in the piston. Observation of the fluorescence signal with a CCD camera was possible via a fixed mirror. 3-pentanone was excited by the beam of a broad-band XeCl-excimer laser with pulse energies of about 250 mJ/pulse (at a laser pulse duration of 17 ns) which was focussed, by means of a cylindrical Galilean telescope, to form a light sheet of 25×1.2 mm.

For the OH detection a tunable KrF-excimer laser (248 nm) similar to the ones described in detail in Sect. 3.1 was used. The LIF signals were separated with suitable filters to distinguish unambiguously between OH and fuel fluorescence. A narrow-band filter (center wavelength 295 nm; FWHM 10 nm) was used to monitor the ($v' = 3$, $v'' = 2$)-fluorescence band of the OH($A^2\Sigma^+$–$X^2\Pi$)-transition. The use of a colored glass filter (Schott WG 345) allowed detection of the fuel tracer fluorescence without any interference from other species. Both filters made sure that no Rayleigh scattered light or stray light was detected. The plane illuminated by the laser light sheet was imaged onto image-intensified CCD cameras equipped with special UV optics. The short laser pulse duration of about 20 ns in combination with short gate times (typically 25 ns) of the image intensifiers provided the complete suppression of the flame luminosity.

Figure 15a shows measured fuel distributions as a function of crank angle (°ca). Inhomogeneities in the mixing phase – before ignition takes place – are clearly resolved. Such inhomogeneities may be wanted in same cases, e.g., to ensure reproducible ignition in stratified systems, but may, in turn, also cause hot spots which can lead to an enhanced NO formation. Ongoing combustion is marked by the disappearance of the fuel LIF signal. For different air/fuel ratios large differences in flame speed and, thus, in the duration of the combustion were observed. Combustion is much slower at fuel-lean equivalence ratios than it is under fuel-rich conditions.

Simultaneously recorded images of fuel and OH radicals show the correlation between fuel consumption and burning or burned areas. For lean and stoichiometric mixtures the OH concentration shows a rather smooth distribution. But there is an obvious change in the OH distribution when switching over to fuel-rich operating conditions. In this case the OH concentration exhibits much more irregular profiles with higher gradients. This cannot be seen in the fuel images. Averaged images (from 64 single frames) of fuel and OH radical distributions were calculated and are depicted in Fig. 15b. An important result here is that the flame burns asymmetrically through the cylinder in all cases. In the region of the (hotter) exhaust valves the flame burns faster. This observation is in agreement with velocity measurements in the cylinder. Since it is known that under very lean conditions the flame extinguishes [74], it would be important not only to visualize the fuel concentration near the spark plug, but also the occurrence and distribution of OH radicals. the correlation between fuel/air mixture, ignition and flame propagating. Such studies would be especially impor-

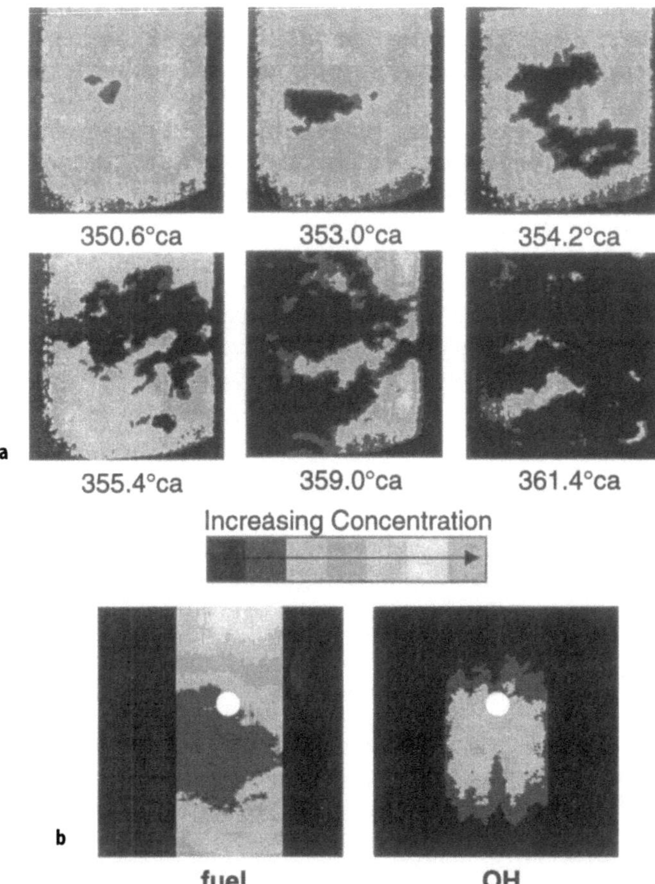

Fig. 15 a, b. Fuel distribution and flame propagation in a four-stroke engine. a Start of combustion visualized by LIF of 5 % 3-pentanone doped to the non-fluorescing fuel (iso-octane) for different detection timings after ignition (at 335° ca). b Distribution of fuel and OH radicals in the same engine (averaged images over 64 single frames). The white dots indicate the position of the spark plug

tant for the development of engines to be operated under ultra-lean conditions. In addition, systematic studies of the engine behavior under extreme running conditions, like cold start, can carried out using the described laser-based methods.

3.5
In Situ Alkali Concentration Measurements in a Fluidized-Bed Coal Combustor

Alkali metal compounds like, e. g., NaCl, and KCl, are unwanted trace species of the coal combustion process. Even at very low concentration levels (down to several ppb), the deposition of alkali vapors can lead to severe damage of reactor parts and downstream equipment in gas turbines [75]. In order to remove these offending species appropriate filter systems have to be developed. The effectiveness of such filters, however, can only be assessed if absolute alkali concentrations can be measured on-line, in situ, and under the appropriate temperature and pressure conditions. Classical collective sampling techniques have frequently been used due to their common availability, but they require sampling/analysis times of more than 1/2 h, in pressurized systems often of several hours. In addition, transient behavior, e.g., upon changes of combustion conditions, cannot be detected in this way.

A suitable optical on-line diagnostics method for detecting gas-phase alkali concentrations is the excimer laser induced fragmentation fluorescence (ELIF) technique [76–78]. In this method alkali compounds (MX) are photodissociated using, e.g., 193 nm laser radiation of a broad-band ArF-excimer laser. In the molecular fragmentation process electronically excited $Na(3^2P)$ and $K(4^2P)$ atoms (M*) can be formed via:

$$MX + h\nu_{Laser} \rightarrow M^* + X \tag{14}$$

Fluorescence which is subsequently emitted from the excited atoms ($M^* \rightarrow M + h\nu_{ELIF}$) can then be detected at wavelengths of $\lambda_{ELIF} = 589$ nm (Na) and $\lambda_{ELIF} = 768$ nm (K), respectively. The relationship between the fluorescence intensity I_{ELIF} and the number density of excited atoms N_{M^*} is given by:

$$I_{ELIF} = C\,(\tau_{eff}/\tau_{rad})\,N_{M^*} \tag{15}$$

For $I_{Laser}\,\alpha(\lambda_{Laser}) \ll 1$, N_{M^*} is related to the initial number density N_{MX} of the parent MX molecules, their optical absorption cross section $\alpha(\lambda_{Laser})$, and the intensity of the photodissociation laser, I_{Laser}, by the following expression:

$$N_{M^*} = N_{MX}\,\phi(M^*)\,\alpha(\lambda_{Laser})\,I_{laser} \tag{16}$$

where $\phi(M^*)$ is the quantum yield for M* formation in the fragmentation process. Insertion of Eq. 16 into Eq. 15 leads to the final expression:

$$I_{ELIF} = C\,(\tau_{eff}/\tau_{rad})\,N_{MX}\,\phi(M^*)\,\alpha(\lambda_{Laser})\,I_{laser} \tag{17}$$

The ratio of the effective fluorescence lifetime τ_{eff} to the radiative lifetime $\tau_{rad} = 1/A$ accounts for fluorescence losses due to quenching of excited alkali atoms by collisions. A denotes the Einstein coefficient of spontaneous emission. In the flue gases of coal combustors, N_2, CO_2, O_2, and H_2O are known to be the most important quenching partners with typical mole fractions of 0.76, 0.14, 005, and 0.04, respectively [78]. Therefore the value of $\tau_{eff} = 1/(A + Q)$ can be determined by calculating the total quenching rate, $Q = \Sigma_i\,q_i$, from the quenching rates q_i of the most important quenching partners. The latter can be obtained from the corresponding quenching cross sections reported in [79]. In Eq. 17, the photodissociation laser intensity, I_{Laser}, could be determined from the measured laser pulse energies and the area of the laser beam. C describes a calibration factor which was determined in separate ELIF experiments using a heatable reference cell to produce well-defined alkali number densities.

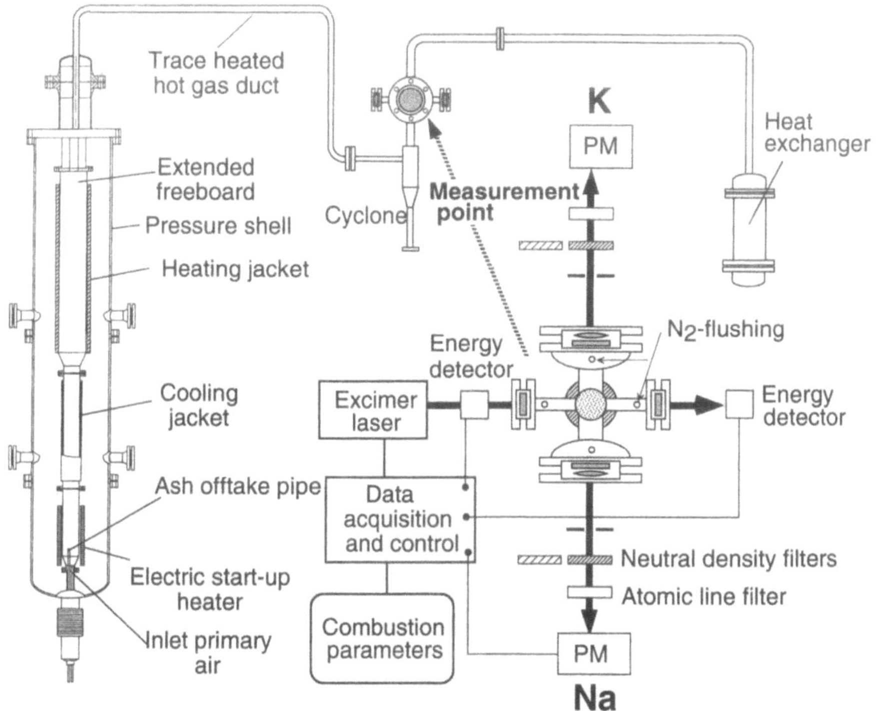

Fig. 16. Pressurized fluidized-bed reactor and ELIF setup for in situ alkali concentration measurements

In situ alkali concentration measurements were carried out in a pressurized fluidized-bed reactor with a rating of about 100 kW at 10 bar. The reactor is depicted in Fig. 16. It includes a lower, bubbling fluidized-bed combustion zone of 1.5 m bed depth and an upper, extended freeboard zone. After leaving the reactor, part of the ash is removed from the flue gas in a cyclone located right in front of the optical access port for the alkali measurements. To allow operation at 10 bar, the reactor head, flue gas pipe, cyclone and optical access were designed using heat-resistant steel. Care was taken to optimize the N_2-flushing of the ELIF laser entrance/exit, and fluorescence exit windows in order to minimize dust deposition.

Figure 17 shows results obtained in experiments performed under various conditions. In each case, the measured temporal history of the potassium (K) and sodium (Na) concentrations is shown together with the corresponding flue gas temperature. In general, concentration levels for both alkali compounds were below 10 ppb for the hard coal, but in the range 300–800 ppb for the lignite. The reason for the observed large difference in alkali concentration for the two coals is their very different mineral composition. Analysis of the ash samples showed a very high proportion (about 30%wt) of alumosilicates in the hard coal, which are known to retain alkali compounds strongly at the tempe-

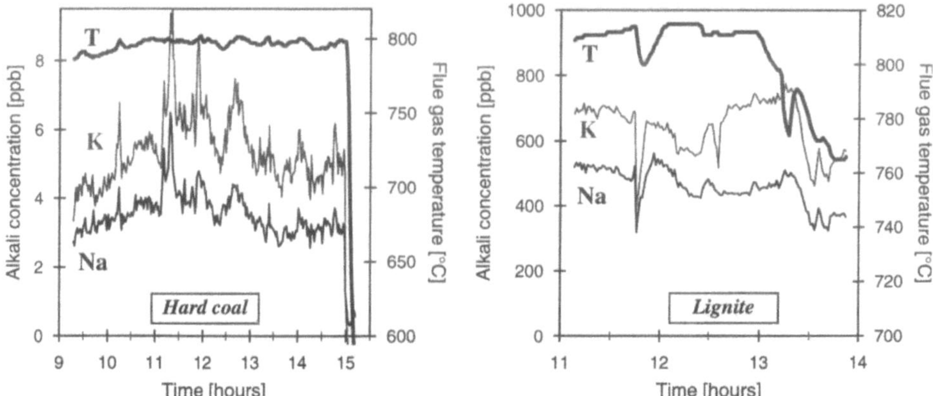

Fig. 17. Sodium (Na) and potassium (K) concentrations in the pressurized fluidized-bed coal combustor, measured with ELIF. The simultaneously recorded flue gas temperature is also shown

ratures of fluidized-bed combustion. In comparison, the lignite ash contains only about 3%wt alumosilicates, even though the amount of ash is about the same (about 5%wt) for the two coals and the proportion of alkali is significantly higher in the hard coal.

Ongoing developments and refinements of the ELIF method will allow reliable and detailed investigations of alkali release and conversion in pressurized coal combustors in actual power plants. Furthermore, the extension of the ELIF method to heavy metal compounds appears feasible.

Acknowledgements. The authors would like to thank Prof. M. Baer (Soreq-NRC, Israel) for stimulating discussions on the $H + D_2$ reaction and Prof. U. Maas (ITV, University of Stuttgart) for providing the data of the simulation calculations shown in Fig. 6. Collaboration with M. Baer was supported by grant No. E1447 of the Bundesministerium für Bildung und Forschung (BMBF, Germany). Financial support from the Deutsche Forschungsgemeinschaft (via SFB 123 and 359) and the State of Baden-Württemberg (TECFLAM, IWR) is also gratefully acknowledged.

References

1. R.P. Wayne, Chemistry of Atmospheres (Clarendon, Oxford, 1985)
2. J. Warnatz, U. Maas, R.W. Dibble, Combustion, (Springer, Heidelberg New York 1996); A. Buschmann, J. Wolfrum, U. Maas, J. Warnatz, Physikalische Blätter 52, 213 (1996)
3. J. Wolfrum, H.-R. Volpp, J. Warnatz, R. Rannacher, (Eds.), Gas Phase Chemical Reaction Systems: Experiments and Models 100 Years after Max Bodenstein, Springer Series in Chemical Physics, Vol. 61 (Springer, Heidelberg New York 1996)
4. F.F. Crim, J. Troe, (Eds.), Unimolecular Reactions, Ber. Bunsenges. Phys. Chem. 101, 309 (1997)
5. J. Wolfrum, V. Sick, K. Kompa, (Eds.), Laser Diagnostics for Industrial Processes, Ber. Bunsenges. Phys. Chem. 97, 1503 (1993)

6. W.C. Gardiner, (Ed.), Combustion Chemistry, (Springer, New York 1994)

7. M. Bodenstein, Z. Phys. Chem. 22, 23 (1897)

8. L. Wilhelmy, Pogg. Ann. physik und Chemie 81, 413 (1850); L. Wilhelmy in Ostwald's Klassiker der exakten Wissenschaften, Vol. 29, (Ed.) W. Ostwald, (Engelmann, Leipzig 1891)

9. S. Arrhenius, Z. phys. Chem. 4, 226 (1889)

10. W. Stiller, Arrhenius Equation and Non-Equilibrium Kinetics (Teubner, Leibzig 1989)

11. I.W.M. Smith, B.R. Rowe, I.R. Sims in [3]

12. H. Eyring, M. Polanyi, Z. phys. Chem. 12, 279 (1931)

13. W.C. McC. Lewis, J. Chem. Soc. 109, 796 (1916); ibid. 111, 457 (1917); 113, 471 (1917)

14. M. A. Eliason, J.O. Hirschfelder, J. Chem. Phys. 30, 1426 (1959); K.E. Shuler in Chemische Elementarprozesse, (Ed.) H. Hartmann, (Springer, Heidelberg New York 1968)

15. R.M. Marshall, J.H. Purnell, J. Chem. Soc. A 1968, 2301 (1968)

16. F. London, Sommerfeldfestschrift (Hirzel, Leipzig, 1928); Z. Elektrochem. 35, 552 (1929)

17. a) D.G. Truhlar, R.E. Wyatt, Ann. Rev. Phys. Chem. 27, 1 (1976); b) M. Baer, (Ed.), Theory of Chemical Reaction Dynamics, Vol I-IV (CRC Press, Boca Raton 1985); c) G.C. Schatz in The Theory of Chemical Reaction Dynamics, (Ed.), D.C. Clary, (Reidel, Dordrecht 1986)

18. B. Liu, J. Chem. Phys. 58, 1925 (1973); P. Siegbahn, B. Liu, ibid. 68, 2457 (1978); D.G. Truhlar, C.J. Horowitz, ibid. 68, 2466 (1978); ibid. 71, 1514 (1979)

19. A.J.C. Varandas, F.B. Brown, C.A. Mead, D.G. Truhlar, N.C. Blais, J. Chem. Phys. 86, 6258 (1987)

20. A. Farkas, Z. phys. Chem. B10, 419 (1930)

21. R.D. Levine, The Farkas Memorial Symposium: Fifty Years of H+H_2 Kinetics, Int. J. Chem. Kinet. 18, 9 (1986); see also J.J. Valentini, D.L. Phillips in Advances in Gas Phase Photochemistry and Kinetics, Vol. 2, (Eds.) M.N.R. Ashfold, J.E. Baggott, (Royal Society of Chemistry, London 1989)

22. H. Buchenau, J.P. Toennies, J. Arnold, J. Wolfrum, Ber. Bunsenges. Phys. Chem. 94, 1231 (1990)

23. J.V. Michael in Advances in Chemical Kinetics and Dynamics, Vol. 1, (Ed.) R. Barker, (JAI Press, Greenwich 1992)

24. S. Kristyan, Computer in Physics, 8, 556 (1994)

25. T.N. Kitsopoulos, M.A. Buntine, D.P. Baldwin, R.N. Zare, D.W. Chandler, Science 262, 1852 (1993)
    L. Schnieder, K. Seekamp-Rahn, J. Borkowski, E. Wrede, K.H. Welge, F.J. Aoiz, L. Bañares, M.J. D'Mello, V.J Herrero, V. Sáez Rábanos, R.E. Wyatt, Science 269, 207 (1995)
    E. Wrede, L. Schnieder, K.H. Welge, F.J. Aoiz, L Bañares, V.J. Herrero, Chem. Phys. Lett. 265, 129 (1997)

26. Y.-S. Wu, A. Kuppermann, Chem. Phys. Lett. 205, 577 (1993); 213, 636 (1993); 255, 105 (1995)
    M.J. D'Mello, D.E. Manolopoulos, R.E. Wyatt, Science 263, 102 (1994)

27. L. Szilard, T.A. Chalmers, Nature 134, 462 (1934)
    J.K. Lee, B. Musgrave, F.S. Rowland, J. Chem. Phys. 32, 1266 (1960);
    R. Wolfgang, Prog. React. Kinetics 3, 97 (1965)

28. R.A. Ogg, R.R. Williams, J. Chem. Phys. 13, 586 (1945)
    G.A. Oldershaw, D.A. Porter, Nature 223, 490 (1969)

29. A. Kuppermann, Isr. J. Chem. 7, 303 (1969)

30. C.R. Quick, J.J. Tiee, Chem. Phys. Lett. 100, 223 (1983)
    K. Kleinermanns, J. Wolfrum, Appl. Phys. B 34, 5 (1984); Chem. Phys. Lett. 104, 157 (1984); J. Chem. Phys. 80, 1446 (1984)

31. S. Koppe, T. Laurent, P.D. Naik, H.-R. Volpp, J. Wolfrum, T. Arusi-Parpar, I. Bar, S. Rosenwaks Chem. Phys. Lett. 214, 546 (1993)
    T. Laurent, P.D. Naik, H.-R. Volpp, J. Wolfrum, T. Arusi-Parpar, I. Bar, S. Rosenwaks, Chem. Phys. Lett. 236, 343 (1995)
    S. Koppe, T. Laurent, P.D. Naik, H.-R. Volpp, J. Wolfrum, Can. J. Chem. 72, 615 (1994) "John Polanyi 65th Birthday Issue"

32. H.-R. Volpp, J. Wolfrum, in [3]
    S. Koppe, T. Laurent, P.D. Naik, H.-R. Volpp, J. Wolfrum, 26th Symp (Int.) on Combustion/
    The Combustion Institute, p. 489, 1996
33. G. Hilber, A. Lago, R. Wallenstein, J. Opt. Soc. Am. B. 4, 1753 (1987)
34. G.W. Johnston, B. Katz, K. Tsukiyama, R. Bersohn, J. Phys. Chem. 91, 5445 (1987)
    J. Park, N. Shafer, R. Bersohn, J. Chem. Phys. 91, 7861 (1989)
35. P.J. Estrup, R. Wolfgang, J. Am. Chem. Soc. 82, 2661 (1960)
    A. Rosenberg, R. Wolfgang, J. Chem. Phys. 2159, 41 (1964)
    W. Stiller, Nichtthermisch aktivierte Chemie (Birkhäuser, Basel 1987)
36. W.J. v. d. Zande, R. Zhang, R.N. Zare, K.G. McKendrick, J.J. Valentini, J. Phys. Chem. 95,
    8205 (1991)
37. C.W.H. Press, B.P. Flannery, S.A. Teukolsky, W.T. Vetterling, Numerical Recipes in C
    (Cambridge University Press, Cambridge 1988)
38. I. Schechter, R. Kosloff, R.D. Levine, J. Phys. Chem. 90, 1006 (1986); R.D. Levine, R.B.
    Bernstein, Molecular Reaction Dynamics and Chemical Reactivity (Oxford University
    Press, Oxford 1987)
    Molekulare Reaktionsdynamik (Teubner, Stuttgart 1991)
39. N.C. Blais, D.G. Truhlar, Chem. Phys. Lett. 102, 120 (1983); J. Chem. Phys. 83, 2201 (1985)
40. D.M. Charutz, I. Last, M. Baer, J. Chem. Phys. 106, 7654 (1997)
41. V. Ebert, R. Hemberger, W. Meienburg, J. Wolfrum in [5]
42. J. Wolfrum, V. Ebert in Optical Measurements: Techniques and Applications, (Ed.) F.
    Mayinger, (Springer, Heidelberg New York 1994)
43. T. Heitzmann, J. Wolfrum, U. Maas, J. Warnatz, Z. Phys. Chem. 188, 177 (1995)
44. P. Andresen, A. Bath, W. Gröger, H. W. Lulf, G. Meijer, J. J. terMeulen, Appl. Opt. 27, 365 (1988)
    J. A. Gray, R. L. Farrow, J. Chem. Phys. 95, 7054 (1991)
    D. E. Heard, D. R. Crosley, J. B. Jeffries, G. P. Smith, A. Hirano, J. Chem. Phys. 96, 4366 (1992)
45. P. M. Doherty, D. R. Crosley, Appl. Opt. 23, 713 (1984)
46. A. Arnold, B. Lange, T. Bouché, T. Heitzmann, G. Schiff, W. Ketterle, P. Monkhouse, J.
    Wolfrum, Ber. Bunsenges. Phys. Chem. 96, 1388 (1992)
47. P. Deuflhard, E. Hairer, J. Zugck, Numerical Mathematics 51, 501 (1987)
48. C. Chevalier, J. Warnatz in [6]
49. I. Gran, U. Maas (personal communication)
50. S.B. Pope, Prog. Energy Combust. Sci. 11, 119 (1986)
51. U. Maas, S.B. Pope, Combust. Flame 88, 239 (1992)
    U. Maas in [3]
52. N. Peters, Prog. Energy Combust. Sci. 10, 319 (1984)
53. G. Damköhler, Z. für Elektrochem. 46, 601 (1940)
54. R. Borghi in Recent Advances in Aeronautical Science (Pergamon, London 1985)
    N. Peters, 21 Symp. (Int.) on Combustion/The Combustion Instiutte, p. 1231 (1986)
55. A. Buschmann, F. Dinkelacker, T. Schäfer, M. Schäfer, J. Wolfrum, 26th Symp. (Int.) on
    Combustion/The Combustion Institute, p. 437, 1996
56. H. Pitsch (personal communication)
57. M. Baum in [3]
58. R. W. Dibble, R. E. Hollenbach, 18th Symp. (Int.) on Combustion/The Combustion
    Institute, p. 1231, 1988
59. M. Schäfer, W. Ketterle, J. Wolfrum, Appl. Phys. B52, 341 (1991)
60. W. Ketterle, M. Schäfer, A. Arnold, J. Wolfrum, Appl. Phys. B 54, 109 (1992)
61. T. C. Chew, R. E. Britter, K. N. C. Bray, Combust. Flame 75, 155 (1989)
62. P. D. Ronney, V. Yakhot, Combust. Sci. Tech. 86, 31 (1992)
63. J. Chomiak, 16th Symp. (Int.) on Combustion/The Combustion Institute, p. 1665, 1977
    Z.-S. She, E. Jackson, S. A. Orszag, Nature 344, 226 (1990)
    I. G. Shepherd, W. T. Ashurst 24th Symp. (Int.) on Combustion/The Combustion Institute,
    p. 485, 1992
64. P. Andresen, G. Meijer, H. Schlüter, H. Voges, A. Koch, W. Hentschel, W. Oppermann, E. W.
    Rothe, Appl. Optics 29, 2392–2404 (1990)

A. Arnold, F. Dinkelacker, T. Heitzmann, P. Monkhouse, M. Schäfer, V. Sick, J. Wolfrum, W. Hentschel, K. P. Schindler, 24th Symp. (Int.) on Combustion/The Combustion Institute, p. 1605, 1992

Th. M. Brugman, R. Klein-Douwel, G. Huigen, E. van Walwijk, J. J. ter Meulen, Appl. Phys. B, 57, 405–410 (1993)

65. A. Bräumer, V. Sick, J. Wolfrum, V. Drewes, R. R. Maly, M. Zahn, SAE Paper No. 952462, Toronto (1995)

66. C. Schulz, B. Yip, V. Sick, J. Wolfrum, Chem. Phys. Lett. 242, 259 (1995)

67. C. Schulz, V. Sick, J. Wolfrum, V. Drewes, M. Zahn, R. Maly, 26th Symp. (Int.) on Combustion/The Combustion Institute, p. 2597, 1996

68. C. Schulz, V. Sick, J. Heinze, W. Stricker, in Laser Applications to Chemical and environmental analysis, Vol. 3, OSA Technical Digest Series, Optical Society of America, 133 (1996)

C. Schulz, V. Sick, J. Heinze, W. Stricker, Appl. Opt. 36, 3227 (1997)

69. A. Orth, V. Sick, J. Wolfrum, R. Maly, M. Zahn, 25th Symp. (Int.) on Combustion/The Combustion Institute, p. 143, 1994

70. F. Großmann, P.B. Monkhouse, M. Ridder, V. Sick, J. Wolfrum, Appl. Phys. B 62, 249 (1996)

71. O.L. Gijzmann, Faraday Trans. II 70, 708 (1973)

P.B. Merkel, D.J. Kearns, J. Chem. Phys. 58, 398 (1973)

72. R. Tait and D. Greenhalgh, Ber. Bunsenges. Phys. Chem. 97, 1619 (1993)

73. A. Arnold, A. Buschmann, B. Cousyn, M. Decker, V. Sick, F. Vannobel, J. Wolfrum, SAE-paper No. 932696, Philadelphia (1993)

S. Seeger, V. Sick, H.-R. Volpp, J. Wolfrum; Isr. J. Chem. 34, 5 (1994)

74. H. Becker, A. Arnold, R. Suntz, P. Monkhouse, J. Wolfrum, R. Maly, W. Pfister, Appl. Phys. B50, 473 (1990)

75. G.A. Osborne, Fuel 71, 131 (1992)

L. Scandrett, R. Clift, Inst. Energy 57, 391 (1984)

76. R.C. Oldenburg, S.L. Baughcum, Anal. Chem. 58, 1430 (1986)

77. K.T. Hartinger, P.B. Monkhouse, J. Wolfrum, H. Baumann, B. Bonn, 25th Symp. (Int.) on Combustion/The Combustion Institute, p. 193, 1994

78. F. Greger, K.T. Hartinger, P.B. Monkhouse, J. Wolfrum, 26th Symp. (Int.) on Combustion/The Combustion Institute, p. 3301, 1996

79. B. Earl, R. Herm, S.-M. Lin, C.A. Mims, J. Chem. Phys. 56, 4568 (1972)

B. Earl, R. Herm, J. Chem. Phys. 60, 4568 (1974)

J.R. Barker, R.E. Weston, Jr., J. Chem. Phys. 65, 1427 (1976)

CHAPTER 12

Fluorescence Techniques for Probing Molecular Interactions in Imprinted Polymers

O. S. Wolfbeis, E. Terpetschnig, S. Piletsky and E. Pringsheim

1
Introduction

There is a growing interest in the design of artificial receptors and enzymes because the modeling of enzymes and receptors is helpful for understanding natural molecular recognition processes. In addition, artificial receptors and catalytic systems can be used to control processes such as the preparation and purification of chemicals, but in particular to design diagnostic tools. There is a growing interest in the design of artificial receptors and enzymes because the modeling of enzymes and receptors is helpful for understanding natural molecular recognition processes. In addition, artificial receptors and catalytic systems can be used to control processes such as the preparation and purification of chemicals, but in particular to design diagnostic tools. Imprinted polymers form a rather recent class of materials with recognition capabilities. The term refers to matrices of organic polymers possessing cavities of specific size, shape and charge and therefore are capable of recognizing molecules fitting into such cavities. Such materials are obtained by polymerizing monomers and cross-linkers in the presence of the molecule to be recognized ("template"). During polymerization, the geometry of the template-monomer complex is maintained in the growing polymer matrix. Removal of the templates leaves cavities displaying a shape and arrangement of functional groups corresponding to those of the template. Table 1 (not exhaustive) gives an overview on materials that are frequently used, on typical analytes („templates") that have been investigated, and on potential fields of applications.

Remarkable progress has been accomplished regarding the design of organic molecules based on low-molecular-weight ring systems such as crown ethers, cryptandes, and cyclodextrins which contain cavities for specific binding and are capable of molecular recognition [13–18]. However, their preparation is tedious and the specificity of the materials sometimes limited. An interesting scheme for preparing specific and artificial receptor sites in cross-linked polymers was introduced in the 1970s [10, 19–23]. In a process called imprinting polymerization, a template-monomer complex is formed and polymerized in the presence of cross-linking agents. The term "template" refers to the analyte which may be a molecular species (such as a drug or a carbohydrate), but could also be a biomolecule such as an antibody or DNA oligomer. During polymerization, the geometry of the template-monomer complex is maintained in the growing polymer matrix. Removal of the templates leaves cavities displaying a shape and arrangement of functional groups corresponding to those of the template (see Fig. 1). This technique has become a simple and straightforward method for preparing synthetic polymers of predetermined selectivity.

Meanwhile, molecularly imprinted polymers (MIPs) have been developed for a broad range of potential applications [20–23, Table 1]. Fluorescence spectroscopy has only seldomly been applied to study either the MIP itself, or the interaction between the MIP and the target analyte. This is surprising because fluorescence not only is extremely sensitive, but also provides unique methods to study the interaction of two species on a molecular level as is evidenced by

Table 1. Materials Commonly Used for Preparation of Molecularly Imprinted Polymers (MIPs), Representative Templates (Analytes), and Typical Applications

Functional monomer	Cross-linker	Template	Application	Ref.
2,6-bis(acrylamido)-pyridine	EDMA[b]	barbiturates	HPLC	[1]
MAA[a]	EDMA[b]	atrazine	chromatography	[2]
2-vinylpyridine and/or MAA[a]	EDMA[b]	amino acid derivatives	chromatography	[3]
4-vinylpyridine	divinylbenzene, styrene	complex between dibenzoylmethane and Cobalt(II)ion	as a stationary phase in HPLC	[4]
MAA	EDMA[b]	pentamidine, benzamide	chromatography	[5]
MAA[a]	pentaerythritol tri- or tetra-acrylate, 2,2-bis (hydroxy-methyl) butanol trimeth-acrylate	amino acid derivates and peptides	as a chiral stationary phase in HPLC	[6]
MAA[a]	trimethylolpropane trimethacrylate	β-adrenergic antagonists	capillary electro-chromatography	[7]
N-acryloylalanine	EDMA[b]	S-propanolol	capillary electro-phoresis	[8]
Allylamine	EDMA[b]	sialic acid	biomimetic receptor	[9]
MAA[a]	EDMA[b]	theophylline, diazepam	antibody mimics	[10]
MAA[a]	EDMA[b]	dansyl-L-phenyl-alanine	sensor	[11]
MAA[a]	EDMA[b]	benzyltriphos-phonium chloride	conductometric chemical sensor	[12]

[a] MAA: methacrylic acid.
[b] EDMA: ethylene glycol dimethacrylate.

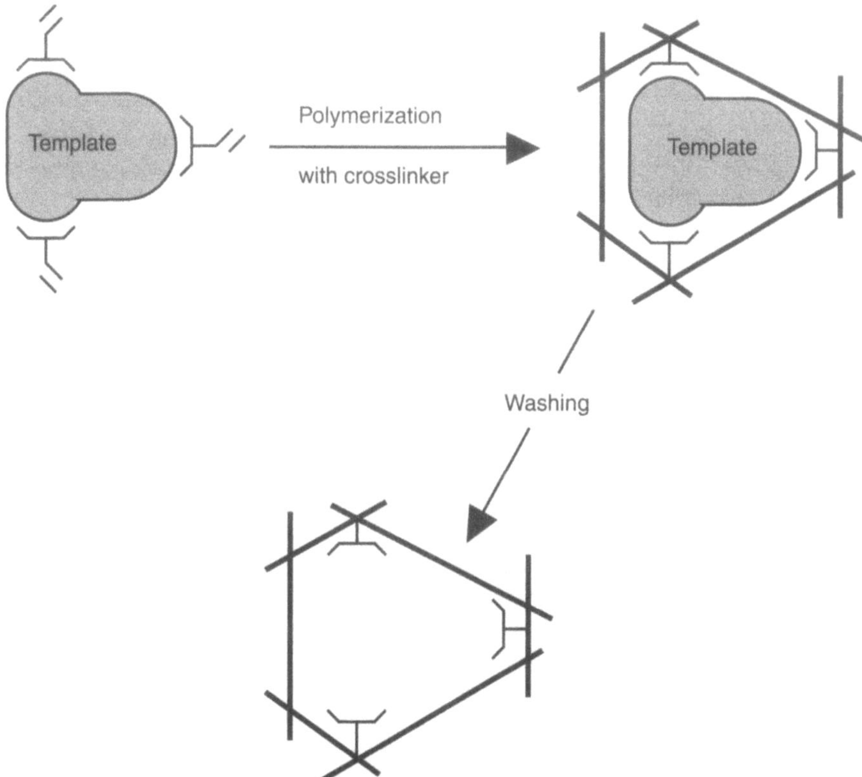

Fig. 1. Scheme of the imprinting polymerization. First, the templates form a weak complex with the monomers. Co-polymerization of this complex with the cross-linker and removal of the template yields an imprinted polymer containing micro-cavities with specifically oriented binding groups which are complementary to the template

the success of fluorescence immunoassays and various methods for DNA analysis including hybridization and sequencing.

The purpose of this article is to summarize the work performed so far using molecularly imprinted polymers in combination with fluorescent techniques. They can be used for (a), the characterization of imprinted polymers, (b), the interaction between template and MIP and (c), the development of fluorescent biosensors.

2
Fluorescence Techniques for Characterization of Imprinted Polymers

The most popular technique for making MIPs is based on bulk polymerization of a mixture of methacrylate functional monomers and cross-linker in a solvent in the presence of a template, followed by grinding and extraction of the print species (the template). A typical monomeric mixture consists of 35–45% cross-linker (divinyl benzene or ethylene glycol dimethacrylate), 2.5% template, 2.5–8% functional monomer (methacrylic acid, ethylene glycol dimethacrylate, vinylpyridine), 50% solvent (chloroform, acetonitrile), and 0.5% free-radical initiator (AIBN). The polymerization typically occurs at 60–80 °C within 8 h. Up to 70–90% of the template can be removed from the imprinted polymers by washing with water or organic solvents. The resulting polymer is a highly cross-linked but porous network with a surface area of 50–300 m^2/g that depends on the polymerization conditions including the solvent, initiator, and applied temperature. Despite the widespread use of macroporous cross-linked polymer networks, its characterization remains elusive. Most data refer to macroscopic properties such as surface area, swelling, and porosity, whilst information on microscopic properties is scarce.

Fluorescence is a powerful tool for the analysis of polymer structure and molecular interactions, as is impressively shown in this book. One of the most rewarding methods for analysis of polymer structures is based on the use of environmentally sensitive fluorescent probes. Among the fluorophores which show such properties, the dansyl probes such as 1 (6-dimethylamino-1-naphthalenesulfonate) are frequently used [24–26]. In polar solvents such as water, its fluorescence emission is weak, but in nonpolar solvent it exhibits strong emission and undergoes a shortwave shift in the fluorescence maximum. It can be excited at around 350 nm, has a large Stokes' shift, its fluorescence is not quenched by oxygen, and the naphthalene ring is small enough to be included into the micropores of MIPs without disturbing the micro-environment.

1

The probe was applied by Shea et al. to investigate highly cross-linked polymers with respect to (a), the solvatation of the polymer chain, (b), diffusion of ionic reagents through the network, and (c), estimation of polymer heterogeneity [27–29]. A variety of cross-linked co-polymers using dansyl probe 1 have been synthesized. An analysis of the solvatochromic shifts of the fluorescence emission of these materials, after being soaked in various solvents, has been found to correlate with the facility with which solvents penetrate the polymer gel of these materials [27]. Similarly, quenching of the dansyl fluorescence was used to evaluate the penetration of ionic reagents into the gel phase of cross-linked polymers [28]. Dansyl probes obviously allow rapid analysis of complex cross-linked networks and the evaluation of the effects of polymerization conditions (such as the degree of cross-linking and the nature of the monomers) on solvent and solute penetrability through the network.

An elegant experiment has been carried out for demonstrating the micro-heterogeneity of cross-linked polymers [29]. The representation of the gel phase of a cross-linked polymer network can be described in terms of two extremes. The one is a completely homogeneous, the other a completely heterogeneous network. The latter type consists of highly cross-linked domains that are interconnected by regions of lower cross-linking. In order to identify which one represents the polymer structure more adequately, a fluorescence study was carried out on polymers based on monofunctional (2) and difunctional (3) dansyl monomers. If totally homogeneous, no difference should be observed in the fluorescence of the resulting polymers.

It was found that the environments of probes 2 and 3 were different. Probe 3 was located (on average) within the less accessible, cross-linked region, whereas probe 2 resided in the more accessible, less cross-linked area. The differences in the microenvironment of the two probes resulted in a solvent-induced shift of the dansyl chromophore (Fig. 2). The values for λ_{cal} were obtained using Eq. 1 which allows the calculation of the fluorescence emission maxima of probes

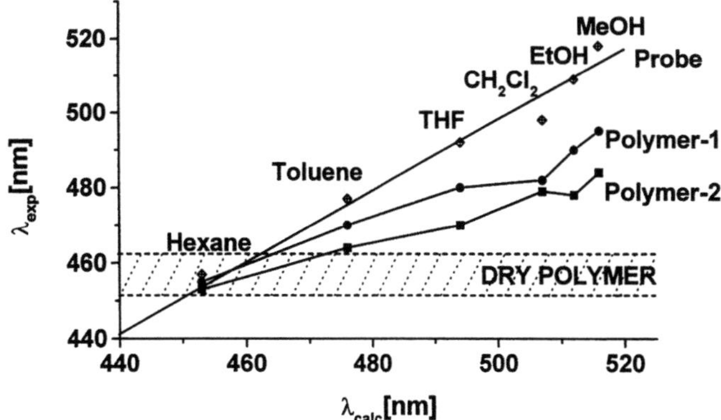

Fig. 2. Plot of calculated versus experimental fluorescence emission of solvent-equilibrated polymers containing probe 2 (polymer 1) and 3 (polymer 2) and probe 2 for different organic solvents

(e.g. **2** or **3**) in any solvent based on the values of π^* (solvent dipolarity/polarizability), α (hydrogen bond donor strength) and β (hydrogen bond acceptor strength).

$$\lambda_{cal} \text{ (nm)} = 53.45\pi^* + 20.48\,\alpha + 9.932\beta + 457.1 \tag{1}$$

In experiments with dansyl-containing functional monomers, it was shown that a heterogeneous model is preferable for representation of highly cross-linked polymer structures. In accordance with this model, imprinted polymers consist of primary particles (so-called *nuclei* or *domains*) grown together during polymerization. The diameter of these particles is from 100 to 200 nm. In between is a system of permanent pores (macropores) with diameters ranging from 20–50 nm [20].

Fluorescent measurements based on dansyl-derivatized polymers were used for the characterization of the environment of binding sites of MIPs. Two polymers, one imprinted by sialic acid (P1) and one not imprinted (K1) were prepared [9]. For the preparation of P1, the polymerization was performed with allylamine, vinylphenylboronic acid and ethylene glycol dimethacrylate in the presence of sialic acid. The mass ratio of the components is approximately 1:1:1. K1 was prepared in the absence of template. The amino functions introduced into the polymers by co-polymerization with allylamine were labelled with dansyl chloride (1), and the fluorescence of the polymers was measured in four different solvents. For comparison, the fluorescence of a blank polymer with adsorbed dansyl glycine (K2) was measured under the same conditions. It was assumed that in the imprinted polymers the dansyl probe is preferably located within the less accessible, hydrophobic cavities (imprints), while in non-specific polymers it should reside in the more accessible area (Fig. 3). When compared to dansylglycine in water solution (where the emission maximum is at 548 nm), the spectra of aqueous suspensions of the dansyl-modified polymers

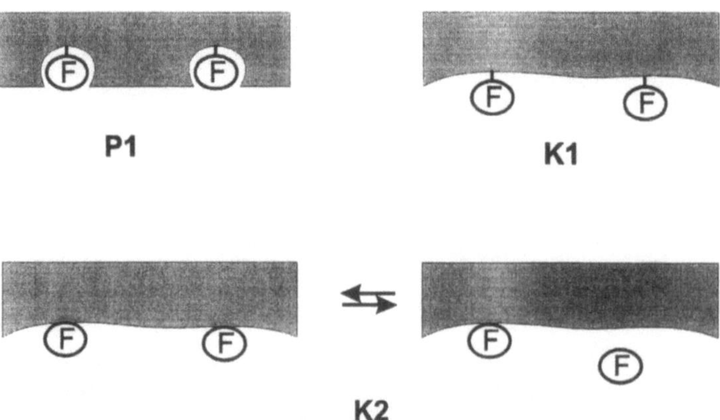

Fig. 3. Difference in the environment of the dansyl probe (F) in imprinted (P1) and non-imprinted polymers (K1 and K2) as revealed by the fluorescence emission maximum. While in K1 the probes are covalently attached, they are adsorbed on the surface of the polymer in K2

(P1 and K1, respectively) were blue-shifted (to 478 and 485 nm, respectively). The data are summarized in Fig. 4.

At the same time, the emission maximum of P1 was at shorter wavelength and its emission intensity higher than that of K1 in all solutions. This indicates that in P1 the dansyl group is located in a more hydrophobic environment than in K1. No difference in fluorescence was detected for non-modified P1, K1 or K2 polymers. The bathochromic shift of the dansyl spectrum of K2 in comparison to those of P1 and K1 can be explained by the fact that the fluorescence of the dansyl group in K2 is composed of both the emission from dansyl groups in solution and from those on the surface of the polymer which is distinctly more hydrophobic.

These observations suggest that the high affinity for a template is provided by imprints formed in the polymer structure by template molecules whose hydrophobicity is higher than that of non-specific binding sites. Electrostatic interactions inside hydrophobic cavities are stronger than those on the outside. In such cases, the binding constant for imprinted polymer-template interaction will be larger than the binding constant for non-imprinted polymer-template interactions. Because of this high total hydrophobicity of polymers made from ethylene glycol dimethacrylate (EDMA), it can also be assumed that the specific interactions between functional monomers and the template are mainly of an electrostatic nature.

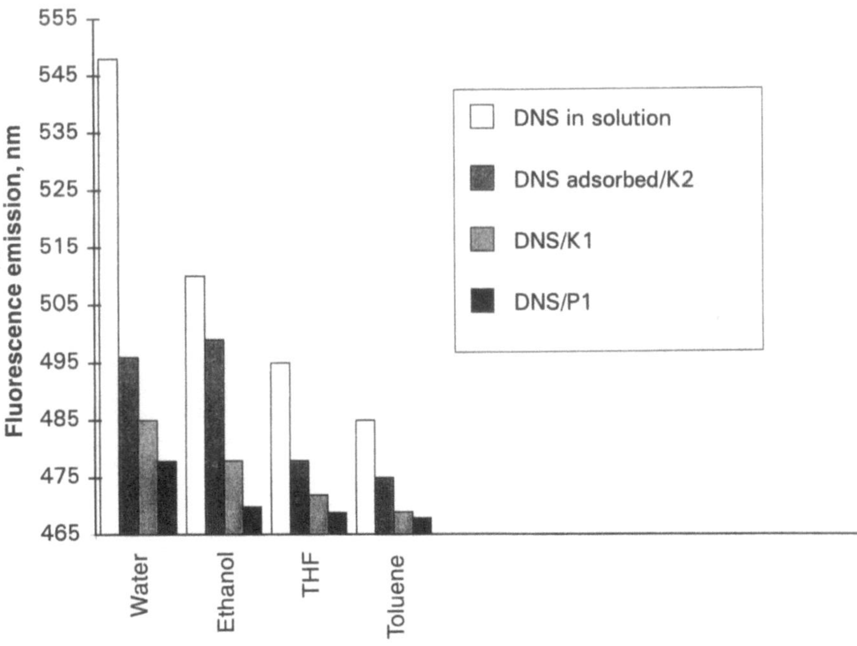

Fig. 4. Emission wavelengths of dansyl probes in different solvents. Pure dansyl-L-Phe, dansyl-L-Phe adsorbed on polymer, dansyl-labelled imprinted polymer, and dansyl-labelled non-imprinted polymer, respectively, were placed in solvents of different polarity and the maxima of the emission spectra recorded at λ_{exc} 355 nm. The top of bars indicates the emission maxima

The template-induced diffusion process of a non-charged solute into a polymer network was studied using ortho-phthalic aldehyde (OPA) and mercaptoethanol (ME). OPA and ME form a fluorescent complex in the presence of the polymer containing primary amino groups, as shown in Scheme 3. Again, an imprinted polymer for sialic acid (P1) and a non-imprinted polymer (K1) using allylamine, vinylphenylboronic acid and ethylene glycol dimethacrylate was prepared [9]. The formation of the fluorescent OPA-ME complex in the absence and presence of sialic acid, glucose, mannose and galactose using polymers P1 and K1 was investgated.

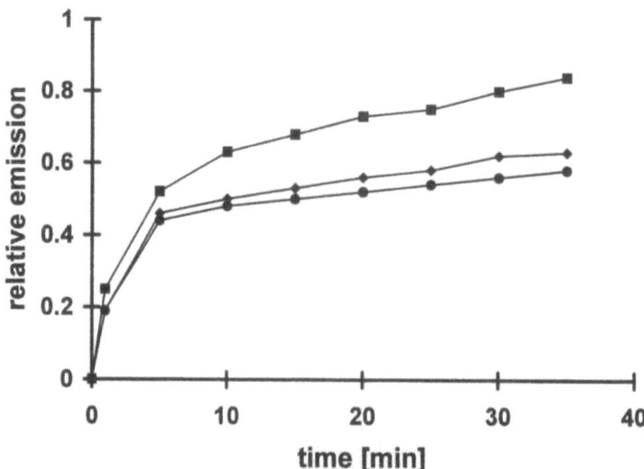

Fluorescence measurements were carried out using a suspension of particles of 1–5 µm diameter. Figure 5 shows the time-dependent changes in the fluorescence intensity due to the formation of the OPA-complex, both in the absence and presence of glucose, mannose and sialic acid. The intensity of P1 is significantly increased in the presence of sialic acid (template), which indicates that more polymer amino groups are available for reaction with the OPA reagent. Glucose and mannose cause smaller intensity changes than sialic acid, which demonstrates the high selectivity of the polymer receptor for the template molecules. Fluorescence emission increases with sialic acid concentrations up to 10 µM, but tends to saturate and even decrease at higher concentrations, which is probably due to the competition of sialic acid and the OPA reagent for the free interaction sites. Importantly, the addition of up to 5 µM concentrations of sugars (sialic acid, glucose or mannose) has no effect on the fluorescence of

Fig. 5. Fluorescence emission of OPA-modified polymer P1 in 100 mM sodium borate buffer of pH 10.0 over time. -■- Fluorescence in presence of 5 µM sialic acid -◆-; fluorescence in presence of 5 µM glucose or mannose; -●- fluorescence in the absence of carbohydrates

the non-imprinted polymers. This detection scheme may therefore be used to selectively detect sialic acid which plays an important role in the diagnosis of malignant diseases.

The observed effects can be explained in terms of electrostatic and conformational interactions between the template (analyte) and the polymer domains containing the selective cavities. This, in turn, is likely to result in a conformational rearrangement of the polymer structure (Fig. 6), thereby affecting the rate of diffusion of organic molecules. Imprinted polymers immersed in buffer show a limited swelling as a result of osmotic pressure, associated with surface ionization. As a result, the volume of polymer domains containing selective cavities increases. Diffusion of molecules through the polymer depends on the diameter and shape of the micropores between these domains and is retarded in the case of larger molecules. Re-formation of weak intermolecular bonds between the template molecules and the polymer during specific sorption is believed to cause exclusion of previously adsorbed ions and water molecules from the cavities, thereby decreasing swelling of the polymeric surface layer.

Fluorescence resonance energy (Förster) transfer (FRET) has been applied to investigate polymers both in solution and in the solid state. FRET is sensitive to intramolecular and intermolecular interactions [30–34]. A typical energy transfer experiment in a polymeric system involves labelling the polymer chains with two different chromophores, one being the donor, the other the acceptor. Absorption of light by the chromophore results in the promotion of an electron to an excited singlet state. This energy may be dissipated nonradiatively via a mechanism known as Förster energy transfer. FRET occurs through induction of a dipole oscillation in the (unexcited) acceptor by the excited-state

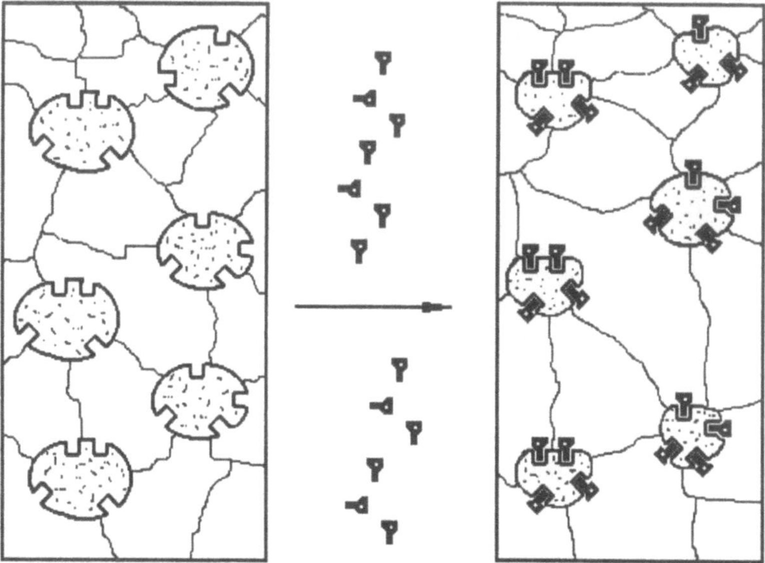

Fig. 6. Schematic drawing of template-induced structural changes occurring in MIPs

donor chromophore. The rate of energy transfer, (k_{ET}), between the chromophores is a function of the sixth power of the distance r between the donor and the acceptor. Because of the strong distance dependence of energy transfer, monitoring the fluorescence intensity in a system of labelled polymers can be considered a valuable parameter to measure the changes in the distance distribution of donor and acceptor moecules which reside at different sites in the imprinted polymer. FRET can therefore be used to investigate the polymer structure on interaction with templates.

In our experiments based on fluorescent resonance energy transfer (FRET) [35], three polymers specific for L-phenylalaninamide (L-Phe-NH$_2$) were prepared in DMF solution using a cocktail of acryloyl derivatives of fluorescein and eosin, methacrylic acid and ethylene glycol dimethacrylate. The first polymer contained only fluorescein (the donor, MIP-F), the second contained only eosin (the acceptor, MIP-E), while the third (MIP-F,E) contained both dyes in the monomer mixture. The emission spectra of these resulting polymers are shown in Fig. 7. It can be seen that there is strong FRET occurring in the double-labelled polymer (MIP-E,F), since the longwave acceptor emission increases whereas the shortwave donor emission decreases when compared to the polymers containing only the donor (MIP-F), or only the acceptor (MIP-E).

The effect of L-Phe-NH$_2$ (template) and its optical antipode on the emission of polymer MIP-E,F is shown in Fig. 8. The changes in the energy transfer efficiency between donor and acceptor can be associated with changes in the structure of the polymer as a result of polymer-template interaction as described above. The induced changes were higher in the case of L-Phe-NH$_2$ when compared to D-Phe-NH$_2$. In the case of the individually labelled polymers (MIP-F and MIP-E) the effect of the template on the emission spectra resulted in a small increase only in the intensities of both the donor and the acceptor. This can be

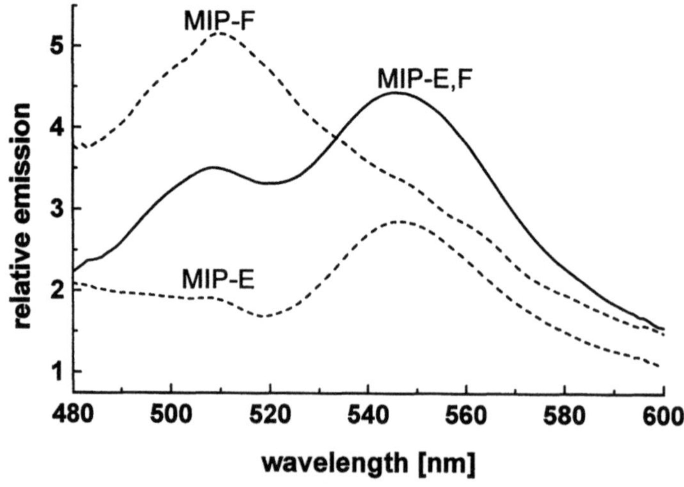

Fig. 7. Relative emission spectra of polymers containing fluorescein (MIP-F) as the donor, eosin (MIP-E) as the acceptor, and both the donor and the acceptor (MIP-E, F)

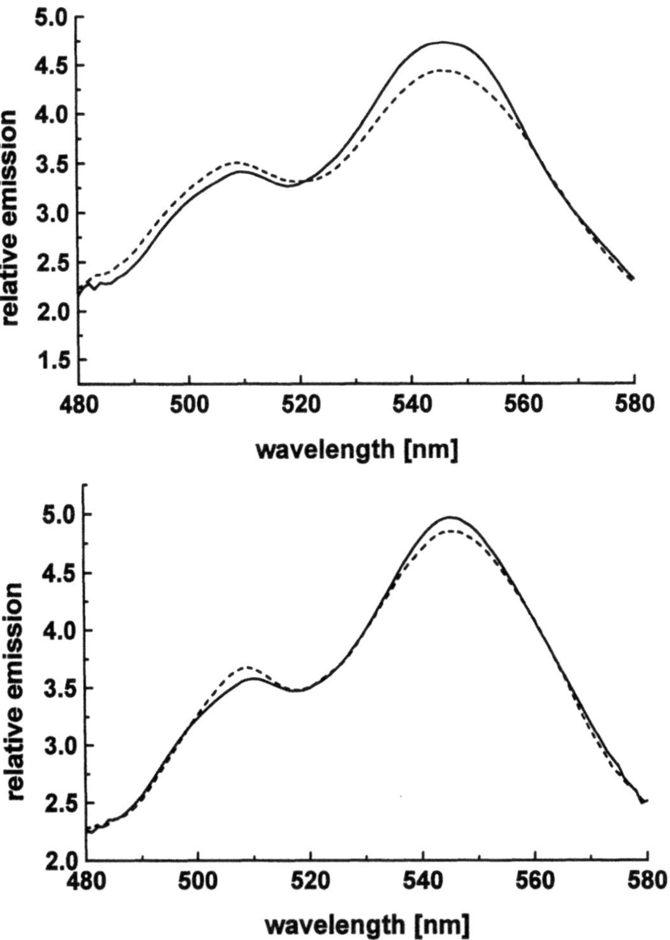

Fig. 8. Spectral changes of donor and acceptor emission in MIP-E,F in the presence of 10 μM concentration of L-phenylalaninamide (upper panel) and D-phenylalaninamide (lower panel)

explained by pH-changes occurring on addition of the template. In contrast to the MIP-E and MIP-F polymer, in the MIP-E,F polymer the specific effect of the template on the energy transfer process can be seen even in strong (100 mM) buffer solution, where pH-dependent effects on the emission spectra are supposed to be negligible. Nevertheless, the template-induced energy transfer in the MIP-E,F (Fig. 8) is small but the energy transfer efficiency and thus the sensitivity of the system can be improved using donor-acceptor pairs with different Förster distances (R_0).

3
Fluorescent Sensors Based on Imprinted Polymers

Molecular sensors undergo a change in their optical properties in response to the presence of chemical species. Hence, molecular recognition plays a major role in sensors. Both biological and abiotic systems can be exploited [36–38]. Since molecular imprints offer the possibility of specific recognition, they are potentially suited for use in sensor technology, and fluorescence is a powerful means for optical detection.

Imprinted polymers can be used in a modified version of fluorescent immunoassays (FIA), with antibodies replaced by imprinted polymer, to result in a so-called fluorescence-imprinted polymer immunoassay (FIPIA). Direct binding of the fluorescent template to the MIP, or competitive binding of free and fluorescent-labelled analyte, can be detected using fluorescence techniques. The imprinting method may also be used for sensor development where the recognition element is replaced by the MIP. Due to the replacement of a rather instable biomolecule, the stability of such sensors may be substantially improved. Sensors based on imprinted polymers have several potential advantages in comparison to sensors based on natural receptors. Thus, the performance of biosensors based on MIPs is not affected by harsh conditions during measurement [11, 39] and they can be operated in water and organic solvents. MIP-based sensors may also be developed for numerous substances, even for those which are not substrates of any natural enzyme or receptor.

By replacing the recognition element of an enzyme-based (or an antibody-based) sensor by a catalytically active (or ligand-specific) MIP, sensor systems can be obtained that are not only specific but distinctly more stable than sensors where the recognition element is protein-based. Three general schemes may be differentiated: The first is based on selective enrichment of fluorescence-labelled templates on, or inside, the polymer. This can be used for signal formation and amplification [40]. A typical example is provided by the binding of dansyl-labelled Phe to a polymer imprinted with Phe-NH$_2$ [11].

The second scheme is based on monitoring of template-induced alterations in the structure of the polymer backbone which acts as a molecular receptor. The change in the tertiary structure of the polymer results in altered fluorescence properties of polymer-bound probes. A typical example is provided by the binding of Phe to a Phe-NH$_2$-imprinted MIP labelled with a donor (fluorescein) and an acceptor (eosine) which results in energy transfer (Fig. 8). Finally, MIPs that have cavities with enzyme-like catalytic activity may be used in place of native enzymes for the design of sensors. Catalytic reactions will occur inside the polymer and this will lead to the formation or elimination of fluorescent species.

Among the numerous fluorescent techniques that can be applied to signal transduction, the following seem to be most suited for monitoring template-MIP interactions:

1) Measurements of the changes in the intensity or decay time of the fluorescence of MIPs, if *fluorescent* functional monomers are being used for the preparation of the polymer [40].

2) Detection of the fluorescence resonance energy transfer (FRET) between a donor-labelled polymer and an acceptor-labelled analyte. The competition between labelled and free analyte for the binding sites takes place on the free imprinted polymer, or on polymer immobilized on the detector surface. Replacement of labelled acceptor molecules by the analyte leads to a change in the emission properties of the donor and are reflected as changes in the intensity or decay time of luminescence.

Competitive binding of free and labelled analyte by an imprinted polymer was applied for the detection of herbicides [40]. The method is based on the competition between fluorescein-labelled triazine and unlabelled target molecules for the specific binding sites in the triazine-imprinted polymer. The system contains methacrylic acid as the functional monomer and shows a high selectivity and sensitivity for triazine in the 0.01–100 mM range (Fig. 9). Compared to the radio-labelled template [10] this method has the advantage of being safer to use.

An optical sensor for dansyl-L-phenylalanine was developed using a dansyl-L-phenylalanine-imprinted polymer and a fiber-optic sensing device [39]. The accumulation of the fluorescent template in the polymer matrix results in a signal increase which can be used to detect 10 mg/l of substrate within 4 h. The sensor is capable of discriminating between D- and L-isomers (Fig. 10).

Monitoring the optical properties of MIPs on interaction with templates was applied for the detection of vapors of organic solvents based on optical changes of a polymer embedded dye. For optical detection, a substituted 3,3-diphenyl-

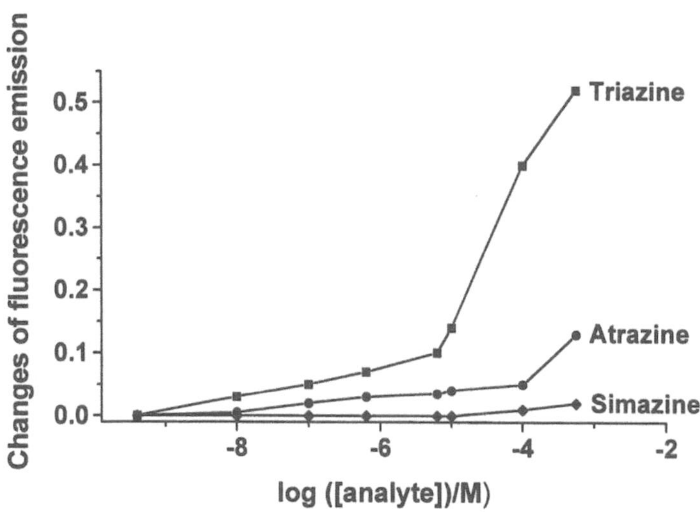

Fig. 9. Effect of herbicides on the adsorption of fluorescein-triazine by a triazine-imprinted polymer. The fluorescent measurements of the supernatant were performed after incubation of 10 mg of polymer in 1 mL of a 25 µM fluorescein-triazine solution in ethanol with herbicide during 3.5 h at room temperature. Changes in fluorescence intensity were detected at 515 nm (λ_{ex} = 336 nm)

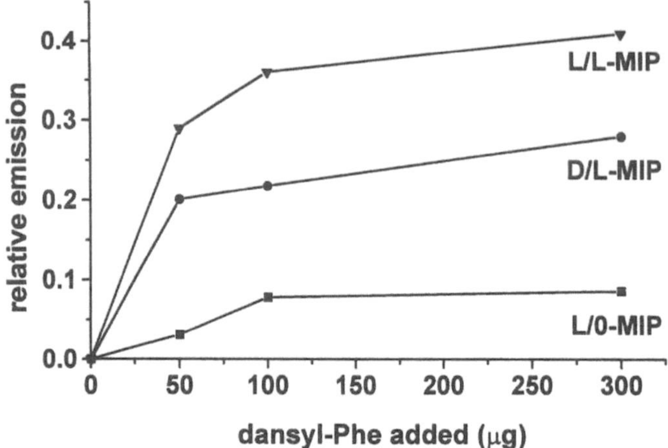

Fig. 10. Fiber-optic measurement of N-dansyl-L-Phe imprinted polymer (L-MIP) and non-imprinted polymer (O-MIP) after overnight incubation with analyte (L- or D-dansyl-Phe)

phthalide was intercalated in an acidic polyurethane matrix. Interaction with solvent molecules reduces the acidity thus changing the absorption properties of the intercalated dye [41].

Recently, we demonstrated the use of displacement chromatography on a molecularly imprinted polymer (MIP) against L-Phe-NH$_2$ as a stationary phase for the detection of Phe-related substances [42]. Methanolic solutions of 4-nitrophenol or Rhodamine B were used as the mobile phase in the detection system and to fill the cavities of the MIP. Subsequently, templates such as L-Phe-NH$_2$, D-Phe-NH$_2$, D- and L-Phe, and -Trp were injected into the mobile phase (Fig. 11). Because of the much higher affinity of these substances for the binding

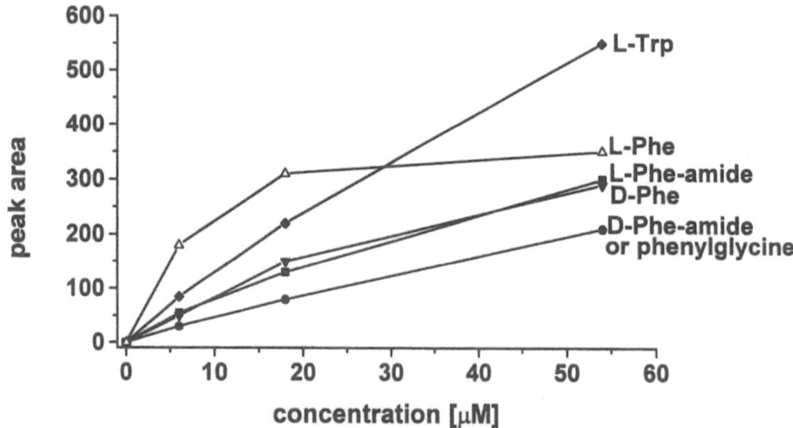

Fig. 11. Dependence of the peak area of dye replaced by various analytes from an L-Phe-amide imprinted polymer. Displacement was detected in methanol containing 10 µM of Rhodamine B

sites, they displaced the dye molecules, resulting in a "displacement peak" whose intensity was proportional to the concentration of the substance and its affinity to the MIP. Thus, we were able to differentiate micromolar concentrations of the optical isomers of Phe and Phe-NH$_2$. The response generated by L-Phe-NH$_2$ was measured in different solvents and found to be 1.3 to 2 times higher compared to that of the respective optical antipode. These experiments demonstrate the potential of the stream-line displacement method of a non-specific dye from a MIP to be used for the development of multisen-sor systems.

A sensing system for the determination of chloramphenicol (CAP), utilizing molecularly imprinting polymers (MIPs) and HPLC has been developed by Levi et al. [43]. The method is based on competitive displacement of a CAP-methyl red dye conjugate from specific binding cavities in an imprinted polymer by the analyte. The best results were obtained using diethylaminoethyl methacrylate as functional monomer and a monomer/template ratio of 2:1. HPLC with a mobile phase containing CAP-MR was used as the detection system and injection of CAP resulted in a proportional displacement of the conjugate. The system offers a selective and rapid method for CAP detection, discriminates between similar molecules and is effective below and above the therapeutic range.

So far the most convincing demonstration of a sensor based on molecular imprinting polymers in combination with optical signal transduction is a fiber optic sensor device in which dansyl-L-Phe-NH$_2$ binds to the polymer placed at the tip of a fiber. The resulting fluorescent signal varies as a function of the concentration of the analyte [11].

4
Future Perspectives

The use of fluorescence in combination with imprinted polymers can open new ways to create sensitive and stable systems for detection of biologically significant analytes. So far, it has mainly been the selectivity and availability of enzymes and antibodies that have promoted their use as recognition systems in bioanalysis. However, the specificity of MIPs for a certain template can be comparable to that of polyclonal antibodies and therefore MIPs may be employed as recognition systems for the development of fluorogenic assays. A major additional aspect is the cost situation. Enzymes and antibodies are much more expensive in comparison to polymers imprinted with the analyte and therefore have the potential of replacing proteins in biosensors technology.

Fluorescent-imprinted polymers with high specificity and sensitivity for the analyte in combination with fiber optics may be used to measure specific analytes under conditions that are not tolerated by sensors based on immobilized antibodies. In addition, such sensors are likely to be more easily sterilized. Hence, the use of molecularly imprinted polymers as biomimicking recognition elements in chemical sensors will allow the production of low-cost sensors with high overall operational stability. Certainly, their application is not limited to absorption-based or luminescence-based detection.

Nevertheless, several problems need to be solved before such sensors will be used in practice. For example, better polymers are needed for imprinting of water-soluble species such as proteins and nucleic acids, and methods need to be developed for polymer deposition on the surface of optical fibers or planar wave guides. In terms of instrumentation, diode lasers are desirable light sources: their light flux is coherent and allows better incoupling into wave guides, and they are inexpensive and small. Diode lasers necessitate, however, labels with longwave absorption and emission which at present are scarce.

Fluorescence analysis is mostly based on measurements of fluorescence intensity of a probe or indicator as a function of the concentration of the analyte. Optical signal quantitation in imprinted polymers is often difficult because the intensity is affected by additional factors such as light scatter or inhomogeneities in polymer structure. One way to overcome this problem is to measure the decay time of luminescence, rather than its intensity. This approach is highly advantageous because measurements of decay time are independent of probe concentration, detector sensitivity, and are less affected by light scattering. Other techniques that may be exploited in MIP-based fluorescent assays include fluorescence resonance energy transfer, quenching or anisotropy.

An approach towards a sensor system based on energy transfer is to introduce a fluorescent donor and a fluorescent acceptor into a MIP using labelled fuctional monomers. The efficiency of energy transfer from donor to acceptor is affected by the spatial changes of the polymer network on binding of the template and can be monitored via measurement of intensity or lifetime in a FRET system. Other approaches may be based on the competition of labelled and unlabelled species for fluorescently labelled MIPs (see Fig. 12). In this case, binding or release of labelled species is related to changes in the energy transfer rate which again affects the lifetime and intensity of the emitted light.

Fluorescence anisotropy is another parameter which is not affected by changes of intensity and/or inhomogeneities in the sensing environment. The physical basis of these measurements is the polarization (anisotropy) of the

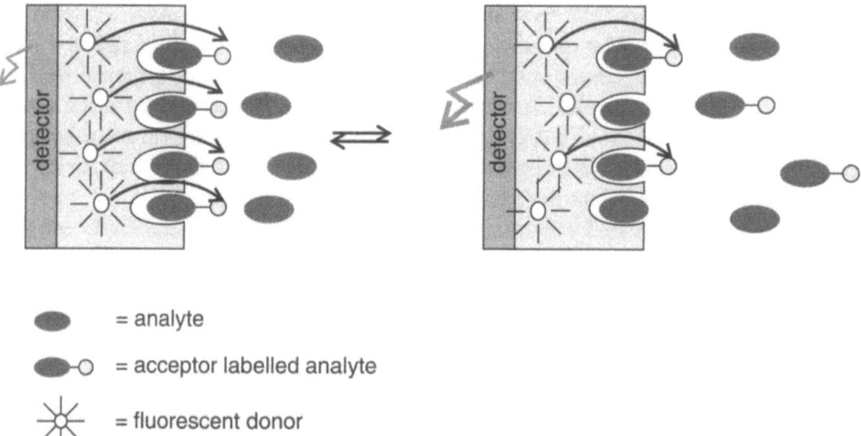

Fig. 12. MIP sensor based on energy transfer from a donor-labelled MIP to an acceptor-labelled template (see the text)

emitted light on excitation with vertically polarized light. The anisotropy reports on the rotational motion of the fluorophore during the lifetime of the exited state and can be used for measurement of receptor ligand interactions or for quantitative measurement of analyte concentrations in so-called fluorescence polarization assays. Artifical receptor molecules like MIPs have the potential to replace expensive natural antibodies and thus enable analysis of antigens under non-physiological conditions, in particular at elevated temperature and high or low pH which cannot be applied using natural receptors.

5
Conclusion

Molecular imprinting is a convenient procedure for preparing synthetic materials with receptor capabilities. The main driving forces for further developments are the low costs of the materials employed and the facile preparation. They can be operated over a range of temperature and pressure as well as in different solvent systems, and are less influenced by pH-effects as compared to natural enzymes and receptors. The successful combination of fluorescence techniques and imprinted polymers will enable the development of sensitive and selective supramolecular devices and tools for analytical applications. Even though MIP sensors are 100–1000 times less sensitive at present (when compared to biosensors) such biomimetic sensors will have their potential uses in the future.

6
Appendix

Since the completion of the manuscript, substantial progress has been made in this area. Thus, Dickert et al. [44] have prepared molecular imprints that bind polycyclic aromatic hydrocarbons (PAHs) with substantial specificity via *van der Waals* interactions. They have exploited this finding for sensing purposes because the MIP was found to be capable of detecting sub-nanomolar concentrations of PAHs via fluorometry using a conventional spectrometer. The group of Powell [45] has demonstrated the applicability of a fluorescently labelled MIP to bind to cyclic adenosine monophosphate (cAMP). The fluorescent probe 4-(N,N-dimethylamino)styryl-N-vinylbenzylpyridinium chloride is polarity-sensitive, and the resulting sensor can detect as little as 0.1 µM concentrations of cAMP but does not "see" the structurally similar cyclic guanosine monophosphate (cGMP). Finally, Wulff [46] has authored a review on the use of imprinting techniques in synthetic polymers along with the options that exist for designing various kinds of chemosensors and biosensors.

References

1. Tanabe T, Takeuchi T, Matsui J, Ikebukuro K, Yano K, Karube I (1995) Recognition of barbiturates in molecularly imprinted copolymers using multiple hydrogen bonding. J Chem Soc, Chem Comm 2303–2304
2. Matsui J, Miyoshi Y, Doblhoff-Dier O, Takeuchi T (1995) A molecularly imprinted synthetic polymer receptor selective for atrazine. Anal Chem 67:4404–4408
3. Ramström O, Andersson LI, Mosbach K (1993) Recognition sites incorporating both pyridinyl and carboxy functionalities prepared by molecular imprinting. J Org Chem 58:7562–7564
4. Matsui J, Nicholls IA, Takeuchi T, Mosbach K, Karube I (1996) Metal ion mediated recognition in molecularly imprinted polymers. Anal Chim Acta 335:71–77
5. Sellergren B (1994) Direct drug determination by selective sample enrichment on an imprinted polymer. Anal Chem 66:1578–1582
6. Kempe M (1996) Antibody-mimicking polymers as chiral stationary phases in HPLC. Anal Chem 68:1948–1953
7. Schweitz L, Andersson LI, Nilsson S (1997) Capillary electrochromatography with predetermined selectivity obtained through molecular imprinting. Anal Chem 69:1179–1183
8. Walshe M, Garcia E, Howarth J, Smyth MR, Kelly MT (1997) Separation of the enantiomers of propanolol by incorporation of molecularly imprinted polymer particles as chiral selectors in capillary electrophoresis. Anal Comm 34:119–120
9. Piletsky SA, Piletskaya EV, Yano K, Kugimiya A, Elgersma AV, Levi R, Kahlow U, Takeuchi T, Karube I, Panasyuk TL, El'skaya AV (1996) A biomimetic receptor system for sialic acid based on molecular imprinting. Anal Lett 29:157–170
10. Vlatakis G, Andersson LI, Muller R, Mosbach K (1993) Drug assay using antibody mimics made by molecular imprinting. Nature 361:645–657
11. Kriz D, Ramström O, Svensson A, Mosbach K (1995) Introducing biomimetic sensors based on molecularly imprinted polymers as recognition elements. Anal Chem 67:2142–2144
12. Kriz D, Kempe M, Mosbach K (1996) Introduction of molecularly imprinted polymers as recogniton elements in conductometric chemical sensors. Sensors Actuat 33B:178–181
13. Vögtle F (1981) Host Guest Complex Chemistry. Springer-Verlag, New York
14. Atwood JJ, Davies JED, MacNicol DD (1984) Inclusion Compounds. Academic Press, San Diego, CA
15. Inoue Y, Gouel GW (1990) Cation Binding by Macrocycles. Marcel Dekker, New York
16. Cram DJ (1988) The design of molecular hosts, guests, and their complexes. Science 240:760–767
17. Cram DJ (1992) Molecular container compounds. Nature 356:29–36
18. Lehn J-M (1988) Supramolecular chemistry – scope and perspectives. Molecules, supermolecules, and molecular devices. Angew Chem Intl Ed Engl 27:89–112
19. Wulff G, Sarhan A (1972) Use of polymers with enzyme-analogous structures for the resolution of racemates. Angew Chem Intl Ed Engl 11:341–344
20. Wulff G (1995) Molecular imprinting in cross-linked materials with the aid of molecular templates – a way towards artificial antibodies. Angew Chem Intl Ed Engl 34:812–1832
21. Piletsky SA, Piletskaya EV, Elgersma AV, Yano K, Karube I, Parhometz YuP, El'skaya AV (1995) Atrazine sensing by molecularly imprinted membranes. Biosensors Bioelectron 10:959–964
22. Piletsky SA, Dubey IYa, Fedoryak DM, Kukhar VP (1990) Substrate-selective polymeric membranes. Selective transfer of nucleic acid components. Biopolim Kletka 6:55–58 (russ)
23. Nicholls IA, Andersson LI, Mosbach K, Ekberg B (1995) Recognition and enantioselection of drugs and biochemicals using molecularly imprinted polymer technology. Trends Biotechnol 13:47–51
24. Kimura K, Arata Y, Yasuda T, Kinosita K, Nakanishi M (1992) Fluorescence quenching measurements of the membrane bound lipid haptens with length spacers. Biochim Biophys Acta 1104:9–14

25. Petrossian A, Owicki J (1984) Interaction of antibodies with liposomes bearing fluorescent haptens. Biochim Biophys Acta 776:217–227
26. Hofmann K, Titus G, Montibeller JA, Finn FM (1982) Avidin binding of carboxy-substituted biotin and analogs. Biochemistry 21:978–984
27. Shea KJ, Sasaki DY, Stoddard GJ (1989) Fluorescence probes for evaluating chain solvation in network polymers. An analysis of the solvatochromic shift of the dansyl probe in macroporous styrene-divinylbenzene and styrene-diisopropenylbenzene copolymers. Macromolecules 22:1722–1730
28. Shea KJ, Stoddard GJ, Sasaki DY (1989) Fluorescence probes for the evaluation of diffusion of ionic reagents through network polymers. Chemical quenching of the fluorescence emission of the dansyl probe in macroporous styrene-divinylbenzene and styrene-diisopropenylbenzene copolymers. Macromolecules 22:4303–4308
29. Shea KJ, Stoddard GJ (1991) Chemoselective targeting of fluorescence probes in polymer networks. Detection of heterogeneous domains in styrene-divinylbenzene copolymers. Macromolecules 24:1207–1209
30. Youn HJ, Terpetschnig E, Szmacinski H, Lakowicz JR (1995) Fluorescence energy transfer immunoassay based on a long lifetime luminescent metal-ligand complex. Anal Biochem 232:24–30
31. Pekcan O, Egan L, Winnik MA, Croucher MD (1990) Energy transfer in restricted dimensions: a new approach to latex morphology. Macromolecules 23:2210–2216
32. Jiang M, Chen W, Yu T (1991) Controllable specific interactions and miscibility in polymer blends: 3. Non-radiative energy transfer fluorescence studies. Polymer 32:984–989
33. Lakowicz JR, Wiczk W, Gryczynsky I, Fishman M, Johnson ML (1993) End-to-end distance distributions of flexible molecules: frequency-domanin fluorescence energy transfer measurements and rotational isomeric state model calculations. Macromolecules 26:349–363
34. Liu G (1993) Determination of intramolecular distance distribution functions using the "spectroscopic ruler". 1. Theoretical feasibility. Macromolecules 26:1144–1151
35. Piletsky SA, Terpetschnig E, Wolfbeis OS (unpublished results)
36. Czarnik AW (1991) In: Schneider H-J, Dürr H (eds) In Frontiers in Supramolecular Organic Chemistry and Photochemistry. VCH: Weinheim, pp 109–122
37. Inouye M, Konishi T, Isagawa K (1993) Artificial allosteric receptors for nucleotide bases and alkali-metal ions. J Am Chem Soc 115:8091–8095
38. Inouye M, Hashimoto K-I, Isagawa K (1994) Nondestructive detection of acetylcholine in protic media: artificial-signalling acetylcholine receptors. J Am Chem Soc 116:5517–5518
39. Kriz D, Mosbach K (1994) Competitive amperometric morphine sensor based on an agarose immobilized molecularly imprinted polymer. Anal Chim Acta 300:71–75
40. Piletsky SA, Piletskaya EV, El'skaya AV, Levi R, Yano K, Karube I (1997) Optical detection system for triazine based on molecular-imprinted polymers. Anal Lett 30:445–455
41. Dickert F, Thierer S (1996) Molecularly imprinted polymers for optochemical sensors. Adv Mater 987–990
42. Piletsky SA, Terpetschnig E, Andersson HS, Nicholls IA and Wolfbeis OS (1999) Towards the Development of Multisensors Based on Molecularly Imprinted Polymers. Application of Nonspecific Fluorescent Dyes for Monitoring Enantio-selective Ligand-Polymer Binding. Anal Chem, in press
43. Levi R, McNiven S, Piletsky SA, Cheong S-H, Yano K, Karube I (1997) Optical detection of chloramphenicol using molecularly imprinted polymers. Anal Chem 69:2017–2021
44. Dickert F, Besenböck H, Tortschanoff M (1998) Molecular imprinting through van der Waals interactions: fluorescence detection of PAHs in water. Adv Mat 10:149–151
45. Turkewitsch P, Wandelt B, Darling GD, Powell WS (1998) Fluorescent functional recognition sites through molecular imprinting. A polymer-based fluorescent sensor for aqueous cAMP. Anal Chem 70:2025–2030
46. Wulff G (1997) Imprinting techniques in synthetic polymers – new options for chemosensors. In: Scheller FW, Schubert F, Fedrowitz J (eds) Frontiers in Biosensorics. Birkhäuser Verlag, Basel, 1997

**Part 3
Fluorescence Probes in Polymers**

Advanced Light Emitting Dyes:

Monomers, Oligomers, and Polymers

Y. Geerts, K. Müllen

1
Introduction

Substances emitting light in response to external stimuli (light, electric current, X-ray...) are known as *luminescent materials* or *luminophores* [1, 2]. Luminophores occupy a special position in materials science because they are generally associated with high-technology applications such as single-molecule spectroscopy [3], fluorescent solar collectors [4], optical switches [5], lasers [6, 7], fluorescent brighteners [8], labelled antibodies [9] and electroluminescent diodes [10–14]. These examples illustrate the diversity of applications and stress the importance of a unified approach toward "organic" and "polymeric" dye stuff chemistry.

This review presents new organic luminophores from our laboratory. Although synthesis forms the core of this work, synthetic details will be kept to a minimum for the non-experts. For the sake of clarity, this chapter is divided into three paragraphs: *monomers, oligomers,* and *polymers* corresponding to three important classes of luminescent organic materials. This classification may appear somewhat arbitrary to the experienced reader since no strict borderline exists between classes of materials. However, each of these classes of light-emitting materials fulfills a particular need and deserves attention in its own right. For an easy access to the body of literature reviewed, four tables summarize the optical properties of the luminophores presented within this review.

Dyes can be distinguished from *pigments* in that they are soluble organic chromophores [1]. Light-emitting monomeric dyes are organic molecules of relatively low molecular mass. They represent the major class of luminophores from the viewpoint of applications and the diversity of chemical structures and properties: i.e. quantum yields of fluorescence and wavelength of emission maxima [1, 2].

Luminescent conjugated oligomers bridge the gap between monomeric dyes and light-emitting polymers [15, 16]. In particular, homologous series of soluble conjugated oligomers present manifold interests. Firstly, they enable clean characterization and studies of geometric and electronic structures of extended monodispersed π-systems [17]. Many optical and electrical properties of conjugated polymers closely correspond to those of oligomers containing only a few numbers of repeat units, and upon examining physical properties as a function of chain length, a reliable extrapolation to the corresponding defect-free polymer is feasible [18]. Secondly, homologous series of soluble conjugated oligomers supply the basic models for testing the fine tuning possibilities of optical and electrical properties by changing the conjugation pathway (topology), varying the geometric demands, incorporating saturated centers into the main chain and inserting electron-withdrawing or -donating groups [17, 19, 20]. Finally, light-emitting oligomers represent functional materials in their own right [16]. From a synthetic viewpoint, however, luminescent conjugated oligomers often present the disadvantage of a more tedious synthesis than their corresponding polymers [16, 17, 21].

Light-emitting polymers, which are higher molecular weight analogues of luminescent oligomers, may suffer to some extent from their polydispersity, their limited solubility, and their structural inhomogeneities [18]. However, they display the decisive physical advantages, over dyes and luminescent oligomers, of forming films of improved optical quality with a lower degree of crystallinity and better three-dimensional stability due to their high mechanical resistance [22–24]. These advantages preferentially qualify luminescent polymers for applications as *light-emitting diodes* [10, 13, 14, 22, 24]. Closely related to the light-emitting polymers, the organic-inorganic hybrid materials, which are obtained by the sol-gel process, involve either the physical or covalent incorporation of organic dyes in an inorganic matrix. These materials combine the advantage of luminescent monomeric dyes: i.e. availability and structural diversity, with the three-dimensional stability of rigid inorganic matrix [6, 25].

In the next three sections, the monomeric, oligomeric and polymeric dyes are presented in detail. In particular, the similarities and the interdependence of these three classes of materials are stressed rather than differences between them.

2
Monomers

Perylene 1, perylene dicarboximide 4 and perylene tetracarboxdiimide 9a are widely used as dyes since they show oustanding chemical, thermal and photochemical stability as well as quantum yields of photoluminescence ($\varnothing_{PL}$) approaching unity (Schemes 1 and 2) [1, 26–28].

Clearly, the perylene derivatives mentioned above present an unusual set of properties which make them important building blocks for new dyes emitting at long wavelength [29–30] or even in the near infrared (NIR) [30–32, 40]. However, before considering the synthesis of perylene based fluorophores emitting in the NIR, one should consider the low processability inherent to chromophores composed of large π-systems. Specifically, perylene derivatives 1, 4 and 9a suffer from a limited solubility in organic solvents and from poor film-forming abilities by spin coating. These drawbacks are successfully circumvented by the introduction of solubilizing substituents at the periphery of the perylene core as depicted for compounds 2a, b, 3a, b, 5 and 8 [33–35].

Furthermore, a particularly efficient way to increase the solubility of perylene dyes is to introduce phenoxy substituents in the bay region as illustrated in Scheme 2 [4, 26, 30]. The $\varnothing_{PL}$ of tetraphenoxy-substituted perylene 9b

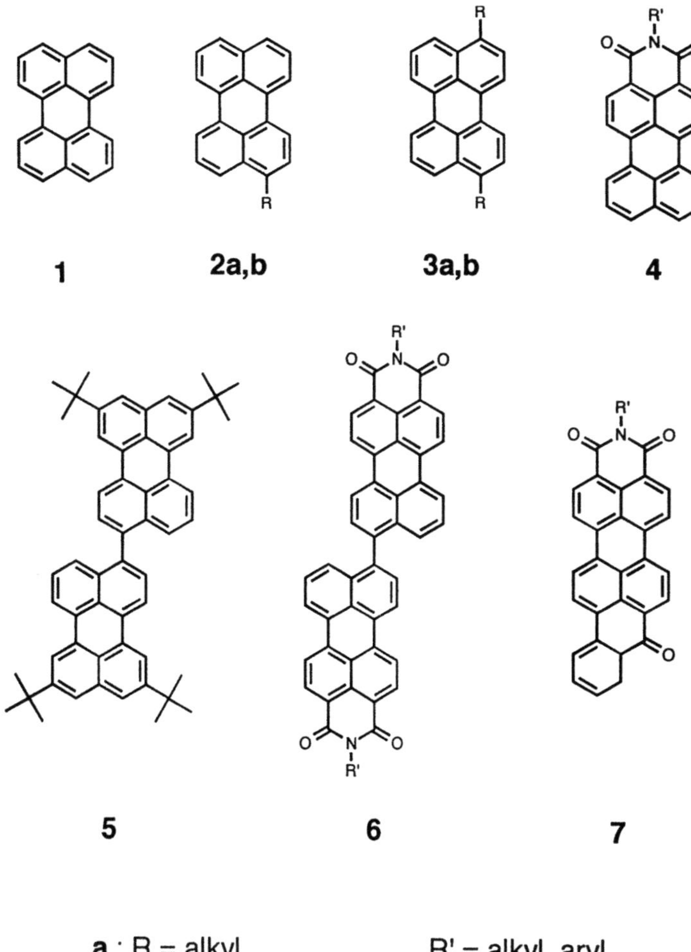

Scheme 1. Perylene-based fluorescent dyes

reaches 96% and is only marginally decreased as compared to that of the corresponding unsubstituted perylene **9a**, $\varnothing_{PL}$=100% (Table 1) [26].

The introduction of substituents on perylene units is also advantageous to fine tune the absorption maximum, λ_{max} (abs.), and emission maximum, λ_{max} (em.), e.g. **2a**, **2b**, **3a** and **3b**, with a λ_{max} (em.) = 450 nm, 464 nm, 458 nm and 485 nm, respectively. A common feature shared by perylene dyes **1–3b** is the small Stokes shift observed (~10 nm) as expected from their rigid structure [33]. When considering perylene dyes **4** and **9a** with larger π-systems, the Stokes shift reaches 23 nm and 14 nm, respectively. Much larger Stokes shifts (~70 nm) are exhibited by biperylenyls **5** and **8** [36]. In particular, in polar solvents, biperylenyl **5** behaves unusually and the lifetime of its excited state in-

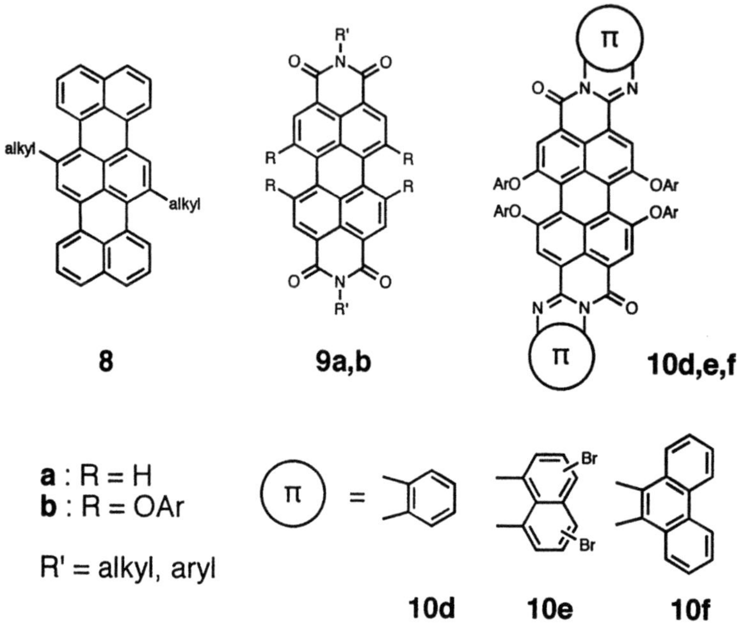

Scheme 2. Terrylene and perylene dyes

creases from a rigid low-temperature matrix to a viscous liquid solvent. In the rigid solvent the emitting state is excitonic and non-polar, whereas at higher temperatures an equilibrium is established between this non-polar state and the polar state with an incomplete charge transfer (CT). Unlike most of the twisted intramolecular charge transfer (TICT), this charge transfer emission is allowed. This emission is related to the excited-state dipole moment indicating an incomplete charge separation, lower than that expected for a TICT state [34].

The development of photochemically and thermally stable dyes absorbing and emitting in the NIR range is particularly challenging because the photochemical inertness and the $\varnothing_{PL}$ of a series of dyes generally decrease when shifting the λ_{max} (em.) to longer wavelength [37, 38]. Another incentive to the synthesis of NIR absorbing and emitting dyes is given by the diversity of attractive applications [39]. A series of long-wavelength absorbing and even emitting dyes, the perylene derivatives **10d–f**, have been prepared starting from tetraphenoxy-substituted perylene tetracarboxydiimide **9b** [29]. The λ_{max} (em.) of luminophores **10d–f** ranges from 661 nm to 768 nm. In a more refined approach λ_{max} (em.) can also be finely tuned by structural variation (Table 1). A second series of dyes which absorb and emit in the NIR range involves the quaterrylenebisdiimides **13a, b**, the terrylenebisdiimides **14a, b**, and the benzoylterrylenebisdiimides **15a, b** [32, 40]. Their synthesis starts with an industrial chromophore, the perylenedicarboximide **4**, and is briefly outlined in Scheme 3. One should notice the simplicity of this synthetic pathway allowing large-scale preparation [32]. The λ_{max} (em.) of luminophore **14b** (**14a**) reaches 707 nm

Table 1. Selected Spectroscopic Data of Fluorescent Dyes

Compound	Absorption λ_{max} (nm)	Emission λ_{max} (nm)	$\varnothing_{PL}$ (%)	Solvent	Ref.
1	435	445	98	dioxane	[33, 57]
2a	443	450	[a]	dioxane	[33]
2b	457	464	[a]	dioxane	[33]
3a	450	458	[a]	dioxane	[33]
3b	476	485	[a]	dioxane	[33]
4	512	535	[a]	CHCl$_3$	[95]
5	448	450	[a]	n-hexane	[34]
	504	522	71	n-propanol	
6	530	600	[a]	CHCl$_3$	[30]
7	566	582	[a]	CHCl$_3$	[96]
8	535	545	[a]	C$_{12}$H$_{12}$	[35]
9a	526	540	100	CHCl$_3$	[4, 26, 30]
9b	576	613	96	CH$_2$Cl$_2$	[4, 26, 30]
10d	629	661	[a]	CH$_2$Cl$_2$	[29, 30]
10e	712	768	[a]	CH$_2$Cl$_2$	[29, 30]
10f	659	680	[a]	CH$_2$Cl$_2$	[29, 30]
13a	762	797[b]	[a]	CH$_2$Cl$_2$	[30, 40]
13b	781	[a]	[a]	CH$_2$Cl$_2$	[30, 40]
14a	650	673	90 ± 10	CHCl$_3$	[32]
14b	664	707	[a]	CHCl$_3$	[32]
15a	676	701	60 ± 10	CHCl$_3$	[32]
15b	687	735	[a]	CHCl$_3$	[32]

[a] Not measured.
[b] Currently under investigation.

(673 nm) whereas that of **15b** (**15a**) is at 735 nm (701 nm) as shown in Table 1. The most striking features of terrylene **14a** and benzoylterrylene **15a** are their extraordinaryly high quantum yields of photoluminescence, $\varnothing_{PL} = 90 \pm 10\%$ and $\varnothing_{PL} = 60 \pm 10\%$, respectively [32]. The photostabilities of **14a, b** and **15a, b** were estimated by exposing chloroform solutions to UV light (l=366 nm) for a prolonged period of time. After one week no significant changes in the extinction coefficients of the solutions of **14a, b** and **15a, b** could be observed demonstrating their remarkable photochemical inertness [32]. The quaterrylenediimides **13a, b** absorb definitively in the NIR window, at 762 nm and 781 nm, respectively [30, 40]. Interestingly enough, preliminary experiments have shown that quaterrylene **13a** emits at λ_{max} (em.) = 797 nm in dilute chloroform solution. A more comprehensive study aimed at the characterization of the steady-state and time-resolved photophysical behavior of **13a, b** is currently underway and will be the subject of a forthcoming publication [31].

A related research topic in which we have been actively involved concerns the combination of dye properties with mesogenic behavior and aggregate formation. The motivation is dual: i) Liquid crystallinity (LC) provides opportunities for materials to be macroscopically oriented and to present anisotropic optoelectronic properties: e.g. photoconductivity [41] and polarized light emission [42]. ii) Conversely, in the liquid and in the solid state, the optoelectronic pro-

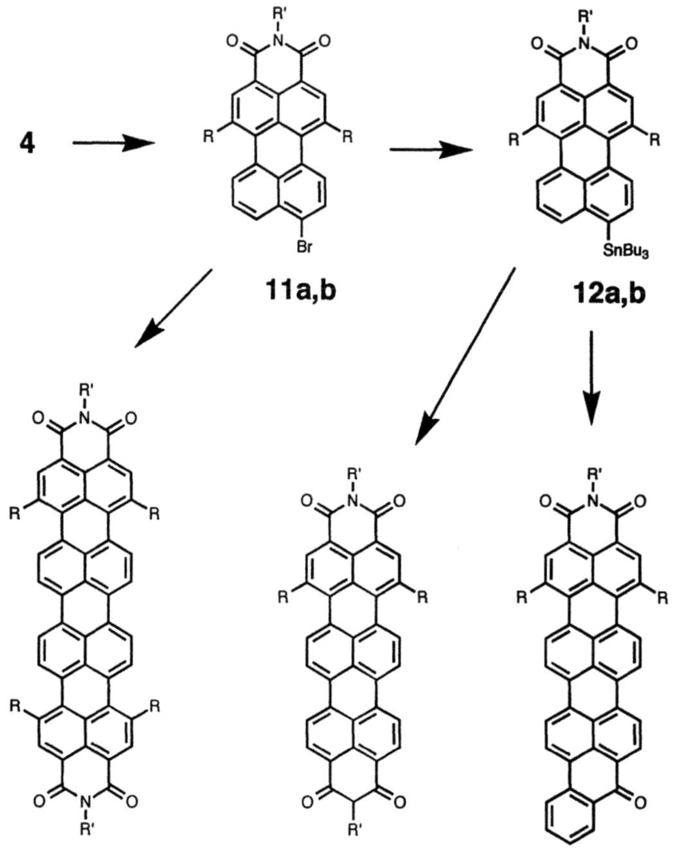

4 → **11a,b** → **12a,b**

13a,b   **14a,b**   **15a,b**

Scheme 3. Synthesis of quaterrylene and terrylene dyes, a = H, b = OAr

perties do not arise from individual molecules but rather are governed by the collective behavior of the chromophores. Hence, it is neccessary to understand and to control the supramolecular packing of a given dye, if its full potential as a functional material is to be estimated [43, 44]. From a structural viewpoint, several phase and aggregate forming dyes are considered: perylenes **16–17** and **22**, coronenes **18**, quinacridones **19–21**, stilbene **23**, and hexa-*peri*-hexabenzo-coronenes **28a, b**.

A common feature shared by the light-emitting dyes **16–19** (Scheme 4) is that they form discotic columnar mesophases [45–48]. This arises from their structure, namely, a rigid aromatic core to which three to six alkyl chains are attached [49]. Discotic mesogens are of special interest for their one-dimensional transport processes such as photoconductivity along the columns [41]. However, discotic mesogens also have the tendency to form stacks in solution.

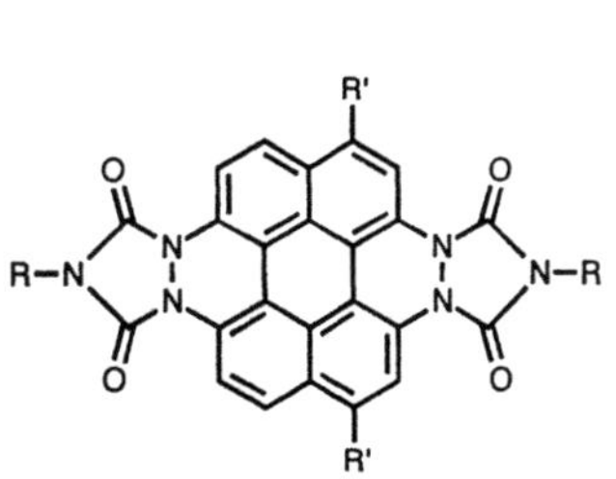

16

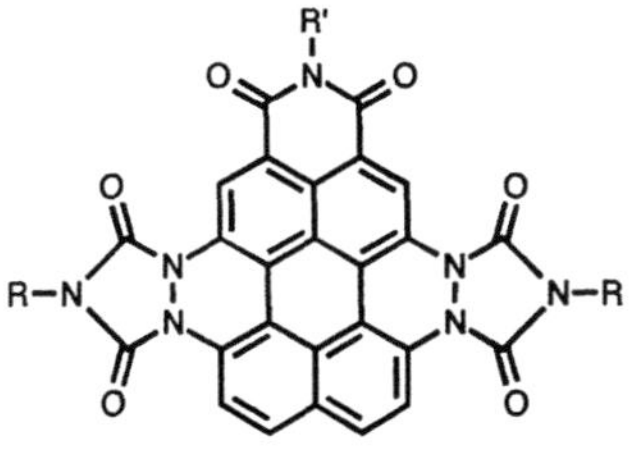

17

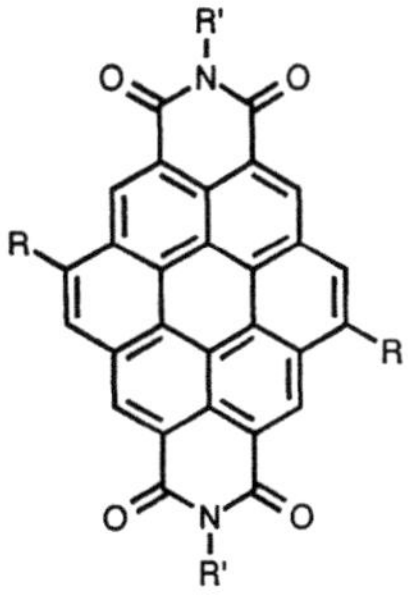

18

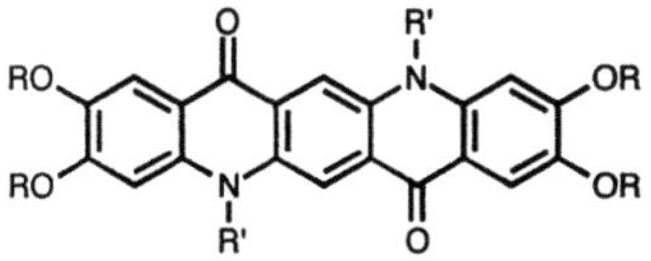

19

Scheme 4. Phase-forming fluorescent dyes, R, R = alkyl

Table 2. Selected Spectroscopic Data of Phase-Forming and Aggregate-Forming Dyes

Compound	Absorption λ_{max} (nm)	Emission λ_{max} (nm)	Solvent	Ref.
16	586[a], 688[b]	606[a], 710[b], 720[c]	MTHF	[50, 45]
17	660[a], 710[a]	768[a]	CHCl$_3$	[46]
18	512[a]	518[a]	dioxane	[48]
19	520	540	CHCl$_3$	[43]
20	526[a], 602[b]	621[b]	toluene	[43, 47]
21	496[a], 562[b] 502[a]	–	toluene	[43, 47]
	502[a]	553[a]	THF	
22	591[a], 592[d]	639	CHCl$_3$	[51]
23	340	540[e]	THF, MTHF [e)	[20]
28a	355, 387[a, f], 436, 443[a, g]	465, 476, 484, 494[a]	n-hexane	[54, 55]
28b	357, 389[a, f], 436, 443[a, g]	465, 476, 484, 494[a], 520[b]	n-hexane	[54, 55]

[a] Monomer in solution.
[b] Aggregate in solution.
[c] Liquid crystalline phase.
[d] In the solid state.
[e] Methyltetrahydrofuran (MTHF) at 77 K.
[f] B$_b$ and L$_a$ transitions, respectively.
[g] L$_b$ transitions.

This phenomenon has been investigated for the perylene dye **16** in solution as a function of concentration by steady-state and time-resolved fluorescence anisotropy [50]. The aim of this study was to bridge the gap between dilute solution and liquid crystalline properties [41]. Specifically, luminophore **16** exhibits a $\lambda_{\max}$ (em.) = 606 nm, at 2×10^{-6} M in 2-methyltetrahydrofuran, emanating from the monomer, whereas at 5×10^{-5} M in the same solvent the $\lambda_{\max}$ (em.) is shifted to 710 nm and originates from aggregates (Table 2). In the LC phase, the $\lambda_{\max}$ (em.) = 720 nm which is close to that of aggregate species in concentrated solution [50]. LC perylene dye **17** exhibits two $\lambda_{\max}$ (abs) at 660 m and 710 nm in dilute solution whereas its $\lambda_{\max}$ (em.) at 768 nm lies in the NIR window [46]. These findings illustrate the importance of aggregate formation on the emission properties of dyes and emphasize the importance of a comprehensive supramolecular characterization.

Interestingly enough, coronene dye **18** absorbs and emits at a shorter wavelength, i.e. $\lambda_{\max}$ (abs.) = 512 nm and $\lambda_{\max}$ (em.) = 518 nm, than the structurally related perylene **9a** [48]. Preliminary results on the emission behavior of LC coronene **18** in the solid-state and in polystyrene (PS) matrix as a function of concentration indicate that the $\lambda_{\max}$ (em.) depends on the conditions of film preparation [48].

The LC quinacridone derivative **19** bearing six solubilizing groups absorbs at 520 nm and emits 540 nm in dilute chloroform solution. The relatively small Stokes shift observed (20 nm) is in agreement with the rigid aromatic structure of quinacridone **19** and with the absence of aggregate formation in dilute solution [47].

Another striking example of the importance of aggregate formation on the optical properties of dyes is illustrated by the quinacridones **20** and **21**, which only differ by the number and the nature of their solubilizing side chains (Scheme 5) [43, 47]. The quinacridones **20** and **21** build large aggregates by hydrogen bonding in low polarity solvents such as toluene as evidenced by various spectroscopic techniques. Interestingly, in this solvent, the aggregate of the quinacridone **20** emits at 621 nm whereas that of quinacridone **21** presents no emission at all. Note that in tetrahydrofuran, i.e. a competing solvent for hydrogen bonding, the dilute solution of the quinacridone **21** emits at 553 nm [43, 47]. This finding indicates that the emitting behavior of the quinacridones **20** and **21** is governed by the supramolecular structure of their aggregates which is induced by their substitution pattern [47]. It was shown by FT-IR spectroscopy that the quinacridone **20** forms tape-like aggregates enabling π–π interactions between chromophores, whereas quinacridone **21** organizes in tubular aggregates where no π-stacking can occur [47].

Perylene **22** aggregates by hydrogen bonding in chloroform solution and in the solid state [51]. Although perylene **22** interacts by hydrogen bonding with its direct neighbors, it does not crystallize but forms a glass as evidenced by X-ray analysis. In dilute solution **22** exhibits a strong fluorescence at 639 nm but no fluorescence could be observed in concentrated solution and in bulk [51, 52]. This can only be attributed to the hydrogen-bonding interactions which offer deactivation channels for the excited state of **22**. Specifically, in this case, no π–π interactions could reasonably be invoked since the perylene cores cannot ap-

Scheme 5. Aggregate-forming fluorescent dyes

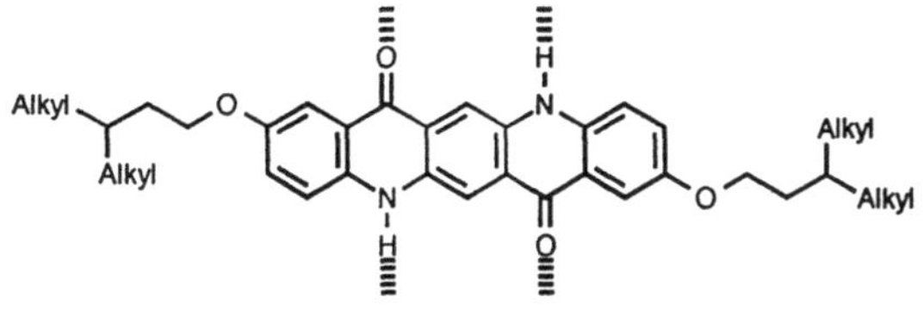

20

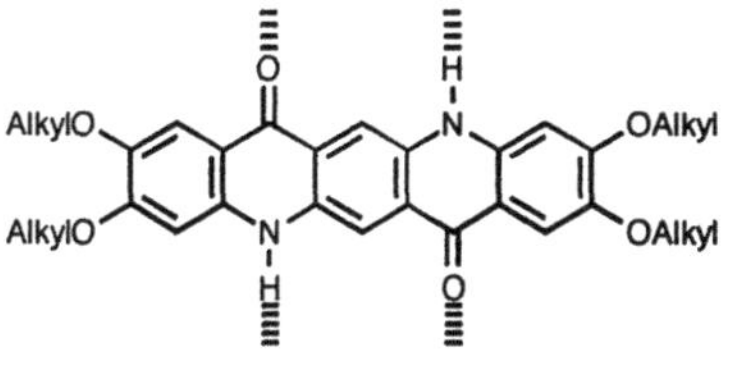

21

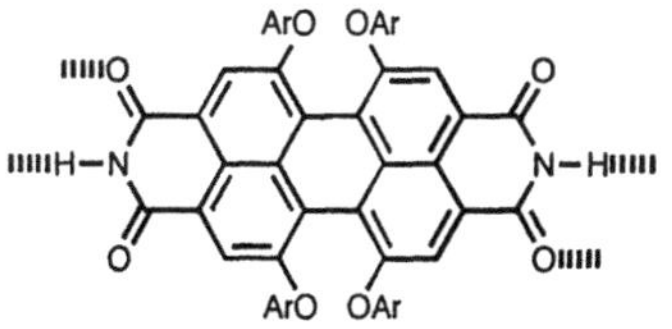

22

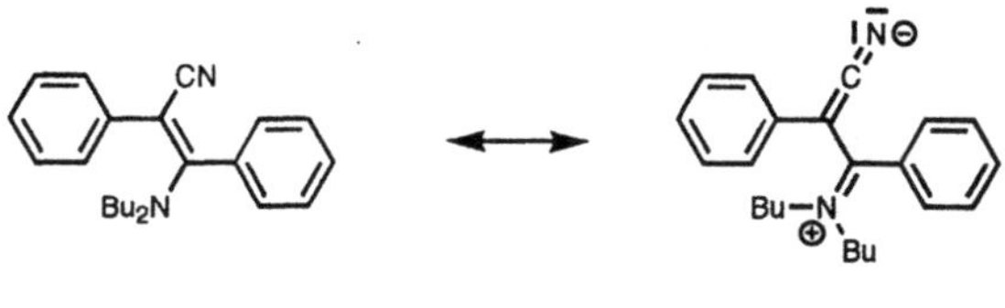

23   **24**

proach each other because of the four phenoxy groups in the *bay-regions* [30]. The specific role of the hydrogen-bonding interactions is also corroborated by the fact that the corresponding *N*-alkyl and *N*-aryl perylenes **9c** emit strongly in bulk.

The previous examples of emitting dyes have stressed the importance of hydrogen – bonding and π-stacking on aggregate formation which directly influence the emitting behavior of the dyes. However, dipolar interactions can also play a marked role on the aggregation and photophysical behaviors of dyes as shown by the α, β-donor-acceptor-substituted stilbene **23**. This compound deserves particular attention for two reasons [20]. Firstly, its synthesis is highly efficient and is based on a newly developed cation-anion coupling (Scheme 6). After deprotonation of benzyl cyanide **25**, coupling with cation **26** and elimination of methyl sulfide, stilbene **23** is obtained in 96% isolated yield [20]. The

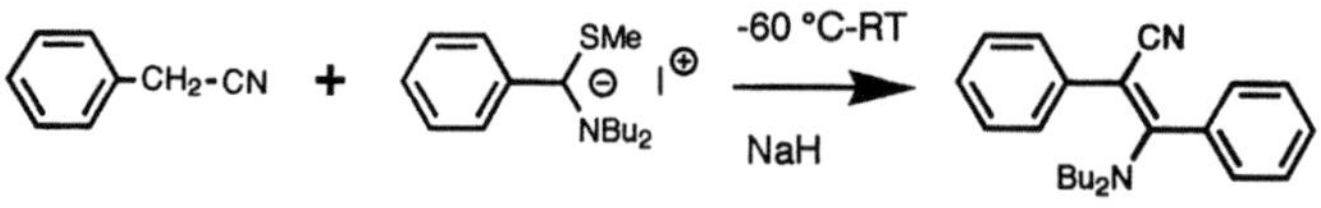

25 + **26** → **23**

$-60\ °C-RT$, NaH

Scheme 6. Synthesis of α,β-donor/acceptor-substituted stilbene **23**

efficiency of this reaction allows the synthesis to be extended to oligomers and even polymers (see below). Secondly, stilbene **23** shows no fluorescence at room temperature. Fluorescence quenching can be explained by participation of the dipolar structure **24**, which permits free rotation about the σ-bond in the actual structure of **23** (Scheme 5). However, a broad fluorescence band is detected for **23** at 540 nm (77 K in methyltetrahydrofuran), for which the size of the Stokes shift is the result of twisting of the sub-units relative to each other [20].

While the emitting behavior of the stilbene **23** documents the importance of the dipolar moment on its aggregation behavior, the effect of higher order dipolar moments should not be overlooked. The alkylated hexa-*peri*-hexabenzocoronenes **28a,b**, which are characterized by the absence of a dipolar moment, illustrate the role of the quadrupolar moment on aggregate formation [53]. Alkylated hexa-*peri*-hexabenzocoronenes **28a,b** are easily synthesized by oxidative cyclodehydrogenation of hexa(alkylphenyl)benzenes **27a,b** in good yields (Scheme 7) [54]. Interestingly, the phase behavior, the aggregate forming character, the solubility and the photophysical properties of **28a,b** are governed by the structure of their alkyl substituents [54, 55]. The bulkiness of the *tert*-butyl substituents prevents the aggregation of **28a** by π–π interactions. Therefore, its photophysical properties reflect that of single molecules in solution. This has been investigated by steady-state and time-resolved fluorescence spectroscopy [55]. Specifically, emission peaks at 465 nm, 476 nm, 484 nm and

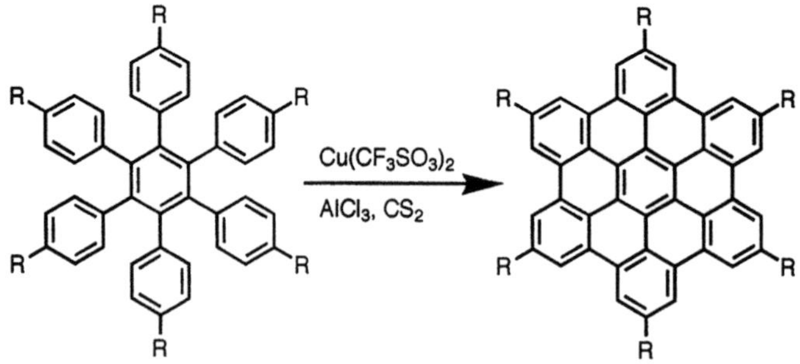

27a,b → **28a,b**

$Cu(CF_3SO_3)_2$, $AlCl_3$, CS_2

Scheme 7. Synthesis of alkylated hexa-peri-hexabenzocoronene derivatives, **a** = *t*-butyl, **b** = *n*-alkyl

494 nm, and an excited state lifetime of 55.6±0.1 ns, were observed. Conversely, the n-alkyl substituents of **28b** do not prevent π-stacking from occuring; consequently, aggregates are formed in solution and the photophysical behavior varies with concentration and solvents [55]. Hexabenzocoronene **28a** shows a $\emptyset_{PL}$ of 6% and the same emission maximum as that of **28b** in dilute solutions (10^{-6} in *n*-hexane). In more concentrated solution a broad emission appears at 520 nm [55].

From the aforementioned examples of phase-forming and aggregate-building dyes, one cannot overestimate the importance of supramolecular interactions in solution and the solid state on the emitting behavior of luminophores. In particular, supramolecular interactions not only enable the tuning of the luminescence intensity [20, 43, 51], but also offer an additional way to bathochromically shift λ_{max} (em.) since aggregates generally emit at longer wavelengths than their corresponding monomers [47, 48, 50].

3
Oligomers

This section is dedicated to oligomeric fluorophores which differ from monomeric light-emitting materials of the previous section, not so much by their higher molecular mass, but rather by their chemical structure which is composed of several repeat units.

Our group has been strongly involved in the preparation of homologous series of soluble conjugated oligomers which has given rise to a large number of spectroscopic studies [16, 18, 21, 56]. Within this section, several oligomers, i.e. oligo(rylene)s **29a–d**, oligo(*p*-phenylenevinylene)s (OPV) **30a–e**, ladder oligo(*p*-phenylene)s (LOPP) **31a–c**, 2,7-oligo(pyrene)s **32a–c**, star-shaped oligo(phenylene)s **33a, b** and rod-like oligo(*p*-phenylene)s **34a, b** are presented and their photophysical properties discussed in the light of their structures (Scheme 8).

The synthesis of soluble oligorylenes **29a–d** has paved the way to the study of their intriguing luminescent properties resulting from their two-dimensional ribbon-type structures [19, 37, 57]. Tetra-*tert*-butylperylene **29a** displays a λ_{max} (abs.) = 439 and λ_{max} (em.) = 446 nm but is also characterized by a small Stokes shift of 7 nm and a $\emptyset_{PL}$ of 94% in dioxane solution [37, 57]. Its fluorescence spectrum in thin solid films is less structured in comparison to the solution spectrum and its Stokes shift increases to 39 nm [36]. The λ_{max} (em.) = 485 nm of **29a** in the solid state contrasts with that of perylene **1,** which exhibits a fluorescence maximum above 600 nm. This particular difference stresses the importance of *tert*-butyl solubilizing groups which prevent the π-stacking of the perylene core of **29a** in the solid state [36]. The rotational motions of perylene **1** and tetra-*tert*-butylperylene **29a** have been investigated by time-resolved fluorescence anisotropy spectroscopy in a series of *n*-alcohols [58]. Perylene **1** can be analyzed according to a biexponential anisotropy model. The high ratio, $D_{pa}/D_{pe} = 16$, between the two diffusion coefficients parallel (D_{pa}) and perpendicular (D_{pe}) to the polarization direction of the excitation light indicates that the molecule can be modelled as a disk but not as an oblate ellipsoid of revolu-

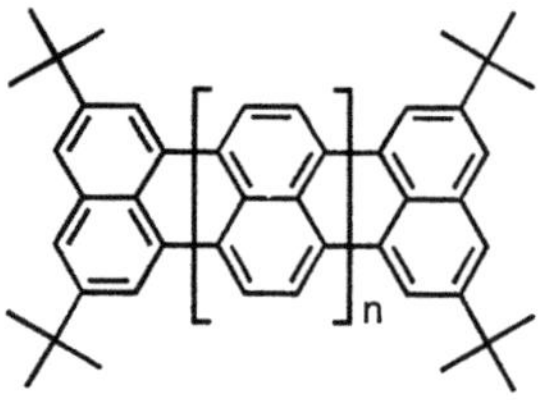

29a-d : n = 0-3

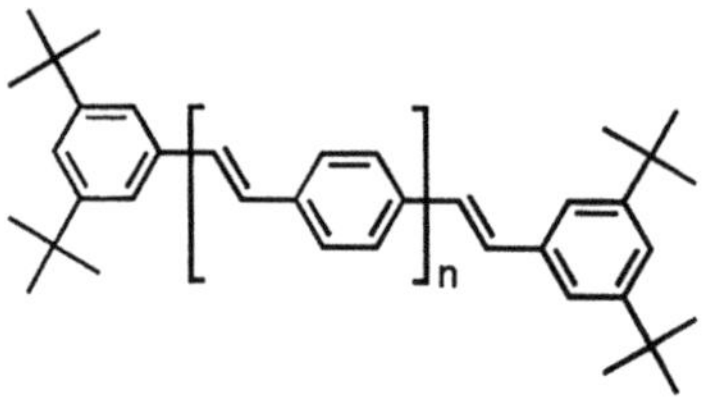

30a-e : n = 1-5

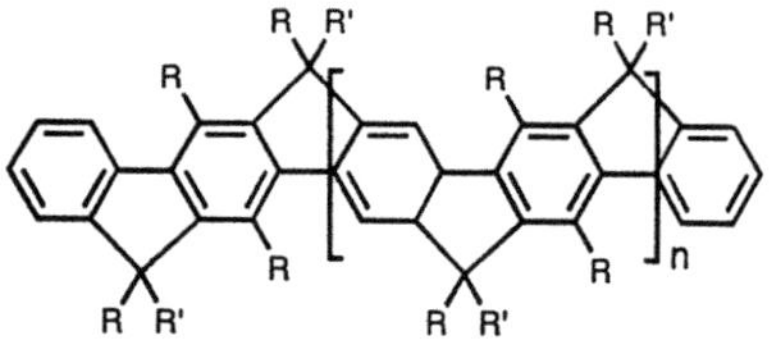

31a-c : n = 0-2

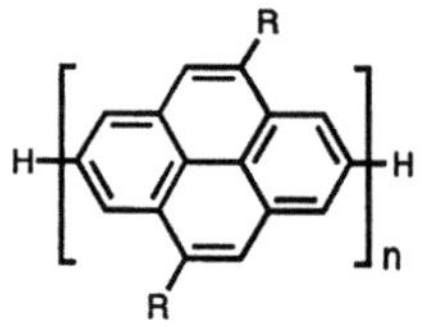

32a-c : n = 0-2

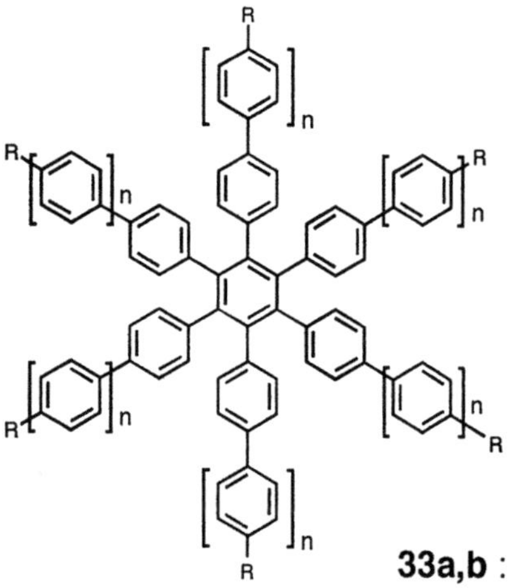

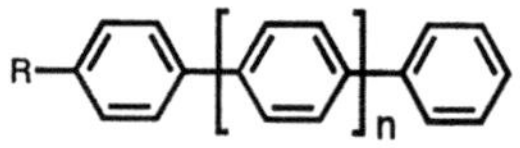

34a,b : n = 1,2

33a,b : n = 2,3

Scheme 8. Fluorescent oligomers, R = alkyl, R′ = aryl

tion. Substituting four bulky *tert*-butyl groups at the positions 2, 5, 8 and 11 of the perylene core, compound **29a**, leads to an anisotropy decay that can be analyzed according to a monoexponential anisotropy model [58].

Tetra-*tert*-butylterrylene **29b** shows a λ_{max} (abs.) = 560 and λ_{max} (em.) = 573 nm. On the other hand, **29b** is also characterized by a small Stokes shift of 13 nm and a $\varnothing_{PL}$ of 70% in dioxane solution [37, 57]. De Schryver et al. have investigated the time-resolved fluorescence anisotropy decay of **29b** in *n*-alcohol and *n*-alkanes [59]. A comparable behavior to that of perylene **29a** (see above) has been observed for **29b**, i.e. it displays anisotropy decays in alcohols and alkanes which are correctly approximated by a monoexponential model. The rotational correlation times of **29b** are proportional to the viscosity of n-alkane and n-alcohol homologous series and are faster in the latter solvent. This is explained in terms of the higher free volume in alcohols than in alkanes [59]. Soluble tetra-*tert*-butylterrylene **29b** is also an extremly valuable luminophore for single-molecule spectroscopy and has given rise to a large body of spectroscopic studies [3]. The interest in single-molecule spectroscopy arises from the fact that the photophysical behavior of the molecules is individually probed. Therefore, the spectral response of the dyes is no longer averaged over a large number of molecules, typically $\sim 10^{14}$ molecules in a conventional fluorescence measurement at a concentration of 10^{-6} mol/l. It can certainly be anticipated that new physical and chemical behavior will be observed using this unprecedented spectroscopic technique [3]. An illustrative example is given by the spectral diffusion of **29b** in an amorphous polymer probed by single-molecule spectroscopy [60]. A broad distribution of line widths was observed giving evidence for the wide variety of locally varying dynamics experienced by the probe terrylene. Interestingly, the distribution of line widths are essentially the same for polyethylene and polyisobutylene matrices leading to the conclusion that the local environment around **29b** is amorphous for both polymers [60].

Tetra-*tert*-butylquaterrylene **29c** absorbs at λ_{max} (abs.) = 660 nm and emits at λ_{max} (em.) = 678 nm. In contrast to perylene **29a** and terrylene **29b**, $\varnothing_{PL}$ reaches only 5% [37, 57]. This observation has led to a comprehensive comparison of the spectrodynamic behavior of terrylene **29b** and quaterrylene **29c** by ultrafast spectroscopy. Terrylene **29b** shows normal excited singlet absorption, nanosecond fluorescence and a long-lived transient absorption assigned to a triplet state whereas quaterrylene **29c** exhibits a new transient absorption which rises for 30 ps, and no associated gain. These results are interpreted, in the case of **29c**, by the ultrafast ($\leqslant 500$ fs) formation of a nonradiant intermediate precursor state of a long-lived reversible photoproduct [61].

The higher homologue, pentarylene **29d,** absorbs at 748 nm (λ_{max}). Unfortunately, the reliable determination of λ_{max} (em.) and $\varnothing_{PL}$ has been hampered by the low solubility of this compound [37, 57].

Clearly, the numerous photophysical studies which have been carried out on the oligo(rylene)s **29a–d** stress the importance of homologous series of soluble conjugated oligomers for obtaining structure-property relationships. Another outstanding oligomeric series is the oligo(*p*-phenylenevinylene) series **30a–e**. These compounds have attracted considerable interest from photophysicists [17, 62, 63] since they are soluble, well-defined oligomeric analogues [21, 64] of the poly(*p*-phenylenevinylene) **37a**, which is one of the most important polymers for third-order nonlinear optics [65] and for electroluminescence [66a]. The absorption and emission maxima of the oligophenylenevinylenes **30a–e** are collected in Table 3. Fluorescence depolarization spectroscopy has been applied to OPVs **30a–c**. This technique shows that **30a–c** are isotropically dispersed in a polystyrene matrix and behave accordingly, i.e. no rotational depolarization within the fluorescence lifetime and no excitation energy transfer occurs [66b]. Solid-state concentration effects on the optical absorption and emission of OPVs **30a–c** and PPV **37a** have been investigated, focussing on the difference between molecules isolated by dispersion in an inert host and concentrated molecular

Table 3. Selected Spectroscopic Data of Light-Emitting Oligomers

Compound	Absorption λ_{max} (nm)	Emission λ_{max} (nm)	$\emptyset_{PL}$ (%)	Solvent	Ref.
29a	439	446, 485[a]	94	dioxane	[36, 37, 57, 59]
29b	560	573	70	dioxane	[36, 37, 57, 59]
29c	660	678	5	dioxane	[37, 57]
29d	748	[b]	[b]	dioxane	[37, 57]
30a	360	420[d] 420[c]	[d]	CHCl$_3$	[64, 66b, 67,
	383[c]	420[c]		solid state	97]
30b	387	455[d]	[e]	CHCl$_3$	[64, 66b, 67]
	416[c]	463[c]		solid state	
30c	403	477[d]	[e]	CHCl$_3$	[64, 66b, 67,
	440[c]	471[c]		solid state	97]
30d	412	–	[e]	CHCl$_3$	[64, 97]
	454[c]	486[c]		solid state	
30e	418	–	[e]	CHCl$_3$	[64, 97]
	466[c]	496[c]		solid state	
31a	335	339	[e]	CH$_2$Cl$_2$	[36, 71]
31b	390	395	[e]	CH$_2$Cl$_2$	[36, 71]
31c	414	419	[e]	CH$_2$Cl$_2$	[36, 71]
32a	344	378[f]	[e]	CH$_2$Cl$_2$	[73, 74]
32b	350	411[f]	[e]	CH$_2$Cl$_2$	[73, 74]
32c	353	422[f]	22[g]	CH$_2$Cl$_2$	[73, 74]
33a	305	360–380[h]	66	CH$_2$Cl$_2$	[75]
33b	316	374[i], 385, 425[j]	93	CH$_2$Cl$_2$	[75]
34a	285	332, 346, 364	99	CH$_2$Cl$_2$	[75]
34b	299	354, 367, 390[i]	96	CH$_2$Cl$_2$	[75]

[a] In the solid state.
[b] Measurements hampered by low solubility.
[c] 0–0 phonon luminescence adapted from ref. 97.
[d] In toluene.
[e] Not measured.
[f] in dioxane.
[g] In tetrahydrofuran.
[h] Broad, fine structure at 77 K.
[i] Shoulder.
[j] Further splitting on lowering the temperature.

films [67]. In OPV polymethylmethacrylate blends, containing a low concentration of oligomer, the PL decay is exponential, independent of both temperature and oligomer length. This implies that the fundamental radiative lifetime of OPVs **30a–c** is essentially independent of the number of phenylenevinylene repeat units. Conversely, there is a substantial red shift of λ_{max} (abs) and λ_{max} (em.) as a function of oligomer length (Table 3). Spin cast oligomer and polymer films have a fast and strongly temperature-dependent PL decay that approaches that of the dilute oligomers at low temperature. The difference in PL decay corresponds to changes in relative PL efficiency [67]. The PL of OPV **30c** thin films has also been investigated and a dramatic PL quenching has been observed upon deposition of a Ca thin film. Notably, a submolecular coverage (0.035 monolayer or 0.1 Å) of Ca on the **30c** surface reduced the PL of a 300 Å film of **30c** by as much as 50% [68]. This work brings some valuable information for the fabrication of electroluminescent diodes (ELD) because Ca is one of the lowest work function materials frequently used as the cathode in ELD. Moreover, OPV **30d** is an oligomer of PPV which, unlike PPV, can be sublimed to form impurity-free thin films in an ultrahigh vacuum environment [68].

Unsubstituted poly(p-phenylene) **37a** (PPP) has always attracted a lot of interest as blue light-emitting material for electroluminescent diodes. This is due to its simple chemical structure, to its high fluorescence in the solid state and to its excellent thermal stability [69, 70]. However, a detailed structure-property analysis of this polymer has always been hampered for several reasons. Firstly, unsubstituted PPP **37a** is rigorously insoluble. Secondly, there is no synthetic route leading to structurally defined high molecular weight PPP. Finally, the π-conjugation is decreased as a result of the mutual distortion of the aromatic rings of about 23° [56]. These reasons have created a definite need for defect-free soluble oligo- and poly (p-phenylene)s. In this paragraph, three examples of oligomers structurally related to PPP are presented, namely, the ladder-type oligo(p-phenylene)s (LOPP) **31a–c**, 2,7-oligo(pyrenylene)s **32a–c**, and star-shaped oligo(phenylene)s **33a, b**. Additionally, four examples of soluble poly-p-phenylenes are given in the next section.

Planar p-phenylene oligomers **31a–c** have been synthesized with a number of repeat units, n = 0–2, corresponding to a number of phenyl rings, x = 3–7, [71]. The planarity of **31a–c** arises from their ladder-type structure and induces a rapid convergence limit of the optical properties on the order of x = 12 ± 2 [56, 71]. The LOPPs **31a–c** exhibit an intense blue photoluminescence in dilute solution, λ_{max} (em.) = 339 nm (n = 0), 395 nm (n = 1) and 419 nm (n = 2), whereas λ_{max} (em.) = 448 nm for the corresponding ladder-type poly(p-phenylene)s **39** which are discussed in the next section. The very small and constant Stokes shift of 5 nm observed for LOPPs **31a–c** documents the rigidity of the conjugated π-system and indicates that little geometric changes between the relaxed ground and the excited state occur [71]. A comparative site selective fluorescence spectroscopic study of LOPP of **31a–c** leads to the conclusion that the entire Stokes shift observed for the polymer **42** is due to spectral diffusion which becomes smaller with improving structural perfection [72].

The 2,7-oligo(pyrenylene)s **32a–c**, although structurally related to oligo(p-phenylene)s, differ completely in their photophysical properties [73, 74]. As a result of molecular orbital coefficients of the frontier orbitals of the bridgehead centers which are almost zero, to a first approximation the pyrene units are electronically decoupled. This new situation is evidenced by the absorption spectra of **32a–c**, where only a weak dependence on the λ_{max} (abs.) appears as a function of the chain length, i.e. λ_{max} (abs.) = 344 nm (monomer), 350 nm (dimer), 353 nm (trimer) [73]. The dependence of the emission spectrum of **32a–c** (dioxane) on the chain length, i.e. λ_{max} (em.) = 378 nm (monomer), 411 nm (dimer), 422 nm (trimer) is observed [74]. Significantly enough, no CT state could be invoked in the case of the dimer **32b** and trimer **32c** since their λ_{max} (em.) remains almost unshifted in solvent of increasing polarity [74].

The star-shaped hexaterphenylylbenzene **33a** and hexaquaterphenylylbenzene **33b** have been synthesized with the aim of studying the behavior of luminophores in a restricted space [75]. Formally, the star-shaped compound **33a** can be regarded as being constructed of three septiphenyl chains with one benzene unit in common. However, the interplanar angle makes an arrangement of six electronically coupled terphenylyl chains a more plausible structure [75]. The comparison of the absorption spectra of **33a,b** with that of *p*-terphenyl **34a** and *p*-quaterphenyl **35b** indicates that the picture of completely independent chromophores is not valid (Table 3). It is also interesting that the $\emptyset_{PL}$ of the star-shaped compounds **33a,b** are only slightly smaller than those of oligo(-*p*-phenylene)s **34a,b**. Moreover, the similarity between the fluorescence spectra of the film and those of the solution excludes the formation of aggregates in the solid state [75]. The above results illustrate the specific role of oligomeric series in the understanding of the photophysical properties of the corresponding polymers [17].

Scheme 9. 9,10-Anthrylene dimers and trimers

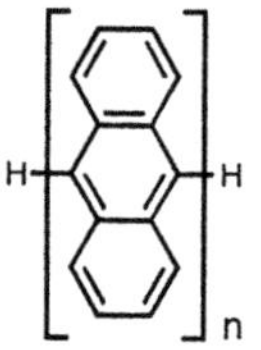

35a,b : n = 2,3

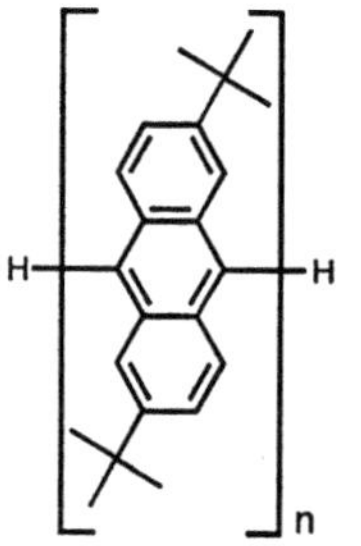

36a,b : n = 2,3

The last example in this section on oligomers deals with the peculiar photophysical behavior of 9,10-anthrylene dimers and trimers **35a,b** and **36a,b** (Scheme 9) which form intramolecular charge separated (CS) states [76]. From the spectroscopic viewpoint, it should be expected that the emission spectra of alkyl-substituted dyes not forming aggregates would not significantly differ from their unsubstituted analogues. However, the dimers and trimers **35a,b** and **36a,b** present a striking example of the converse behavior. Specifically, the existence of a CS state and hence the emission maxima are controlled by the substitution pattern of the anthrylene core [76]. The emission maxima of the unsubstituted anthrylene dimers **35a** and trimers **35b** behave accordingly and are more Stokes shifted in a polar solvent (DMF) than in a less polar solvent (cyclohexane). This dependence of the emission maxima upon solvent polarity is less pronounced for the *tert*-butyl-substituted anthrylene dimers **36a** and trimers **36b**. This observation is explained in terms of a lower availability of the CS state for the substituted anthrylenes **36a,b**, in comparison to the unsubstituted anthrylenes **35a,b**, due to their smaller solvation energy, as a result of their larger Onsager radii [76].

4
Polymers

Polymeric dyes presented in this section contrast from the aforementioned monomeric and oligomeric dyes by their larger molecular mass, their polydispersity, and by their film-forming ability.

The poly(*p*-phenylene)s **37a, b** [77], the poly(*p*-phenyleneethylinene) (PPE) **38** [cw], poly(*p*-phenylenevinylene)s **39a, b** [78] represent three main classes of conjugated light-emitting polymers. Their spectroscopic data are collected in Table 4. Evidently, the current importance of PPP, PPE, and PPV in science and technology is related to the discovery of the electroluminescence of PPV and to the prospect of fabricating low-cost large-area light sources [66a, 78a, 79]. One should also mention that some basic questions regarding the photophysics of these polymers remain unanswered [78b]. One issue directly related to light emission concerns the quantum yield of photoluminescence of PPV. By definition, the quantum yield of photoluminescence, $\emptyset_{PL}$, is given by the product of the quantum yield of fluorescence, $\emptyset_f$, times the quantum yield of exciton formation, $\emptyset_{exc}$, i.e. $\emptyset_{PL} = \emptyset_f \emptyset_{exc}$. Until now, it has generally been considered that $\emptyset_{exc} = 1$ and thus $\emptyset_{PL} = \emptyset_f$. This assumption seems, however, not to hold in all cases and a $\emptyset_{exc} = 0.1 - 0.2$ has recently been measured [78b]. However, our contribution as synthetic chemists to the field of light-emitting polymers is to synthesize soluble, film-forming and defect-free structures to enable photophysical studies on defined materials [15, 21, 18, 80–83]. Another important aspect of our work is the development of new light-emitting polymers **40–48** which deviate from the conventional PPP, PPE and PPV structures (Schemes 10 and 11).

A first example is the α,β-donor-acceptor-substituted poly(*p*-phenylene-vinylene) **40** which is structurally related to PPV but which differs completely in its light-emitting properties [84]. In contrast to PPV **39b**, which emits strongly green light in solution [85, 86], the α,β-donor-acceptor-substituted PPV **40** shows only a weak emission between 400 nm and 500 nm. This is remarkable since its corresponding monomer, the α,β-donor-acceptor-substituted stilbene **23**, displays no emission at all at room temperature (see above) [20]. In both monomer **23** and polymer **40**, the quenching of the fluorescence is related to the degree of rotational freedom about the central vinylene bond, which possesses a certain σ-character due to the donor and acceptor substituents. This rotational freedom opens channels for the non-radiative deactivation of the excited states. In the case of the polymer **40**, aggregation in solution via dipolar interactions imposes to some extent an immobilization of the vinylene units and thus enables a weak emission [52].

Interest in the poly(*p*-phenylene) composed of cyclophane units **41**, arises from the combination of chirality with the emission properties of the PPP backbone [87]. Since the torsion angle between neighboring aromatic units in PPP strongly depends on the mutual steric hindrance of the substituents attached, the isotactic and the syndiotactic structures will display different optical and electronic properties to those of the atactic one. The atactic PPP **41** possessing its absorption maximum at 332 nm and featuring blue light emission, λ_{max} (em.) = 344 nm, represents an attractive material for blue light emitting diodes [87].

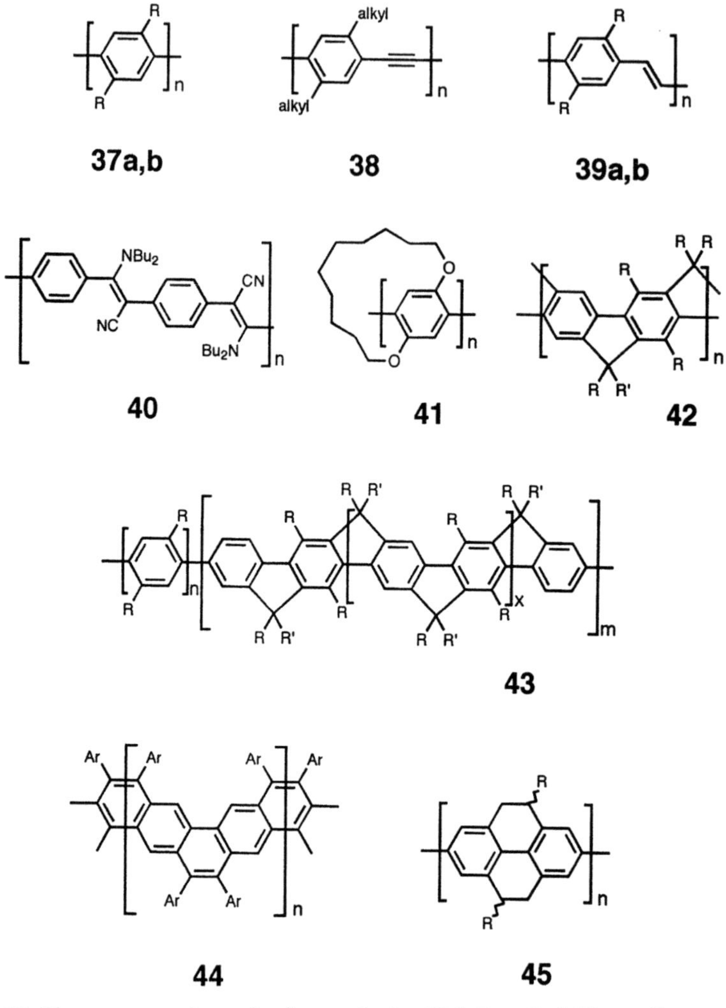

Scheme 10. Fluorescent conjugated polymers (a: R = H, b: R = alkyl, R′ = aryl)

The ladder-type poly(p-phenylene) (LPPP) **42**, the PPP copolymer **43**, and their corresponding oligomers **31 a – c** (mentioned in the former section) have attracted considerable attention from photophysicists [10, 14, 69, 88, 89]. This relates, first of all, to their defect-free planar structure, to their excellent solubility, to their good film-forming abilities and to their high $\varnothing_{PL}$ (see Table 4) which make them ideal candidates for blue light emitting diodes [10, 13, 21, 25, 70, 90]. The LPPP **42** displays an intense emission, λ_{max} (em.) = 447 nm, in solution [71]. The PL maximum is shifted to ~ 560 nm in the solid state and the electroluminescence (EL) peaks at about ~ 630 nm [44]. These discrepancies between the solution and the solid state and between PL and EL can be traced to the formation of aggregates as evidenced by femtosecond pump probe experi-

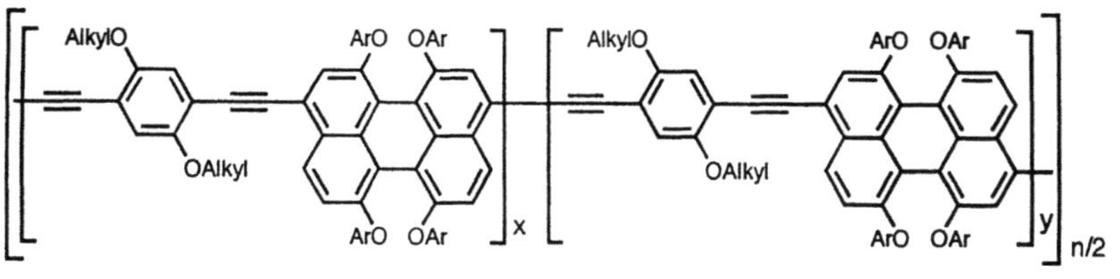

46

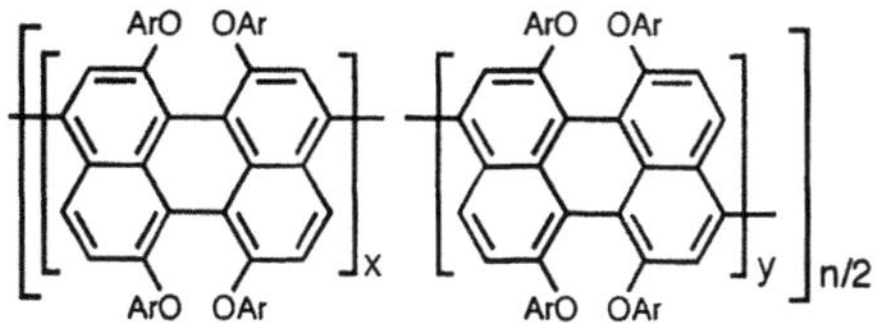

47

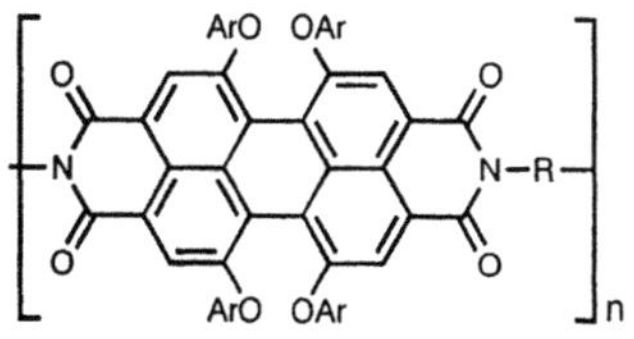

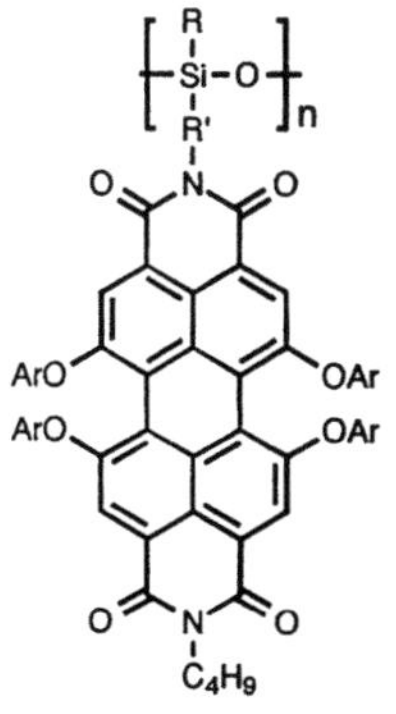

48a,b

49

Scheme 11. Fluorescent polymers containing perylene subunits, **48a**: R = alkyl, **48b**: R = aryl, **49**: R, R′ = alkyl

ments [89] and femtosecond luminescence up-conversion spectroscopy [88]. Consequently, excitons migrate to lower energy traps (aggregates) and emit at lower frequency [14, 44, 88]. A significant decrease in the yellow part of the emission depending on the monomer ratio is achieved for PPP copolymer **43**, namely λ_{max} (em.) = 430 nm, 440 nm, 448 nm, and 452 nm for a molar ratio of copolymer n/m = 70:30, 60:40, 50:50 and 40:60 [13].

The ladder-type polymer **44** is a new type of angular polyacene which in comparison to linear [n] acenes have a wider band gap [91]. The photoluminescence in solution of **44** is characterized by the appearance of a sharp and structured emission band centered around 478 nm. For films of **44**, in addition to the described blue emission, a shoulder appears in the yellow region of the spectrum which can be assigned to the formation of aggregates [91].

The soluble poly(p-phenylene) containing tetrahydropyrene repeating units **45** differs from alkyl-substituted PPP **37b** in that the solubilizing alkyl substi-

Table 4. Selected Spectroscopic Data of Fluorescent Polymers

Compound	Absorption λ_{max} (nm)	Emission λ_{max} (nm)	$\O_{PL}$ (%)	Solvent	Ref.
37a	~ 340	~ 420, 460	[a]	[b,c]	[69, 70]
37b	310	400	[a]	$C_{12}H_{12}$	[83]
38b	388	428	[a]	$C_{12}H_{12}$	[83]
39a	~ 400	~ 505 ~ 540	7–15	[b,c]	[65, 78]
39b	397–404	532	47	[b,c]	[78, 85, 86]
40	386	400–500	[a]	$CHCl_3$	[52]
41	332	415	[a]	CH_2Cl_2	[87]
42	438	447 ~560[b] ~630[d]	40–100, 30[e]	CH_2Cl_2	[10, 11, 12, 14, 44, 71, 88, 89, 98]
43	335, 370, 390, 415, 450[f]	430, 440, 448, 452[b,g]	[a]	CH_2Cl_2	[13, 14]
44	431	478	[a]	CH_2Cl_2	[91]
45	385	425[h] 457[c]	[a]	$C_{12}H_{12}$	[92]
46	552	597	60	$CHCl_3$	[11, 12, 24]
47	443	473	[a]	$CHCl_3$	[24, 30]
48a	589	615	[a]	$CHCl_3$	[22]
48b	592–608	613–622	[a]	$CHCl_3$	[22]

[a] Not measured.
[b] Photoluminescence in the solid state.
[c] In the solid state.
[d] Electroluminescence in the solid state.
[e] In solution and in the solid state, respectively.
[f] See text.
[g] Molar ratio of copolymer n/m = 70:30; 60:40; 50:50 and 40:60.
[h] In tetrahydrofuran.

tuents do not induce a torsion about the aryl–aryl single bonds. The fluorescence spectrum measured in tetrahydrofuran solution shows an emission maximum at 425 nm, while in the solid state, the maximum is shifted to 457 nm. These emission characteristics suggest the use of **45** as material for ELD [92].

Up to now, we have considered light-emitting conjugated polymers which absorb and emit at relatively short wavelength in comparison to the monomeric dyes presented in the first section. A new situation arises from the presence of large chromophoric units such as perylene fragments in the main chain of conjugated polymers. A typical example of such structures is given by the poly(perylene-co-diethynylbenzene) (PPDB) **46** (Scheme 11). This polymer deserves particular attention due to recent discoveries in the field of ELD [11, 12, 24]. In solution, PPDB **46** shows λ_{max} (abs.) = 552 nm and emits red light with λ_{max} (em.) = 597 nm. Its emission spectrum in films is comparable to that in solution with an emission maximum around 595 nm but with a long wavelength emission tail, resulting from the formation of a weakly bound interchain excited state which also appears in the solid state. The $\O_{PL}$, which is low in the solid state

(0.6%) due to interchain quenching effects, reaches a value of 60% in solution [12]. A new situation arises when low amounts of PPDB **46** are blended with LPPP **42**. Specifically, for a doping concentration of about 1 wt% of PPDB **46** in the PPDB/LPPP blend, an almost quantitative energy transfer from LPPP **42** to PPDB **46** occurs. As a result, the PL and EL of the polymer blend are dominated by the red PPDB emission. This observation is likely related to the good compatibility of both rigid-rod polymers, i.e. no phase separation in thin films could be observed within the resolution (>10 nm) of an atomic force microscope [12]. Interestingly enough, the maximum of EL quantum efficiency, $\emptyset_{EL} = 1.6\%$ is reached for a blend containing as little as 0.2 wt% of PPBD in LPPP. Additionally, the use of polymethylmethacrylate as a charge carrier blocking material results in the partial hindrance of energy transfer from the blue to the red and allows the fabrication of white light-emitting diodes [12]. It is anticipated that these white ELD will find a considerable number of applications, especially as back lighting in liquid crystal displays [12].

Independent chromophoric units can also be introduced in the main chain and side chain of polymers. In these cases the polymer chain serves only as carrier and brings the advantage of film formability and mechanical resistance. The covalent incorporation of dyes in polymers prevents their migration which is often observed for the physical dispersion of dyes in polymer matrix [93]. In addition, the covalent incorporation of dyes allows them to be homogeneously distributed. Dye migration into the polymer matrix is often encountered in industrial applications. The prevention of dye migration is essential to ensure, for example, that a plastic bottle does not contaminate its contents [93].

There exists, of course, a large body of examples of dyes covalently attached to polymers. We will focus on derivatives of perylene dyes and, in particular, on perylene **1** and perylene tetracarboxdiimide **9b**. These two compounds with their $\emptyset_{PL}$ approaching unity have prompted numerous studies aimed at their incorporation into polymer structures to combine the intrinsic film-forming properties of polymers with the desirable photophysical properties of perylene dyes [23]. An example of perylene-containing polymers is given by the polyperylene **47**, which is exclusively composed of tetraphenoxy-substituted perylene repeating units. It absorbs at λ_{max} (abs.) = 443 nm and emits at λ_{max} (em.) = 473 nm in chloroform solution [24, 30]. The absorption and emission characteristics of **47** are close to those of its constituitive tetraphenoxy-substituted perylene monomers indicating that for steric reasons little electronic interactions occur between repeating units [24, 30]. Several perylene-based polyimides **48a, b**, depicted in Scheme 11, have been synthesized [22, 24]. The λ_{max} (em.) of **48a, b** ranges from 615 to 622 nm depending on the nature of the spacer connecting the perylene tetracarboxdiimide units (Table 4). There are also examples of side-chain perylene polymers such as the functionalized polysiloxane **49** [23, 94]. Not surprisingly, the λ_{max} (em.) of **49** remains comparable to the one of main chain polymer **48**.

A novel strategy for the incorporation of dyes into mechanically resistant materials is the fabrication of inorganic/organic hybrids **50**, composed of a silicon-based sol/gel glass as an inorganic matrix and a molecularly dispersed organic emitter (Scheme 12) [6]. Using this method, the molecular dispersion of

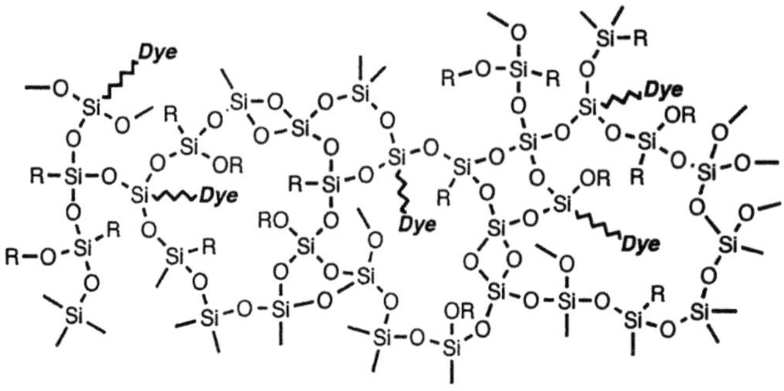

50

Scheme 12. Inorganic/organic hybrid systems containing fluorescent dyes

the organic chromophore was obtained via the formation of covalent bonds between the silicon-based matrix and the perylene-based dye molecule, as depicted in Scheme 12 [25]. Sol/gel techniques also offer the possibility of investigating the photophysical properties and photochemical stability of dyes imbedded into rigid silica cages. It is currently debated that a rigid silica environment may reduce intramolecular motions and rearrangements and thus enable the enhancement of photostability and fluorescence capability of dye molecules [6].

It can be concluded from this section that, inspired by conventional PPP, PPE and PPV structures, and also by the intrinsic photophysical behavior of organic dyes, a large variety of light-emitting polymers has been developed. In particular, these polymers have not only contributed to a better understanding of the photophysics of conjugated polymers, but also qualify as functional materials for ELD applications.

5
Summary and Outlook

We have presented in this review three different classes of new light-emitting dyes: monomers, oligomers, polymers, in a unified approach toward "organic" and "polymeric" dye stuff chemistry. We have focussed, in particular, on the preparation of highly photoluminescent defect-free structures. These structures have attracted wide interest among photophysicists and materials scientists. This is largely documented by the numerous fundamental spectroscopic and photophysical studies carried out on our materials. It should not be overlooked, on the other hand, that the properties of our materials make them ideal candidates for applications, especially in the field of light-emitting diodes.

Acknowledgements. The work described in this review article would not have been possible without the help of several spectroscopists and physicists, in particular Prof. Armstrong, Prof. Basché, Prof. Bässler, Prof. Brauchle, Prof. Brédas, Dr. Bubeck, Prof. DeSchryver, Prof. Friend, Prof. Heinze, Prof. Leising, Dr. Mahrt, Dr. Meerholz, Prof. Okada, Prof. Rettig, Dr. Tang, and Prof. Zerbi. The financial support of BASF AG and of the Bundesminister für Forschung und Bildung are also gratefully acknowledged. The help of Prof. Tolberg for correcting this manuscript is also aknowledged. Our cordial thanks are due to those members of the group who have contributed to the reviewed results and those names that appear in the references.

References

1. Zollinger, H.; „Color Chemistry" VCH Verlagsgesellschaft, Weinheim, 1987
2. Krassovitskii, B.M.; Bolotin, B.M.; „Organic Luminescent Materials", VCH Verlagsgesellschaft, Weinheim, 1988
3. Basché, T.; Moerner, W.E.; Orrit, M.; Wild, U.P.; „Single-Molecule Optical Detection, Imaging and Spectroscopy" VCH Verlagsgesellshaft, Weinheim, 1997, and references cited therein
4. Seybold, G.; Wagenblast, G.; Dyes and Pigments, 1989, 11, 303
5. O'Neil, M.P.; Niemczyk, M.P.; Svec, W.A.; Gosztola, D.; Gaines III, G.L.; Wasielewski, M.R.; Science, 1992, 257, 63
6. Burgdorff, C.; Löhmannsröben, H.G.; Reisfeld, R.; Chem. Phys. Lett. 1992, 197, 358
7. Gvishi, R.; Reisfeld, R.; Burshtein, Z.; Chem. Phys. Lett. 1993, 213, 338
8. Christie, R.M.; Polym. Int. 1994, 34, 351
9. Mayer, A.; Neuenhofer, S.; Angew. Chem. Int. Ed. Engl. 1994, 33, 1044
10. Tasch, S.; Niko, A.; Leising, G.; Scherf, U.; Appl. Phys. Lett. 1996, 68, 1090
11. Tasch, S.; Hochfilzer, C.; List, J.W.E.; Leising, G.; Geerts Y.; Schlichting, P.; Rohr, U.; Scherf, U.; Müllen, K.; Phys. Rev. B. submitted
12. Tasch, S.; Hochfilzer, C.; List, J.W.E.; Leising, G.; Geerts Y.; Schlichting, P.; Rohr, U.; Scherf, U.; Müllen, K.; Appl. Phys. Lett. submitted
13. Huber, J.; Müllen, K.; Salbeck, J.; Schenk, H.; Scherf, U.; Stehlin, T.; Stern, R.; Acta Polym. 1994, 45, 244
14. Conwell, E.; Trends Polym. Sci. 1997, 5, 218
15. Klärner, G.; Müller, M.; Morgenroth, F.; Wehmeier, M.; Sozka-Guth, T.; Müllen, K.; Synth. Met. 1997, 84, 297
16. Geerts, Y.; Klärner, G.; Mullen, K.; Hydrocarbon Oligomers In Electronic Materials: The Oligomer Approach, Müllen, K. and Wegner, G. (eds.), VCH, Weinheim, 1997
17. Electronic Materials: The Oligomer Approach, Müllen, K. and Wegner, G. (eds.), VCH, Weinheim, 1997
18. Baumgarten, M.; Bunz, U.; Scherf, U.; Müllen, K.; Organic Synthesis and Materials Science In Molecular Engineering for Advanced Materials, Becher, J. and Schaumburg, K. (eds.), Kluwer Academic Publishers, 1995
19. Müllen, K.; Pure & Appl. Chem. 1993, 65, 89
20. Klärner, G.; Former, C.; Yan, X.; Richert, R.; Müllen, K.; Adv. Mater. 1996, 8, 932
21. Scherf, U.; Müllen, K.; Synthesis, 1992, 23
22. Dotcheva, D.; Klapper, M.; Müllen, K.; Macromol. Chem. Phys. 1994, 195, 1905
23. Müllen, K; Quante, H.; Benfaremo, N.; „Perylene Polymers" in „Polymeric Materials Encyclopedia", Salamone, J.C. ed., CRC Press, Boca Raton, 1996
24. Quante, H.; Schlichting, P.; Rohr, U.; Geerts, Y.; Müllen, K.; Macromol. Chem. Phys. 1996, 197, 4029
25. Geerts, Y.; Keller, U.; Scherf, U.; Schneider, M.; Müllen, K.; Am. Chem. Soc., Polym. Prep. 1997, 38, 315
26. Rademacher, A.; Märkle, S.; Langhals, H.; Chem. Ber. 1982, 115, 2927
27. Graser, F.; Hädicke, E.; Liebigs. Ann. Chem. 1980, 1994; 1984, 483
28. Nagao, Y.; Misono, T.; Dyes Pigm. 1984, 5, 171
29. Quante, H.; Geerts, Y.; Müllen, K.; Chem. Mater. 1997, 9, 495
30. Quante, H.; Doktor Arbeit, Mainz, 1995

31. Geerts, Y.; Quante, H.; Platz, H.; Mahrt, R.; Hopmeier, M.; Böhm, A.; Müllen, K.; J. Mater, Chem., in press

32. Holtrup, F.O.; Müller, G.R.J.; Quante, H.; De Feyter, S.; De Schryver, F.C.; Müllen, K.; Chem. Eur. J. 1997, 3, 219

33. Schlichting, P.; Rohr, U.; Müllen, K.; Liebigs Ann./Recueil 1997, 395

34. Dobowski, J.; Grabowski, Z.; Paeplow, B.; Rettig, W.; Koch, K.H.; Müllen, K.; New. J. Chem. 1994, 18, 525

35. Anton, U.; Göltner, C.; Müllen, K.; Chem. Ber. 1992, 125, 2325

36. Schmidt, A.; Armstrong, N.R.; Göltner, C.; Müllen, K.; J. Phys. Chem. 1994, 98, 11780

37. Koch, K.H.; Müllen K.; Chem. Ber. 1991, 124, 2091

38. Müllen, K.; Scherf, U.; Makromol. Chem., Macromol. Symp. 1993, 69, 23

39. Fabian, J.; Nakazumi, H.; Matsuoka, M.; Chem. Rev. 1992, 92, 1197

40. Quante, H.; Müllen, K.; Angew. Chem. Int. Ed. Engl. 1995, 34, 1323

41. van de Craats, A.; Warman, J.; Müllen, K.; Geerts, Y.; Brandt, J.-D.; Adv. Mater. 1998, 10, 36

42. Cimrovà, V.; Remmers, M.; Neher, D.; Wegner, G.; Adv. Mater. 1996, 8, 146

43. Keller, U.; Müllen, K.; De Feyter, S.; De Schryver, F.C.; Adv. Mater. 1996, 8, 490
Keller, U.; Müllen, K.; Geerts, Y.; unpublished results

44. Grüner, J.; Wittmann, H.F.; Hamer, P.J.; Friend, R.H.; Huber, J.; Scherf, U.; Müllen, K.; Moratti, S.C.; Holmes, A.B.; Synth. Met. 1994, 67, 181

45. Göltner, C.; Pressner, D.; Müllen, K.; Spiess, H.W.; Angew. Chem. Int. Ed. Engl. 1993, 32, 1660
Pressner, D.; Göltner, C.; Spiess, H.W.; Müllen, K.; Ber. Bunsenges. Phys. Chem. 1993, 97, 1362

46. Müller, G.R.J.; Ph.D. Thesis, Mainz, 1996
Müller, G.R.J, Meiners, C.; Geerts, Y.; Enkelmann, V.; Müllen, K.; J. Mater. Chem. 1998, 8, 61

47. Keller, U.; Doktor Arbeit, Mainz, 1996

48. Gross, M.; Meerholz, K.; Bräuchle, C.; Schlichting, P.; Rohr, U.; Müllen, K.; Angew. Chem. Int. Ed. Engl. 1998, 37, 1435

49. Collings, P.J.; Hird, M.; Introduction to Liquid Crystals, Taylor and Francis, London, 1997

50. Biasutti, M.A.; De Feyter, S.; De Backer, S.; Dutt, G.B.; De Schryver, F.C.; Ameloot, M.; Schlichlting, P.; Müllen, K.; Chem. Phys. Lett. 1996, 248, 13

51. Geerts, Y.; Platz, H.; Quante, H.; Meiners, C.; Müllen, K.; unpublished results

52. Herrmann, A.; Geerts, Y.; Keller, U.; Müllen, K.; unpublished results

53. Hunter, A.C.; Sanders, J.K.M.; J. Am. Chem. Soc. 1990, 112, 5525
Williams, J.H.; Acc. Chem. Res. 1993, 26, 593

54. Herwig, P.; Kayser, C.W.; Müllen, K.; Spiess, H.W.; Adv. Mater. 1996, 8, 510

55. Biasutti, M.A.; Rommens, J.; Vaes, A.; De Feyter, S.; De Schryver, F.C.; Herwig, P.; Müllen, K.; Bull. Soc. Chim. Belg., in press

56. Grimme, J.; Kreyenschmidt, Uckert, F.; Müllen, K.; Scherf, U.; Adv. Mater. 1995, 7, 292

57. Bohnen, A.; Koch K.-H.; Lüttke, W.; Müllen, K.; Angew. Chem. Int. Ed. Engl. 1990, 29, 525

58. De Backer, S.; Negri, M.R.; De Feyter, S.; Bhaskar Dutt, G.; Ameloot, M.; De Schryver, F.C.; Müllen, K.; Holtrup, F.; Chem. Phys. Lett. 1995, 233, 538

59. De Backer, S.; Bhaskar Dutt, G.; Ameloot, M.; De Schryver, F.C.; Müllen, K.; Holtrup, F.; J. Phys. Chem. 1996, 100, 512

60. Tittel, J.; Kettner, R.; Basché, Th.; Bräuchle, C.; Quante, H.; Müllen, K.; J. Lumin. 1995, 64, 1

61. Meyer, Y.H.; Plaza, P.; Müllen, K.; Chem. Phys. Lett. 1997, 264, 643

62. a) Meerholz, K.; Gregorius, H.; Müllen, K.; Heinze, J.; Adv. Mater. 1994, 6, 671. b) Schenk, R.; Gregorius, H.; Müllen, K.; Adv. Mater. 1991, 10, 492. c) Brendel, P.; Grupp. A.; Mehring, M.; Schenck, R.; Müllen, K.; Huber, W.; Synth. Met. 1991, 45, 49. d) Barth, S.; Bässler, H.; Wehrmeister, T.; Müllen, k.; J. Phys. Chem. 1997, 106, 321

63. a) Tian, B.; Zerbi, G.; Schenck, R.; Müllen, K.; J. Phys. Chem. 1991, 95, 3191. b) Tian, B.; Zerbi, G.; Schenck, R.; Müllen, K.; J. Phys. Chem. 1991, 95, 3198. c) Cornil, J.; Beljonne, D.; Shuai, Z.; Hagler, T.W.; Campbell, I.; Bradley, D.D.C; Brédas, J.L.; Sprangler, C.W.; Müllen, K.; Chem. Phys. Lett. 1995, 247, 425

64. Schenck, R.; Gregorius, H.; Meerholz, K.; Heinze, J.; Müllen, K.; J. Am. Chem. Soc. 1991, 113, 2634

65. Mathy, A.; Ueberhofen, K.; Schenk, R.; Gregorius, H.; Garay, R.; Müllen, K.; Bubeck, C.; Phys. Rev. B 1996, 53, 4367

66. a) Borroughes, J.H.; Bradley, D.D.C.; Brown, A.R.; Marks, R.N.; Friend, R.H.; Burn, P.L.; Holmes, A.B.; Nature 1990, 347, 539. b) Hennecke, M.; Dammerau, T.; Müllen, K.; Macromolecules 1993, 26, 3411

67. Heller, C.M.; Campbell, I.H.; Laurich, B.K.; Smith, D.L.; Bradley, D.D.C.; Burn, P.L.; Ferraris, J.P.; Müllen, K.; Phys. Rev. B 1996, 54, 5516

68. Choong, V.; Park. Y.; Gao, Y.; Wehrmeister, T.; Müllen, K.; Hsieh, B.R.; Tang, C.W. Appl. Phys. Lett. 1996, 69, 1492

69. Grem, G.; Leising, G.; Synth. Met. 1993, 55-57, 4105

70. Grem, G.; Leditzky, G.; Ulrich, B.; Leising, G.; Adv. Mater. 1992, 4, 37

71. Grimme, J.; Scherf, U.; Macromol. Chem. Phys. 1996, 197, 2297

72. Pauck, T.; Bässler, H.; Grimme, J.; Scherf, U.; Müllen, K.; Chem. Phys. 1996, 210, 219

73. Kreyenschmidt, M.; Baumgarten, M; Tyutyulkov, N.; Müllen, K.; Angew. Chem. Int. Ed. Engl. 1994, 33, 1957

74. Kreyenschmidt, M.; Ph.D. Thesis, Mainz, 1996

75. Keegstra, M.A.; De Feyter, S.; De Schryver, Müllen, K.; Angew. Chem. Int. Ed. Engl. 1996, 35, 774

76. Nishiyama, K.; Honda, T.; Okada, T.; Reis, H.; Baumann, W.; Müller, U.; Müllen, K.; J. Phys. Chem. 1998, 102, 2936

77. a) Schlüter, A.D.; Wegner, G.; Acta Polymer. 1993, 44, 59. b) Tour, J.M.; Adv. Mater. 1994, 6, 190. c) Kovacic, P.; Jones, M.B.; Chem. Rev. 1987, 87, 357

78. a) Friend, R.H.; Denton, G.J.; Halls, J.J.M.; Harisson, N.T.; Holmes, A.B.; Köhler, A.; Lux, A.; Moratti, S.C.; Pichler, K.; Tessler, N.; Towns, K.; Wittmann, H.F.; Solid State Comm. 1997, 102, 249. b) Rothberg, L.J.; Yan, M.; Papadimitrakopoulos, F.; Galvin, M.E.; Kwock, E.W.; Miller, T.M.; Synth. Met. 1996, 80, 41. c) Obrzut, J.; Karasz, F.E.; J. Chem. Phys. 1987, 87, 2349

79. Pope, M.; Kallmann, H.; Magnante, P.; J. Chem. Phys. 1963, 38, 2042

80. Weiss, K.; Michel, A.; Auth, E.M.; Bunz, U.H.F.; Mangel, T.; Müllen, K.; Angew. Chem. Int. Ed. Engl. 1997, 36, 506

81. Bunz, U.; Francke, V.; Klapper, M.; Uckert, F.; Müllen, K.; Polymer Synthesis with Organometallic Intermediates In Proceedings of the Fifth Symposium on Organic Synthesis via Organometallics in Heidelberg, Friedr. Vieweg & Sohn, Braunschweig/ Wiesbaden, 1997

82. Garay, R.O.; Baier, U.; Bubeck, C.; Müllen, K.; Adv. Mater. 1993, 5, 561

83. Mangel. T.; Eberhardt, A.; Scherf, U.; Bunz, U.H.F.; Müllen, K.; Macromol. Rapid Commun. 1995, 16, 571
Weiss, K.; Michel, A.; Auth, E.M.; Bunz, U.H.F.; Mangel, T.; Müllen, K.; Angew. Chem. Int. Ed. Engl. 1997, 36, 506

84. Klärner, G.; Former, C.; Martin, K.; Räder, J.; Müllen, K.; Macromolecules, 1998, 31, 3571

85. Staring, E.G.; Demandt, R.C..; Braun, D.; Rikken, G.L.; Kessener, Y.A.; Venhuizen, T.H.; Wynberg, H.; ten Hoeve, W.; Spoeltra, K.J.; Adv. Mater. 1994, 6, 935

86. Sonoda, Y.; Kaeriyama, K.; Bull. Chem. Soc. Jpn. 1992, 65, 853

87. Huber, J.; Scherf, U.; Macromol. Rapid Commun. 1994, 15, 897

88. Mahrt, R.F.; Pauck, T.; Lemmer, U.; Siegner, U.; Hopmeier, R.; Hennig, R.; Bässler, H.; Göbel, E.O.; Haring Bolivar, P.; Wegmann, G.; Kurz, H.; Scherf, U.; Müllen, K.; Phys. Rev. B. 1996, 54, 1759

89. Pauck, T.; Hennig, R.; Perner, M.; Lemmer, U.; Siegner, U.; Mahrt, R.F.; Scherf, U.; Müllen, K.; Bässler, H.; Göbel, E.O.; Chem. Phys. Lett. 1995, 244, 171

90. Scherf, U.; Müllen, K.; Adv. Polym. Sci. 1995, 123, 1

91. Chmil, K.; Scherf, U.; Acta Polymer. 1997, 48, 208
Chmil, K.; Scherf, U.; Makromol. Chem., Rapid Commun. 1993, 14, 217

92. Kreyenschmidt, M.; Uckert, F.; Müllen, K.; Macromolecules 1995, 28, 4577

93. Christie, R.M.; Polym. Int. 1994, 34, 351
94. Beck, K.H.; Etzbach, K.-H.; Schmidt, H.-W. EP 0 422 535A1, 1990. Etzbach, K.-H; Beck, K.H.; Wagenblast, G. EP 0 422 538A1, 1990
95. Feiler, L.; Langhals, H.; Polborn, K.; Liebigs Ann. 1995, 1229
96. Holtrup, F.O.; Müller, G.R.J.; Uebe, J.; Müllen, K.; Tetrahedron, 1997, 53, 6847
97. Schmidt, A.; Anderson, M.L.; Dunphy, D.; Wehrmeister, T.; Müllen, K.; Adv. Mater. 1995, 7, 722
98. Scherf, U.; Müllen, K.; Makromol. Chem., Rapid Commun. 1991, 12, 489

CHAPTER 14

Fluorescence Probes in Polymers and Liquid Crystals:
Complex Macromolecular Chain Dynamics (Proposal from the Far East)

H. Ushiki

1
Prologue

In our previous book [1], I wrote the following: "At the point of nihility without time nor space, an explosion of dense matter marked the origin of the universe. Immediately after this accident, a fireball of radiation at a unbelievably high temperature formed in a tiny volume. Many photons were produced by collisions between quarks and anti-quarks. This is the Big-Bang theory of the genesis of the universe. It is not unusual for books on the concept of light to begin with a quotation taken from the Bible on the creation of the world. Indeed, the relationship between mankind and light has a long history and it has been strongly connected to all aspects of life, from antiquity to the modern ages. Man has always been careful to distinguish light from heat. Light is always given a much greater importance throughout the ages. Light is a crucial tool in many religions. Men do not speak badly of light. Note for instance that light is used in the Japanese word Goraiko describing the admiration of the sun, while heat is included in an expression describing hell, Shakunetsu-Jigoku, the scorching heat in hell. Hence it comes as no surprise that man has always been fascinated by the multiple properties of light".

In this report, I have summarized the review of "Fluorescence Probe Methods", together with my invited lecture entitled "Complex Macromolecular Chain Dynamics" given at the 5th International Conference on Methods and Applications of Fluorescence Spectroscopy in Berlin. On this occasion, I would like not only to present an overview of fluorescence probe techniques, but also to explain the meanings of these methods in various scientific research fields. Therefore I will try to discuss the position of fluorescence probe methods based on a viewpoint of post-modernism [2]. The important key words in the former stage of this review are "Classification Based on Structure for Macromolecular Chain Dynamics", "PLASMA (Perfective Laboratory Automation System for Macromolecular Analysis)", "Fluorescence Probe Method", "Time-Resolved Fluorescence Depolarization Method", "Dynamics of Morphological Formation Process", and "Power Law Analysis". On the other hand, in the later stage of this review, I will briefly discuss the phenomena of macromolecular transition, micellar formation, volume phase transition of gels, and structural transition of liquid crystal based on the measurement of fluorescence probe methods.

Owing to limited space in this review, I will not be able to explain in detail the concept of "Complex Macromolecular Chain Dynamics" based on a post-modernism viewpoint. Therefore, I would like to be so bold as to present many figures as schematic diagrams based on a transparency style in order to clarify the easy understanding of the concept of complex chemical physics even though some referees have pointed out that these figures are too densely packed. In the near future, I will arrange the concept of "Classification Based on Structure for Macromolecular Chain Dynamics" from a post-modernism viewpoint as a book subtitled "Proposal from the Far East".

2
Classification Based on Structure for Macromolecular Chain Dynamics

2.1
Space-Time Concept

I would like to start with an explanation of the relationship between natural hierarchy and dynamics. All materials are formed of a combination of elements and the relationships between the dynamics of these elements. These elements are distributed into various classes of nature, and the classes form a natural hierarchical structure from a microscopic class (atoms and molecules) to a macroscopic class (condensates). In each class, there is a special movement. For example, the schematic diagram of class organization based on a microscope photograph of *Amoeba Limax* is shown in Fig. 1. Using an operation of microscopic degree, we can understand the class organization of nature, from *Amoeba Limax* of microscope level, organella of light scattering level, biomembrane and phosphatidyl choline of molecular level, a methylene chain of functional group level, and to wave function of molecular orbital level. In spatial class organization of condensed matters, we can easily understand it by dividing the classification into six levels (naked eye, microscope, light scattering, molecular, functional group, molecular orbital levels). We know well a natural hierarchy based on space, and also of course a natural hierarchy based on time. As the concept of class organization in nature connects with the recent research theme of 'complex system', 'complex fluid', 'supramolecules', 'cluster', etc., it is very important to discuss the fundamental principle of the relationship between each class of nature in order to clarify the mechanism of structural formation processes in condensed matters. In particular, it is necessary to discuss it from the viewpoint of dynamics in order to understand the vector developments, reaction → relaxation → transition → structural formation, in condensed matters.

In recent years, new knowledge about phase diagrams and fluctuation behavior at the critical points of micellar solution, liquid crystals, polymers, condensates, gels, etc. has been discussed by many authors, especially in order to clarify the mechanism of morphological formation processes of biomacromolecules. Many researchers think that it is very important to understand the physical and chemical properties of a near critical point based on both a theoretical and experimental viewpoint because some long range interaction forces

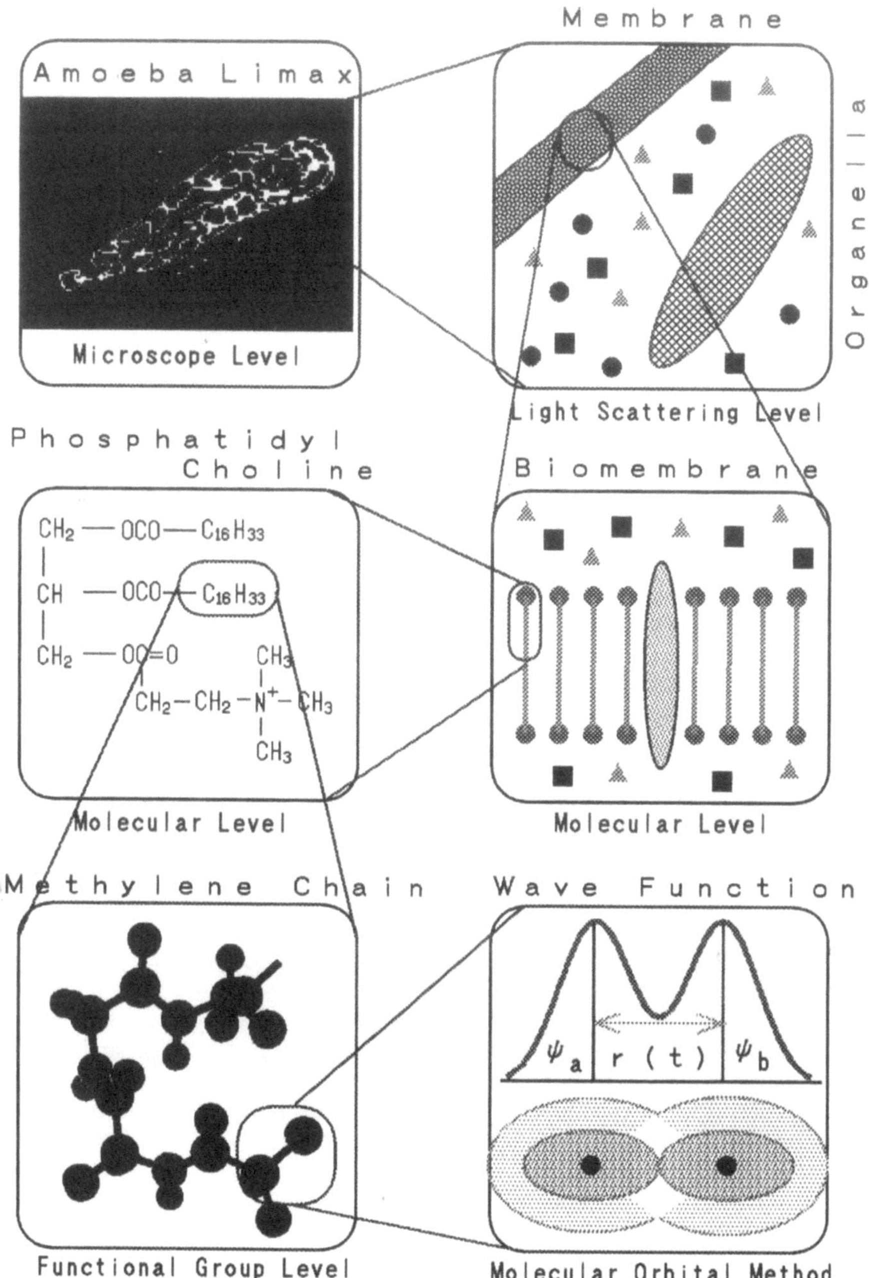

Fig. 1. Schematic diagram of class organization in nature

will be produced in this system when the concentration fluctuations increase as a near critical point is approached. Thus much research on the dynamic behavior of phase at a near critical point has been fundamentally discussed from the viewpoint of mutual diffusion processes based on thermodynamics. But, in order to bind both phenomena of molecular dynamic behavior at a near critical point and a morphological formation process, I think that the key viewpoint of this research is a self-diffusion process of material itself, that is, it is very important to measure directly the self-diffusion processes of the element in each class and to analyze the mechanism of the relationship between self-diffusion and mutual diffusion processes at a near critical point.

For the purpose of discussion about movement of elements in each class, the image of macromolecules is the best example because a macromolecule has a large number of motional degrees of freedom, and the particular motional behavior of each element is distributed into each class from a macroscopic class to a microscopic one. As its simple example, the schematic diagram of a macromolecular space-time image is shown in Fig. 2. Macromolecular dynamic processes hold many motions. The horizontal axis is space, and the vertical one is relaxation time. We know that there are many movements, for example, two state motions, local motions, segmental motions, cooperative diffusions, repetitive motions, etc., so I would like to demonstrate the fact that it is important to understand the existence of the relationship between space and time in various scientific phenomena with macromolecules. Therefore the concept of class organization with a space-time image in nature is very important in order to discuss "Complex Macromolecular Chain Dynamics".

As an experimentalist handling spectroscopic techniques, I can easily obtain the measurement image of five class organizations. for example, naked eye, microscope, light scattering, molecular, functional group levels in macromolecules based on various optical spectroscopic techniques [3]. Generally speaking, the naked eye level and microscope level are included in the macroscopic class, the light scattering level in the mesoscopic class, and the molecular and functional groups levels in the microscopic class. At naked eye and microscope levels, a typical measurement technique is CCD video graphics analysis. At the light scattering level, it is static and dynamic two-dimensional light scattering analysis. At the molecular level, it is fluorescence molecular probe analysis and in the functional group level, it is by using a two-dimensional image Raman microscope. Therefore it will be necessary to measure the dynamic behavior of macromolecules of all levels in order to discuss the topic "Complex Macromolecular Chain Dynamics".

2.2
Non-Uniformity

In order to understand the dynamic behavior of an element in each class of nature, the strongest tool is thought to be statistical mechanics. This research field has the role of clarifying the relationship between various macroscopic properties of materials based on our everyday life and various microscopic properties of atoms and molecules that make up materials. Historically speaking,

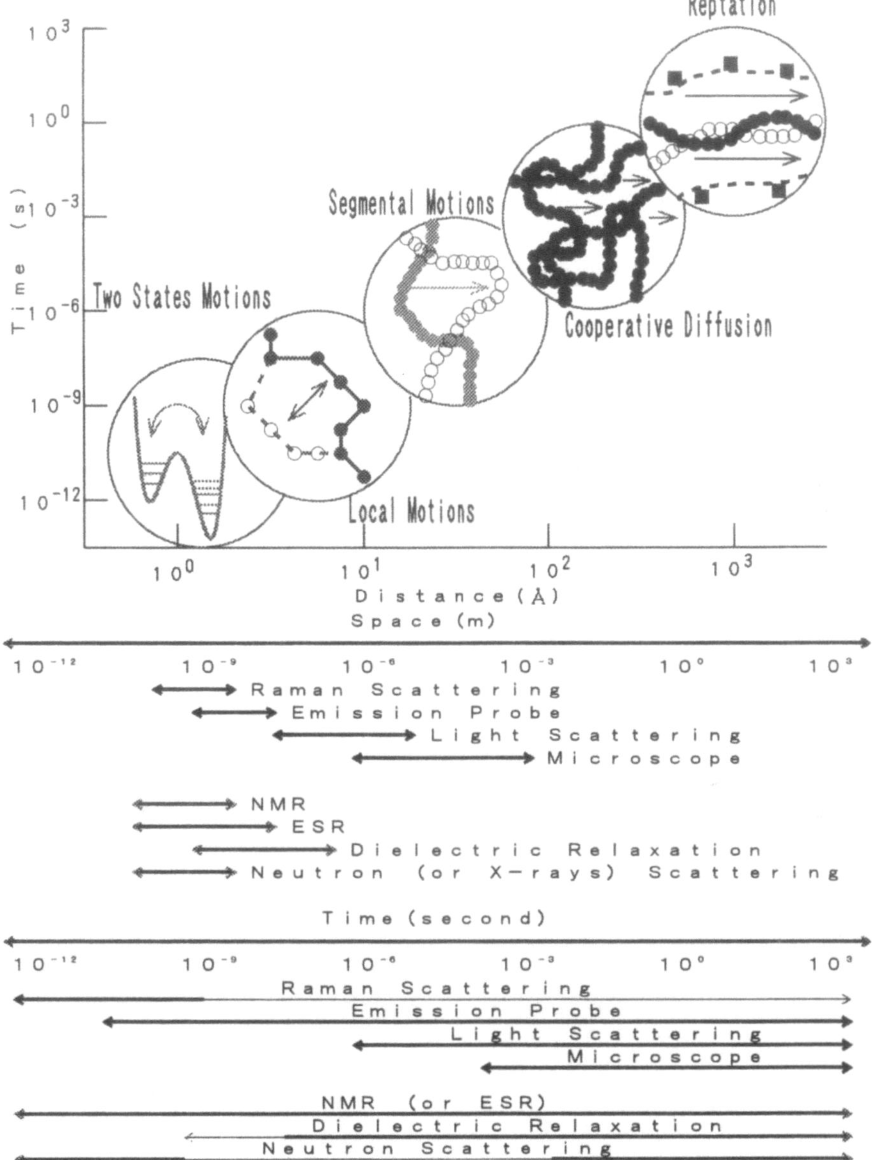

Fig. 2. Schematic diagram of space-time image in macromolecules

based on an "order" concept in the field of statistical mechanics, the main research theme has gradually changed from periodical to non-periodical structures via the random structure of the material system. This developmental process means an extension to the "order" concept. In recent years, many researchers have begun to discuss a new structure, non-uniformity or non-periodical rather than periodical or random structures of material system. The discussion about this new structure will be inevitably attributed to the viewpoint of above-mentioned class organization in nature. Many researchers have recently become interested in the mechanism of comparative weak molecular interactions [3]. The cause of macroscopic behavior in various materials can be ascribed to five molecular interactions classified into van der Waals forces, saturating chemical bonds, affinity between compounds, Coulomb forces, and topological interactions. The morphological formation processes of condensates are performed by these molecular interactions.

Here let's briefly consider the non-uniformity of space. The schematic diagram of non-uniformity in condensed matters is shown in Fig. 3. At first, the property of non-uniformity can be classified into microscopic and macroscopic classes. In the microscopic class, the origin of non-uniformity is classified into replaceable, topological, and structural types. Therefore we can obtain the outline of the relationship between microscopic and macroscopic classes in non-uniformity of condensed matters by means of these classified types. Until now we have understood the mechanism of non-uniformity using the effective medium approximation method, but it was difficult to understand the topological

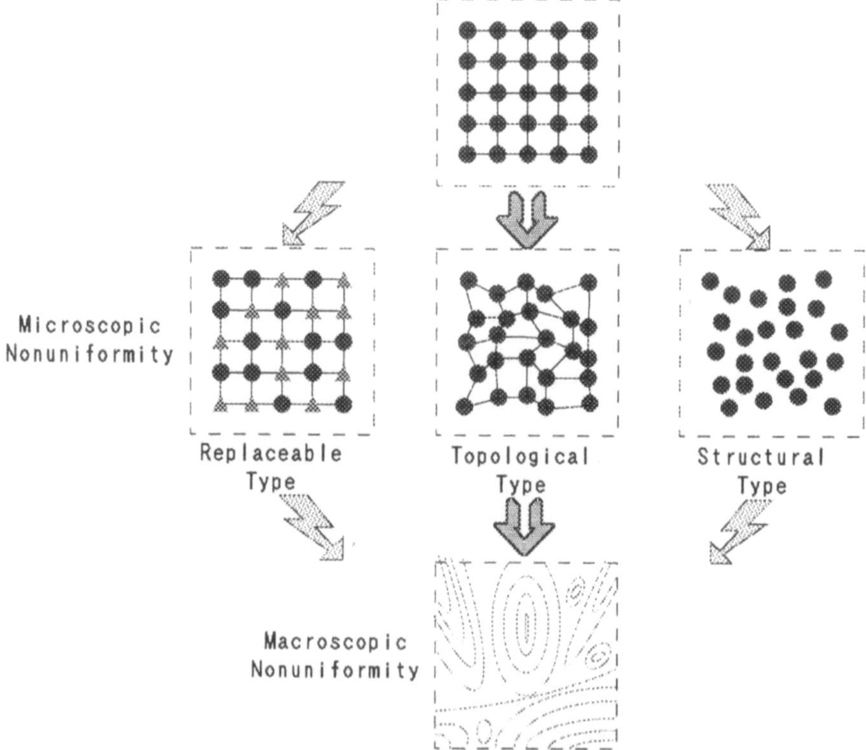

Fig. 3. Schematic diagram of non-uniformity in condensed matters

Table 1. Properties of Classified Types at the Microscopic Level

Microscopic non-uniformity	Long range order	Short range order
replaceable type	maintained	almost maintained
topological type	breakdown	almost maintained
structural type	breakdown	partly maintained

type based on comparative weak molecular interaction. The properties of these types in microscopic non-uniformity are shown in Table 1.

Therefore the interaction in the topological type has an ambiguity (network structure) which cannot be adequately approximated by general functions. In the past using the approximation method for non-uniformity of condensed matters, there was a precondition for a ergodic hypothesis. The space averaging can be replaced by an ensemble one. Recently it has become important to discuss the dynamic behavior of materials in non-uniformity media. As a result, it is important to discuss the approximation method of non-ergodic processes. Therefore we must find a new method with an excellent calculation technique for averaging in a non-uniform field in order to pick up some important information in the microscopic class [3].

I would like to propose that we consider the non-uniformity system from a different viewpoint and, as the next stage, I will try to explain the relationship between renormalization group, fractal, and percolation at a near critical point of condensed matters. The schematic diagram is shown in Fig. 4. In the upper part of this figure, we suppose that the amount of the component represented by the triangles gradually increases as following order from bottom to upper sides in microscopic class (left side). Hence we suppose that macroscopic degree gradually increases as following order from left to right sides at each state. The operation of macroscopic degree using the renormalization group method is given by Eq. 1:

$$X = f_b(Y) = f_b f_a(Z) = f_{ba}(Z) \tag{1}$$

where Z, Y, X are states in the microscopic, mesoscopic, and macroscopic class, and f_a, f_b, and f_{ba} are operators of macroscopic transformation, respectively. We can easily imagine the relationship between microscopic and macroscopic classes based on Fig. 4. In the upper part of Fig. 4, we cannot really judge a critical point in the microscopic class (left side), but we can easily find a critical point in the macroscopic class (right side). In other words, the state at the critical point is not influenced by a macroscopic transformed operation using the renormalization group method. I demonstrate the fact that it is important not to change at a critical point with operation of macroscopic degree, that is, there is a self-similarity at a critical point. We will be able to extend this idea with a macroscopic transformed operation to a percolation phenomenon. If we take out the all-square components from the upper diagram in Fig. 4, we will obtain the bottom diagram in Fig. 4. In a similar manner to above, we cannot judge a critical percolated point in the microscopic class (left side), but we can easily

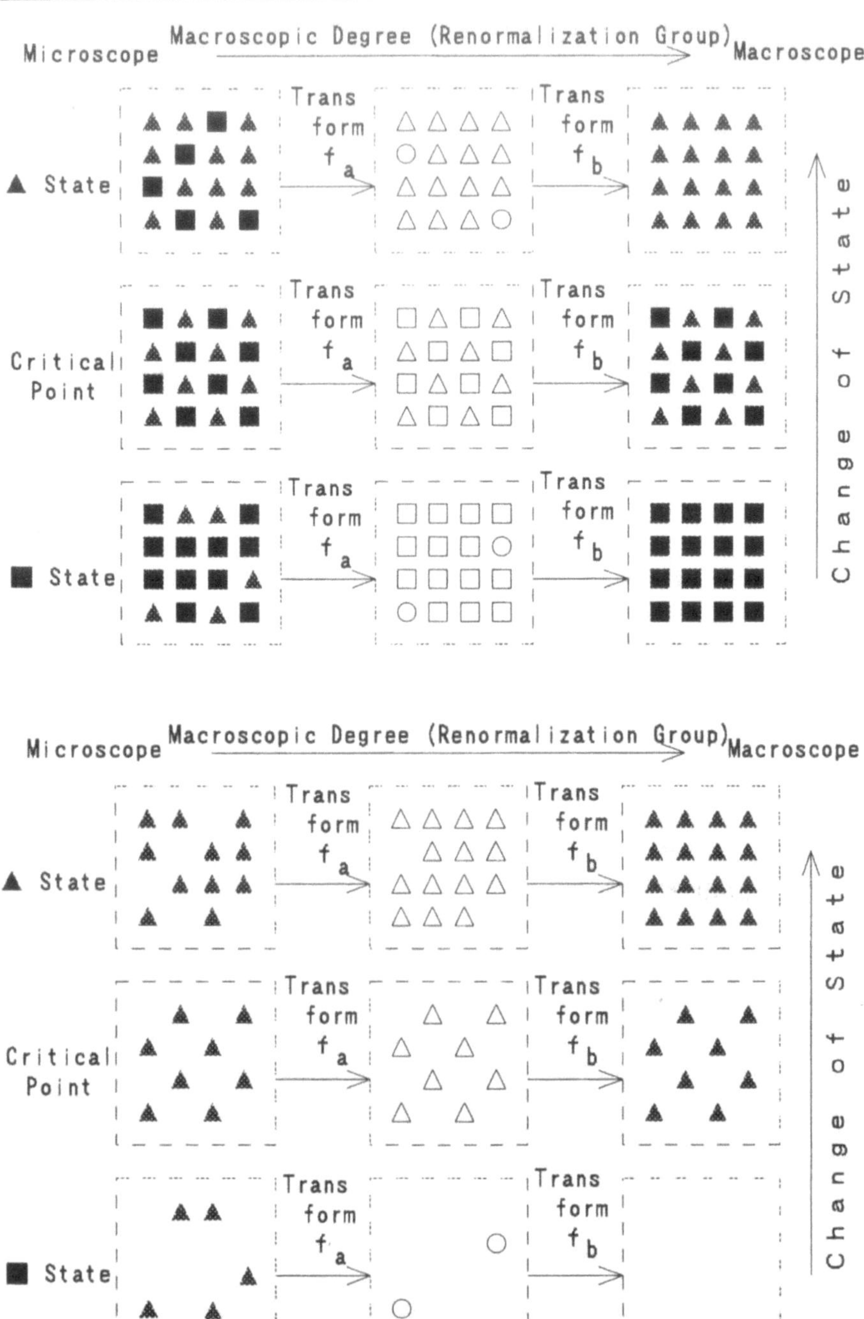

Fig. 4. Schematic diagram of relationship between renormalization group, fractal, and percolation at critical point

find a critical percolated point in the macroscopic class (right side). Hence we will be able to define fractal dimension at a critical point because the fractal definition is constant with a macroscopic transformed operation using the renormalization group method, that is, it is self-similarity. Therefore I think that there is a relationship between renormalization group, percolation, fractal, and critical point from the viewpoint of class organization in condensed matters. As the macroscopic conditions directly correspond to the microscopic ones at a critical point, many researchers are interested in critical phenomena in order to discuss the relationship between macroscopic and microscopic states. I would like to emphasize that it is important to study the dynamic behaviors of various elements at a near critical point in order to discuss the essence of "Complex Macromolecular Chain Dynamics".

Next I would like to consider the non-uniformity system from a different viewpoint and briefly think about the relationship between fractal structure and power law. In order to understand this relationship, we can introduce three excellent reports based on the discussion of relaxation time of reactants in condensed matters [4–6]. In the past, we knew that the response functions of complex condensed matters would show various non-exponential functions. In particular, in the field of dielectric relaxation measurement, many discussions had taken place about relaxation functions for various dynamic modes of condensed matters. For example, Debye type (Eq. 2), Cole-Cole type (Eqs. 3 and 4), Davidson-Cole type (Eq. 5), and Williams-Watts type (Eq. 6) functions were reported. Their response functions are given by the following equations:

$$\phi(t) = \frac{1}{\tau} \exp\left(-\frac{t}{\tau}\right) \tag{2}$$

$$\phi(t) = \frac{a}{\tau\Gamma(1+a)} \left(\frac{t}{\tau}\right)^{-(1-a)} \quad \left(\frac{t}{\tau}\right) \ll 1 \tag{3}$$

$$\phi(t) = \frac{a}{\tau\Gamma(1+a)} \left(\frac{t}{\tau}\right)^{-(1+a)} \quad \left(\frac{t}{\tau}\right) \gg 1 \tag{4}$$

$$\phi(t) = \frac{a}{\tau\Gamma(\beta)} \left(\frac{t}{\tau}\right)^{-(1-\beta)} \exp\left(-\frac{t}{\tau}\right) \ll 1 \tag{5}$$

$$\phi(t) = \frac{\delta}{\tau} \left(\frac{t}{\tau}\right)^{-(1-\delta)} \exp\left(-\left(\frac{t}{\tau}\right)^{\delta}\right) \tag{6}$$

where τ and $\Gamma(x)$ are the relaxation time and gamma functions, respectively. However, the discussion about their relaxation functions had a weak point. It is very difficult to understand the physical meaning of α, β, and δ as a constant, that is, their relaxation functions are empirical ones. Therefore many researchers in this field have recently become interested in the discussion about the meaning of the power constant of hyperbolic and stretched exponential functions. At the beginning, de Gnenne [5] and Evesque [6] theoretically discussed the relationship between relaxation time and fractal structure using bimolecular reaction field. The schematic diagram of the relationship between structure, relaxation time, and power law based on their reports is shown in the bottom graph in Fig. 5. They foretold the phenomenon that bimolecular reaction kinetics in porous glass would show a stretched exponential type function

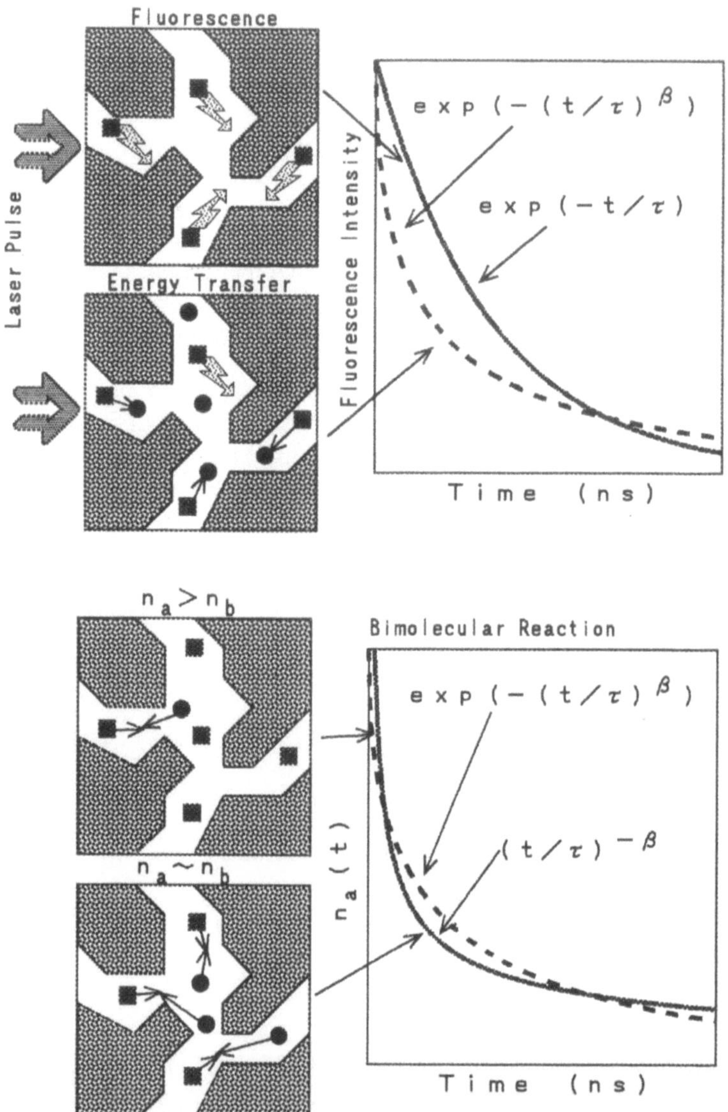

Fig. 5. Schematic diagram of the relationship between structure, relaxation time, and power law

(Williams-Watts type) or hyperbolic type function (Cole-Cole type). Here it was very important to find out the physical meaning of β connected with the fractal structure of porous glass. Following this, Even et al. [4] experimentally discussed the relationship between relaxation time and fractal structure using a singlet-singlet energy transfer process from Rhodamine B to malachite green in porous glass. The schematic diagram based on their report is shown in the upper graph in Fig. 5. They found that the fluorescence decay curve of Rhodamine B in porous glass with quenchers showed a stretched exponential type function, and they discussed the relationship between the constant β and the fractal dimension of its porous glass structure.

From my considerations about the space-time concept and the non-uniformity system, I would like to propose that the power law analysis could be used as a strong tool in order to understand some of the dynamic behavior of condensed matters in each level. In the future development of "Complex Macromolecular Chain Dynamics" from the viewpoint of post-modernism, I would expect it to be very important to clarify each relationship between a natural class organization with a space-time image, a macroscopic transformed operation with a renormalization group method, various critical and percolation processes, and power law analysis with the Laplace transformation.

3
PLASMA (Perfective Laboratory Automation System for Macromolecular Analysis)

We have made a new apparatus that I have named PLASMA (Perfective Laboratory Automation System for Macromolecular Analysis) [7]. The aim of this apparatus was to the investigation of our main theme entitled "Classification Based on Structure for Macromolecular Chain Dynamics". The concept of this apparatus has the aspect that my laboratory is a single apparatus, and PLASMA generally consists of a personal computer network with many original electric circuits and program software. Our final image of PLASMA's thoughts corresponds to the fact that time-resolved computer graphics of macromolecular dynamics can be presented by our new analytical methods with microscope CCD graphics, two-dimensional static and dynamic light scattering, various spectra (absorption, action, emission, and emission anisotropy ratio), various time-resolved spectra (emission and emission anisotropy ratio), various correlation spectra (fluorescence and IR), two-dimensional imaging (ultra high speed CCD video camera and Raman scope with microscope), etc.

The schematic diagram of PLASMA IV is shown in Fig. 6. PLASMA IV consists of five level divisions for optical spectroscopic measurements as follows:

(1) Naked Eye Level (video graphics analysis)
(2) Microscope Level (various microscope analysis with CCD graphics and spectroscopy)
(3) Light Scattering Level (two-dimensional static and dynamic light scattering analysis)

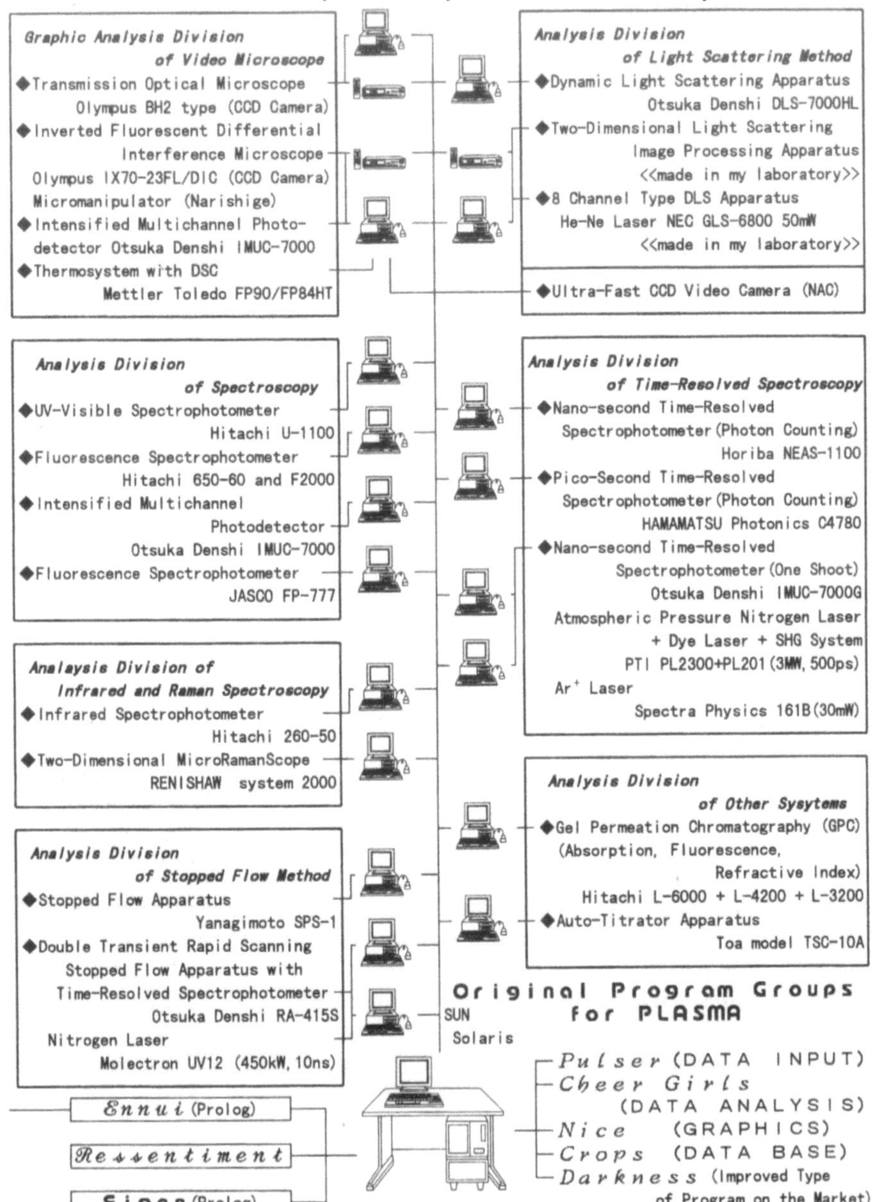

Fig. 6. Schematic diagram of PLASMA IV

(4) Molecular Level (various spectroscopic analysis with absorption, action, emission, and depolarization spectra, pico- and nanoseconds time-resolved spectroscopic techniques)
(5) Functional Group Level (image Raman scope with microscope)

PLASMA is controlled by about two hundred of our original program softs with about three thousands lines. Our original program softs of PLASMA can be divided into five divisions as follows:

(1) Division of Data Input (called "Pulser")
(2) Division of Data Analysis (called "Cheer Girls")
(3) Division of Graphics (called "Nice")
(4) Division of Data Base (called "Crops")
(5) Division of Improved Type of Program on the Market (called "Darkness")

In the future, our computer system of PLASMA will gradually expand from Turbo PASCAL on MS-DOS to Delphi on WINDOWS and C++ on UNIX. I would like to emphasize that it is important to use MS-DOS in order to control various apparatus because we cannot easily get in to WINDOWS and UNIX. I dislike complicated OS.

3.1
Experimental Spectroscopic Techniques for Naked Eye Level

The optical spectroscopic measurement for naked eye level in PLASMA is shown in Fig. 7. In this section, I will try to explain this measurement technique using the discussion based on the temperature-induced volume phase transition behavior of poly(N-isopropylacrylamide) (PNIPA) gels [8].

An average PNIPA gel cannot shrink quickly at a critical temperature (about 34 °C) because its transition phenomenon needs some time in order to discharge water from the PNIPA gel. In this work, temperature-induced volume phase transition processes of PNIPA gels were measured by a high vision CCD video camera, and these monitored graphics were analyzed by our original software system. We proposed our original measurement technique in order to analyze the volume changes and whitening processes based on computer graphics, and briefly discussed the relationship between shrinking and whitening processes in temperature-induced volume phase transition behavior of PNIPA gels. The samples (diameter: 2–6 cm, thickness: 0.1–0.5 mm) were synthesized and put into our original cell based on the experimental apparatus in Fig 7. In order to estimate the shrinking and whitening processes of PNIPA gels at near critical temperature, the obtained CCD video pictures were transformed into numerical values in a V-RAM system. The light intensities on V-RAM (RGB: 320 × 200 dots) were divided into 0–255 values based on the imaginary V-RAM system exploited by us originally [9]. As an example, the graphics of contour lines for PNIPA gel forms (T = T_c + 1.0 °C, diameter: 42.0 mm, thickness: 0.40 mm) photographed by a high-vision CCD video camera are shown in Fig. 9 together with the graphics of a two value system. The distribution map of light intensity within the closed curve line on the graphics of the two value system is

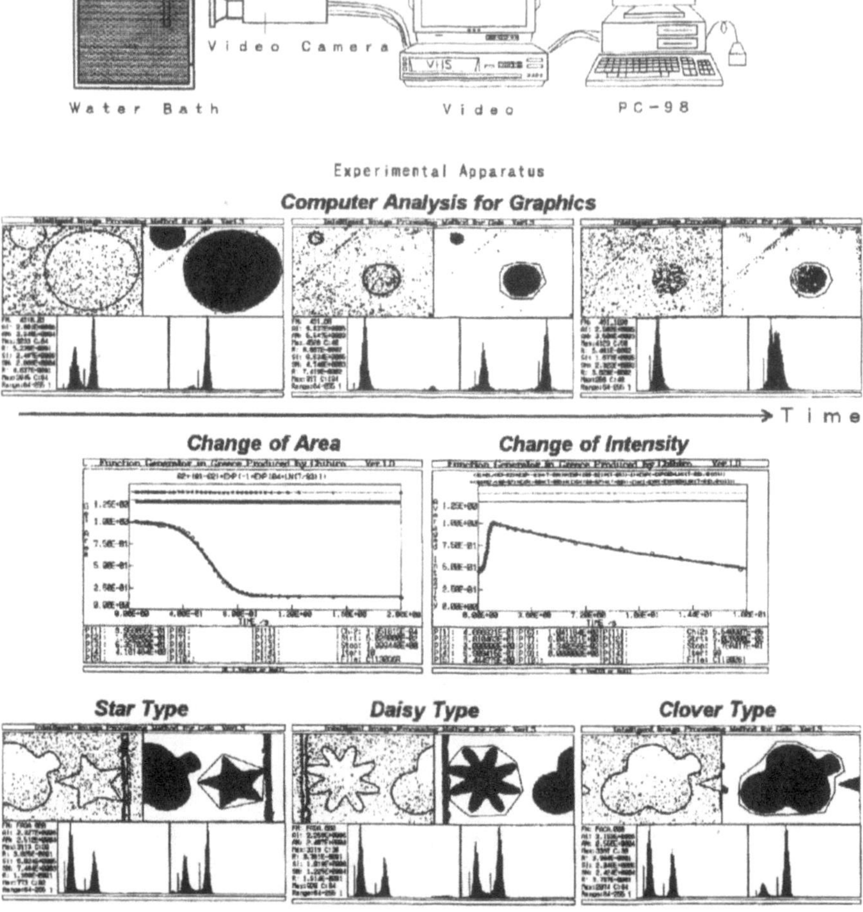

Fig. 7. Optical spectroscopic measurement at the naked eye level

shown on the right side of the bottom position of each of the graphics shown in the upper position of Fig. 7. The plotted graphs for time (min) vs. normalized gels area, S, and normalized integrated light intensity, I, in the shape change process of PNIPA gels at a near critical temperature are shown in the middle of Fig. 7.

We can always estimate every curve by the use of our PLASMA software system. We can choose functions of the type in Eq. 7 as plots for time vs normalized gels area, S, in various functions.

$$S = S_\infty + (S_0 - S_\infty) \exp\left(-\left(\frac{t}{\tau}\right)^\beta\right) \tag{7}$$

where S_0, S_∞, and τ are normalized gel areas at initial and final stages, and the averaged time of collapse of the skin layer in the shape change process of PNIPA gels, respectively. In the case of the stretched exponential function, the physical meaning of parameter β is very important, and the β will mean the sharpness of the skin layer collapse. On the other hand, we can obtain the function as in Eq. 8 as plots for time vs. normalized integrated light intensity, I, based on the continuous reaction equation in the field of kinetics, that is, we assume that the whitening process proceeds in both the skin layer and the inner part of PNIPA gels.

$$I = W_{skin}\left(B_0 + B\exp\left(-\left(\frac{t}{\tau}\right)^\beta\right)\right) + W_{inner}\left(B_0 + B\left(1 + \exp\left(-\left(\frac{t}{\tau}\right)^\beta\right)\right)\right) \tag{8}$$

$$W_j + A_j\left(\frac{k_{1j}}{k_{2j} - k_{1j}}\right)\exp\left(-k_{2j}t\right)\left(\exp\left((k_{2j} - k_{1j})t\right) - 1\right) \tag{9}$$

where k_{ij} is the rate constant in the whitening process of volume phase transition of PNIPA gels. The suffix i means the whitening process, 1, and the transparent process, 2, and the suffix j means the part of the skin layer and inner part, respectively. Using this measurement technique of PLASMA, various complex phenomena in naked eye level can be transformed into some phenomelogical functions via numerical values. We are now trying to discuss the mechanism of shape effect at the volume phase transition process of PNIPA gels using our original technique. The monitored CCD video pictures of star, daisy, and clover types of PNIPA gels at a near critical temperature are shown at the bottom of Fig. 7. Can you point out which shape of PNIPA gels will shrink fastest at a near critical temperature?

3.2
Experimental Spectroscopic Techniques for Microscope Level

The optical spectroscopic measurement for microscope level in PLASMA is shown in Fig. 8. In this section, I would like to try to explain this measurement technique using a discussion about the judgment of the living or dead state of an amoeba under a microscope.

In Fig. 8, microscope graphics on the left- and right-hand sides show the living and the dead state of Amoeba Limax, respectively. Hence each computer graphic measured by a microscope CCD video camera in the upper, middle, and bottom positions shows three-dimensional graphics for transmittance of light, fractal dimensional analysis, and two dimensional Fourier transformational analysis for the microscope image, respectively. It is very difficult to judge the living or dead state based only on numerical values of the shape of an amoeba in these graphics. Therefore I have devised two-dimensional fractal analysis and two dimensional Fourier transformational analysis methods based on the

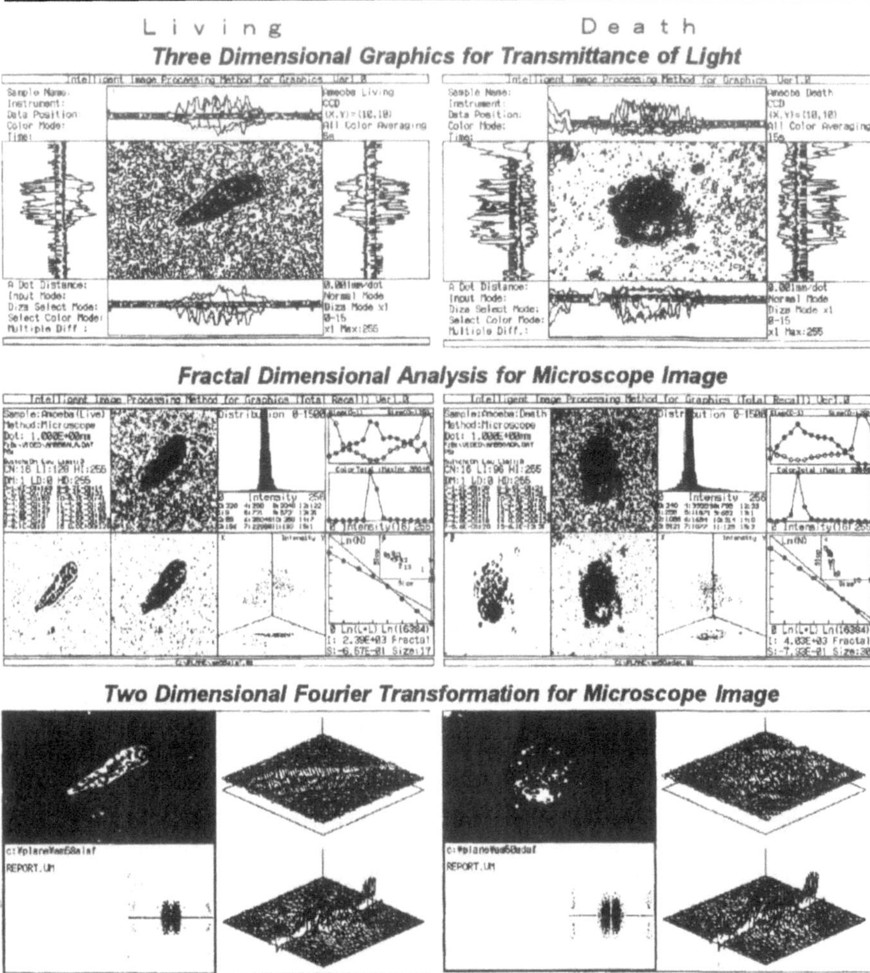

Fig. 8. Optical spectroscopic measurement at the microscope level

transmittance of light in microscope photographs in order to easily distinguish the dead state from the living state of an amoeba. In our analysis, we could not distinguish the dead state from the living state of an amoeba using the two dimensional Fourier transformational analysis method., but we were able to differentiate between these states of an amoeba using the two-dimensional fractal analysis method. Here we were able to obtain the values of fractal dimension D = 1.304 (living state) and D = 1.586 (dead state), that is, the distribution of nutritious materials within a dead cell differed from that within a living cell of an amoeba. In other words, the movement of nutritious materials within dead and living cells of an amoeba show white and pink noise behavior, respectively. Gods know the meaning of these absolute numerical values!

3.3
Experimental Spectroscopic Techniques for Light Scattering Level

The optical spectroscopic measurement for light scattering level in PLASMA is shown in Fig. 9. In this section, I will try to explain this measurement technique using the discussions based on two dimensional light scattering analysis of critical nonionic micellar solution immediately after shear quench [9–10] and

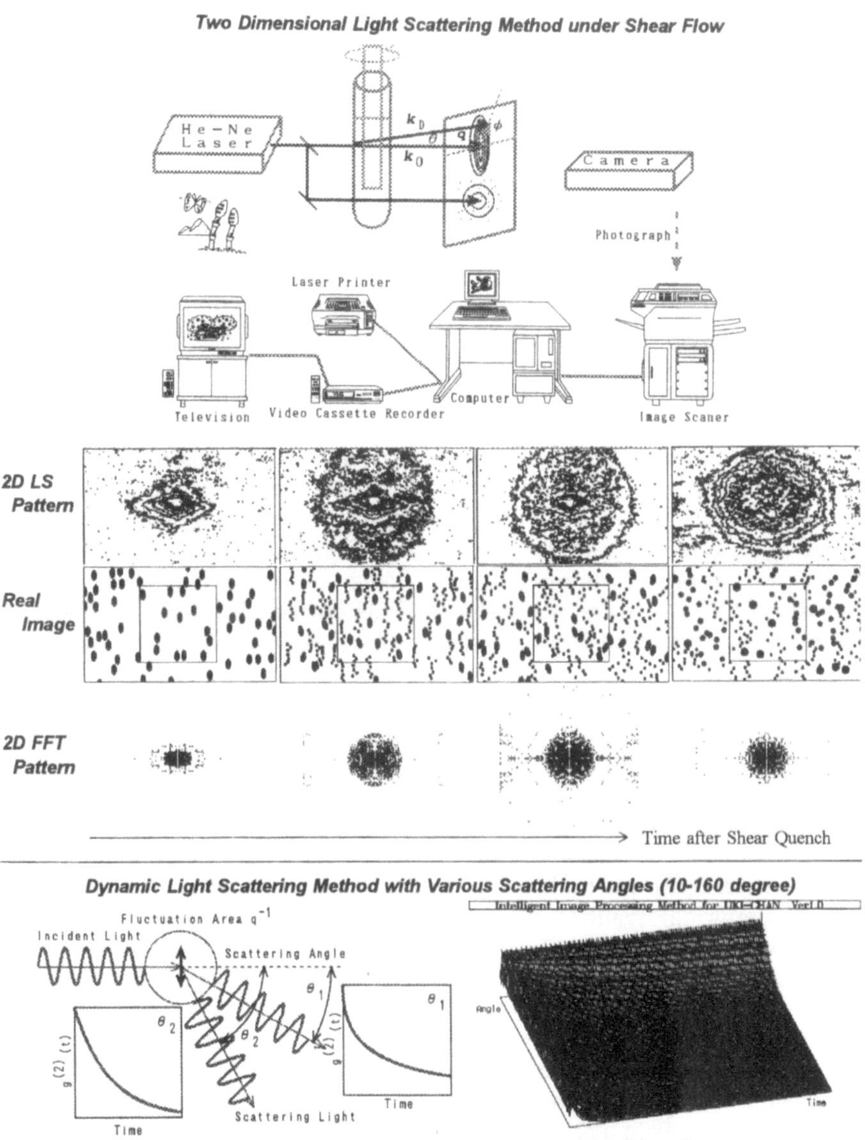

Fig. 9. Optical spectroscopic measurement at the light scattering level

dynamic light scattering analysis at various scattering angles (10–160°) of aqueous PNIPA solution at a near critical temperature [11].

The sample was aqueous 2.2 wt% tetraethylene glycol n-dodecyl ether ($C_{10}E_4$: $C_mH_{2n+1}O(CH_2CH_2O)_nH$, m = 4, n = 4). The apparatus system with two dimensional light scattering –video type image processing analysis method for critical fluid under shear flow – is shown in the upper part of Fig. 9. The incident light (He-Ne laser) is scattered in a quartz sample cell with a rotational viscometer, and the scattering light can be visualized by the use of projection onto a screen. The two dimensional patterns of critical fluid after shear quench are monitored by a high vision CCD video camera. These video picture data are transformed into digital signals as graphic computer information using our original program software (Nice). The graphics data on V-RAM in a personal computer are calculated by many of our original program software (Cheer Girls). The four graphics for two dimensional light scattering – video type image processing analysis (two dimensional light scattering pattern, real image, two dimensional Fourier transformational pattern) in critical $C_{10}E_4$/water mixtures after shear quench ($\gamma = 20\ s^{-1} \to 0\ s^{-1}$) at T = T_c + 18 mK are shown in the middle of Fig. 9. These graphics data are at t = 3 s, 22 s, 33 s, and 158 s after shear quench in the following order from left to right, respectively. It is well-known that the degree of ellipticity on two dimensional light scattering pattern increases with increasing shear rate. It is very important to discuss the difference of two dimensional light scattering patterns in the range of t_{after} = 10–50 s after shear quench because these two dimensional light scattering patterns obviously show the morphological behavior of $C_{10}E_4$/water domains in the process as it comes to an equilibrium. According to the time-dependence of the two dimensional light scattering pattern in critical fluid after shear quench shown in Fig. 9, the ellipsoidal form on two dimensional light scattering pattern is held until about t_{after} = 13 s, and the new pattern generates like a rough road for about 18–33 s. This result demonstrates the fact that the new pattern appears when the shear rate of the solution as a whole decreases below a certain rate. The final two dimensional light scattering patterns can be attained to a spherical one via a sunny-side up type. Although it is very difficult to transform from a two dimensional light scattering pattern to a two dimensional real image, I propose a new excellent technique called "Image Production" based on matching table dictionary between a real image and the two dimensional Fourier transformational pattern in order to clarify the morphological formation process of critical $C_{10}E_4$/water mixture after shear quench. In this experimental system, the important qualities of two dimensional light scattering – image processing analysis are to find out the model functions of orientation, form and distribution, and broken pieces of domains. With a change in the orientational distribution of droplets, the two dimensional Fourier transformed image pattern changes in the following sequence: ellipse type → horizontal butterfly type → circle type → vertical butterfly type → circle type. Using our "Image Production" technique, we were finally able to obtain the real images shown in the middle of Fig. 9. In a previous paper [9–10], I proposed a new method that used a two dimensional light scattering–video type image processing analysis with a semi-empirical self-consistent field method which operates as follows:

(1) measurement of two dimensional light scattering photographs
(2) data transformation from CCD video pictures to imaginary V-RAM system
(3) analysis of elementary functions (Guinier, Zimm, and Debye plots)
(4) analysis of Onuki's function [12]
(5) analysis of Hashimoto's function [13]
(6) (Image Production ↔ two dimensional Fourier transformed check process) repeat
(7) two dimensional real image.

From this analysis, we were able to determine that a weak stretched sphere of oil was torn to pieces immediately after shear quench, and that these small oil spheres gradually gathered together again.

Until now the dynamic behavior of condensed matters at a near critical point has been measured by the dynamic light scattering method. In order to obtain much information about dynamic behavior at a near critical point, I would like to suggest a new excellent technique, i.e. a two dimensional dynamic light scattering method using an ultra high speed CCD video camera. As a pre-experimental method for this new technique, we have proposed a dynamic light scattering image processing method [11]. The concentration of PNIPA was kept at less than 0.1 % at the final step of preparation, and the sample was set up in the thermostat with an accuracy of about 10 mK. The recorded 151 $g^{(2)}$ (t) decay curve data for dynamic light scattering of an aqueous PNIPA solution at a near critical temperature were converted into our personal computer system in PLASMA. The 151 decay curves of dynamic light scattering (light scattering angle: 10–160°) of the aqueous PNIPA solution were calculated by our original program software system "Cheer Girls" based on the non-linear least squares method with Marquardt's method. The three dimensional graphics for X/Y/Z = time (μs)/light scattering angle/$g^{(2)}$ (t) intensity of aqueous PNIPA solution at a near critical temperature is shown at the right-hand side of the bottom part of Fig. 11. In our experiment we were able to establish that all $g^{(2)}$ (t) decay curves were fitted by a stretched exponential function. If the relationship between parameters (β and τ values) of stretched exponential function and various light scattering angles is examined in detail, we are able to understand the relationship between size (space) and dynamic property (time) in condensed matters at a near critical point.

3.4
Experimental Spectroscopic Techniques for Molecular Level

The optical spectroscopic measurement for molecular level in PLASMA is shown in Fig. 10. In this section, I will try to explain this measurement technique using the discussion based on an image processing stationary spectroscopic method for the structural phase transition behavior of liquid crystals [14].

Using a general fluorescence spectrophotometer as shown at the top of Fig. 10, we can easily obtain spectral data based on excitation spectra, emission spectra, emission anisotropy ratio spectra, and static light scattering spectra

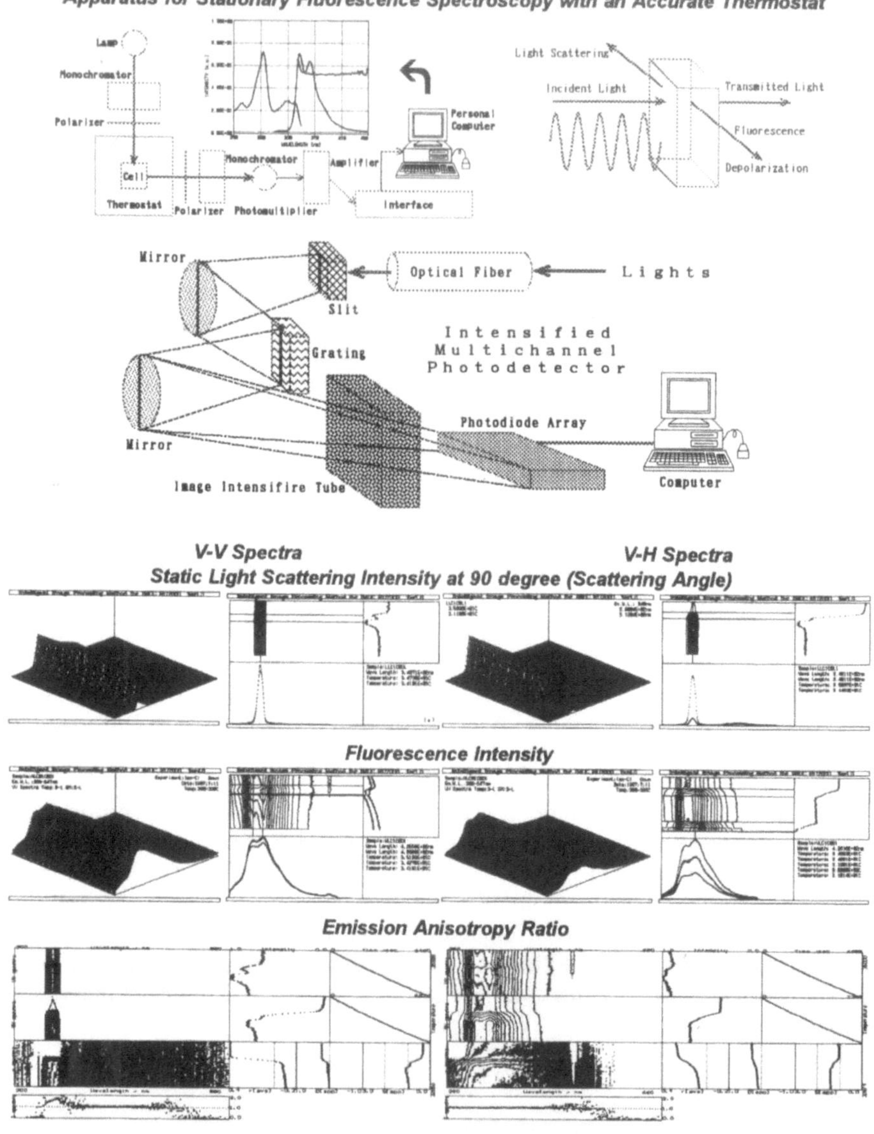

Fig. 10. Optical spectroscopic measurement at the molecular level

(scattering angle: 90°). As we have recently obtained an apparatus of an intensified multichannel photodetector shown in the middle of Fig. 10, we can now easily obtain a large amount of spectra data at various strictly controlled temperatures using our original thermocontroller (resolution temperature: approx. 3 mK, heating rate: 0.3 mK/s). The sample was 9,10-diphenylanthracene (DPA) doped in cholesteryl oleyl carbonate (COC). The liquid crystal phase transition

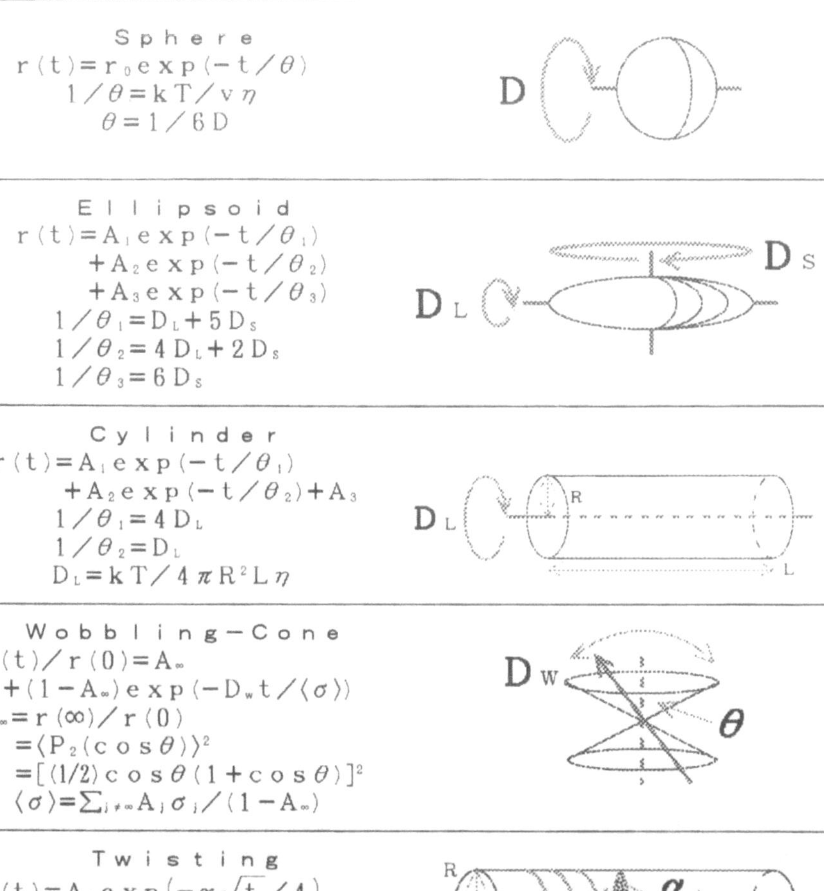

Fig. 11. Model functions for simple rotational motions. (r: emission anisotropy ratio, θ: fluorescence rotational correlation time, k: Boltzmann constant, T: absolute temperature, V: volume of spherical body, η: solvent viscosity, D: rotational diffusion coefficient, α: angle between long axis of cylinder and emission transition moment of a chromophore, R: radius of cylinder, L: length of cylinder, Y: molecular elasticity)

temperatures of COC are known to be at about 17 °C (smectic A phase → cholesteric phase) and 34 °C (cholesteric phase → isotropic phase) [15]. The three dimensional and contour graphics for static light scattering spectra of COC at 90 °C, and fluorescence and emission anisotropy ratio spectra of DPA in COC from 31° to 35 °C are shown at the bottom of Fig. 10. In the spectral graphs for static light scattering and fluorescence shown at the bottom of Fig. 10, VV (left side) and VH (right side) spectra mean parallel-parallel and parallel-perpendicular directions of incident and emitted light, respectively. The anisotropy ratio spectra for static light scattering and emission are shown in the left- and right-hand sides of the bottom of Fig. 10, respectively. In these three dimensional graphics, the X, Y, and Z axes are wavelength (nm), temperature from 31–34 °C, and each intensity, respectively.

It is very difficult to measure and analyze the structural phase transition behavior of a liquid crystal using the fluorescence probe method because the experimental condition of liquid crystals creates some problems based on the muddiness and the anisotropy of domains. However, our measurement technique may be able to overcome these problems. Using this technique, the structural phase transition behavior may be clarified based on the measurement of each dynamic orientation process of the domains and the molecules in liquid crystals.

In experimental spectroscopic techniques of molecular level, the time-resolved emission spectroscopic technique is very renowned as a motional fluorescence probe method. The time-resolved fluorescence depolarization method, in particular, is an excellent technique for clarifying the mechanism of molecular or macromolecular dynamic behavior in condensed matter. Therefore, I would like to explain the time-resolved fluorescence depolarization method more fully in a later section.

3.5
Experimental Spectroscopic Techniques for Functional Group Level

Until now there has been no effective experimental data available using an image Raman scope with a microscope in order to discuss "Complex Macromolecular Chain Dynamics". Now we have research plans for phase separation processes in polymer alloy, volume phase transition processes of gels, morphological change of micelle, etc. In the near future, we hope to start the study of complex fluids using optical spectroscopic measurements for the functional group level. I hope that postdoctoral students in my laboratory will take an active part in this research technique.

4
Overview of Fluorescence Probe Methods

The fluorescence probe method is an excellent technique in many research fields of physics, chemistry, biology, and medicine for understanding the microscopic behavior of condensed matter. This method has some useful properties such as ultra high sensitivity (clean), molecular labeling (molecular spy), ultra wide dynamic range (strong experimental tools), many measurement channels (many informations), etc. In the near future, this method may become more developed than the NMR method in scientific research fields, but at the moment it suffers from problems such as complex data, little generality, much difficulty, etc.

After light is absorbed in molecules, excited molecules relax their energy via various photophysical and photochemical reaction processes [1]. Therefore it is by using the molecular probe method that we find out well our objective physical amount by using these relaxation processes skillfully. Various interaction processes between radiation and molecular fields are well-known as one-photon processes (absorption and emission), two-photon processes (Rayleigh scattering, Raman scattering, and two-photon absorption), three-photon processes [second harmonic generation (SHG) and hyper-Raman scattering], and four-photon processes [coherent anti-Stokes Raman scattering (CARS)] [1]. On the other hand, there are various photophysical processes (fluorescence, thermal deactivation, phosphorescence, S-S and T-T energy transfers, excimer and exciplex formations, etc.) and photochemical processes (photoisomerization, photodimerization, photocyclization, photoaddition, photosubstitution, photoscission, photooxidation, photoionization, photopolymerization, etc.) of excited state molecules after the absorption process [1]. From my above-mentioned viewpoint, I would like to emphasize that it is important to take into account the photophysical and photochemical processes of molecules after the absorption process when using the general fluorescence probe method.

In our previous book [1], we proposed that the fluorescence probe techniques could be classified into three methods, a micro-environmental fluorescence probe based on twisted internal charge transfer, a motional fluorescence probe based on triplet-triplet energy transfer or excimer formation or fluorescence depolarization, a and micro-structural fluorescence probe based on excimer formation or intramolecular charge transfer of macromolecules.

4.1
The Micro-Environmental Fluorescence Probe Method

In the initial stages of the molecular probe method, 1-anilinonaphthalene-8-sulfonate (ANS) has become very well-known as a fluorescence probe. The emission spectra, emission energy, quantum yield, and $E_T(30)$ of 1-anilino-naphthalene-8-sulfonate (ANS) change in various organic solvents [16–17]. The peak wavelength of ANS gradually red-shifts with increasing solvent polarity, and the fluorescence intensity of ANS gradually decreases with the increasing solvent polarity. Using these properties of ANS, we can discuss the micro-environmental state around a probe molecule.

The photophysical mechanism of ANS is explained as follows [1]. The electronic interaction between the non-bonding electron on the alkylamino group and the p-electron molecular orbital on the aromatic ring depends on the angle between the C_A and N atoms of the ANS molecule, that is, the fluorescence spectra of ANS depend on the rotational motion around the C_A-N bond. Therefore, the fluorescence spectrum of ANS is strongly influenced by solvent polarity and viscosity based on an electronic interaction of molecular orbitals. In recent years, this sort of chromophore has been called a twisted intramolecular charge transfer (TICT) probe.

4.2.
Motional Fluorescence Probe Methods

We can discuss the local motions and self-diffusion processes around a probe by the use of a motional fluorescence probe. This probe technique is divided into three methods, singlet-singlet and triplet-triplet energy transfers, excimer formation processes, and fluorescence depolarization. In this section, I will explain the motional fluorescence probe method based on energy transfer and excimer formation processes.

There is a triplet-triplet energy transfer process between benzil B (donor) and anthryl A (accepter) groups. Horie and Mita synthesized some polystyrenes having a chromophore (anthryl or benzil group) at the end or center position of the polymer chains. They were first able to measure the DP (degree of polymerization)-dependence on the diffusion-controlled polymer-polymer reaction rate in various solutions using this technique. They measured the triplet lifetime of PS-B in various concentrations of PS-A solution, and obtained the quenching constant k_q between PS-B and PS-A using the Stern-Volmer relationship [18]. Hence the DP-dependence on the diffusion-controlled intramolecular end-to-end reaction rate in various solutions was measured based on delayed fluorescence technique by Ushiki, Horie, and Mita [19].

Up until about ten years ago, the excimer formation and energy migration processes between polymer side chains were studied by many researchers. In the epoch of the oil shock, many researchers in the field of photo-science had probably considered making use of sunlight as the energy source in daily life. By the use of this probe method, the conformational change behavior of polymer side chains had been studied in detail by a few researchers [20]. The relationship between conformational change and configuration of polystyrene oligomers had been clarified by the experimental measurement of time-resolved excimer formation processes. From these results, the schematic diagram of excimer formation processes in each configuration of polystyrene oligomers was analyzed. Hence the DP-dependence on the diffusion-controlled intramolecular end-to-end reaction rate in various solutions was measured based on excimer formation processes by Yamamoto's group and Winnik's group [21].

It is well-known that the strongest tool in motional fluorescence probe methods is a time-resolved fluorescence depolarization method in order to investigate molecular and macromolecular dynamic behavior. In a later section, I will briefly comment on this method.

4.3
The Micro-Structural Fluorescence Probe Method

The micro-structural fluorescence probe method is an excellent technique when we discuss the molecular orientational state in an amorphous solid phase. As a result, this probe method has often been used for the research of polymer alloy.

Chromophore groups (for example, naphthyl, pyrenyl, carbazolyl, etc.) attached to polymer side chains are able to form various sorts of excimer states (for example, intramolecular adjacent excimer forming site, intramolecular non-adjacent excimer forming site, and intermolecular excimer forming site). Frank has measured the ratio of dimer to monomer emission intensities of naphthyl groups in poly(2-vinylnaphthalene) (P2VN) or poly(acenaphthylene) (PAN)-various polymer mixtures [22]. In their plotted graphs for the emission intensity ratio between dimer and monomer of the naphthyl group vs the gap of the Hildebrand solubility parameter between guest (P2VN or PAN) and host (various polyalkyl methacrylates) polymer, these ratios for dimer to monomer emission intensities of naphthyl groups increase with the increasing absolute values for the gap of the Hildebrand solubility parameter between guest and host polymers. This experimental result demonstrates that a guest polymer can solute as a macromolecular level in host polymers at near $\sigma_{host} - \sigma_{guest} = 0$. In other words, their experiment proves that the Flory–Huggins thermodynamics theory can be applied to the microscopic class (molecular level).

5
Trend of the Time-Resolved Fluorescence Depolarization Method

It is well-known that the time-resolved fluorescence depolarization technique is the most excellent technique among fluorescence probe methods in order to discuss "Complex Macromolecular Chain Dynamics" based on molecular level. In this section, I would like to discuss the trend of this method within the limited space available.

The excited transition probability of a molecule which absorbs incident light is shown as a function of the angle α between directions of its absorption transition moment and electric vector of incident light, that is, $E(\alpha) = E_0 \cos^2 \alpha$. Therefore the fluorescence which is emitted from molecules with an opto-anisotropic field ought to be observed as remarkably polarized light. However, we cannot observe this polarized light in a general optical system. The origin of fluorescence depolarization phenomena was classified into four categories: the gap between absorption and emission transition moments, intermolecular energy transfer, Brownian motion, and non-uniformity of fluorescent molecular orientation. Consequently it is by the fluorescence depolarization method that we measure the degree of fluorescence depolarization and analyze four physical factors based on the origin of fluorescence depolarization phenomena in this experimental system. If there is not intermolecular energy transfer in an experimental system, we can extract only Brownian motional effect from the obtained experimental emission anisotropy ratio data. In recent years, this technique has been given attention in order to measure the Brownian motion of molecules and macromolecules as this method has some useful properties, such as high sensitivity, molecular labeling, and real time measurement.

As we can easily obtain the recent apparatus for measuring time-resolved fluorescence depolarization, time-correlated single photon counting and time-wavelength correlated streak scope methods, compared with about ten years ago, much research on "Complex Macromolecular Chain Dynamics" will be reported in the near future based on the time-resolved fluorescence depolarization technique.

5.1
Epoch of Establishment of Relationship Between Fluorescence Depolarization and Brownian Motion

When Perrin proved the existence of molecules by the use of the fluorescence depolarization technique, the study of the micro-Brownian motion of a molecule had started. If we suppose that a molecule having an angle ω between the absorption and emission transition moment rotates through an angle δ based on Brownian motion, the emission anisotropy ratio r is given by Eqs. 10 and 11.

$$r = Z \text{ (depolarization term of inner origin)}$$
$$\text{(depolarization term of outer origin} \tag{10}$$

$$r = Z \left(\frac{3\cos^2 \omega - 1}{2} \right) \left(\frac{\overline{3\cos^2 \delta - 1}}{2} \right) \tag{11}$$

Without a Brownian motional effect, the value of the emission anisotropy ratio r exists in $0.4 \geqq r_0 = Z$ (depolarization term of inner origin) $\geqq -0.2$ as the angle ω can change within $0 - \pi/2$. With a Brownian motional effect, the time-dependence on the emission anisotropy ratio of a rigid sphere is given by Eq. 13 via Eq. 12 based on the Brownian motion of a rigid sphere:

$$\overline{\cos^2 \delta}\, \frac{1}{3} + \frac{2}{3\rho} \exp\left(-\frac{3t}{2}\right) \tag{12}$$

$$r(t) = r \exp\left(-\frac{t}{\theta}\right) \qquad \theta = \frac{\rho}{3} = \frac{1}{6D} \tag{13}$$

where ρ, θ, and D are the relaxation time of rotation, the fluorescence rotational correlation time, and the rotational diffusion coefficient, respectively. Therefore the emission anisotropy ratio induced by stationary light is shown by Eq. 15 using $1/\theta = kT/V\eta$:

$$r = \frac{\displaystyle\int_0^\infty I(t)\, r(t)\, dt}{\displaystyle\int_0^\infty I(t)\, dt} = \frac{\displaystyle\int_0^\infty I_0 \exp\left(-\frac{t}{\tau}\right) r_0 \exp\left(-\frac{t}{\theta}\right) dt}{\displaystyle\int_0^\infty I_0 \exp\left(-\frac{t}{\tau}\right) dt} \tag{14}$$

$$\frac{1}{r} = \frac{1}{r_0}\left(1 + \frac{\tau}{\theta}\right) = \frac{1}{r_0}\left(1 + \frac{kT}{V\eta}\right) \tag{15}$$

where k, T, V, and η are the Boltzmann constant, the absolute temperature, the volume of the spherical body of the molecule, and the solvent viscosity, respectively. Eq. 15 is known as the Perrin-Weber plots [23].

5.2
Epoch of Establishment of the Time-Resolved Fluorescence Depolarization Method

After the experimental measurement technique for time-resolved fluorescence depolarization had been devised, researchers in the field of biophysics began to discuss non-exponential emission anisotropy ratio decay curves. Between 1960 and 1980, many models (sphere, ellipse, and cylinder) with double or triple exponential type function emission anisotropy ratio decay curves were proposed by many authors [24]. These model functions are listed in Fig. 11.

5.3
Epoch of Discussion About Restricted Rotational Motion

In the decade 1970–1980, research of biomolecular dynamics was further extended by a number of investigators using ERS and NMR techniques. As biomacromolecules have complicated structures, their biomacromolecular chain dynamics processes are very complex. Naturally, the measured time-resolved emission anisotropy ratio decay curves of chromophores in condensed biomolecules showed various non-exponential functions [25]. In this time, the researchers in the field of biophysics picked up some important information about intramolecular motions of biomacromolecules from these non-exponential function type emission anisotropy ratio decay curves. As a result, some models (wobbling-cone and twisting) for emission anisotropy ratio decay curves based on restricted rotational motion of a chromophore in a biomembrane and twisting motion of DNA were proposed. These model functions are listed in Fig. 11.

5.4
Epoch of Discussion About Local Motion Polymer Chain Segment

The discussion about local motions of polymer main chains using time-resolved emission anisotropy ratio decay curves was first ignited by Monnerie's group [26]. The time-resolved emission anisotropy decay curves of a chromophore incorporated into a polymer main chain are very complex, that is, these curves exhibit non-exponential functions. Between 1970–1990, many researchers in the field of polymer chemical physics became interested in the discussion about the segmental motion of a polymer due to the fact that Morawetz proposed the existence of crankshaft motions in a polymer [27]. Hence the researchers in the field of polymer dynamics were influenced by some new techniques based on the developments of ESR and NMR methods for biomacromolecules. After it became possible to discuss crankshaft motions of polymer main chains, research about time-resolved emission anisotropy decay curves of a chromophore incorporated into a polymer main chain were experimentally reported by the groups of Monnerie, Yamamoto, Ushiki, and Ediger.

In this epoch, the experimental researchers for segmental motions of polymer main chains were interested in finding an exact model based on an autocorrelation orientational function of polymer chains. The model functions for the local motion of polymer chain segments are listed in Fig. 12, where WC, WH, IR, HH, GDL, VJGM, JS, BY, and Woessner are the wobbling-cone model, the Weber and Helfand model, the single exponential model, the Hall and Helfand model, the Viovy, Monnerie, and Brochon model, the Valeur, Jarry, Geny, and Monnerie model, the Jones and Stockmayer model, the Bendler and Yaris model, and the Woessner model [23, 25, 26, 28], respectively. However, it was very difficult to find the best fit function in many proposed model functions for local motion of polymer chain segments. In this discussion about polymer chain dynamics, there was a very serious problem in which experimentalists had to estimate many parameters in proposed model functions from

Basic Function: $(1-A)\exp(-t/T_1)\cdot B+A$

	A	B
WC	$0\sim1$	1 
WH	$0\sim1$	$\exp(-t/T_2)\,I_0(t/T_2)$ 
IR	0	1 
HH	0	$\exp(-t/T_2)\,I_0(t/T_2)$ 
GDL	0	$\exp(-t/T_2)\{I_0(t/T_2)+I_1(t/T_2)\}$ 
VJGM	0	$\exp(t/T_2)\,\mathrm{erfc}(t/T_2)^{1/2}$ 
JS		$0.578\exp(-t/T_1)+0.422\exp(-t/T_2)$
BY		$\dfrac{1}{2}\sqrt{\pi}\,\dfrac{\mathrm{erfc}\sqrt{t/T_1}-\mathrm{erfc}\sqrt{t/T_2}}{\sqrt{t/T_2}-\sqrt{t/T_1}}$

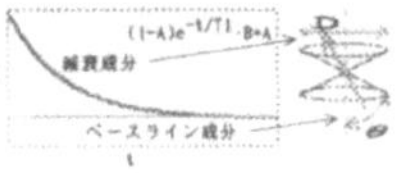

Woessner

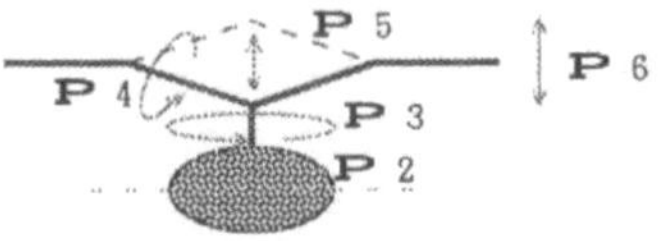

$$r(t)=P_1\{A_1+A_2\exp(-t/P_3)+A_3\exp(-t/P_3)\}\exp(-t/P_4)\{0.578\exp(-t/P_5)+0.422\exp(-t/P_6)\}$$
$$A_1=(1/4)(3\cos^2 P_2-1)^2,\quad A_2=(3/4)\sin^2 P_2,\quad A_3=(3/4)\sin^4 P_2$$

Fig. 12. Model functions for local motion of polymer chain segment. (A: pre-factor, T: relaxation time, $I_i(t/T)$: Bessel function, $\mathrm{erfc}(t/T)$: error function)

obtained data based on time-resolved emission anisotropy ratio decay curves of a chromophore incorporated into a polymer main chain. But theorists could easily introduce many parameters into these theoretical functions. As the result of this course of study, Monnerie's, Yamamoto's, Ushiki's and Ediger's groups had a difficult task trying to judge a best fit model function based on the time-resolved fluorescence depolarization method. I remember that at this time it was very important to understand the application limit of time-resolved fluorescence depolarization method.

5.5
New Epoch

5.5.1
Flow Chart of Fluorescence Probe Measurement in PLASMA

As mentioned above, we knew that it was very difficult to analyze the local motion of the polymer main chain by the use of only the χ^2 value between an obtained time-resolved emission anisotropy decay curve and a model function. From our experience during the previous discussion about polymer dynamics based on the time-resolved fluorescence depolarization probe method, I was able to establish a universal research process for the fluorescence probe measurement in PLASMA. In my laboratory, we analyzed the fluoresence and depolarization decay curves based on the flow chart of fluorescence measurement in PLASMA. In order to obtain more information about the time-resolved fluorescence probe method, I propose a new analysis method with a computer system. This method is performed by the following process:

(1) sample
(2) measurement of absorption spectra
(3) measurement of excitation spectra
(4) measurement of emission spectra
(5) measurement of emission anisotropy ratio spectra
(6) measurement of time-resolved emission spectra
(7) measurement of time-resolved emission anisotropy ratio spectra
(8) analysis of various χ^2-maps
(9) discussion [29].

5.5.2
χ^2-*Map Method in PLASMA*

In order to estimate the validity of the proposed model function on the basis of the relationship between our measured time-resolved emission anisotropy ratio decay curves and model functions, we propose a new calculation technique, named the "χ^2-map" method. Here I would like to explain the technique briefly. The number-average molecular weight, the molecular-weight distribution, and the content of the carbazolyl group of copoly(styrenevinylcarbazole) [P(St-Cz)] are $M_n = 2.5 \times 10^4$, $M_w/M_n = 1.84$, and 7.0% unit/polymer, respectively. The macromoleculer dynamics of a polymer side chain of the sample [P(St-Cz) in polyoxyethylene(PEO)/dichloroethane(DCE) = 7:3 mixture] were analyzed along the above-mentioned flow chart of fluorescence probe measurement in PLASMA. The time-resolved fluorescence and emission anisotropy ratio decay curves are shown in the upper part of Fig. 13. As each dynamic mode of the polymer side chain had been theoretically studied by Woessner as shown in

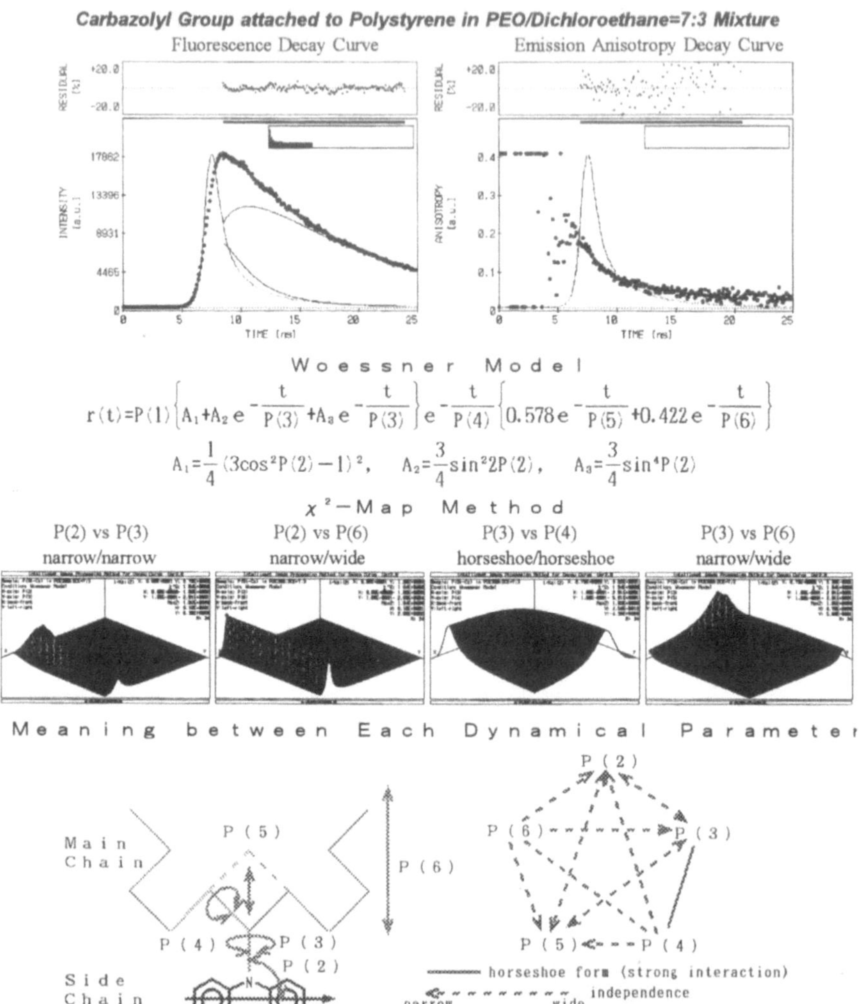

$$r(t)=P(1)\left\{A_1+A_2\,e^{-\frac{t}{P(3)}}+A_3\,e^{-\frac{t}{P(3)}}\right\}e^{-\frac{t}{P(4)}}\left\{0.578\,e^{-\frac{t}{P(5)}}+0.422\,e^{-\frac{t}{P(6)}}\right\}$$

$$A_1=\frac{1}{4}(3\cos^2P(2)-1)^2,\qquad A_2=\frac{3}{4}\sin^2 2P(2),\qquad A_3=\frac{3}{4}\sin^4 P(2)$$

Fig. 13. Schematic diagram of χ^2-map method for local motions of polymer side chains

Fig. 13, we adopted this model in order to separate each dynamic mode in the vicinity of the polymer side chain, where parameters P(1), P(2), P(3), P(4), P(5), and P(6) are the pre-factor, the angle between the excited transition dipole moment and the rotational axis of the chromophore attached to a polymer side chain, the rotational relaxation time of a polymer side and main chains, the relaxation time of the three-bond crankshaft and the tetrahedral lattice motions of a polymer main chain, respectively. The image for the physical meaning of each parameter in Woessner's model is shown on the left-hand side of the bottom position of Fig. 13. The fluorescence rotational correlation times of each dynamic mode of a polymer chain are calculated from our original program software system using the non-linear least-squares method with deconvolution based on Marquardt's method. Finally, the χ^2 values of 100x100 points are calculated in the vicinity of a stable point against a trial function, and the χ^2-map is shown in the computer graphics on a monitored vision calculated by our original PLASMA program. The χ^2-map for P(i) vs. P(j) is shown in the middle of Fig. 21. In these graphics, the Z-axis represents the reciprocal of the χ^2 values. For example, in the χ^2-map of P(2) vs. P(6), the Z-axis value is large within a very narrow region on the X-axis [P(2)] but the Z-axis value is large within a very wide region on the Y-axis [P(6)]. In other words, the result shows the fact that the P(6) parameter is meaningless but the P(2) parameter has a physical meaning in the relationship between our experimental data and the Woessner model function. On the other hand, the form of χ^2-map of P(3) vs. P(4) represents a horseshoe type. The result shows the fact that it is difficult to separate into two dynamic parameters in the relationship between our experimental data and the Woessner model function. The summary of the χ^2-map for each P(i) vs. P(j) is shown at the bottom of Fig. 13. All arrows gather at the dynamic parameters P(2), P(3), and P(5) in our experimental system. Consequently, the result shows that our measured time-resolved emission anisotropy ratio decay curves of P(St-Cz) in PEO/DCE = 7:3 mixture can be expressed by only three dynamic parameters P(2), P(3), and P(5) in the Woessner model function. By the use of our calculation technique, we can judge the best fit model function based on the time-resolved fluorescence depolarization method with a few restricted conditions.

5.5.3
Stretched Exponential and Hyperbolic Model Functions

At the next stage, we began to again determine an exact model for the auto-correlation orientational function of polymer chains because we were not able to give up the discussion of the local motion of polymer main chains based on the time-resolved fluorescence depolarization method. We tried to apply our χ^2-map method to the analysis of time-resolved emission anisotropy ratio decay curves of α, ω-dianthrylpolystyrene (APSA) and monoanthrylpolystyrene (PS-APS) in ethyl acetate/tripropionin = 6:94 mixture. Using our χ^2-map method, these experimental decay curves could be estimated by many theoretical models, single exponential, double exponential, JS, VJGM, HH, GDL, WH, WC, BY, KWW (Kohlraush, Williams, and Watts: stretched exponential), and hyperbolic functions. Here we determined that the auto-correlation functions based on time-resolved emission anisotropy ratio decay curves of APSA and PSAPS were given by a hyperbolic function as shown in Eq. 16 [30]:

$$r(t) = A \left(-\frac{t}{\tau} \right)^{-\beta} \tag{16}$$

In order to clarify the parameters τ and $\beta = (T - T_r)/T_r$, we proposed a new model, the "Intermittent Motion" model, for the local motion of polymer chains based on a fractal-time random walk theory. According to our model, the meaning of these parameters showed that τ and T_r are the effective orientational relaxation time and the characteristic activation energy of the polymer segments, respectively. In our new model, we predicted that the decay functions of the emission anisotropy ratio of a chromophore attached to a polymer main chain would transfer from a stretched exponential function to a hyperbolic function with an increase in its polymer concentration.

5.5.4
Fractal-Time Random Walk Theory

In this section, I would like to briefly discuss the "Intermittent Motion" model for the local motion of a polymer chain based on the fractal-time random walk theory. We presume that the orientational relaxation process of a polymer segment is caused by the reaction of a migrating solvent molecule in solution as shown in Eq. 17:

$$A + B \rightarrow A' + B' \tag{17}$$

where A, B, A', and B' are a polymer segment at an initial direction, a solvent molecule before reaction, a relaxed polymer segment, and a solvent molecule after reaction, respectively. Therefore the number of A, n_A is given by Eq. 18 as a function of time t:

$$\frac{dn_A}{dt} = - R_0 D n_A n_B \tag{18}$$

where n_j, R_0, and D denote the density of species i, the reaction radii, and the diffusion coefficient ($D = D_A + D_B$), respectively. Here we assume that n_A as a function of time is equivalent to the orientational auto-correlation function r (t) of the polymer segments.

$$n_A (T) = r (t) \tag{19}$$

The mean square distance of migration $\langle L^2 (t) \rangle$ of B is assumed to be given by Eq. 20 based on Tuac [31].

$$\langle L^2 (t) \rangle = Dt \tag{20}$$

If we presume that the orientation of a polymer segment relaxes after the segment is activated by the reaction with solvent molecules above the threshold E_c, the state density of the polymer segments is written as in Eq. 21:

$$\varrho (E) = \varrho_c \exp \left(- \frac{E - E_c}{k_B T_r} \right) \tag{21}$$

where E, ϱ_c, and k_B are the potential energy of a polymer segment, the state density at E_c, and the Boltzmann constant, respectively. T_r ($= E_a/k_B$) corresponds to the characteristic activation energy of polymer segments. According to Orenstein [32], the interval time t_w for the orientational relaxation of a segment at energy E can be expressed as in Eq. 22:

$$\frac{1}{t_w} = \nu_{ph} \exp\left(-\frac{E - E_c}{k_B(T_r - T_r)}\right) \tag{22}$$

where ν_{ph} and T indicate the frequency of the photon and the temperature, respectively. On rearranging Eq. 22, we can obtain:

$$-\frac{E - E_c}{k_B} = (T - T_r)\ln\frac{1}{t_w \nu_{ph}} \tag{23}$$

Eq. 21 is rearranged as follows:

$$\varrho(E) = \varrho_c\left(-\frac{1}{t_w \nu_{ph}}\right)^{\frac{T - T_r}{T_r}} \tag{24}$$

Here the distribution function $\psi(t_w)$ of the interval times for relaxation is related to the state density $\varrho(E)$ as shown in Eq. 25:

$$\varrho(E)\,dE \sim \psi(t)\,dt_w \tag{25}$$

On integrating both numbers of Eq. 25 and substituting Eq. 24 into Eq. 25, we obtain Eq. 26:

$$\psi(t) \sim t_w - 1^{-\frac{T - T_r}{T_r}} = t_w - 1 - \gamma \tag{26}$$

When $\psi(t_w)$ is expressed by Eq. 26, the mean square distance of migration $\langle L^2(t)\rangle$ of a species B is given by Eq. 27 [33].

$$\langle L^2(t)\rangle \sim t^\gamma \tag{27}$$

Substituting the above equation into Eq. 20, D is expressed as a function of time:

$$D \sim t^{\gamma - 1} \tag{28}$$

Therefore Eq. 18 is transformed as

$$\frac{dn_A}{dt} = -\frac{\xi_0}{t^{1-\gamma}}n_A n_B \tag{29}$$

where ξ_0 is the minimum radii of reaction between A and B at T. The representation of our theory is shown at the bottom of Fig. 14. According to Tauc [31], we can solve Eq. 29 under two conditions of $n_A \sim n_B$ and $n_A \ll n_B$. If we should solve Eq. 29 with the conditions of $n_A \sim n_B$ and $n_A \ll n_B$, we can obtain n_A as hyperbolic and stretched exponential functions of time t, respectively.

$$n_A(t)\frac{n_{A0}}{1 + \dfrac{\xi_0 n_{A0}}{\gamma}t^\gamma} \sim \frac{\gamma}{\xi_0}t^{-\gamma} \tag{30}$$

$$n_A(t) = n_{A0}\exp\left(-\frac{\xi_0 n_{B0}}{\gamma}t^\gamma\right) \tag{31}$$

where n_{A0} and n_{B0} are the density of n_A and n_B at $t = 0$, respectively. Finally, we can apply Eqs. 30 and 31 and obtain two relationships between the parameters in this manner as follows,

$$\beta = \gamma \tag{32}$$

$$\xi_0 \sim \frac{\beta}{r_0\,\tau_{\mathrm{eff}}^{\beta}} \tag{33}$$

5.5.5
Intermittent Motion of Polymer Chain

In our theoretical analysis, we obtained a power law type orientational auto-correlation function for polymer chain dynamics . In both our experimental and theoretical studies , the results are equivalent to saying that the primary orientational relaxation process of polymer segments is an "Intermittent Motion", that is, the orientational relaxation of a segment starts and stops at intervals. If our theoretical analysis is right, a crossover behavior from Eq. 31 to 30 based on the orientational auto-correlation function of the local motion of a polymer chain will be observed with the decreasing solvent/segment ratio (B/A).

In order to prove this crossover behavior, the emission anisotropy ratio decay curves of anthryl groups end-capped polystyrene (APSA) in various polystyrene (PS)/dibutylphthalate (DBP) mixtures were measured by a time-resolved single photon counting technique in PLASMA. The time-resolved fluorescence and emission anisotropy ratio decay curves of APSA in 20% PS/DBP solution are shown at the top of Fig. 14. The obtained emission anisotropy ratio decay curves of APSA are estimated by our original computer analysis method based on Wahl's argolism and the χ^2-map method. The χ^2-maps of HPB and KWW models are shown in the middle of Fig. 14. In this analysis, we were able to establish that the time-resolved emission anisotropy ratio decay curves of APSA in dilute and concentrated polystyrene solutions are best-fitted by KWW (stretched exponential) and HPB (hyperbolic) functions, respectively. The half-width values on χ^2-maps for KWW and HPB functions in each polystyrene concentration are listed in the middle of Fig. 14. Consequently we can confirm the fact that there is a crossover behavior between KWW and HPB functions. We could prove our theoretical model for local motion of polymer chain, the "Intermittent Motion" model, based on the fractal-time random walk theory.

Now a new epoch based on "Complex Macromolecular Chain Dynamics" from a viewpoint of post-modernism is becoming known to us. I have a clear hunch of what will happen in the near future. For the new epoch to come, the concept of probe methods as a spy will be further extended not only by fluorescence techniques but also some new techniques in each level from the microscopic class to the macroscopic class via the mesoscopic class, and it will be a very excellent tool for discussing "Dynamics in Morphological Formation Processes" having representations of bio-mechanics and intelligent systems based on the post-modern viewpoint [2].

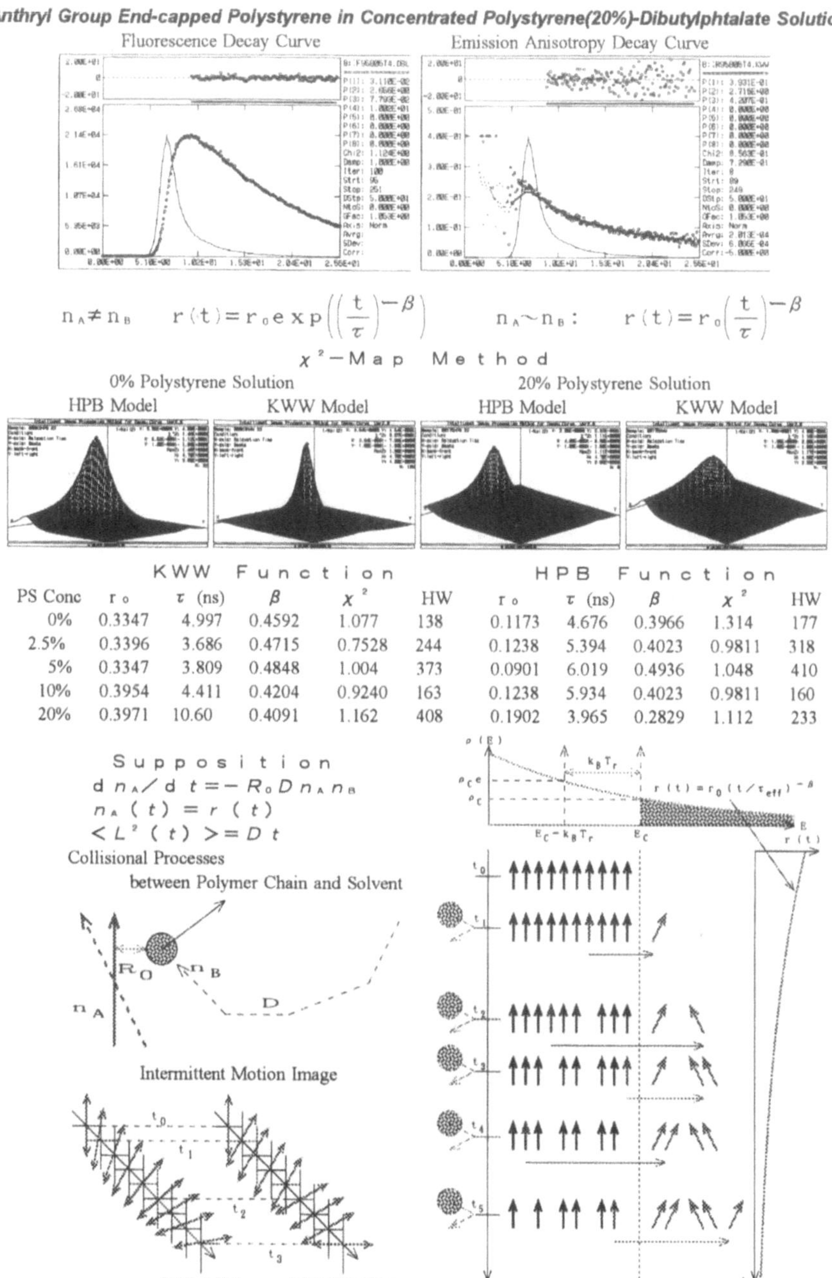

$$n_A \neq n_B \qquad r(t) = r_0 \exp\left(\left(\frac{t}{\tau}\right)^{-\beta}\right) \qquad n_A \sim n_B: \qquad r(t) = r_0\left(\frac{t}{\tau}\right)^{-\beta}$$

PS Conc	\multicolumn{5}{c}{KWW Function}	\multicolumn{5}{c}{HPB Function}
	r_0	τ (ns)	β	χ^2	HW	r_0	τ (ns)	β	χ^2	HW
0%	0.3347	4.997	0.4592	1.077	138	0.1173	4.676	0.3966	1.314	177
2.5%	0.3396	3.686	0.4715	0.7528	244	0.1238	5.394	0.4023	0.9811	318
5%	0.3347	3.809	0.4848	1.004	373	0.0901	6.019	0.4936	1.048	410
10%	0.3954	4.411	0.4204	0.9240	163	0.1238	5.934	0.4023	0.9811	160
20%	0.3971	10.60	0.4091	1.162	408	0.1902	3.965	0.2829	1.112	233

$$d\,n_A / d\,t = -R_0 D\, n_A n_B$$
$$n_A(t) = r(t)$$
$$< L^2(t) > = D\, t$$

Fig. 14. Schematic diagram for intermittent motion of polymer main chains

6
Overview of Fluorescence Probe Techniques in Complex Fluid Science

Initially, the fluorescence probe technique was exploited in order to analyze a very small amount of biomolecules in the field of biochemistry. After that, this method was further extended to obtain chemical and physical information about micro-environment, motion, and micro-structure in the vicinity of a molecular probe in many scientific fields. In particular, this technique was very effective in the fields of biophysics, polymer science, solution chemistry, amorphous solid state chemical physics, statistical mechanics, etc. because these fields had to discuss various complicated states of materials from the viewpoint of the microscopic class. This technique, which had some useful properties such as ultra high sensitivity, molecular labeling, an ultra wide dynamic range, and many measurement channels, was a very excellent tool for clarifying the relationship between the various macroscopic properties of materials and the microscopic properties of atoms and molecules making up materials in condensates. The fluorescence probe method has recently been adopted to investigate the mechanism of complex fluid and condensed matters.

6.1
Macromolecular Transition Phenomena Measured by Fluorescence Probes

Since early times, fluorescence probe methods have usually been applied to study polymer chemistry and polymer physics. In the decade between 1960–1970, researchers using this method in the field of polymer science obtained some hints about condensed matters from the fruitful results in the fields of biochemistry and biophysics. After that, much research using fluorescence probe methods in the field of polymer science was reported. In particular, there was an explosion in the amount of research for photo-performance polymer materials [34]. In this section, I will briefly discuss the macromolecular transition behavior measured by fluorescence probe methods.

In macromolecular science, helix-coil and coil-globule transition phenomena and, of course, protein folding processes as well, are generally well-known by many researchers in various scientific fields. The helix-coil transition behavior of proteins, polypeptides, and DNA were studied by light scattering, dielectric relaxation, ESR, and NMR methods. In the field of the fluorescence probe method, the helix-coil transition behavior of poly(γ-benzyl-L-glutamate) (PBLG), DNA, and poly(glutamic acid) was reported [25, 35]. The emission anisotropy ratio of a chromophore attached to the N-terminal group of PBLG in various mixed organic solvents was measured, and the mechanism of the helix-coil transition behavior and the aggregation process of PBLG analyzed. On the other hand, the time-resolved emission anisotropy ratio decay curves of a chromophore interchelated DNA molecule were measured, and the mechanism of the temperature- and the pH-dependence on helix-coil transition behavior was analyzed. Naturally, the mechanism of the structural change of various proteins was studied by the use of time-resolved fluorescence and emission anisotropy ratio decay curves of a chromophore attached to SH and amino groups of this protein. The mechanism of muscles from the viewpoint of the molecular level has been studied using fluorescence probe techniques [36].

The coil-globule transition behavior of polystyrene and poly(N-isopropylacrylamide) (PNIPA) chains was studied by theory, light scattering methods, and fluorescence probe methods [37]. In the first stage, the phenomenon of the coil-globule transition was foretold by a theorist in the field of statistical mechanics. After that, the coil-globule transition behavior of polystyrene in cyclohexane under θ-temperature was measured by the light scattering method, and the polymer-induced coil-globule transition phenomena of polystyrene in various PEO/benzene mixtures were determined by the fluorescence probe method. On the other hand, the temperature-induced coil-globule transition behavior of PNIPA in water was further extended by many researchers after the volume phase transition phenomena of PNIPA gels had been determined by Tanaka. Recently, a few researchers have asked the question as to whether the coil-globule transition phenomenon of a polymer exist. Therefore, the coil-globule transition phenomenon of a polymer has been carefully studied in a new system with surface-active agents based on both the methods of light scattering and the fluorescence probe.

Until now, it has been very difficult to measure directly the protein folding processes by the use of the fluorescence probe method because this phenomenon is over in a few seconds. In recent years, the studies of macromolecular folding processes based on photo-reaction have been reported as protein folding models. The processes of intramacromolecular cross linkage reactions of polystyrene having anthryl and eosinyl groups have been measured by both the photodimerization and the fluorescence probe methods [38]. In the field of polymer science, much research into macromolecular morphological formation dynamics will be performed in the future using the fluorescence probe method.

I think that sooner or later the fluorescence probe method with a light scattering tool will be an effective technique in order to clarify the mechanism of some macromolecular transition phenomena based from the viewpoint of the microscopic and mesoscopic classes.

6.2
Micellar Formation Phenomena Measured by Fluorescence Probes

The behavior of lipid bilayer, micellar (surface-active agents), and phase separation was mainly studied by static and dynamic light scattering methods. In recent years, research into this behavior has made rapid progress based on fluorescence probe methods because many researchers in this field have begun to be interested in the dynamic mechanism of molecular condensed processes in the mesoscopic class. However, it may be difficult to measure some emission behavior of chromophores in a cloudy solution system.

At the first stage, the structure and dynamics of lipid bilayer as a biomembrane model were investigated by the fluorescence probe technique [39]. The self-diffusion coefficients of a chromophore in various lipid bilayers were estimated by time-resolved emission anisotropy ratio and excimer emission decay curves. After that, the formation processes of various phase diagrams of surface-active agents/water mixtures were studied by excimer formation processes, fluorescence depolarization, and singlet-singlet and triplet-triplet energy transfer processes [40]. Hence, recently, the researchers for liquid-liquid phase separation processes have adopted both the fluorescence probe technique and the light scattering method [41]. However it is not easy to measure emission anisotropy ratios of a chromophore in a cloudy solution system at a near critical temperature.

6.3
Volume Phase Transition Phenomena of Gels Measured by Fluorescence Probes

After Tanaka had discovered the phenomenon of solvent- and temperature-induced volume phase transition of gels, some researchers tried to measure the static and dynamic properties under this transition process monitored by fluorescence probe methods in the field of polymer science [42]. Firstly, the fluorescence, excimer emission, and emission anisotropy ratio spectra of a chromophore incorporated into poly(acrylamide) gels were measured in various acetone/water mixtures. In this experimental system, many researchers were interested in solvent effects at volume phase transitions of gels. Following on from this, a spectroscopic measurement with an accurate thermostat around room temperature was established. The temperature-induced volume phase transition behavior of gels were popularly discussed based on the measurements of fluorescence, excimer emission, and emission anisotropy ratio spectra of a chromophore attached to PNIPA gel chains from the viewpoint of the molecular level.

On the other hand, much work on the volume phase transition behavior of gels was reported using the NMR method. Every researcher thinks that the NMR method is a very strong tool for clarifying the volume phase transition process of gels but the discussion based on NMR method differs from that based on the fluorescence probe method. The data measured by fluorescence probe and NMR methods are based on molecular and functional group levels, respectively. In the near future, it will be very important that we have the viewpoints of all levels of class organization in nature in order to clarify the mechanism of the volume phase transition of gels.

6.4
Structural Transition Phenomena of Liquid Crystal Measured by Fluorescence Probes

The structural phase transition behavior of liquid crystals has mainly been studied by light scattering and DSC methods. In liquid crystals, it is very difficult to measure some optical spectra because there are problems associated with liquid crystals spectroscopically. There are essentially both properties of cloud state (strong light scattering) and molecular orientation (optical anisotropy) in a liquid crystal system. Initially [43], the research of liquid crystals was performed using the fluorescence probe method based on excimer formation processes. In particular, the polymer liquid crystal systems were studied using the fluorescence probe method because this system was barely cloudy. Hence the formation process of polymer lyotropic liquid crystals could be estimated by the fluorescence probe method based on the emission anisotropy ratio because this experimental system was also clear.

Recently, we proposed a new optical measurement technique in order to estimate the structural phase transition behavior of liquid crystals. Our new technique in PLASMA consists of DSC, a microscope photomonitor, microscope CCD video graphics, depolarized emission spectra, emission anisotropy ratio spectra, and static light scattering with our original accurate thermostat. Our objective is that the structure phase transition behavior of liquid crystals is simultaneously measured and analyzed by the above-mentioned six pieces of apparatus. In our technique, I have tried to classify the obtained experimental data according to each level of class organization of nature, that is, I have tried to divide the obtained experimental information into both effects of molecular orientation and domain shape because the obtained experimental data for the structure phase transition behavior of liquid crystals will certainly contain both molecular orientation and domain shape effects.

In this final section, I will briefly discuss the structural phase transition behavior of cholesteryl oleyl carbonate (COC) measured by our new technique. The DSC and transmittence intensity curves of COC with a heating rate of 5°C/min under a polarized microscope are shown at the top of Fig. 15. The liquid crystal phase transition temperatures of COC are well-known at about 17°C from the smectic A phase to the cholesteric phase and 34°C from the cholesteric phase to the isotropic phase [15]. These curves demonstrate that there are two transition points, smectic A/cholesteric phase transition at about 22 °C and cholesteric/isotropic phase transition at about 35°C, respectively. The CCD video pictures at the smectic A phase, the smectic A/cholesteric phase transition point, and the cholesteric phase are shown at the bottom of Fig. 15, where the left- and right-hand side of the graphs are analyzed by the two dimensional Fourier transformation and the fractal dimension method based on light and shade, respectively. Each two dimensional Fourier transformational pattern shows generally the imaginary two dimensional light scattering one. According to the left-hand side graphs of Fig. 15, the change of light scattering patterns can be ascribed to the effect of the domain shape at each liquid crystal phase. We were able to easily obtain the values of the total averaged fractal dimension, $D = 1.99$ (smectic A phase), $D = 1.87$ (smectic A/cholesteric phase transition point), and $D = 1.71$ (cholesteric phase), from the right-hand side graphs of Fig. 15. These numerical values will be related to the effect of the domain shape at each liquid crystal phase.

The three-dimensional graphics for VV emission spectra (180 lines) of 2-anilinonaphthalene-6-sulfonate (ANSA) doped in COC at 18–24°C and 32–38°C are shown on the left side of the upper and middle positions of Fig. 16, respectively. Hence the contour graphics for VV emission, VH emission, and emission anisotropy ratio spectra of ANSA doped in COC at 18–24°C and 32–38°C are shown on the right side of the upper and middle positions of Fig. 16, respectively. Our apparatus for emission spectra can easily measure a large amount of spectra data at various strictly controlled temperatures using our original ther-

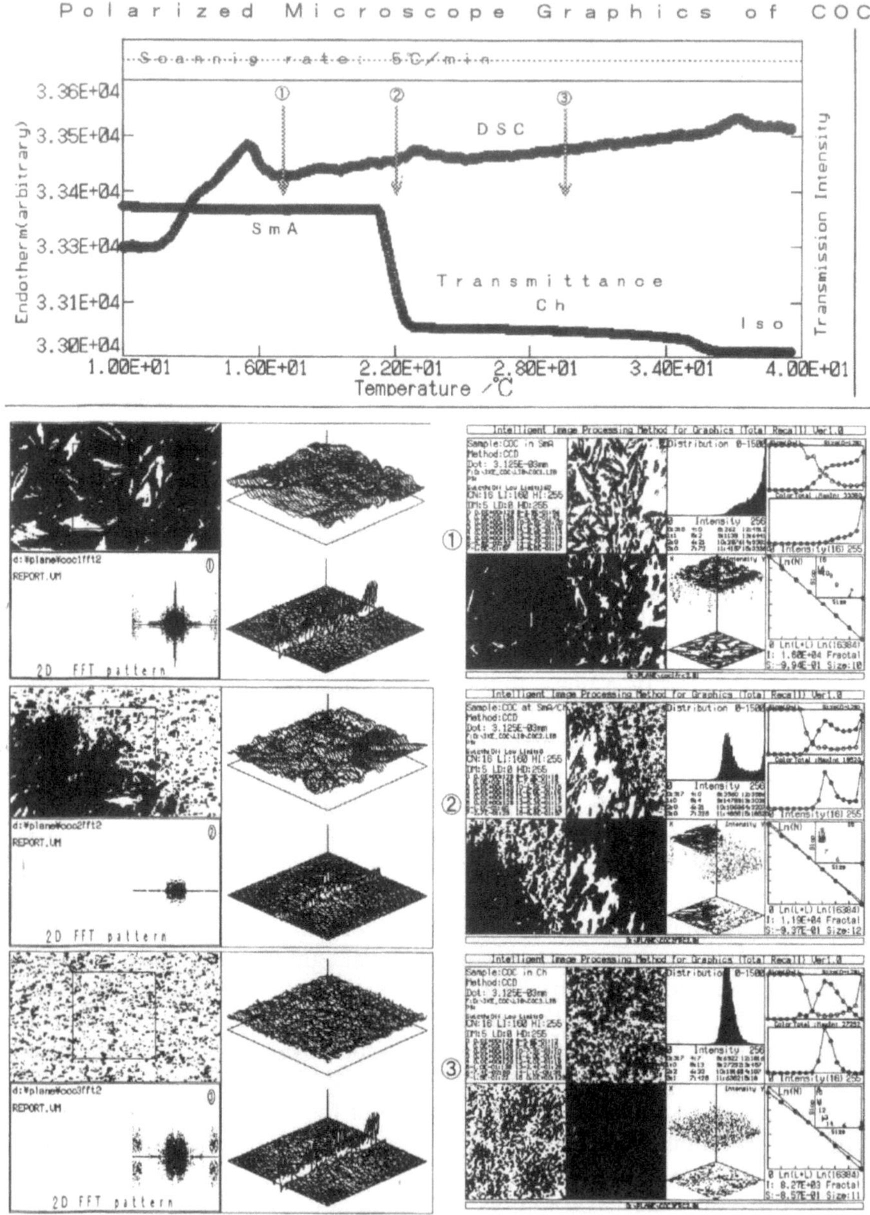

Fig. 15. Schematic diagram for microscope CCD video pictures in structural phase transition of COC

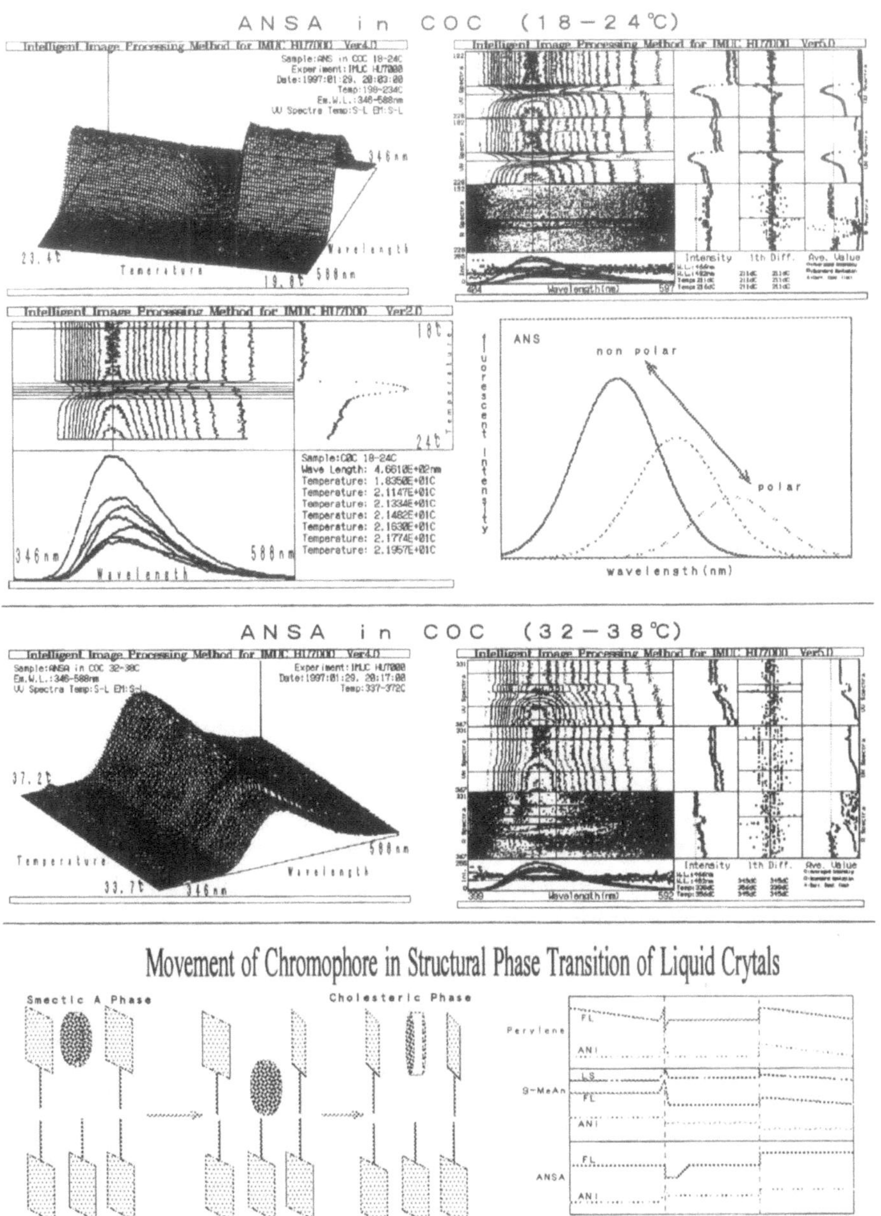

Fig. 16. Schematic diagram for emission spectra of ANS in structural phase transition of COC

mocontroller (resolution temperature: about 3 mK, heating rate: 0.3 mK/s). These graphics show the fact that the intensities of VV emission, VH emission, and emission anisotropy ratio spectra of ANSA doped in COC sharply change at each near critical temperature. However, it is very dangerous to discuss the structural phase transition behavior of COC based only on intensity data of VV emission, VH emission, and emission anisotropy ratio spectra of ANSA doped in COC as the intensity data of these spectra surely hold both the effects of molecular orientation and domain shape.

The three-directional graphics for VV emission spectra of ANSA doped in COC at 18 – 24 °C is shown on the left-hand side of the upper position of Fig. 16. ANSA is well-known as a twisted intramolecular charge transfer (TICT) probe. The fluorescence spectra of this probe are strongly influenced by solvent polarity and viscosity based on an electronic interaction of molecular orbitals. The schematic diagram of fluorescence spectra of a TICT probe is shown on the right-hand side at the top of Fig. 16. The three-directional graphics show the fact that the peak position of the fluorescence spectra of ANSA doped COC red-shifts and blue-shifts in the range of about 0.8 °C at near the smectic A/cholesteric phase transition temperature. From this experimental result, it can be presumed that an ANSA molecule will slide and come back along a COC molecule at the smectic A/cholesteric phase transition point because the direction of the COC molecules will change slightly all together at the smectic A/cholesteric phase transition point. The image of this process is shown on the left-hand side at the bottom of Fig. 16.

In research using optical spectroscopic measurements, it is very difficult to analyze the dynamic behavior of condensed matters in muddy samples. But we usually had to try to measure various spectra of muddy samples in order to clarify "Complex Macromolecular Chain Dynamics" in the field of complex chemical physics. Therefore it will be very important to analyze the dynamic behavior of condensed matters based on all levels of class organization in nature because the phenomenon of muddiness is mainly caused by light scattering in themesoscopic level.

7
Epilogue

In this review, I have tried to discuss not only an overview of the fluorescence probe technique but also the meanings of this method in the scientific research field from a viewpoint of post-modernism. Here I would like to make several conclusions about methods and applications of optical spectroscopy in the field of complex chemical physics. My views on "Classification Based on Structure for Macromolecular Chain Dynamics", "PLASMA", "Dynamics of Morphological Formation Process", and "Power Law Analysis" are closely related to the development of the fluorescence probe method about ten years ago. Therefore my discussion for optical spectroscopic measurement holds the possibility of manufacturing some new experimental and theoretical techniques based on a fluorescence application method in the field of complex chemical physics. For a new epoch to come, I can imagine that new optical spectroscopic techniques will consist of various simultaneous measurements with all levels of class organization in nature, naked eye level, microscope level, light scattering level, molecular level, functional group level, and molecular orbital level. Therefore I believe that our original concept of PLASMA will grow to act as the go-between between the old and the new period in the field of 'METHODS AND APPLICATIONS OF FLUORESCENCE SPECTROSCOPY'. My first remark about the development of complex chemical physics is the fact that the viewpoint of simultaneous measurement technique with all levels of class organization in nature is very important.

As my second remark in this review, it is of great importance not to keep research ideas only to oneself. We need practically to work on the problems that still await solutions together with everybody else based on post-modern thought. Here I would like to introduce our designs for new measurement techniques in PLASMA that are now going on.

(1) Establishment of measurement and analysis for dynamic light scattering at all scattering angles
(2) Establishment of measurement and analysis for phase transition behavior within a quasi-biocell using a micromanipulator
(3) Establishment of simultaneous measurement and analysis for two dimensional polarized dynamic light scattering and two dimensional fluorescence depolarization in an anisotropy state by the use of an ultra-high-speed CCD video camera
(4) Establishment of measurement and analysis for time-resolved spectroscopy with stopped flow apparatus.

In a new epoch, it will be most important that we train many clever young people in this area of research, that is, we ought to discuss some educational problems for young researchers. We have no time to exchange scientific ideas.

The third remark is some thoughts about complex chemical physics. I feel that the framework of complex chemical physics has not yet been established. The age in which the space-time concept has been discussed only in fields of high energy physics and astronomy have finished. In our everyday life, it is a matter of course that various phenomena in each level of class organization in nature closely connect with the space-time concept. There has been a serious problem in how we should have a viewpoint within a scientific research framework. It is important that we learn how to think about this framework of a scientific field. For example, I do expect that the field of complex chemical physics will develop in the near future if the concept of "Complex" in this research field only has a meaning of complexity. I think that the meaning of "Complex" in the field of complex chemical physics will connect with the concept of "Phase". In next century, I feel that new scientific research will develop around our understanding of the concept of "Phase" with the expansions of various scientific research fields.

Acknowledgements. I gratefully acknowledge that a large amount of support for our research is generously provided by Ministry of Education, Science, Sports, and Culture in Japan.

References

1. K. Horie, H. Ushiki, 'Hikari-Kinou-Bunshi-no-Kagaku', Kodansha Scientific (1992) in Japanese; K. Horie, H. Ushiki, F.M. Winnik, 'Interactions of Light and Materials – Fundamental and Practical Aspects of Photo- and Opto-sciences', Kodansha-VCH (1997) in press

2. H. Ushiki, F. Tsunomori, Bull. Fac. Gen. Ed. Tokyo Univ. Agr. Tech., 26, 109 (1989) in Japanese; H. Ushiki, K. Ushiki, Bull. Fac. Gen. Ed. Tokyo Univ. Agr. Tech., 27, 105 (1990) in Japanese
   H. Ushiki, F. Tsunomori, Bull. Fac. Gen. Ed. Tokyo Univ. Agr. Tech., 29, 85 (1992) in Japanese; H. Ushiki, F. Tsunomori, K. Hamano, Bull Fac. Gen. Ed. Tokyo Univ. Agr. Tech, 30, 35 (1993) in Japanese; H. Ushiki, F. Tsunomori, K. Hamano, Bull Fac. Gen. Ed. Tokyo Univ. Agr. Tech, 31, 115 (1994) in Japanese

3. H. Ushiki, K. Hamano, 'Spectroscopic Study on Structural Formation Dyamics of Condensed Phase Based on Phase Transition Phenomena', Scientific Research Reports Supported by Grant-in Aid from Ministry of Education, Science, Sports, and Culture in Japan (1996) in Japanese: H. Ushiki, 'Establishment of New Material Estimation Based on Computer Analysis of Photo-multi-functional Measurements', Scientific Research Report Supported by Grant-in Aid from Ministry of Education, Science, Sports, and Culture in Japan (1997)

4. U. Even, K. Rademann, J. Jortner, N. Manor, R. Reisfeld, J. Luminescence, 31/32, 634 (1984)

5. P.G. de Genne, J. Chem. Phys., 76, 3316 (1982)

6. U. Evesque, J. Physique, 44, 1217 (1983)

7. H. Ushiki, Rep. Prog. Polym. Phys. Jpn., 35, 419 (1992)

8. H. Ushiki, C. Hashimoto, J.Rouch, Rep. Prog. Polym. Phys. Jpn., 39, 179 (1996); H. Ushiki, C. Hashimoto, J. Rouch, Rep. Prog. Polym. Phys. Jpn., 40, (1997) in press

9. H. Ushiki, F. Tsunomori, K. Hamano, Rep. Prog. Polym. Phys. Jpn., 37, 463 (1994)

10. H. Ushiki, F. Tsunomori, K. Hamano, Rep. Prog. Polym. Phys. Jpn., 35, 431 (1992); H. Ushiki, F. Tsunomori, S. Kashimori, K. Hamano, Rep. Prog. Polym. Phys. Jpn., 36, 377 (1993)

11. H. Ushiki, Y. Fukase, J. Rouch, Rep. Prog. Polym. Phys. Jpn., 39, 135 (1996)

12. A. Onuki, K. Yamazaki, K. Kawasaki, Annals Physics, 131, 217 (1981)

13. T. Takebe, K. Fujioka, R. Sawaoka, T. Hashimoto, J. Chem. Phys., 93, 5271 (1990)

14. H. Ushiki, K. Honda, J. Rouch, Rep. Prog. Polym. Phys. Jpn., 40, (1997) in press; H. Ushiki, Y. Katsumoto, J. Rouch, Rep. Prog. Polym. Phys. Jpn., 40 (1997) in press

15. H. Ushiki, K. Honda, F. Tsunomori, K. Hamano, Rep. Prog. Polym. Phys. Jpn., 38, 111 (1995)

16. L. Stryer, J. Mol. Biol., 13, 483 (1965)

17. H. Dodik, E. Kosower, J. Phys. Chem., 81, 50 (1977)

18. K. Horie, I. Mita, Polymer J., 8, 227 (1976)
    K. Horie, I. Mita, Polymer J., 9, 201 (1977); I.Mita, K.Horie, J. Macromol. Sci. Rev, Macromol. Chem. Phys., C27, 108 (1987)

19. H. Ushiki, K. Horie, I. Mita, Polymer J., 13, 191 (1981); K. Horie, W. Schnabel, I. Mita, H. Ushiki, Macromolecules, 14, 1422 (1981)

20. H. Itagaki, K. Horie, I. Mita, Macromolecules, 20, 2774 (1987); H. Itagaki, K. Horie, I. Mita, Macromolecules, 22, 2520 (1989); H. Itagaki, K. Horie, I. Mita, Progr. Polym. Sci., 15, 361 (1990)

21. M.A. Winnik, T. Redpath, D.H. Richard, Macromolecules, 13, 328 (1980); T. Kanaya, K. Goshiki, M. Yamamoto, Y. Nishijima, J. Am. Chem. Soc., 104, 3580 (1982); C. Cuniberti, A. Perico, Prog. Polym. Sci., 10, 271 (1984)

22. C.W. Frank, M.A. Gashgari, Ann. New York Acad. Sci., 366, 387 (1981)

23. F. Perrin, Ann. Phys. (Paris), 12, 169 (1929)
    G. Weber, Adv. Protein Chem., 8, 415 (1953)

24. T. Tao, Biopolymers, 8, 609 (1969)
    Ph.Wahl, New Tech. Biophys. Cell Biol., 2, 233 (1975)

25. K. Kinoshita, S. Kawamoto, A. Ikegami, Biophys., 20, 289 (1977)

M.D. Barkley, B.H. Zimm, J. Chem. Phys., 70, 2991 (1979); G. Lipari, A. Szabo, Biophys. J., 30, 489 (1980)

26. B. Valeur, L. Monnerie, J. Polym. Sci., Polym. Phys., 14, 11 (1976); T. Sasaki, M. Yamamoto, Y. Nishijima, Macromolecules, 21, 610 (1988); M.D. Ediger, Ann. Rev. Phys. Chem., 42, 225 (1991); F. Tsunomori, H. Ushiki, Rep. Prog. Polym. Phys Jpn., 36, 357 (1993)

27. H. Morawetz, I.Z. Steinberg, eds., Ann. New York Acad. Sci., 366 (1981)

28. D.E. Woessner, J. Chem. Phys., 36, 1 (1962); A.A. Jones, W.H. Stockmayer, J. Polym. Sci. 15, 847 (1977); J.T. Bendler, R. Yaris, Macromolecule, 11, 650 (1978); C.K. Hall, E. Helfand, J. Chem. Phys., 77, 3275 (1982); T.A. Weber, E. Helfand, J. Phys. Chem., 87, 2881 (1983); J.L. Viovy, L. Monnerie, J. Brochon, Macromolecules, 16, 1845 (1983)

29. H. Ushiki, F. Tsunomori, Rep. Prog. Polym. Phys. Jpn., 36, 369 (1993)

30. F. Tsunomori, H. Ushiki, Polymer J., 28, 576 (1996); F. Tsunomori, H. Ushiki, K. Horie, Polymer J., 28, 582 (1996); F. Tsunomori, H. Ushiki, Polymer J., 28, 588 (1996); H. Ushiki, F. Tsunomori, Rep. Prog. Polym. Phys. Jpn., 39, 127 (1996); H. Ushiki, Y. Katsumoto, F. Tsunomori, J. Rouch, Rep. Prog. Polym. Phys. Jpan., 39, 131 (1996)

31. J. Tuac, Festkorperprobleme, 22, 85 (1982)

32. J. Orenstein, M. Kastner, D. Monroe, J. Non-Cryst. Solid, 35–36, 951 (1980)

33. H. Takayasu, 'Fractal Kagaku', Asakura, Tokyo 1987 in Japanese

34. G.G. Guilbault, 'Practical Fluorescence', Marcel Dekker, New York (1973); G. Smet, Adv. Polym. Sci., 50, 17 (1983); J.E. Guillet, 'Polymer Photophysics and Photochemstry', Cambridge Univ. Press, Cambrige (1985); D. Phillips, ed, 'Polymer Photophysics', Chapman and Hall, New York (1985); M.A. Winnik, ed, 'Photophysical and Photochemical Tools in Polymer Science', NATO ASI series Vol 182, Dordrecht: Reidel (1986); C.E. Hoye, J.M. Torkelson, 'Photophysics of Polymers', ACS Symposium Series 358 (1987); H. Ushiki, K. Horie, 'Influence of Molecular Structure on Polymer Photophysical and Photochemical Properties', edited by N.P. Cheremisinoff, Handbook of Polymer Science and Technology, Marcel Dekker Inc., (1989); H.J. Schneider, H. Durr, J.M. Lehn, ed., 'Frontiers in Supramolecular Organic Chemistry and Photochemistry', VCH (1991)

35. Ph. Wahl, J. Paoletti, J.B. LePeque, Proc. Natl. Acad. Sci. USA, 65, 417 (1970); D.P. Millar, R.J. Robbins, A.H. Zewail, Proc. Natl. Acad. Sci. USA, 77, 5593 (1980); H. Ushiki, I. Mita, Polymer J., 13, 837 (1981); I. Ashikawa, K. Jr. Kinoshita, A. Ikegami, Y. Nishimura, M. Tsuboi, K. Watanabe, K. Iso, J. Biochem., 22, 6018 (1983); H. Ushiki, I. Mita, Polymer J., 16, 751 (1984) H. Ushiki, I. Mita, Polymer J., 17, 1297 (1985); H. Ushiki, F. Tanaka, I. Mita, Polymer J., 22, 827 (1986); H. Ushiki, F. Tsunomori, Rep. Prog. Polym. Phys. Jpn., 35, 435 (1992); H. Ushiki, F. Tsunomori, Rep. Prog. Polym. Phys. Jpn., 36, 365 (1992); H. Ushiki, F. Tsunomori, Rep. Prog. Polym. Phys. Jpn., 36, 369 (1993)

36. Y. Sasaki, F. Tsunomori, T. Yamashita, K. Horie, K. Kohama, H. Ushiki, Rep. Prog. Polym. Phys. Jpn., 37, 723 (1994); Y. Sasaki, F. Tsunomori, T. Yamashita, K. Horie, H. Ushiki, R. Ishikawa, K. Kohama, J. Biochem., 116, 236 (1994)

37. T. Tanaka, Phys. Rev. Lett., 40, 820 (1978); T. Tanaka, Sci. Am., 244, 124 (1981); H. Ushiki, F. Tanaka, Eur. Polym. J., 21, 710 (1985); F. Tanaka, H. Ushiki, J. Chem. Phys., 84, 5925 (1986); H. Ushiki, F. Tanaka, S. Sakanoue, Eur. Polym. J., 23, 323 (1987); F. Tanaka, H. Ushiki, Macromolecules, 21, 1041 (1988); F.W. Winnik, Polymer, 31, 2125 (1990); A. Matsuyama, F. Tanaka, J. Chem. Phys., 94, 781 (1991); Th. Binkert, J. Oberreich, M. Meewes, R. Nyffenegger, J. Ricka, Macromolecules, 24, 5806 (1992); H. Ringsdorf, J. Simon, F.M. Winnik, Macromolecules, 25, 5353 (1992); H. Ushiki, S. Kashimori, F. Tsunomori, Rep. Prog. Polym. Phys. Jpn., 37, 467 (1994); H. Ushiki, F. Tsunomori, rep. Prog. Polym, Phys, Jpn., 38, 107 (1995); K. Ono, K. Ueda, T. Sasaki, S. Murase, M. Yamamoto, 29, 1584 (1996)

38. T. Torii, H. Ushiki, K. Horie, Polymer J., 24, 1057 (1992)
    T. Torii, H. Ushiki, K. Horie, Polymer J., 25, 173 (1993)

39. T. Pullerits, A. Freiberg, Chem. Phys., 149, 409 (1991); E. John, F. Jahnig, Biophys. J., 60, 319 (1991); X. Han, R.W. Gross, Biophys. J., 63, 309 (1992); P.J. Tummino, A. Gafni, Biophys. J., 64, 1580 (1993); J.G. Ray, P.K. Sengupta, Chem. Phys. Lett., 230, 75 (1994); T. Parasassi, A.M. Giusti, M. Raimond, E. Gratton, Biophys. J., 68, 1895 (1995); W.D. Niles, F.S. Cohen, Rev.

Sci. Instrum., 66, 3527 (1995); J.Y.A. Lehtonen, P.K.J. Kinnunen, Biophys. J., 68, 525 (1995); J.L. Thomas, K.A. Borden, D.A. Tirrell, Macromolecules, 29, 2570 (1996)

40. F. Candau, R.H. Ottewill, ed., 'Scientific Methods for the Study of Polymer Colloids and Their Applications', Kluwer Acad. Pub. (1988); F. Quina, E. Abuin, E. Lissi, Macromolecules, 23, 5173 (1990); A. Yekta, J. Duhamel, P. Brochard, H. Adiwidjaja, M.A. Winnik, Maromolecules, 26, 1829 (1993); A. Yekta, J. Duhamel, H. Adiwidjaja, P. Brochard, M.A. Winnik, Langmuir, 9, 881 (1993); O. Soderman, M. Jonstromer, J. van Stam, J. Chem. Soc. Fraday Trans., 89, 1799 (1993); M.I. Viseu, S.M.B. Costa, J. Chem. Soc. Fraday Trans., 89, 1925 (1993); O. Anthony, R. Zana, Macromolecules, 27, 3885 (1994); P. Hansson, M. Almgren, Langmuir, 10, 2115 (1994); A. Yekta, B. Xu, J. Duhamel, H. Adiwidjaja, M.A. Winnik, Maromolecules, 28, 956 (1995); I. Astafieva, K. Khougaz, A. Eisenberg, Macromolecules, 28, 7127 (1995); T.J. Martin, S.E. Webber, Macromolecules, 28, 8845 (1995); P.B. Bisht, K. Fukuda, S. Hirayama, J. Chem. Phys., 105, 9349 (1996); E. Alami, M. Almgren, W. Brown, J. Francois, Macromolecules, 29, 2229 (1996); E. Feitosa, W. Brown, P. Hansson, Macromolecules, 29, 2169 (1996); E. Alami, M. Almgren, W. Brown, Macromolecules, 29, 5026 (1996); R. Walter, J. Ricka, Ch. Quellet, R. Nyffenegger, Th. Binkert, Macromolecules, 29, 4019 (1996); T. Fujieda, K. Ohta, N. Wakabayashi, S. Higuchi, J. Colloid Interface Sci., 185, 332 (1997); E.B. Abuin, E.A. Lissi, A. Aspee, F.D. Gonzalez, J. Varas, J. Colloid Interface Sci., 186, 332 (1997); C.G. Blanco, M.M. Velazquez, S.M.B. Costa, P. Barreleiro, J. Colloid Interface Sci., 189, 43 (1997); K. Akiyoshi, S. Deguchi, H. Tajima, T. Nishikawa, J. Sunamoto, Macromolecules, 30, 857 (1997)

41. B. Wandelt, D.J.S. Birch, R.E. Imhof, R.A. Pethrick, Polymer, 33, 3558 (1992); J. Zhang, L.L. Lee, J.F. Brennecke, J. Phys. Chem., 99, 9268 (1995); F. Tsunomori, H. Ushiki, K. Hamano, Rep. Prog. Polym. Phys. Jpn., 38, 99 (1995); F. Tsunomori, H. Ushiki, K. Hamano, Chem. Phys. Lett., 252, 121 (1996)

42. J. Parreno, I.F. Pierola, Polymer, 31, 1768 (1990); Y. Hu, K. Horie, H. Ushiki, Macromolecules, 25, 6040 (1992); Y. Hu, K. Horie, H. Ushiki, F. Tsunomori, T. Yamashita, Macromolecules, 25, 7324 (1992); Y. Hu, K. Horie, T. Torii, H. Ushiki, X. Tang, Polymer J., 25 123 (1993); Y. Hu, K. Horie, H. Ushiki, T. Yamashita, F. Tsunomori, Macromoelcules, 26, 1761 (1993); Y. Hu, K. Horie, H. Ushiki, Polymer J., 25, 651 (1993); Y. Hu, K. Horie, H. Ushiki, F. Tsunomori, Eur. Polym. J., 29, 1365 (1993); F. Fujimoto, Y. Nakajima, M. Kashiwabara, H. Kawaguchi, Polym. Internat. 30, 237 (1993); H. Ushiki, K. Minomo, F. Tsunomori, A. Murata, Rep. Prog. Polym. Phys. Jpn., 38, 153 (1995)

43. H. Ushiki, F. Tsunomori, Rep. Prog. Polym. Phys. Jpn., 36, 369 (1993); T. Torii, T. Yamashita, H. Ushiki, K. Horie, Eur. Polym. J., 29, 465 (1993); H. Hashimoto, M. Hasegawa, K. Horie, T. Yamashita, H. Ushiki, I. Mita, J. Polym. Sci. Part B: Polym. Phys., 31, 1187 (1993); H. Ushiki, K. Honda, K. Hamano, Rep. Prog. Polym. Phys. Jpn., 38, 111 (1995); H.W. Huang, K. Horie, T. Yamashita, M. Sone, M. Tokita, J. Watanabe, Y. Maeda, Macromolecules, 29, 3485 (1996); G. Mao, J. Wang, S.R. Clingman, C.K. Ober, J.T. Chen, E.L. Thomas, Macromolecules, 30, 2556 (1997); A.E. Huber, C. Viney, Macromolecules, 30, 2662 (1997); H. Ushiki, K. Honda, J. Rouch, Rep. Prog. Polym. Phys. Jpn., (1997) in press

CHAPTER 15

Fluorescence Method for Monitoring Gelation and Gel Swelling in Real Time

Ö. Pekcan, Y. Yılmaz

Abstract. Gelation by free-radical cross-linking copolymerization (FCC) of methyl methacrylate (MMA) and ethylene glycol dimethacrylate (EGDM) was studied by real-time monitoring of pyrene (Py) intensity using the steady-state fluorescence (SSF) method. Samples with various cross-linker densities and toluene contents were prepared and gelations were observed at various temperatures. The percolation theory was employed to interpret the results. For all samples the gel fraction and the weight average degree of polymerization exponents β and γ were measured and found to be around 0.42 and 1.67, respectively. SSF measurements were then performed in order to study the swelling processes in gels prepared by FCC of MMA and EGDM at 75 °C for various toluene contents. After drying the gels, swelling and desorption experiments were performed in toluene at 50 °C by real-time monitoring of Py fluorescence intensity. The Li–Tanaka equation was employed to produce the swelling parameters. Cooperative diffusion coefficients (D_c) were measured and found to be around 10^{-6} cm^2/s for gels swollen in toluene.

1
Introduction

Gelation is the phase transition from a state without a gel to a state with a gel, i.e. gelation involves the formation of an infinite network. One particular approach to gelation theory is the percolation model [1]. It can be explained simply as follows: Monomers are thought to occupy the sites of a periodic lattice, and between the two nearest neighbors of the lattice sites a bond is formed randomly with probability p. The conversion factor p is the ratio of the actual number of bonds which have been formed between monomers at a given moment to the maximally possible number of such bonds. There is, in general, a sharp phase transition at some intermediate critical point $p = p_c$ where an infinite cluster starts to appear: The gel phase for p above p_c, the sol phase for p below p_c. Such simple gelation theories often make the assumption that the parameter p alone determines the behavior of the gelation process, though p may in turn depend on temperature, concentration of monomers, and time t [1]. In the case of chemical or irreversible gelation, a bond once formed is not easily broken again. Critical phenomena take place in the region in which p is very close to or the same as the gel point p_c. Thus if p is asymptotically close to p_c, critical behavior is observed. Of course, any real experiment can never reach this purely mathematical limit, but one can try to come as close as possible to asymptopia [1].

Some physical quantities can be described as a power law at the critical point during the sol-gel phase transition. For example, gel fraction G, and weight average degree of polymerization DP_w, can be written in the following forms [1]:

$$G = B(p - p_c)^\beta, \quad (p \to p_c^+) \tag{1}$$

$$DP_w = C(p_c - p)^{-\gamma}, \quad (p \to p_c^-) \tag{2}$$

where β and γ are the critical exponents, and asymptotic proportionality factors B and C are referred to as the critical amplitudes. At $p = p_c$, G and DP_w diverge. Since an infinite macromolecule appears for p above p_c but not for p below p_c, it is likely that some average molecular weight or degree of polymerization diverges, if p approaches p_c from below (denoted as $p \to p_c^-$). One can obtain the exponents β and γ using double logarithmic plots of the measured quantities according to Eqs. 1 and 2, respectively. Such a log-log plot reveals that data should be particularly accurate near the gel point. In particular, a small shift in p_c results in a large shift in the critical exponent [1]. To our knowledge so far, no one has yet reached the experimental state of the art to determine the β and γ exponents accurately for the sol-gel transition. However, in this paper, we introduce a novel technique to determine β and γ exponents quite accurately.

The fluorescence and phosphorescence intensities of aromatic molecules are effected by both radiative and non-radiative processes [2]. If the possibility of perturbation due to oxygen is excluded, the radiative probabilities are found to be relatively independent of environment and even of molecular species. Environmental effects on non-radiative transitions which are primarily intra-molecular in nature are believed to arise from a breakdown of the Born–Oppenheimer approximation [3] The role of the solvent in such a picture is to add the quasi-continuum of states needed to satisfy energy resonance conditions. The solvent acts as an energy sink for rapid vibrational relaxation which occurs after the rate-limiting transition from the initial state. Years ago, Birks et al. studied the influence of solvent viscosity on the fluorescence characteristics of pyrene solutions in various solvents and observed that the rate of monomer internal quenching is effected by solvent quality [4]. Weber et al. reported the solvent dependence of energy trapping in phenanthrene block polymers and explained the decrease in fluorescence yield with the static quenching, caused by the solvent-induced trapping states [5]. As the temperature of the liquid solution is varied, the environment about the molecule changes and much of the change in absorption spectra and fluorescence yields in solution can be related to the changes in solvent viscosity. A matrix that changes little with temperature will enable one to study molecular properties themselves without changing environmental influence. Poly-(methylmethacrylate) (PMMA) has been used as such a matrix in many studies [6]. Recently, we reported the viscosity effect on low frequency, intramolecular vibrational energies of excited naphthalene in swollen PMMA latex particles [7].

In this work, these properties of an aromatic molecule were used to monitor the sol-gel phase transition in free-radical cross-linking copolymerization using in situ fluorescence techniques. Gelation was monitored with the fluorescence intensity variation against time. The β and γ exponents were determined from these curves.

The swelling properties of chemically cross-linked gels can be understood by considering the osmotic pressure vs the restraining force [8–12]. The total free energy of a chemical gel consists of bulk and shear energies. In fact, in a swollen gel the bulk energy can be characterized by the osmotic bulk modulus K, which is defined in terms of the swelling pressure and the volume fraction of polymer at a given temperature. On the other hand, the shear energy which keeps the gel in shape can be characterized by shear modulus G. Here shear energy minimizes the non-isotropic deformations in the gel. The theory of kinetics of swelling for a spherical chemical gel was first developed by Tanaka et al. [13] where the assumption is made that the shear modulus, G is negligible compared to the osmotic bulk modulus. Later, Peters et al. [14] derived a model for the kinetics of swelling in spherical and cylindrical gels by assuming non-negligible shear modulus. Recently Li and Tanaka [8] have developed a model where the shear modulus plays an important role which keeps the gel in shape due to coupling of any change in different directions. This model predicts that the geometry of the gel is an important factor, and swelling is not a pure diffusion process.

Several experimental techniques have been employed to study the kinetics of swelling, shrinking and drying of chemical and physical gels among which are neutron scattering [15], quasi-elastic light scattering [14], macroscopic experiments [9] and in situ interferometric measurements. Using the fluorescence technique, a pyrene derivative was employed as a fluorescence probe to monitor the polymerization, ageing and drying of aluminosilicate gels [16], where peak ratios in emission spectra were monitored during these processes. The volume phase transition of poly(acrylamide) gels were monitored by fluorescence anisotropy and lifetime measurements of dansyl groups [17]. Recently, we reported in situ observations of the sol-gel phase transition in free-radical cross-linking copolymerization using the steady state fluorescence (SSF) technique [18, 19].

In this work, swelling parameters of gels formed by free-radical cross-linking copolymerization (FCC) of MMA and EGDM were reported. Py was used as a fluorescence probe to monitor swelling and desorption processes in real time

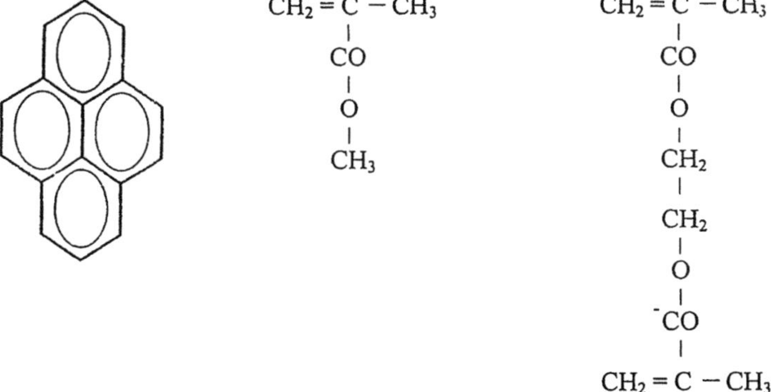

during in situ fluorescence experiments. Desorption curves are subtracted from the swelling curves to obtain the pure swelling behavior of these gels. The chemical structures of Py, MMA and EGDM are given below, respectively.

2
Experimental

2.1
Probing Gelation by Fluorescence

Here, at first, sol-gel transition in free-radical cross-linking copolymerization of MMA and EGDM was probed by using the steady-state fluorescence technique. The radical copolymerization of MMA and EGDM was performed in bulk or in toluene solutions at different temperatures in the presence of 2,2′-azobisisobutyronitrile (AIBN) as an initiator. Py was used as a fluorescence probe to detect the gelation process, where below p_c, MMA, linear and branched PMMA chains act as an energy sink for the excited Py but above p_c, the PMMA network provides an ideal, unchanged environment for the excited Py molecules.

Table 1. Experimentally measured β, γ and t_c values during the gelation of the samples prepared with various cross-linker densities and toluene contents[a]

Sets	Gels	EGDM $\times 10^{-3}$ (vol%)	Toluene (vol%)	T ± 2 (°C)	t_c (s)	β	γ
Set 1	1	5	-	75	2183	0.430	1.403
	2	7.5	-	75	1435	0.414	1.719
	3	10	-	75	1396	0.433	1.730
	4	12.5	-	75	1302	0.423	1.749
	5	15	-	75	1252	0.445	1.770
	6	20	-	75	1521	0.426	1.775
	7	30	-	75	873	0.420	1.704
Set 2	8	2.5	-	85	1609	0.432	1.582
	9	5	-	85	1395	0.468	1.713
	10	7.5	-	85	926	0.430	1.427
	11	25	-	85	791	0.396	1.789
Set 3	12	5	-	70	3072	0.408	1.815
	13	5	-	75	2392	0.420	1.536
	14	5	-	85	1239	0.427	1.588
Set 4	15	10	0.075	75	2065	0.367	1.493
	16	10	0.100	75	1714	0.432	1.772
	17	10	0.130	75	1775	0.428	1.789
	18	10	0.140	75	2725	0.360	0.530
	19	10	0.150	75	1874	0.238	1.771
	20	10	0.180	75	2879	0.349	0.994
	21	10	0.230	75	3035	0.201	0.854

[a] AIBN Content was kept 0.26 wt% for all samples.

Naturally, from these experiments one may expect a drastic increase in the fluorescence intensity, I, of Py around the gel point.

EGDM has been commonly used as a cross-linker in the synthesis of polymeric networks [20]. Here, for our use, the monomers MMA (Merck) and EGDM (Merck) were freed from the inhibitor by shaking with a 10% aqueous KOH solution, washing with water and drying over sodium sulfate. They were then distilled under reduced pressure over copper chloride. The initiator, AIBN (Merck), was recrystallized twice from methanol. The polymerization solvent, toluene (Merck), was distilled twice over sodium.

Steady-state fluorescence measurements were performed using a Model LS-50 spectrometer from Perkin Elmer, equipped with a temperature controller. All measurements were made at a 90° position and slit widths were kept at 2.5 mm.

Mainly, four different sets of experiments were carried out: in the first set; AIBN (0.26 wt%) was dissolved in MMA and this stock solution was divided and transferred into round glass tubes of 15 mm internal diameter for fluorescence measurements. Different samples were prepared with various EGDM contents for bulk polymerization at 75 °C. In the second set; by using stock solution, samples were again prepared with various EGDM contents for bulk polymerization at 85 °C. In the third set; various samples with the same EGDM contents were used for polymerization at different temperatures. In the last set of experiments; samples were prepared with various toluene contents using the stock solution of the first experimental set and measurements were carried out with these samples at 75 °C during solution polymerization in toluene, where the EGDM content was kept at 0.01 vol%. Details of the samples for all sets are listed in Table 1. Py concentrations were taken as 4×10^{-4} M for all samples in these

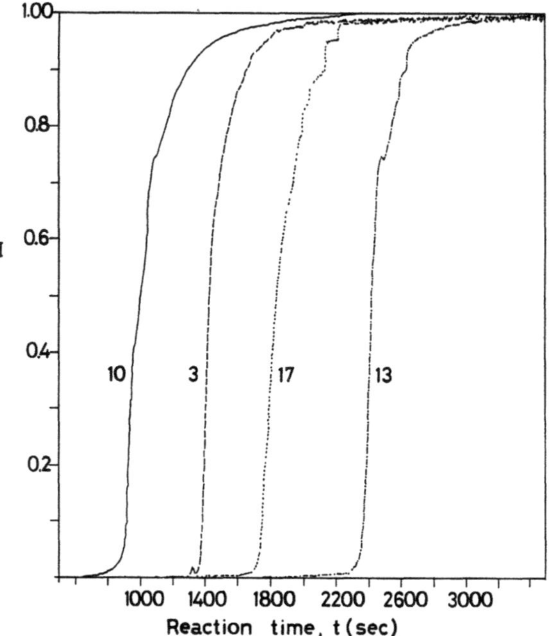

Fig. 1. Plots of P_y intensity I vs reaction time t during polymerization. The numbers *3, 10, 13* and *17* correspond to samples in experimental sets 1, 2, 3 and 4 in Table 1

experimental sets. Samples were deoxygenated by bubbling nitrogen through for 10 min and then radical copolymerization of MMA and EGDM was performed in the fluorescence accessory of the spectrometer. The Py molecule was excited at 345 nm during in situ experiments and variation in fluorescence emission intensity, I, was monitored with the time-drive mode of the spectrometer, by staying at the 395 nm peak of the Py spectrum. No shift was observed in the wavelength of the maximum intensity of Py and all samples were kept at their transparency during the polymerization process. Scattering light from the samples was also monitored during gelation experiments and no serious variation was detected at 345 nm intensity. Normalized Py intensities vs reaction times are plotted in Fig. 1 for samples from each set. Gelation curves in Fig. 1 represent asymptotic behaviors, which give evidence of typical critical phenomenon [1, 21].

2.2
Probing Gel Swelling by Fluorescence

All measurements were made at the 90° position and slit widths were kept at 2.5 mm. In situ swelling and slow release experiments were both performed in

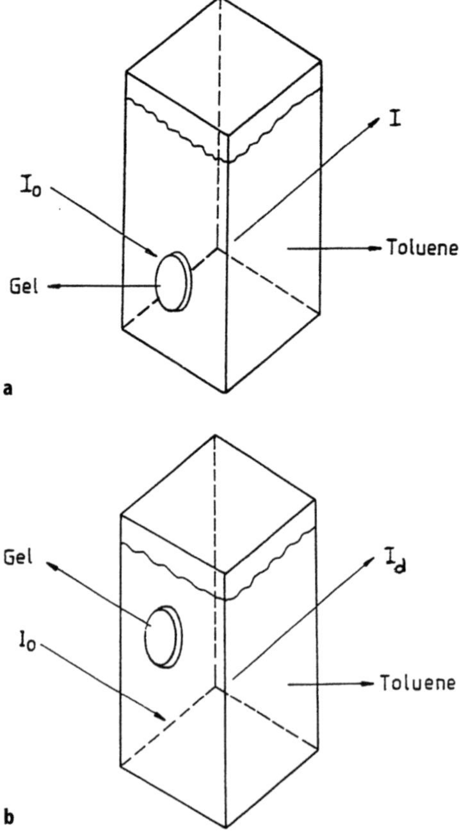

Fig. 2a, b. Fluorescence cell in LS-50 Perkin Elmer spectrometer. Monitoring of a gel swelling b desorption processes. I_0 and I_d are the excitation and emission intensities at 345 and 395 nm, respectively

a 1×1 cm quartz cell at 50 °C. Gel samples were attached to one side of a quartz cell by pressing the disk with thin steel wire. The quartz cell was filled with toluene. This cell was placed in the spectrometer and the fluorescence emission was monitored at a 90° angle so that during the swelling experiment only the gel was illuminated by the excitation light. During the desorption experiments, gel samples were shifted slightly to an upward position so that only the cell with toluene was illuminated by the excitation light. Here the fluorescence emission from Py molecules was monitored which were desorbed from the swelling gels. Fig. 2a, b presents the fluorescence cell and the gel positions in swelling and slow desorption experiments, respectively. In both experiments disk-shape gels were used which were dried and cut from the cylindrical gels obtained from FCC. The radii of these disk-shape gels were around 7.7 mm.

During the swelling and slow release experiments the cell was illuminated with 345 nm excitation light and the Py intensity, I, was monitored at 395 nm using the „time-drive" mode of the spectrometer. In the swelling experiment no shift was observed in the wavelength of maximum intensity of Py and gel samples were kept at their transparencies during the experiments. Typical fluorescence spectra of a gel before and after swelling are shown in Fig. 3.

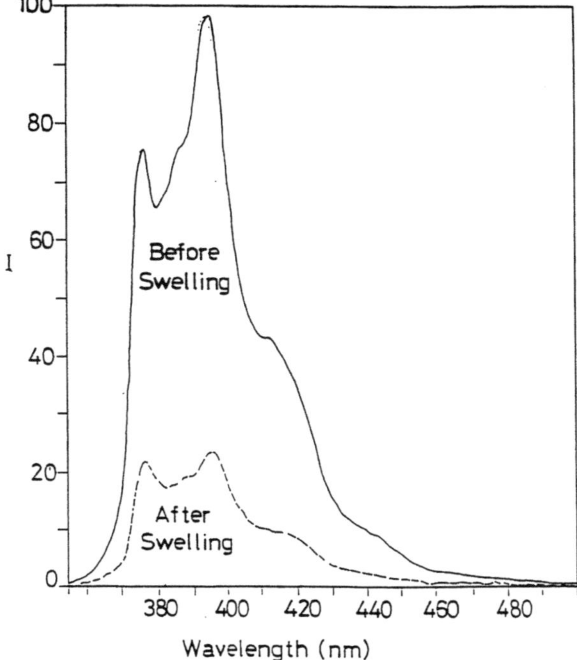

Fig. 3. Emission spectra of pyrene (P_y) taken before and after the swelling process is complete. P_y is excited at 345 nm

3
Results and Discussion

3.1
Gelation

If one assumes that the reaction time, t, for FCC is proportional to the probability p about the close proximity of the percolation threshold, then the above results can be quantified as follows. Above t_c the fluorescence intensity I monitors the growing gel fraction G, and below t_c the weight average degree of polymerization DP_w. Then Eqs. 1 and 2 can be written in the following forms:

$$I = A\,(t - t_c)^\beta, \quad t \rightarrow t_c^+ \tag{3}$$

$$I = D\,(t - t_c)^{-\gamma}, \quad t \rightarrow t_c^- \tag{4}$$

Here, the critical time t_c corresponds to the gel point p_c and A and D are the new critical amplitudes. The value t_c can be found from considering the first derivative of the experimentally obtained I–t curve with respect to t. In Fig. 4a we

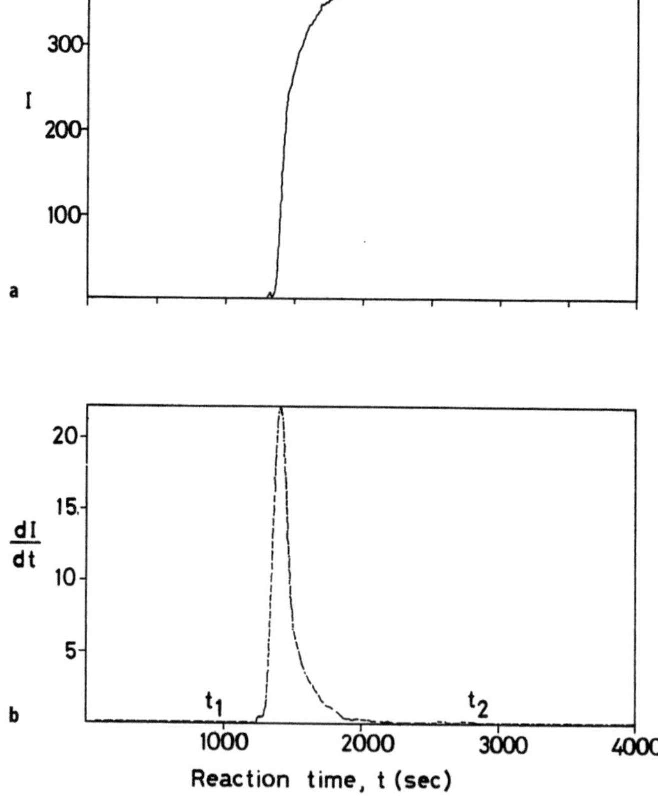

Fig. 4a, b. Determination of t_c: plots of **a** P_y fluorescence intensity I and **b** its first derivative dI/dt against reaction time t. t_c was chosen at P_c

present the I–t curve, and in Fig. 4b, we plot the derivative, dI/dt against t for Sample 3 in Fig. 1. The maximum in the dI/dt curve corresponds to the inflection point in the curve I(t). The conversion from time (t) to the probability (p) scale is effected by taking the end points of the time interval (t_1, t_2) over which the dI/dt curve differs from zero, to correspond to $p = 0$ and $p = 1$, as shown in Fig. 4b. Assuming a linear relationship between t and p, the position of the critical point, t_c, on the time axis is found from $p_c = (t_c - t_1)/(t_2 - t_1)$, where $p_c = 0.248$ is the percolation threshold found for the simple cubic lattice from Monte Carlo simulation [15]. The measured t_c values are listed in Table 1. It can be seen that t_c values decrease as EGDM content is increased at a given temperature (Set 1, 2) which indicates that gelation happens faster in high EGDM samples. For the samples at constant EGDM content, increasing temperature causes faster gelation (Set 3). For a given temperature and EGDM content, t_c decreases as toluene content is increased [21] (Set 4). Since t_c is directly proportional to the bond formation probability p, in the presence of toluene the behavior of t_c can be explained by the site bond percolation model [21], i.e. gel formation is retarded by toluene molecules, occupying the sites of the three dimensional lattice.

Below t_c, since the Py molecules are free, they can interact with and be quenched by sol molecules; as a result I decreases. In this region the fluorescence intensity originates from the clusters embedded in the monomeric medium. Each of these clusters are formed from polymer molecules and I is proportional to the weight average degree of polymerization. However, above t_c, since most of the Py molecules are frozen in the EGDM network, the intensity I yields very large values. The plots of $\log I = \log A + \beta \log(t - t_c)$ and $\log I = \log D + \gamma \log(t_c - t)$, for Sample 3 shown in Fig. 1 are presented in Fig. 5, in the critical region [1] of $10^{-2} < 1|1 - t/t_c| < 10^{-1}$ above and below t_c, respectively.

The critical exponents β and γ were determined from the slopes of the curves in Fig. 5 and are listed in Table 1, together with other β and γ values for the

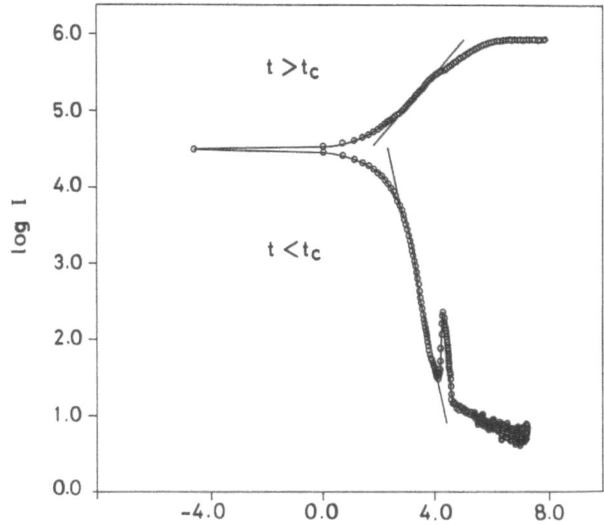

Fig. 5. Log-log plots of Eqs. 3 and 4 for the data given in Fig. 1. The $10^{-2} < |1 - t/t_c| < 10^{-1}$ region above and below t_c was chosen for the best fit to obtain β and γ values

corresponding samples for all the experimental sets. Average β and γ values were found to be 0.421 ± 0.005 and 1.673 ± 0.03, respectively, which are independent of the EGDM contents and temperature. These values are in good agreement with the values found from the bond percolation theory [23, 24], which confirms the choice of t_c as inflection point.

From Table 1, one can conclude that in the sets labelled 1 and 2 the critical times t_c are shifted to smaller values as the EGDM content is increased. Furthermore, the experimental sets 4 and 3 show that, at a given cross-linker content, gel formation is retarded with rising dilution but is accelerated as the temperature is increased. However, in all these sets the β and γ values vary around 0.42 and 1.7, respectively, independent of the EGDM contents and temperature, as is to be expected from bond percolation theory [1]. All of these results are in accordance with the theory of gelation [25] and measured exponents β and γ are consistent with the universal exponents in bond theory [1]. Here we have to note that increasing toluene content increases the site bond percolation during gelation, i.e. sites start to be occupied by toluene molecules. This is the reason why in the last four samples (18–21) in Table 1, set 4 had difficulty forming a gel, which also produced strange β, γ pairs. However, the first three samples (15–17) present meaningful β, γ pairs, most probably due to lower toluene content.

3.2
Swelling Kinetics

In swelling experiments, continuous volume transitions are expected which should result in a continuous decrease in the Py intensity I during swelling. Here one may expect that as solvent uptake (W) increases, desorption of Py molecules from the swollen gel increases, as a result Py intensity from the gel decreases. On the other hand, during desorption experiments one should expect an increase in I, due to increasing amount of Py molecules which are released into toluene in the cell.

Li and Tanaka showed that the kinetics of swelling and shrinking of a polymer network or gel obey the following relationship [8]:

$$\frac{W(t)}{W_\infty} = 1 - \sum_{n=1}^{\infty} B_n e^{-t/\tau_n} \tag{5}$$

where $W(t)$ and W_∞ are the swelling or solvent uptake at time t and at infinite equilibrium, respectively. Here B_n represents a constant related to the ratio of shear modulus G and longitudinal osmotic modulus, M, which is defined by the combination of shear and osmotic bulk moduli as [11, 12] $M = 4/3\, G + K$. τ_n is the swelling rate constant. In the limit of large t or if τ_1 is much larger than the rest of τ_n, all high-order terms ($n \geq 2$) in Eq. 5 can be neglected, then Eq. 5 becomes

$$\frac{W(t)}{W_\infty} = 1 - B_1 e^{t/\tau_1} \tag{6}$$

Here B_1 is given by the following relationship [8]:

$$B_1 = \frac{2(3-4R)}{\alpha_1^2 - (4R-1)\,3-4R)} \tag{7}$$

where $R = G/M$ and α_1 is given as a function of R, i.e.:

$$R = \frac{1}{4}\left[1 + \frac{\alpha_1 J_0(\alpha_1)}{J_1(\alpha_1)}\right] \tag{8}$$

Here, J_0 and J_1 are the Bessel functions. In Eq. 6, τ_1 is related to the collective cooperative diffusion coefficient D_c of a gel disk at the surface and given by the relationship [8]:

$$\tau_1 = \frac{3a^2}{D_c\alpha_1^2} \tag{9}$$

where a represents the half of the disk thickness in the final infinite equilibrium state which can be experimentally determined.

Py intensities, I_p vs swelling time are plotted in Fig. 6 for the gels prepared by toluene polymerization. The numbers in Fig. 6 correspond to gel samples listed in Table 2. It is interesting to note that solvent uptake is much faster in loosely formed gels (5 and 6) than densely formed gels (1).

The swelling curves in Fig. 6 were obtained during in situ florescence experiments described in Fig. 2a, where, at the beginning, all Py molecules are in the gel and I_{0s} is obtained. After solvent penetration starts, some Py molecules are was-

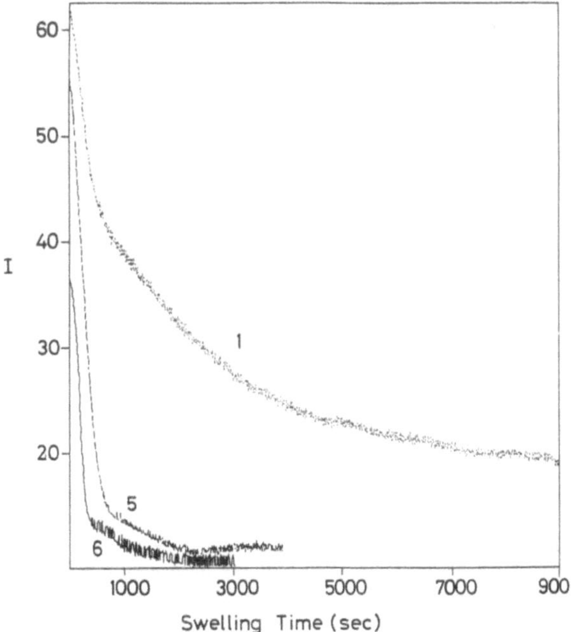

Fig. 6. Total pyrene intensity, I, vs slow release time for the gel samples listed in Table 2. The gel in the cell was illuminated at 345 nm during swelling measurements. Data for the plot were obtained using the time-drive mode of the spectrometer

Table 2. Swelling parameters of disk-shape gels[a]

Gels	1	2	3	4	5	6
Toluene vol%	0.10	0.13	0.15	0.20	0.23	0.30
τ_1 (s)	3062	3631	1538	1545	1021	962
B_1	0.87	0.86	0.59	0.67	0.74	0.78
α_1	1.33	1.40	2.16	2.0	1.77	1.64
$D_c \times 10^{-6}$ (cm²/s)	8.62	10.32	6.67	6.34	12.37	16.08
a (cm)	0.12	0.15	0.12	0.11	0.11	0.11

[a] τ_1 and B_1 values were obtained from Eq. 12 by linear regression of the swelling data in Fig. 11.

hed out from the swollen part of the gel into the cell, as a result the Py intensity, I_s, from glassy gel decreases as the swelling time increases. At the equilibrium state of swelling, the Py intensity from glassy gel reaches the $I_{\infty s}$ value where the solvent uptake by swollen gel is W_∞. The cartoon representation of these swelling stages are shown in Fig. 7, where the intensity from the desorbed Py molecules are presented by I_d. The relationship between solvent uptake W and fluorescence intensity I_s from the glass part of the gel is given by the following relationship:

$$\frac{W}{W_\infty} = \frac{I_{0s} - I_s}{I_{0s} - I_{\infty s}} \tag{10}$$

Since $I_{0s} \gg I_{\infty s}$, then Eq. 10 becomes:

$$\frac{W}{W_\infty} = 1 - \frac{I_s}{I_{0s}} \tag{11}$$

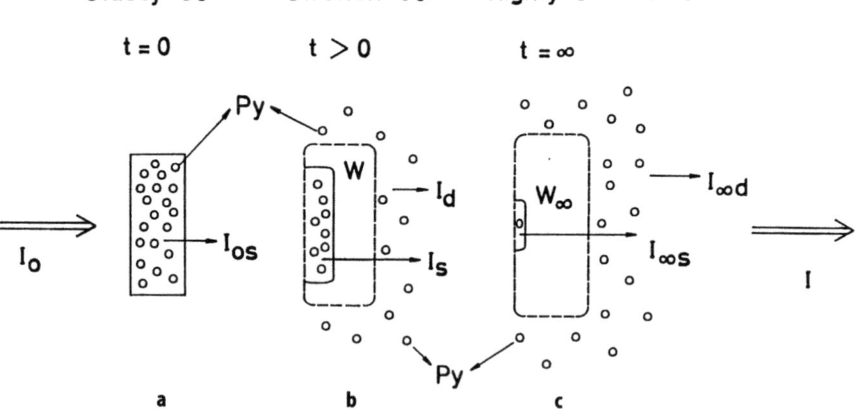

Fig. 7 a–c. Cartoon representation of the swelling processes in the gel during solvent uptake. Fluorescence intensities from P_y molecules are also presented. **a** Gel before swelling, I_{0s} is the fluorescence intensity from glassy gel at t = 0. **b** Swollen gel. I_s and I_d represent the fluorescence intensities from glassy gel and desorbed P_y molecules at t > 0. W is the solvent uptake. **c** Highly swollen gel. $I_{\infty s}$ and $I_{\infty d}$ are the fluorescence intensities at t = ∞. W_∞ is the solvent uptake at t = ∞

This relationship predicts that as W increases, I_s decreases. Here, it is assumed that Py molecules are completely washed out from the swollen part of the gel. Combining Eq. 11 with Eq. 6, the following useful relationship can be obtained:

$$\ln (I_s/I_{0s}) = \ln B_1 - t/\tau_1 \tag{12}$$

If one thinks that the fluorescence intensity curves in Fig. 6 originate only from the gels, then Eq. 12 has to be obeyed by the data. The digitized form of the data in Fig. 6 are plotted in Fig. 8 using Eq. 12, where we fail to observe the linear relationship in the swelling curves. This is not surprising, because during the swelling experiments desorbing Py molecules also contribute to the fluorescence intensity which prevents the observation of pure swelling curves as shown in Fig. 7. In fact the data in Fig. 6 present the total Py intensity, I curves, during in situ swelling experiments, which are presented at times by the following relationships:

$$t = 0, I_0 = I_{0s} + I_{0d}$$

$$t > 0, I = I_s + I_d \tag{13}$$

$$t = \infty, I_\infty = i_{\infty s} + I_{\infty d}$$

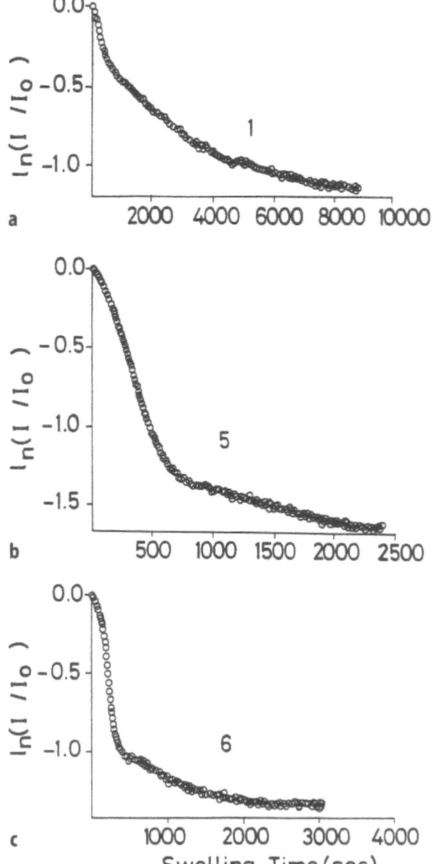

Fig. 8a–c. Logarithmic plot of the digitized data in Fig. 6. **a, b** and **c** represent the data for the gel samples marked with 1, 5 and 6 in Fig. 6 and in Table 2

Fig. 9. P_y intensity, I_d, from desorbing P_y molecules vs desorption time for the gel samples listed in Table 2. The cell was illuminated at 345 nm during desorption measurements. Data were obtained using the time-drive mode of the spectrometer

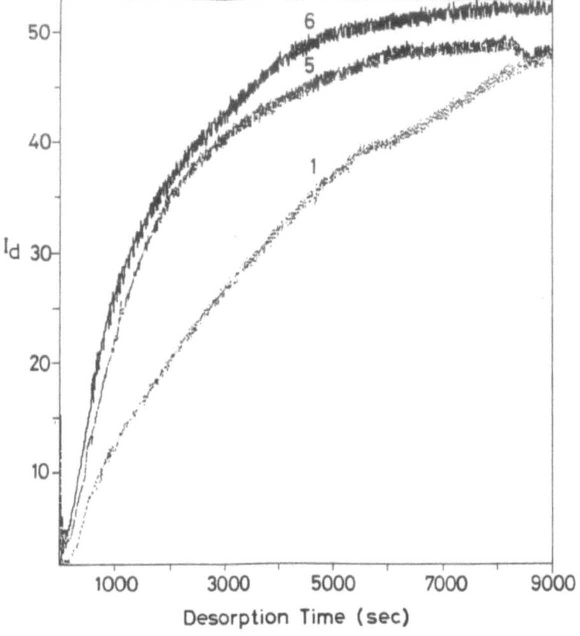

Fig. 10. P_y intensity, I_s, from the glassy part of the gel which is obtained by subtracting the data in Fig. 6 from the data in Fig. 9

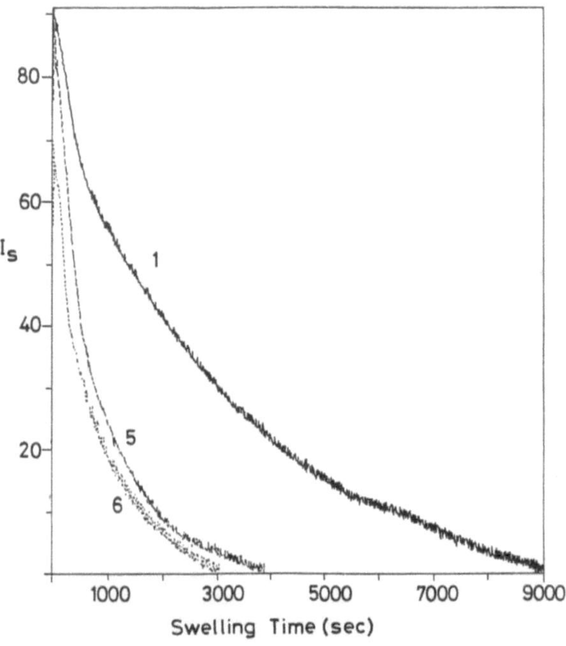

where I_d is the Py intensity from the desorbing Py molecules as shown in Fig. 7. Plots of I_d vs desorbing time for three different gels are shown in Fig. 9 which are obtained from the experiments performed according to Fig. 1b. In Fig. 9, I_d increases as the desorbing time increases for all the gel samples. Here numbers represent the gels given in Table 1 and in Fig. 6. Since I_d is directly proportional to the number of Py molecules in toluene, the behavior of the intensity curves in Fig. 6 suggest that Py molecules are desorbed much faster from densely formed gels (5 and 6) then loosely formed gel (1).

In order to produce the pure swelling intensity (I_s) curves, data in Fig. 6 are subtracted from the data in Fig. 9 according to Eq. 3 and plotted in Fig. 10. In order to see the correctness of the pure swelling curves in Fig. 10, data were digitized according to Eq. 12 and plotted in Fig. 11, where nice linear relationships are obtained except at very short and long time regions. Long time deviations are explained by the saturation of solvent uptake. Short time deviations may correspond to fast relaxation processes in the gel at the early swelling stage. Linear regression of curves in Fig. 11 at intermediate time region provides us

Fig. 11. Logarithmic plot of the digitized data of Fig. 10, which obey the Eq. 12. **a**, **b** and **c** represent the data for the gel samples marked with 1, 5 and 6 in Fig. 10 and Table 2

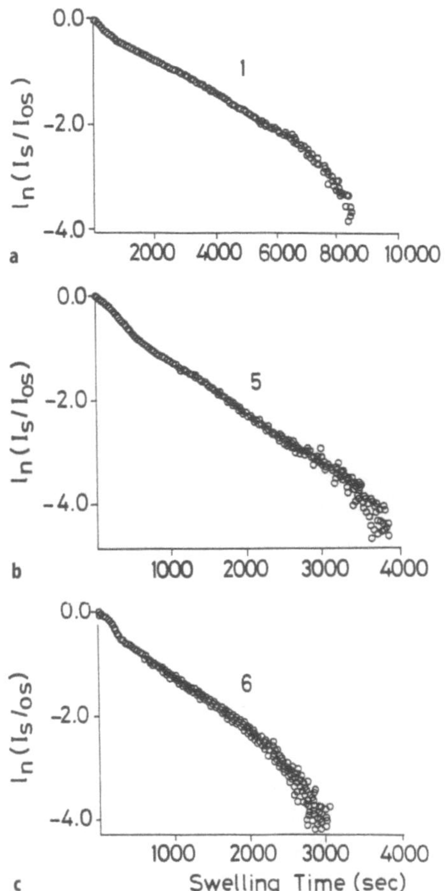

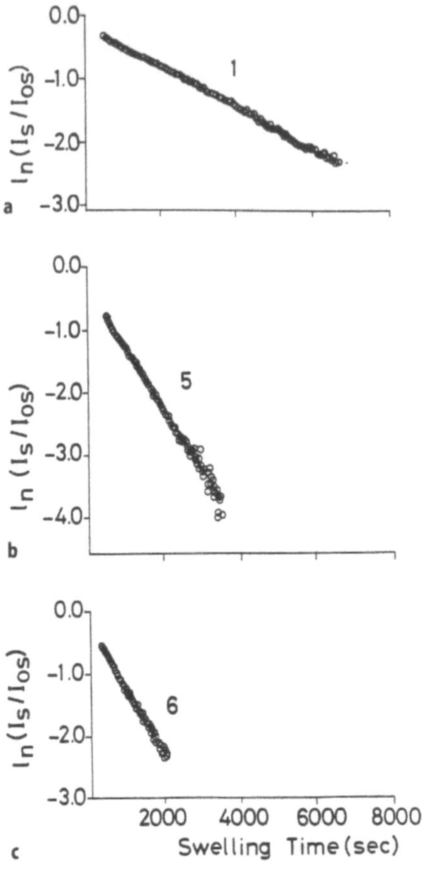

Fig. 12. Linear regressions of the data presented in Fig. 11 at the intermediate time regions. B_1 and τ_1 values were obtained from the intersections and the slopes of the plots in **a, b** and **c** for the gels marked with 1, 5 and 6 in Fig. 11

with B_1 and τ_1 values from Eq. 12. Fits are shown in Fig. 12. Taking into account the dependence of B_1 on R one obtains R values and from α_1-R dependence α_1 values were produced [1]. Then using Eq. 9, cooperative diffusion coefficients, D_c, were determined for these disk-shape gels. Experimentally obtained parameters, τ_1 and B_1, together with α_1 and D_c values are summarized in Table 2, where a values are also presented for each gel. Here one should notice that measured D_c values generally present larger numbers for loosely formed gels than densely formed gels, which states that faster cooperative diffusion occurs in densely formed gels due to high shear. As expected τ_1 values in Table 2, however, show reverse dependence on D_c values, i.e. loosely formed gels have smaller τ_1 values than densely formed gels. This result can be understood by realizing the fact that loosely formed gel has more vacant spaces, which can take more solvent easily and needs less time to reach its fully swollen state. Here one has to notice that τ_1 values are more consistent than D_c values. At this stage of our work we are unable to answer this question and more work needs to be done.

References

1. D. Stauffer, A. Coniglio, M. Adam, Gelation and Critical Phenomena. Advances in Polymer Science. Springer-Verlag, Berlin, Heidelberg 74, 44 (1982)
2. L.J. Kropp, R.W. Dawson, "Fluorescence and Phosphorescence of Aromatic Hydrocarbons in Poly(methylmethacrylate)" Molecular Luminescence International Conference, Ed. E. C. Lim. (1969), New York, Benjamin
3. M. Bixon, J. Jortner, J. Chem. Phys. 48, 715 (1968)
4. J. B. Birks, M. D. Lumb, I. H. Munro, Proc. Roy. Soc. A 277, 289 (1964)
5. K. Kamioka, S. E. Weber, Y. Morishima, Macromolecules 21, 972 (1988)
6. P. F. Jones, Siegel, J. Chem. Phys. 50, 1134 (1969)
7. Ö. Pekcan, J. Apply. Polym. Sci. 57, 25 (1995)
8. Y. Li, T. Tanaka, J. Chem. Phys. 92(2), 1365 (1990)
9. M. Zrinyi, J. Rosta, F. Horkay, Macromolecules 26, 3097 (1993)
10. S. Candau, J. Baltide, M. Delsanti, Adv. Polym. Sci. 7, 44 (1982)
11. E. Geissler, A. M. Hecht, Macromolecules 13, 1276 (1980)
12. M. Zrinyi, F. Horkay, J. Polym. Sci., Polym. Phys. Ed 20, 815 (1982)
13. T. Tanaka, D. Filmore, J. Chem. Phys. 20, 1214 (1979)
14. A. Peters, S.J. Candau, Macromolecules 21, 2278 (1988)
15. J. Bastide, R. Du0plessix, C. Picot, S. Candau, Macromolecules 17, 83 (1984)
16. J.C. Panxviel, B. Dunn, J.J. Zink, J. Phys. Chem 93, 2134 (1989)
17. Y. Hu, K. Horie, H. Ushiki, F. Tsunomori, T. Yamashita, Macromolecules 25, 7324 (1992)
18. Ö. Pekcan, Y. Yılmaz, O. Okay, Chem. Phys. Lett., 229, 537 (1994)
19. Ö. Pekcan, Y. Yılmaz, O. Okay, Polymer (in press)
20. O. Okay, Ç. Gürün, J. Appl. Polym. Sci 46, 421 (1992)
21. Ö. Pekcan, Y. Yılmaz, O. Okay, Polymer 37, 2049 (1996)
22. Ö. Pekcan, Y. Yılmaz, O. Okay, Chem. Phys. Lett. 229, 537 (1994)
23. D. Stauffer, Phys. Reps. 54, 1 (1979), and in: int. Conf Disordered Systems and Localization, C. di Castro (ed), Springer Lecture Notes in Physics 149, 9 (1981)
24. L. Hoshen et al., J. Phys. A12, 1285 (1979)
25. P. J. Flory, Principles of Polymer Chemistry, Cornell University, Ithaca NY. 1953

Photophysical Studies Provide Thermodynamic Insights into Block-Copolymer Micelle Formation in a Selective Solvent

C. W. Frank, D. Y. Ylitalo

1
Introduction

A large number of theoretical [1–4] and experimental efforts [5–16] have been directed toward describing the micellization of block copolymers in solvents selective for one of the blocks. For appropriate conditions of concentration, temperature, and pressure, spherical micelles may be formed from aggregates of approximately 100 block copolymer molecules in which the shell or corona is composed of the block having favorable enthalpic interactions with the solvent, and the core is composed of the block having unfavorable solvent interactions. In the present study, we are interested in two aspects of the micellization phenomenon: (i) the effect of solvent size on the equilibrium between the individual block copolymer chain (the unimer) and the micelle, as interpreted through the closed-association model, and (ii) the extent to which homopolymer chains of the core block material may be solubilized in the micelle core. Both of these issues involve thermodynamic questions that can be addressed by use of photophysical experimental techniques, the first taking advantage of intrinsic excimer fluorescence from the core block and the second incorporating a fluorescently labeled, free probe molecule that partitions between the micelle core and the surrounding solvent.

In the following, we present highlights of the theoretical and experimental literature related to the solvent influence on the unimer/micelle equilibrium, our first topic of inquiry. Of particular theoretical interest is the work of Munch and Gast [4] who followed the mean-field treatment of Leibler et al. [2] to examine the effect of solvent size on the critical micelle concentration (cmc) and micelle aggregation number. For ratios of copolymer molecular weight to solvent molecular weight between 20 and 50, they observed a decrease in both the cmc and the aggregation number with increasing solvent size. They explained the drop in cmc as being a result of the decreased solubility of the block copolymer in the larger solvent, where the entropy of mixing is reduced. The lower background concentration of free chains that accompanies the lower cmc is then equilibrated with a more dilute micellar phase and a lower aggregation number. While their calculations do not approach the range of solvent molecular weights in general use, the trends with solvent size are interesting for comparison with experimental data.

Several experimental studies have been published on block copolymer micelles in alkane solvents of varying sizes. Stacy and Kraus [6] observed a decrease in aggregation number for several styrene-butadiene block copolymer micelles as the solvent became better for the butadiene block, for example, from hexane to heptane to dodecane. Somewhat later, Bahadur et al. [7] studied the styrene-isoprene block copolymer micelle system in several alkanes and concluded that the hydrodynamic radius was most strongly affected by the enthalpic interactions between the solvent and the shell-forming isoprene block. As the solvent quality for isoprene was improved, the hydrodynamic radius increased, reflecting the increased stretching of the isoprene chains. Finally, Candau et al. [8] measured the ^{1}H- and ^{13}C-NMR spectra of poly(styrene-b-ethylene propylene) (PSPEP) copolymers as a function of temperature in octane, decane, dodecane, and hexadecane. At low temperatures, they observed no motion of the styrene units, indicating a viscous, possibly glassy core. As the temperature was increased, the mobility of the phenyl groups increased, indicating possible core plasticization. For temperatures between 373 and 408 K, the mobility of the styrene units approached that of polystyrene in a good solvent, indicating that the micelles had completely broken up. The temperature at which this occurred increased as the solvent size increased.

While solubilization of hydrocarbons by surfactant micelles in aqueous solution has been the subject of much research [17–21], the analogous phenomenon of homopolymer solubilization in the core of block copolymer micelles, the subject of our second inquiry, is less well characterized. Over two decades ago, Skoulious et al. [22] showed that micelles of AB block copolymers with a core of aggregated A blocks can solubilize homopolymers of A provided that they are shorter than the core A block. Since that time, several researchers [23–25] have noted the solubilization phenomenon in such systems but no effort has been made to quantify the equilibrium constant for homopolymer solubilization. This quantity can be important for understanding the behavior of a block copolymer system in which small amounts of homopolymer remain from the synthesis. Additionally, in cases where a homopolymer of the same chemical structure as the micelle core block is solubilized, the response of the homopolymer to changes in environment may reflect some of the same thermodynamic forces that are acting on the micelle core block itself.

In this paper, we follow the approach of Yeung and Frank [10] and Ylitalo and Frank [25] and apply excimer photophysics to quantitatively describe the effect of alkane solvent size on the thermodynamics of micellization of PSPEP block copolymers in heptane, dodecane, and hexadecane. We also develop a methodology for calculating the partition coefficient for a homopolymer partitioning between a block copolymer micelle core and the solvent phase. In particular, we measure the partitioning of a pyrene doubly end-tagged polystyrene (Py-PS-Py) between the PSPEP core and a heptane solvent phase. In addition to varying concentration and temperature, the standard parameters, we present data on the effect of pressure on the partitioning. This is relevant to the application of such block copolymers as additives to lubricating oil that is subjected to high pressures in actual use.

2
Experimental

2.1
Materials and Preparation

Poly(styrene-*b*-ethylene propylene) (PSPEP) block copolymer was kindly provided to us by Shell Development Co. PSPEP is a diblock copolymer prepared by anionic block copolymerization of styrene and isoprene followed by complete hydrogenation of the isoprene. Thus, one block consists of an all-carbon backbone with phenyl rings on alternate carbons; the second block effectively consists of repeating units composed of a two carbon "ethylene" unit and a three carbon "propylene" unit with the associated pendent methyl group. The distinctly different chemical character of the two blocks – aromatic for the polystyrene portion and aliphatic for the ethylene-propylene portion – give rise to the selective solvent micellization behavior. Number-average molecular weights of the styrene and ethylene-propylene blocks were 35000 and 61000, respectively; the overall polydispersity was 1.02, as determined by GPC. Before use, the copolymer was subjected to several cycles of precipitation from tetrahydrofuran into methanol or acetone to remove any traces of polystyrene homopolymer, if present, and added stabilizers. The solvents used were spectro-grade *n*-heptane (Burdick and Jackson), *n*-dodecane (Kodak) and *n*-hexadecane (Burdick and Jackson), all non-solvents for polystyrene but good solvents for poly(ethylene propylene). Each solvent was checked for fluorescent impurities and filtered through a 0.02 µm filter before use. Stock solutions of polymer in solvent were made and then transferred into cylindrical quartz cuvettes for degassing. Each sample was subjected to five freeze-pump-thaw cycles to remove dissolved oxygen and then vacuum sealed at less than 2×10^{-5} Torr. Since oxygen is an efficient quencher of polystyrene fluorescence, its removal was necessary to increase the fluorescence signal and, in particular, to minimize the differential quenching of the excimer fluorescence relative to the monomer. The sealed tubes were heated at 423 K for 6 h to ensure complete dissolution of the copolymer and then slowly cooled to room temperature over 8 h. This thermal treatment produced a reproducible, equilibrium micelle structure, from which no appreciable changes in micelle size or fluorescence spectra were observed after storing at room temperature for several weeks.

Probe chains for the core solubilization study were polystyrene homopolymer, doubly end-tagged with pyrene (Py-PS-Py), with molecular weights of 2000, 4000, 8000 and 32000; their synthesis has been described elsewhere [26]. The end-labeling reaction was performed by coupling the living polystyryl anion (from an anionic polymerization) with 1-bromobutylpyrene, leading to direct carbon-carbon attachment to the polystyrene chain. This approach is preferred to the common alternative method of esterification of hydroxy-terminated polystyrene with pyrene butyric acid because the esterification route introduces a polar group that could modify solvation behavior. Solutions were prepared in heptane such that the number of micelles was five times the number of probe chains to avoid the formation of intermolecular excimers within the micelle core. Each sample was subjected to the same freeze-pump-thaw cycles and thermal equilibration as described above for the pure block copolymer solution. No changes were observed in the fluorescence spectra of the probe molecules in heptane before and after the thermal treatment.

2.2
Fluorescence Spectroscopy

Intrinsic fluorescence measurements on the pure PSPEP micellar solutions were conducted in a modular spectrofluorimeter that has been described elsewhere [27]. Fluorescence measurements were taken at 2 to 7 K intervals with 45–60 min equilibration time between data points. Similar experiments were performed with 20- and 90-min equilibration times; in each case, the fluorescence data agreed within experimental error, indicating that the 45-min equilibration time is certainly sufficient to establish thermodynamic equilibrium. Experiments were also run by decreasing the temperature from above the critical micelle temperature (cmt), and no change in the measured cmt was observed. Spectra were corrected for instrument response and to remove the solvent Raman scattering peak. The corrected spectra were then transformed to an energy scale and resolved into monomer and excimer peaks. The shape of the excimer peak was assumed to be that of sec-butyl benzene in heptane with an allowance for spectral blue shift as the temperature was increased. The excimer peak was assumed to be Gaussian on an energy scale. A Levenberg-Marquardt non-linear regression routine was used to deconvolve the excimer and monomer peaks, and goodness-of-fit was based on inspection of the weighted residuals and analysis of variance.

Fluorescence measurements of the probe-containing samples at atmospheric pressure were performed on a SPEX Fluorolog 212 spectrofluorimeter, which has been described previously [26]. Emission spectra were taken with a 343 nm excitation wavelength and emission monochromator band pass of 3.6 nm. Excitation spectra were taken at 376 nm and 510 nm for monomer and excimer emission, respectively. The excitation monochromator band pass was also set at 3.6 nm. Fluorescence measurements at elevated pressures were made on the modular spectrofluorimeter [27]. Pressure was maintained in a cubic stainless steel optical pressure cell in which sapphire windows occupied two adjacent window ports for fluorescence excitation and emission monitoring. This cell and the pressure-generating system have been described in detail elsewhere [27].

2.3
Dynamic Light Scattering

Dynamic light scattering (DLS) experiments were performed on a system that has been described previously [9]. The major components of this system include a Lexel model 95–2 2 W argon-ion laser with the 514.5 nm line selected, a Brookhaven Instruments BI-200 goniometer, and a Brookhaven BI2030 136-channel correlator. DLS measurements at 293 K were carried out at with a 90°

scattering angle. Data from the correlator were collected by an IBM AT computer, and a cumulant analysis was performed. The data accepted based on baseline agreement were then analyzed for the average micelle size and size distribution by the CONTIN inversion program provided by Provencher.

2.4
Viscometry

Intrinsic viscosity measurements were made in an Ubbelohde capillary viscometer immersed in a water bath at 293 K. Flow times were in excess of 200 s, so that no kinetic corrections were necessary.

3
Results and Discussion

3.1
Micelle-Free Chain Equilibrium

We begin by characterizing micelle structure using standard techniques. Table 1 summarizes our results on the hydrodynamic radius, intrinsic viscosity, and aggregation number of the micelles formed in each solvent. Although these measurements were made at 293 K, subsequent photophysical measurements were performed at elevated temperatures. The reason for this is that we wanted to measure the properties of the micelles only, while at higher temperatures, a large number of free chains exist in equilibrium with the micelles. The aggregation number may be obtained from Eq. 1 [15]

$$[\eta] = \frac{10\pi R_{\mathrm{h}}^3 N_{\mathrm{AV}}}{3M_{\mathrm{m}}} \tag{1}$$

where M_{m} is the micellar mass, $[\eta]$ is the intrinsic viscosity, and N_{AV} is Avogadro's number. The decrease in aggregation number with increasing solvent size is consistent with the predictions of Munch and Gast [4].

If a closed-association model is assumed to apply to the block copolymer micellization, an increase in temperature will shift the micelle equilibrium towards freely dispersed chains. Moreover, the core is expected to imbibe some of the alkane solvent, as suggested by the work on the PSPEP/octane system by Candau et al. [8], who employed NMR techniques sensitive to the PS core block and observed a swelling of the core as the temperature was increased. Both of

Table 1. Solvent Dependence of Micelle Sizes

Solvent	R_{h}/nm	$[\eta]$/cm³/g	N_{agg}
heptane	46.8	65	103
dodecane	46.8	70	96
hexadecane	43.3	64	83

these effects will lead to a change in the chemical environment of the PS block component, that is, from one rich in aromatic PS segments to one with an increasing aliphatic solvent content. Fluorescence techniques are particularly well suited for studying aromatic ring-containing polymer systems because of the possibility of intra- or intermolecular excimer formation occurring. (Given the importance of this technique, we provide a brief tutorial on the photophysics of aromatic vinyl polymers in the Appendix.) The excimer-forming site is expected to be sensitive to such a change in the local environment because stabilization of the phenyl ring overlap constitutes a delicate balance between steric repulsive forces and resonance attractive forces. Indeed, significant variation in the spectral characteristics of the excimer fluorescence from poly(1-vinylnaphthalene), poly(2-vinylnaphthalene), and their model compounds, has been observed in a homologous series of alkyl benzene solvents [28].

The sensitivity of excimer fluorescence to local chemical environment was the basis for the initial work of Yeung and Frank [10] in which the temperature dependence of the PS excimer band-width and band-position provided information on the changes in intermolecular interactions within and around the PS block of the PSPEP block copolymer in heptane solvent. At a temperature dependent on the solution concentration, they observed an increase in the slope of the excimer band-width plotted against inverse temperature and proposed that this point represented the critical micelle temperature (cmt). Their subsequent thermodynamic analysis based on this proposal was consistent with other results from the literature, providing verification of the technique. In a recent study [25], we found the excimer band-position to be a more reliable measure of the cmt at elevated pressures. In that work, the excimer band-position decreased with increasing temperature below the cmt, reflecting the shift in the micelle-free chain equilibrium towards free chains. At the cmt, the band-position reached a minimum value similar to that of PS in heptane and then increased with the same temperature dependence as PS in heptane at higher temperatures. In the present study, we present both band-width and band-position data for heptane and hexadecane solvents, but we will use the former set for the cmt determination in order to compare with the earlier results [10] in dodecane. In most instances, the agreement between the break in the band-width data and the minimum in the band-position is within 2 to 4 K.

Figures 1 and 2 show the dependence of the excimer band-width on inverse temperature for various concentrations of PSPEP in heptane and hexadecane, respectively. All of the concentrations examined are higher than the cmc at room temperature [10], so that the overwhelming majority of the PSPEP chains will be located in micelles. The band-width is relatively insensitive to concentration at room temperature, and the initial values are all approximately 5000 cm^{-1} in heptane and 4900 cm^{-1} in hexadecane; as a result, the data in Figs. 1 and 2 have been vertically shifted for clarity. The most striking observation, which applies to both solvents, is the dramatic change in the slope of the band-width data at a particular temperature, which is attributed to the cmt [10]. Previous dynamic light scattering experiments for PSPEP in n-heptane [9] have shown that the first-order autocorrelation function was single exponential for PSPEP concentrations between 0.01 and 0.5%. For larger concentrations, a bimodal distribution of re-

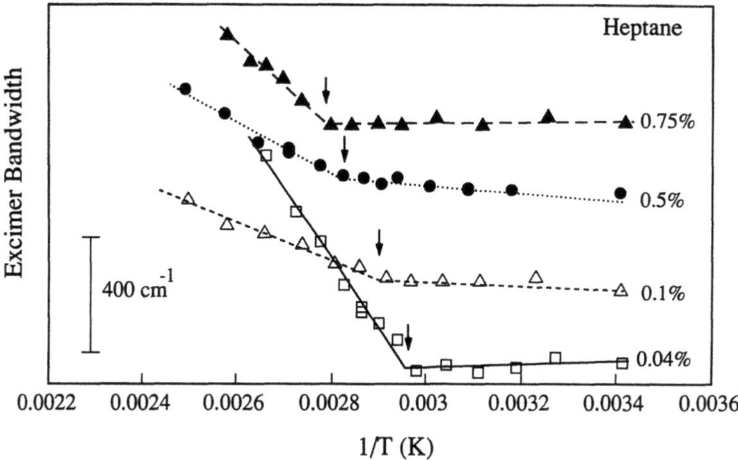

Fig. 1. Inverse temperature dependence of the relative excimer band-width (scale shown in cm^{-1}) of the PSPEP block copolymer in heptane at the concentrations (wt%) shown. Data have been shifted vertically for clarity, and the cmt values are denoted by the *arrows*

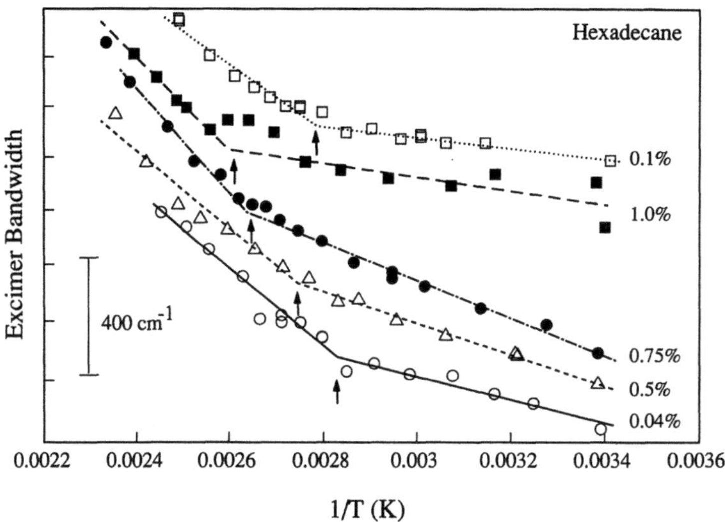

Fig. 2. Inverse temperature dependence of the relative excimer band-width (scale shown in cm^{-1}) of the PSPEP block copolymer in hexadecane at the concentrations (wt%) shown. Data have been shifted vertically for clarity, and the cmt values are denoted by the *arrows*

laxation times was observed. This may reflect micelle-micelle interactions and, thus, the photophysics of excimer formation within the core might be expected to be altered. However, all of the thermodynamic conclusions are based on data obtained for PSPEP concentrations of 0.5% or less.

The physical basis for this slope change is the alteration of the environment of the PS excimer-forming sites as the equilibrium shifts toward free chains. More specifically, the nearest neighbors change from PS repeat units in the micelle core to include an increasing content of alkane solvent molecules. The fact that there is no abrupt change in band-width suggests that the micelle breakup may be gradual with contributions to the observed fluorescence coming from both free chains as well as ones still within micelles. Upon complete dissolution of the micelle, the immediate environment of the excimer-forming site will be enriched in solvent, although there could be contributions from intramolecular phenyl rings due to the formation of unimolecular micelles. In addition, whereas the band-width initially remains constant with increasing temperature in heptane, it increases modestly in hexadecane. This difference must be associated with variations in the PS conformation or environment in the micelle core for the two solvent systems.

Corresponding band-position results for the same spectra are shown in Figs. 3 and 4. Initial values of the excimer band-position at 293 K are independent of composition over the range of 0.04 to 0.5 wt%; the data have been vertically shifted for clarity. As temperature is increased, the band-position passes through a minimum, attributed to the cmt [10], and the values obtained are in good agreement with the band-width data. Since the potential-energy surface for the phenyl-phenyl interactions in PS is unknown, it is difficult to provide an

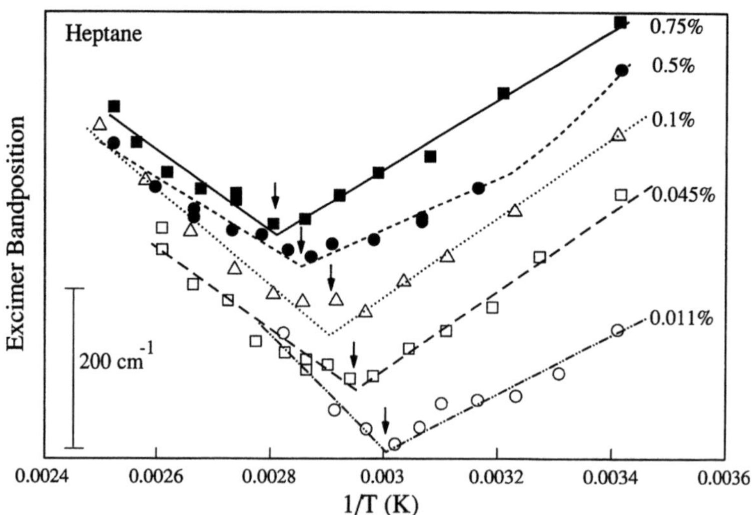

Fig. 3. Inverse temperature dependence of the relative excimer band-position (scale shown in cm^{-1}) of the PSPEP block copolymer in heptane at the concentrations (wt%) shown. Data have been shifted vertically for clarity, and the cmt values are denoted by the *arrows*

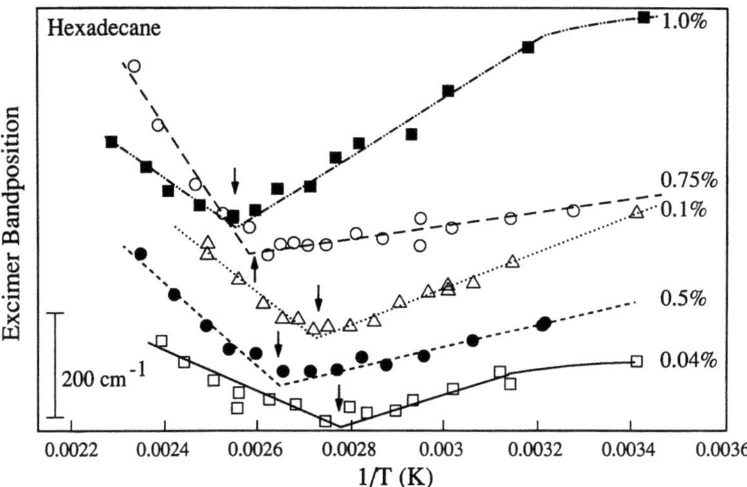

Fig. 4. Inverse temperature dependence of the relative excimer band-position (scale shown in cm^{-1}) of the PSPEP block copolymer in hexadecane at the concentrations (wt%) shown. Data have been shifted vertically for clarity, and the cmt values are denoted by the *arrows*

energetic explanation for the decrease in excimer band-position with decreasing temperature. Nevertheless, a previous study [25] of a low molecular weight PS homopolymer, which is soluble in the alkane solvent, has shown a comparable intrinsic temperature dependence of the band-position. Thus, we expect that at the highest temperature where the micelles are totally disrupted the excimer-forming sites will reflect an environment that is comparable to low molecular weight PS, with the possibility of some modification if unimolecular micelles are formed. Superimposed on this electronic-based decrease in band position with increasing temperature is then a morphology-based increase that occurs when the micelle is formed at lower temperatures.

Since the solution concentration is equal to the critical micelle concentration at the critical micelle temperature [10], data from Figs. 1 and 2 may be used to estimate the standard free energy, ΔG^0, enthalpy, ΔH^0, and entropy, ΔS^0, of micellization using the method of Price [11]. Assuming the closed association process for micellization to apply, Price describes an equilibrium between free chains, A_1, and micelles of aggregation number n, A_n, with:

$$K_n = [A_n]/[A_1]^n \tag{2}$$

where K_n is the equilibrium constant for micellization. ΔG^0 is then given by:

$$\Delta G^{\circ}_n = RT \ln [A_1] - \frac{RT}{n} \ln [A_n] \tag{3}$$

The second term can be neglected, given the large aggregation number, as indicated in Table 1; thus, $[A_1]$ can be taken as the cmc. With these simplifications:

$$\Delta G^{\circ}_n = RT \ln (cmc) \tag{4}$$

Price then calculated ΔH^0 with the van't Hoff equation:

$$\Delta H_n^o = - RT^2 \frac{d\ln(cmc)}{dT} \tag{5}$$

which can be integrated to give:

$$\ln(cmc) = \frac{\Delta H^o}{RT} + C \tag{6}$$

where C is a constant of integration.

We may determine ΔH^0 from the slopes of the plots of $\ln(cmc)$ against inverse temperature, shown in Fig. 5, and, with the use of Eq. 4, we can then calculate ΔG^0 and ΔS^0 at 298 K. The summary of these values in the three solvents is shown in Table 2. The modest driving force for micellization is due to the slight excess of the negative standard enthalpy of micellization over the standard entropy of micellization.

The major contribution to ΔH^0 is the result of replacing PS-solvent contacts with lower energy PS-PS and solvent-solvent contacts. This process is analogous to polymer demixing, and the enthalpy of micellization is analogous to the negative enthalpy of mixing of polymer and solvent. In the simple Flory-Huggins treatment, the enthalpy of mixing is proportional to the Flory-Huggins χ parameter, described by:

$$\chi = \beta + \frac{V_1}{RT}(\delta_1 - \delta_2)^2 \tag{7}$$

where β is a non-combinatorial entropy term, V_1 is the molar volume of the solvent, and δ_1 and δ_2 are the solubility parameters for the solvent and polymer, respectively.

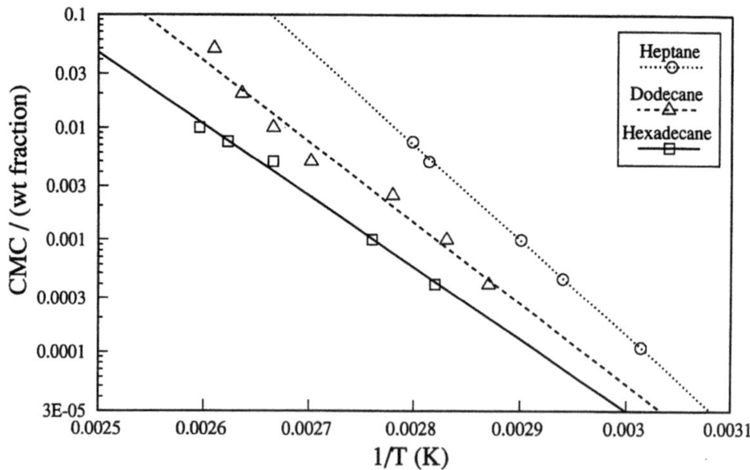

Fig. 5. Semi-log plot of critical micelle concentration (cmc) and inverse temperature for PS-PEP in heptane ($\cdots\bigcirc\cdots$), dodecane ($---\triangle---$), and hexadecane ($—\square—$). Dodecane data are from ref [10]

Table 2. Solvent Dependence of Micellization Thermodynamics at T = 378 K

Solvent	δ (cal/cm)$^{3/2}$	ΔG^0 (kJ/mol)	ΔH^0 (kJ/mol)	$T\Delta S^0$ (kJ/mol)
heptane	7.4	-6	-160	-154
dodecane	7.8	-13	-138	-125
hexadecane	8.05	-16	-122	-106

Figure 6 shows the standard enthalpy of micellization plotted against the solubility parameter difference multiplied by the solvent molar volume. The linear dependence indicates that the solvent dependence of ΔH^0 is adequately captured by the solubility parameter change in different solvents.

The solvent size effect on the micellization thermodynamics is observed through the solvent dependence of ΔS^0 shown in Table 2. As the solvent size is increased, the entropy of mixing of the free chains is reduced, thereby decreasing the entropy lost upon micellization. In the absence of solubility parameter effects, this results in a decrease of the cmc as the solvent size is increased, in agreement with the results of Munch and Gast [4]. These forces are opposed by the decrease in the χ parameter for PS with the larger alkane solvents, which tends to increase the cmc.

Despite the decrease in aggregation number from heptane to dodecane, the hydrodynamic radius of the micelles formed in each solvent is the same, as shown in Table 1. This is a result of the small solubility parameter difference

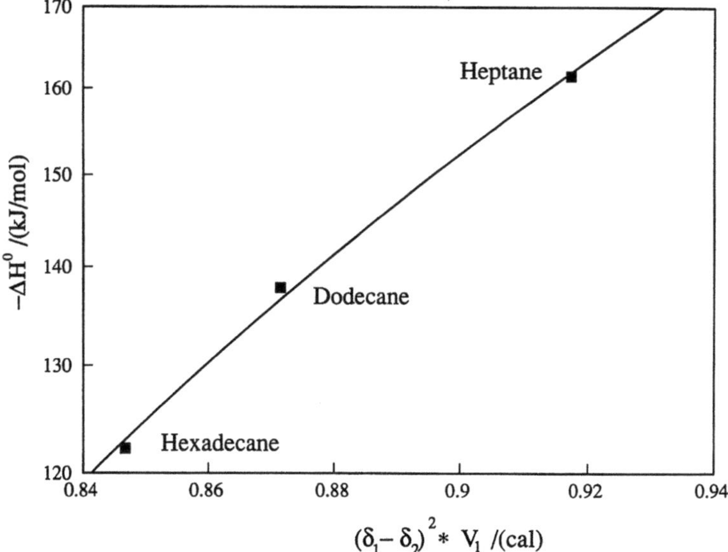

Fig. 6. Standard enthalpy of micellization for PSPEP in the three alkane solvents as a function of solubility parameter difference

between dodecane ($\delta = 7.8$ (cal/cm^3)$^{1/2}$) [29] and the ethylene-propylene block ($\delta \sim 7.8$ (cal/cm^3)$^{1/2}$) [30]. Whitmore and Noolandi [3] have calculated the thickness of the micelle shell as a function of solvent molecular weight, assuming athermal interactions between the solvent and the corona block of the polymer. They observed that the corona chains became more stretched as the solvent size decreased, indicating an increased swelling of the corona by the solvent. Our results may indicate contributions from both solubility parameter differences as well as swelling differences due to solvent-size effects. Between heptane and dodecane, it is apparent that the effect of the enthalpic interactions is most important, while between heptane and hexadecane the effect of mixing entropy in the shell becomes more important.

3.2
Homopolymer Partitioning Between Core and Solvent

Our second application of photophysics to block copolymer micelle thermodynamics involves study of the partitioning of homopolymer between the micelle core and surrounding solvent. To determine the relative amount of solubilized chains, we elected to use intramolecular excimer formation due to end-to-end cyclization of a telechelic probe molecule. The Py-PS-Py probe molecule allows dispersed and solubilized chains to be easily distinguished because excimer fluorescence occurs readily in solution but infrequently, if at all, in the dense micelle core. The relative amount of excimer fluorescence can then be easily related to the number of chains residing in the solvent phase.

To determine an appropriate probe molecular weight for the partitioning experiments, we dissolved probe chains with molecular weights from 2000 to 32000 in a 2% PSPEP micelle solution. Figure 7 shows the I_e/I_m ratio as a function of probe molecular weight. Two phenomena will influence the results – one related to the inherent ability of the chains to cyclize and the other to the partitioning behavior. Diffusion-controlled cyclization theory [31] predicts a $M^{-\gamma}$ dependence of the rate constant for cyclization, where γ is slightly above or below 1.5 depending on whether non-draining or free-draining coils are assumed. Higher molecular weight chains will cyclize at a slower rate and, therefore, exhibit less excimer fluorescence. Since the solubility of the homopolystyrene decreases in the solvent phase with increasing molecular weight, we expect a higher fraction of chains to reside in the micelle core where cyclization will be hindered. We then compare the I_e/I_m values in the micelle solution to that in pure heptane solution. For the lowest molecular weight probes, the addition of PSPEP micelles has little effect on the I_e/I_m ratio, indicating either that cyclization occurs in the micelle core or that little solubilization occurs; in either case, the probe is not useful for a study of the partitioning. At the other extreme, very little cyclization was observed from the 32000 M_w probe molecule; thus, we selected the 8000 M_w Py-PS-Py for our studies. With this probe, cyclization occurs readily in heptane but is greatly diminished in the micelle solution.

Figure 8 shows the I_e/I_m ratio for the 8000 M_w Py-PS-Py probe as a function of PSPEP concentration at atmospheric pressure and 293 K. For reference, the value in pure heptane is also shown. It is significantly larger than when added

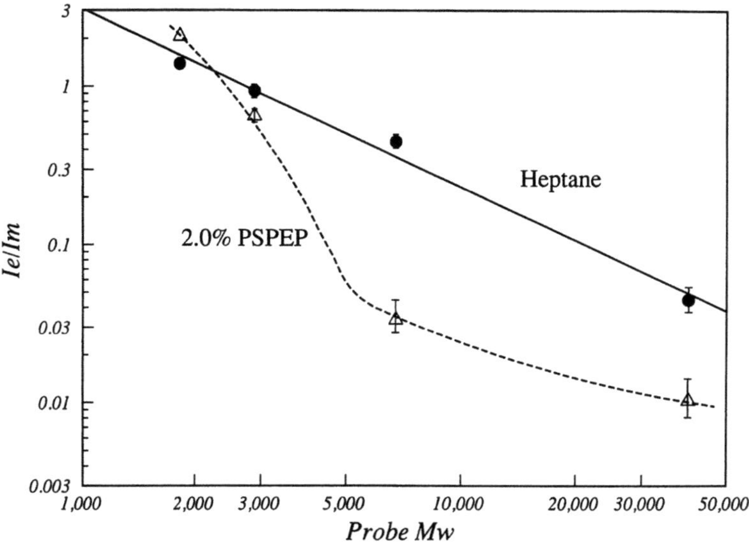

Fig. 7. Py-PS-Py molecular weight dependence of the I_e/I_m ratio in heptane solution (●) and 2% PSPEP solution (△)

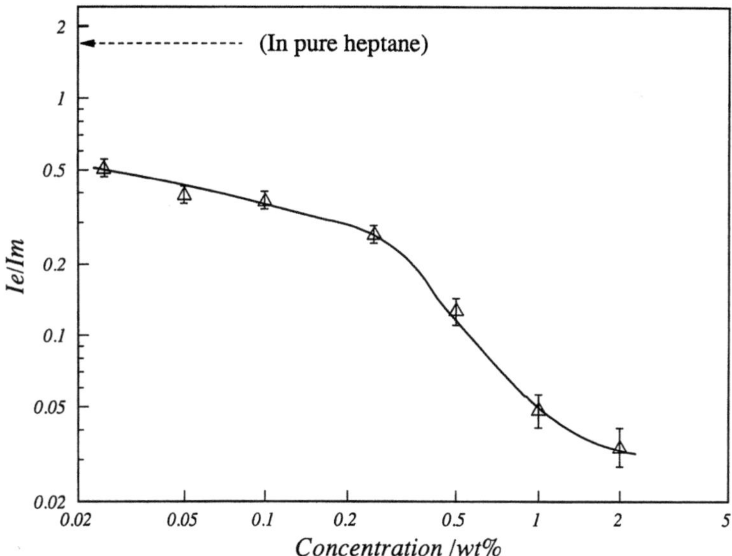

Fig. 8. PSPEP concentration dependence of the I_e/I_m ratio for the 8000 Py-PS-Py probe. The arrow indicates the value for the probe in pure heptane

to the PSPEP solution, indicating substantial incorporation of the probe in the micelles. An increase in PSPEP concentration produces a further decrease in I_e/I_m, indicating an increase in the relative amount of solubilized probe chain.

The substantial decrease in I_e/I_m for PSPEP concentrations of 0.5% or greater may be associated with the micelle-micelle interactions observed from dynamic light scattering [9]. To quantify the partitioning of the Py-PS-Py probe, we assume that no cyclization occurs within the micelle core and that cyclization in the dispersed phase occurs at the same rate as in pure solvent. We then may write I_e/I_m as a sum of contributions to give:

$$\frac{I_e}{I_m} = \left(\frac{I_e}{I_m}\right)_{hep} \left(1 + \frac{I_{m,mic}}{I_{m,hep}}\right)^{-1} \tag{8}$$

where the (mic) or (free) subscripts indicate solubilized or free chains, respectively. The I_e/I_m ratio in heptane can be measured independently, and we can then relate $I_{m,mic}/I_{m,free}$ to the ratio of solubilized to free chains. To do this we must determine the relationship between monomer intensity and the number of chains in each phase. We begin with the probabiliity of eventual monomer emission, M, [32] defined by Eq. 9:

$$\frac{I_e}{I_{m,}} = \frac{Q_e}{Q_m} \left(\frac{1}{M} - 1\right) \tag{9}$$

where Q is the quantum efficiency for either excimer or monomer. In the micelle core, where I_e/I_m vanishes, M is equal to 1. In the dispersed phase, we can calculate M from the value of I_e/I_m in pure heptane. With these values, we relate:

$$\frac{I_{m,mic}}{I_{m,hep}} = \left(\frac{Q_{m,mic}}{Q_{m,hep}}\right) \left(\frac{M_{mic}}{M_{hep}}\right) \left(\frac{N_{mic}}{N_{hep}}\right) \tag{10}$$

where N is the number of chains in either the micelle core or the dispersed phase; this leads to the fraction of solubilized chains X_m:

$$X_m = \frac{N_{mic}}{N_{mic} + N_{hep}} \tag{11}$$

It has been shown previously in complexation and adsorption experiments [33, 34] that the wavelengths of the peak maxima in the pyrene excimer and monomer excitation spectra shift as the pyrene is transferred from hydrophilic to hydrophobic regions. Because excimer formation can only result after a single pyrene chromophore has been excited, the excitation spectra for monomer and excimer in a homogeneous solution should be identical. If, however, excimer formation occurs from a subset of the entire monomer population, as would be the case if excimers form only in the dispersed phase, the excitation peak position would be shifted for excimer and monomer. Figure 9 shows excitation peak position data for monomer and excimer for the 8000 Mw probe chain in PSPEP solutions. In pure solvent, the excitation peak positions are identical, but as more chains enter the micelle core, the monomer position shifts to longer wavelengths, indicating some stabilization of the pyrene excited state.

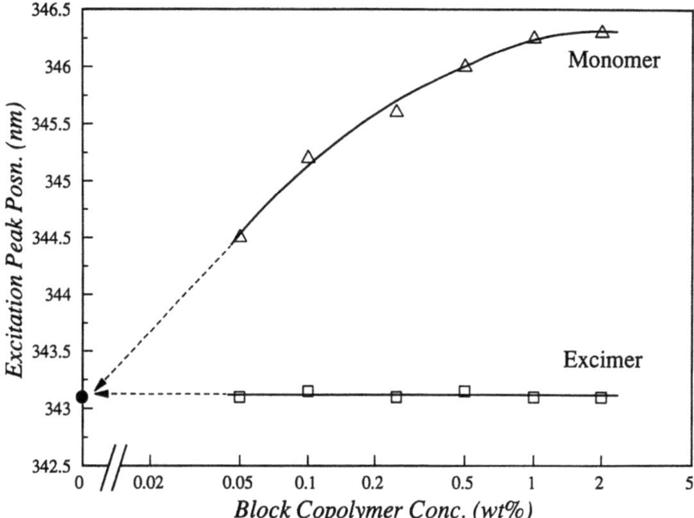

Fig. 9. PSPEP concentration dependence of the pyrene excimer ($\square$) and monomer ($\triangle$) excitation peak positions. The *solid circle* indicates the value for both monomer and excimer in pure heptane

At the same time, the excimer excitation position remains constant with increasing concentration, and at the same value as observed in pure heptane. These data indicate that, although the Py-PS-Py probe chains are becoming increasingly solubilized with higher micelle concentrations, excimer formation occurs only in the heptane-like dispersed phase. Therefore, no cyclization occurs in the micelle core.

The calculated fractions of solubilized probe molecule are shown as a function of PSPEP concentration in Fig. 10 and as a function of temperature for selected PSPEP concentrations in Fig. 11. Figure 10 also includes data obtained from the monomer excitation peak position. These values were obtained by determining the peak shift at each concentration relative to the amount of shift between the peak positions in pure heptane and in a 2 wt% PSPEP solution with a 32 000 M_w probe chain. We consider this ratio to represent the fraction of chains solubilized in the micelle core. The good agreement between these two results gives us confidence in our calculation of the solubilized fraction.

From the X_m data and the analysis of Hayase [21], we can calculate the partition coefficient, K_p, defined as the ratio of the mole fraction of probe chains solubilized in the micellar cores to the mole fraction in the continuous phase. We arrive at the following relationship:

$$K_p = \frac{Y_{PS}}{X_{PS}} \tag{12}$$

$$Y_{PS} = \frac{100 X_m}{680} \tag{13}$$

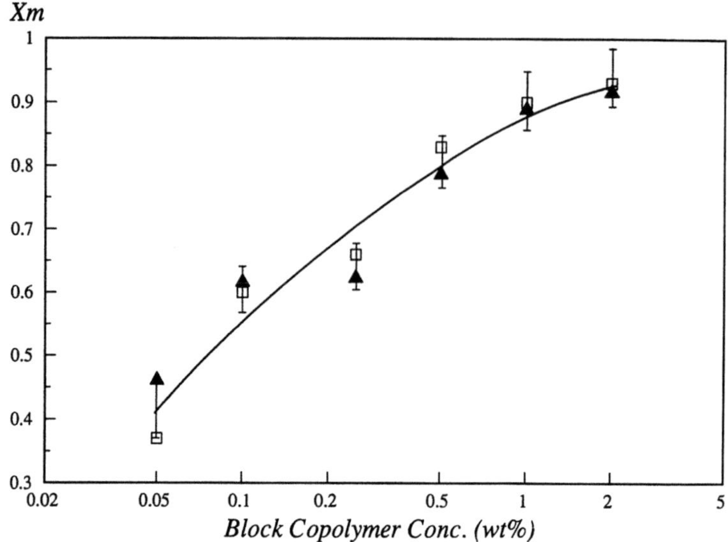

Fig. 10. PSPEP concentration dependence of X_m, the fraction of Py-PS-Py probe chains solubilized in the micelle core using the I_e/I_m (▲) and excitation position (□) calculations

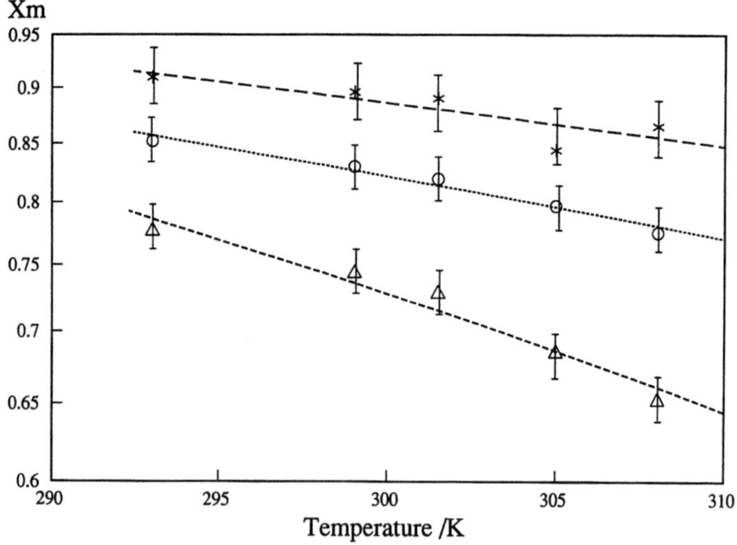

Fig. 11. Temperature dependence of the solubilized fraction, X_m, in 0.5% PSPEP (*), 1% PSPEP (O), and 2% PSPEP (△) at atmospheric pressure

$$X_{PS} = \frac{(1 - X_m)\,A}{B - CMC + (1 - X_m)\,A} \tag{14}$$

where A is the total probe concentration in mol/liter and B is the total PSPEP concentration in mol/liter. The cmc values are obtained from the previous work of Yeung and Frank [10]. Figure 12 depicts the concentration dependence of K_p above the cmc, calculated from both the excitation and the I_e/I_m data. At low PSPEP concentration, the partition coefficient is larger than the value at higher concentrations, indicating a higher propensity for the probe chains to solubilize in the micelle core. The leveling of Kp for PSPEP concentration may be related to corona interpenetration which would decrease the driving force for core solubilization.

The standard free energy of partitioning, Δ_p^0, can be calculated by Eq. 15:

$$\Delta G_p^0 = -\,RT \ln K \tag{15}$$

Using the value of $\ln K_p$ to be 13.8 for concentrations between 0.1 and 2% in Fig. 12, we find that ΔG_p^0 is -34 kJ/mol at 298 K. The temperature dependence of the free energy allows us to calculate the standard enthalpy of solubilization, ΔH_p^0, with the van't Hoff equation, which leads to Eq. 16

$$\ln K_p = -\,\frac{\Delta G_p^0}{RT}. \tag{16}$$

The plot of $\ln K_p$ versus $1/T$ yields a straight line with a slope of ΔH_p^0. At high temperatures, the assumption that there is no excimer formation within the micelle core may break down as the core becomes more fluid and the ratio of mi-

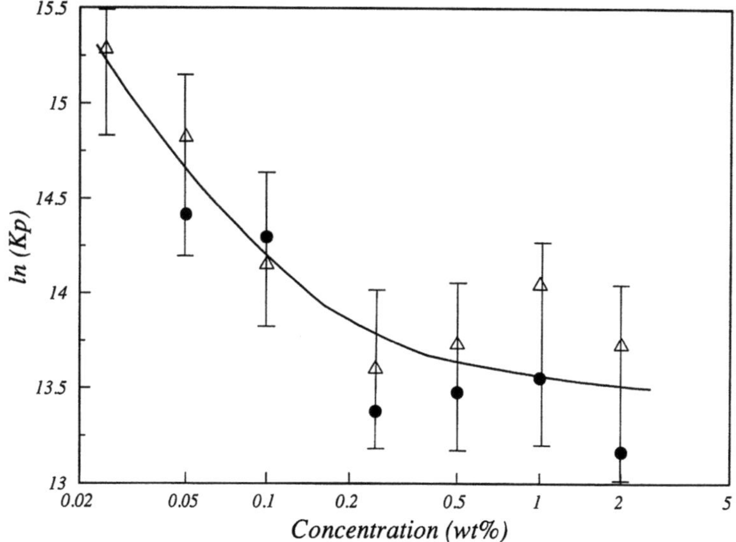

Fig. 12. PSPEP concentration dependence of K_p, the partition coefficient for the 8000 Mw probe chain as calculated from excitation data (●) and I_e/I_m data (△)

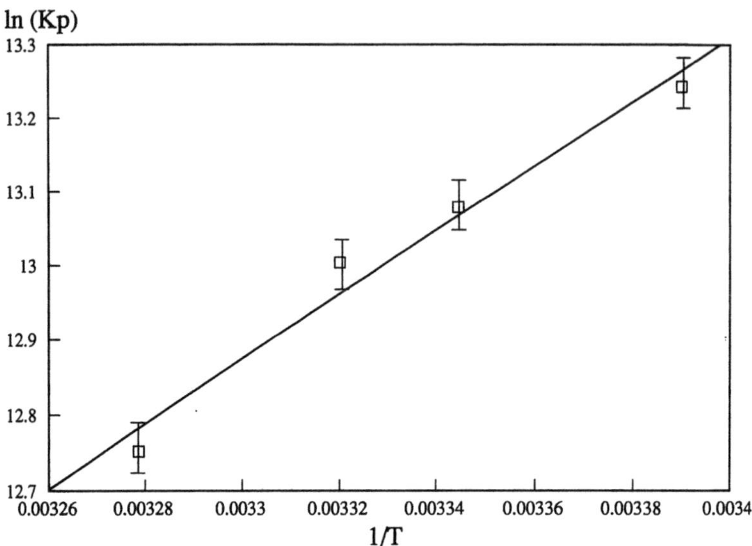

Fig. 13. Temperature dependence of the partition coefficient, K_p, at atmospheric pressure

celles to probe chains increases. Therefore, we restrict ourselves to data in the narrow range between 293 and 308 K, with the results plotted in Fig. 13. From the slope, we obtain $\Delta H_p^0 = -40$ kJ/mol.

In our previous study on the same block copolymer [10], we found that the standard enthalpy of micellization was -160 kJ/mol. We can compare the two results if we normalize these values by the molecular weights of the probe molecule (8000) and that of the PS block in the PSPEP (35000); we thus obtain $\Delta H_p^0 = -175$ kJ/mol. The relatively close agreement indicates that the thermodynamics of solubilization mimic the enthalpic portion of micellization. This is because the micellization enthalpy stems mainly from the transfer of the PS core block from the solvent phase into the micelle core. The entropic contribution plays a much greater role in micellization than in solubilization, as can be seen by comparison of the standard entropy values. For solubilization, $T\Delta S^0 = -6$ kJ/mol, whereas the micellization data from Table 2 gives -154 kJ/mol.

Similar experiments can be performed at elevated pressures, although slight modifications must be made in the data analysis to account for changes in cyclization rates as the viscosity is increased at higher pressures. Additionally, the excitation peak positions for both monomer and excimer will shift with increasing pressure due to the changes in refractive index and density of the solvent. With these complications in mind, we can easily determine the pressure dependence of X_m for the Py-PS-Py probe chain in PSPEP micelle solution, with the results shown in Fig. 14. The increase in pressure forces probe chains out of the micelle core and into the dispersed phase. At pressures above 100 MPa, this effect seems to have saturated. The agreement between the excitation position and the I_e/I_m data is good at the higher concentration but not as good at the

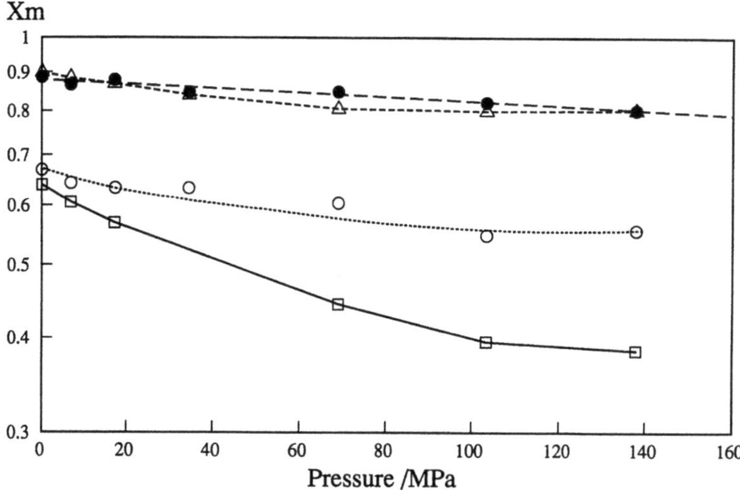

Fig. 14. Pressure dependence of the solubilized fraction of probe chains in 0.1% PSPEP solution with I_e/I_m ($\square$) and excitation position ($\bigcirc$) and 0.5% PSPEP solution with I_e/I_m ($\triangle$) and excitation position ($\bullet$)

0.1% concentration. This may be the result of the difference in environment between the 0.1% PSPEP micelle core and the 0.5% PSPEP micelle core as a function of pressure, e.g. due to increased solvent incorporation in the core at the lower concentration.

It is also possible to calculate K_p at elevated pressures, with the results shown in Fig. 15. The values for both concentrations are in excellent agreement and exhibit a decrease in K_p with increasing pressure. Since K_p is directly related to ΔG_p^0 as shown in Eq. 15, this indicates a decrease in the thermodynamic driving force for core solubilization as the pressure is increased. As indicated by the data at atmospheric pressure, the dominant component of the free energy for solubilization is the large negative enthalpy gained by replacing PS-heptane contacts by PS-PS contacts in the micelle core. This process is analogous to a reverse mixing process for the PS chains, so we can use relationships for the enthalpy of mixing to gain insight into the solubilization process.

The effect of pressure on solvent quality for a polymer solution has been addressed by Flory in his equation-of-state theory [35, 36] and by Patterson in his corresponding states theory for polymer solutions [37, 38]. In each case, the authors realized the importance of relative free volumes between the solvent and solute. In polymer solutions where the solvent free volume is vastly greater than that of the polymer, an increase in pressure will decrease the free volume difference, typically leading to enhanced solubility. We have previously used Patterson's corresponding states treatment to calculate the effect of pressure on the PS-heptane χ parameter [25]. We have found that as the pressure is increased, χ decreases quickly at lower pressures and then levels off at higher pressures. This corresponds to a decrease in the magnitude of K_p with an initial in-

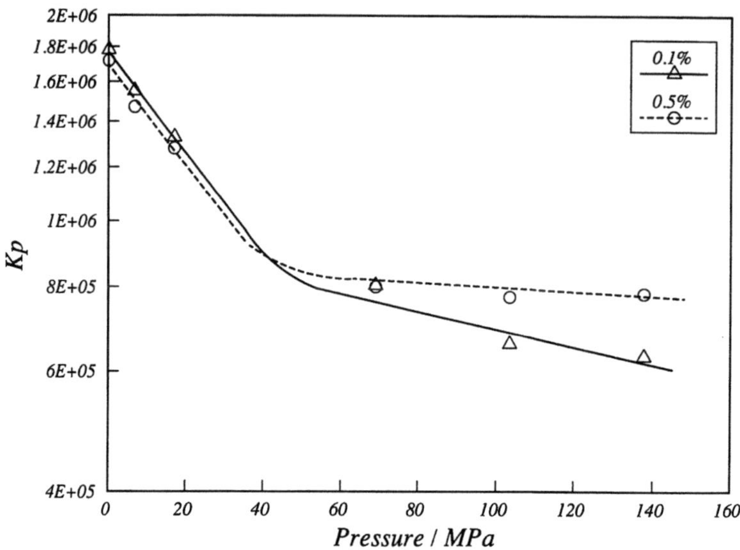

Fig. 15. Pressure dependence of the partition coefficient, K_p, for 0.1% PSPEP ($\triangle$) and 0.5% PSPEP (o)

crease in pressure and then a leveling off at higher pressures, which is what is observed in Fig. 15. Thus, the primary influence of pressure on solubilization is through the χ parameter.

Summary

In an extension of earlier work [9, 10, 26], we have provided further evidence that the photostationary state spectral parameters of excimer band-width and excimer band-position are quite useful for monitoring the chemical environment of excimer-forming sites in aryl-containing block polymer systems. The free-chain/micelle equilibrium is an excellent system for application of this photophysical tool because of the dramatic change in morphology. As the size of the solvent is increased in the PSPEP/alkane system, the cmc decreases, in agreement with the mean-field calculations [4]. The decrease in cmc with increasing solvent size is accompanied by a decrease in micelle aggregation number. Comparisons of the standard enthalpy and entropy of micellization indicate that the solvent size effect is captured by the entropic contribution, while the energetic interactions are captured by the solvent solubility parameter difference.

In addition, we have presented an approach for quantitatively measuring the extent to which a fluorescently labeled homopolymer partitions between a block copolymer micelle core and the dispersed phase. The excellent agreement between the excimer-to-monomer ratio and the excitation position approach indicates that photophysical techniques allow accurate measurement of the partition coefficient. Furthermore, the good agreement of the enthalpy of solubilization with the enthalpy of micellization is an indication that the polystyrene-heptane contacts dominate the enthalpic term in micellization. As the pressure is increased, the decrease in the partition coefficient corresponds to an increase in solvent quality of heptane for polystyrene, thereby reducing the driving force for solubilization. This is similar to micellization at higher temperatures.

Appendix. Photophysics of Aromatic Vinyl Polymers
Since an understanding of the photophysics of aromatic vinyl polymers is essential to an appreciation of its application to study the thermodynamics of block copolymer micelles in selective solvents, we present a brief overview of some of the highlights in this Appendix. We begin with consideration of a generic Jablonski diagram (Figure A1) that shows the relationships between the singlet ground electronic state of an isolated aromatic ring (S_0), the first excited electronic state (S_1) and the excimer ($^1D^*$). For generality, this diagram also shows the lowest lying triplet state (T_1), but that does not concern us for the PSPEP system of interest. Absorption of radiation by the aromatic chromophore, typically in the ultraviolet region of the spectrum, could lead to a variety of vibronic excitation states, denoted by the levels n = 0, 1, 2 ..., as evidenced by a structured absorption spectrum. Following absorption, rapid internal conversion to the lowest vibrational state occurs such that radiative emission to the ground electronic state is from n = 0 in S_1; a structured emission spectrum of the aromatic chromophore, termed monomer emission, arises because the radiative transition could be to any one of several vibronic levels in the ground electronic state S_0.

Several other processes could also occur. For example, internal conversion, indicated by the dashed line, is a non-radiative process that can also take place between the excited and ground singlet states of the isolated aromatic chromo-

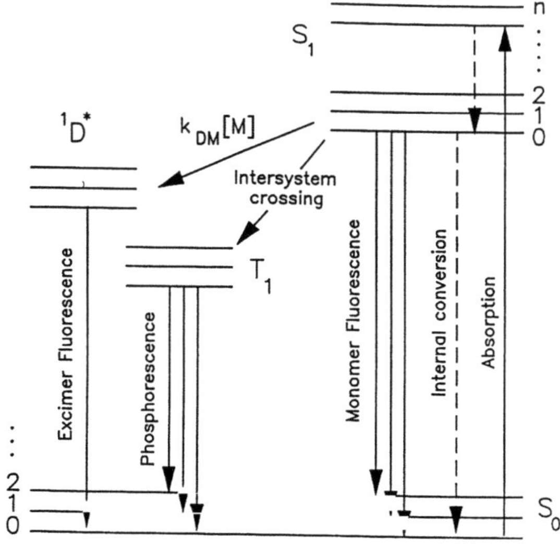

Fig. A1. Schematic Jablonski diagram illustrating absorption, radiative and nonradiative emission processes in an excimer-forming system

phore. Intersystem crossing involves a transition between states of different multiplicity, and is not important for PSPEP. A final possibility is that the excitation be localized on an excimer-forming-site (EFS), which is a coplanar sandwich formed by two identical aromatic chromophores, such as the phenyl rings of the polystyrene within PSPEP. The process by which this excitation is localized at an EFS is complex and can involve segmental rotational motion of one unexcited phenyl ring into the vicinity of a second, excited one or energy migration among individual, isolated phenyl rings through a dipole-dipole process until the excitation is trapped at a lower energy EFS. Energy migration is always a possibility in high chromophore density systems, such as in the core of the PSPEP block copolymer micelle, whereas the rotational „sampling" of the EFS requires some degree of segmental mobility. Equation 9 lumps all of the photophysical processes into the quantity M, which is the probability of eventual monomer emission, as opposed to any of the other processes denoted in Fig. A1.

An essential feature of the photophysical experiments described in this paper is that we have discovered that the spectral characteristics of the photostationary excimer fluorescence are sensitive to the local environment of an EFS. This can be understood qualitatively with reference to Fig. A2, which is another

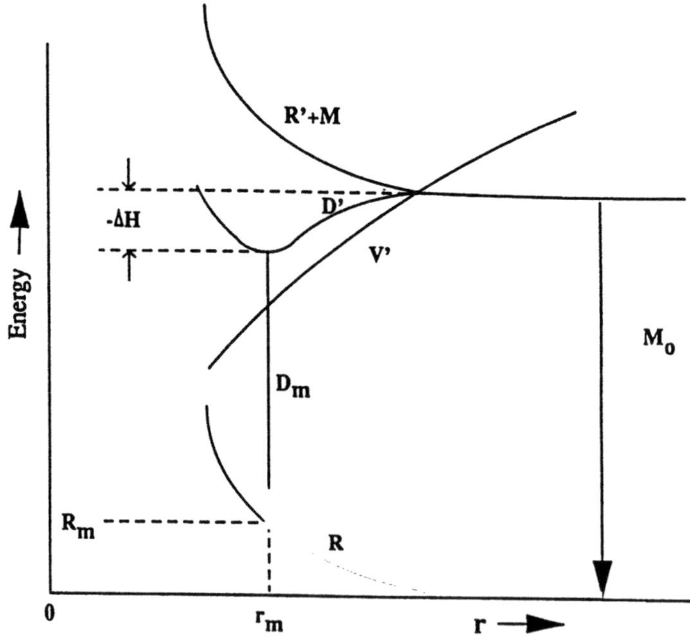

Fig. A2. Schematic energy level diagram illustrating excimer stabilization for a particular combination of repulsive and attractive potentials
R = repulsive potential in ground state; R′ = repulsive potential in excited state; V′ = excimer interaction potential; D′ = resultant excimer potential; M_0 = 0–0 molecular transition; r_m = equilibrium separation of excimer; D_m = peak excimer fluorescence transition; –ΔH = excimer bindung energy.

energy level diagram with the emphasis on the relative positions of the non-excited aromatic chromophore and the excited aromatic chromophore. At very large distances of separation, the excited chromophore is effectively isolated and emission occurs with energy M_o. As the two chromophores approach each other, two types of intermolecular interaction become important: (i) a repulsive potential (represented by R in the ground state or R' in the excited state; typically, these are assumed to be of the same form), which reflects the steric repulsion and (ii) an attractive interaction (V') in the excited state, which is associated with a configuration interaction between a charge transfer interaction and a neutral excitation resonance. Depending on the nature of the aromatic groups, the balance of the repulsive and attractive interactions may lead to a minimum in the energy of the excited dimer (excimer) or EFS complex; if so, stabilization occurs and it will be possible to observe an emission from the excimer. Note that the emission from the EFS occurs from the approximately parabolic minimum to a repulsive ground electronic state, represented by the potential R. The ground state is totally repulsive because both chromophores are in their ground states and are interacting purely with van der Waals repulsive forces; this causes the excimer emission band to be featureless.

A typical photostationary state emission spectrum from PSPEP is shown in Fig. A3. The monomer emission occurring from the isolated phenyl rings is the poorly resolved shoulder at high energy and the predominant emission band is that due to the excimer. The excimer band has been deconvoluted from the monomer through use of a nonlinear Marquardt regression analysis, as described in the Experimental section. Although the data are shown here on a wavelength scale, all of the deconvolutions are performed after the raw data were converted to an energy scale, for which we expect the excimer emission band to be symmetric. The reported values of excimer band-width and band-position

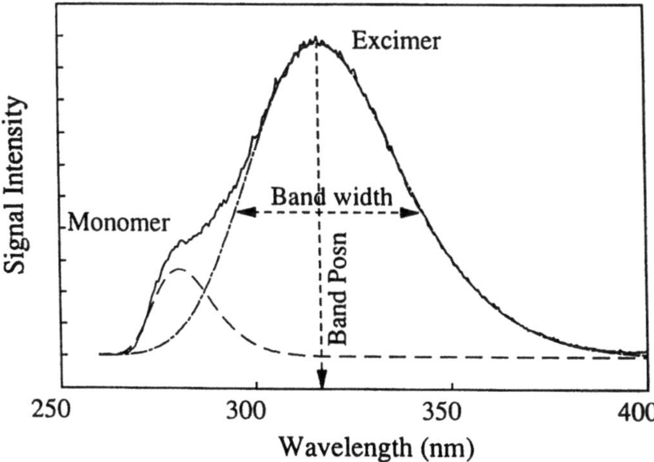

Fig. A3. Typical emission spectrum for PSPEP micelle solution illustrating the deconvolution of the excimer and monomer emission bands

result from this deconvolution. It is also possible to obtain values of I_D/I_M from integration of the areas under the excimer and monomer emission bands.

The environmental sensitivity of the excimer emission arises from the details of the position and shape of the minimum in the D′ stabilized state and the position and steepness of the repulsive ground state R. The energy of the excimer emission band, of course, is represented by D_m and may be shifted by a variation in the attractive excimer interaction potential V′. Of more interest for the experimental results on PSPEP described in this chapter is the consequence of a change in the steepness of the repulsive ground state potential; an increase in steepness will lead to a broader excimer emission band whereas a decrease in steepness will lead to a narrower excimer band. One might expect that the nature of the first coordination sphere around an EFS would have a strong influence on the nature of the repulsive potential, e.g. an EFS located in a largely aromatic environment such as the core of the PSPEP micelle might be expected to have a less repulsive potential than an EFS located in an alkane solvent environment. It does not seem possible to provide any more quantitative insight than this because of the complexity of the conformational energy surface.

Acknowledgements. We would like to thank the NSF Polymers Program and Shell Development Company for their financial support of this project. We also thank Dale Handlin of Shell Development Co. for the polymer samples and polymer characterization work.

References

1. Noolandi, J., Hong, K.M., *Macromolecules 1983, 16,* 1443
2. Leibler, L., Orland, H., Wheeler, J.C., *J. Chem. Phys. 1983, 79,* 3550
3. Whitmore, M.D., Noolandi, J., *Macromolecules 1985, 18,* 657
4. Munch, M.R., Gast, A.P., *Macromolecules 1988, 21,* 1360
5. Tuzar, Z., Kratochvil, P., *Adv. Coll. Int. Sci. 1976, 6,* 201
6. Stacy, C.J., Kraus, G., *Polymer Eng. Sci. 1977, 17,* 627
7. Bahadur, P., Sastry, N.V., Marti, S., Riess, G., *Coll. Surf. 1985, 16,* 337
8. Candau, F., Heatley, F., Price, C., Stubbersfield, R.B., *Eur. Poly. J. 1984, 20,* 685
9. Yeung, A.S., Frank, C.W., *Polymer 1990, 31,* 2089
10. Yeung, A.S., Frank, C.W., *Polymer 1990, 31,* 2101
11. Price, C., Kendall, K.D., Stubbersfield, R.B., Wright, B., *Poly. Commun. 1983, 24,* 326
12. Mandema, W., Emeis, C.A., Zeldenrust, H., *Makro. Chem. 1979, 180,* 2163
13. Higgins, J.S., Dawkins, J.V., Maghami, G.G., Shakir, S.A., *Polymer 1986, 27,* 931
14. Bednar, B., Devaty, J., Koupalova, B., Kralicek, J., Tuzar, Z., *Polymer 1984, 25,* 1178
15. Tuzar, Z., Plestil, J., Konak, C., Hlavata, B., Sikora, A., *Makromol. Chem. 1983, 184,* 2111
16. Tuzar, Z., Kratochvil, P., *Makromol. Chem. 1972, 160,* 301
17. McBain, M.E.L., Hutchinson, E., *Solubilization,* Academic Press, New York, 1955
18. Schick, M.J., ed., *Nonionic Surfactants,* Marcel Dekker, New York, 1967
19. Shinoda, K., Nakagawa, T., Tamamushi, B., Isemura, T., eds., *Colloidal Surfactants,* Academic Press, New York, 1963
20. Hayase, K., Hayano, S., *J. Coll. Int. Sci. 1978, 63,* 446
21. Hayase, K., Hayano, S., *Bull. Chem. Soc. Japan. 1977, 50,* 83
22. Skoulious, A., Helffer, P., Gallot, Y., Selb, J., *Makromol. Chem. 1971, 148,* 305
23. Krause, S., *ACS Polymer Prepr. 1970, 11,* 568
24. Ikada, Y., Horii, F., Sakurada, I., *J. Poly. Sci., Poly. Chem. Ed. 1973, 11,* 27

25. Ylitalo, D. Y., Frank, C.W., *Polymer 1996, 37,* 4969
26. Tang, W., Hadziioannou, G., Smith, B., Frank, C.W., *Polymer 1988, 29,* 1718
27. Fitzgibbon, P.D., Frank, C.W., *Macromolecules 1981, 14,* 1650
28. Frank, C.W., *Organic Coatings and Plastics Chemistry 1981, 45,* 433
29. Brandrup, J., Immergut, E.H., eds., *Polymer Handbook* 1975, Wiley, New York
30. Frensdorff, H.K, *J. Appl. Poly. Sci. 1973, 17,* 1101
31. Wilemski, G., Fixman, M., *J. Chem. Phys. 1979, 60,* 866, 878
32. Semerak, S.N., Frank, C.W., *Adv. Poly. Sci. 1983, 54,* 31
33. Hemker, D.J., Garza, V., Frank, C.W., *Macromolecules 1990, 23,* 4411
34. Char, K., Frank, C.W., Gast, A.P., *Langmuir 1989, 5,* 1335
35. Eichinger, B.E., Flory, P.J., *Trans. Far. Soc. 1968, 64,* 2035, 2066
36. Flory, P.J., Orwoll, R.A., Vrij, A., *J. Am. Chem. Soc. 1964, 86,* 3507
37. Patterson, D., *Macromolecules 1969, 2,* 672
38. Gaeckle, D., Patterson, D., *Macromolecules 1972, 5,* 136

**Part 4
Applications of Fluorescence Spectroscopy in Biology**

Fluorescence Microscopy and the Reactions of Single Molecules

G. Pilarczyk, S. Monajemashi, C. Hoyer, V. Uhl, K. O. Greulich

1
Fluorescence Microscopy vs Fluorescence Solution Studies

The majority of applications of fluorescence reported in this book deal with fluorescence in solution. In many of them high spatial resolution is not required and therefore detectors for extreme sensitivity or extreme temporal accuracy can be used. However, for a number of applications in chemistry, biology, and medicine, images from microscopes are required and their resolution should ideally be close to the diffraction limit. Modern cameras allow high spatial resolution and high sensitivity at moderate temporal resolution: Single photon counting cameras with a sensitivity of better than a nanolux achieve the sensitivity of individual photomultipliers. Cooled CCD cameras are one to two orders of magnitude less sensitive (10 to 100 nanolux) but have a better spatial resolution and are easier to handle. The temporal resolution is poor as compared to photomultipliers or avalanche photodiodes. Fortunately, a large number of applications in chemistry, biology, and medicine do not need higher temporal resolution. In the present contribution we will report on a few such applications: single molecule detection and handling as well as the observation of single molecule reactions in the fluorescence microscope. This microscope will also be used to look into individual cells and to study the distribution of their chromosomes and finally to detect how calcium waves direct the coordinated beating of the heart and heart tissues.

2
CCD Cameras and Pitfalls in Quantitative Fluorescence Microscopy

The crucial element in modern fluorescence microscopes is the camera. During processing the image of the object is distorted to some extent. Such distortions are not problematic in standard imaging, but have to be taken very seriously when the data of an image have to be used for detailed measurements, for example, when spectral, spatial, or temporal information has to be quantified. Thus, for *quantitative* fluorescence microscopy a large variety of possible pitfalls have to be considered. Some of them are shortly addressed in the following. Readers who are not interested in such quantitative aspects may skip these details and continue with the general chapter on fluorescence microscopy.

2.1
Spatial Distortions of Moving Objects

When objects in motion are observed, the imaging system superimposes its temporal characteristics onto that of the object. This problem is most pronounced with scanners, but also to some extent with CCD cameras. A second effect, „blooming", increases the temporal and spatial width of a signal. In other words, it blows up the real size of the object. Cameras which suppress blooming are often of no help since they artificially shrink the size of the moving object.

2.2
Distortions of the Intensity Information by the Camera

An additional widespread tool in advanced CCD-camera technology is the automatic black level compensation. Here the camera adjusts the darkest parts of the picture to the lowest digital channel. This compensation disturbs the information about the absolute signal intensity of the specimen. If the information is needed it is necessary to guide a part of the outcoming signal to a photomultiplier and to calibrate the CCD signals this way.

2.3
Distortions of the Spectral Information by Color CCD Cameras

Color CCD cameras are problematic in fluorescence microscopic applications. The picture's color is collected through three different color filters. The transmission curves of the filters overlap one another and are very broad. Thus it is not possible to reconstruct with certainty which wavelength of the specimen's signal represents which color in the picture.

2.4
Conversion of Light Power (lux) into Intensity (W/cm²)

When selecting a camera for high sensitivity, problems occasionally occur, since the sensitivities are given in nanolux or microlux and the real interest is in the minimum intensity in W/cm^2. Below we give an example for calculating the number of photons impinging on a CCD-camera pixel at a certain light power. This may be helpful if one wants to measure the light power of the specimen and wants to estimate if this signal can be recorded by a CCD camera with a sensitivity given in lux. We have made the following simplifications. The light was taken to have a wavelength of 500 nm. Estimations using the unit steradian were based on the unit sphere. The size of a camera pixel was taken as $9 \times 9 \ \mu m^2$.

2.4.1
Calculation of the Number of Photons Impinging on a Pixel

Intensity of a black emitter during the emission of 1 candela (cd):

$$I = \sigma \times T^4$$

where σ: Stefan/Boltzmann constant, T: absolute temperature of the emitter, under these conditions 2042.5 K

Put in:

$$I = 5.67 \times 10^{-8} \, \frac{W}{m^2 K^4} \times (2.042 \times 10^{-3} \, K)^4$$

Simplify:

$$I = 9.858 \times 10^5 \, \frac{W}{m^2} \approx 10^6 \, \frac{W}{m^2} \approx 10^2 \, \frac{W}{cm^2}$$

Definition of the unit of illumination (illuminance):

$$1 \, Lux = \frac{1 \, Lumen}{m^2} = \frac{1 \, cd \times sr}{m^2} = 10^2 \, \frac{W}{cm^2} \times \frac{1}{60}$$

The value of a complete steradian (sr) equals:

$$[\Omega] = 4\pi \times sr = steradian \, (r)$$

For the circle with radius 1:

$$sr = 4\pi \times 1$$

It follows:

$$1 \, Lux = 10^2 \, \frac{W}{cm^2} \times \frac{1}{60} \, cm^2 \times sr$$

Put in:

$$1 \, Lux = 10^2 \, \frac{20 \, W}{m^2} = 2 \times 10^{-3} \, \frac{W}{cm^2}$$

The value of a pixel on a CCD cameras chip with 9×9 µm equals:

$$(9 \times 10^{-4} \, cm)^2 = 8.1 \times 10^{-9} \, cm^2$$

The power on this pixel generated by the application of 1 Lux equals:

$$I = 2 \times 10^{-3} \, \frac{W}{cm^2} \times 8.1 \times 10^{-9} \, cm^2 = 1.6 \times 10^{-11} \, W$$

Assume for the photons that a wavelength of 500 nm is used.

Put in:

$$1 \, Lux = 1.6 \times 10^{-11} \, Ws = 1.6 \times 10^{-11} \times 2.5 \times 10^{18} \, \frac{photons}{pixel \times s} = 5 \times 10^9 \, \frac{photons}{pixel \times s}$$

Under the illuminance of 1 Lux using light with a wavelength of 500 nm 5×10^9 photons per second are applied to a CCD camera's pixel of 9×9 µm.

3
Fluorescence Microscopy

A typical fluorescence microscope contains a lamp for excitation of the fluorescence dye and a special filter with high transmission values for light emitted by this dye. In the case of epi-fluorescence, the light of a mercury vapor lamp (HBO) is directed via an excitation filter, a beam splitting mirror, and the objective into the object plane. The emitted light is guided through the objective and a blocking filter to the ocular or to a camera. The full potential of fluorescence microscopy can only be exploited when the optical, biochemical and physiological properties of the fluorescent dyes are optimized and fit with the light source and the optical filters. In the following sections, some examples are given to stress this point.

3.1
Improved Signal-to-Noise Ratio and Physiological Selectivity

In fluorescence microscopy the signal-to-noise ratio is much better compared to visualization under phase contrast or bright-field imaging. The reason is that in contrast to other methods the visualized object emits a light signal itself and is embedded into a non-emitting background.

The experimentator can not only select for the structure in the sample but also for the physiological function by a suitable choice of dye. There are a large number of commercially available dyes with affinities to different biological structures, so it is possible to localize parts of the subcellular machinery exclusively with no interference with structures in the direct neighborhood. This method can be used to circumvent the microscope's restriction in localizing structures below the (Abbe) resolution limit.

3.2
In Situ Measurement of Ionic Movement with Subcellular Resolution

The application of fluorescence microscopy for the observation of living objects, single cells as well as small organisms, provides advantages over comparative methods such as the electrophysiologic ones. The movement of ions in single cells and small cellular aggregates can be measured without capacitative restrictions. Using digital cameras the temporal resolution is comparable to the electrophysiologic methods but more than one cell can be measured at the same time. With appropriate dyes (such as di-4-anepps or di-8-anepps which have potential dependent fluorescence yields), the electric potential of the plasma membrane of a single cell can be visualized. In contrast to the direct measurements of the electric state using microelectrodes the fluorescence approach offers the possibility of visualizing local differences in the potential of one cell.

3.3
Problems in Measuring the Ionic Composition of Living Objects

While optical measurement of ions in cells gives a number of answers it also causes a number of problems. The composition of compounds inside a living system makes it very difficult to measure ion concentrations. The substance of interest has a very low concentration, mostly in the nanomolar to micromolar range and cannot be purified. The pollution is in the range of around 30–60% of the total mass of the specimen. The probe volume is very small, mostly in the range of some picoliters. The interesting temporal resolution of the measurement makes it necessary to automatize the measurements. The selectivity of the ion indicator is low and influenced by many ions which are of no interest for the measurement. Most of the introduced fluorescent substances influence the system or are themselves toxic. The stability of the fluorescent probes, especially the optic stability, is low. Nothing is known about the chemical composition of the local environment and its interaction with the fluorescent probe. The macroscopic definitions of concentration, electrical charge, and others are not suitable in the range of just some hundred molecules in the observed volume. A comparison of the system under conditions of observation with the untouched system is not possible in any case.

4
Optical Tweezers for the Handling of Individual DNA Molecules

The fluorescence microscope can not only be used as an analytical tool in biology, chemistry and medicine, but also as a micromanipulating instrument. If a continuous wave laser is coupled into the microscope, the laser can be used for manipulation of biological objects. This setup is called optical tweezers (see, for example, Ashkin et al. 1986, Ashkin 1997). The basis for optical tweezers is the pressure exerted by light. This is not only used today to trap and cool atoms or molecules (see Physics Nobel prize 1997) but also for larger particles such as microspheres of several micrometers in diameter or living biological cells (G. Leitz 1995, G. Leitz 1994, S. Monajembashi et al. 1997, K. O. Greulich and G. Weber 1992).

In some single molecule experiments it is mandatory to handle the molecules before their dynamics can be studied in the fluorescence microscope or reactions of them with other molecules can be observed. While a micromechanical handling is in principle conceivable, it is certainly inpractical in experimental routine. Therefore, optical tweezers have found their place in some branches of single molecule research, particularly when long, extended macromolecules or macromolecular complexes such as microtubules, actin, or DNA are involved.

5
A First Application: Single Molecule Handling and Enzyme Reactions

The handling of single molecules is of increasing importance in life sciences, particularly in single molecule DNA analysis. Critical steps of such an analysis are isolation, preparation and handling of individual DNA molecules. It is not possible to isolate and handle a single molecule solely by dilution since inter- and intramolecular adhesion cause aggregation of large numbers of molecules. In order to overcome this problem DNA molecules are bound in suspension to polystyrene microbeads (see below) and handled by hydrodynamic flow, electric fields and optical tweezers. In detail, such a bead with a DNA molecule is trapped by optical tweezers, the electrical field is switched on and the negatively charged DNA molecule is stretched and pulled to the positive electrode. In the experiments described below, lambda phage DNA (48 Mbp, 16 μm long) has been bound to polystyrene beads one micrometer in diameter. The crucial step is the *preparation* of the single molecules.

5.1
Single-Molecule Preparation

The molecules have to be labeled with biotin. In order to increase the specificity the labeling reaction was performed in two steps. Firstly, terminal transferase was used to bind a poly(A)-tail to the ends of the lambda DNA. Secondly, an oligo(T)-biotin molecule was hybridized to the poly(A)-tail. Then the streptavidin covered beads were added to the solution in large excess over the DNA molecules and bound to the DNA. Due to the fact that both ends of the DNA molecule are labeled with biotin, one could expect that two beads, one at each end, could be found. Actually this happens very rarely. An explanation may be that the molecule is often disrupted in the reaction process. This would also explain the observation of stretched molecules which are often shorter than the theoretical maximum length. To observe the molecule, the DNA was stained with the fluorescent dye SYBR Green. An intensified CCD camera connected to an image processing software was able to detect and enhance the signals.

5.2
DNA Molecules and Enzyme Reactions

The endonuclease ApaI is used to restrict the DNA while it is bound to the bead and stretched in the electric field. This enzyme has only one restriction site on the lambda phage DNA. That means that after a successful cutting two fragments of different length should result and be visible. One fragment bound to the bead and the other free part pulled to the positive electrode.

It is important that the reaction is done under observation and for the starting point to be controlled. For a restriction reaction, the enzyme needs Mg^{2+}-ions. If these ions are complexed with a caged compound like DM-Nitrophen, the reaction is inhibited. Only after photolysis of the complex the are Mg^{2+}-ions liberated and the ApaI endonuclease activated which restricts the lambda phage DNA into the two specific fragments.

Why use SYBR Green? This dye is not very common and, in principle, one could use any DNA-sensitive dye. The difference is only the excitation wavelength because this is of special interest if one wants to photolyze a caged compound. In our case the DM-Nitrophen can be irreversibly destroyed with UV light, since the DNA should be observed and handled before the reaction starts and the excitation wavelength of the fluorescent dye should be a long way from that of the caged compound. Only if this prerequisite is satisfied, can be one certain not to activate the enzyme. Figure 1 shows the enzymatic digestion of an individual DNA molecule by ApaI.

5.3
Validation

It has still to be shown that the filamentous structure attached to the bead indeed represents a single DNA molecule. In spite of the fact that the stoichiometry of the coupling reaction should guarantee that only one DNA molecule is bound to one bead, aggregation of several molecules might occur. However the reaction itself gives hints to the number of individual molecules in a filament. One has to observe the fragments which result from the restriction reaction with ApaI (keep in mind: one restriction site – two fragments).

Three outcomes are possible: no restriction at all, restriction with more than the expected fragments or restriction with the correct number of expected fragments.

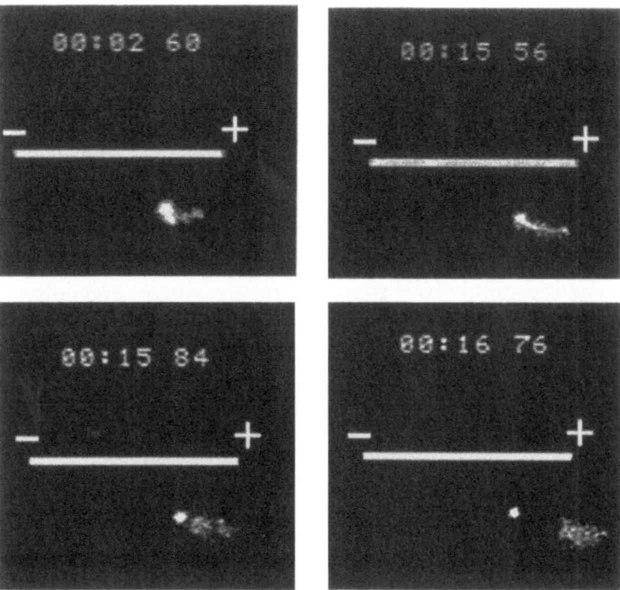

Fig. 1. Enzymatic cutting of a single DNA molecule. At time 02 60 a microbead with a DNA molecule is shown. The microbead has previously been transported into the visual field and is now held in position by the optical tweezers. The scale bar is 30 µm. "+" and "–" indicate the direction of the electric field. At time 15 56 the DNA molecule is stretched by the field. At 15 84 the DNA molecule is cut by the restriction endonuclease ApaI and at 16 76, after the DNA molecule is released, blurring of the DNA molecule is recognizable due to increased thermal motion

In the first and second case we could discard or repeat the experiment. In the third case there are still two possibilities. The first and simple one is to accept that just one molecule was bound to the bead. The second way is to wonder if it could be possible that two or more molecules were bound to the bead in the same way. That would mean very close to each other so that the restriction was done by the enzyme at the same time and therefore only the expected two fragments would be observed.

To eliminate this situation, the complicated two-step binding reaction was introduced as explained previously. The tailing reaction always produces tails of different lengths, between 50 and 150 bp. The hybridization of the oligo(T)-biotin to this poly(A)-tail will therefore rarely be at the same position. That means the length of the two hypothetical DNA tails bound to the bead will be different and so will the position of the restriction sites. Not to forget additionally that even the orientation of the molecule can be reversed due to the fact that it is biotin labeled on both ends.

We assume that a restriction reaction starts now at the different positioned restriction sites, and we also assume that the restriction will really work. What could happen? The only reasonable answer is: nothing. Nothing, because of the already mentioned inter- and intramolecular adhesion. The fragments bind together and collapse. No drift of a fragment in the electric field is observed. Obviously, this is a useful and reproducible method for binding a single molecule to a bead and for performing single molecule enzyme reactions under observation and control.

6
Observation of Enzyme Reactions on a Molecular Level

The detection of single molecules brings chemical analysis into the range below concentrations of zeptomol (10^{-21} mol). Therefore, single molecule detection in liquid environment has been the focus of attention for the last few years (Xue and Yeung 1995, Hoyer et al. 1996, Funatsu et al. 1995, Xu and Yeung 1997, Tokunaga 1997). However, monitoring single molecule reactions of biological relevance, such as enzyme reactions, is in most cases still difficult. There have been several approaches to the subject published, for example, Xue and Yeung examined single molecule reactions with the enzyme lactate dehydrogenase (LDH-1) by capillary electrophoresis. Most importantly, result differences in the chemical reactivity among the individual lactate dehydrogenase molecules could be detected (Xue and Yueng, 1995). Hoyer et al. showed the cutting of individual fluorescence labelled DNA molecules by a restriction endonuclease using light microscopy (Hoyer et al, 1996, see above in this article).

In the following, a method is developed that allows the observation of a single molecule reaction by using the autofluorescence of the reaction's product. The enzyme lactate dehydrogenase (LDH-1) catalyzes the reaction of lactate and nicotinamide adenine dinucleotide (NAD^+) to pyruvate and the reduced form of nicotinamide adenine dinucleotide (NADH) and H^+. NADH exhibits a fluorescence band with maximum at 460 nm which can be excited with the 365 nm emission of a highpressure mercury arc lamp, for example, the fluorescence illumination of a microscope.

6.1
Experimental Procedure

Two picoliters (2×10^{-12} l) of enzyme solution are injected with a microinjection unit into a drop of substrate solution placed on a slide in an inverted microscope. The enzyme solution and the substrate are prepared with 20 mM Tris buffer [Tris(hydroxymethyl)-aminomethane] and are adjusted to pH 9.1. At pH 9.1, LDH-1 shows maximum activity for catalysis of the lactate to pyruvate conversion. To increase the viscosity of the substrate solution polyethylene glycol (PEG, Av. Mol. Wt.: 10000) is added in a proportion of 60%. For the substrate solution the concentrations are as follows: lactic acid 3 mM, NAD^+ 1 mM. The concentration of the enzyme solution is 1 unit per 1000 ml. The different viscosities of the enzyme solution (no PEG) and the substrate solution (60% PEG) lead to the formation of a bubble of enzyme solution in the drop of substrate. Under typical injection conditions such a bubble has a diameter of about 20 micrometers and is stable for 20 min.

From the following approximate calculations one can assume that only few enzyme molecules (in the order of 100) are injected. The injected volume is 2 pl of enzyme solution.

This means 2×10^{-12} units of enzyme as the concentration of the enzyme solution is 10^{-3} unit/ml. One unit reduces 1 micromol of pyruvate per minute from definition (37 °C, pH 7.5). This leads to a maximum reaction rate of $v_{max} = 1.7 \times 10^{-8}$ mol s^{-1}. According to the theory of Michaelis–Menten, $v_{max} = k \cdot [E]$ and one gets $[E] = 1.7 \times 10^{-11}$ mol, where k is the turnover number of the enzyme (k = 1000 s^{-1} for LDH) and $[E]$ is the concentration of active centers of the enzyme (Stryer, 1988). This means that one unit contains 1.7×10^{-11} mol active centers of enzyme or, in other words, one unit is equal to 10^{13} enzyme molecules. Since 2×10^{-12} units of enzyme are injected, the injected volume of enzyme solution in the drop of substrate contains approximately 20 active enzyme molecules.

After injection of the enzyme solution areas with increasing fluorescence intensity from the reaction's product, NADH become visible in the microscope. Fluorescence excitation source is the fluorescence illumination lamp of the inverted microscope with 365 nm excitation filter set (Zeiss Axiovert microscope with 63×/1.3 oil immersion objective and HBO 50 W fluorescence illumination lamp, Carl Zeiss Jena, Germany). The microscope is equipped with a 420–490 nm band pass emission filter. This filter set produces lower background intensity than the frequently used DAPI filter set for other fluorescence microscopy applications with NADH as chromophore. The fluorescence image is recorded every 30 s with an intensified CCD camera and stored on the hard disk of a PC [image intensifier: Soliton, Gilching, Germany; CCD (Dalsa) and PC image acquisition board: Stemmer, Puchheim, Germany]. The exposure time on the CCD depends on the gain of the image intensifier and is in this work 100 ms at maximum.

The use of other camera equipment such as a cooled CCD or intensified video camera is possible in principle. Nevertheless, a cooled CCD, in spite of its usually better spatial resolution, is not recommended because of the lower detection sensitivity. The use of a cooled CCD would result in exposure times of typically 30 s. Since the reaction's product NADH is very delicate to photoinduced oxidation fluorescence illumination of more than 5 s is not acceptable. For this reason permanent video recording is also not possible.

To reduce photoinduced oxidation of NADH in this work, the excitation source is mechanically shuttered when no image is recorded. The actual fluorescence illumination time for one image acquired with the camera is 1 s. This illumination time is of negligible influence on the NADH fluorecence intensity.

6.2
Observation of the Enzyme Reaction

The injection of enzyme solution leads to the formation of a bubble of enzyme solution in the highly viscous substrate. At the border of the bubble where the enzyme solution is in contact with the substrate the production of NADH takes place. When the concentration of the enzyme solution is low enough, discrete and unequally distributed zones of increasing NADH fluorescence intensity and different size can be observed after enzyme injection. In the following these zones are referred to as reaction zones. Higher concentrations of enzyme solution lead to a continuous distribution of NADH fluorescence all over the bubble of enzyme solution within one second. For enzyme concentrations lower than 10^{-3} unit/ml no fluorescence has been detected so far.

Figure 2 shows the development of the reaction zones in the bubble of enzyme solution in the time after one injection. The time between the pictures is one minute. The reaction zones seem to move irregularly along the border of the bubble because of Brownian movement or thermal diffusion. The fluorescence images are analyzed with the computer program NIH-image.

The variation in the fluorescence intensity in the reaction zones is used to follow the progress of the reaction. Therefore, the fluorescence intensity in three chosen reaction zones is plotted vs time (lower part of Fig. 2). The time zero can be extrapolated via the intersection of the straight lines with the axis at fluorescence zero. The increase of the fluorescence intensity monitors the reaction rate i. e. the NADH production. The slopes are for zone A 2.6 a. u./min, for zone B 3.3 a. u./min and for zone C 6.1 a. u./min. The NADH fluorescence in reaction zone C starts approx. 5 min later than in zone A and B. This indicates that the enzyme reaction is delayed in zone C. The reason could be a longer travelling time of the enzyme or enzyme cluster from the interior to the border of the bubble where it makes contact with the substrate.

The discrete zones of NADH fluorescence can be interpreted as clouds of the reaction product (NADH) around single enzyme molecules. The unequal distribution of the fluorescent zones shows that the enzyme molecules are not equally distributed in the solution. They seem to form small clusters or to exist as single molecules in the middle of the reaction zones. From the kinetics of the enzyme reaction in one reaction zone the number of individual lactate dehy-

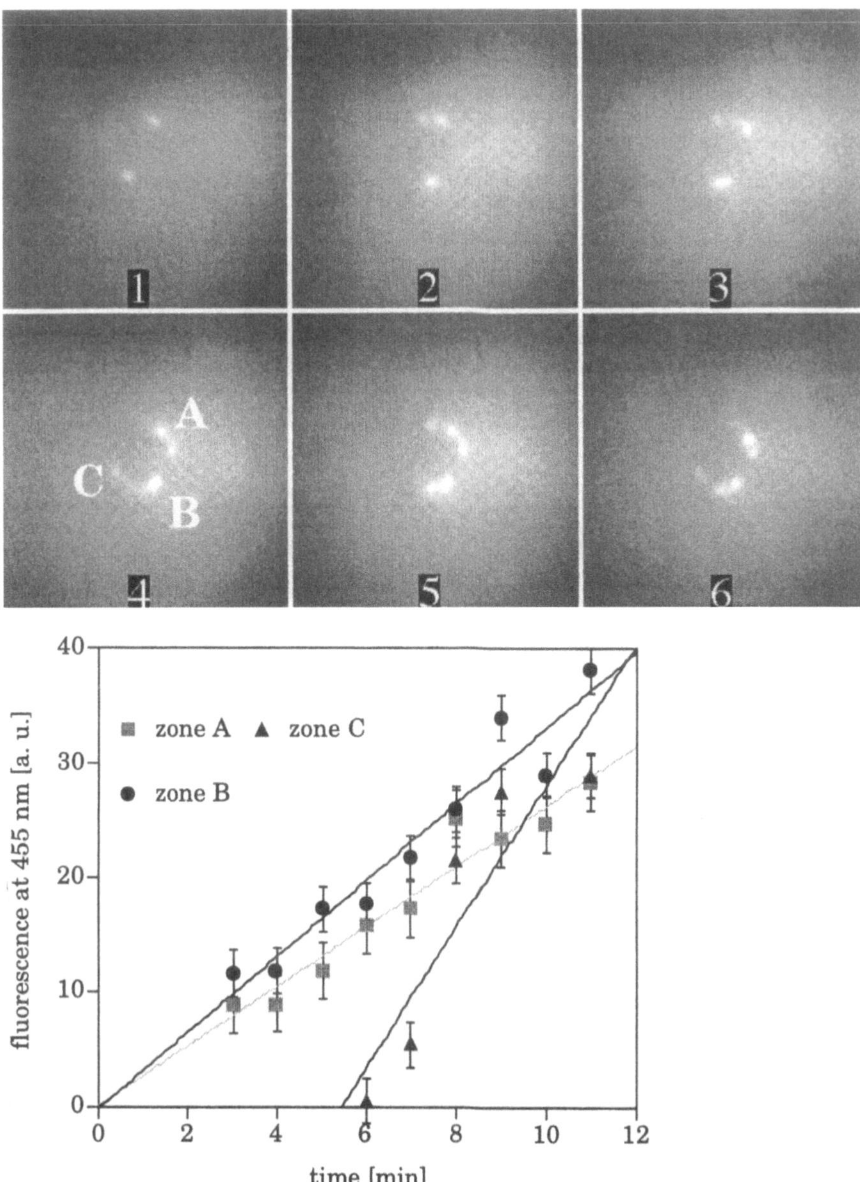

Fig. 2. Development of the reaction zones in the bubble of enzyme solution for one injection. The time between the measurements is 1 min. The reaction zones seem to move along the border of the bubble. The fluorescence intensity is measured in 3 reaction zones in the enzyme bubble as a function of time. The slopes are for zone A 2.6 a.u./min, for zone B 3.3 a.u./min and for zone C 6.1 a.u./min. The start of reaction zone C is delayed by approx. 5 min. The time zero can be extrapolated via intersection of the straight lines with the axis at fluorescence zero

drogenase molecules in the zone can be estimated. The slopes of zone A and zone B differ by 26.9%. The ratio is 1.269 or approximately 5:4. Thus one may speculate that zone B contains 5 enzyme molecules and zone A 4 enzyme molecules. In other words, each LDH-1 molecule contributes with 0.65 a.u./min to the total reaction rate. Then zone C would contain 9–10 molecules. The total number of observed molecules would then be 18–19, quite close to the expected value of 20 (see calculations above).

This is, however, only one limiting case. When one allows for somewhat larger errors, zone A and B might represent the same number of enzyme molecules with an average rate of 0.295 ± 0.035 a.u./min. Zone C represents approximately the double value. Thus zone A and B contain n molecules and zone C 2n molecules where n can range from 1 to 5. If n is larger than 1 the question is how this quite improbable distribution of molecules is generated. When n is smaller than 5 it is to explain where the missing molecules are. At the present time no answer to these questions can be given, and further measurements on this topic are being prepared.

7
Fluorescence In Situ Hybridization

The experiments described above allow fluorescence studies on individual molecules. In those experiments the molecules were prepared and studied as individual molecules. Fluorescence microscopy can not only be used on a single molecule basis, but also for the study of DNA on a single cell basis.

Nucleic acid hybridization provides the possibility of detecting DNA or RNA sequences that are complementary to any isolated nucleic acid, such as a viral genome or a cloned DNA sequence. For example, the cloned DNA is fluorescently labeled. This fluorescence DNA is then used as a probe for hybridization to complementary DNA or RNA sequences. Nucleic acid hybridization can be used to detect homologous DNA or RNA sequences not only in cell extracts, but also in chromosomes or intact cells, a procedure called in situ hybridization. In situ hybridization (ISH) has become a powerful and versatile tool for the detection and localization of nucleic acid sequences within cell or tissue preparations. The technique is capable of giving a high degree of spatial information in locating specific DNA or RNA sequences either within individual cells or chromosomes, or to a small sub-population of cells within a tissue sample.

High-resolution localization of genes on chromosomes is possible with spatially correlated introduction of fluorescent DNA probes. This method is generally referred to as fluorescence in situ hybridization (FISH). Fluorescence in situ hybridization to metaphase chromosomes is a direct and precise method and allows the mapping of a cloned gene to a locus defined by a chromosome band. In this case, the hybridization of fluorescent probes to specific cells or subcellular structures is analyzed by microscopic examination. For example, labeled probes can be hybridized to intact chromosomes in order to identify the chromosomal regions that contain a gene of interest. Fluorescence in situ hybridization can also be used to detect specific mRNAs in different types of cells within a tissue. FISH has become a useful tool in the medical field and in fundamental research.

In situ hybridization may be used, for example, to identify sites of gene expression, to analyze the tissue distribution of transcription, or to identify and localize viral infection or for chromosome mapping.

Two types of FISH can be distinguished: the direct and indirect method. In the direct method, the fluorochrome is directly bound to the probe. In the indirect method the probe contains an element (e.g. biotin or digoxigenin) with a specific affinity to another element which is fluorochrome conjugated (e.g. avidin-FITC). Another indirect FISH protocol is based on the principle of hybridizing a specific DNA probe which contains chemically modified nucleotides (i.e. digoxigenin-dUTP) to the chromosome's DNA. In a second step a fluorescently labeled antibody recognizes this probe.

7.1
Technical Aspects of the Optimization of Fluorescence In Situ Hybridization

Many problems in FISH concern the accessibility of the sample. The accessibility is influenced by the way the biological material is fixed, the denaturation method (only for DNA targets) and the size of the probe.

For FISH on metaphase spreads of chromosomes or on interphase nuclei, one can use fresh blood lymphocytes or cultured cell lines which are routinely fixed with methanol/acetic acid. In contrast, cells or tissue sections are fixed with (para)formaldehyde. There are a number of standard methods that can be used to introduce the fluorescent nucleotides into DNA probes (see also other chapters in this book). The most common way is to use nick translation, where the DNA template is first nicked with DNaseI and then the gaps are closed up via DNA polymerase I where the fluorescent nucleotide is introduced.

The major factor that limits the sensitivity of FISH is the amount of non-specific background and thus the conditions for hybridization and washing are critical for successful results. An important factor for optimal hybridization is the melting temperature (T_m) of the hybrids which is influenced by the nature of the probe and target (RNA-RNA-hybrids or DNA-DNA hybrids), by the length of the probe and by the composition of the hybridization/washing solution. Formamide is frequently used, since it decreases T_m, allowing lower hybridization temperatures to be used to achieve a high stringency.

7.2
Investigating the Architecture of Chromosomal DNA Domains
in the Cell Nuclei with Fluorescence In Situ Hybridization

Using FISH an important discovery about interphase chromosomes was made. Probes that stain or "paint" whole chromosomes show that individual chromosomes occupy restricted subcompartments or domains in the mammalian interphase nucleus, rather than being dispersed throughout the nuclear volume (Cremer et al. 1988, Lichter et al. 1988). FISH was applied to analyze fundamental processes of the formation of chromosomal aberration and to quantify radiation-induced chromosomal changes using whole chromosome painting probes.

The spatial correlation and the form of the selected chromosomes (#2, #7, #9, #X) in the interphase nucleus was investigated by fluorescence in situ hybridization (FISH). We used a conventional epifluorescence microscope (Zeiss, Axioskop) and a cooled, low-background and high-resolution CCD-camera system (Photometrics München, MAC200-A, Germany) equipped with an image processing software. The whole chromosome painting probes were stained either with Cy 3 or Cy 5. Since lymphocytes are not synchronized, interphase nuclei in different cell cycle phases could be found on the cover slide.

Interphase in situ experiments with human lymphocytes and whole chromosome painting probes for chromosomes #7, #9 and # X have shown that chromosome domains are highly compact and clearly separated from each other and arranged in pairs of homologues. The homologue chromosomes of-

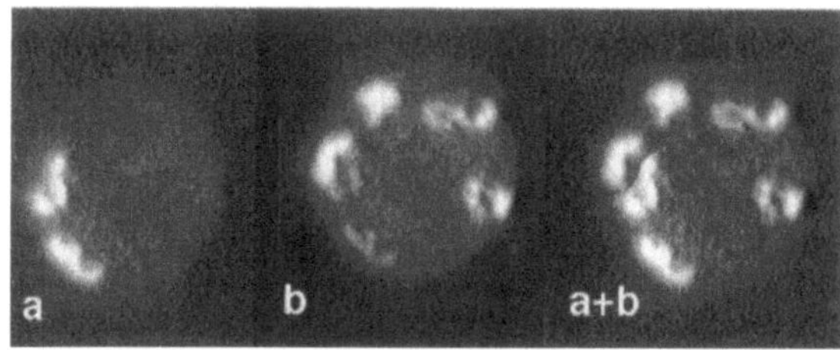

Fig. 3. Chromosomes #7, #9 and #X, probably in a cell cycle state close to mitosis, in (a) chromosome #9, in (b) chromosomes #7 and #X and in (a+b) the combination of both images (a) and (b) are demonstrated. The chromosomes are still compact, nearly in the metaphase, and are pairwise arranged

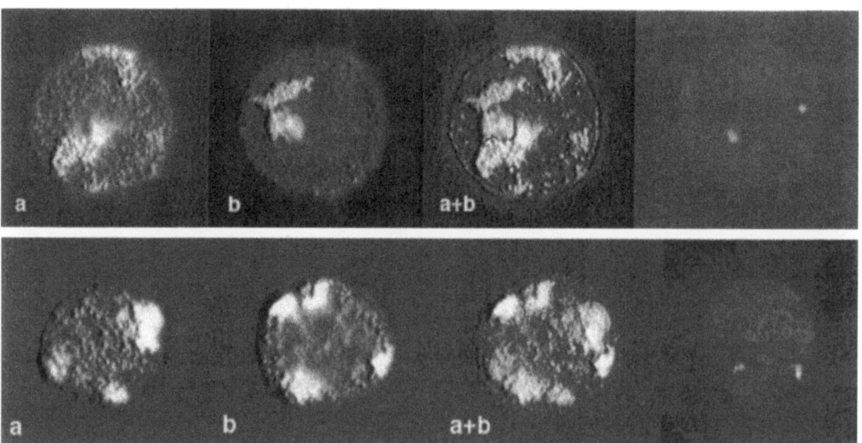

Fig. 4. Dynamics of chromosome territories during the cell cycle. Chromosomes #2, #9 (Cy 5: *a*) and #7, #X (Cy 3:*b*) and α-satellite (centromere) of chromosome 4 (*right hand* with two spots), in G$_2$-phase

ten appear on a circle in the exterior area of the nucleus. These nuclei probably represent the arrangement of chromosomes #7, #9, and #X in a cell cycle phase perhaps immediately after mitosis, as demonstrated in Fig. 3.

In most cases, however, the domains are bigger and packed more loosely. They fill much larger regions of the cell nucleus. The arrangement of interphase nuclei was investigated using multicolor FISH on four selected chromosomes (#2, #9, #7, #X) in unsynchronized lymphocytes from peripheral blood. We only checked the nuclei derived from cells in the G_2-phase. α-Satellite DNA (high repetitive DNA from the centromere region of chromosomes) can be used as a cell cycle marker. The cell cycle phase of the nuclei will be identified if, simultaneously, an α-satellite probe is hybridized. At the G_1-phase (early cell cycle phase) one hybridization signal and at the G_2-phase (after the duplication of the DNA) two hybridization signals appeared (Haaf et al. 1997). Figure 4 shows the dynamics of chromosome territories in the G_2-phase.

The positioning of chromosomes is related to cell cycle state. This relationship is visible in the non-random distribution of chromosomal subdomains such as centromeric and telomeric regions (Ferguson and Ward 1992, Vourch et al. 1993). The shape of these domains is irregular and highly variable and depends on the cell cycle phase and the cell type. The variable character of the chromosomal domains implies that systematic patterns of chromosomal organization, if they exist, can only be derived from a statistical analysis of a large number of observations.

8
Fluorescence in Living Cells: Calcium Waves Organize Heart Beating

In the in situ hybridization experiments of the previous section we looked into fixed, i.e. dead, cells. The target was the DNA of a single cell. Since each cell has only two sets of DNA molecules, in situ hybridization was still, at least in principle, a single molecule technique. In the following studies we will observe bulks of molecules.

8.1
Preservation of Coupling Capacity During Microscopic Preparation

The coordination of the contraction of the entire heart is based on the behavior of the single heart cell and its cooperation with the direct neighborhood. The single cell has an intrinsic periodicity in the regulation of the contraction. Inside the differentiated tissue bulk all cells of the muscle are coupled electrically. The resistance of this coupling is comparatively low and in the range of around 1 MΩ. For comparison, the resistance of a native lipid bilayer with embedded proteins is in the range of $1-10$ GΩ. So it is possible to interpret the bulk tissue as a series of coupled oscillators (Niggli and Lipp 1995). This oscillation is visible in small aggregates of cardiac myocytes derived from mammalian or avian embryonic hearts. After the enzymatic disaggregation of the heart's ventricle and plating to Petri dishes or Petriperm dishes (a Petri dish with a thin foil instead of a massive bottom), the cells form small electrically coupled groups with a behavior similar to the behavior of the bulk ventricle (Rabkin et al. 1995). These cells can be observed in a microscope with high spatial and temporal resolution. This way the coordination processes of an entire heart can be simulated under microscopic control.

8.2
Observation of Coupled Oscillation by the Visualization of Cytosolic Metabolites

The best way to observe the intracellular as well as the intercellular processes in the system described above is to look for a small cytosolic metabolite that performs the same oscillation as the entire system. This metabolite should be fluorescent itself (like NAD^+, NADH, or NADPH) or it should be localizable via a fluorescent dye.

8.3
Cytosolic Calcium Oscillations

A candidate with the latter properties is the bivalent cation calcium. Its cytosolic concentration oscillates simultaneously with the cycling of the cardiac myocyte (Pertsov et al., 1993). This is performed by the cell through a set of ion selective channels for the rise of the cytosolic calcium concentration and ion selective pumps for removing the calcium ions from the cytosol. The cell has cytoplasmic organelles with the capability to store and to resequester the cytosolic calcium from the cytosol. Herein the mitochondria (long term calcium homeostasis) and the sarcoplasmic reticulum (short term regulation of the cycling) are of special importance. In addition it is possible to transport calcium ions to the extracellular space. Many of the involved calcium transporting proteins are regulated by calcium itself. So the rise, as well as the decrease, in the concentration can be regulated by internal feedback processes. This means in practice that for the activation of a cardiac myocyte a short stimulus is sufficient and the cellular system performs an independent cycle of excitation and refractoriness. A part of this cycle generates a signal that excites the neighboring cells. So a single stimulus at a restricted region inside an aggregate of coupled cardiac myocytes will result in a wave of excitation over the whole electrically coupled population. Visualizing the transient rises of the cytosolic calcium concentration during a cycle of excitation one can follow this wave with microscopic methods and can estimate the parameters of this oligocellular oscillator (Lipp and Niggli, 1996).

8.4
Influencing Heart Beating with Drugs

In addition to the observation of the cytosolic calcium transients it is of great importance to manipulate the concentration of this metabolite inside the living cell. This is of special interest because mismatches in the whole heart's calcium based excitability and calcium homeostasis can lead to failures in the coordination of the electrical as well as the mechanical behavior. A wide set of cardiac diseases is related to this phenomenon. The cytosolic calcium level can be manipulated indirectly by stimulating the calcium channels or calcium pumps. Many readers perform such a manipulation every morning. "coffee" and "black tea" contain the alkaloid "caffeine" which has an influence on the calcium channel protein in the sarcoplasmic reticulum of heart cells, the ryanodine receptor. After incorporation it binds to this channel and increases its probability to be in an open state (Ishide et al., 1984). So the stimulus for the excitation of the cardiac myocytes decreases and the beating frequency is increased: The heart is beating faster.

There is another group of drugs which are applied to regulate the excitability of the heart by influencing the cytosolic calcium concentration. These drugs bind to the plasmalemma based calcium channel proteins like the dihydropyridine receptor (also known as the L-type calcium channel) and regulate the entering of the calcium from the exterior to the cytosol. These drugs have two influences on the cardiovascular system: firstly, they regulate the overall excitability of the heart muscle and secondly, they influence the contractile state of the blood vessels. They de-excite the muscle layer around these blood vessels and in this way the blood pressure is decreased. This reduction of the heart's after-load prevents the hypertrophy of the heart muscle.

8.5
Direct Manipulation of the Cytosolic Calcium Concentration

All these substances are widely used in the cardiological experiments, but these experiments all fail in one sense: it is not possible to measure the effect of these drugs in a group of heart cells in a non-homogeneous physiologic state. To apply it to the partial ischemic heart muscle is a good simulation but here the behavior of the single cell is not visible. To apply it to cultures of small re-aggregated cell groups improves the ability to observe the region of interest but now the ischemic condition can be applied only to the culture in its entirety. So far it has not been possible to make a selective stimulation of only a few cells inside an electrically coupled and coordinated behaving group of ventriculocytes.

8.6
Diazo-2 as a UV-Sensitive Calcium Switch

We are able to influence the cytosolic calcium level in only a few cells which are in tight physiologic contact with their surrounding cells. All the other cell are not influenced. So we can generate and visualize a situation very similar to the cardiac infarctous situation: A more or less small part of the heart undergoes a physiological transition into a state with decreased excitability while the rest of the heart muscle is only marginally influenced (Harken, 1993). In the patient's heart this mostly takes place following an under-oxygenation of parts of the ventricle because of two reasons. Either the heart is swollen by increasing the size of its cells, or the blood flow through the vessels is decreased following a sedimentation of the calcium salts of cholesterols. Both situations have a similar effect. The under-oxygenation stimulates a change in the cellular basal calcium level. This makes it possible for the influenced cells to receive and to process the incoming signals for the contraction. In situations like these the overall coordination of the heart breaks down.

8.7
Reaction of the Coupled Oscillators to the Caged Compounds

We simulate situations like these under microscopic control of the spatial and temporal behavior of the cytosolic calcium level. We introduce the physiologically inactive precursors of calcium buffers into the cytosol. They are not active because chemical protecting groups block their action. These protective groups are connected to the molecule via links which can be broken with ultraviolet light (Adams et al., 1989). So, by application of a moderate UV-pulse, it is possible to switch on (or to switch off, depending on the compound used) the buffering behavior of this substance. In Fig. 5, this principle is shown by the application of the UV-labile compound diazo-2. This compound binds in the uninfluenced state calcium with a very low affinity (Mulligan and Ashley, 1989). After illumination with UV light (around 350 nm wavelength) a diazonium group is cleaved from the molecule and the calcium affinity increases significantly. The observable situation in the experiment is that the cytosolic calcium concentration inside the influenced cell decreases immediately, while the physiological state of the directly coupled cells remains unaffected.

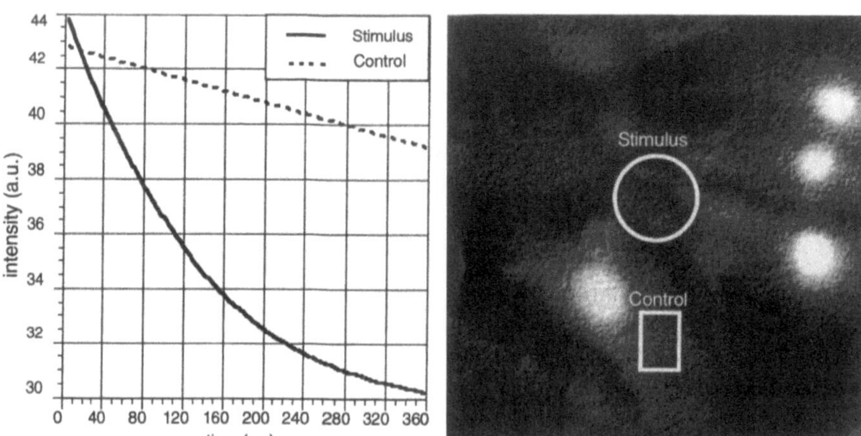

Fig. 5. The decrease in the intracellular calcium concentration via the photolysis of the UV-activated calcium buffer diazo-2 under influence of the light spot is shown. The spot influences the cells in the encircled region and leads to a decrease as displayed in the graph on the left (*lower line*, "Stimulus"). The loss of intensity through bleaching is measured in an unaffected cell as indicated by the rectangular mark (*upper line*, "Control")

Failures like the ones generated via the activation of UV-sensitive calcium buffers have true effects in the coordination of coupled contracting cell groups. A mismatch in such a cell group is measured and displayed in Fig. 5.

8.8
Fluorescence Imaging of the Action of Pacemaker Cells

In a system of coupled oscillators the frequency of the entire system is driven by the member with highest internal frequency (the pacemaker). The pacemaker has to feed the stimulus to the system via a process which is based on tight physiological contact of the cells. If the contact decreases because of physiological or morphological bottlenecks the stimulus reaches the receiving cell group with a temporal delay.

Such a bottleneck and the resulting delay can be seen in Fig. 6 and the graph on the left-hand side. The triggering cell group (the small one on the left side in the picture) is the pacemaker of the entire system. But the trigger has to trespass the tight cytosolic bridge between the pace-making cell and the bigger cell group. This relatively large cell group is pumped by the smaller triggering cell group. The conductive bottleneck following to a tight physiologic bridge and the unbalanced relationship between the small physiological pacemaker and its much bigger following group cause a delay in the coupling of the signal. Vice versa when the big cell group receives the signal, translates it and sends it to the environment the situation is slightly different. Now the big cell group pumps the smaller one which is now in the receiving state. The signal from this group is high enough to compensate for the restrictions given by the physiologic bottleneck. Here no delay can be observed.

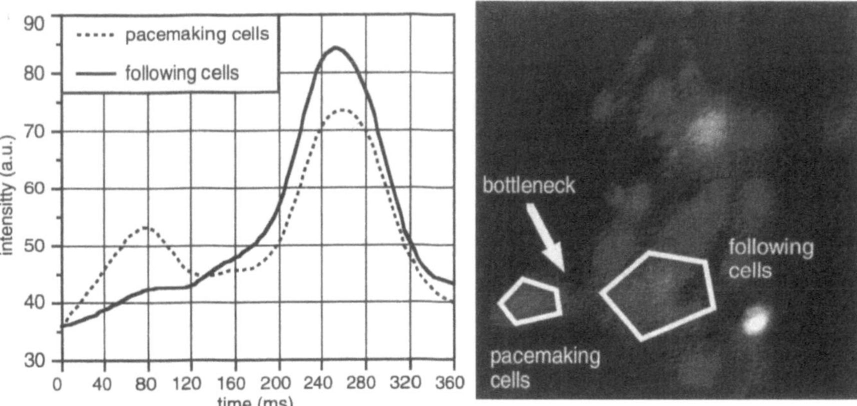

Fig. 6. The spread of an exciting calcium wave from a small cell group left-hand side to a bigger one at the right of the picture is shown. The transient changes in the calcium concentration in both cell groups are shown in the graphs. Here the dotted line gives the calcium concentration of the smaller cell group and the solid line that of the bigger group

8.9
Too High Pacemaker Frequencies Cause Stuttering (Fibrillation) of the Heart

If a pacemakers beats with a very high frequency the pumping capacity of the heart is lost. It is not possible to influence the heart any more via stimulation of the primary trigger of the heart, the sinus node (Starobin et al., 1996). The heart has its own automatic behavior caused by this ventricle based pacemaker. Figure 7 shows such a pacemaker found in a small group of coordinated contracting cardiac myocytes.

A single cell at the bottom of the tissue islet has nearly two times the internal frequency compared to the surrounding. This frequency is based on a small area in the cytosol with a higher excitability than the rest of the cell. This higher excitability is reasoned by a higher ability to release calcium from internal stores and thus by a higher local cytosolic calcium concentration. These regions are called "hot spots" and it is presumed that they are very widespread in the subcellular regulation and compartimentation. In the graph left of the picture, the frequencies of the pacemaker in comparison to the following system are displayed. The pacemaker frequency is directly taken from the cells' "hot spot" region. The frequency of the follower can be taken from any point all over the surrounding cell group. It can be clearly seen that the follower reacts to the second peak of the triggering cell. Perhaps the first impulse is too small because its amplitude is lower than that of the second bigger pulse. As a measure of the stimulating power of a cell, one can take the integral of its singular transients. A measure of its excitability would be the basic level of the calcium concentration and the slope of the ascending flank of the calcium transients. In contrast the slope of the descending flank is a measure of the proteins which remove the

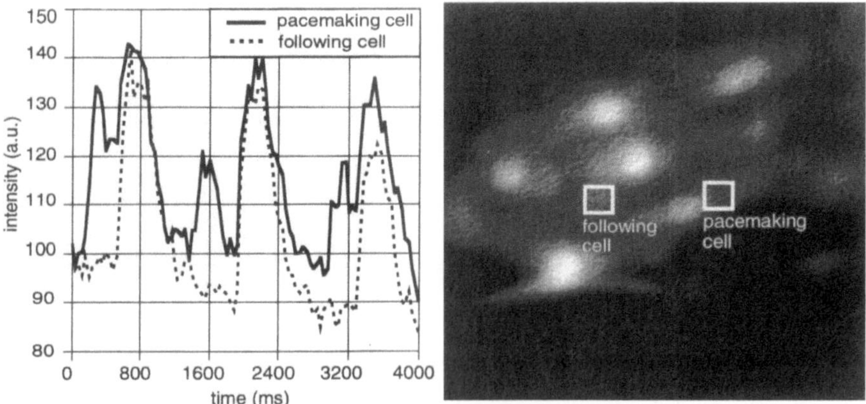

Fig. 7. The pacemaking function of a cell with a higher beating frequency compared to the neighboring cells can be seen here. The pacemaking cell's cytosolic calcium concentration makes more transient rises than that of the surrounding cells. Not every transient of this pacemaker is high enough to stimulate the surrounding cells. Only if the threshold of the surrounding cells is crossed they develop their own excitation

calcium from the cytosol. So they are a measure of the ability of the cell to regulate the basic calcium concentration in the cytosol between two of the excitations.

9
Conclusion

Classical fluorescence microscopy, due to the availability of sensitive cameras, can now be used to study single molecule reactions in solution as well as the biological and physiological details of single cells. This means that biological objects can be studied as molecular or cellular individuals.

References

S.R. Adams, J.P.Y. Kao, Kao, R.Y. Tsien, Biologically useful chelators that take up Ca^{2+} upon illumination, J. Am. Chem. Soc., 111, 7957–7968, 1989

A. Ashkin, J.M. Dziedzic, J.E. Bjorkholm, S. Chu, Observation of a single beam gradient trap for dielectric particles, Optics Letters, 11, 288–290, 1986

A. Ashkin, Optical trapping and manipulation of neutral particles using lasers, Proc. Natl. Acad. Sci. 94, 4853–4860, 1997

T. Cremer, P. Lichter, J. Borden, D.C. Ward, L. Manuelidis, Detection of chromo- some aberrations in metaphase and interphase tumor cells by in situ hybridization using chromosome-specific library probes, Hum. Genet. 80, 235–246, 1988

M. Ferguson, D.C. Ward, Cell cycle dependent chromosomal movement in premitotic human T-lymphocyte nuclei Chromosoma 101, 557–565, 1992

T. Funatsu, Y. Harada, M. Tokunaga, K. Salto, T. Yanagida, Imaging of single fluorescent molecules and individual ATP turnovers by single myosin molecules in aqueous solution, Nature 374, 555–559, 1995

K.O. Greulich, G. Weber, The light microscope on its way from an analytical to a preparative tool, J. Microscopy 167(2), 127–151, 1992

T. Haaf, Analysis of replication timing of ribosomal RNA genes by fluorescence in situ hybridization, DNA and Cell Biology 16, 341–345, 1997

A.H. Harken, Operation bei Herzjagen nach Infarkt, Spekt. d. Wiss., pp. 64–69, 1993

C. Hoyer, S. Monajembashi, K.O. Greulich, Laser manipulation and UV induced single molecule reactions of individual DNA molecules, J. Biotechnology 52, 65–73, 1996

N. Ishide, H. Watanabe, T. Takishima, Effects of high-K low-Na, Diltiazem, Caffeine and Ba on spontaneous fluctuating contractile activity in rat cardiac muscles, Tohoku J. Exp. Med. 144, 331–341, 1984

G. Leitz, K.O. Greulich, E. Schnepf, Displacement and return movement of chloro-plasts in the marine dinophyte Pyrocystis noctiluca. experiments with optical tweezers, Bot. Acta 17, 90–94, 1994

G. Leitz, E. Schnepf, K.O. Greulich: Micromanipulation of statoliths in gravity sensing Chara rhizoids by optical tweezers. Planta 197(2), 278–288, 1995

P. Lichter, T. Cremer, J. Borden, L. Manuelidis, D.C. Ward, Delineation of individual human chromosomes in metaphase cells by in situ suppression hybridization using recombinant DNA libraries, Hum. Genet. 80, 224–234, 1988

P. Lipp, E. Niggli, Submicroscopic calcium release signals as fundamental events of excitation – contraction coupling in guinea-pig cardiac myocytes, J. Physiol., 492 (1), 31–38, 1996

S. Monajembashi, C. Hoyer, K.O. Greulich, Laser microbeams and optical tweezers convert the microscope into a versatile microtool ,Microscopy and Analysis 97(1), 7–9, 1997

I. P. Mulligan, C.C. Ashley, Rapid relaxation of single frog skeletal muscle fibres following laser flash photolysis of the caged calcium chelator, diazo-2, FEBS Lett., 255(1), 196–200, 1989

E. Niggli, P. Lipp, Subcellular features of calcium signalling in the heart muscle: what do we learn?, Cardiovasc. Res. 29, 441–448, 1995

A.M. Pertsov, J.M. Davidenko, R. Salomonsz, W. T. Baxter, J. Jalife, Spiral waves of excitation underlie reentrant activity in isolated cardiac muscle, Circ. Res. 72, 631–650, 1993

G. Pilarczyk, K.O. Greulich, The combined use of UV-labile calcium chelators and calcium sensitive dyes in a microscope with two light sources influencing different regions in a group of coordinated contracting cardiac myocytes, Optical Biopsies and Microscopic Techniques, Proc., SPIE 3197, 1997

S.W. Rabkin, N. Freestone, G.A. Quamme, Nifedipine does not alter the increased cytosolic free magnesium during inhibition of mitochondrial function in isolated cardiac myocytes, Gen. Pharmac. 25 (7), 1483–1491, 1994

J.M. Starobin, Y.I. Zilberter, E.M. Rusnak, C.F. Wavelet formation in excitable tissue: the role of wavefront-obstacle interactions in inititating high-frequency fibrillatory-like arrhythmias, Biophys. J. 70, 581–594, 1996

L. Stryer, Biochemistry, Chap. 8, W. H. Freeman and Co., New York, 1988 (German edition Spektrum Akad. Verlag, Heidelberg, Berlin, New York, 1991)

M. Tokunaga, K. Kitamura, K. Saito, A.H. Iwane, T. Yanagida, Single-molecule imaging of fluophores and enzymatic reactions achieved by objective-type total internal reflection microscopy, Biochemical and Biophysical Research Communications 235, 47–53, 1997

C.Vourc'h, D. Taruscio, A.L. Boyle, D.C. Ward, Cell cycle dependent distribution of telomeres, centromeres and chromosome-specific subsatellite domain in the interphase nucleus of mouse lymphocytes, Exp. Cell Res. 205, 142–151, 1993

X.-H. Xu, E. Yeung, Direct measurement of single molecule diffusion and photode composition in free solution, Science 275, 1106–1109, 1997

Q. Xue, E. Yeung, Differences in the chemical reactivity of individual molecules of an enzyme, Nature 373, 681–683, 1995

CHAPTER 18

Solvent Relaxation in Biomembranes

M. Hof

1
Introduction

Fluorescence spectroscopy has considerably contributed to the present picture of the structure and function of biomembranes consisting predominantly of phospholipids. Among the fluorescence techniques employed such as quenching [1], energy transfer [2, 3], lifetime (distributions) [4–6], and excimer formation [7, 8], the determination of fluorescence anisotropy [4] has certainly been the dominating fluorescence method in studies of biological and model membranes. Fluorescence polarization studies, however, exhibit some major limitations: the experimentally determined steady-state and time-resolved anisotropies characterize the motional restrictions of the 'reporter' molecule itself and give therefore only *indirect* information about the dye environment, with the consequence that, if the probe is bound covalently to the lipid [e. g., 1-(4-trimethylammoniumphenyl)-6-phenyl-1,3,5-hexatriene; TMA-DPH], this attachment may dominate the recorded depolarization behavior. The membrane order parameters obtained from freely mobile probes like 1,6-diphenyl-1,3,5-hexatriene (DPH) result from a broad distribution of localization within the hydrophobic interior, the detailed characterization of which reveals inherent ambiguities [9]. Moreover, the anisotropy technique is limited to the characterization of the hydrophobic bilayer interior, since the anisotropy of head-group labelled phospholipids appears to be rather insensitive to environmental changes [4]. These limitations do not restrict the application of the solvent relaxation method, the principles of which and their applications in membrane studies will be presented in this review. It will be shown that the solvent relaxation technique makes it possible to observe *directly* the viscosity and polarity changes in the vicinity of the probe molecule with well-defined location in both membrane domains. Surprisingly, only a very low number of fluorescence investigations of biomembrane structure are based on this technique and exploit its advantages compared to other fluorescence methods. This review presents an overview of a set of membrane probes with well-defined locations reaching from the outermost interface region to the hydrophobic interior and the time-resolved analysis of their solvent relaxation behavior. The validity and sensitivity of the solvent relaxation method in membrane studies is demonstrated for variations of temperature, membrane curvature, lipid composition as well as by membrane binding studies of calcium ions and peripheral membrane proteins.

The results achieved through the compilation of appropriate probes might suggest a more intensive application of this approach in biological membrane studies.

2
Basic Principles of Solvent Relaxation

The fluorescent probes used in solvent relaxation studies are characterized by a large change in the dipole moment $\Delta\mu$ ($\Delta\mu = \mu(S_1) - \mu(S_0)$) upon electronic excitation from the ground state S_0 to the excited state S_1. Since the time scale of the electronic transition is much shorter than that of nuclear motion, the excitation causes an ultra-fast change in the probe's charge distribution but does not affect the position or orientation of the surrounding solvent molecules. The solvent molecules are, thus, forced to adapt to the new situation, and start to reorient themselves in order to find an energetically favored position with respect to the excited dye. The dynamic process starting from the originally created non-equilibrium Franck–Condon state (F) and gradually establishing a new equilibrium in the excited state (R) is called solvent relaxation (SR). This relaxation red-shifts the probe's emission spectrum continuously from the emission maximum frequency of the Franck–Condon state ($v(0)$ for t = 0) to the emission maximum of the fully relaxed R-state ($v(\infty)$ for t = ∞). Since a more polar solvent leads typically to a stronger stabilization of the polar R-state, the overall shift Δv ($\Delta v = v(0) - \text{n}(\infty)$) increases with increasing solvent polarity for a given change of the solute's dipole moment $\Delta\mu$. The accurate mathematical description of this relationship depends on the choice of the dielectric solvation theories [10–17]. The fundamental 'dielectric continuum solvation model' [15–17] predicts a linear proportionality between Δv and a dielectric measurement of the solvent polarity F for a large variety of solvents [18], where F is a function of the static dielectric constant ε_0 and the optical refractive index n of the solvent (F = $[(\varepsilon_0 - 1)/(\varepsilon_0 + 2)] - [(n^2 - 1)/(n^2 + 2)]$; for illustration see Fig. 9 in [18]). Deviations from this linearity are found for a few aromatic solvents, indicating that F is a poor measure of the nuclear polarizability of aromatic solvents. However, for dielectric relaxation in phospholipid/water systems it can be assumed that the Δv/F-proportionality is valid, and that changes in Δv directly reflect polarity changes in the dye environment, giving the first major information provided by the solvent relaxation technique in membrane studies.

The second information obtainable from the measurement of time-dependent spectral shifts is based on the fact that the SR kinetics is determined by the mobility of the dye environment. The response of solvent molecules to the electronic rearrangement of the dye is fastest in the case of water: more than half of its overall solvation response occurs under 55 femtoseconds (fs) [19]. If the dye is located in a viscous medium, like phospholipid membranes, the solvent relaxation takes place on the nanosecond (ns) time scale [20]. In vitrified solutions, on the other hand, the dye may fluoresce before solvent relaxation towards the R-state is completed [21]. The time evolution of the solvation energy relaxation process can be described by means of the normalized spectral response function or so-called "correlation function" $C(t)$:

$$C(t) = (\nu(t) - \nu(\infty))/\Delta\nu \qquad (1)$$

The development of ultrafast spectroscopic methods with fs-resolution has led to an accurate determination of $C(t)$ for a large variety of polar solvents and has allowed the physical interpretation within the framework of existing theoretical models. Summarizing the experimental results of several research groups [18, 19, 22, 23], the solvents at ambient temperatures respond with a fast inertial (librational) motion in the range between 50 and 500 fs, depending on the selected solvent. After this initial period of solvation response, which was shown to be a Gaussian function of time, the diffusion of the solvent molecules, occurring on the pico- to subnanosecond time scale leads to further solvation energy relaxation towards the R-state. Although the diffusive contribution to $C(t)$ is often described by (multi)exponential decays, the application of "distributions of decay times" analysis indicates that the diffusive relaxation might be better described by a distribution of relaxation times than by discrete exponential relaxation processes [18]. The fs-experimental results are in good agreement with recently presented theories [10–14], which basically represent further refinements of the dielectric continuum model introduced by Bakshiev and co-workers [15, 16], especially by taking into account the molecular nature of the solvent.

3
Fluorescence Approach to Solvent Relaxation

3.1
Time-Resolved Emission Spectra and Solvent Relaxation Times

The literature on the theoretical and experimental aspects of time-resolved fluorescence of polar molecules in polar solvents goes back over thirty years [15, 16, 24, 25]. Although there have been several attempts to simplify the characterization of the SR process [25, 26], the determination of the normalized spectral response function $C(t)$ is certainly the most general and most precise way to characterize the time course of the solvent response. The $C(t)$ function can be determined either by the "single wavelength method" [27] or, as most frequently used, by "spectral reconstruction" [28]. This latter method requires a minimum of *a priori* assumptions and involves a reconstruction of time-resolved emission spectra (TRES) from wavelength-dependent time-resolved decay data. It has been used for obtaining the results presented in this work and is outlined here.

Fluorescence decay data are collected at regular intervals spanning the fluorescence emission spectrum. The individual wavelength-dependent decays are fitted to a sum of exponentials using an iterative reconvolution technique until a fit, which is mathematically satisfactory, is obtained [29]. It is important to note that the purpose of these fits is to represent the "pure" (i.e. undistorted by the excitation pulse width and the time-dependence of the detecting electronics) fluorescence decay curve at each single wavelength and not to provide a physical interpretation of the decays. The TRES at a given time t, $S(\lambda; t)$, is obtained by the fitted decays, $D(t; \lambda)$, by relative normalization to the steady-state spectrum, $S_0(\lambda)$, as follows:

$$S(\lambda; t) = (D(t; \lambda) \times S_0(\lambda))/ \int_0^\infty D(t; \lambda)\, dt \qquad (2)$$

Typical results of such spectral reconstruction are illustrated in Fig. 2a,b. The examination of the shape of the TRES can give information about the nature of the investigated relaxation process. The existence of isoemissive points [30], for example, indicates that the examined excited state process is not continuous, but rather the existence of an equilibrium between two excited states, like excimer or exciplex formation. After conversion from a wavelength representation to one linear in frequency, the TRES are fitted by the empirical "log-normal function" [31], which gives a realistic picture of broad, asymmmetric electronic emission bands [28, 32]. Should the log-normal model fail to fit the data, excited state processes other than pure continuous solvent relaxation may be considered. From the fitted spectra the emission maximum frequencies $v(t)$, the total Stokes shift Δv, and the full width at half maximum of the TRES ("half width") (Fig. 7) are usually derived. Since $v(t)$ contains both information about polarity (Δv) and viscosity of the reported environment, the spectral shift $v(t)$ is normalized to the total shift Δv. The resulting "correlation functions" $C(t)$ (Eq. 1) describe the time course of the solvent response (Fig. 4) and allow for comparison of the SR-kinetics and, thus, of relative microviscosities, reported from environments of different polarities.

Although the fs-results encourage a physical interpretation of the shape and time course of the $C(t)$, a user of the solvent relaxation technique has to be aware that experimental factors like the signal-to-noise ratio, the excitation pulse width, or the dynamics of the detectors influence the quality of the $C(t)$ data. In some cases a fitting by physical models might then appear as an over-interpretation of such data. As described above, the $C(t)$ function results from several mathematical operations treating measured fluorescence decays and emission spectrum, and experimental errors can multiply. When applying the SR-technique in anisotropic media like membranes it has to further be considered that the dye might probe a distribution of microenvironments and $C(t)$ might then represent a continuous superposition of several solvent responses. In all investigations on membrane labels done by our group employing a sub-nanosecond "Single Photon Counting" apparatus [33–39], the solvent response appears to be rather complex and cannot be satisfactorily and consistently described by (multi)exponential relaxation models. In order to characterize the overall time scale of the solvation response, we omit therefore data fitting and use an (integral) average relaxation time:

$$\langle \tau_t \rangle \equiv \int_0^\infty C(t)\, dt \qquad (3)$$

3.2
Red-Edge Excitation Spectroscopy

Laboratories which are not equipped with a time-resolved fluorescence spectrometer, but with steady-state absorption and emission spectrometers, can still obtain qualitative information about viscosity and polarity of the probed environment by exploiting the solvent relaxation process. This application of steady-state fluorimetry is termed 'Red-Edge Excitation Spectroscopy' (REES) and it is limited to those cases where the dipolar relaxation process is comparable or slower than the fluorescence process. It uses the fact that the wavelength of the emission band maximum of polar fluorophores in motionally restricted media, such as in very viscous solutions [40, 41], glass-like matrices [21] or membranes [20], shifts to longer wavelength by shifting the excitation wavelength toward the red-edge of the absorption band [42]. Since the observed shift should be maximal if the solvent relaxation is much slower than the fluorescence, and it should be zero if SR is fast and, thus, independent of the excitation wavelength of the entire fluorescence originating from the relaxed R-state, it can serve as an indicator of the mobility of the probe surrounding [40, 41]. Usually, red-edge excitation shift values range from several up to 40 nanometers (nm) depending on the chosen solute and solvent system. A recently reported unique large shift of 76 nm for a hemicyanine dye in "AOT"-micelles indicates an ionic rather than a dielectric relaxation process [43]. The red-edge excitation shift is especially useful when using dyes, the absorption and fluorescence maxima of which hold linear correlations with the polarity of low-viscosity solvents [21, 28], because then the probed polarity as well as the hypothetical emission maximum of the fully relaxed R-state can be estimated from the absorption maximum. Since this work is focused on the time-resolved approach towards solvent relaxation in membranes, the readers are referred to three review articles on steady-state REES that, to a great extent, describe biological applications of this technique [20, 40, 41].

4
Solvent Relaxation Probes in Phospholipid Membranes

4.1
Overview

The most popular solvent relaxation probes in biophysical studies of phospholipid membranes are certainly the amino-substituted naphthalene derivatives. Representatives of this family are the negatively charged 1,8-ANS (1-anilinonaphthalene-8-sulfonate) [44], 2,6-ANS (2-anilinonaphthalene-6-sulfonate) [44], 2,6-TNS (2-(p-toluidinylnaphthalene)-6-sulfonate) [34, 45, 46], the neutral probes NPN (N-phenyl-1-naphthylamine) [47], Dansyl Lysin (N-e-(5-dimethylaminonaphthalene-1-sulfonyl)-L-lysine) [48], Prodan (6-propionyl-2-(dimethylamino)naphthalene) [35, 49, 50], Laurdan (2-dimethylamino-6-lauroylnaphthalene) [51, 52], and the positively charged probe Patman (6-palmitoyl-2-[[2-(triethylammonium)ethyl]methylamino]naphthalene chloride) [35, 53].

The chemical structures of these probes illustrate the general principle of constructing a probe with a large change of the dipole moment upon excitation: the delocalized π-electron system (naphthalene) is substituted with an electron-donating group (substituted amino groups in all the above listed cases) and -accepting group (variations of acyl or sulfonato groups). In recent membrane studies it has been further shown that the fluorescence characteristics of the family of NBD (7-nitrobenz-2-oxa-1,3-diazol-4-yl) labelled lipids are also significantly influenced by solvent relaxation of the probed environment [20, 54]. The above listed labels are supposed to be located at or very close to the head group region of the phospholipid bilayer.

For the systematic characterization of structural changes in both membrane domains, however, a set of membrane labels is desirable with defined location reaching from the outermost interface region to the hydrophobic interior of the membrane. Since the time-dependent Stokes shift Δv depends both on the solvent polarity and on the change in the solute's dipole moment $\Delta\mu$, detection of polarity differences at different positions within the bilayer requires that the chromophore of the used compounds is identical. In the following sections a set of 7 compounds based on two different chromophores is presented, which permits the determination of polarity and viscosity changes at defined positions along the axis perpendicular to the membrane plane. The series of n-(9-anthroyloxy) fatty acids (n-AS, where n = 2, 6, 9, 12, 16) (Fig. 1a) serve as the pro-

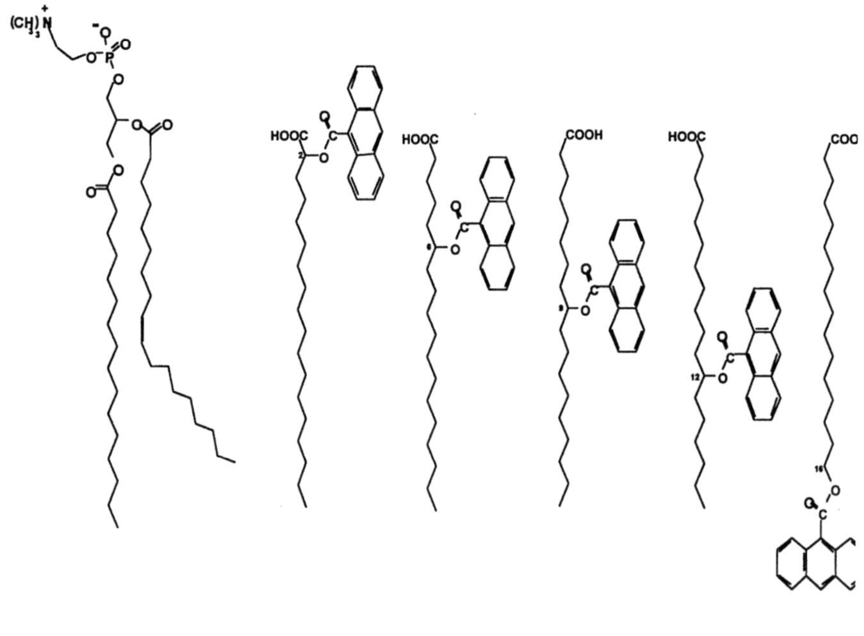

Phosphatidylcholin   2-AS   6-AS   9-AS   12-AS   16-AP

Fig. 1a. Molecular structures of n-anthroyloxy fatty acids (n-AS) and their schematic positions relative to phosphatidylcholine

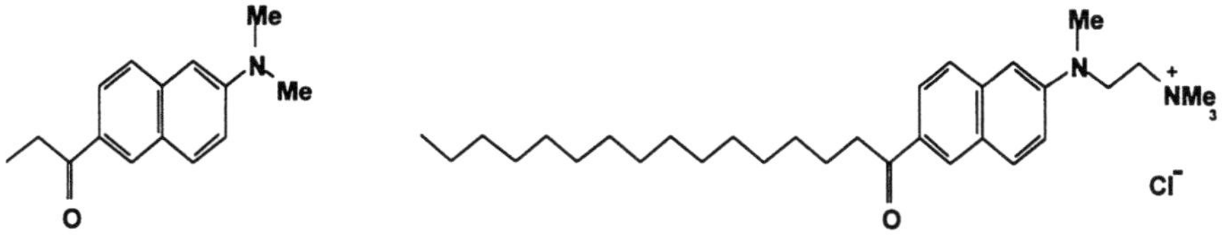

Fig. 1b. Molecular structures of Prodan (*left-hand side*) and Patman

bes for the hydrophobic backbone, the interface region is probed by Prodan and its long alkyl chain derivative Patman (Fig. 1b).

4.2
n-AS Dyes Probing the Hydrophobic Backbone

The better the probed environment is defined, the more valid the information about viscosity and polarity that can be gained by the solvent relaxation method. On that score the n-anthroyloxy fatty acids (n-AS) constitute a unique set of fluorescent dyes with the great advantage that a common chromophore can be used, which is covalently attached at different positions (n = 2, 6, 9, 12, 16) along the acyl chain of the fatty acid (stearic acid for n = 2 – 12; palmitic acid for n = 16). The n-AS probes are known to insert into the membrane with the stearoyl chains parallel to the phospholipid acyl chains with the long axes of the anthroyl ring on average almost perpendicular to the plane of the membrane [55]. The chromophore shows little motion around the ester linkage and it is positioned at a defined depth of the bilayer as confirmed by both ^{1}H NMR [56] and quenching studies using Ca^{2+} and dimethylamine [57]. Although the photophysics of 9-anthroic esters has been intensively studied [58–62], their potential use as SR probes, however, has been overlooked. The observed wavelength dependence of the fluorescence decay behavior has been attributed to an intramolecular conformational change from the initially excited Franck–Condon state to the state with equilibrium geometry characterized by a more planar orientation of the aromatic ring relative to the carboxyl group [58]. In a recent study [37] we have shown that this picture holds for the n-AS dyes in non-polar solvents, but that intermolecular, continuous solvent relaxation dominates the wavelength-dependent emission behavior in polar environments. While the total Stokes shifts Δv in highly viscous solvents like paraffin oil evoking from the intramolecular relaxation process are small and independent of the fluorophore position (in nm: 7–10), the n-AS dyes show much larger Δv values increasing within the series 16-AP < 12-AS < 9-AS < 6-AS < 2-AS (Figs. 2a,b, and 3; Table 1), when incorporated in phosphatidylcholine small unilamellar vesicles (PC-SUV). With 2-AS, a Stokes shift of 39 nm is observed within the first 20 ns after excitation. After this time, the relaxation is essentially complete for all dyes, as may be seen in Fig. 3 which summarizes the detected increases of the emission maxima as a function of time after excitation. Apparently, 2-AS, which is located closest to the membrane/water interface, probes the most polar en-

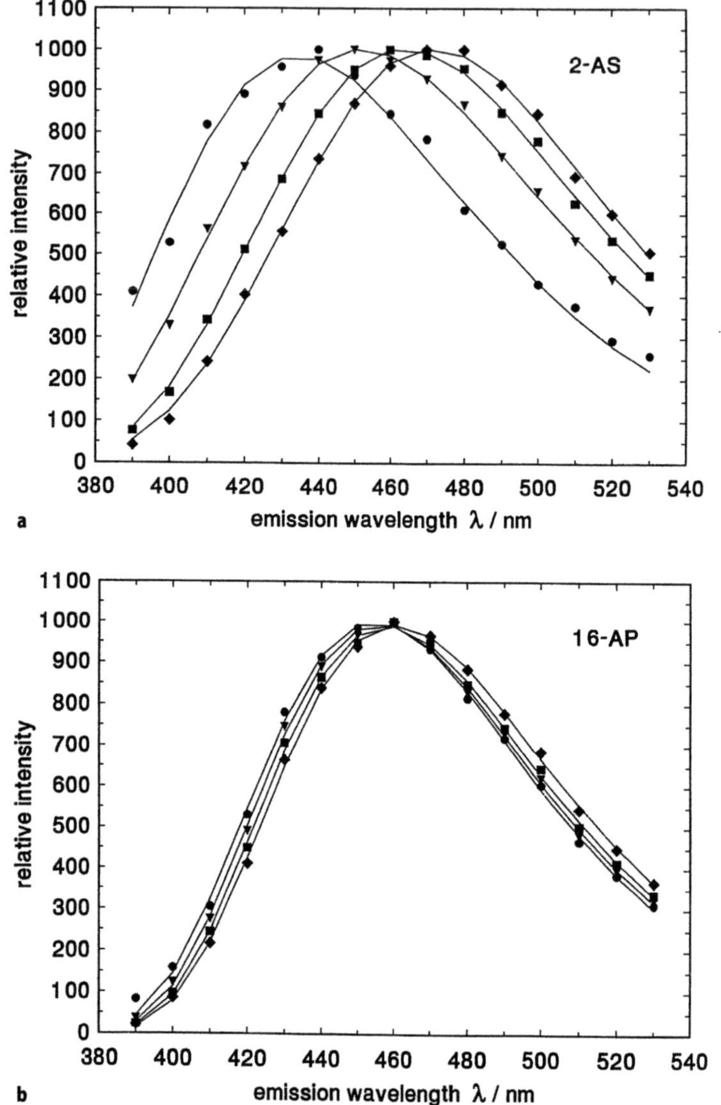

Fig. 2a, b. Time-resolved emission spectra for the 2-AS (**a**) and 16-AP (**b**) in PC-SUV at 25 °C. The respective chromophores are indicated inside the plots. Spectra are shown at 0.2 ns (*circles*), 2.0 ns (*triangles*), 5.0 ns (*boxes*) and 20.0 ns (*diamonds*) after excitation

vironment. The Stokes shift of 6 nm observed for 16-AP is comparable to those detected in non-polar, viscous solvents, and is probably entirely due to intra-molecular relaxation and indicates the absence of water molecules close to the center of the bilayer. The presented trends illustrate the Δv/solvent polarity relationship and give further evidence for the idea of a polarity gradient perpendicular to the membrane/water interface. Moreover, the results motivate the use

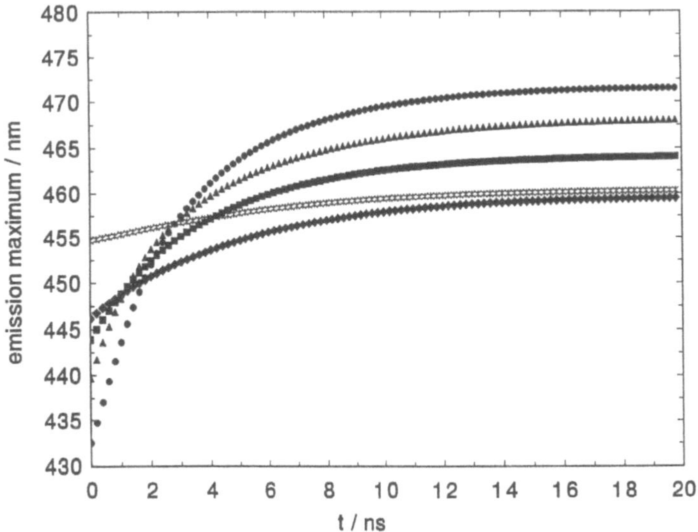

Fig. 3. Time course of the emission maxima (in nm) as a function of time after excitation of the n-AS in PC-SUV at 25 °C. *Circles*: 2-AS; *triangles*: 6-AS; *boxes*: 9-AS; *diamonds*: 12-AS; *asteriscs*: 16-AP

of this method for detecting externally induced polarity changes within the hydrophobic backbone of the bilayer.

In Fig. 4 the corresponding correlation functions C(t) for the n-AS dyes in PC-SUV are presented. The rate of the relaxation process decreases with increasing depth of incorporation of the fluorophore (Table 1). This trend contradicts the "microviscosity" determined by steady-state anisotropy measurements [63], which yield a viscosity decrease from the head-group region to the terminal methyl groups of the phospholipid bilayers. In order to rationalize this contradiction it has to be considered that the solvent relaxation process is connected with the properties of the *polar* solvent molecules surrounding the dye. Anisotropy, on the other hand, monitored by fluorescence probes like DPH is determined by motional restrictions of the probe imposed by the hydrophobic phospholipid acyl chains. Thus, the slower SR reports an increasing visco-

Table 1. Stokes Shifts Δv (in nm) and Mean Relaxation Times $\langle \tau_r \rangle$ (in ns) for the n-AS Dyes in PC-SUV at 25 °C

n	$\langle \tau_r \rangle$ (ns)	Δv (nm)
2	2.7	39
6	3.0	27
9	3.6	20
12	4.8	13
16	5.9	6

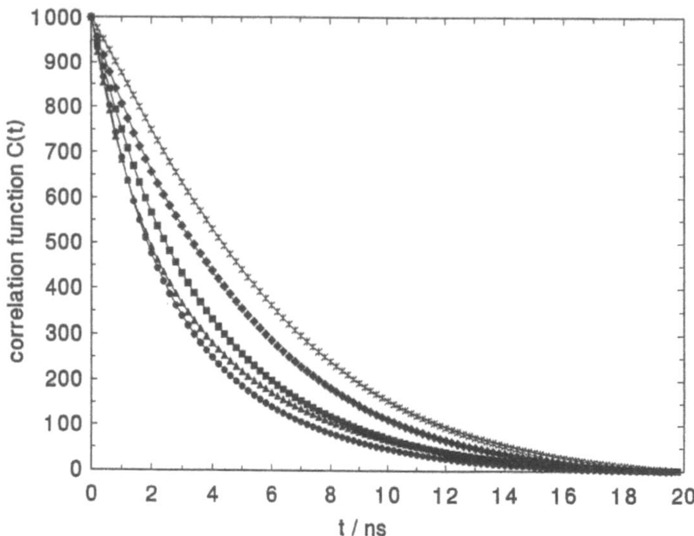

Fig. 4. Correlation functions C (t) calculated according to Eq. 1 for the n-AS dyes in PC-SUV at 25 °C. *Circles*: 2-AS; *triangles*: 6-AS; *boxes*: 9-AS; *diamonds*: 12-AS; *asteriscs*: 16-AP

sity of the water present in the membrane bilayer when going deeper into the membrane. Physiologically induced changes in membrane structure, however, may change the local viscosity of the aqueous solvation shell hydrating the acyl chain domain, which can be characterized by the SR method presented here.

4.3
Hydrophilic Head-Group Region

In search of a set of commercially available head-group labels possessing the same chromophore, but located at different distances from the membrane/ water interface, one certainly ends up with the dyes Prodan, Laurdan and Patman. The chromophore of all three dyes is a 6-acyl-2-amino-substituted naphthalene. Different substitution at the donor (2-amino) and acceptor (6-acyl) group are supposed to lead to a different localization of the chromophore within the phospholipid head-group region. In a recent contribution [35] the relaxation behavior and the relative localization of Prodan and Patman have been defined by fluorescence spectroscopy as well as by high resolution ¹H NMR measurements. It was concluded that Patman is incalated deeper between the lipid acyl chains than Prodan, sensing a less polar and more restricted environment. Prodan, on the other hand, is due to the lack of the acyl chain located very close to the membrane/water interface. As a consequence of its deeper localization in the membrane, Patman probes a considerably slower continuous solvent relaxation dynamic compared to Prodan, which is shown for DMPC (1,2-dimyristol-sn-glycero-3-phosphatidylcholine)-SUV for two temperatures in Fig. 5. Apparently, the monitored mobility decreases towards the hydropho-

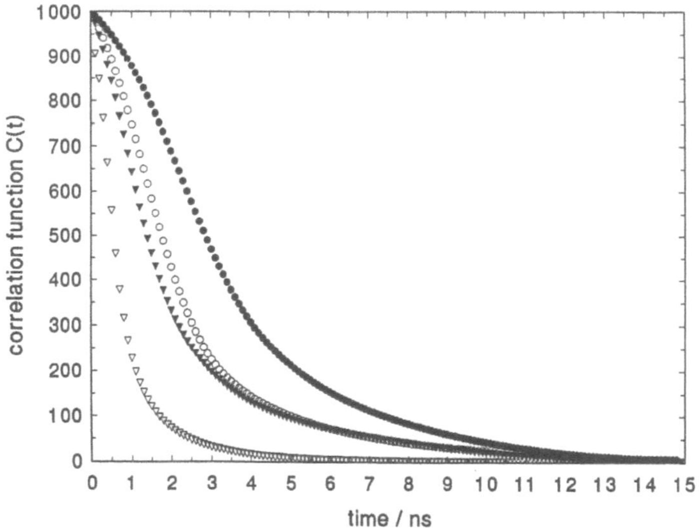

Fig. 5. Correlation functions C(t) calculated according to Eq. 1 for Patman (*circles*) and Prodan (*triangles*) in DMPC-SUV at 30° (*filled symbols*) and 43 °C (*open symbols*)

bic backbone within the head-group region. The localization of Laurdan in membranes has already been the subject of detailed fluorescence investigations [51, 52] and it can be concluded that Laurdan is incorporated deeper than Prodan. A comparison with Patman, however, is still missing. In the applications of the SR technique presented here, Patman and Prodan are employed for the determination of the phospholipid head-group mobility.

5
Structural Changes Monitored by the Solvent Relaxation Method

5.1
Content of Negatively Charged Lipids in the Absence and Presence of Calcium

Although biological membranes contain only relatively small amounts of anionic phospholipids, their presence is essential for many physiological processes on the cellular level. The blood coagulation process, in which phosphatidylserine (PS) directly regulates several proteolytic enzyme activations, may serve as an example (replacement of the positively charged choline in PC by the neutral serine residue yields a head-group charge of – 1 in PS). Since the activities of so-called "vitamin K dependent proteins" [coagulation factors II (prothrombin), VII, IX, X, and proteins C, S, Z] require a calcium intermediated peripheral binding to the PS containing membrane surface, understanding of the protein binding mechanism anticipates structural characterization of the interfacial region. Thus, the labels Prodan and Patman have been used to study the influence of PS and calcium ions on the phospholipid head-group mobility in PS/PC-SUV

[36]. Patman monitors a slight deceleration of the head-group mobility with increasing PS-content in the absence of Ca^{2+} (Table 2) suggesting a lower flexibility of the PS head-group. This conclusion is in agreement with NMR studies which show that the rotational mobility of the PS head-group is lower than that of phosphatidylglycerol, which is, on the other hand, less mobile than the PC head-group [64, 65]. In contrast to pure PC vesicles, the addition of 3 mM Ca^{2+} leads to a considerable deceleration of the solvent relaxation in PS-containing vesicles as observed by both dyes (Table 2). This effect increases with increasing PS-content. Thus, the deceleration of the head-group mobility with increasing PS-content is much more pronounced when Ca^{2+} is present. These results indicate a tighter phospholipid head-group packing with increasing PS-content and suggest a bridging of PS molecules by Ca^{2+} within the plane of the membrane leading to a decrease in lipid mobility. On the molecular level the observed tighter head-group packing can be explained by a coordinative Ca-PS binding. A 1:1 calcium/PS complex would only lead to an increase of the molecular mass of the serine head-group and cannot explain the observed large increase in the mean relaxation times with increasing PS-content. A much more likely explanation is the formation of $Ca(PS)_x$-complexes ($x > 1$). Considering the calcium coordination number six, and the presence of two negative charges in the PS head-group, x should be two [66].

5.2
Binding of Vitamin K Dependent Proteins

The molecular mechanism of the interaction of the vitamin K dependent protein prothrombin with negatively charged membrane surfaces has been intensively investigated by means of fluorescence spectroscopy [67–70]. A clear mechanistic picture, however, is still missing. One reason for this might be that some of the fluorescence techniques used such as pyrene-excimer [67] and DPH anisotropy studies [68] probe only the hydrophobic backbone, and do not detect protein-induced changes in the membrane head-group organization. Recently, we compared protein-induced changes in the SR kinetics monitored by Patman and Prodan with changes in the DPH steady-state anisotropy for the binding of prothrombin to PS-containing membranes [36, 71]. The SR times $\langle \tau_r \rangle$

Table 2. Mean Relaxation Times $\langle \tau_r \rangle$ (in ns) of Patman and Prodan in PC/PS-SUV in the Presence and Absence of Ca^{2+} at 25 °C

PS-content (%)	Prodan without Ca^{2+}	Prodan with 3 mM Ca^{2+}	Patman without Ca^{2+}	Patman with 3 mM Ca^{2+}
0	1.0	1.1	2.1	2.0
10	1.1	1.1	2.0	2.0
20	1.1	1.2	2.0	2.3
30	1.1	1.3	2.2	2.6
40	1.2	1.3	2.3	2.8
50	1.0	1.5	2.5	3.5

in SUV composed of PC/PS 80:20 increase by 100% and 30% detected by Prodan and Patman, respectively, at protein concentrations leading to >90% membrane surface coverage [72]. The observed increase in membrane order under similar conditions reported by DPH, on the other hand, is relatively small (<5%) [71]. The binding of this peripheral protein rigidifies considerably the phospholipid head-group region, but leads only to a very small increase in the packing density of the hydrocarbon region of the bilayer. The observed deceleration of the solvent relaxation process is larger for Prodan, which is bound near the surface of the membrane, than for Patman which is located closer to the hydrocarbon region. Apparently, the binding of the proteins predominantly affects the outermost region of the membrane, where the amino and the carboxyl group of the serine head-group are supposed to be located. These observations exclude hydrophobic protein–membrane interaction and favor the formation of a coordination complex by the PS head-group, calcium ions and the Gla residues of the proteins [73]. In further experiments we are using the high sensitivity of the presented SR approach for comparative binding studies of the factors II and Xa with their N-terminal protein fragments [71], so as to give an answer to the question if and by which mechanism binding sites outside the N-terminal end contribute to the binding of the entire proteins [68, 69, 74].

5.3
Substitution of Diacyl- by Diether-Lipids

Although phospholipids containing ether-linked alkyl or 1-alkenyl chains have been widely found in various types of cells, their biological function is still unknown [75]. One way to learn about their biological function is to investigate the potential change in physico-chemical properties of membranes by enhanced level of ether phospholipids, since the absence of a hydration bond partner may change the molecular packing in the head-group domain. This approach provides the motivation to exploit the SR of Prodan and Patman for the characterization of the head-group mobilities of vesicles containing monoether lipids [50, 76] as well as asymmetric [38] and symmetric diether [39] lipids. In summary, it was concluded that the substitution of the acyl by ether linkages leads to considerably faster solvent relaxation as monitored by both fluorophores. This finding will be demonstrated by the solvent relaxation characteristics of Patman in SUV composed of the symmetric diacyl-(1,2-dipalmitoylphosphatidylcholine) (DPPC) and diether-(1,2-dihexadecylphosphatidylcholine) (DHPC) [39]. The Δv dependence observed for this system gives direct evidence that the ether linkage allows more efficient water penetration in the glycerol region. The results illustrate, moreover, the usefulness of the characterization of the half-widths of the reconstructed TRES for the interpretation of the determined $\langle \tau_r \rangle$ and Δv.

The head-group mobility of DPPC- and DPHC-SUV was monitored at two temperatures one below (20° and 40 °C) and the other above (52 °C) the phase transition temperatures Tm (41° and 43 °C, respectively). Table 3 shows that the total time-dependent Stokes shifts Δv in DPPC-SUV below Tm are relatively small (5 and 17 nm for 20° and 40 °C, respectively), whereas the shifts in DHPC

Table 3. Stokes Shifts Δv (in nm) and Mean Relaxation Times $\langle \tau_r \rangle$ (in ns) for Patman in DPPC- and DHPC-SUV[a]

Temp (°C)	DPPC $\langle \tau_r \rangle$ (ns)	Tm = 41 °C Δv (nm)	DHPC $\langle t_r \rangle$ (ns)	Tm = 43 °C Δv (nm)
20	n.d.	5	2.6	31
40	3.7	17	2.0	43
52	1.2	49	< 0.7	>31

[a] $\langle \tau_r \rangle$ for DPPC at 20°C was no determined because of the small shift Δv. Tm is the lipid phase transition temperature.

at these temperatures are considerably larger (31 and 43 nm, respectively). Apparently, the ether linkage leads to a more polar environment and, thus, in the given phospholipid/water system, to a more efficient water penetration. The correlation functions C(t) (Fig. 6) and the determined $\langle \tau_r \rangle$ (Table 3) confirm for all temperatures the above stated conclusion that the head-group domain probed by Patman is more mobile in the case of ether compared with acyl membranes. It has to be stressed that C(t) for DPPC at 20 °C may be insignificant because of its small shift, but it is included in Fig. 6 in order to visualize the slow relaxation behavior in this system. The very slow solvent relaxation in DPPC below Tm manifests itself as well in the time course of the half-widths (Fig. 7).

If the fluorescence lifetime is long enough to allow complete relaxation to the relaxed R-state, the half-width of the time resolved emission spectra (TRES) first increases to a maximum usually at times comparable to $\langle \tau_r \rangle$ and then decreases with time [18, 77–79]. This regular behavior is found for Patman probing DHPC-SUV below Tm as well as for the DPPC system above Tm. The continuous increase observed for DPPC-SUV below Tm indicates that Patman fluoresces before SR is complete. The opposite extreme is observed for DHPC above Tm: the maximum in the half-width is already present at very short times after

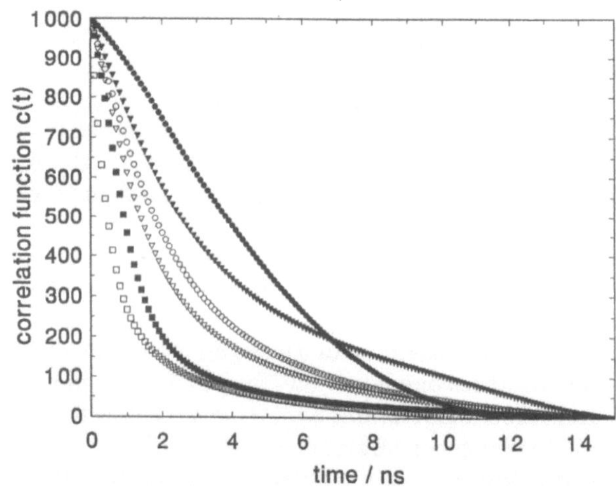

Fig. 6. Correlation functions C(t) calculated according to Eq. 1 for Patman in DPPC-SUV (*filled symbols*) and DHPC-SUV (*open symbols*) at 20 ° (*circles*), 40° (*triangles*), and 52 °C (*boxes*)

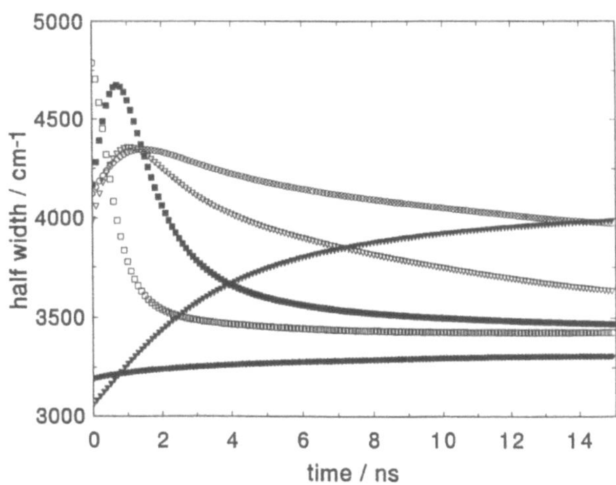

Fig. 7. Full width at half maximum of the TRES ('half width') as a function of time after excitation for Patman in DPPC-SUV (*filled symbols*) and DHPC-SUV (*open symbols*) at 20° (*circles*), 40° (*triangles*), and 52 °C (*boxes*)

excitation. Together with the determined, suprising drop in the determined $\Delta\nu$ for DHPC at 52 °C to 31 nm it can be concluded that a major part of the SR occurs much faster than the time resolution of our experiment, and, thus, this fast part of the SR cannot be detected with the given time resolution.

5.4
Membrane Curvature

Since several membrane processes like binding of peripheral proteins [80], production of thrombin [81] or membrane fusion [82] are influenced by the membrane curvature, its effect on the phospholipid head-group mobility has been determined by the SR behavior of TNS [34]. The probe, which is known to adsorb at the lipid-water interface, has been incorporated into two types of well-defined sized PC-vesicles, characterized by average mass-weighted radii of 11 and 125 nm, respectively. Over the examined temperature range (1–40 °C), the small vesicles showed significantly shorter SR times than their larger counterparts (e.g., 1.3 ns compared to 2.1 ns at 5 °C). Although this work does not take advantage of the well-defined head-group probes Patman and Prodan and involves a simplified "single-wavelength" analysis [26] for the determination of the SR times, the results clearly demonstrate an increase of the head-group mobility with an increase in the membrane curvature.

6
Summary and Outlook

The unique advantage of using solvent relaxation probes in membrane studies is that the determined time dependence of the Stokes shift displays directly the viscosity and polarity properties of the solvent system. In anisotropic media, like biomembranes, a well-defined, uniform probe localization is crucial for further interpretation of the information extracted. Considering these aspects, the

SR studies of Patman and Prodan [35], as well as of the n-AS dyes [37], serve as the basis for viscosity and polarity probing at different positions in both membrane domains. The solvent relaxation technique is shown to detect viscosity and polarity changes in the bilayer due to temperature, membrane curvature, and lipid composition variations as well as due to the binding of external ions and proteins. The defined localization of the mentioned dyes provides the possibility to investigate selectively the protein interaction with different domains of the bilayer. Together with the demonstrated high sensitivity of this technique for the binding of peripheral proteins, the application of the SR method might lead to a deeper mechanistic understanding of protein–membrane interactions than is achievable by standard fluorescence techniques.

From the methodological point of view it is certainly possible to further improve the SR technique in the form presented here. A higher time-resolution of the 'Single Photon Counting' experiment used as well as that from a time-resolved experiment's independent estimation of the $t = 0$ spectrum and its emission maximum n (0) [18, 83], can lead to a more precise determination of the Stokes shifts $\Delta\nu$ and correlation functions C (t). This might allow, such as in the case of the fs-results in low viscous solvents, a physical interpretation of the shape of C (t). An interesting aspect could be – in analogy to the micellar situation [43] – the finding of different "kinds of water" present in the phospholipid bilayer. Moreover, application and synthesis of other compounds combining defined probe location with a large change in charge distribution upon excitation should be further developed. Examples of chromophores with extremely large charge relocalization are hemicyanine monopolic dyes [84] and piperidine-bridged electron donor–acceptor systems [85].

Acknowledgements. The author would like to express his gratitude to Drs. S. Vajda and R. Hutterer as well as to Profs. F.W. Schneider and V. Fidler, by especially emphasizing the diligence and skill in the experiment of Dr. R. Hutterer. Financial support from the "Deutsche Forschungsgemeinschaft", "Fonds der Chemischen Industrie", and of the Czech Grant Agency (GACR 203/95/0755) are acknowledged. Figs. 2–5 are reprinted from [35] and [37] with kind permission of Elsevier Science, Sara Burgerhartstraat 25, 1055 KV Amsterdam, the Netherlands, and Figs. 6 and 7 from [39] with kind permission of Plenum Publishing Corporation, 233 Spring Street, NY 10013 New York, USA.

References

1. Eftink MR (1992) In: Lakowicz JR (ed) Topics in Fluorescence Spectroscopy: Principles, Plenum Press, New York, p 53
2. Van der Meer BW, Coker G, Chen SYS (1994) Resonance Energy Transfer Theory and Data, VCH Verlag, Weinheim
3. Hutterer R, Krämer K, Schneider FW, Hof M (1997) Chem Phys Lipids, in press
4. Stubbs CD, Williams BW (1992) In: Lakowicz JR (ed) Topics in Fluorescence Spectroscopy: Biochemical Applications, Plenum Press, New York, p 231
5. Bernsdorff C, Wolf A, Winter R, Gratton E (1997) Biophys J 72:1264
6. Brand K, Hof M, Schneider FW (1991) Ber Bunsenges Phys Chem 95:1511

7. Duportail G, Lianos P (1996) In: Rosoff M (ed) Vesicles, Marcell Dekker, New York, p 296
8. Hutterer R, Haefner A, Schneider FW, Hof M (1997) In: Slavik J (ed) Fluorescence Microscopy and Fluorescence Probes Vol. 2, Plenum Press, New York, in press
9. Van der Heide UA, van Ginkel G, Levine YK (1996) Chem Phys Lett 253:118
10. Bagchi B, Oxtoby SD, Fleming GR (1984) Chem Phys 86:257
11. Rips I, Klafter J, Jortner J (1988) 89:4288
12. Rips I, Klafter J, Jortner J (1988) 88:3246
13. Friedman HL, Raineri FO, Hirata F, Perng PC (1995) J Stat Phys 78:239
14. Raineri FO, Friedman HL, Perng PC (1994) Chem Phys 183:187
15. Bakshiev NG (1964) Opt Spectosc (USSR) 16:446
16. Mazurenko YT, Bakshiev NG (1970) Opt Spectosc (USSR) 28:490
17. Liptay W (1974) In: Lim EC (ed) Excited States, Academic Press, New York, Vol 1, p 129
18. Horng ML, Gardecki JA, Papazyan A, Maroncelli M (1995) J Phys Chem 99:17320
19. Jimenez R, Fleming GR, Kumar PV, Maroncelli M (1994) Nature 369:471
20. Mukherje S, Chattopadhyay A (1995) J Fluorescence 5:237
21. Hof M, Lianos P (1997) Langmuir 13:290
22. Zhang H, Jonkman AM, van der Meulen, Glasbeek M (1994) Chem Phys Lett 236:587
23. Bingemann D, Ernsting NP (1995) J Chem Phys 102:2691
24. Ware WR, Lee SK, Brant GJ, Chow PP (1971) J Chem Phys 54:4729
25. Lakowicz, JR (1983) Principles of Fluorescence Spectroscopy, Plenum Press, New York, p 217
26. Gonzalo I, Montoro T (1985) J Phys Chem 89:1608
27. Nagarajan V, Bearly AM, Kang TJ, Barbara PF (1987) J Chem Phys 86:3183
28. Maroncelli M, Fleming, GR (1987) J Chem Phys 86:6221
29. Hof M, Schleicher J, Schneider FW (1989) Ber Bunsenges Phys Chem 93:1377
30. Wehry EL, Rogers LB (1966) In: Hercules E (ed) Fluorescence and Phosphorescence Analysis, Interscience, New York, 125
31. Siano DB, Metzler DF (1969) J Chem Phys 51:1856
32. Castner EW Jr, Maroncelli M, Fleming GR (1987) J Chem Phys 86:1090
33. Hutterer R, Schneider FW, Perez N, Ruf H, Hof M (1993) J Fluor 3:257
34. Hof M, Hutterer R, Perez N, Ruf H, Schneider FW (1994) Biophys Chem 52:165
35. Hutterer R, Schneider FW, Sprinz H, Hof M (1996) Biophys Chem 61:151
36. Hutterer R, Hof M (1996) In: Slavik J (ed) Fluorescence Microscopy and Fluorescence Probes, Plenum Press, New York, p 232
37. Hutterer R, Schneider FW, Lanig H, Hof M (1997) Biochim Biophys Acta 1323:195
38. Hutterer R, Schneider FW, Fidler V, Grell E, Hof M (1997) J Fluor 7:161S
39. Hutterer R, Schneider FW, Hof M (1997) J Fluor 7:27
40. Demchenko AP (1992) In: Lakowicz JR (ed) Topics in Fluorescence Spectroscopy: Biochemical Applications, Plenum Press, New York, p 65
41. Demchenko AP (1994) Biochim Biophys Acta 1209:149
42. Galley WC, Purkey RM (1970) Proc Natl Acad Sci USA 67:1116
43. Hof M, Lianos P, Laschewsky A (1997) Langmuir 13:2181
44. Slavik J (1982) Biochim Biophys Acta 694:1
45. Lakowicz JR, Thompson RB, Cherek H (1983) Biochim Biophys Acta 815:295
46. Demchenko AP, Shcherbatska NV (1985) Biophys Chem 22:131
47. Matayoshi ED, Kleinfeld AM (1981) Biochim Biophys Acta 644:233
48. Epand RM, Leon BTC (1992) Biochemistry 31:1550
49. Chong PL (1988) Biochemistry 27:399
50. Sommer A, Paltauf F, Hermetter A (1990) Biochemistry 29:11134
51. Parasassi T, De Stasio G, d'Ubalso A, Gratton E (1990) Biophys J 66:120
52. Parasassi T, Di Stefano M, Loiero M, Ravagnan G, Gratton E (1994) Biophys J 66:763
53. Lakowicz JR, Bevan DR, Maliwal BP, Cherek H, Balter A Biochemistry 22:5714
54. Chattopadhyay A, Mukherje S (1993) Biochemistry 31:3804
55. Jones SR, Willing RI, Thulborn KR, Sawyer WH (1979) Chem Phys Lipids 24:11
56. Podo F, Blasie JK (1972) Proc Natl Acad Sci USA 69:1032

57. Thulborn KR, Sawyer WH (1978) Biochim Biophys Acta 511:125
58. Matayoshi ED, Kleinfeld AM (1981) Biophys J 35:215
59. Werner TC, Hercules DM (1969) J Phys Chem 73:2005
60. Werner TC, Mathews T, Soller B (1976) J Phys Chem 80:533
61. Werner, T.C. (1976) In: Wehry EL (ed) Modern Fluorescence Spectroscopy Vol 2, Plenum Press, New York, p 277
62. Berberan-Santos MN, Prieto MJE (1991) J Phys Chem 95:5471
63. Blatt E, Ghigginio KP, Sawyer WH (1981) J Chem Soc Faraday I 77:2551
64. Seelig J (1978) Biochim Biophys Acta 515:105
65. Browning JL, Seelig J (1980) Biochim Biophys Acta 19:1262
66. Feigenson GW (1989) Biochemistry 28:7453
67. Jones ME, Lentz BR (1986) Biochemistry 25:567
68. Tendian SW, Lentz BR (1990) Biochemistry 29:6720
69. Tendian SW, Lentz BR, Thompson NL (1991) Biochemistry 30:10991
70. Pearce KH, Hof M, Lentz BR, Thompson NL (1993) J Biol Chem 268:22984
71. Hutterer R, Hermens WTh, Wagenvoord R, Hof M, in preparation
72. Cutsforth GA, Whittaker RN, Lentz BR (1989) Biochemistry 28:7453
73. Comfurius P, Smeets EF, Willems M, Bevers EM, Zwaal RFA (1994) Biochemistry 33:10319
74. Lecompte MF, Miller IR (1980) Biochemistry 19:3434
75. Paltauf F (1983) In: Mangold HK, Paltauf F (eds) Ether Lipids: Biochemical and Biomedical Aspects, Academic Press, New York, p 309
76. Hermetter A, Prenner E. Loidl J, Kalb E, Sommer A, Paltauf F (1993) In: Wolfbeis OS (ed) Fluorescence Spectroscopy, Springer Verlag, Berlin, p 149
77. Richert R, Wagener A (1991) J Phys Chem 95:10115
78. Maroncelli M (1993) J Mol Liq 57:1
79. Vincent M, Galley J, Demchenko AP (1995) J Phys Chem 99:14923
80. Abott AJ, Nelsestuen GL (1987) Biochemistry 26:7994
81. Giesen PLA, Willems GM, Hermens WTh (1993) J Biol Chem 266:1379
82. Lentz BR, McIntyre GF, Parks DJ, Yates JC, Massenburg D (1992) Biochemistry 31:2642
83. Fee RS, Maroncelli M (1994) Chem Phys 183:235
84. Fromherz P (1995) J Phys Chem 99:7185
85. van Stokhum IHM, Scherer T, Brouwer AM, Verhoeven JW (1994) J Phys Chem 98:852

Luminescent Lanthanide Chelates for Improved Resonance Energy Transfer and Application to Biology

P. R. Selvin

1
Introduction to Resonance Energy Transfer

Fluorescence resonance energy transfer (FRET) is a technique widely used to measure the distance between two points which are separated by approximately 10–80 Å. Lanthanide-based RET (LRET) is a recent modification of the technique with a number of technical advantages, yet relies on the same fundamental mechanism – subject to careful interpretation of various terms. A number of excellent reviews on FRET have been written [14, 21–28]. A recent review of LRET has also appeared [9], as well as a summary of lanthanide luminescence [29] and its application to bio-assays [30], so here we provide only a brief summary of the relevant theory.

In FRET, an excited fluorescent donor molecule attached at one point in the biomolecule transfers energy to an acceptor molecule attached at a second site, through a non-radiative dipole-dipole coupling which is inversely proportional to the sixth-power of the distance between the two dyes. The excited-state donor can decay to the ground state via energy transfer or via other processes (such as fluorescence and non-radiative mechanisms) and the efficiency of energy transfer, E, is the energy transfer rate divided by all rates which de-excite the donor:

$$E = k_{et}/(k_{et} + \Sigma k_i) = 1/(1 + \Sigma k_i/k_{et}) \tag{1}$$

where Σk_i is the sum of all rates that de-excite the donor other than those due to energy transfer. Because k_{et} depends on R^{-6}:

$$E = 1/(1 + R^6/R_0^6) \tag{2}$$

where R_0 is the distance at which the efficiency of energy transfer is 50% ($\Sigma k_i/k_{et} = 1$). R_0 depends on the spectral properties and relative orientation of the donor and acceptor (discussed below).

E can be measured because the donor's intensity and excited-state lifetime are reduced in proportion to the amount of energy transfer:

$$E = (1 - I_{DA}/I_D) = 1 - \tau_{DA}/\tau_D \tag{3}$$

where I_{DA}, τ_{DA}, are the donor's intensity and excited-state lifetime in the presence of an acceptor, and I_D, and τ_D are the same parameters in the absence of an acceptor. Energy transfer also leads to an increase in the acceptor's intensity, called sensitized emission. E can therefore be determined by measuring the increase in fluorescence of the acceptor due to energy transfer and comparing this to the residual donor emission:

$$E = [I_{AD}/q_A]/[I_{DA}/q_D + I_{AD}/q_A] \tag{4}$$

where I_{DA} is the integrated area under the donor emission curve in the presence of the acceptor, I_{AD} is the integrated area of the sensitized emission of the acceptor (i.e., not including the fluorescence due to direct excitation of the acceptor) and q_i is the quantum yield for the donor or the acceptor.

Hence, by measuring donor and/or acceptor emission, one can calculate E and, given R_o, the distance can then be inferred:

$$R = R_o (1/E - 1)^{1/6} \tag{5}$$

R_o is typically 20–55 Å for organic dyes (although some new far-red and infrared dyes can have R_os up to 70 Å [31]), and 50–75 Å when using lanthanides as donors [18]. It can be shown [23] that:

$$R_o = (0.211) (J q_D n^{-4} \kappa^2)^{1/6} \text{ (in Å)} \tag{6}$$

$$J = \int \varepsilon_A(\lambda) f_D(\lambda) \lambda^4 d\lambda / \int f_D(\lambda) d\lambda \quad \text{in M}^{-1} \text{ cm}^{-1} \text{ nm}^4 \tag{7}$$

where J is the normalized spectral overlap of the donor emission (f_D) and acceptor absorption (ε_A in units of $M^{-1}cm^{-1}$ where M is units of mol/liter), q_D is the quantum efficiency (or quantum yield) for donor emission in the absence of acceptor (q_D = number of photons emitted divided by the number of photons absorbed), n is the index of refraction (1.33 for water; 1.29 for many organic molecules) and κ^2 is a geometric factor related to the relative orientation of the transition dipoles of the donor and acceptor and their relative orientation in space. κ^2 can vary anywhere from 0 to 4 if the donor and acceptor are rigid and polarized, is limited to between 1/3 and 4/3 if either donor or acceptor rotate rapidly or have isotropic emission (such as lanthanides in solution), and is 2/3 if both the donor and acceptor rapidly rotate during the donor's excited-state lifetime.

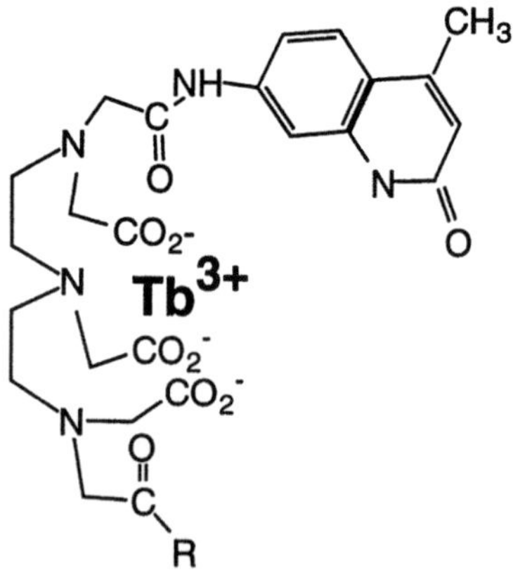

Fig. 1. Diethylenetriaminepentaacetate (DTPA) chelate of terbium or europium with covalently attached antenna molecule, carbostyril 124 (cs124). The antenna absorbs the excitation light (λ_{exc} = 337 nm) and transfers energy to the lanthanide, which can then transfer energy to an acceptor or luminesce. The antenna is necessary because the lanthanides have extremely weak absorption cross sections. The chelate binds the lanthanide tightly ($K_b > 10^{14} M^{-1}$), excludes water from the primary coordination sphere of the lanthanide which would otherwise quench luminescence, and allows for site-specific modification through a reactive group, R

2
Introduction to LRET and Its Theory

We and others have introduced a modification of the standard FRET technique, which we call luminescence (or lanthanide-based) resonance energy transfer (LRET). In LRET, the donor is a luminescent lanthanide (terbium or europium) complex instead of the conventional organic-based fluorophore. A lanthanide complex (chelate) which we have frequently used is shown in Fig. 1. As mentioned, the mechanism of energy transfer to an organic acceptor from a lanthanide donor, or from an organic-based donor, is the same – subject to careful interpretation of various terms – and hence the conventional dipole-dipole theory of Förster is applicable. At the same time, LRET offers many technical advantages over FRET which enable new biological systems to be studied.

The reason FRET and LRET have the same mechanism is because emission, or the electric field established by an excited-state lanthanide, arises primarily from electric dipole transitions within the atom, just as transitions in an organic fluorophore – transitions which lead to fluorescence and energy transfer – also arise from electric dipole (E1) transitions. Hence the electric field surrounding the donor, whether that be a lanthanide or an organic-based fluorophore, looks dipolar in nature, i.e., it decays as R^{-3} at distances much less than the wavelength of emitted light. Energy transfer depends on the square of the electric field (see [14] for a review) and so in both cases depends on $1/R^6$.

A priori, the electric dipolar nature of lanthanide emission is surprising since it arises from $4f - > 4f$ electronic transitions, and hence E1 transitions are formally forbidden by parity. However, a small admixture of 5d states enables the E1 mechanism. The semi-forbidden nature of the transitions, and the fact that they arise from inner-shell electrons which are generally well-shielded from the environment, leads to long (millisecond) excited-state lifetimes and sharply-spiked emission spectra (Fig. 2). It is important to note, however, that magnetic dipole (M1) transitions do exist and, indeed, the europium $^5D_0 - > {}^7F_2$ transition around 590–600 nm is magnetic dipolar [29, 32]. The $^5D_4 - > {}^7F_5$ transition of terbium around 546 nm may have some M1 character, although the extent is unclear because terbium photophysics are rather complicated. Magnetic transitions produce only a feeble electric field, and hence cannot transfer energy [33]. Yet higher order terms, such as quadrupole terms, are not significant in solution. Therefore, when using Förster's theory of dipole-dipole interaction with the lanthanides, the spectral overlap term must exclude any emission lines which arise from magnetic dipole transitions.

A second term which must be treated with caution is the quantum yield term in the calculation of R_o when using lanthanides. This is discussed in detail elsewhere [9], although the conclusion is straightforward. Two "quantum yields" are sometimes discussed with lanthanide chelates, the "total" quantum yield (q_T) of the chelate, and the lanthanide quantum yield (q_{Ln}). q_T is the probability of lanthanide emission given that the *chelate-complex* absorbs a photon. Recall that chelate-complex absorption occurs primarily due to the antenna molecule. q_{Ln} is the probability of lanthanide emission given that the *lanthanide* is excited.

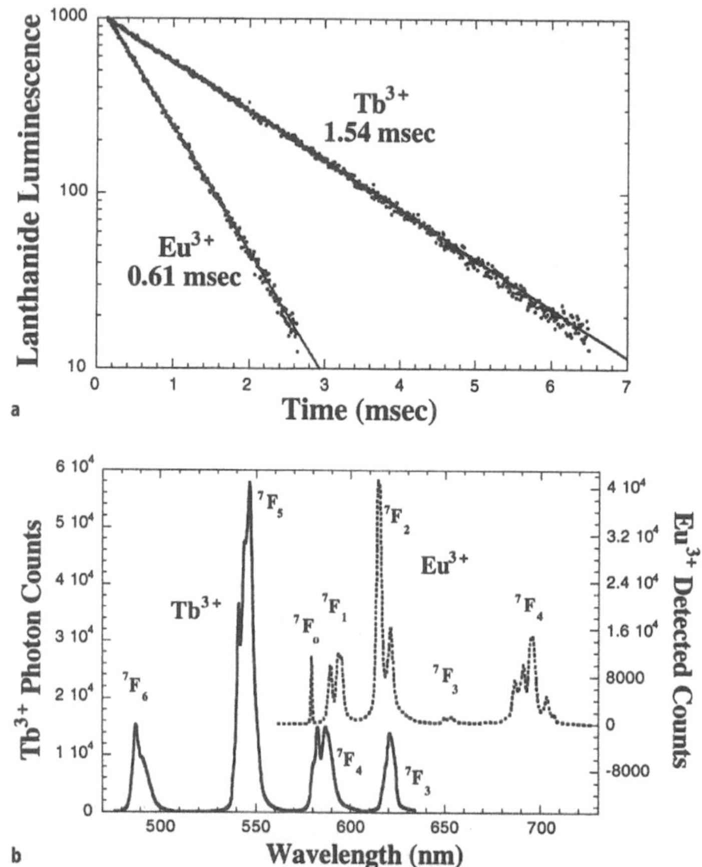

Fig. 2a, b. Excited state lifetime (a) and emission spectra (b) of Tb^{3+} – and Eu^{3+} – DTPA-cs124 in H_2O at pH 7.0. For emission spectra [Tb^{3+}] at 0.25 µM and Eu^{3+} at 2 µM. Lifetime measurements acquired in 10 s and normalized. Spectral data acquired with CCD integrated for 3 s with 600 g/mm grating, entrance slit of 0.1 mm, chopper delay of 145 µs. Excitation is approximately 5 µJ/pulse, 40 Hz pulse-rate at 337 nm (see instrumentation section, below). Emission for Tb^{3+} is $^5D_4 \rightarrow {}^7F_J$; for Eu^{3+} it is $^5D_0 \rightarrow {}^7F_J$

It is q_{Ln}, the lanthanide quantum yield, which counts in resonance energy transfer. Why? The efficiency of energy transfer is related to relative rates of energy transfer from the *excited lanthanide* to the acceptor vs. other processes such as non-radiative decays (Eq. 1). This is exactly related to q_{Ln} – in fact energy transfer can be described as a decrease in the donor (or lanthanide) quantum yield in the presence of acceptor: fewer photons are emitted by the donor in the presence of an acceptor because some of the energy that would go into emission is taken up by the acceptor. In contrast, if the antenna of the chelate absorbs light, but does not transfer energy to the lanthanide, the lanthanide is simply not excited and that process is irrelevant in terms of energy transfer, although q_T decreases. q_T is related to q_{Ln} simply by the efficiency of (internal) energy transfer from antenna to chelate (q_{int}): $q_T = q_{Ln} \times q_{int}$.

With the proper understanding that the terms in R_0 are the lanthanide quantum yield and the spectral overlap due only to electric dipole transitions, one can then use the standard equations in Förster's theory in LRET.

3
Advantages of LRET

We have shown that the use luminescent lanthanide atoms as donors – namely terbium or europium, placed in the appropriate chelate – and organic-based acceptors yields many technical advantages over conventional FRET [9, 12–14, 18]. (See Fig. 3) Mathis has developed a similar lanthanide-based system where he uses a allophycocyanine (a large protein with many chromophores and high quantum yield) as the acceptor [10, 11]. An example which highlights these advantages is shown in Fig. 3, where energy is transferred from a europium chelate attached at one end of a DNA oligomer to an organic dye (called Cy5) at the other end.

Specifically, the advantages are:

- The efficiency of energy transfer using lanthanides as donors is considerably higher compared to organic-based donors, leading to larger signals and longer measurable distances. The value of R_0 is typically 50–70 Å with current chelates and conventional organic-based dye acceptors, and as high as 90 Å with allophycocyanin as the acceptor. In addition, R_0 can be optimized to the particular system by varying the amount of H_2O/D_2O because D_2O increases the lanthanide quantum yield [34], thereby changing R_0. Lastly, the lanthanide quantum yield is insensitive to pH over a wide range (at least pH 5–9) and, hence, if the acceptor's absorbance does not change these R_0s are maintained over a wide pH range.
- In LRET, because donor emission is generally unpolarized, energy transfer is primarily dependent on the distance between donor and acceptor, and not on their relative orientation, leading to more precise distance measurements. With the donor emission unpolarized, κ^2 is constrained to between 1/3 and 4/3, the extreme cases corresponding to a completely rigid acceptor perpendicular or parallel to the radius vector, respectively. If one then assumes $\kappa^2 = 2/3$, as is often done, then at most the distance inferred is in error by $\pm 12\%$ [21].
- Because of the donor's unusually long lifetime (ms), LRET lifetime measurements can be made with high precision: lifetime measurements avoid problems of concentration determination and allow analysis of heterogeneous mixtures. The long lifetime also means it is likely that the donor's and acceptor's orientation is unpolarized because they have a millisecond to reorient, rather than the nanoseconds typical of organic lifetimes.
- In LRET, a 50- to 100-fold improvement in signal to background of the sensitized emission compared to conventional FRET has been achieved, largely through the reduction of background. In particular, the acceptor's sensitized

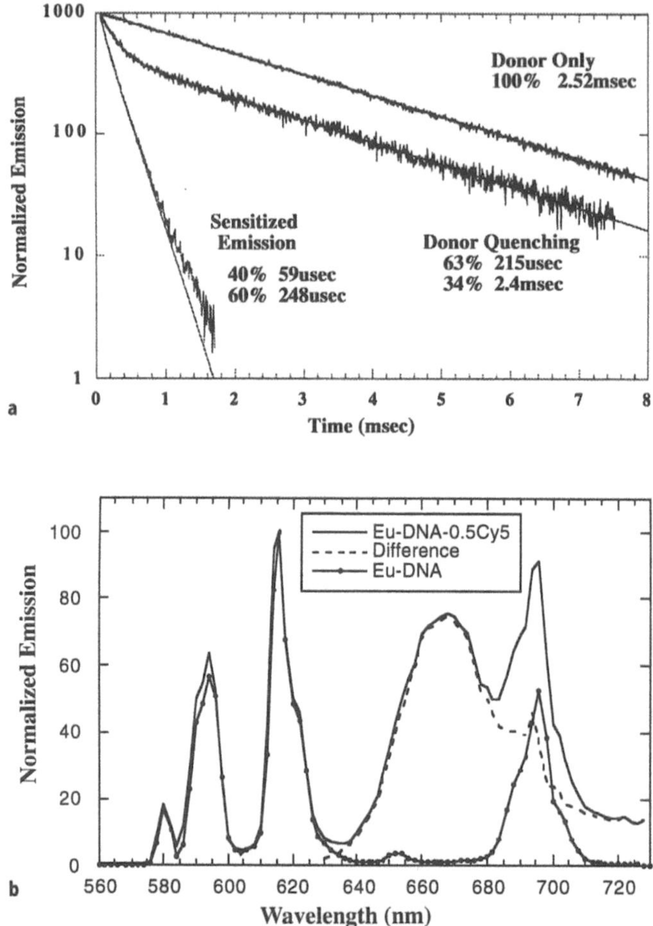

Fig. 3a, b. Lifetime data (**a**) and delayed-emission spectra (**b**) and for Eu-DTPA-cs124 transferring energy to Cy5 (adapted from [12]). The europium donor is bound to one end of a DNA oligomer (10 base-pairs in length) and the Cy5 acceptor is bound to the opposite end of a complementary DNA oligomer which can bind the donor-strand. Eu in the absence of acceptor has a lifetime of 2.5 ms. When sub-stoichiometric amounts of acceptor-DNA are added (approximately 0.5 acceptors per donor) and allowed to bind to the donor strand, the donor lifetime becomes bi-exponential, with the short-lifetime component corresponding to donor-acceptor complexes, and the long lifetime corresponding to those donors without an acceptor strand. The lifetime of the sensitized emission (at 668 nm) from the donor-acceptor pairs can be measured and the 248 μs lifetime is in good agreement with the 215 μs lifetime found from the donor decay. (The 69 μs component in the sensitized emission is due to detector ringing.) (**b**) The emission spectra is collected after a ≈ 100 μs delay following the excitation pulse to eliminate any prompt fluorescence (see *implementation of LRET,* above). Any acceptor emission is therefore due to energy transfer, and such sensitized emission is clearly evident. If the D-A curve is subtracted from the D-only curve, the difference has the expected Cy5 emission shape and the average amount of energy transfer can be calculated based on Eq. 4

emission can be measured with no background, either from donor luminescence or from direct excitation of the acceptor. Specifically, the emission of
the acceptor due to excitation by the excitation light, which has a nanosecond
lifetime, is discriminated against using pulsed excitation and time-delayed
detection. The donor emission is discriminated against by wavelength filtering – donor emission is highly spiked, with regions of darkness (see Fig. 2)
– and, with the appropriate choice of acceptor, acceptor fluorescence is in this
dark region. The improvement in signal to background enables long
distances to be measured (where the signal is weak), and easier handling of
incompletely labeled samples (because excess acceptor or donor do not contribute background).

- LRET, the lifetime of the acceptor decay due to energy transfer, i.e., sensitized
emission, can be measured (see "sensitized emission" lifetime curve in
Fig. 3a). This decay follows the donor's lifetime because it is being "fed" by
the long-lived donor, which is the rate-limiting step. As soon as the acceptor
receives a quanta of energy from the donor, it will, within its nanosecond lifetime, emit that energy as an emission photon. The ensemble decay of many
such acceptor molecules is a signal at the acceptor's emission wavelength
which decays with the donor lifetime in the donor-acceptor complex. The
sensitized emission lifetime is insensitive to concentration, *and arises only
from the fully labeled donor-acceptor complex.* Hence if there is incomplete
labeling, with some biomolecules labeled only with donor or only with acceptor, these do not contribute to the sensitized emission signal. This significantly increases the number and type of samples which can be looked at
with resonance energy transfer since complete labeling is often not possible
and leads to difficulties in conventional FRET.

4
Implementation of LRET

4.1
Instrumentation

The instrumentation has been described in detail elsewhere [9]. In summary,
solutions of chelated terbium or europium (see Figs. 1 and 4) are placed in a
quartz cuvette and excited with a pulsed nitrogen laser (337 nm, 5 ns pulse-
width, 40 Hz repetition rate) for typically 400–1600 pulses. The lanthanide
emission at right-angles to the excitation path is focused on the entrance slit of
a dual-emission port imaging spectrometer which has a mechanical chopper in
front of the slit, synchronized to the excitation pulse to eliminate prompt fluorescence. For time-resolved measurements, light within the spectrometer is
steered to the emission port containing a cooled gallium-arsenide photomultiplier tube operating in photon-counting mode and a multichannel scalar with
2 µs time-resolution. For these measurements the chopper is typically not used
and prompt fluorescence is discriminated against digitally. For time-delayed
emission spectra, the chopper is used to trigger the laser such that emission less

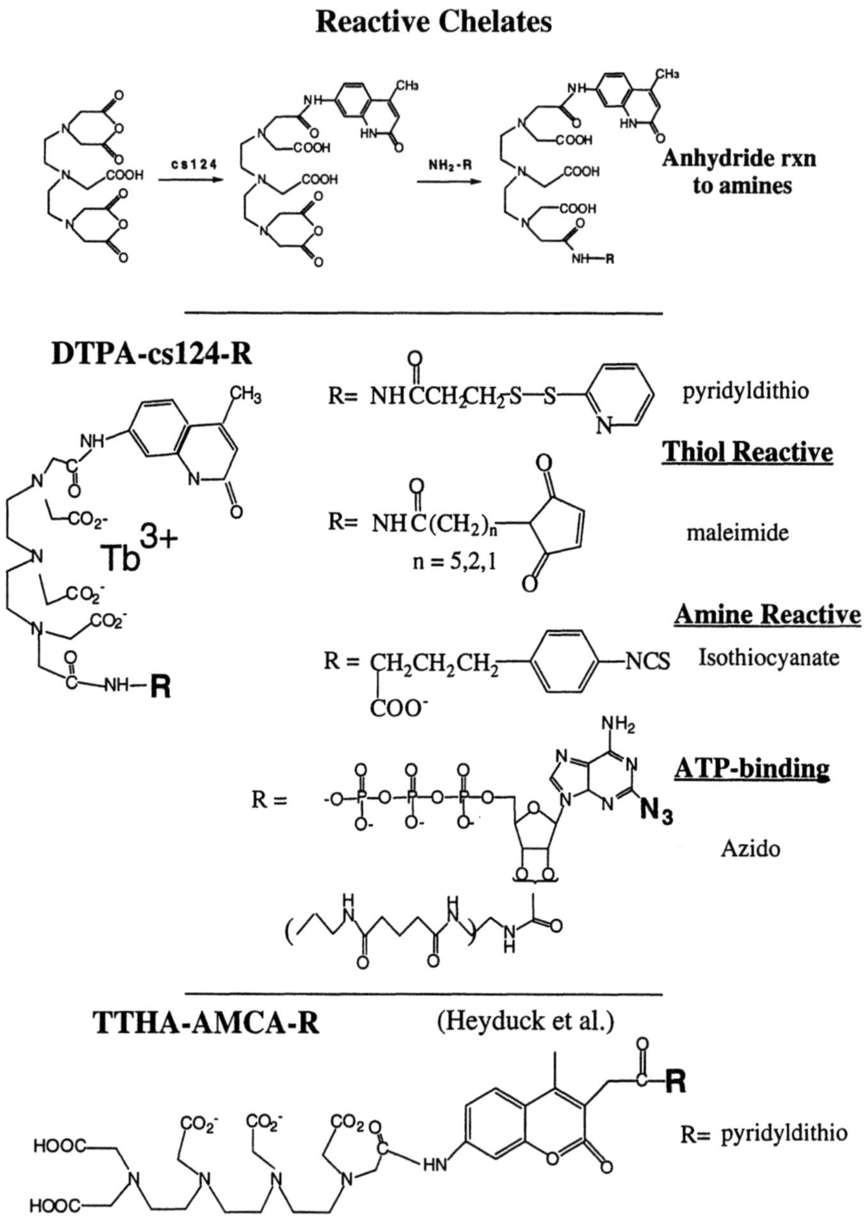

Reactive Chelates

Fig. 4. Reactive polyaminocarboxylate chelates. The anhydride or isothiocyanate reacts with amines; pyridyldithio or maleimides react with thiols (cysteines on proteins), and the azido-ATP analog will photoreact around an ATP binding site. The azido-ATP analog has two forms of different linker length

than 100 µs after the excitation pulse is blocked. The remaining luminescence (donor emission and sensitized emission) is routed to the spectrometer port containing a liquid-nitrogen cooled charge-coupled device (CCD) connected to a personal computer.

4.2
Reactive Chelates

An important element in LRET is the site-specific attachment of donor and acceptors. The acceptor technology, because it uses conventional dyes, is well developed. For donors, a number of bipyridine-based lanthanide chelates with reactive groups have been synthesized (reviewed in [30]). We and others have made reactive forms of luminescent lanthanide polyaminocarboxylate chelates for site-specific attachment to amines [4, 35, 36] and to thiols [15, 36, 37] and most recently to ATP-binding sites [38] (see Fig. 4). The polyaminocarboxylate chelates are significantly simpler to synthesize than the bipyridine chelates.

To make the polyaminocarboxylate chelates reactive, the anhydride-based reaction, because it contains all commercially available components, is the most readily available to researchers without chemistry expertise (see top of Fig. 4). We have developed procedures for using it with DNA which has been end-labeled with a primary amine [35]. Root has used it successfully to label antibodies [17]. However, the anhydride reaction is far from ideal: it is sensitive to hydrolysis by water and therefore generally requires a large excess, it can lead to a mixture of products, including cross-linking and various side reactions, and the anhydride is not reactive to thiols. We and others have therefore generated the reactive chelates shown in Fig. 4.

5
Biological Applications

LRET has recently been applied by ourselves to the study of actomyosin – the two proteins, actin and myosin – which form the bulk of muscle tissue and are responsible for muscle contraction [19, 20, 38]. Root has recently used LRET to study the interaction of the proteins dystrophin and actin in the muscle cell [17], and Heyduk and co-workers have used LRET to study protein-induced DNA bends [16]. Here we discuss each of these examples.

5.1
Protein-Induced DNA Bends

Heyduk and co-workers used LRET to measure the bending of DNA induced by a class of proteins known as high-mobility-group (HMG) proteins. The bending of DNA is essential to the packing of DNA into chromosomes, and is often also involved in transcriptional regulation. Except for the histone proteins, which bend DNA into sharply curved structures called nucleosomes, HMG proteins are the most abundant proteins associated with chromatin. One subset of HMG proteins binds to DNA with high affinity at specific sequences, and a second

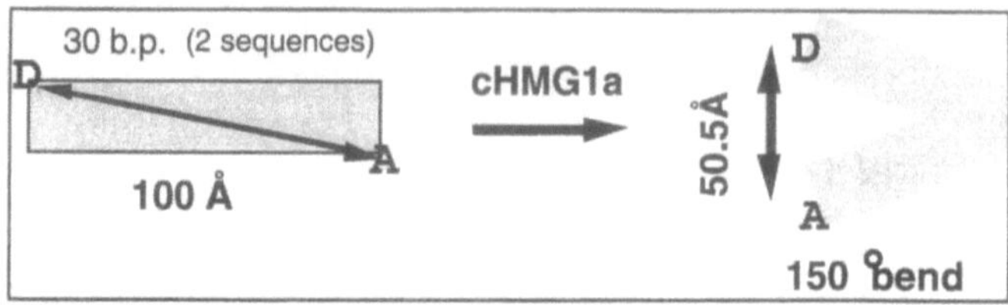

Fig. 5. Schematic representation of donor-acceptor labeled DNA which is bent by the addition of a DNA-binding HMG protein. (Adapted from [16])

subset binds with relatively little specificity (aside from binding to A-T rich regions). A number of the sequence-specific HMG proteins have been shown to sharply bend DNA when bound to their high-affinity site, but before the Heyduk paper, only indirect evidence existed that sequence non-specific HMG proteins bent DNA.

Heyduk and co-workers used LRET to measure whether two particular HMG proteins, called cHMG1a and HMGI, bend DNA upon binding at a non-specific sequence [16]. Furthermore, they mutated cHMG1a, in one case removing the C-terminal negatively charged region (now called cHMG1a/102) and the other removing both a positively and a negatively charged region (now called cHMG1a/84). The former causes an increase in the binding constant of the protein for DNA and the latter causes a decrease in affinity. Whether removal of these charged regions affected the protein's ability to bend DNA was unclear until their work.

The authors end-labeled complementary 30-base pair strands of DNA, one with a europium donor chelate, and the other with the acceptor dye Cy5. The idea is that in the absence of bends, the DNA adopts its normal B-form configuration and is therefore straight. The donor-acceptor distance is then approximately 100 Å (3.4 Å/base along the helix axis, and secondarily, a helix diameter of approximately 15 Å). This results in a small amount of energy transfer. Upon addition of a protein that bends the DNA, the ends come considerably closer, resulting in greater energy transfer, and the donor-acceptor distance and angle of DNA bend can then be calculated (see Fig. 5). The 30-mer was A-T rich, but did not contain a high-affinity sequence for HMG proteins, and a second 30-mer of different non-high-affinity sequence was also used to ensure that the results were not unique to the sequence(s) chosen.

The chelate was Eu-DTPA-7-aminomethylcoumarinacetate attached to DNA through the coumarin's 3-position which contained either an NHS ester for reaction with amine-modified DNA, or a pyridyldithio group for reaction with thiol-modified DNA (See Figs. 4 and 5). The Cy5-NHS acceptor was coupled to 5′-amino-modified DNA of complementary sequence and, after purification, the two strands were hybridized. This donor-acceptor pair has an R_0 of 55 Å in aqueous solution. Addition of D_2O, which leads to a higher quantum yield for europium emission, increases this R_0 and, indeed, the authors used this to optimize R_0 for the particular distance they were interested in.

Representative donor and sensitized emission decay curves are shown in Fig. 6 and the results are summarized in Table 1. (Fig. 6 data is from a 30-mer

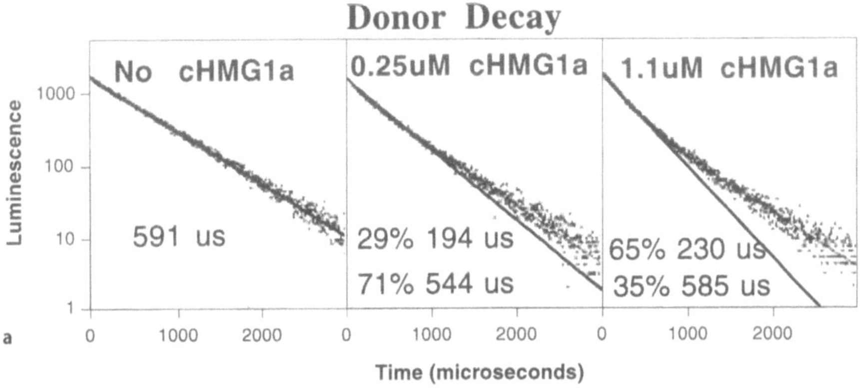

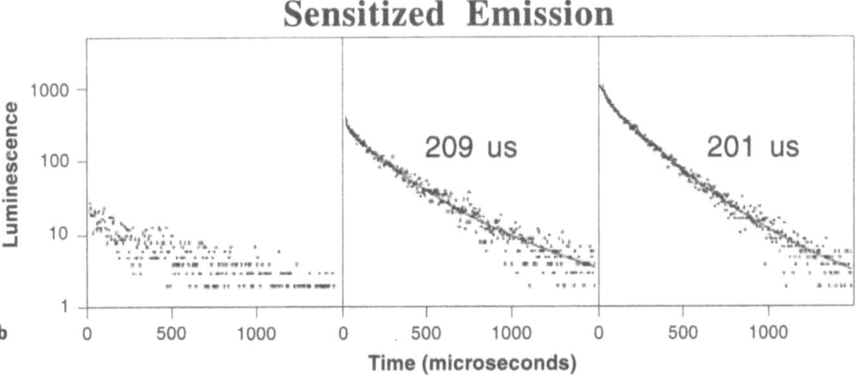

Fig. 6a, b. Europium donor lifetime (**a**) and sensitized emission lifetime (**b**) with increasing amount of HMG protein added to the donor-acceptor labeled DNA. Both figures indicate significant energy transfer is induced by the binding of the protein. (Adopted from Figs. 3 and 4 [16].)

end-labeled using a 17-atom linker between donor and the 5′-end of the DNA, whereas Table 1 data are from a donor attached via a shorter 6 atom linker, and hence the lifetimes are slightly different). In the absence of cHMG1a, the donor decay in the donor-acceptor (D-A) labeled DNA duplex is single exponential and long. Addition of protein to a donor-only duplex DNA, a control experiment, does not significantly change the donor lifetime. Addition of cHMG1a to the D-A sample leads to a bi-exponential donor decay, with an ≈ 200 μs component and a ≈ 580 μs component. Titration with increasing amounts of cHMG1a does not change the two time constants but increases the relative amplitude of the short-time component (Fig 6a). The straightforward interpretation is that the protein is binding in sub-stoichiometric amounts, but, when it is bound, it causes an increase in the amount of energy transfer, and hence a decrease in donor-lifetime. Note that the donor decay in the presence of protein is quite easy to fit quantitatively to a two exponential model, thereby detecting the heterogeneity present in the sample.

Table 1. Europium Lifetime and Energy Transfer for Donor-Acceptor Labeled DNA with and Without Bound Proteins and Their Mutant Forms. [16]

Sample	Lifetime μs	Energy transfer F	Distance Å
Donor	602.9 ± 2		
Donor+cHMG1a	616.7 ± 7		
Donor-acceptor	586.4 ± 3	0.027 ± 0.01	100.0 ± 10
Donor-acceptor+cHMG1a	226.0 ± 22	0.63 ± 0.04	50.3 ± 1.5
Donor + cHMG1a/102	614.0 ± 1		
Donor-acceptor+cHMG1a/102	233.0 ± 23	0.62 ± 0.04	50.7 ± 1.5
Donor + cHMG1a/84	606.0 ± 4		
Donor-acceptor+cHMG1a/84	243.9 ± 47	0.60 ± 0.08	51.4 ± 3.0

DNA Bending by HMG Proteins.
cHMG1a/102: (–) charged domain removed. K_b up.
cHMG1a/84 : (–) & (+) domains removed. K_b down.

The signal from the DNA with bound protein can be further isolated from DNA without protein by looking at the lifetime of the sensitized emission of the acceptor around 668 nm (Fig. 6b). In the absence of protein there is very little signal (left-hand panel, Fig. 6b) because europium is silent at this wavelength, and the direct acceptor emission is very short lived. As protein is added, the intensity of the sensitized emission increases (because there is a higher population of protein-bound DNA) but the lifetime remains constant (middle and right-hand panel of Fig 6b). The agreement between the sensitized emission lifetime and the short-component of the donor lifetime confirms that the short-component is due to energy transfer (and not some trivial quenching of the donor).

Because lanthanides have a very long excited-state lifetime, significant *inter*molecular energy transfer can occur, arising from diffusion during the ms donor lifetime. Intermolecular energy transfer can also arise from non-specific aggregation. To confirm that the 200 µs component arose from intramolecular energy transfer, the authors mixed equimolar amounts of double-stranded donor-only DNA and double-stranded acceptor-only DNA at 240 nM, twice the concentration used in the protein-DNA experiments and added protein. The lifetime was long and single exponential, further confirming that the 200 µs component is due to intramolecular bending of DNA by the protein.

The distance corresponding to the 200 µs component can be calculated from Eq. 3, using a calculated R_0 of 55 Å. One issue that arises in determining absolute distances is the orientation factor, κ^2, in R_0. Europium emission in solution is expected to be unpolarized. However, we have recently shown that the 616 nm europium in a crystal is highly polarized [39] and hence one must measure the europium polarization to make sure it is unpolarized in solution. Furthermore, Heyduk found that Cy5 is highly polarized (anisotropy = 0.28) and so there is some uncertainty in R_0, and hence the absolute distance determined – probably in the order of 10%.

Once the distance is determined, one can then infer the angle of the DNA using the model in Fig. 5. There are subtleties associated with how far the donor and acceptor are from the helix axis and the tether lengths, and these are discussed in their manuscript. They report a distance of 51 Å, corresponding to a bending angle of 150°.

When the experiment is repeated using mutants of cHMG1a, very similar results are found (Table 1), confirming that these mutants also bend DNA, despite their altered binding efficiency. A different class of HMG protein, called cHMGI, was found to bend the DNA only a minor amount. Finally, as a side point, note in Table 1 that the accuracy in lifetime determination, particularly when the signal is mono-exponential, is extremely high, such that lifetime quenching of a few percent can be measured, corresponding to a measurable distance of almost $2R_o$, or 100 Å in this case.

5.2
Molecular Association: Dystrophin and Actin

Root used LRET to detect the association between actin and the protein dystrophin inside a muscle cell [17]. When absent, dystrophin leads to muscular dystrophy, a disease characterized by the progressive atrophy of muscle. Dystrophin is present in the inner muscle cell membrane and is believed to stabilize muscle fibers by binding to actin filaments and ultimately creating a bridge to the extracellular matrix (Fig. 7). This model requires the direct association (close proximity) between dystrophin and actin, and there is significant in vitro evidence for this. Since it is known that some proteins bind in vitro but not in vivo, Root asked the question whether this association is present in the muscle cell.

Root used thin (20 µM) tissue sections of muscle cells and specifically labeled dystrophin with monoclonal antibodies labeled with Tb-DTPA-cs124. To label the antibodies, he used the dianhydride of DTPA, mixed with cs124 (see Fig. 4). Actin was stained with phalloidin-tetramethylrhodamine. This pair leads to a R_o of 59 Å. A large R_o is necessary to get reasonable energy transfer because of the large size of antibodies (≈ 100 Å). If the donor-antibody on dystrophin is in close proximity, roughly within R_o, of the acceptor-labeled actin, then energy transfer should be observed. If the distance is large ($\gg R_o$), then no energy transfer is expected.

In fixed tissues donor intensity decreased by 40%, and in unfixed tissue by 60%, in the presence of an acceptor, indicating significant energy transfer and therefore a close association between anti-dystrophin antibodies and actin (Fig. 8). Sensitized emission (using delayed detection to eliminate prompt acceptor fluorescence) was also observed. In addition, Root measured donor and sensitized emission lifetime measurements (fitting to a single exponential) and found that the sensitized emission lifetime (at 568 nm) was similar, but somewhat shorter, than the donor lifetime (at 547 nm) (Fig. 9). This indicates that most, but not all, of the anti-dystrophin antibodies had an acceptor nearby. (A comparison of donor lifetime to sensitized-emission lifetime gives information about the distribution of donor-acceptor complexes: donor-emission measure-

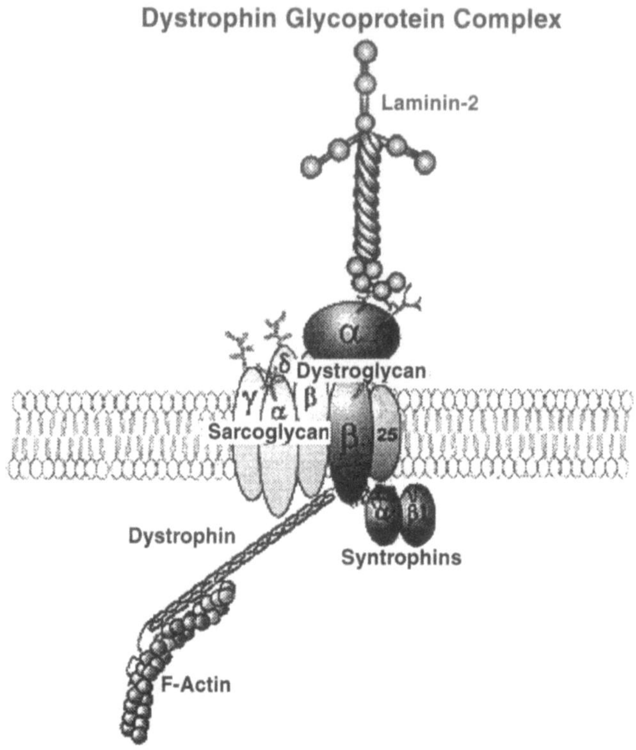

Fig. 7. Schematic arrangement of actin anchoring to membrane and intermembrane proteins via the protein dystrophin. Such a model requires direct interaction between dystrophin and actin. (Fig. from Kevin Campbell, Univ. of Iowa)

ments are weighted towards those transferring less energy, and sensitized emission measurements are weighted towards those transferring more energy. For example, if some of donors have no acceptors nearby, this will lengthen the average donor lifetime, whereas it does not affect the sensitized emission lifetime because this latter signal arises only from those complexes which can transfer energy).

Root also looked at the spatial distribution of the anti-dystrophin antibodies by performing LRET in a microscope. Examining 90 μm diameter regions of the tissue sections, dual-labeled samples yielded a sensitized emission lifetime of 0.7 ± 0.1 but donor emission lifetimes were more variable, ranging from 0.7 ms to 1.4 ms. Root suggested that the association of dystrophin for actin may have micro-heterogeneity within the cell.

Finally, Root compared the detection of molecular co-localization using LRET (Root called his method SEIRET – sensitized emission immuno-resonance energy transfer) with the conventional technique of immunofluorescence colocalization. In immunofluorescence colocalization, the association of two objects are inferred by staining them with different dyes and looking for

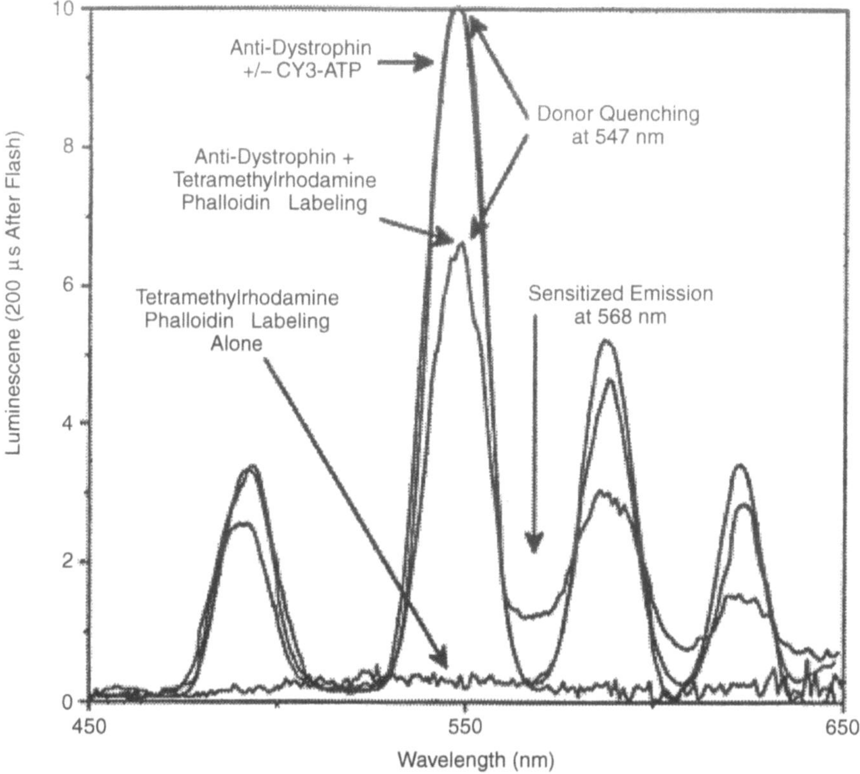

Fig. 8. Delayed emission spectra showing energy transfer from Tb-labeled dystrophin antibodies to rhodamine-labeled actin in a skeletal muscle fiber tissue section. The decrease in Tb intensity at 546 nm and increase in sensitized emission of the acceptor at 568 nm confirms energy transfer. Energy transfer from dystrophin to fluorescently labeled ATP did not show any energy transfer, despite their co-localization in the membrane (see text). From [17]

spatial overlap of the fluorescence from the two dyes in the microscope. This technique, while widely used, is limited in spatial resolution by conventional diffraction (sub-micron resolution), and so cannot differentiate between a true molecular association (nm scale) and nearby-binding (sub-micron). For this purpose, Root stained muscle sections with the terbium-labeled anti-dystrophin antibody, and with ATP bound to the dye Cy3, which is spectrally very similar to tetramethylrhodamine. Both were found in the cell periphery by immunofluorescence microscopy, but no energy transfer was found (Fig. 8), indicating that they were in the same vicinity but not molecularly interacting. Hence, the combination of using lanthanides as donors, which can produce significant energy transfer even with antibodies, and the extra spatial resolution of resonance energy transfer, yields a much more accurate picture of molecular interactions than conventional immunofluorescence microscopy.

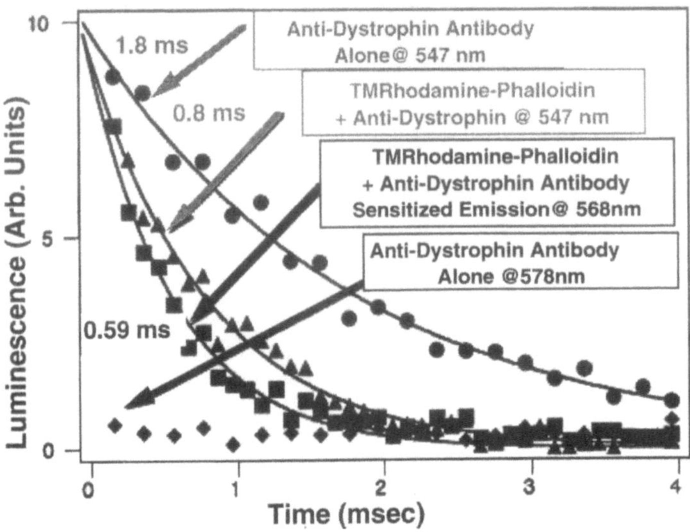

Fig. 9. Lifetime measurements on unfixed muscle cells (see also Fig 8). A decrease in terbium lifetime and the presence of sensitized emission confirm energy transfer. The sensitized emission lifetime is somewhat shorter than the donor lifetime, indicating some heterogeneity (see text). Adapted from [17]

5.3
Molecular Mechanism of Muscle Contraction

Muscle contraction and a variety of subcellular motion in eucaryotes (cells with nuclei) are caused by the interaction of two proteins, myosin and actin. (For a general reference, see [40]). Each protein polymerizes to form filaments, and it is the relative sliding of these filaments with respect to each other which leads to muscle contraction and motion. The sliding is caused by conformational changes in myosin "heads" which protrude from the thick filament, bind and ultimately pull the thin filament (Fig. 10). This model is often called the "tilting oar" or "swinging crossbridge" model of muscle mechanics because the myosin head is thought to act like an oar, undergoing a rotation or "powerstroke" which sweeps the thin filament past the thick filament. The energy for this work is derived from the chemical compound ATP (adenosine triphosphate). More specifically, the conformational change within the myosin head is believed to be a relative motion of the myosin head's two domains after the hydrolysis of ATP into ADP and the release of inorganic phosphate: the light chain domain, also called the neck, is thought to rotate with respect to the catalytic (or globular) domain, made of a so-called heavy chain (see Figs. 10 and 11). The light chain domain consists of two polypeptides of approximately 20 kDaltons in mass, the Essential Light Chain (ESL) and the Regulatory Light Chain (RLC), both of which are folded onto an alpha helix of the heavy chain which extends from the globular domain (see Fig. 11). The rotation of the light chain domain moves the

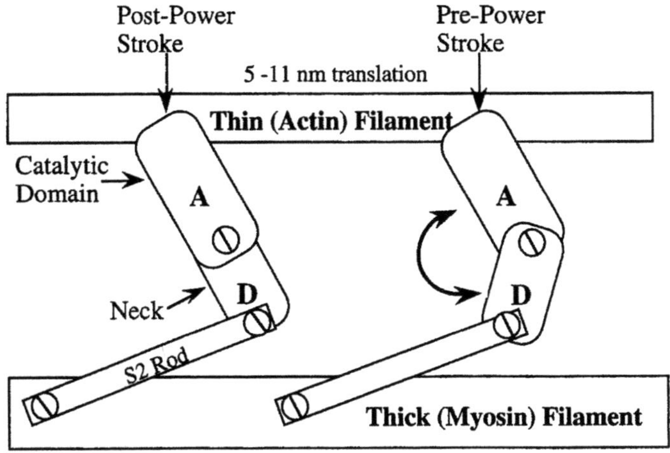

Fig. 10. Schematic view of muscle contraction. Contraction is caused by conformational changes within the two domains of the myosin head, which leads to a relative sliding of the thin (or actin) filament with respect to the thick (or myosin) filament. The energy is derived from the splitting of Adenosine Triphosphate (ATP). It is generally believed that the neck region of the myosin head rotates through approx. 45°, while the catalytic domain is rigidly attached to the thin filament. This leads to an approximately 5 nm translation of the thin filament. Such a change in orientation between the neck and catalytic domain should change the distance between a donor and acceptor placed on each part, and hence should be detectable by luminescence resonance energy transfer

catalytic domain which, because it is strongly bound to the actin at this point in the catalytic cycle, translates the thin filament.

A number of workers have attached spectroscopic probes to the light chains in an effort to observe this rotation during the powerstroke and, with a recent notable exception [41], the best that has been achieved is the detection of a very small (3°) rotation [42]. A roughly 45° rotation is expected if the tilting oar model is to account for the approximately 5 nm movement of the actin filament per myosin conformational change and ATP hydrolysis.

An alternative approach to rotationally sensitive spectroscopic probes is to infer a neck swing by measuring the distance between specific sites in the light chain region and the catalytic domain. As the angle between these regions changes, so too should the distance between particular sites. Recent FRET work measured distances from the essential light chain to the catalytic domain but did not detect a neck swing [43]. These measurements, however, are technically difficult. In general, FRET is not sensitive to the relatively large distances involved (70 Å or more), and is particularly problematic with myosin because complete labeling of the protein is difficult. Furthermore, labeling the regulatory light chain is better than labeling the essential light chain since the former is further from the fulcrum and hence will lead to larger distance changes if a neck swing occurs (Fig. 11).

We have recently shown that LRET is capable of measuring the relatively large distances between the regulatory light chain and globular domains, even

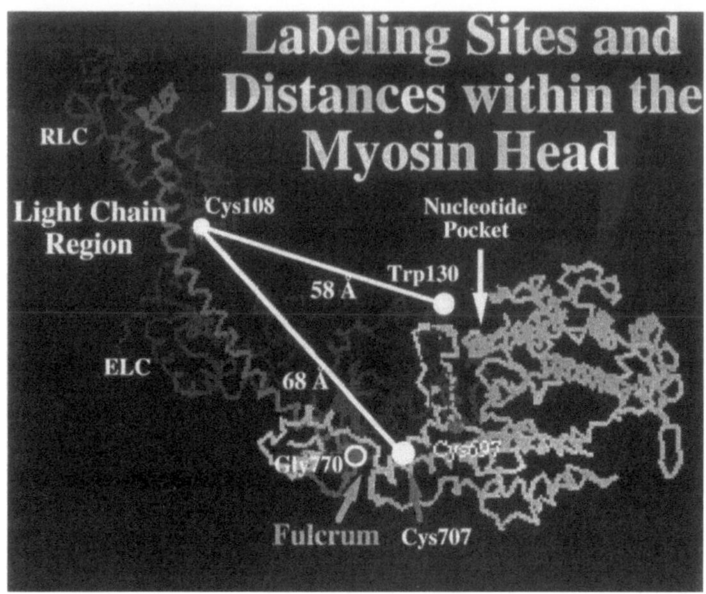

Fig. 11. Labeling sites and their distances. The globular domain contains a nucleotide binding region and an actin binding region. (Actin binds at the right-hand side of the figure.) We have labeled Trp 130 and Cys 707 in the globular domain, as well as Cys 108 on the regulatory light chain. The two light chains bind to an alpha-helix of the heavy chain which extends from the globular region. The light-chain domain is believed to undergo a rotation about a fulcrum, likely near Gly 770. Distances are from the crystal structure of Rayment [44]

when they are not fully labeled with donor and acceptor [9, 19, 20]. Here we present two sets of data: energy transfer from the RLC to a point in the globular domain (cys 707, also called SH1) believed to be near the putative fulcrum; and energy transfer from the active (ATP-binding) site to the regulatory light chain. The latter is particularly important because the donor placed at the nucleotide site (see Fig. 11) does not interfere with the myosin's enzymology (ability to utilize free ATP) and is far from the fulcrum.

5.3.1
Regulatory Light Chain to SH1

Labeling. Skeletal regulatory myosin has many cysteines, including two on the regulatory light chain. To specifically label a unique site on the RLC, chicken gizzard RLC, which has a unique cysteine at position 108, was purified and labeled with Tb-DTPA-cs124-[epsilon-maleimido-propylhyzadine] (see maleimide chelate with n = 2 in Fig. 4). SH1 of skeletal heavymeromyosin (HMM) was labeled with tetramethylrhodamine iodoacetamide (TMRIA), which is known to react specifically with SH1, even in the presence of other cysteines in myosin. HMM is a proteolytic fragment of myosin which is fully functional enzymatically and can translate actin in solution, and is used in these experiments because it is more soluble under our conditions than whole myosin. The labeled gRLC was exchanged into HMM by adding a 10–20-fold excess of gRLC-TMRIA over myosin, heating to 34 °C in the absence of magnesium, and then cooling on ice and adding back in magnesium (which stabilizes the light chain binding). Labeling SH1 with the acceptor TMRIA follows standard procedures and simply involves mixing the reactive dye with myosin and then removing excess unreactive dye after a few hours.

Results. Figure 12a show donor lifetime in the absence and presence of acceptor. The donor-only sample is bi-exponential, with the predominant population having a long (1.85 ms) lifetime. The short component, about 10% of the chelates, likely arises from a population which interacts with the protein in such a way that the terbium lifetime is quenched. In the presence of acceptor on the catalytic domain, the lifetime becomes tri-exponential. The most important term is the 1.47 ms term. This lifetime arises from an energy transfer and indicates 21% energy transfer (1–1.47/1.85) or a distance of 73 Å using a calculated R_o of 59 Å. Because the acceptor is quite rigid (anisotropy = 0.34), there is an uncertainty of ± 10% in this distance due to the orientation factor in R_o. The distance of 73 Å is in excellent agreement with the crystal structure, where the distance between the α-carbons of SH1 and the equivalent of cys 108 is 68 Å [44].

There are two addition lifetimes besides the 1.47 ms component. The 60 μs lifetime term of the terbium decay in the presence of acceptor is not of physiological interest: it arises from detector ringing due to the burst of the nanosecond prompt fluorescence of the acceptor. The 0.48 ms componnt is more complex. It may arise partly from the quenched donors (those 10% of donors with a short, 0.47 ms lifetime, even in the absence of acceptor – see previous paragraph) and partly from some donors which are undergoing significant energy transfer: 1–0.48/1.85 = 74% energy transfer = 44 Å between donor and acceptor. Whether this is due to non-specific labeling of acceptor or an alternative conformation of myosin is under investigation.

Figure 12b shows the same samples with the addition of polymerized actin. The donor-only lifetime decay is unchanged, while the donor-acceptor lifetime increases from 1.47 ms to 1.54 ms. This indicates that actin causes a decrease in energy transfer of 4.5%, which has an uncertainty of ± 1.0% on multiple samples. This difference could be caused either by a rotation of SH1, or an increase in distance, from 73 to 77 Å, or a combination of the two. Addition of ADP to actomyosin (data not shown) causes an increase in energy transfer of 4%. This is likely due to a rotation of TMR at SH1 because it has been shown via dichroism experiments in fibers that ADP causes a 20–30° rotation of TMR-SH1 when myosin is bound to actin.

5.3.2
Active Site to Regulatory Light Chain

Figure 13a shows lifetime and sensitized emission of a myosin head covalently labeled at the active site with the short-linker version of the azido-ATP-chelate

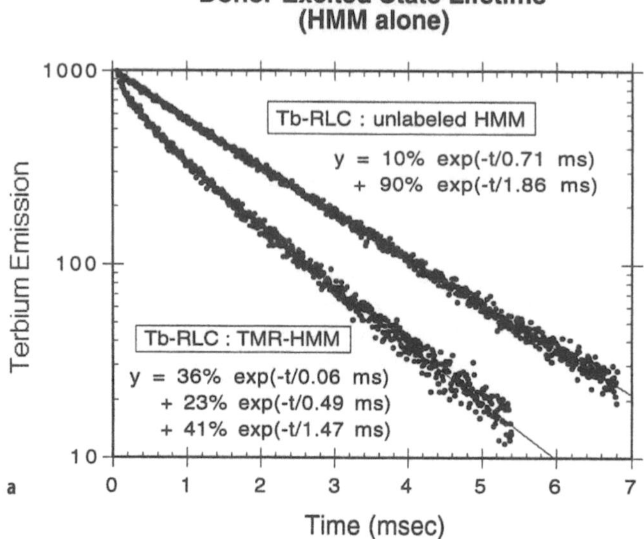

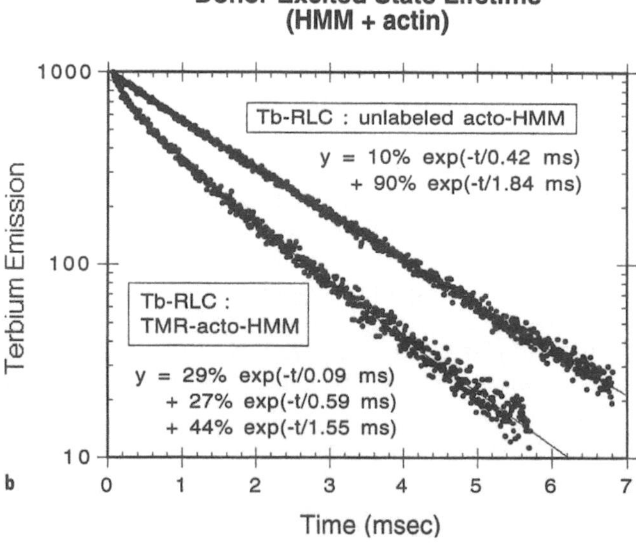

Fig. 12a, b. Energy transfer from a terbium donor placed on the regulatory light chain to a rhodamine acceptor place at SH1 on the globular domain of myosin, in the absence (a) and presence (b) of polymerized actin. The donor-only signal is predominantly single exponential and 1.85 ms, which is reduced to 1.47 ms or 1.54 ms in the presence of acceptor without and with actin, respectively. Hence, in the absence of actin, 20.5% energy transfer is measured (1 – 1.47/1.85) corresponding to a distance of 73 Å between donor and acceptor. Addition of actin decreases this energy transfer to 15.5%, likely due either to a rotation or translation of the acceptor at SH1

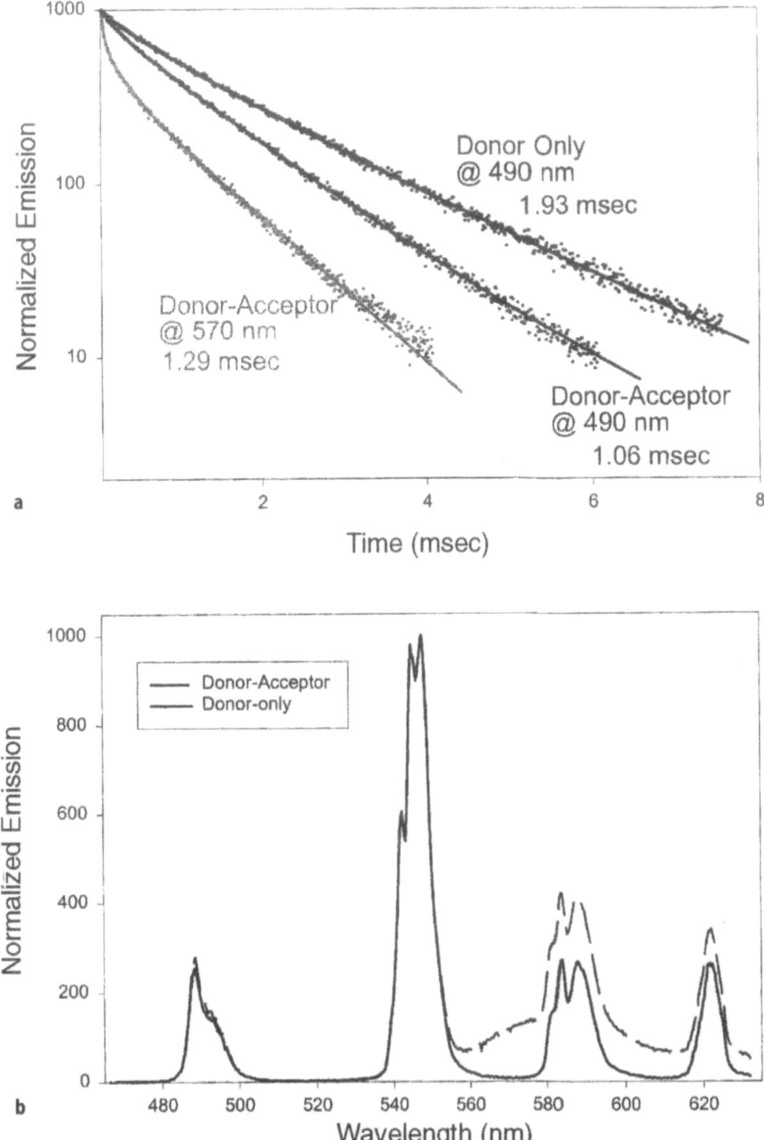

Fig. 13a, b. LRET on myosin: from the active site of the catalytic domain to the regulatory light chain. (**a**) Lifetime measurements showing long, 1.9 ms lifetime of donor-only is reduced to 1.06 ms in the presence of acceptor. The acceptor receives energy from the donor and re-emits light (1.29 ms), with a lifetime-decay which equals the shortened-donor lifetime. The difference between the 1.29 and 1.06 ms is not statistically significant for this particular measurment. (**b**) Time-delayed spectral measurements, showing donor-only (solid curve) and donor acceptor (dashed-line) curves. The increase in emission in the 560–600 nm region is due to emission of the acceptor after it has received energy from the donor. Because the spectrum is taken by collecting light beginning 150 μs after the excitation pulse, all direct emission of the acceptor (nanosecond lifetime) is eliminated. See text for more details.

(see Fig. 4). The donor-labeling procedure is somewhat involved and discussed in the original manuscript [38]. The acceptor TMRIA is placed on gRLC and exchanged into myosin. Both the donor lifetime, sensitized emission lifetime, and emission spectra indicate significant energy transfer. The donor-only has a lifetime of 1.93 ms, which is reduced to 1.06 ms in the presence of acceptor, indicating 45% energy transfer or a distance of 62 Å (R_o = 60 Å) which agrees well with the 58 Å found in the crystal structure. (The 490 nm curve of the donor-acceptor sample shown in Fig. 13a contains, in addition to the 1.06 ms component, a 15% component which remains at 1.93 ms due to incomplete exchange of the acceptor-labeled RLC into the myosin.) Energy transfer is confirmed by measuring the sensitized emission lifetime at 570 nm, which is primarily 1.29 ms, in reasonable agreement with the 1.06 ms found by donor lifetime. (There is also a shorter component in the sensitized emission whose origin is under investigation.) Qualitatively, the presence of energy transfer is very clear from the shape of the emission spectra in Fig. 13b: the emission of the acceptor in the 560–620 nm range is all due to energy transfer since the fluorescence is collected only after a 140 µs delay, which eliminates all prompt fluorescence.

These experiments show that LRET is capable of measuring distances between the catalytic domain and regulatory light chain, even if labeling of the protein is not complete. The precision in the measurements is high, which means that relatively small changes in energy transfer, corresponding to subtle conformational changes, can be detected. Indeed, we have now detected changes in distance between the light chain domain and the catalytic domain upon binding of different nucleotides, and these results will be presented elsewhere [20].

6
Lifetime- and Color-Tailored Fluorophores

6.1
Overview

FRET has also recently been used to generate new DNA-binding heterodimeric fluorophores with improved *spectral* (but not temporal) properties. [45, 46]. Here we show that LRET can be used in a similar manner to generate a new class of DNA-binding heterodimeric fluorophores with improved *spectral* and *temporal* properties. We call these dyes Lifetime- and Color-Tailored Fluorophores (LCTFs) because they potentially have tunable excited-state lifetimes, ranging from 0.1–300 µs, and simultaneously tunable emission color, ranging from 500–750 nm. The technology is very new and to date only rudimentary proof-of-principle experiments have been accomplished. However, the potential, should the technology become practicable, is significant. LCTFs potentially quadratically increase the number of fluorophores simultaneously detectable – relying on both color (spectral) discrimination and temporal discrimination (see Fig. 14). For example, LCTFs with five different colors, each having five (or more) widely differing lifetimes, may be possible. Each LCTF can also potentially bind to a unique DNA sequence and, by a combination of tem-

Fig. 14. Schematic view of multiple labels created with LCTFs. Lifetime- and Color-Tailored Dyes potentially quadratically increase the number of labels, and hence signals, which can be simultaneously detected. This is because multi-color fluorescence can be combined with multi-lifetime fluorescence

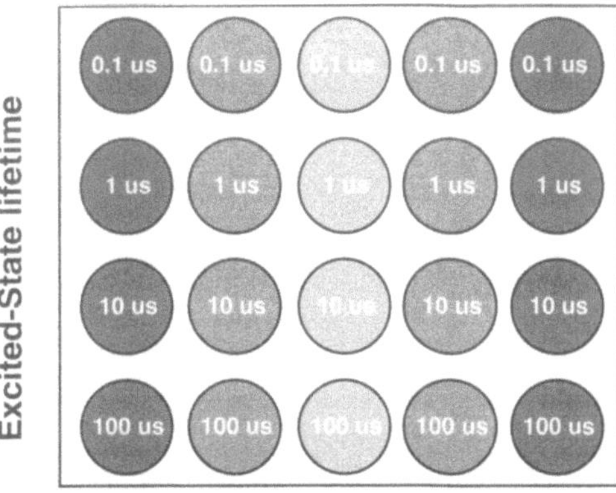

poral and spectral resolution, can be independently analyzed. Hence on a single sample, potentially 25 DNA sequences can be analyzed. In addition, the microsecond to millisecond lifetime of these dyes is long compared to autofluorescence but short compared to common saturation rates. Hence the signal intensity should not be adversely affected by the relatively long lifetime, but the background, often due to autofluorescence with nanosecond lifetime, should be readily discriminated against. The LCTF technology has not yet been applied to a particular biological system but, should it become feasible, the ability to detect multiple signals on a single sample would, for example, significantly speed up the detection of genetic changes during carcinogenesis via fluorescence in situ hybridization [47, 48].

6.2
Introduction to FRET-Based Dyes

An example of using FRET to generate new DNA-binding dyes is shown in Fig. 15. This work has been pioneered by Glazer and Mathies, as well as scientists at Molecular Probes Inc. [49–54]. In Fig. 15, the donor is fluorescein and the acceptor is a dye called TAMRA. These dyes are an efficient energy transfer pair. The DNA serves as a scaffold to place the donor and acceptor close enough together – significantly less than R_0 – such that most of the donor's energy is transferred to the acceptor, but far enough apart so that strong coupling between the dyes, which can alter their fluorescent properties, does not occur. The DNA also serves as a reactive moiety, enabling coupling to DNA of complementary sequence.

The heterodimeric complex is excited near the donor's absorption maximum, so the absorbance is dominated by the donor and, with a large amount of energy transfer, the emission is dominated by the acceptor. (Recall for future reference that the acceptor emission arising from energy transfer, as opposed to direct excitation, is called "sensitized emission".) The total Stokes shift is there-

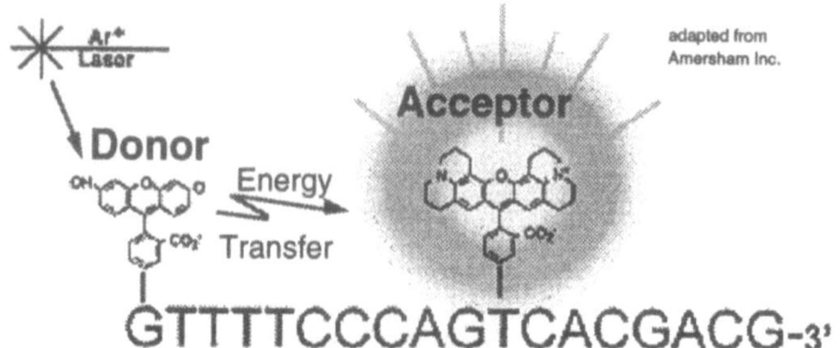

Fig. 15. Heterodimeric DNA dyes. By choosing the appropriate pair which can undergo significant energy transfer, new dyes with the absorption property of one dye and the emission property of the second dye can be generated. These lead to improved sensitivity over single dyes. (Fig. adapted from Amersham Inc.)

fore unusually large and by choosing different acceptors with the same donor, a single excitation wavelength can be used to excite dyes with varying emission wavelengths. These factors increase sensitivity by discrimination against excitation light and Raman scattering, and enable detection of several probes on a single sample.

6.3
The Potential of Using Lanthanides

The same principle can be applied using LRET, where now the donor is a luminescent lanthanide complex and the acceptor is a conventional organic fluorophore (see Fig. 16). As long as a large amount of energy transfer occurs, the emission is dominated by the acceptor – the emission color is therefore determined by the choice of acceptor, and the Stokes shift is very large ($\lambda_{exc} = 337$ nm for lanthanide donors shown here). In addition, the lifetime of the complex can potentially be made anywhere from 0.1–300 ms. These lifetimes are long compared to autofluorescence and, because they're adjustable, temporal discrimination can isolate and detect many probes independently on single sample.

The ability to tailor the emission lifetime arises from the lanthanides' unusually (millisecond) lifetime (in the absence of acceptor). The lifetime, τ, of the donor and sensitized emission in a complex such as the one in Fig. 16 is:

$$\tau = \tau_0\,(1 - E) = \tau_0\,/[1 + (R_0/R)^6]$$

where τ_0 is the lanthanide lifetime in the absence of acceptor (0.5–2.6 ms depending on lanthanide and chelate). The distance between donor and acceptor to achieve this lifetime is $R = R_0\,([\tau_0/\tau] - 1)^{-1/6}$. With $\tau_0 = 1.5$ ms, for example, if the donor and acceptor are placed such that the energy transfer is 90%, then the sensitized emission (and donor emission) will decay with a lifetime of 150 µs. The distance for 99% energy transfer is $0.465\,R_0$, with a resulting lifetime of 15 µs. For Tb-chelate as donor, and tetramethylrhodamine as acceptor [13], where $R_0 = 60$ Å, this corresponds to 28 Å. A probe with a 1.5 µs lifetime can

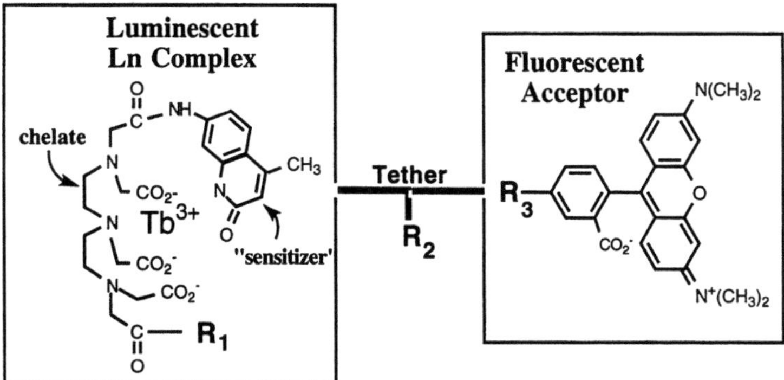

Fig. 16. Structure of a particular lifetime- and color-tailored fluorophore. The lanthanide chelate donor is excited, transfers energy to the fluorescent acceptor which then emits light with a color determined by the particular acceptor and with a lifetime determined by the donor's inherent lifetime and amount of energy transfer. By varying the distance and/or R_o between the two, the energy transfer and hence lifetime of the complex can be tailored as desired

potentially be achieved by placing the donor and acceptor $0.316\,R_o$ apart for 99.9% energy transfer, or 19 Å for Tb-DTPA-cs124 and tetramethylrhodamine. Similarly, a probe with a 150 ns lifetime can be achieved by placing the Tb-chelate and TMR a distance of 13 Å apart, assuming $R_o = 60$ Å. Importantly, because the R_os are large when using lanthanides, these distances are expected to be beyond the strong-coupling limit (typically about < 10 Å) where Dexter exchange [33] occurs and it becomes difficult to predict emission properties.

6.4
Proof of Principle

The full range of energy transfer and lifetimes has yet to be explored. However, Fig. 17 shows that variable excited-state lifetimes, albeit over a limited range, have been generated by LRET. The decay curves in this figure are all at 570 nm, where tetramethylrhodamine emits but terbium does not. By varying the distance between the donor and acceptor, the sensitized emission lifetime changes. We have also used fluorescein (emission 525 nm) as an acceptor with terbium, and have used Cy5 (maximal emission 668 nm) as an acceptor with europium. In the latter case, we have achieved over 90% energy transfer. Figure 18 shows two such constructs where they have similar lifetimes (250–270 µs) but different emission colors.

These examples are proof of the LCTF principle. However, these constructs are double-stranded DNA oligomers labeled with donor and acceptor at the 5′ ends of complementary strands (see Fig. 19, left-hand-side), and hence have no reactive part. In order to make an LCTF which is capable of reacting with a larger piece of DNA, a single DNA oligomer, labeled 5′–3′ or internally, needs to be made (Fig. 19). By varying the distance and choice of donor and/or acceptor,

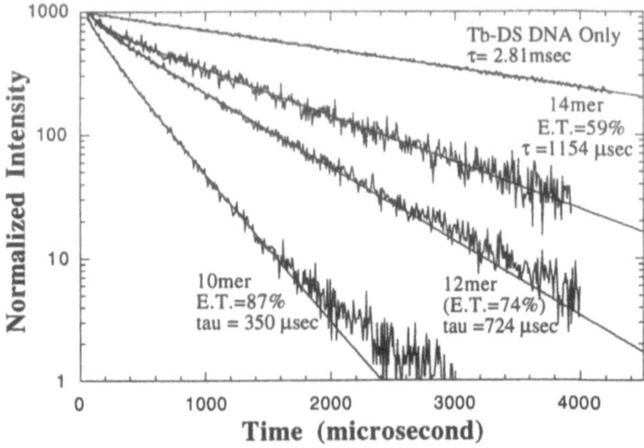

Fig. 17. Lifetime-tailored dyes. The lifetime of the sensitized emission can be tailored by altering the distance between donor and acceptor. Energy transfer from terbium to tetramethylrhodamine, separated by different 10, 12, or 14 base pairs of DNA, yields a sensitized emission lifetime at 570 nm of 0.35, 0.72 and 1.15 ms, respectively. The donor lifetime (546 nm) in the absence of acceptor is 2.81 ms (top curve). (Data from [13])

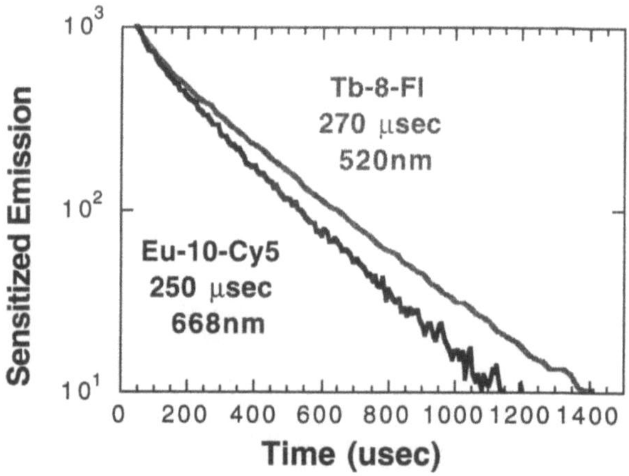

Fig. 18. Color-tailored dyes with fixed lifetime. Just as the lifetime can be tailored (see Fig. 17), so too can the emission color be varied by using different acceptors while keeping the lifetime constant. The terbium and fluorescein construct is separated by 8 base pairs and emits in the green with a lifetime of 0.27 ms. The europium and Cy5 construct is separated by 10 base pairs (and is in D_2O) and emits in the red with a lifetime of 0.25 ms

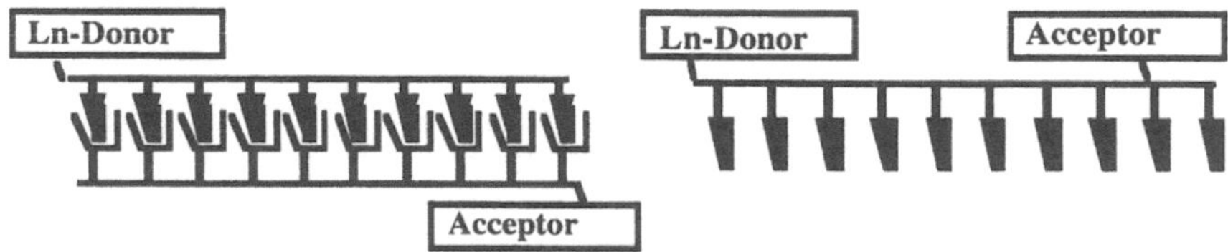

Fig. 19. Two methods of attaching a donor and acceptor via DNA. Only the doubly labeled single-stranded complex (*right-hand side*) is a true LCTF since it can react with a target DNA via the unpaired base

based on ours the desired emission lifetime and spectrum can be generated, and by choosing the appropriate base-sequence, this LCTF can serve as a probe for a specific sequence on a larger piece of DNA. Using our isothiocyanate-based chelates [35], the construction of a single-stranded LCTF should be straightforward.

There are, however, several drawbacks associated with using lanthanides as donors in LCTFs. First is that their absorbance is relatively weak. Carbostyril 124, for example, used in our chelates, has an extinction coefficient of about 12000, and of this roughly 25 % of the energy is transferred to terbium – hence the effective absorption coefficient is approximately 3000. Second is the possibility of a direct transfer of energy from the donor's antenna chromophore (the carbostyril 124) to the acceptor, bypassing the lanthanide. This would decrease the intensity of the long-lived emission of the acceptor since the resulting acceptor emission would be short-lived (nanosecond) and hence not be of interest. Both of these potential problems are minimized by designing antenna molecules which have large absorbance and have efficient transfer of energy to the lanthanide. Extinction coefficients up to 80000 M^{-1} cm^{-1} in ethanolic solution have been made [55], and a greater understanding of the mechanism of energy transfer from antenna to lanthanide will help increase this rate and efficiency, thereby minimizing problems of short-circuiting [56–59]. Choosing acceptors whose absorption has minimal overlap with the antenna fluorescence, or finding antenna molecules which are efficient at transferring energy to lanthanides yet have poor fluorescence, further decreases the short-circuiting effect. Yet a completely different alternative would be to use other metal-ligand complexes as donors [60].

6.5
Lifetime- and Color-Tailored Fluorophores – Conclusion

Although the development of LCTFs is in its infancy, should this technology become practicable, the implications are significant. Heterodimeric fluorophores in which both the emission spectra and lifetime can be tuned will be possible. Such fluorophores would potentially increase quadratically the number of detectable signals in a single sample, relying on both multi-color fluorescence and multi-lifetime fluorescence for discrimination. If spectral discrimination can resolve five dyes, and temporal discrimination can resolve five lifetimes, then the number of signals one could look at in a single sample would be 25. Combinatorics, which is now used in multi-color fluorescence [61], could further increase this number. The instrumentation for temporal resolution would also likely be simplified since the megahertz modulation necessary for resolving nanosecond dyes [62] becomes kilohertz instead. Furthermore, LCTFs, because they have relatively long lifetimes, would enable efficient discrimination against prompt emission, whether from Raman or autofluorescence (which tends to arise from organic molecules and hence have a nanosecond lifetime). LCTFs also have the advantage over simply using lanthanides alone in that many emission colors are possible and their lifetime is sufficiently short such that saturation is not a significant problem. (Lanthanide emission of a millisecond limits the emission rate to 1000 photons/s/molecule, whereas 10–100 kHz is frequently desired and can potentially be achieved with LCTFs.)

7
Conclusion

Luminescence resonance energy transfer has a number of technical advantages over conventional FRET. These advantages include efficient energy transfer (large R_os) which depends primarily on the donor-acceptor distance and not their orientation (restricted range of κ^2), the ability to accurately measure energy transfer because of the lanthanide's long lifetime and excellent signal to background of sensitized emission. These factors enable resonance energy transfer to be used in biological systems which were previously inaccessible using conventional FRET. These include large protein complexes such as actin and myosin, as well as antibodies, and protein-DNA systems.

Acknowledgements. The author thanks Roger Cooke, Elise Burmeister Getz, Handong Li, Ming Xiao, and Ralph Yount, for collaborative work using LRET on actomyosin; and Tomasz Heyduk and Douglas Root for helpful discussions. This work was supported in part by National Institutes of Health grant AR44420 to PRS.

References

1. Yu, H., E.P. Diamandis. (1993) Ultrasensitive time-resolved immunofluorometric assay of prostate-specific antigen in serum and preliminary clinical studies. Clin. Chem. 39: 2108–14
2. Oser, A., W.K. Roth, G. Valet. (1988) Sensitive non-radioactive dot-blot hybridization using DNA probes labelled with chelate group substituted psoralen and quantitative detection by europium ion fluorescence. Nucleic Acids Res. 16:1181–1196
3. Oser, A., M. Collasius, G. Valet. (1990) Multiple End Labeling of Oligonucleotides with Terbium Chelate-Substituted Psoralen for Time-Resolved Fluorescence Detection. Anal. Biochem. 191:295–301
4. Saha, A.K., K. Kross, E.D. Kloszewski, D.A. Upson, J.L. Toner, R.A. Snow, C.D.V. Black, V.C. Desai. (1993) Time-Resolved Fluorescence of a New Europium Chelate Complex: Demonstration of Highly Sensitive Detection of Protein and DNA Samples. J. Am. Chem. Soc. 115:11032

5. Hemmilä, I., S. Dakubu, V.-M. Mukkala, H. Siitari, T. Lovgren. (1984) Europium as a label in time-resolved immunofluorometric assays. Anal. Biochem. 137:335–343

6. Seveus, L., M. Vaisala, I. Hemmila, H. Kojola, G.M. Roomans, E. Soini. (1994) Use of Fluorescent Europium Chelates as Labels in Microscopy Allows Glutaraldehyde Fixation and Permanent Mounting and Leads to Reduced Autofluorescence and Good Long-Term Stability. Microscopy Res. and Technique 28:149–154

7. Marriott, G., M. Heidecker, E.P. Diamandis, Y. Yan-Marriott. (1994) Time-Resolved Delayed Luminescence Image Microscopy Using an Europium Ion Chelate Complex. Biophys. J. 67:957–965

8. Stryer, L., D.D. Thomas, C.F. Meares. 1982. Diffusion-Enhanced Fluorescence Energy Transfer, In Ann. Rev. of Biophys. Bioeng., ed. L. J. Mullins, pp. 203–222. Palo Alto, CA: Annual Reviews, Inc

9. Selvin, P.R. (1996) Lanthanide-based resonance energy transfer. IEEE J. of Selected Topics in Quantum Electronics: Lasers in Biology 2:1077–1087

10. Mathis, G. (1995) Probing molecular interactions with homogeneous techniques based on rare earth cryptates and fluorescence energy transfer. Clin. Chem. 41:1391–1397

11. Mathis, G. (1993) Rare Earth Cryptates and Homogeneous Fluoroimmunoassays with Human Sera. Clin. Chem. 39: 1953–1959

12. Selvin, P.R., T.M. Rana, J.E. Hearst. (1994) Luminescence resonance energy transfer. J. Am. Chem. Soc. 116:6029–6030

13. Selvin, P.R., J.E. Hearst. (1994) Luminescence energy transfer using a terbium chelate: Improvements on fluorescence energy transfer. Proc. Natl. Acad. Sci, USA 91: 10024–10028

14. Selvin, P.R. 1995. Fluorescence Resonance Energy Transfer, In Methods in Enzymology, ed. K. Sauer, pp. 300–334. Orlando: Academic Press

15. Heyduk, E., T. Heyduk. (1997) Thiol-reactive luminescent Europium chelates: luminescence probes for resonance energy transfer distance measurements in biomolecules. Anal. Biochem. 248:216–227

16. Heyduk, E., T. Heyduk, P. Claus, J.R. Wisniewski. (1997) Conformational Changes of DNA Induced by Binding of Chironomus High Mobility Group Protein 1a (cHMG1a). J. Biol. Chem. 272:19763–19770

17. Root, D.D. (1997) In Situ molecular association of dystrophin with actin revealed by sensitized emission immuno-resonance energy transfer. Proc. Natl. Acad. Sci., USA 94

18. Li, M., P.R. Selvin. (1995) Luminescent lanthanide polyaminocarboxylate chelates: the effect of chelate structure. J. Am. Chem. Soc. 117:8132–8138

19. Getz, E.B., R. Cooke, P.R. Selvin. (in press) Luminescence Resonance Energy Transfer Measurements on Myosin. Biophys. J.

20. Xiao, M., H. Li, E.B. Getz, R. Cooke, R.G. Yount, P.R. Selvin. 1998. Luminescence Resonance Energy Transfer Measurements from the Active Site of Myosin. Presented at the Biophysical Society, Kansas City, MO 1998

21. Stryer, L. (1978) Fluorescence Energy Transfer as a Spectroscopic Ruler. Ann. Rev. Biochem. 47:819–846

22. Fairclough, R., H., C. Cantor, R. 1978. The use of Singlet – Singlet Energy Transfer to Study Macromolecular Assemblies, In Methods in Enzymology, ed., pp. 347–379

23. Cantor, C.R., P.R. Schimmel. 1980. Biophysical Chemistry. San Francisco: W. H. Freeman and Co

24. Herman, B. (1989) Resonance Energy Transfer Microscopy. Meth. Cell Bio. 30:219–243

25. Coker, G., III, S.Y. Chen, B.W. van der Meer. 1994. Resonance Energy Transfer: VCH Publishers, Inc

26. Clegg, R.M. (1995) Fluorescence Resonance Energy Transfer. Curr. Op. Biotech. 6: 103–110

27. Clegg, R.M. 1996. Fluorescence Resonance Energy Transfer, In Fluorescence Imaging Spectroscopy and Microscopy, ed. X. F. Wang, B. Herman, pp. 179–251: John Wiley & Sons, Inc

28. dos Remedios, C.G., P.D.J. Moens. 1998. Fluorescence resonance energy transfer – applications in protein chemistry, In Resonance Energy Transfer, ed. D. L. Andrews, A. A. Demidov. Chichester: John Wiley and Sons

29. Bunzli, J.-C.G. 1989. Luminescent Probes, In Lanthanide Probes in Life, Chemical and Earth Sciences, Theory and Practice, ed. J.-C. G. Bunzli, G. R. Choppin, pp. 219–293. New York: Elsevier

30. Sammes, P.G., G. Yahioglu. (1996) Modern Bioassays using Metal Chelates as Luminescent Probes. Modern Bioassays using Metal Chelates as Luminescent Probes 13:1–28

31. Bastiaens, P.I.H., T.M. Jovin. 1998. Fluorescence Resonance Energy Transfer (FRET) Microscopy, In Cell Biology: A Laboratory Handbook, ed. J. E. Celis, pp. 136–146. New York: Academic Press

32. Drexhage, K.H. (1970) Monomolecular Layers and Light. Sci. Amer. 222:108–119

33. Dexter, D.L. (1953) A Theory of Sensitized Luminescence in Solids. J. Chem. Phys. 21: 836–850

34. Horrocks, W.D., Jr., D.R. Sudnick. (1979) Lanthanide Ion Probes of Structure in Biology. Laser-Induced Luminescence Decay Constants Provide a Direct Measure of the Number of Metal-Coordinated Water Molecules. J. Am. Chem. Soc. 101:334–350

35. Li, M., P.R. Selvin. (1997) Amine-reactive forms of a luminescent DTPA chelate of terbium and europium: Attachment to DNA and Energy Transfer Measurements. Bioconjugate Chem. 8:127–132

36. Takalo, H., V.-M. Mukkala, H. Mikola, P. Liitti, I. Hemmila. (1994) Synthesis of Europium(III) Chelates Suitable for Labeling of Bioactive Molecules. Bioconjugate Chem. 5:278–282

37. Selvin, P.R., J. Chen. (1998) Thiol-reactive luminescent lanthanide chelates. manuscript in preparation

38. Li, H., J. Grammer, R. Cooke, P.R. Selvin, R.G. Yount. (1998) Synthesis and Spectral Characterization of a Photoaffinity ATP Analog Containing a Luminescent Lanthanide Chelate. manuscript in preparation

39. Vereb, G., E. Jares-Erijman, P.R. Selvin, T.M. Jovin. (submitted) Time and spectrally resolved imaging microscopy of lanthanide chelates. Biophys. J.

40. Stryer, L. 1995. Biochemistry. San Francisco: W.H. Freeman

41. Baker, J.E., I. Brust-Mascher, S. Ramachandran, L.E.W. LaConte, D.D. Thomas. (in press) A Large and Distinct Rotation Of The Myosin Light Chain Domain Upon Muscle Contraction. Proc. Nat. Acad. Sci. USA

42. Irving, M., T.S. Allen, C. Sabido-David, J.S. Craik, B. Brandmeier, J. Kendrickjones, J.E.T. Corrie, D.R. Trentham, Y.E. Goldman. (1995) Tilting of the light-chain region of myosin during step length changes and active force generation in skeletal muscle. Nature 375: 688–691

43. Smyzynski, C., A.A. Kasprzak. (1997) Effect of nucleotides and actin on the orientation of the light chain-bindig domain in myosin subfragment 1. Biochemistry 36:13201–13207

44. Rayment, I., W.R. Rypniewski, K. Schmidt-Base, R. Smith, D.R. Tomchick, M.M. Benning, D.A. Winkelmann, G. Wesenberg, H.M. Holden. (1993 A) Three-dimensional structure of myosin subfragment-1: A molecular motor. Science 261:50–57

45. Ju, J., C. Ruan, C.W. Fuller, A.N. Glazer, R.A. Mathies. (1995) Fluorescence energy transfer dye-labeled primers for DNA sequencing and analysis. Proc. Nat. Acad. Sci. USA 92: 4347–51

46. Ju, J., A.N. Glazer, R.A. Mathies. (1996) Energy transfer primers: A new fluorescence labelling paradigm for DNA sequencing and analysis. Nature Medicine 2:246–249

47. Bergerheim, U.S., K. Kunimi, V.P. Collins, P. Ekman. (1991) Deletion mapping of chromosomes 8, 10, and 16 in human prostatic carcinoma. Genes Chromosomes Cancer 3: 215–20

48. Cher, M.L., Ito, T., Weidner, N., Carroll, P.R., Jensen, R.H. (1995) Mapping of regions of physical delection on chromosome 16q in prostate cancer cells by fluorescence in situ hybridization (FISH). J. Urology 153:249–254

49. Glazer, A.N., L. Stryer. (1983) Fluorescent Tandem Phycobiliprotein Conjugates: Emission Wavelength Shifting by Energy Transfer. Biophys. J. 43:383–386

50. Rye, H., D. Yue S; Wemmer, M. Quesada, R. Haugland, R. Mathies, A. Glazer. (1992) Stable fluorescent complexes of double-stranded DNA with bis-intercalating asymmetric cyanine dyes: properties and applications. Nucleic Acids Res 20:2803–12

51. Benson, S.C., R.A. Mathies, A.N. Glazer. (1993) Heterodimeric DNA-binding dyes designed for energy transfer: stability and applications of the DNA complexes. Nucl. Acids Res. 21:5720–5726
52. Benson, S.C., P. Singh, A.N. Glazer. (1993) Heterodimeric DNA-binding dyes designed for energy transfer: synthesis and spectroscopic properties. Nucl. Acids Res. 21:5727–5735
53. Glazer, A.N., H.S. Rye. (1992) Stable Dye-DNA Intercalation Complexes as Reagents for High-Sensitivity Fluorescence Detection. Nature 359:859–861
54. Glazer, A., R. Mathies. (1997) Energy-transfer fluorescent reagents for DNA analyses. Curr Opin Biotechnol 8:94–102
55. Yamada, S., F. Miyoshi, K. Kano, T. Ogawa. (1981) Highly Sensitive Laser Fluorimetry of Europium(III) with 1,1,1-Trifluoro-4-(2-Thienyl)-2,4-Butanedione. Anal. Chim. Acta 127:195–198
56. Weissman, S.I. (1942) Intramolecular energy transfer: the fluorescence of complexes of europium. J. Chem. Phys. 10:214
57. Crosby, G.A., R.E. Whan, R.M. Alire. (1961) Intramolecular energy transfer in rare earth chelates: the role of the triplet state. J. Chem. Phys. 34:743
58. Abusaleh, A., C. Meares. (1984) Excitation and De-Excitation Processes in Lanthanide Chelates Bearing Aromatic Sidechains. Photochem. & Photobiol. 39:763–769
59. Kirk, W.R., W.S. Wessels, F.G. Prendergast. (1993) Lanthanide-Dependent Perturbations of Luminescence in Indolylethylenediaminetetraacetic Acid-Lanthanide Chelate. J. Phys. Chem. 97:10326–10340
60. Lakowicz, J.R. 1997. Long Lifetime Metal-Ligand Complexes as Probes in Biophysics and Clinical Chemistry, In Methods in Enzymology, ed. L. Brand, M. L. Johnson
61. Ried, T., A. Baldini, T.C. Rand, D.C. Ward. (1992) Simultaneous visualization of seven different DNA probes by in situ hybridization using combinatorial fluorescence and digital imaging microscopy. Proc. Natl. Acad. Sci. 89:1388–1392
62. Gadella, T.W.J., T.M. Jovin, R.M. Clegg. (1993) Fluorescence Lifetime Imaging Microscopy (FLIM) – Spatial Resolution Of Microstructures On The Nanosecond Time Scale. Biophysical Chem. 48:221–239

Part 5
Fluorescence Techniques in Medicine –
a Challenge for the Future

CHAPTER 20

Fluorescence Lifetime Imaging Microscopy

B. Herman, X. F. Wang, P. Wodnicki, A. Perisamy, N. Mahajan, G. Berry, G. Gordon

1
Introduction

The marriage of fluorescence microscopy with digital imaging has resulted in an expansion in our understanding of cellular function. Currently, most fluorescence microscopic imaging is performed as a measurement of emission or excitation intensity. These types of measurements have several limitations: a) intensity-based fluorescence imaging can be difficult to quantify; b) fast (in the order of nano- to picosecond) dynamic events in cell physiology and biology cannot be studied; c) the emitted fluorescence from biological samples is often complex and represents contributions from a number of molecular species which cannot be individually analyzed; and d) autofluorescence and background fluorescence limit the sensitivity of detection. In contrast, the measurement of fluorescence lifetimes does not suffer from these limitations. Fluorescent lifetimes possess the added benefits of being independent of local intensity, concentration and photobleaching of the fluorophore. Fluorophores with similar spectra may have significant differences in their lifetimes, and the same fluorophore may display distinct lifetimes in different environments. Because fluorescent lifetimes are not affected by scattering, measurements of fluorescent lifetimes provide more sensitive and quantitative information from complex structures. Very recently, the combination of the sensitivity of fluorescence lifetimes to environmental parameters with highly sensitive rapidly gatable image intensifiers has allowed fluorescent lifetimes to be monitored in a spatial manner in single living cells. This technology has been termed Fluorescence Lifetime Imaging Microscopy (FLIM). The reason that fluorescence lifetime imaging is useful is based on the fact that most biological systems are not homogeneous, but rather exist as 3-D spatially heterogeneous structural and functional complexes. Fluorescence lifetime imaging provides useful and unique information in terms of deciphering the 3-D spatially heterogeneous structural and functional activities of cells and tissues.

2
Wide Field FLIM

There are two methods commonly employed to measure fluorescent lifetimes: time-domain pulse methods [1, 2] and frequency-domain or phase-resolved methods [3]. Time-domain lifetime measurements employ pulsed excitation and the fluorescent lifetime is determined directly from the fluorescence signal or by photon counting detection. Frequency-domain (phase-resolved) lifetime measurements utilize sinusoidally modulated light as an excitation source and lifetimes are calculated from the phase shift or modulation depth of the fluorescence emission signal. It is possible to obtain fluorescence lifetime images by a combination of lifetime determination techniques (time- or frequency-domain) with high speed two-dimensional detectors or scanning techniques such as mechanical stage scanning, laser beam scanning, or electronic (e.g. image dissector, streak-camera) scanning methods.

In wide field FLIM, parallel image detection using a high speed 2-D gated image intensifier with gating times from sub-nanosecond to a few nanoseconds is employed [4–6]. Using these types of detectors, the decay time resolution attainable is limited by the gate time and gating jitter of the detector. In the FLIM developed in our laboratory, a picosecond pulsed light source consisting of a mode-locked YAG laser, a dye laser and a cavity-dumper provide picosecond pulses with tunable wavelengths from the UV to the IR region at rates of single shot to 76 MHz (Fig. 1). Picosecond excitation pulses are delivered to a fluorescence microscope through a multimode optical fiber system which is focused to illuminate the entire field of view through a $40 \times$ Fluor (oil, N.A. = 1.3) lens. Time-resolved fluorescence microscopic images are obtained using a high speed gated multichannel plate (MCP) image intensifier operating at a maximal frequency of 10 kHz. The image acquired on the gated MCP image intensifier is focused at unity magnification onto a slow-scan cooled CCD camera. Time-resolved images can be accumulated on the CCD chip by repeated pulsed light excitation. Following the production of an image of sufficient signal to noise, the image is read out to a computer. For weak fluorescence emission, the target integration in the CCD is effective at enhancing signal-to-noise ratios (S/N) of detected images. In a given situation, the available fluorescence light level determines the integration time required to obtain acceptable signal-to-noise ratios. Cooling the CCD to reduce dark current to negligible levels allows very long integration times.

2.1
Acquisition and Processing of Fluorescent Lifetime Images

Lifetimes are calculated at rapid rates from multigate data using the ratio method (Fig. 2). To obtain lifetime images, the gated MCP is turned on for a very brief interval (i.e. 2 ns) at a particular time interval (t_1) after the exciting pulse for a set duration (ΔT). The acquired image from the gated MCP is sent to the slow-scan cooled CCD camera (which is continuously left on). This process is repeated a large number of times (i.e. 10 000) at this same t_1 with the emitted intensity from the gated MCP being continuously accumulated on the CCD. The final image on the CCD is then read out, the time of onset of the gate window with respect to the excitation pulse (t_2) is temporally shifted and the whole process is repeated.

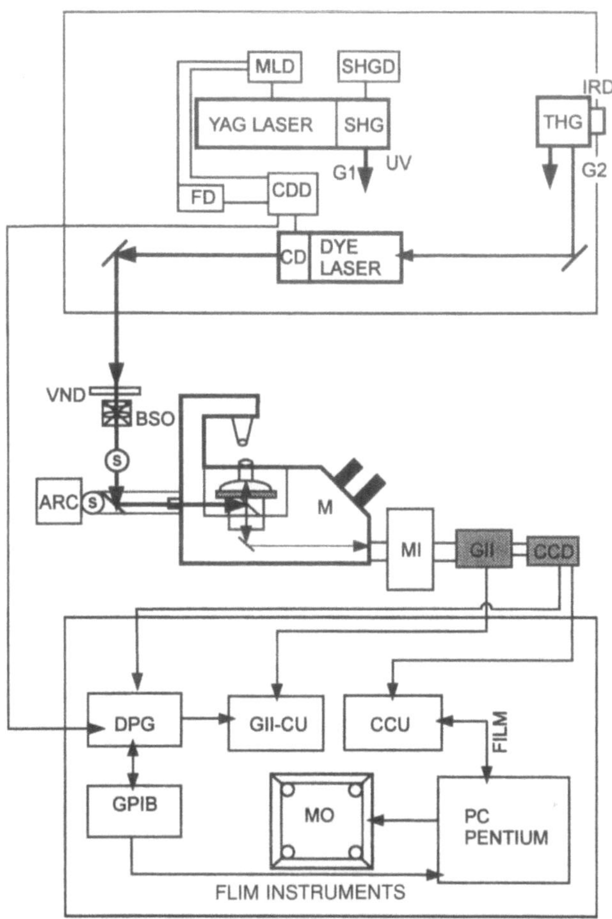

Fig. 1. FLIM combines fluorescence lifetime detection and imaging techniques. It is possible to measure fluorescence lifetime images by the combination of lifetime determination (time- and frequency-domain) techniques with high speed two-dimensional detectors and scanning techniques such as mechanical stage scanning, laser beam scanning, or electronic (e.g. image dissector, streak-camera) scanning methods. Abbreviations are as follows: inverted microscope (*M*); variable neutral density filter (*VND*); beam steering optics (*BSO*); shutter (*S*); frequency divider (*FD*); digital delay pulse generator (*DPG*); single stage microchannel plate gated image intensifier (*GII-CCD*); general purpose interface bus (*GPIB*); cavity dumper driver unit (*CDD*); mode locker driver-*MLD*; second harmonic generator (*SHG*) and the driver (*SHGD*); third harmonic generator (*THG*); cavity dumper (*CD*); arc lamp (*ARC*); multi image module (*MI*); monitor (*MO*); gated image intensifier control unit (*GII-CU*); camera control unit (*CCU*)

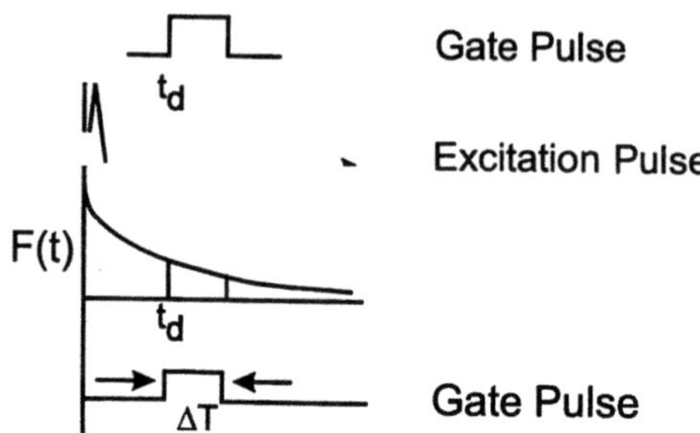

Fig. 2. The principle of operation the gated image intensifier used for time-resolved FLIM. At some time (t_D) after the excitation pulse, a gate pulse of duration ΔT is delivered to the gated image intensifier. Thus, for the duration of ΔT, the intensifier accumulates photons from the sample. Each image accumulated on the intensifier during ΔT is then read onto an integrating CCD camera and this process is repeated (using the same t_D and ΔT), until the image integrated on the CCD is of sufficient signal-to-noise. A second image is then acquired but with a different t_D, enabling the use of a simple ratioing of the two images to determine the lifetime

Rapid calculation of fluorescent lifetimes is performed using a simple, compact, easy to implement algorithm. A single exponential decay of the fluorescence signal excited by a short duration pulsed light source can be written as:

$$F(t) = A \exp(-t/\tau)$$

Since the decay of fluorescence is detected at two different delay times (t_1 and t_2) with the gate width ΔT, the fluorescence lifetime (τ) can be calculated directly using only four parameters: the gated fluorescence signals, D_1 and D_2, and t_1 and t_2:

$$\tau = (t_2 - t_1)/\text{natural} \log(D_1/D_2)$$

3
Confocal FLIM (CFLIM)

Conventional microscopes create images with a depth-of-field at high power of 2–3 µm. Since the resolving power of optical microscopes is about 0.2 µm, superimposition of detail within this plane of focus obscures structural detail that would otherwise be resolved. In addition, light from out of focus planes creates diffuse halos around objects of study. In contrast, confocal microscopes create optical sections which are 0.5 µm thick. Confocal microscopes reject light from out-of-focus planes, providing images with much higher resolution and image content. Confocal microscopy can be thought of as a computerized axial tomography (CAT) scanner for cells and tissues. In principle, CFLIM can be based on either time- or frequency-domain lifetime methods, although time-domain methods suffer from the drawback that a high laser repetition rate is needed to acquire an image in a reasonable time since each pixel will require many measurements to generate the decay curve. Conversely, frequency-domain methods allow the use of a conventional continuous wave (CW) laser and an external high frequency light modulator to produce short pulses of light for excitation. In the frequency-domain method, high frequency modulated laser excitation (using an inexpensive and compact CW laser coupled with external modulation of the laser output) provides the possibility for high temporal resolution confocal FLIM.

3.1
Description of Confocal FLIM System and Data Analysis

Figure 3 is a schematic representation of a confocal FLIM, and is based on a laser scanning confocal microscope (LSCM) and frequency-domain lifetime detection equipment. High frequency (Mhz-Ghz) sinusiodally modulated light from a CW laser excites the sample and the fluorescence lifetimes are determined from phase shifts or amplitude demodulation factors. When the sample in a single point under study is excited with time-dependent sinusoidally modulated light E(t) can be represented as follows:

$$E(t) = A\,[1 + Me \cos(\omega t)]$$

where A is the dc intensity component of the exciting light, $\omega = 2\pi f$ where f is the modulation frequency, and Me is the ratio of the amplitude of the ac intensity to the dc intensity component. For a single component sample with an exponential decay, the resulting wavelength dependent fluorescence, F(t), will have frequency f but will be phase shifted by an angle F which is unique to the particular fluorescent species:

$$F(t) = F_0\,[1 + Mf \cos(\omega t - F)]$$

where F_0 is the average fluorescence intensity. The phase shift F and the modu-

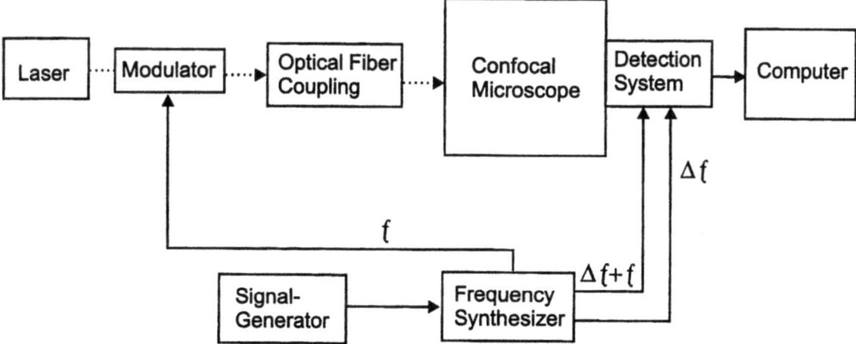

Fig. 3. Schematic diagram of confocal FLIM based on the use of phase-resolved fluorescence lifetime detection and laser scanning confocal microscope

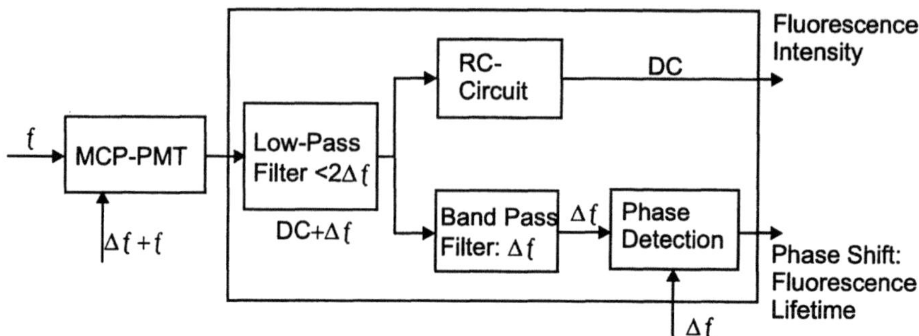

Fig. 4. Functional description of high speed heterodyne phase detection circuitry

lation factor M are related to the monoexponential fluorescence lifetime τ by the following relationships:

$$\tan F = \omega\tau$$

$$M = Mf/Me = 1/[1 + (\omega\tau)^2]^{1/2}$$

Measurements of fluorescent lifetimes by frequency modulation have been extensively employed for time-resolved fluorescence spectroscopy and require only one phase shift parameter to be measured [7, 8]. The use of the phase shift F for lifetime determination is advantageous because an electronic circuit can detect it as it occurs, which allows real time CFLIM. Real time CFLIM is made possible by the use of heterodyne detection, where high-frequency signals are transformed into low frequency signals which contain the same phase and modulation information as the original high-frequency signal [9]. The heterodyne detection circuit in the PMT is shown in Fig. 4. It is also possible, using high-speed phase shift detection circuitry, to measure lifetimes very rapidly at each pixel allowing real time CFLIM. If resolution of multicomponent lifetimes is required, data must then be obtained at several different modulation frequencies and more complicated image processing software using global data analysis techniques is required [10].

4
Two-Photon FLIM

While confocal fluorescence microscopy substantially improves the spatial detail in images of 3-D objects, it is still limited, especially when examining dynamic processes in living cells, due to phototoxicity and photobleaching of the fluorophore during observation. This effectively limits the excitation intensity that can be used and can result in low signal levels precluding observation of dynamic events in cells and tissues. These problems can be minimized with two-photon excitation. Two-photon excitation microscopy provides the advantages of greatly reduced photobleaching and photodamage, greater detection efficiency (due to the fact that the emission light does not have to be descanned), and deeper penetration into tissue specimens [11]. Two-photon (multiphoton) microscopy is made possible by the very high local instantaneous intensity provided by a combination of diffraction-limited focusing of a single laser beam in the microscope and the temporal concentration of 100 femtosecond pulses generated by a mode-locked laser. Excitation intensity varies as the square of the distance from the focal plane. Thus, the probability of two-photon absorption outside the focal region falls off with the fourth power of the distance along the optical axis (z direction). This effectively results in fluorophore excitation occurring only at the point of focus (where it is needed) eliminating unnecessary photobleaching and phototoxicity. The z axis resolution of two-photon FLIM is equal to that of a confocal microscope, but there is no out-of-focus photobleaching or photodamage as is the case with confocal microscopy, and no attenuation of the excitation beam by out-of-focus absorption. Deeper tissue penetration is made possible because: a) the excitation source is not attenuated by absorption by fluorescent molecules above the plane of focus; b) the longer excitation wavelengths used in two-photon excitation suffer less Raleigh scattering; and c) the fluorescence signal is not degraded as much by scattering from within the sample.

4.1
Two Photon FLIM Data Acquisition and Processing

Two-photon FLIM is a relatively new technology and has only been used in a limited number of situations. In two-photon FLIM, time-gated detection is employed. The emitted fluorescence resulting from a femtosecond excitation pulse is collected in two "windows", each of which is delayed relative to the excitation pulse. Unlike wide-field FLIM, where each gate window is detected individually, in two-photon FLIM, single photon counting is used to acquire the fluorescence intensity for every laser pulse in n windows (i.e. 2–8), sequentially, allowing detection of multiple photons per excitation pulse [12]. For monoexponential decay, the ratio of the windows is related to the lifetime in the following manner:

$$\tau = -\Delta t / \text{natural log } (I_b/I_a)$$

where Δt is the time between the beginning of the windows and I_a and I_b are the integrated fluorescence intensity in windows a and b, respectively.

5
Biomedical Applications of FLIM

FLIM is advantageous to use in the following situations. 1) In multiparameter imaging, spectral discrimination is essential for accurate identification of labeled structures. Spectral overlap of the absorption and emission of the different dyes used limit the number of spectral components that can be resolved. However, because fluorescent molecules with similar spectra often display distinct lifetimes, multiparameter imaging of cellular structures is possible [13]. Thus, the number of analytical parameters which can be examined is increased by FLIM even at the same wavelength [14]. 2) FLIM can discriminate autofluorescence from living cells from fluorophore fluorescence on the basis of distinct lifetime differences leading to increased contrast and sensitivity. 3) The fluorescence lifetimes of various probes are sensitive to numerous chemical and environmental factors such as pH, oxygen, and Ca^{2+}. Thus, FLIM can directly image the structural environment immediately surrounding the probe. 4) FLIM can also provide quantitative information regarding rotational and translational dynamics of cellular constituents and microviscosity in living cells through measurements of time-resolved emission anisotropy [15]. 5) FLIM can also be used to quantify the interaction, binding or association of two types of molecules using fluorescence resonance energy transfer (FRET) spectroscopy [16–18]. FRET FLIM is particularly useful in examining temporal and spatial changes in the distribution of fluorescent molecules in the living cells by monitoring changes in the donor lifetime in the presence and absence of the acceptor.

5.1
Multiparameter Imaging

Currently available technology for DNA sequencing [19–21] is based on the spectral emission characteristics of the individual dideoxynucleotides labeled with different fluorescent probes in the gel during or after electrophoretic separation. Detection of fluorescence requires having the requisite sensitivity to detect the intensity from each of the reporter molecules coupled to the dideoxynucleotides as well as sufficient spectral resolution to be able to distinguish A, C, T and G. In the current instruments used for DNA sequencing, the emission spectra of the four different dyes substantially overlap. This overlapping spectral interference results in uncertainty in the DNA sequence [22, 23]. Moreover, the speed of existing DNA sequencing instruments is slow because of limitations in the methods used to scan the gel. To overcome the constraints of the current instruments used for DNA sequencing (in terms of sensitivity, spectral resolution and speed), the combination of multiple capillary electrophoresis (CE) and two-dimensional (2-D) FLIM can be used for automated DNA sequencing (Fig. 5). Using FLIM, spatial patterns of different bands with similar spectra and which are physically overlapping can be analyzed. Such 2-D parallel fluorescence lifetime image detection without moving parts results in very high speed DNA sequencing (at >1 base/s per lane) because the 2-D image device "looks at" all the capillaries at all times.

5.2
Discrimination of Fluorophore vs Autofluorescence

A common problem affecting essentially all studies employing fluorescence microscopy, whether in living cells, fixed tissues, or clinical samples, and which serves to effectively decrease the signal-to-noise ratio of detection, is autofluorescence. Autofluorescence can originate from solvents, solutes, cell and tissues, fixatives and the optical components of the microscopic system. Autofluorescence typically has an intensity equivalent to 102–104 fluorescein molecules [24] and the lifetime of autofluorescence and other nonspecific fluores-

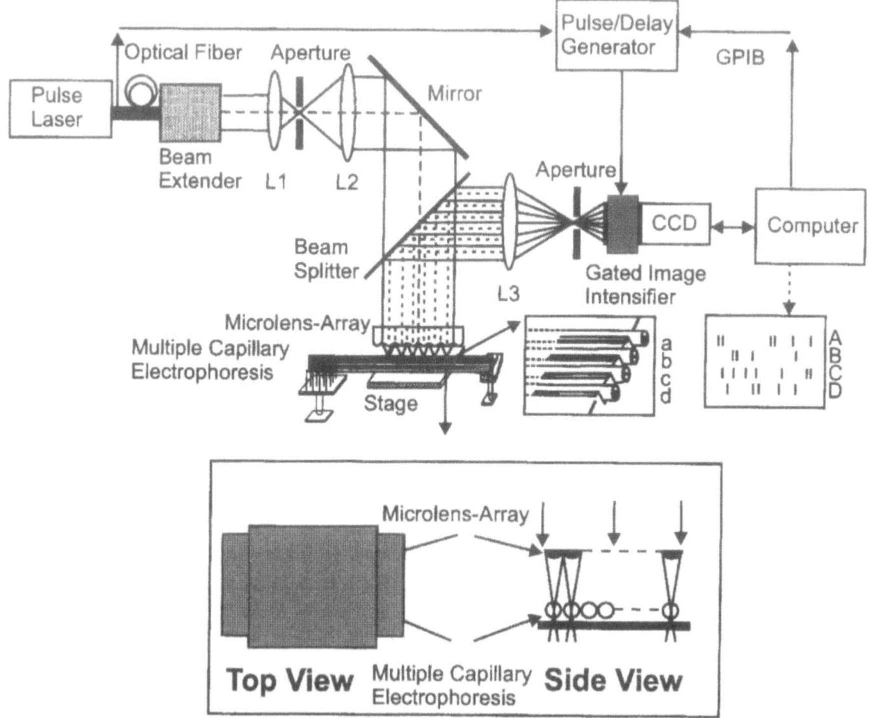

Fig. 5. Principle of DNA sequencing using FLIM and multiple capillary electrophoresis (CE) techniques

cence is in the range of 1–10 ns. One application of FLIM is to separate autofluorescence from fluorophore fluorescence based on a difference in their respective lifetimes [24]. In principle, if the target to be detected in a specimen could be labeled with a probe possessing a long lifetime (relative to the lifetime of the autofluorescence), following excitation with a short pulse of light, the lifetime of the fluorophore fluorescence could be separated from the lifetime of the autofluorescence. This would lead to a substantial increase in detection sensitivity (Fig. 6). In practice, fluorophore fluorescence emitted from the long lifetime probe is measured after a delay time when the background fluorescence (autofluorescence) has completely decayed. The ability to discriminate between fluorophore fluorescence and autofluorescence and other nonspecific fluorescence suppresses background by two orders of magnitude, providing sensitivity equivalent to radioimmunoassay [25].

Several long lifetime probes are commercially available. Historically, lanthanide chelate labels have been employed because of their long lifetime (several orders of magnitude longer than autofluorescence) and large Stokes shift. As an example, the phosphor yttriumoxisulfide activated with europium ($Y_2O_2S:Eu$) emits maximally at 620 nm and has a half-life of 700 µs, possesses

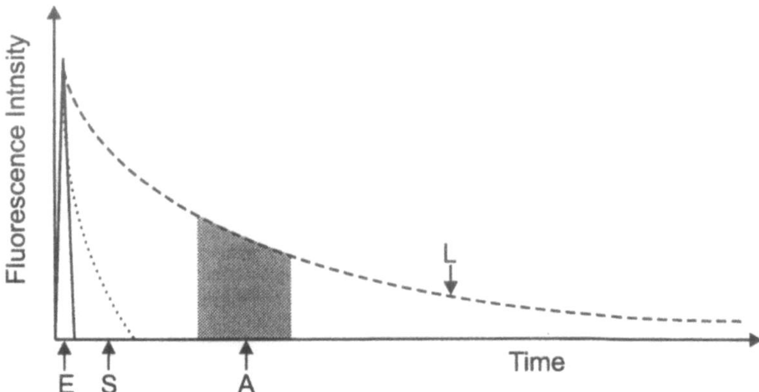

Fig. 6. Theory of long-lifetime FLIM

strong luminescence, is very photostable (relative to Fluorescein or Rhodamine), and is not significantly influenced by pH or temperature. Labeling is achieved by covalently binding antibodies to the surface of the probe. Other long lifetime probes include Zn_2MnAs and $ZnS:Ag$, which emit in the green and blue respectively, thus permitting simultaneous multiparameter measurements. Figure 7 illustrates the detection of specific fluorescence from a Eu^{3+} chelate derivatized fluorescent in situ hybridization (FISH) probe directed against human Papillomavirus (HPV) 16 DNA. Using conventional fluorescence imaging (A), both the signals from the HPV 16 DNA as well as autofluorescence and other nonspecific signals are observed. However, taking advantage of the long lifetime of the Eu^{3+} chelate and employing time-resolved FLIM, the autofluorescence and other nonspecific fluorescence has been removed resulting in a significant increase in the sensitivity of detection.

5.3
Ion Imaging

The role of various ions (e.g. Ca^{2+}, H^+, Na^+, Mg^{2+}, K^+, etc.) as important signaling molecules is well established. The ability to observe dynamic changes in these ions in individual cells has been made possible by the development of fluorophores whose absorption and/or emission are sensitive to changes in the levels of these ions [27, 28]. Ion sensitive probes currently are of two types – "ratiometric", where inverse changes in the intensity at two different wavelengths of excitation or emission are reflective of the ion concentration, and "non-ratiometric", where an increase or decrease in intensity is used to report ion levels in cells. Ratiometric probes have the advantage of being able to correct for differences in path length and probe concentration, but like their non-ratiometric counterparts are difficult to calibrate in situ [29, 30]. In addition to being diffi-

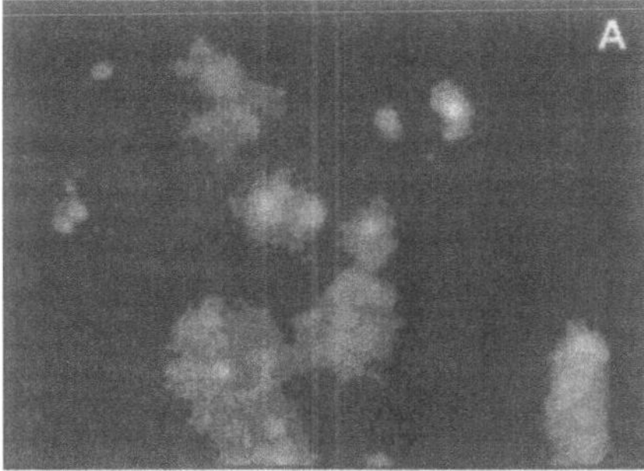

Fig. 7A, B. Fluorescence in situ hybridization (FISH) images for human Papillomavirus (HPV) DNA 16. (**A**) fluorescence image detected by conventional fluorescence detection method. The image includes the signal from HPV 16 DNA and, at the same time, other autofluorescence and nonspecific signals. (**B**) Fluorescence image detected by time-resolved FLIM using long lifetime probe (Eu^{3+} chelate). The detection sensitivity has been significantly improved by removing autofluorescence and other background

cult to calibrate, intensity based measurements of ion concentrations can be affected by other factors in the fluorophore's local environment, which may or may not lead to a measurable change in intensity.

However, fluorescent lifetime measurements of ion concentrations do not suffer from these same drawbacks. The principle of the use of fluorescent lifetime measurements for assessing ion concentrations is based on the fact that the lifetime of the ion-bound form of the fluorophore differs from the lifetime of the free form of the probe. Each ion-sensitive fluorophore will have two lifetimes, that of the ion-free form and the ion-bound form. As the concentration of the ion changes, the amplitudes (relative contribution of each lifetime to the observed lifetime), but not the measured lifetime of the ion- bound and ion-free forms of the fluorophore, changes. Ion concentrations then can be determined by changes in the respective amplitudes of the lifetimes of the ion-bound and ion-free forms of the probe.

For quantitative estimation of ion concentrations using lifetime measurements to be useful, the lifetime of the ion-bound and ion-free forms must remain constant independent of the probe environment. Studies have been recently published which provide data demonstrating this to be the case [31–33]. Sanders and colleagues have demonstrated that fluorescent lifetime pH calibration curves of carboxyl SNAFL-1 (Molecular Probes) using buffers of different pH, or in situ clamping of intracellular and extracellular pH using nigericin, gave identical calibration curves, whereas these same calibration curves obtained by ratio imaging were different. These investigators also demonstrated that Ca^{2+} levels can be determined in cells using Calcium Green (154719–40–1 Glycine, N-[2-[2-[2-[bis(carboxymethyl)amino]-5-[[2′,7′-dichloro-3′,6′-dihydroxy-3-oxospiro[isobenzofuran-1(3H),9′-[9H]xanthene)-5-yl]carbonyl]amino]-phenoxy]ethoxy]phenyl]-N-(carboxymethyl)-, hexapotassium salt) or Fluo-3 ($C_{36}H_{45}Cl_2N_7O_{13}$). Like carboxyl SNAFL-1 (151898–26–9 spiro[7H-benzo[c]xanthene-7,1′(3'*H*)-isobenzofuran]-ar′-carboxylic acid, 3,10-dihydroxy-3′-oxo), calibration of these probes was environment-insensitive. In another study, the effect of hydrophobicity, protein concentration and potential interaction of cellular constituents on the Ca^{2+}-sensing capabilities of Calcium Crimson $(C_{53}H_{49}K_4N_5O_{16}S_2)$, using both intensity-based and fluorescent lifetime analyses, was examined. As shown in Fig. 8, at constant Ca^{2+}, the hydrophobicity, and the concentration of protein or cellular constituents, all had a substantial affect on Calcium Crimson's accuracy in reporting Ca^{2+} based on measurements of the Calcium Crimson intensity. However, very little change in the lifetime of Calcium Crimson occurred as a function of protein or cellular constituent concentration or hydrophobicity. These findings indicate that the lifetime of Calcium Crimson is relatively insensitive to environmental factors, and suggests

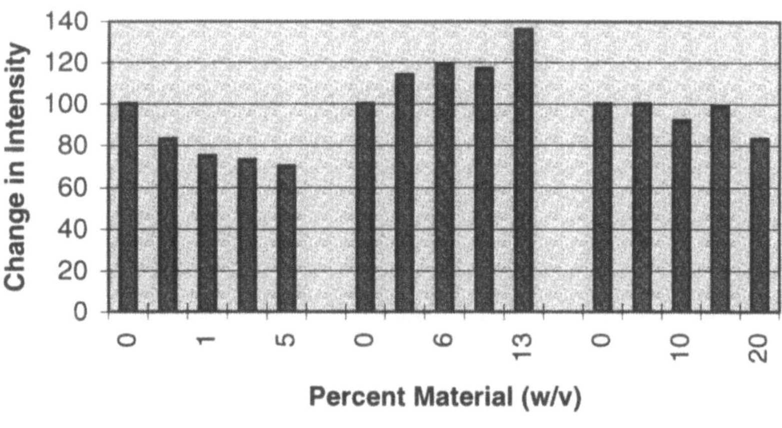

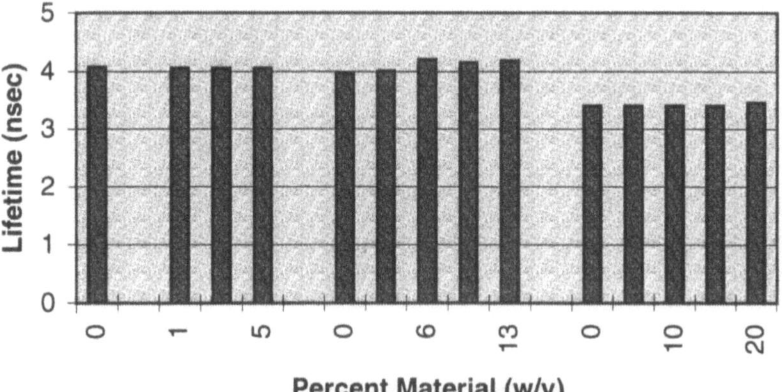

Fig. 8. Comparison of the effects of probe environment on Ca^{2+} quantitation using Calcium Crimson intensity (*top*) and lifetime (*bottom*) measurements. *Left set of bars* [0–5% bovine serum albumin (BSA)], *middle set of bars* [0–13% ethanol, (ETOH)], *right set of bars* (0–20% cell extract)

that calibration of Calcium Crimson in situ, like carboxyl SNAFL-1 (Molecular Probes) can be accomplished simply by the use of buffers. These results suggest straightforward calibration procedures in buffer might suffice for quantitating ion concentrations in situ when employing measurements of fluorescent lifetime.

5.4
FRET FLIM

FRET has been used as a spectroscopic ruler to examine the interactions between two different appropriately labeled fluorescent objects. FRET is a process by which a fluorophore (donor) in an excited state may transfer its excitation energy to a neighboring chromophore (acceptor) nonradiatively through dipole–dipole interactions. For FRET to occur, the donor molecules' fluorescence emission spectrum must overlap the absorbance spectrum of a fluorescent acceptor molecule. If the donor- and acceptor-labeled objects are close enough, then the energy of excitation will be transferred from the donor to the acceptor via a nonradiative dipole–dipole interaction. Because this energy transfer occurs via a dipole–dipole interaction, FRET decreases in proportion to the inverse sixth power of the distance separating the donor and acceptor fluorophores and is effective at measuring separation of the donor- and acceptor-labeled molecules when they are within 10–100 Å of each other [33, 34]. FLIM-based measurement of FRET allows the donor–acceptor distance to be more precisely measured. In addition, a FRET signal corresponding to a particular field (or pixels) within a microscope image provides additional information beyond the microscopic limit of resolution down to the molecular scale. The ability to measure interactions and distances of molecules provides the principal and unique benefits of FRET for microscopic imaging [35–37]. FRET imaging is particularly useful in examining *temporal* and *spatial* changes in the distribution of fluorescently conjugated biological molecules in living cells.

The rate of energy transfer (KT) or the energy transfer efficiency (ET) are both related to the lifetime of the donor in the presence or absence of the acceptor:

$$K_T = (1/\tau_D)\,(R_0/r)^6$$

when R_0 the critical Forster distance, the donor-acceptor separation for which the transfer rate equals the donor de-excitation rate in the absence of acceptor, τ_D the lifetime of the donor in the absence of the acceptor and r, the distance separating the donor and acceptor molecules E_T is related to r by:

$$r = R_0\,(1/E_T - 1)^6 \quad \text{where } ET = \tau_{DA}/\tau_D$$

where τ_{DA} is the lifetime of the donor in the presence of the acceptor. Thus, by measuring the lifetime of the donor with and without acceptor, one can calculate the distance between donor and acceptor.

The application of FLIM to the measurement of FRET is just beginning. Measurement of FRET using FLIM has been applied to studies of growth factor receptor dimerization, endosomal fusion, and the interaction of cellular proteins [38–41]. Studies of epidermal growth factor (EGF) receptors have been carried out using frequency-domain lifetime imaging microscopy and FRET [38]. EGF receptor clustering during signal transduction was monitored and a stereochemical model for EGF-dependent activation of the EGF receptor tyrosine kinase has been proposed. Time-resolved FRET imaging has also been applied to study the extent of membrane fusion of individual endosomes in single cells [39]. Using time-domain fluorescence lifetime imaging microscopy (FLIM) and FRET, the extent of fusion, and the number of fused and unfused endosomes, were clearly visualized and quantitated. FRET imaging has also been undertaken to determine whether the human Papillomavirus (HPV) 16 E6 protein and the tumor suppressor p53 protein are in close (≤ 50 Å) physical proximity to each other [40]. The results of these studies documented for the first time that HPV 16 E6 is physically associated with p53 in cervical carcinoma cells. Most recently, the interaction of two regulators of apoptosis (programmed cell death) has been investigated using FRET FLIM in combination with mutant Green Fluorescent Proteins (GFP) as donor and acceptor [41].

6
Other Applications of FLIM

In addition to these studies, there exist a number of potential applications of FLIM. FLIM can also be used for the quantitative imaging oxygen concentration of cells and tissue. Using phosphorescence quenching and FLIM, quantitative oxygen distribution has been measured [42]. By quantitatively detecting oxygen levels, localization of tumors and evaluation of their state of oxygenation has been determined [43]. The fluorescence lifetime of the electron carrier nicotinamide adenine dinucleotide (NADH) increases after binding to proteins, and, using FLIM, it should be possible to image the free and protein-bound NADH in cells [44]. As evidence accumulates that cellular membrane structure, composition and function are temporally heterogeneous, FLIM should provide a sensitive technique capable of obtaining data relative to both the dynamics and heterogeneous nature of membrane components related to plasma membrane fluidity, transportation and fusion [45, 46]. Another application is to use FLIM in combination with near-field fluorescence microscopy allowing visualization of molecular information beyond the diffraction limit. Since fluorescent molecules are limited at one detection point of the near-field, conventional CW excitation in one small spot generates substantial photobleaching and photodamage. The use of pulsed light excitation for FLIM will also have the added advantage of diminishing photobleaching and photodamage while still being able to obtain lifetime information. By scanning at each point, 2-D dynamic near-field lifetime imaging can be undertaken, which should result in enhanced image contrast.

Acknowledgements. The work described here was supported by grants from the American Cancer Society, American Heart Association, Gustavus and Louise Pfeiffer Foundation, National Institutes of Health, National Science Foundation, North Carolina Biotechnology Center, UNC Research Council and the Whitaker

Foundation. We also wish to gratefully acknowledge support from Hamamatsu Photonic Systems, Olympus Co., and ISS Inc. Much of the work described in this review would not be possible without the substantial contributions of many of my colleagues. While the list is long, particularly important contributions were made by David Albertini, Joseph Beechem, Richard Berlin, Jim DiGuiseppi, Pamela Diliberto, Salvatore Fernandez, Fred Fay, Kathryn Florine-Casteel, Hans Gerritsen, Charles R. Hackenbrock, Ken Jacobson, Dennis Koppel, John J. Lemasters, Steve Lockett, Fred Maxfield, Dave Piston, Deborah K. Smith, Ted Salmon, Ken Spring and D. Lansing Taylor.

References

1. J. N. Demas, Excited State Lifetime Measurements Academic Press, New York, (1983)
2. D. V. O'Connor, D. Phillips, Time Correlated Single Photon Counting Academic Press, New York, (1984)
3. Lakowicz, J. R., Principles of Fluorescence Spectroscopy, Plenum Press, New York, (1983)
4. Wang, X. F., Periasamy, A., Coleman D. M., Herman, B. Fluorescence lifetime imaging microscopy: instrumentation and applications. Critical Review in Analytical Chemistry. 23, 369–395 (1992)
5. Wang, X. F., Periasamy, A., Gordon, J., and Herman, B. Fluorescence lifetime imaging microscopy (FLIM) and its applications. SPIE Proc. on Time-Resolved Laser Spectroscopy in Biochemistry 2137, 64–77 (1994)
6. Periasamy, A., Wang, X. F., Wodnicki, P., Gordon, G., Kwon, S., Diliberto, P., Herman, B., High-speed fluorescence microscopy: lifetime imaging in the biomedical sciences, JMSA, 1, 13–23 (1995)
7. Lakowicz, J. A., Szmacinski, H., Nowaczyk, K., Johnson, M. L., Fluorescence lifetime imaging. Analytical Biochemistry. 202, 316–330 (1992)
8. Lakowicz, J. A., Szmacinski, H., Nowaczyk, K., Johnson, M. L., Fluorescence lifetime imaging of calcium using Quin-2. Cell Calcium. 13, 131–147 (1992)
9. Wang. X.F., Periasamy, A., Wodnicki, P., Gordon, G., Herman, B. Time-resolved fluorescence lifetime microscopy (FLIM): instrumentation and biomedical applications. in Fluorescence Spectroscopy and Microscopy., eds. Herman, B. Wang, X.F. John Wiley & Sons, Inc., New York, pp. 313–350, (1996)
10. Gadella, T. W. J., Clegg, R. M., Jovin, T. M., Fluorescence lifetime imaging microscopy: pixel-by-pixel analysis of phase-modulation data, Bioimaging, 2, 139–158 (1995)
11. Xu C. Zipfel W. Shear JB. Williams RM. Webb WW. Multiphoton fluorescence excitation: new spectral windows for biological nonlinear microscopy. Proc. Natl. Acad. Sci. 93, 10763–10768 (1996)
12. Sanders, R., Draaijer, A., Gerritsen, H.C., Houpt, P.M., Levine, Y.K. Quantitative pH imaging in cells using confocal fluorescence lifetime imaging. Anal. Biochem. 227, 302–308 (1995)
13. McGown, L. B., Fluorescence lifetime filtering, Anal. Chem. 61, 839 A (1989)
14. Wang, X. F., Uchida, T., Maeshima, M., Minami, S., Fluorescence pattern analysis based on time-resolved ratio method, Appl. Spectrosc. 45, 560 (1991)
15. Verkman-A-S. Armijo-M. Fushimi-K., Construction and evaluation of a frequency-domain epifluorescence microscope for lifetime and anisotropy decay measurements in subcellular domains, Biophys-Chem. 40(1). P 117–25 (1991)
16. T. Forster, Ann. Physik., 2, 55–75 (1948)
17. Stryer, L., Fluorescence energy transfer as a spectroscopic rules, Annu. Rev. Biochem. 47, 819 (1978)
18. Dewey, T. G. (Ed.) Biophysical and Biochemical Aspects of Fluorescence Spectroscopy, Plenum Press, New York, (1991)

19. Collins, F. D. Galas. A new five-year plan for the U.S. human genome project. Science. 262, 43–46 (1993)

20. Smith, L. M., J.Z. Sanders, R.J. Kaiser, P. Hughes, C. Dodd, C.R. Connell, L.E. Hood. Fluorescence detection in automated DNA sequence analysis. Nature. 321, 674–679 (1986)

21. Van Ranst, M., Fiten, P., G. Opdenakker. Comparison of three non-isotopic automated DNA sequence analysis systems. Clinical Chemistry. 38, 465 (1992)

22. Ansorge, W., Sproat, J., M. Zenke. Automated DNA sequencing: Ultrasensitive detection of fluorescent bands during electrophoresis. Nucleic Acids Res. 15, 4593–4603 (1987)

23. Chang, K. R.K. Force. Time-resolved laser-induced fluorescence study on dyes used in DNA sequencing. Appl. Spectroc. 47, 24–29 (1993)

24. Periasamy, A., Siadat-Pajouh, M., Wodnicki, Wang, X. F., Herman, B., Time-gated fluorescence microscopy for clinical imaging, Microscopy and Analysis, March, 33–35 (1995)

25. Mottram P, Soderback: Applications of lanthanide chelates for time-resolved fluoroimmunoassay. Am. Clin. Lab. 9, 5:34–38 (1990)

26. van Gijlswijk RP. Zijlmans HJ. Wiegant J. Bobrow MN. Erickson TJ. Adler KE, Tanke HJ. Raap AK. Fluorochrome-labeled tyramides: use in immunocytochemistry and fluorescence in situ hybridization. Journal of Histochemistry & Cytochemistry. 45(3):375–382 (1997)

27. Tsien, R. Y., Fluorescence imaging creates a window on the cell, Chem. Engineer. News, 72 (29), 34–44 (1994)

28. Tsien, R. Y., Fluorescence probes of cell signaling, Ann. Rev. Neurosci., 12, 227–253 (1992)

29. Diliberto, P. A., Wang, X. F., B. Herman. Confocal imaging of [Ca^{2+}] in living cells. in A Practical Guide to the Study of Ca^{2+} in Living Cells, Methods in Cell Biology, Academic Press, pp. 243–262 (1994)

30. Lakowicz, J. A., Szmacinski, H., Johnson, M. L., Calcium imaging using fluorescence lifetimes and long-wavelength probes, J. Fluorescence, 2, 47–62 (1992)

31. Sanders, R., Gerritsen, H.C., Draaijer, A., Houpt, P.M., Levine, Y.K. Fluorescent lifetime imaging of free calcium in living cells. Bioimag. 2, 131–138 (1994)

32. Herman, B., Wodnicki, P., Kwon, S., Periasamy, A., Gordon, G.W., Mahajan, N., Wang, X.F. Recent developments in monitoring calcium protein interactions in cells using fluorescence lifetime microscopy. J. Fluores. 7, 85–91 (1997)

33. Selvin, P.R. Fluorescence resonance energy transfer. Methods in Enzy. 246, 300–334 (1995)

34. Gadella, T.W.J.J., Jovin, T.M. Clegg, R.M. Fluorescence lifetime imaging microscopy (FLIM): spatial resolution of microstructures on the nanosecond time scale. Biophys. Chem. 48, 221–239 (1994)

35. Herman, B., Resonance energy transfer microscopy, Methods in Cell Biology Vol. 30, eds. D. L. Taylor, Y-L Wang, Academic Press, New York, (1989)

36. Jovin, T. M., Arndt-Jovin, D. FRET microscopy: digital imaging of fluorescence resonance energy transfer, in Cell Structure and Function by Microspectrofluorometry, Academic Press, San Diego, (1989)

37. Wu, P. Brand, L. Review: Resonance energy transfer: methods and applications. Anal. Biochem. 218, 1–13 (1994)

38. Gadella, T.W.J.J., Jovin, T.M. Clegg, R.M. Fluorescence lifetime imaging microscopy (FLIM): spatial resolution of microstructures on the nanosecond time scale. Biophys. Chem. 48, 221–239 (1994)

39. Oida, T., Sato, Y., Kusumi, A. Fluorescence lifetime imaging microscopy (flimscopy): methodology development and application to studies of endosome fusion in single cells. Biophy. J. 64, 676–685 (1993)

40. Liang, X. H., Volkmann, M., Klein, R., Herman, B., Lockett, S.J. Co-localization of the tumor suppresser protein p53 and human Papillomavirus E6 protein in human cervical carcinoma cell lines. Oncogene 8 2645–2652 (1993)

41. Mahajan, N. Herman, B. Alterations in the molecular interaction of Bcl-2 and Bax during apoptosis assessed using fluorescence resonance energy transfer (FRET) microscopy and green fluorescent protein (GFP)-Bax and blue fluorescent protein (BFP)-Bcl-2 expressing proteins. Micro. & Microanal. 3, 135–136 (1997)

42. Shonat, R. D., Wilson, D. F., Riva, C. E., Pawlowski, M., Oxygen distribution in the retinal and choroidal vessels of the cat as measured by a new phosphorescence imaging method, Applied Optics, 31, 3711–3718 (1992)
43. Wilson, D. F., Cerniglia, G. J., Localization of tumors and evaluation of their state of oxygenation by phosphorescence imaging, Cancer Research, 52, 3988–93 (1992)
44. Lakowicz, J. A., Szmacinski, H., Nowaczyk, K., Johnson, M. L., Fluorescence lifetime imaging of free and protein-bound NADH. Proc. Natl. Acad. Sci. USA. 89, 1271–1275 (1992)
45. Thevenin-B-J. Periasamy-N. Shohet-S-B. Verkman-A-S., Segmental dynamics of the cytoplasmic domain of erythrocyte band 3 determined by time-resolved fluorescence anisotropy: sensitivity to pH and ligand binding. Proc. Natl. Acad. Sci. USA. 91(5).1741–1745 (1994); Bicknese-S. Periasamy-N. Shohet-S-B. Verkman-A-S., Cytoplasmic viscosity near the cell plasma membrane: measurement by evanescent field frequency-domain microfluorometry. Biophys-J. 65, 1272–1282 (1994)

Injection Based Heterogeneous Fluorescence Immunoassays

J. N. Miller, M. Evans, M. T. French, D. A. Palmer

1
Introduction

Following the introduction almost 40 years ago of radioimmunoassays [1], research began quite soon into the use of other non-isotopic labels in similar assays. Since the emission of fluorescence from solutions and surfaces can be detected with high sensitivity, reasonable selectivity and perfect safety, it was not surprising that fluorescent labels, already familiar in cytology, were amongst the first such alternatives to be studied [2], and in the last thirty years fluorescence immunoassays have been widely researched and much used in clinical and environmental chemistry [3]. Interest in them is now greater than ever with the recognition that analogous methods can be used to study other receptor binding interactions, for example, in the high throughput screening of compounds for a specific pharmacological activity.

While several types of fluorescence immunoassay are available [3], the most common type relies upon the competitive binding principle shown in the following scheme:

$$X + X^* + Ab = X \cdot Ab + X^* \cdot Ab$$

Here X is the analyte whose concentration is to be determined, and Ab represents specific anti-X antibodies. The X^* molecules are derivatives of X (or occasionally analogues of X) containing a covalently linked fluorescent or fluorigenic label. The assay protocol is arranged so that the combined levels of X and X^* are in excess over the amount of available antibody. There is thus a *competition* between X and X^* for a limited number of antibody binding sites. As the concentration of X in the sample increases, the X^* molecules are displaced from such binding sites, and the levels of $X^* \cdot Ab$ fall. The calibration curve for the assay is thus a plot of the amounts of $X^* \cdot Ab$ formed (often at equilibrium, but kinetic assays are also feasible) against the concentration of X, and the analysis is essentially the determination of $X^* \cdot Ab$ levels. It is assumed here that antibodies (polyclonal or monoclonal) are the reagents but alternative receptors, including synthetic ones, are sometimes used.

All such fluorescence antibody/receptor assays can be classified into two main types. In many instances the emission properties (intensity, wavelengths, polarisation etc.) of a fluorescent label covalently linked to an antigen (or antibody) do not change significantly when the labelled molecule binds to a specific antibody (or antigen). As in radioimmunoassays a physical separation step is thus necessary before the extent of binding can be measured. Such methods are said to be *heterogeneous*. In other cases, however, the fluorescence emission properties of the fluorophore labelled molecule undergo sufficiently large changes on, for example, antigen-antibody binding for the latter interaction to be studied without a separation step being necessary. These *homogeneous* assays offer improved ease of use, especially when inexperienced staff are involved, and apparently simpler automation in the study of numerous samples than heterogeneous methods. Elegant homogeneous methods utilising the principles of fluorescence polarisation changes [4], the intensity changes accompanying resonance energy transfer [5], and steric effects on the hydrolysis of fluorigenic enzyme substrates [6], were all made commercially available some years ago.

These and other homogeneous assays nonetheless suffer from the crucial disadvantage that, because they include no separation step, all the endogenous sample components are still present when the final fluorescence measurement is made. Many sample matrices such as blood serum/plasma have a significant "autofluorescence" which (along with scattered light signals, which are particularly troublesome when fluorophores with small Stokes shifts are used) generate a background signal which adversely affects the limits of detection accessible [7]. These limits of detection are often several orders of magnitude worse than those of the same fluorophore in pure solution.

Users of fluorescence immunoassays are thus apparently faced with a choice between homogeneous methods which are simple to perform but may lack sensitivity in real applications, and heterogeneous ones which, because endogenous components are removed in a separation step, can achieve excellent sensitivities but are harder to perform and automate. Many approaches to the solution of this problem have been attempted. The most widely explored are (i) the use of long-wavelength (> 650 nm) fluorescence labels, since autofluorescence and scattered light signals in this region are much reduced and modern solid state devices provide convenient and low-cost instrument components [8]; and (ii) the use of lifetime discrimination, especially via the incorporation of enhanced lanthanide luminescence in the labeling technology [9], to discriminate between the (narrow bandwidth) millisecond lifetime label emission and the (broad) nanosecond autofluorescence background. In each case, however, the properties of the labels make assay development far from straightforward.

Just over 20 years ago the automatic analysis technique known as *flow injection analysis* was introduced [10]. Its underlying principles are the use of non-segmented liquid flow in conditions of controlled dispersion (sample spreading), but by using its intrinsic simplicity and flexibility it has proved to be adaptable to a great variety of assays of varying complexity. Work in this laboratory speedily showed that fluorescence detection could be readily adapted to it [11], and it was successfully applied in an automatic energy transfer immunoassay [12] and to other receptor-binding studies [13]. In recent years it has been shown that this method can also automate heterogeneous techniques so successfully as to provide a very realistic approach to precise and sensitive high throughput fluorescence immunoassays [14]. This review summarises its major benefits, characteristics and applications. With only a few exceptions it is based on a search of the literature from 1985 onwards, but in practice most of the published work dates from the 1990s.

2
Experimental Aspects

The basis of the heterogeneous fluorescence immunoassays described here is that after the antigen/hapten–antibody reaction has occurred, the antibody bound and unbound fractions are separated on some form of solid-phase reactor incorporated into the flow injection analysis manifold. In most cases the bound fraction is eluted from the reactor using a change of conditions (pH, ionic strength, etc.), and the reversal of this change regenerates the reactor prior to the injection of the next sample. Such immunoassays can take many forms, involving off-line, on-line and on-reactor incubation, a range of possible fluorescent labels, a variety of formats and chemistries for the immunoreactor itself, fluorescence measurements on the reactor-bound or (more rarely) unbound fractions, and so on. It is a tribute to the flexibility of the flow injection approach that it can readily accommodate these numerous variants, usually by means of quite simple and robust flow manifold designs.

In most of the methods reviewed here, samples of a few tens or hundreds of µL are injected into the flow system (Fig. 1) by a simple low-pressure rotary valve, and are carried through flow tubing of 0.5–1.0 mm internal diameter at volume flow rates of a few mL min^{-1} (i.e. linear flow rates of ca. 50–100 mm s^{-1}), the carrier stream being impelled by peristaltic or syringe pumps or sometimes by gas pressure. Solid-phase reactors incorporating immunologically active stationary phases perform the necessary separation, their volumes often being 0.5 mL or less: it was recognised in early work that these reactors do not have a seriously damaging effect on the overall dispersion in the flow injection system. (In one application developed in our laboratory a reactor containing ca. 45 µL of protein A bonded to controlled pore glass particles produced an overall sample dispersion of ca. 7 at a flow rate of 1 mL per minute. The same flow manifold used at the same flow rate but without the reactor produced a dispersion about half as great). Fluorescent-labeled fractions are usually detected in conventional or purpose built fluorescence spectrometers incorporating simple flow cells: in view of the significant dispersion of the samples before they reach the flow cell, the very low volume cells used in h.p.l.c. are generally not requir-

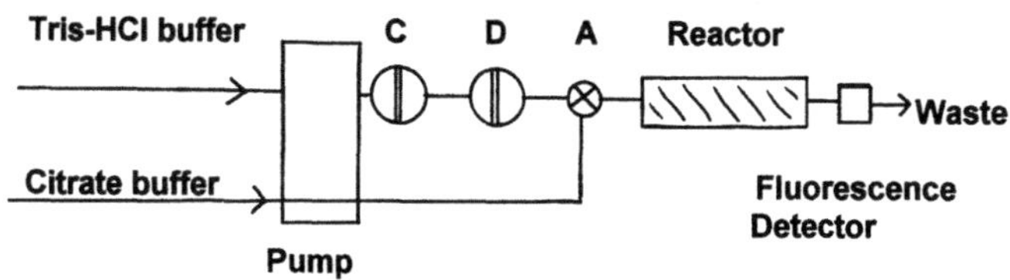

Fig. 1. Flow injection system

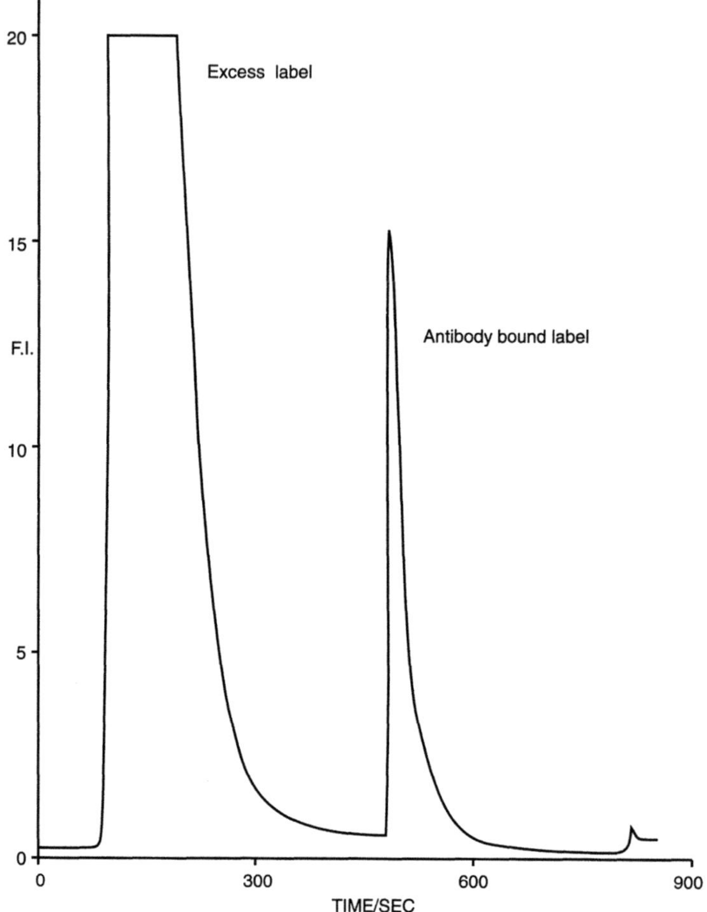

Fig. 2. Detector response of the flow injection system

ed – volumes of 20 µL or even more are quite adequate. The flow cell should be designed so that sudden changes in tubing diameter do not cause unnecessary increases in dispersion, and as always efficient illumination of the sample and efficient light collection optics maximise sensitivity. When a conventional solid-phase reactor is used (see below), the detector output shows two peaks for each injected sample, the first corresponding to the material not bound to the immunoreactor, the second being the bound peak eluted by a change in the reactor conditions. The height or area of the latter is normally measured (Fig. 2). Some applications of flow injection analysis with fluorescence detection of the solid (reactor) phase itself have also been reported: in principle this allows detection of the bound fluorescent molecules before their elution.

3
Fluorescent Labels

Since heterogeneous fluorescence immunoassays involve a separation step, it might be thought that there are no special requirements for a suitable fluorescent label beyond the conventional desire for high sensitivity and stability, and easy labeling of antigens or antibodies. In practice, however, there are further requirements relating to the change of elution conditions required to separate the bound labeled materials from the solid-phase reactor. In many cases this step involves separating the antigens/haptens from antibodies immobilised on the reactor, or separating immune complexes from a protein A or protein G reactor (see below). In each case this involves the use of low pH (2–3) eluting buffers, so fluorophores which are strongly quenched in acid solution are not ideal. It would of course be possible to restore the eluted fraction to near-neutral pH by means of an extra fluid flow delivering a correcting alkaline buffer downstream from the reactor, but this extra complexity is undesirable, especially as it is now possible to use label fluorophores which have a sensitivity comparable to that of fluorescein derivatives, but are more robust towards pH effects [15]. In practice, despite the known problems of a low emission intensity at low pH, and a reduced quantum yield after conjugation to analytes or antibodies, fluorescein and derivatives such as 5- and 6-carboxyfluorescein have been by far the most commonly used fluorophores. In some instances they have been used in methods where its susceptibility to pH changes is not a serious drawback, for example, in systems where the immunoreactor is constantly renewed or in liposome-based methods (see below). Rhodamine and related dyes, such as tetramethylrhodamine and Texas Red (a sulforhodamine 101 derivative), have also been used, as have a europium chelate, Lucifer Yellow (an aminophthalimide usually linked to proteins etc *via* a vinyl sulfone group), the methoxypyrenetrisulfonic acid derivative Cascade Blue, and fluorescent proteins such as phycoerythrin. A further variant utilises an enzyme such as β-galactosidase as the label, and a fluorigenic substrate (umbelliferone galactoside) as the detection reagent.

The use of fluorophore-laden liposomes as labels in heterogeneous immunoassays is an important technical advance which has been led by Durst and co-workers [16]. Since numerous fluorescent molecules can be incorporated into a single liposome (ca. 10^5 molecules in a liposome of ca. 135 nm diameter), an instant signal amplification effectively takes place when the antibody-bound antigen labeled liposomes are disrupted (by a detergent), rather than the kinetic amplification which results from the use of an enzyme and a fluorigenic substrate. Moreover, covalent linkage of the fluorophore to the analyte molecule is not necessary, and the susceptibility of the fluorophore to pH is not an issue: a wider range of fluorophores might thus be practicable. Finally, the detergent used as release agent might have the additional bonus of enhancing the emission of certain fluorophores. Some alternative indirect labeling methods utilising the [strept]avidin-biotin interaction have been studied in our laboratory. The best approach (though it does not involve a signal amplification if the analyte is a small molecule) utilises biotin as the immediate analyte label (biotinylation of numerous small molecules and macromolecules is simple) with fluorophore-labeled avidin as a universal marker: if this approach is used, with the antibodies in solution and a protein A or G reactor to bind them, then the same reactor and the same fluorescent label can be used for *any* single analyte [17].

As in all immunoassays it is necessary to minimise non-specific binding effects, such as the binding of labeled but antibody-unbound fractions to protein A or G columns, or the binding of labeled or unlabeled analyte molecules to non-specific sites on antibody based reactors.. The choice of fluorophore may be important in this context. Work in our laboratory suggests that such effects must be separately investigated for each reactor material and each fluorophore. There are even cases where (as would be hoped) the analyte does not bind to a protein A or G reactor on its own, the label alone also does not, but the labelled analyte (undesirably) does!

Long-wavelength and near-IR fluorophores have attracted much attention because of the low background signals at their emission wavelengths. In one sense this advantage is minimal in heterogeneous assays, as the intrinsic separation step will remove most of the autofluorescent interferences (this also applies to lanthanide based fluorescent labels which seek to minimise any background using lifetime discrimination). Such labels remain of interest, however, as their excitation wavelengths are well matched to widely available diode laser light sources (for example the cyanine dye Cy 5 is well suited to excitation by a 670 nm laser), thus offering the real possibility of compact, even portable, detectors for flow injection immunoassays.

Finally, it should be noted that dual or multiple assays are possible, i.e. determining two or more fluorophores simultaneously, thus detecting analytes using a number of non-cross-reacting immunoassays (see below). Such situations obviously make extra demands on the fluorophores used if they are to be readily distinguished by the detector. Purely spectroscopic approaches to such multi-component assays include the use of rapid scanning spectrometers to secure well resolved synchronous fluorescence spectra [18] from the different fluorophores, and the resolution of lanthanide based labels with different millisecond lifetimes [19]. Both these methods have also been suggested for other multi-component fluorescence applications such as nucleic acid sequencing. Chemometric methods for resolving overlapping spectra or lifetime decay curves also offer promising approaches.

4
Solid-Phase Reactors

A successful solid-phase reactor for use in a flow injection immunoassay must be based on a suitable base material with appropriate mechanical and flow properties, along with minimal non-specific adsorption effects, and a specific bonded phase which is also robust and can be used repeatedly. In practice much early work used agarose based stationary phases, but the increasing interest in high speed assays has encouraged the use of more rigid particles such as controlled pore glass and, more recently, perfusion chromatography media [20].

The latter, which have benefited many types of separation, allow much higher flow rates and thus faster assays. Other types of reactor stationary phase include magnetic particles which can be concentrated at one point by the use of a switchable electro-magnet then released after elution of the bound molecules [21].

The choice of the specific activity incorporated into the reactor depends on the immunoassay protocol desired. In many assays for a single analyte, immobilised antibodies have been used: coupling of antibodies to a range of solid phases is achieved by a range of established methods, and the antigen-antibody complexes are readily dissociated by a reduction in pH. Reversal of this pH change renews the antibody activity, a process that can be recycled many times, though experience suggests that antibody based reactors are not as stable in repeated use as those based on antibody-binding proteins. In a few cases, an antigen/hapten has been immobilised on the solid-phase reactor rather than the antibodies.

Much more common has been the use of protein A or protein G as the basis of the immunoreactors. These proteins bind antibodies at sites distinct from the antigen-binding sites, and are thus convenient for use in a wide range of assays. Protein A was originally isolated from *S. aureus*, but is now widely available as a recombinant product [22]. It has several domains which bind (human) IgG at the CH_2 and CH_3 domains of the latter antibody. Strong binding is also obtained with mouse, rabbit and pig antibodies, but the binding to the frequently used bovine, goat and sheep antibodies is unfortunately much weaker. Binding to the different IgG sub-classes of any one species is not equal: for example, binding to human IgG3 is much weaker than the binding to IgG1, IgG2 and IgG4. Protein A is remarkably stable towards changes in pH, a crucial advantage in heterogeneous immunoassays. In many assays the size of the reactor is chosen so that its antibody-binding capacity is several milligrams, while in practice the amount of antibody used in any single assay will be only micrograms. This excess capacity, coupled with the ruggedness of protein A itself, allows a single reactor to be used for thousands of assays before it needs replacement.

A number of authors have used reactors based on protein G rather than protein A. The former protein is isolated from the cell walls of group G *Streptococci* and binds to antibodies from a greater range of species than protein A. It is also available in wild and recombinant forms and is thus a promising alternative to protein A. The binding of some antibodies to protein G is so strong that dissociation of the antibody-protein G complex is difficult or slow even at very low pH. This may cause problems of sample carryover and/or shorter reactor lifetimes in immunoassays. Recombinant products which combine the binding specificities of proteins A and G are now also available. The proteins are also available labeled with a variety of fluorophores: these products may provide additional means of detecting antibody-antigen interactions when the fluorescence of the reactor itself is monitored.

Reactors based on proteins A or G can clearly be used for a range of assay types, and are especially valuable in multi-analyte assays where a general antibody-binding column is required. Incubation of sample analyte, labeled analyte and antibody can take place off-line, with the flow injection manifold used only to bind and measure the antibody fraction; the same reaction can take place on-line, i.e. with the sample and reagents injected together or in rapid succession, or it is possible to inject the antibodies first and allow them to bind to the protein A or G, with the immune reaction subsequently taking place on the surface of the solid-phase reactor.

Since, as noted above, the use of antibody-binding proteins involves the use of a pH change to elute the bound molecules, thus restricting the range of fluorophores that can easily be used with maximum sensitivity, other stationary phase chemistries have been studied. In particular the thiophilic gels or T-gels [23] show some promise. These affinity media contain a thioether group and a sulfone group in close proximity and efficiently bind antibodies from many species (and apparently very few other proteins) in conditions of high ionic strength (ca. 0.5 M) of a lyotropic salt such as potassium sulfate, elution involving merely a lowering of the ionic strength (to e.g. 0.1 M) without change of pH. Such reactors deserve further study, though non-specific binding may be a greater problem than with other columns.

Despite the feasibility of using protein A/G reactors for hundreds or thousands of samples without loss of binding capacity, studies in the laboratory of Ruzicka, one of the pioneers of flow injection analysis, have developed a system that presents a fresh reactor surface to each sample to be analysed [24]. The key component of the system is a "jet ring cell" in which suspended particles, for example, antibody-coated agarose beads, are retained by the flowing carrier stream against a plane optical surface in an annular space observed by the detector system. Subsequent injection of the sample, with the direction of the flow maintained, allows fluorescence changes in the cell to be detected using a fluorescence microscope or a fibre optic attachment to a conventional spectrometer. Reversal of the flowing stream removes the particles prior to the sequential injection of the next batch of beads and the next sample. Very good repeatability and no carry-over are advantages claimed for this arrangement, though a reasonable degree of uniformity of the particle size is apparently necessary. It is clearly not limited to fluorescence detection methods.

5
Applications

This section summarises a selection of recent applications of flow injection based heterogeneous fluorescence immunoassays, emphasising the range of possible analytes as well as the great variety of procedures available. Other critical features include the fluorophores and reactors used, any unusual or noteworthy experimental steps, and the analytical figures of merit that result.

The commonest assays use protein A (immobilised on a variety of stationary phases) in the crucial solid-phase reactor. In a typical example [25], theophylline was determined at the $\mu g\ ml^{-1}$ level using on-line incubation with antibodies and a fluorescein-theophylline conjugate. The protein A was immobilised on controlled-pore glass, and the column eluted with citrate buffer, pH 3.5. Fluorescein is heavily quenched at this pH, but its residual fluorescence was sufficient for this assay. An improved assay [26] used similar reagents, but with the important difference being that the protein A reactor was placed in the light path of the fluorescence detector, thus allowing the protein A bound fractions to be observed directly, prior to washing with acid buffer. This system offered a detection limit of 0.3 ng ml^{-1} at a sampling rate of 10–12 h^{-1}. The same research group described the determination of adrenocorticotropic hormone (ACTH) in the range 0.2–10 mg L^{-1} using Lucifer Yellow VS as the label [27]. This fluorophore is noticeably less fluorescent than fluorescein, but its resistance to pH changes is good. A similar system was used by Palmer et al. [28] to determine serum transferrin. Protein A immobilised on a perfusion chromatography medium has been used to determine testosterone, with the sulforhodamine dye Texas Red as the label, and serum phenytoin levels with Rhodamine as the label. In the latter case a detection limit of less than 1 ng ml^{-1} was obtained, with a total analysis time of only ca. 3.5 min [29]. The drug Wellferon has recently been determined in a similar way with the long-wavelength label Cy5 as the fluorophore. The perfusion columns were found to last for at least 5 000 assays and the fully automated method could be used at concentrations as low as 10 ng ml^{-1} (M. Evans, to be published). Cassidy et al [30], by using sequential addition immunoassay (the antibody, the test sample, then a known amount of the analyte antigen are added to the column in turn) avoided the use of a fluorescent label altogether. A sequential addition method has also been used with magnetic solid-phase particles [31].

Protein G as an alternative stationary phase was used in conjunction with reversed phase chromatography to determine serum transferrin with a precision (RSD) of 6% in the range 5–10^5 ng [32]. A similar dual column technique, using Texas Red as the label, was used to determine serum levels of human growth hormone [33]. A single protein G column sufficed to allow the detection of insulin at 50 pg ml^{-1}. In this case the elution pH was 2.5, the RSD 4%, and the lifetime of the agarose based protein G columns was again many hundreds of samples [34]. A recent report describes the determination of steroids in the sub-pM range using an enzyme label and fluorescence detection. This application was unusual in that the fraction of the immunoassay incubation mixture *not* bound to the protein G column was determined, a suitable fluorigenic substrate being injected via a merging stream downstream from the reactor [35].

The potential of thiophilic gels, previously used mostly for preparative work, was shown by Palmer and Miller [36], who demonstrated good results for the determination of albumin as a model protein and theophylline as a representative small molecule. Immunoassays in which the analyte was bound to the solid phase have been exemplified by the work of Gunaratna and Wilson [37], who immobilised the anti-cancer drug α-(difluoromethyl)ornithine and determined it at the attomole level in a non-competitive assay in which excess of the reagent, peroxidase-labeled monovalent antibody, was determined fluorimetrically. In a rather similar format Wortberg et al. used monoclonal antibodies directed towards triazine herbicides and labelled with a fluorescent Eu(III) complex. An affinity column containing the herbicides on oxirane activated acrylic beads was saturated with these labeled antibodies, and the fraction of the labeled molecules that were displaced by the sample analyte were measured [38].

The elegant approach involving the use of liposomes labeled with the analyte and loaded with fluorescent molecules has been used to determine a number of analytes. Both theophylline and its antibodies have been analysed, the results for serum theophylline levels, using immobilised anti-theophylline antibodies, correlating very well with a fluorescence polarisation immunoassay [39]. Application of this method to environmental problems has been demonstrated by the determination of the herbicides alachlor and imazethapyr. In the latter case good results were obtained in the analysis of spiked tap and pond water samples at concentrations of 1–100 ng ml^{-1} [40].

Two final examples show that this area of study is still developing rapidly and offers many promising advantages. A recent paper has shown some early applications of the renewable immunoactive surface approach of Ruzicka's group. The rate constants for an albumin reaction with anti-albumin antibodies compared favorably with the results obtained using the established surface plasmon resonance technique, but the renewable surface technique was simpler, more widely applicable, and less prone to interferences [41]. In the authors' laboratory, two analytes (albumin and transferrin) have recently been determined simultaneously with the aid of synchronous fluorescence spectra of the labels Cascade Blue and fluorescein, respectively, and the work is being extended, with success, to triple assays.

References

1. Yalow RS, Berson SA (1959) Nature 184:1648
2. Tengerdy RP (1963) Anal Chem 35:1084
3. Hemilla IA (199 1) Applications of Fluorescence in Immunoassays. John Wiley, New York
4. Dandliker WB, Feigen GA (1961) Biochem Biophys Res Commun 5:299
5. Ullman EF, Schwarzberg M, Rubenstein KE (1976) J Biol Chem 251:4172
6. Burd JF, Carrico RJ, Fetter, MC, Buckler RT, Johnson RD, Boguslaski RC, Christner JE (1977) Anal Biochem 77:56
7. Soini E, Hernmila IA (1979) Clin Chem 25:353
8. Miller JN, Brown MB, Seare NJ, Summerfield S (1993) In: Fluorescence Spectroscopy Wolfbeis OS (ed), New Methods and Applications, p 189
9. Leif RC, Thomas RA, Yopp TA, Watson BD, Guarino VR, Hindman DFK Lefkove N, Vallarino M (1977) Clin Chem 23:1492
10. Ruzicka J, Hansen EH (1988) Flow Injection Analysis, 2nd edn. John Wiley, New York
11. Braithwaite JI, Miller JN (1979) Anal Chim Acta 106:395
12. Lim CS, Miller JN, Bridges JW (1980) Anal Chim Acta 114:183
13. Miller JN, Abdullahi GL, Sturley HN, Gossain V, McCluskey PL (1986) Anal Chim Acta 179:81
14. Miller JN (1989) Anal Proc 26:317
15. Hauglund RP (1996) Handbook of Fluorescence Probes. Molecular Probes Inc, Eugene, Orgeon
16. Locascio-Brown L, Plant AL, Durst RA (1988) Anal Chem 60:792
17. Khokhar MY (1993) PhD Dissertation, Loughborough University
18. Stevenson CL, Johnson RW, Vo-Dinh T (1994) Biotechniques 16:1104
19. Xu Y-Y, Pettersson K, Blomberg K, Hemmila 1, Mikola H, Lovgren T (1992) Clin Chem 38:2038
20. Afeyan NB, Gordon NF, Mazsaroff I, Varady L, Fulton SP, Yang YB, Regnier FE (1 1990) J Chromatogr 519:1

21. Forrest GC (1977) Ann Clin Biochem 14:1
22. Palmer DA, French MT, Miller JN (1994) Analyst 119:2769
23. Porath J, Maisano F, Belew M (1985) FEBS Lett 185:306
24. Ruzicka J, Pollenia CH, Scudder KM (1993) Anal Chem 65:3566
25. Fernandez-Hernando P, Miller JN (1991) J Pharm Biomed Anal 9:1121
26. Rico CM, Fernandez MP, Gutierrez AM, Concepcion PC, Camara C (1995) Analyst 120:2589
27. Martinesteban A, Fernandez P, Perezconde C, Gutierrez AM, Camara C (1996) An Quim 92:37
28. Palmer DA, Xuezhen R, Fernandez-Hernando P, Miller JN (1993) Anal Lett 26:2543
29. Palmer DA, Evans M, Miller JN, French MT (1994) Analyst 119:943
30. Cassidy SA, Janis LJ, Regnier FE (1992) Anal Chem 64:1973
31. Pollema CK, Ruzicka J, Christian GD, Lernmark A (1992) Anal Chem 64:1356
32. Janis LJ, Regnier FE (1989) Anal Chem 61:1901
33. Riggin A, Regnier FE, Sportsman JR (1991) Anal Chim Acta 249:185
34. Khokhar MY, Miller JN, Seare NJ (1994) Anal Chim Acta 290:154
35. Kronkvist K, Lovgren U, Svenson J, Edholm LE, Johansson G (1997) J Immunol Methods 200:145
36. Palmer DA, Miller JN (1995) Anal Chim Acta 303:223
37. Gunaratna PC, Wilson GS (1993 3) Anal Chem 65:1152
38. Wortberg M, Middendork C, Katerkamp A, Rump T, Krause J, Cammann K (1994) Anal Chim Acta 289:177
39. Locascio-Brown L, Plant AL, Chesler R, Kroll M, Ruddel M, Durst RA (1993) Clin Chem 39:386
40. Lee M, Durst RA (1996) J Agric Food Chem 44:4032
41. Willumsen B, Christian GD, Ruzicka J (1997) Anal Chem 69:3482

CHAPTER 22

Microfluorometry of Cellular and Subcellular Processing in CNS Cells

W. Müller, S. Schuchmann, A. V. Egorov, T. Gloveli, K. Bittner

1
Introduction

Understanding the behavior of biological cells is a primary target in biomedical research aimed at the development of new therapeutic strategies against a variety of diseases. It has become clear with traditional biochemical and physiological methods that the processing and storage of information, i.e. the behavior of biological cells, can be explained only by models hypothesizing a highly organized subcellular organization of molecular signaling. In order to understand this complex cellular organization correctly, it is of great importance to study it in a state as undisturbed as possible. Fluorescence techniques have turned out to be invaluable tools in these studies because of minimal external interference together with a high temporal and spatial resolution that has allowed the study of many important subcellular signaling events. It has been demonstrated, particularly in nervous system cells, that the performance of information processing and storage relies heavily on functional cellular compartmentation. Individual cortical pyramidal neurons communicate through an estimated 10000 to 30000 synaptic connections on their dendrites respective axon collateral terminals.

Important functions of the central nervous system (CNS) are the analysis of sensory information with respect to a wide variety of aspects as well as storage and retrieval of information including associative reasoning. These functions are achieved by massive parallel information processing and storage by the synaptic cell–cell contacts in a layered network of neurons. Through adequate synaptic connections, certain neurons become active by specific characteristics of sensory input. This means that these neurons fire action potentials that propagate along axons to the next layer of synaptic connections. In addition, electrochemical coupling controls the metabolic state of a cell and can initiate long lasting changes in specific synaptic connections. Such processes, e.g. long-term-potentiation (LTP) of synaptic connections in specific brain areas, are dependent on intracellular changes of free Ca^{2+} and other molecules in neurons and, presumably, in glial cells, too. It has been infered that synaptic changes underly learning and memory by favoring certain spatio-temporal activity patterns in a neuronal-synaptic network. The probability of "remembrance" of such an activity state, i.e. transient rebirth of this state, inherently depends on the similarity of this state to pre-remembrance activity caused by specific sensory input. These features may also explain, at least in part, associative reasoning. Disturbance of the basic signaling mechanisms up to cell death may underlay problems of memory and intelligence in diseases like Alzheimer's and Down's syndrome. A better understanding of information processing in the CNS has contributed to progress in artificial intelligence and may allow novel therapeutic interventions against neurological and psychiatric disorders.

In this paper we will give a brief overview of approaches utilizing fluorescent techniques. We will recapitulate some of the results with respect to cell function in development, plasticity as well as in pathophysiological conditions with particular reference to studies that have been of interest to our laboratory.

Targets: In the CNS one of the natural targets to choose first were electrical signals that have long been recognized as substantial signaling events in this organ in respect to its cells. Since this topic will be presented elsewhere in this book, it is not discussed in this chapter. The next event in cellular signal transmission depends on increases in intracellular free Ca^{2+} that is an ubiquitous second messenger in synaptic release of transmitters and in gating of Ca^{2+}-dependent ion channels, enzymes, transcription factors and muscle contraction. Because of this high importance of Ca^{2+}, the relatively long lifetime of Ca^{2+}-signals and the technical ease of monitoring it inside cells successfully, the number of laboratories and publications addressing cellular Ca^{2+}-signals has grown enormously in recent years. Besides interest in Na^+-, Cl^-- or K^+-changes inside cells, monitoring of enzyme activities, e.g. protein kinase A, protein synthesis, intracellular routing of proteins, release of neurotransmitters, endocytosis, oxidative state, glutathione and free fatty acids all appear to be important processes in physiological and pathophysiological cellular functions and hence are attractive targets in biomedical research.

2
Basic Considerations

For application of fluorescence techniques to the study of cellular signaling the following basic points have been found useful to consider. First, the temporal and spatial resolution dt and dx^*dy^*dz at a defined signal level are determined by the relationship:

$$dt^*dy^*dx^*dz = 1/k$$

where the constant k is proportional to the intensity of emitted light, i.e. to the product of excitation intensity I_λ, indicator concentration C, quantum efficiency Q and absorption coefficient E_λ. This means that doubling spatial resolution in 2 or 3 dimensions requires a four- or eight fold increase in photons collected, either by an increase in the integration time, i.e. temporal resolution dt, or by increasing I_λ or C. Figure 1 demonstrates the basic configuration for microfluorometry and imaging (Fig. 1 A). The emitted light is imaged onto the detector and the signal is either spatially subdivided by the pixels of a CCD chip or spatially integrated by a photodiode. In the confocal setup, an individual "point" is illuminated at a time and only confocal light passes the emission pinhole to be measured by the detector (Fig. 1 B). In this case, the spatial resolution

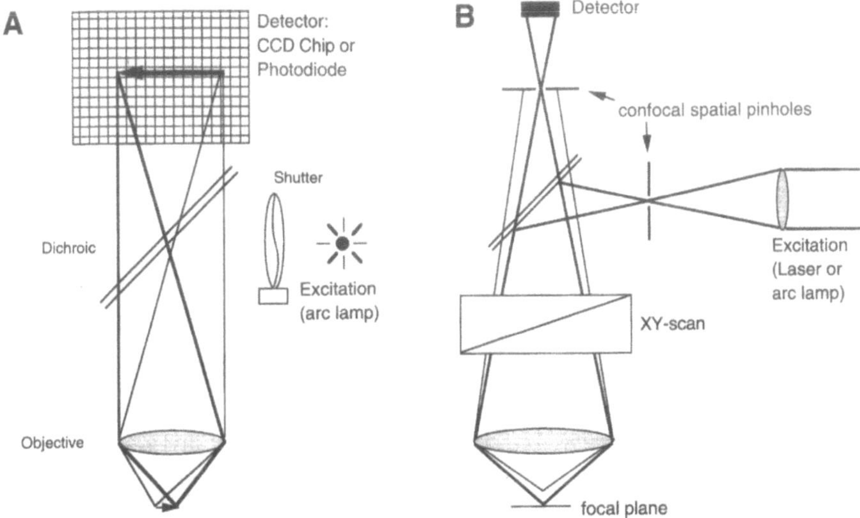

Fig. 1. Scheme of fluorometry using either an epifluorescence microscope with a charge-coupled device (CCD) or a photodiode as a detector (A) or a confocal epifluorescence microscope with "point" excitation and an xy-scan device (B). In standard imaging or fluorometry, the excitation light is reflected by a dichroic mirror and focused by the objective onto the probe. The longer wavelength emission is imaged onto a 2D array of photodetectors (CCD) or collected integrally by a single photodiode (A). In confocal microscopy, spatial pinholes reduce unwanted excitation outside the point under study and reject nonfocal emissions in the detection pathway. In this way lateral resolution as well as optical sectioning in the z-direction is improved. A 2D image is obtained by sequential scanning of an area through light deflection with galvanic mirrors or acousto-optical devices (XY-scan). When two-photon excitation is used, the signal can be improved by removing the pinhole in front of the detector without degrading spatial resolution. This is due to the well confined spot of excitation (dependence of excitation on the square of the intensity) and, hence, origin of scattered emitted photons

(or "point" size) is defined by the size of the pinholes. I_λ can be effectively increased using laser illumination, e.g. in confocal microscopy. This is at the cost of increased photodamage, a process seriously limiting the work with living samples. The downside of an increase in the indicator concentration is an increased disturbance of the system under study, e.g. by enhanced buffering and diffusion of the target molecule. This point is of particular importance when the dye concentration is much higher than the target concentration, as is usually the case with targets with rather low concentration like $[Ca^{2+}]_i$ (see below). Other points to be considered are indicator characteristics like a spectroscopic shift with binding of the target molecule that allows quantitative rather than qualitative measurements and fluorescence changes due to interaction with cellular proteins or viscosity (Poenie 1990; Stehno, Perez et al. 1995). The problems associated with this latter point may be overcome, at least in part, by exploiting fluorescence lifetime changes, as discussed elsewhere in this book (Herman).

2.1
Excitation Light Sources

With respect to the basic measurement requirements, equipment specifications are important to consider. With respect to the excitation light source, high resolution requires high intensity. Mercury and xenon arc lamps are inexpensive sources offering a wide spectrum of wavelengths to choose from. To achieve a high intensity at the probe under study, the arc spot is imaged by the microscope objective onto the probe, i.e. the power density but not the total power of the arc at the respective wavelength is crucial. The relative power density can be evaluated roughly as the product of total power and the relative intensity at the wavelength needed divided by the xy area of the arc. This makes the HBO 100 and XBO 75 arc lamps with a small arc size the choice over stronger models. The strong mercury lines of the HBO lamp can eventually be exploited for extra power but may also be hard to suppress when not wanted because of disturbance of the measurement. When more power is needed, the light intensity from a mercury short arc light bulb can be boosted up to about 100 times its steady-state value for a period of about 1 ms (Denk 1997). Wavelengths are usually selected by filter wheels with dichroic bandpass filters, a split light path with two filters or two monochromators or a scanning monochromator. The latter devices allow wavelength switching in the low ms range.

Alternatively, lasers may be used for very high intensity excitation. Drawbacks, in addition to photodamage, are a very limited selection of wavelengths and the cost. Because of these points lasers are used primarily in confocal microscopy where energy density of arc lamps is insufficient.

2.2
Detectors

With respect to photon detection, low fluorescence light levels from small areas make efficient photon detection a must. Therefore, quantum efficiency of the detector at the emitted wavelengths is an important parameter. Front illuminated charge coupled devices (CCDs) are highly efficient detectors in the infrared but exhibit quantum efficiencies of only about 20% at 500 nm and even below at shorter wavelengths. Therefore, silicon intensified (SIT) analog video cameras are frequently used. This technology is quite affordable due to the use of mass products and in principle allows imaging at video rate (30 or 25 frames/second for the US and Europe, respectively). For quantitative analysis of data, images are usually transferred to a computer via a video frame grabber card. In low light applications, the required high amplification of light intensity results in significant noise problems. Averaging of a number of subsequent frames reduces noise at the cost of speed. Due to the amplification by the intensifier, the dynamic bandwidth of 8 bits is usually reduced down to 4 or 5 bits, thereby limiting the range of brightness levels across a cell that can be measured simultaneously. Analog data can be stored at low cost on a video recorder.

Digital CCD cameras have been used for long time in astronomy and work by integrating the photons on the CCD chip for an arbitrary time prior to read out. This allows a more effective improvement in the signal to noise ratio simply by prolonging the exposure time. To suppress thermal noise the CCD chip is cooled thermoelectrically by 20° to 200 °C below ambient temperature. In addition, slow read out keeps extra noise to a minimum. Cameras are available with dynamic bandwidths varying from 8 to 16 bits. We favor 12 bit systems that in our hands are sufficient for simultaneous recording from bright neuronal somata and fine and therefore dim spines and distal dendrites without problems of detector saturation or insufficient signal from the dim parts of neurons. The temporal characteristics of many biological problems that have been addressed in recent years have led to CCD cameras offering both analog and digital modes, as well as to a gradual increase in digital read out rates up to 4 MHz. Many models allow in addition an increase in speed by i) binning pixels (at the cost of spatial resolution), ii) reading out of subarrays or iii) frame transfer, i.e. a fast shift of the image into a storage part of the chip to be read out during the exposure of the following image. With these techniques, imaging rates for subarrays up to 100 Hz have been achieved, under the condition, of course, of light levels allowing sufficiently short exposure times.

The low quantum efficiency of CCDs in the visible light region is largely due to the gates on the frontside of the chip, hindering the transmittance of photons. To overcome this problem, CCDs have been thinned from the backside by etching and are used in illumination from the backside. Using this technique, a quantum efficiency of some 80 % at 500 nm, i.e. a fourfold signal compared to front illuminated CCD chips, has been achieved. This either results in significantly improved signal over noise or allows the integration or exposure time to be cut. While reduced exposure is beneficial in avoiding photodamage, a significant increase in temporal resolution is achieved only when in frame transfer mode the read out time from the storage section of the chip is sufficiently brief, i.e. not much longer than the exposure time. This depends on the number of pixels read out in relation to the light level. A rather long shift-time of back illuminated CCD chips needs to be taken into account in such considerations.

Another solution to the problem of low transmittance of photons through the gates is by on chip lenses that guide the photons around the gates. High quantum efficiency together with in line parallel read out has made possible full frame image rates of up to 300 Hz, although until now image quality appears to be still significantly inferior.

Confocal microscopy has gained a considerable share of the market because it achieves up to approximately twofold better lateral resolution and better sectioning in the longitudinal axis, compared with conventional light microscopy (Fig. 1 B). The 2D image is obtained by scanning a corresponding area through light deflection with mirrors or acousto-optical devices. High intensity illumination, using lasers, and a detector with high quantum efficiency allow rapid acquisition of 2D images. Recent systems can work at 10 to 20 Hz for subarrays and some at considerably higher rates (e.g. 240 Hz). When a 1D spatial profile is sufficient, line scan modes achieve even higher time resolution.

By avoiding problems of optical alignment of two detectors, the confocal principle is, by the way, inherently well suited to collect emissions at two wavelengths, separated by a dichroic mirror behind the emission pinhole, simultaneously. This can be used with dyes like Indo-1 for ratiometric measurements. Ratiometric confocal measurements may also be achieved by using a combination of two dyes, one with an increase and one with a decrease in emission with binding to molecules of the same species. Due to the lasers being available at low cost, a dye combination of Fluo-3 and Fura Red has been used for Ca^{2+}-measurements. To exclude artifacts due to a separation of the two dyes, coupled variants (chemical complexes of two dyes) have been made available.

A serious problem, caused by the high photon density, is photodamage, that often limits the useful recording time to a few minutes. In this respect, the possibility of two-photon excitation needs to be mentioned. Fluorescent dyes can also be excited by two photons with half the energy when these two photons are sufficiently coincident in time and space. With two-photon excitation at the doubled wavelength, the phototoxicity problem appears to be drastically reduced. Another important point in two-photon excitation is that sectioning in the longitudinal axis is considerably improved by the dependence of excitation on the square of the intensity of the excitation light. This allows an improvement in the signal by removing the pinhole in front of the detector (cf. Fig. 1 B) without degrading spatial resolution, because scattered emitted photons originate exclusively from the well confined spot of excitation. In addition, infrared excitation results in better penetration of the excitation light into the tissue, extending the spatial range of recording in brain slice as well as in vivo brain (Denk 1994; Yuste and Denk 1995; Svoboda, Denk et al. 1997). Drawbacks of confocal microscopy, besides photodamage and the limited selection of excitation wavelengths, are its very high cost, especially when dyes need to be excited in the UV range or when two-photon excitation is used.

Spatial resolution comparable to that of confocal laser scan microscopy at considerably lower cost can be achieved with conventional digital imaging in combination with numerical image deconvolution ("deconvolution confocal microscopy"). This method determines the point spread function of the imaging system to remove the imperfect part of the optical projection from the image (Fay, Carrington et al. 1989; Isenberg, Etter et al. 1996). Because the algorithms used are calculation intensive, deconvolution takes in the order of a minute per image with a workstation computer. Therefore, this technique is currently better suited for off-line analysis of data or imaging of rather slow signals.

When, compared with confocal microscopy, even thinner layers need to be resolved, the evanescent wave formed by a laser beam experiencing total reflection at the glass cytosol interface has been successfully used for excitation. With this technique, transport, docking and secretion of transmitter peptides fused to green fluorescent protein have been directly visualized in living chromaffin cells (Lang, Wacker et al. 1997; Steyer, Horstmann et al. 1997).

3
Calcium Measurements

For conventional microfluorometry and microfluorometric imaging, Fura-2 has become the most popular dye because of these advantages:

1) high fluorescence efficiency
2) good photostability
3) Ca^{2+}-binding constant well suited for intracellular Ca^{2+}-levels
4) good selectivity for Ca^{2+} over disturbing ions (Mg^{2+}, Na^+, H^+)
5) a strong shift of the excitation spectrum with binding of Ca^{2+} allows ratiometric measurements with high resolution
6) easy loading of cells by membrane permeable esters or by iontophoretic injection through a microelectrode or by diffusion in whole cell patch clamp recording
7) a good separation of excitation and emission wavelengths.

Drawbacks that are sought to be overcome or by designing new dyes are:

1) significant buffering of Ca^{2+}-increases
2) Ca^{2+}-gradients are suppressed because of enhanced diffusion
3) not suited to monitoring very high Ca^{2+}-concentrations ($>2\ \mu M$)
4) excitation in the UV region requires special optics and can be harmful to the preparation
5) excitation by UV-lasers is extremely expensive.

Problems 1 to 2 reflect the situation that the free Ca^{2+}-concentration is rather low in relation to the indicator concentration that is required to obtain a sufficient signal. Together with the high affinity of Fura-2 for Ca^{2+} and the amplitude of Ca^{2+}-increases, this results in large amounts of Ca^{2+} bound to the mobile indicator. Accordingly, lower affinity dye variants and coupling to slow dextrans can be used to reduce buffering and enhanced diffusion. Using the dye as a chelator competing with endogenous Ca^{2+} buffers, endogenous buffers and their properties can be evaluated and absolute Ca^{2+}-fluxes estimated by the fluorescence signal at a single Ca^{2+}-dependent excitation wavelength (Neher 1995).

The necessity to excite in the UV region considerably limits the selection of optics. UV excitation is also a considerable handicap in confocal microscopy because it requires an expensive UV laser or even two for excitation ratiometric measurements. When caged compounds that are activated by UV light are used to manipulate the cells, a dye with an excitation outside the band of absorbance of the caged compound has to be used.

Because of the unparalleled advantages of Fura-2, strong demand by physiologists has led to new technical developments, in particular long distance water immersion optics with good near-UV transmission. These long distance water immersion objectives are important for the effective use of the upright microscope configuration (cf. Müller and Connor 1993). For in vivo brain imaging there is no alternative to the upright configuration and for slice work it is preferred because good resolution requires selection of cells close to the surface of the slice facing the objective and selective loading with dye through a microelectrode. This selective loading is difficult to achieve in an inverted configuration where the microelectrode would have to be passed through the slice without much visual control (Regehr and Tank 1990). The long distance design is important for the manipulation with microelectrodes although it limits the numerical aperture of the objectives.

Imaging of cell cultures in an inverted microscope setup leaves more space for micromanipulation and allows the use of high numerical aperture objectives. Although the upright microscope does not offer these benefits, it can be used not only for slices but also for most experiments on dissociated cultures. One advantage is that problems with additional optical elements like plastic culture dishes are nonexistent. Dissociated cultures are easy to load by incubation with the membrane permeable acetoxymethyl ester of Fura-2 and allow simultaneous recording from many cells, facilitating the study of cell to cell variability and of responses that occur only in a small fraction of cells (Müller and Swandulla 1995; Erdmann, Müller et al. 1996). In a study on the generation of rhythmic activity in hypothalamic synaptic networks in culture, this allowed us to screen large numbers of cells for primary bursters and to study the synchronous burst activity of these cells under various experimental conditions, e.g. reduced burst frequency during reduction of synaptic excitation by a glutamate receptor antagonist (Müller and Swandulla 1995).

Because indicator concentrations largely exceed the concentration of elevated free Ca^{2+}, e.g. 100 μM as opposed to 1 μM, diffusion of Ca^{2+} along a gradient is massively potentiated by binding to the indicator (Müller and Connor 1992). This situation makes it difficult to demonstrate transient gradients in small volumes like spines, as demonstrated in Fig. 2 (Müller and Connor 1992; Eilers, Augustine et al. 1995; Yuste and Denk 1995; Egorov and Müller 1997). Since very high peak concentrations of Ca^{2+}-signals are usually extremely localized and decay very rapidly, high speed recording from small volumes and hence a good signal over background is again needed. Demonstration of strong Ca^{2+}-gradients, despite enhanced breakdown of Ca^{2+}-gradients due to facilitated diffusion, has been addressed either by low affinity dyes and/or high speed imaging (Llinas, Sugimori et al. 1992; Eilers, Callewaert et al. 1995) or by the use of high molecular weight Fura-2-dextran that diffuses considerably slower (Müller, Naraghi et al. 1995). While high speed imaging cannot restore the physiological spread of Ca^{2+}, the problem with Fura-2-dextran is its considerably reduced brightness, requiring extra efforts to achieve sufficient signal over background at the recording speed required. Using a low affinity approach, Llinas and colleagues succeeded in demonstrating hot spots of high Ca^{2+} underneath the presynaptic membrane of the squid giant synapse (Llinas, Sugimori et al. 1992). These hot spots are supposed to mediate fast release of transmitter. Petrozzino et al. showed with the low affinity Ca^{2+}-dye mag-Fura-5 that up to 40 μM Ca^{2+} peak concentrations are reached in postsynaptic spines during tetanic stimulation (Petrozzino, Pozzo Miller et al. 1995). Low affinity indicator variants have been found useful not only for monitoring very high concentrations of free calcium, but also for reducing buffering and enhanced diffusion in the case of rather moderate Ca^{2+}-signals.

In confocal microscopy ratiometric Ca^{2+}-imaging has made use of the emission shifted Ca^{2+}-dye Indo-1. Since this dye is known for considerable photobleaching and requires expensive UV optics and lasers for excitation, most confocal studies have concentrated on qualitative single wavelength measurements using Fluo-3 or the spectroscopically similar but brighter indicator Ca-green. Both can be excited with the 488 nm line of a standard Argon/Krypton laser.

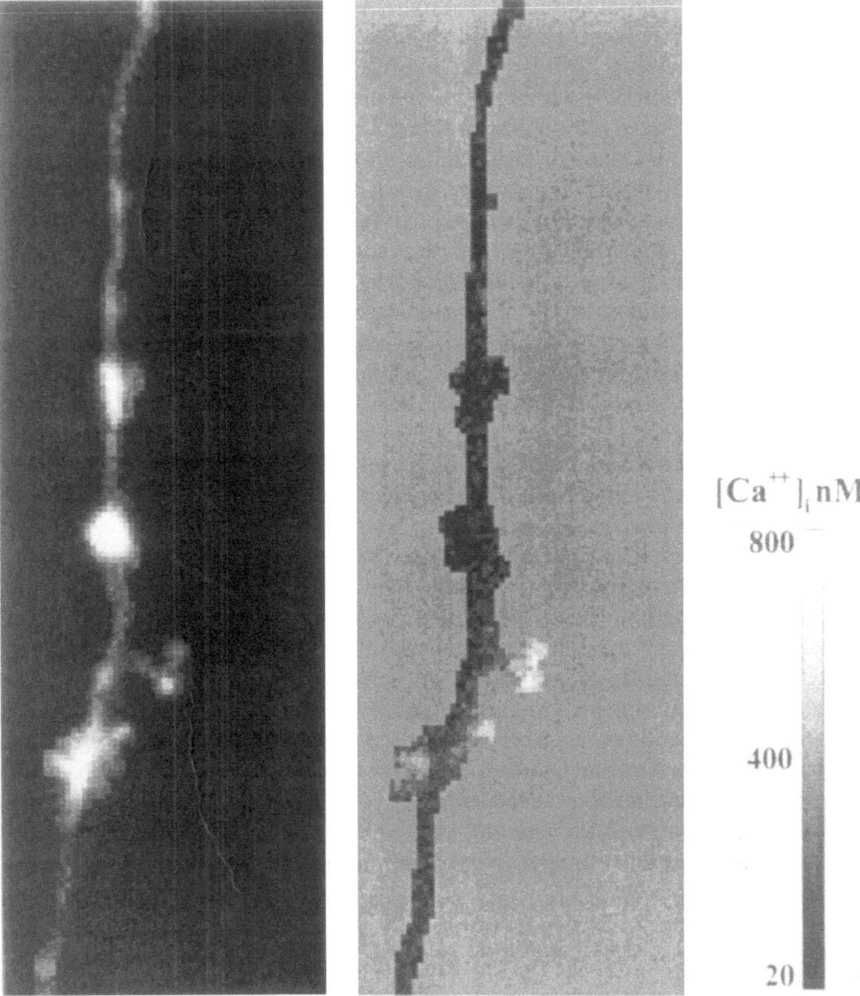

Fig. 2. Raw fluorescence image and gray scale coded ratio image of a rat hippocampal CA1 pyramidal neuron dendrite in brain slice demonstrates sustained increased Ca^{2+}-levels in spines projecting laterally. These sustained Ca^{2+}-levels presumably indicate enhanced Ca^{2+}-influx that may contribute to synaptic potentials as well as to neurodegenerative processes. The cell was selectively injected with the pentapotassium salt of Fura-2 during intracellular sharp microelectrode recording

With confocal imaging, Ca^{2+} in dendrites and spines has been successfully imaged in Purkinje cells in the cerebellar slice. However, spine-specific Ca^{2+}-increases have not been found with conventional confocal microscopy but could be demonstrated by two-photon confocal laser scan microscopy (Denk, Sugimori et al. 1995; Eilers, Augustine et al. 1995). The increased resolution of two-photon confocal laser scan microscopy in slice tissue allowed the characterization of the hippocampal pyramidal neuron dendritic spine to be confirmed and extended as an independent functional unit with respect to Ca^{2+}-dependent processing (Müller and Connor 1991; Yuste and Denk 1995). Presumably because of the high power density of confocal illumination, photo-toxicity/photodamage is a serious problem. Photodamage appears to be largely overcome by two-photon confocal excitation of Ca-green (Svoboda, Denk et al. 1997), most likely due to the change in wavelength though absorption of infrared light by tissue water has to be considered (Denk, Piston et al. 1995).

Cellular Ca^{2+}-signaling exhibits further complexity with respect to a variety of sources for increases in intracellular free Ca^{2+}. For Ca^{2+}-influx across the membrane, amplitudes of Ca^{2+}-increases appear to reflect primarily the surface to volume ratio. Therefore, strong Ca^{2+}-increases in voluminous compartments may indicate release of Ca^{2+} from intracellular Ca^{2+}-stores. Figure 3 demonstrates such an unusual strong somatic Ca^{2+}-increase in an entorhinal cortex pyramidal cell in brain slice during stimulation with the cholinergic agonist carbachol (CCh). Such somatic Ca^{2+}-increases have been shown to contribute to the control of gene expression in the nucleus and may therefore be important for long term cellular adaptive changes.

To address Ca^{2+}-changes in subcellular organelles, a variety of approaches has been used. Actually, dyes targeted to the cytosol are often partially segregated into organelles. This is especially a problem with the membrane permeable acetoxymethyl ester form of dyes and results in a mixed signal making interpretation of results very difficult (Almers and Neher 1985; Connor 1993). Even injected dye may be compartmentalized by active transport mechanisms (Al-Mohanna, Caddy et al. 1994). Dextran-conjugated dyes appear to be largely resistant to sequestration. Loading cells with a low concentration (e.g. 1 µM) of the membrane permeable acetoxymethyl ester of Fura-2 improves the relationship between cytosolic dye and cellular esterase activity, thereby reducing the cytosolic concentration of membrane permeable dye. This appears to re-

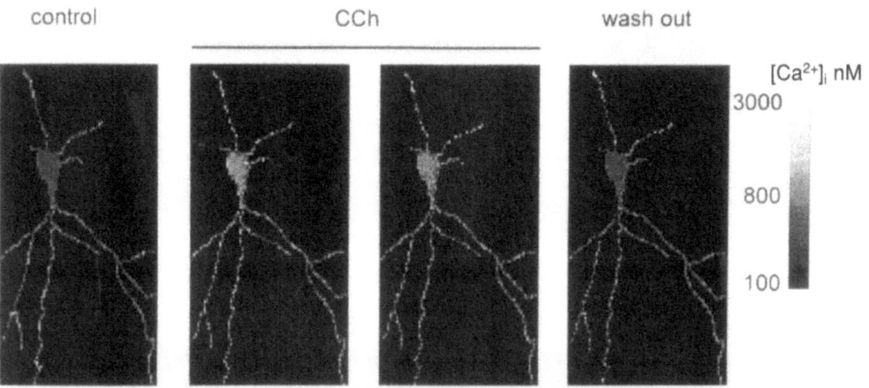

Fig. 3. Gray scale coded ratio images of a rat entorhinal cortex pyramidal neuron visualize intracellular free Ca^{2+} concentration in control, during superfusion of carbachol (CCh) and during wash out of CCh. Note the unusually strong response of somatic free Ca^{2+}

duce organellar loading relative to the cytosol. On the other hand, loading with a high concentration (e.g. 10 µM) of membrane permeable dye appears to improve loading of intracellular organelles. After permeabilization of the plasmamembrane, cytosolic dye can be removed and organellar Ca^{2+} is reported by Ca^{2+}-dye trapped there. Using this method we could demonstrate excessive ATP-driven organellar Ca^{2+}-accumulation in astrocytes that is supposed to be of pathophysiological importance for neurodegenerative damage in the trisomy 16 mouse model of Down's syndrome (Fig. 4, Müller, Heinemann et al. 1997). For qualitative, i.e. non-ratiometric monitoring of Ca^{2+} in mitochondria, the AM ester of rhod-2 has been successfully used in T lymphocytes as well as in hippocampal cultures (Hoth, Fanger et al. 1997; Müller and Kudo, unpublished observations).

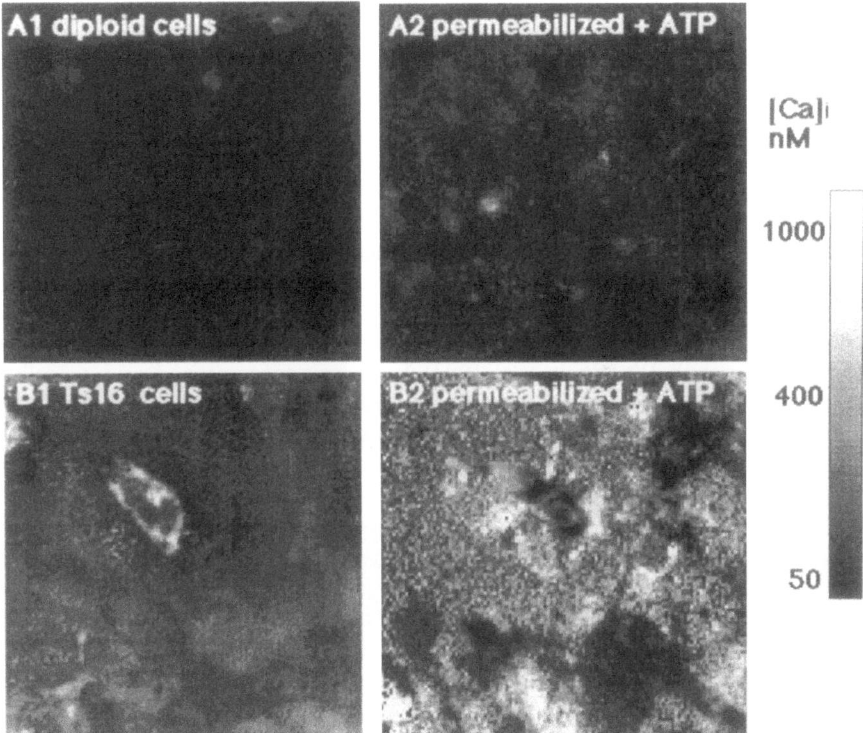

Fig. 4. Gray scale coded ratio images of Fura-2/AM loaded hippocampal astrocytes in dissociated cell culture from diploid and littermate trisomy 16 mouse embryos. The trisomy 16 mouse is a model for Down's syndrome due to genetic homology between mouse chromosome 16 and human chromosome 21. A1 and B1 demonstrate increased resting Ca^{2+}-levels in intact trisomic astrocytes. After permeabilization of the plasmamembrane with digitonin cytosolic Fura-2 was washed away, resulting in a fluorescence signal from dye trapped in intracellular organelles. Panels A2 and B2 demonstrate that addition of ATP results in moderate accumulation of Ca^{2+} in organelles from diploid cultures but a massive accumulation of Ca^{2+} in organelles from trisomic cells

Most likely because of a net positive charge, rhod-2 is taken up into mitochondria by a membrane potential driven uptake. The marriage of molecular
and imaging techniques gave birth to another innovative approach. Rizzuto and
colleagues used cellular sorting sequences to obtain specific accumulation of
Ca^{2+}-indicators in cellular compartments such as mitochondria or the nucleus
(Pozzan, Rizzuto et al. 1994). This idea has been extended by Miyawaki et al.,
who constructed a Ca^{2+}-sensor based on the Ca^{2+}-binding protein calmodulin,
a calmodulin binding peptide and two variants of green fluorescent protein
(GFP). Binding of Ca^{2+} to calmodulin makes calmodulin wrap around the binding peptide and increases fluorescence resonance energy transfer between the
flanking GFPs. The resulting spectroscopic shift allows ratiometric measurements. Using appropriate localization signals in addition and expressing the
construction in HeLa cells, Ca^{2+}-dynamics has been visualized in the cytosol,
nucleus and endoplasmic reticulum, where it was determined as varying between 60 to 400 µM at rest (Miyawaki, Llopis et al. 1997).

4
Measurements of Other Ions and Molecules

Compared to Ca^{2+}, the subcellular importance of other intracellular second
messengers appears to be considerably less interesting, as judged from the
number of imaging studies addressing their dynamics and function. Of course,
protons are long known for their importance in many biochemical reactions
and pH has been measured with ion-selective microelectrodes or fluorescence
imaging (Bright, Fisher et al. 1987). There are only very few observations of subcellular cytosolic pH gradients, probably due to a high concentration of mobile
buffers (Grinstein, Woodside et al. 1993). In *Ascaris* spermatozoa, an intracellular pH gradient could be demonstrated that separated areas of assembly and di-
sassembly of the major sperm protein cytoskeleton underlying motility.
Acidification of 1 to 2 pH units has been found in subcellular organelles like the
Golgi apparatus by using a targeted indicator or in the endosomal-lysosomal
pathway by endocytosis of labeled medium, (Overly, Lee et al. 1995; Kim,
Lingwood et al. 1996). In cerebellar astrocytes, glutamate receptor agonists as
well as noradrenaline cause increases in $[H^+]_i$ and $[Ca^{2+}]_i$ or only in $[Ca^{2+}]_i$. In
this case, the changes in $[H^+]_i$ are apparently not simply due to a competition of
protons and Ca^{2+} for common binding sites (Brune and Deitmer 1995).

With respect to neurological and psychiatric diseases, neurodegenerative
mechanisms which possibly contribute to the progress of these diseases attract
particular interest. Besides Ca^{2+}, that appears to have a trigger and modulatory
role, reactive oxygen species, particularly superoxide (O_2^-) radicals and its superoxide dismutase catalyzed product hydrogen peroxide, have been implicated
in the pathogenesis of excitotoxic cell damage and death (e.g. Behl, Davis et al.
1994). Superoxide radicals are produced in the metabolism of free fatty acids
and in the respiratory chain in the mitochondria. Therefore these two pathways
are of particular interest. It could be demonstrated by non-fluorescence techniques, that Ca^{2+}-increases caused by NMDA or metabotropic glutamate receptors
effectively trigger liberation of arachidonic acid (Dumuis, Sebben et al. 1988;

Dumuis, Pin et al. 1990). Using a fluorescent indicator of free fatty acids, i.e. a conjugate of a fatty acid binding protein and the polarity sensitive acrylodan fluorophore named ADIFAB (Mol. Probes, Eugene, Oregon, US) we had some success in monitoring free fatty acids in serum free mouse cerebellum culture during stimulation with glutamate and glycin. Figure 5 demonstrates a gradual decease of the ADIFAB fluorescence signal during stimulation in the vicinity of four selected neurons and after exogenous addition of arachidonic acid (AA).

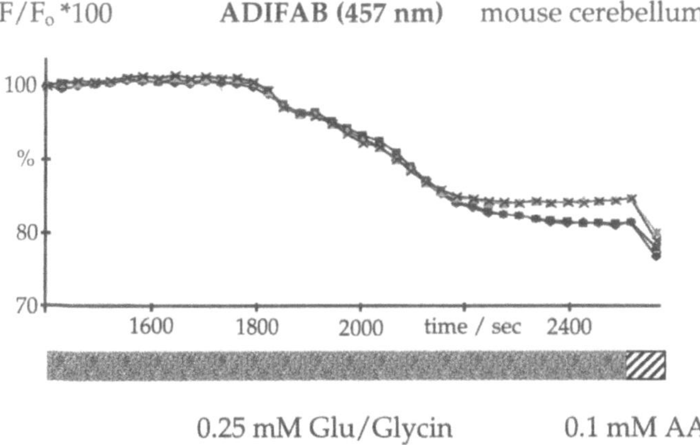

Fig. 5. Plot of fluorescence of the extracellularly applied free fatty acid indicator ADIFAB versus time from areas around four selected neurons in culture demonstrate a gradual decease of the ADIFAB fluorescence signal during stimulation with glutamate (Glu)/glycin and after exogenous addition of arachidonic acid (AA)

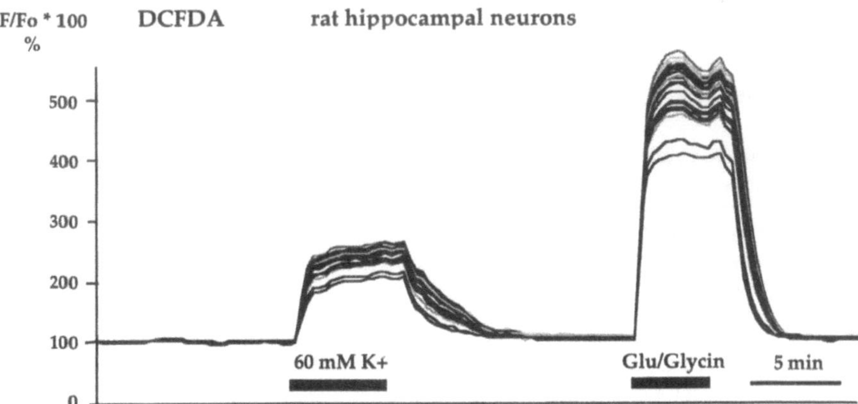

Fig. 6. Plot of fluorescence of the extracellularly superfused indicator dichlorodihydrofluorescein diacetate (DCFDA), that increases its fluorescence after oxidation through reactive oxygen intermediates, demonstrates that stimulation with glutamate (Glu)/glycin is considerably more effective than high potassium in production of reactive oxygen intermediates in the vicinity of selected rat hippocampal neurons in culture

Mitochondrial function, i.e. polarization, superoxide generation, NAD(P)H/NAD(P)$^+$ and flavoprotein, have been assessed by rhodamine 123, dihydrorhodamine 123 or dichlorodihydrofluorescein diacetate (DCFDA) and autofluorescence, respectively. All these parameters are affected by a rise in cytosolic calcium, consistent with a close coupling between Ca^{2+} and the mitochondrial function. To monitor production of free oxygen radicals and H$_2$O$_2$, we used an extracellular dye like DCFDA, that increases its fluorescence after oxidation. Figure 6 shows measurements from the vicinity of hippocampal neurons in culture during superfusion of saline containing the dye and either a high concentration of potassium (K$^+$) or glutamate/glycin for stimulation. Stimulation with glutamate/glycin is considerably more effective than high potassium in increasing the production of H$_2$O$_2$. Both H$_2$O$_2$ responses can be blocked by intracellular buffering of Ca^{2+} by BAPTA.

5
Outlook

Fluorescence imaging studies of second messengers and cellular metabolites have led to a rapid growth in our understanding of neuronal signal processing in health and disease. Ongoing progress in instrumentation, design of fluorescent indicators and molecular biology will sustain the continued growth of our knowledge of the central nervous system. With respect to fluorescent dyes, ratiometric dyes that allow absolute calibration are most wanted for the study of molecules that can today only be evaluated qualitatively.

Acknowledgements. We thank K. Berlin and A. Düerkop for excellent technical assistance. The work reported herein was supported by the Deutsche Forschungsgemeinschaft, Bonn (Heisenberg-Stipendium to W.M., Mu 809/6–1 and –2 and SFB 507-C4) and by the Charité, Berlin, Germany.

References

Al-Mohanna FA, Caddy KW, Bolsover SR (1994) The nucleus is insulated from large cytosolic calcium ion changes. Nature 367:745–750

Almers W, Neher E (1985) The Ca signal from fura-2 loaded mast cells depends strongly on the method of dye-loading. Febs Lett 192:13–18

Behl C, Davis JB, Lesley R, Schubert D (1994) Hydrogen peroxide mediates amyloid beta protein toxicity. Cell 77:817–827

Bright GR, Fisher GW, Rogowska J, Taylor DL (1987) Fluorescence ratio imaging microscopy: temporal and spatial measurements of cytoplasmic pH. J Cell Biol 104:1019–1033

Brune T, Deitmer JW (1995) Intracellular acidification and Ca^{2+} transients in cultured rat cerebellar astrocytes evoked by glutamate agonists and noradrenaline. Glia 14:153–161

Connor JA (1993) Intracellular calcium mobilization by inositol 1,4,5-trisphosphate: intracellular movements and compartmentalization. Cell Calcium 14:185–200

Denk W (1994) Two-photon scanning photochemical microscopy: mapping ligand-gated ion channel distributions. Proc Natl Acad Sci USA 91:6629–6633

Denk W (1977) Pulsing mercury arc lamps for uncaging and fast imaging. J Neurosci Methods 72:39–42

Denk W, Piston DW, Webb W (1995) Two-photon molecular excitation in laser-scanning microscopy. Handbook of biological confocal microooscopy. New York, Plenum Press 2, ed 445–458
Denk W, Sugimori M, Llinas R (1995) Two types of calcium response limited to single spines
  in cerebellar Purkinje cells. Proc Natl Acad Sci USA 92:8279–8282
Dumuis A, Pin JP, Oomagari K, Sebben M, Bockaert J (1990) Arachidonic acid released from
  striatal neurons by joint stimulation of ionotropic and metabotropic quisqualate receptors. Nature 347:182–184
Dumuis A, Sebben M, Haynes L, Pin JP, Bockaert J (1988) NMDA receptors activate the
  arachidonic acid cascade system in striatal neurons. Nature 336:68–70
Egorov A, Müller W (1997) Muscarinic potentiation on stimulation-induced calcium in-
  creases in dendrites and spines of CA1 pyramidal neurons in rat hippocampal slice.
  Pflüger's Arch 433:R 71
Eilers J, Augustine GJ, Konnerth A (1995) Subthreshold synaptic Ca^{2+} signalling in fine den-
  drites and spines of cerebellar Purkinje neurons. Nature 373:155–158
Eilers J, Callewaert G, Armstrong C, Konnerth A (1995) Calcium signaling in a narrow soma-
  tic submembrane shell during synaptic activity in cerebellar Purkinje neurons. Proc Natl
  Acad Sci USA 92:10272–10276
Erdmann S, Müller W, Bahrami S, Vornehm SI, Mayer H, Bruckner P, von der Mark K,
  Burkhardt H (1996) Differential effects of parathyroid hormone fragments on collagen
  gene expression in chondrocytes. J Cell Biol 35:1179–1191
Fay FS, Carrington W, Fogarty KE (1989) Three-dimensional molecular distribution in single
  cells analyzed using the digital imaging microscope. J Microsc 153:133–149
Grinstein S, Woodside M, Waddell TK, Downey GP, Orlowski J, Pouyssegur J, Wong DC,
  Foskett JK (1993) Focal localization of the NHE-1 isoform of the Na+/H+ antiport: assess-
  ment of effects on intracellular pH. Embo J 12:5209–5218
Hoth M, Fanger CM, Lewis RS (1997) Mitochondrial regulation of store-operated calcium
  signaling in T lymphocytes. J Cell Biol 137:633–648
Isenberg G, Etter EF, Wendt GM, Schiefer A, Carrington WA, Tuft RA, Fay FS (1996)
  Intrasarcomere $[Ca^{2+}]$ gradients in ventricular myocytes revealed by high speed digital
  imaging microscopy, Proc Natl Acad Sci USA 93:5413–5418
Kim JH, Lingwood CA, Williams DB, Furuya W, Manolson MF, Grinstein S (1996) Dynamic
  measurement of the pH of the Golgi complex in living cells using retrograde transport of
  the verotoxin receptor. J Cell Biol 134:1387–1399
Lang T, Wacker I, Steyer J, Kaether C, Wunderlich I, Soldati T, Gerdes HH, Almers W (1997)
  Ca^{2+}-triggered peptide secretion in single cells imaged with green fluorescent protein and
  evanescent-wave microscopy. Neuron 18:857–863
Llinas R, Sugimori M, Silver, RB (1992) Microdomains of high calcium concentration in a presynaptic terminal. Science 256:677–679
Miyawaki A, Llopis J, Heim R, McCaffery JM, Adams JA, Ikura M, Tsien RY (1997) Fluorescent
  indicators for Ca^{2+} based on green fluorescent proteins and calmodulin. Nature
  388:882–887
Müller TH, Naraghi M, Neher E (1995) Calcium gradients observed in chromaffin cells under
  conditions of low dye-dependent mobility. Pflüger's Arch 429:R24
Müller W, Connor JA (1991) Dendritic spines as individual neuronal compartments for synaptic Ca^{2+} responses Nature 354:73–76
Müller W, Connor JA (1992) Ca^{2+} signalling in postsynaptic dendrites and spines of mammalian neurons in brain slice. J Physiol Paris 86:57–66
Müller W, Connor JA (1993) High resolution microfluorometry of Ca^{2+} signalling in dendrites
  and spines of central neurons. Jpn J Physiol 43:S131–137
Müller W, Heinemann U, Schuchmann S (1997) Impaired Ca-signaling in astrocytes from the
  Ts16 mouse model of Down's syndrome. Neurosci Lett 223:81–84
Müller W, Swandulla D (1995) Synaptic feedback excitation has hypothalamic neural net-
  works generate quasirhythmic burst activity. J Neurophysiol 73:855–861
Neher E (1995) The use of Fura-2 for estimating Ca buffers and Ca fluxes. Neuropharma-
  cology 34:1423–1442

Overly CC, Lee KD, Berthiaume E, Hollenbeck PJ (1995) Quantitative measurement of intra-organelle pH in the endosomal- lysosomal pathway in neurons by using ratiometric imaging with pyranine. Proc Natl Acad Sci USA 92:3156–3160

Petrozzino JJ, Pozzo Miller L, Connor JA (1995) Micromolar Ca^{2+} transients in dendritic spines of hippocampal pyramidal neurons in brain slice. Neuron 14:1223–1231

Poenie M (1990) Alteration of intracellular Fura-2 fluorescence by viscosity: a simple correction. Cell Calcium 11:85–91

Pozzan T, Rizzuto R, Volpe P, Meldolesi J (1994) Molecular and cellular physiology of intracellular calcium stores. Physiol Rev 74:595–636

Regehr WG, Tank DW (1990) Postsynaptic NMDA receptor-mediated calcium accumulation in hippocampal CA1 pyramidal cell dendrites. Nature 345:807–810

Stehno BL, Perez TC, Clapham DE (1995) Diffusion across the nuclear envelope inhibited by depletion of the nuclear Ca^{2+} store. Science 270:1835–1838

Steyer JA, Horstmann H, Almers W (1997) Transport, docking and exocytosis of single secretory granules in live chromaffin cells. Nature 388:474–478

Svoboda K, Denk W, Kleinfeld D, Tank DW (1997) In vivo dendritic calcium dynamics in neocortical pyramidal neurons. Nature 385:161–165

Yuste R, Denk W (1995) Dendritic spines as basic functional units of neuronal integration. Nature 375:682–684

CHAPTER 23

Fluorescence Diagnosis in the Border Zone of Liver Tumors

J. Beuthan, O. Minet

1
Introduction

For several years laser-induced fluorescence spectroscopy (LIF) has been a valuable tool in medical diagnosis detecting the relative distribution of endogenous chromophors [1–15]. In the case of the reduced form of Nicotinamide Adenine Dinucleotide, NADH, the fluorescence signal reveals the activity status of the observed cells because NADH takes part in some essential features of the cellular energy metabolism (see for example [16] and [17]). This coenzyme takes part in biochemical reactions within living cells, mainly the respiratory cycle, as was discovered, in principle, by O. Warburg in 1931.

For the differential diagnosis of metabolic disturbance and for the quantitative assessment of intraoperative sites or transplants, it is indispensable, in most cases to know the quantitative values of NADH concentration in the particular region.

The general problem arising from this is fluorescence measuring in a strongly scattering medium. The fluorophores are embedded in this medium which is characterized by the following main optical parameters: absorption coefficient μ_a, scattering coefficient μ_s and anisotropy factor g. When detecting similar intrinsic fluorescences on the surface of the medium (e.g. tissue surface or in the bioreactor), it is possible that the fluorescence intensity values measured are different with different optical parameters.

The potential for clinical application using a time-resolved fluorescence spectroscopy was for instance qualitatively demonstrated as follows [9]: Having performed the routine clinical needle biopsy at an astrocytoma, the profile of the NADH level was mapped using the already established biopsy channel. The tip of an optical fiber was introduced as a sensor into a small catheter marked at different positions. Catheter and fiber were inserted into the biopsy channel and moved gradually to map the brain region of interest. This type of measurement provides new possibilities for the intraoperative monitoring of the metabolism in neurosurgery using time-resolved laser-induced fluorescence spectroscopy. For instance, Fig. 1 shows the result of the measurement for a liver tumor. Within the liver tumor at positions 3 to 7, the NADH concentration level is significantly lower. A qualitative mapping of the tumor-dependent metabolic state was possible.

Nevertheless, the possibilities of the diagnostics in medicine grow rapidly with the quantitative interpretation of the measurements. One possible ap-

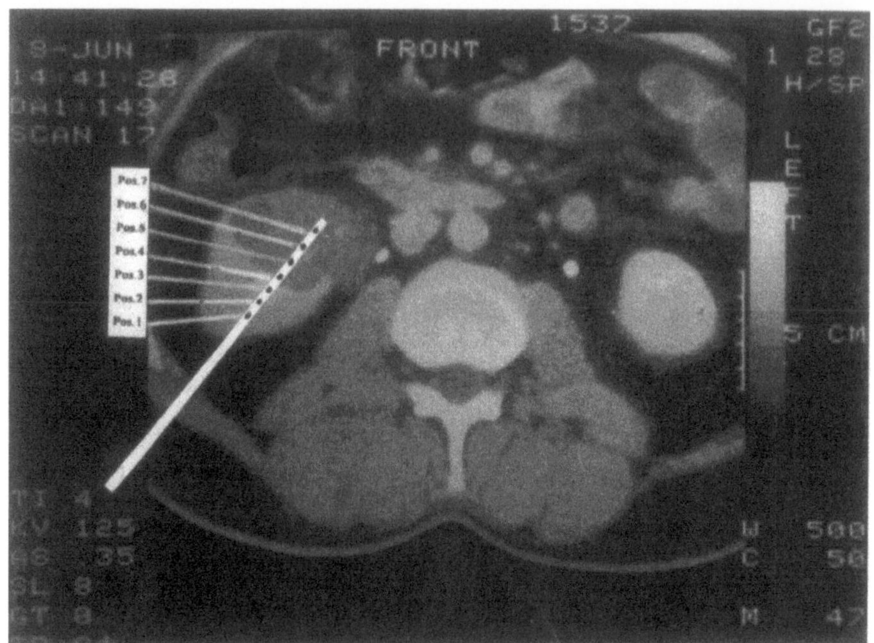

Direction of biopsy (kidney tumor)

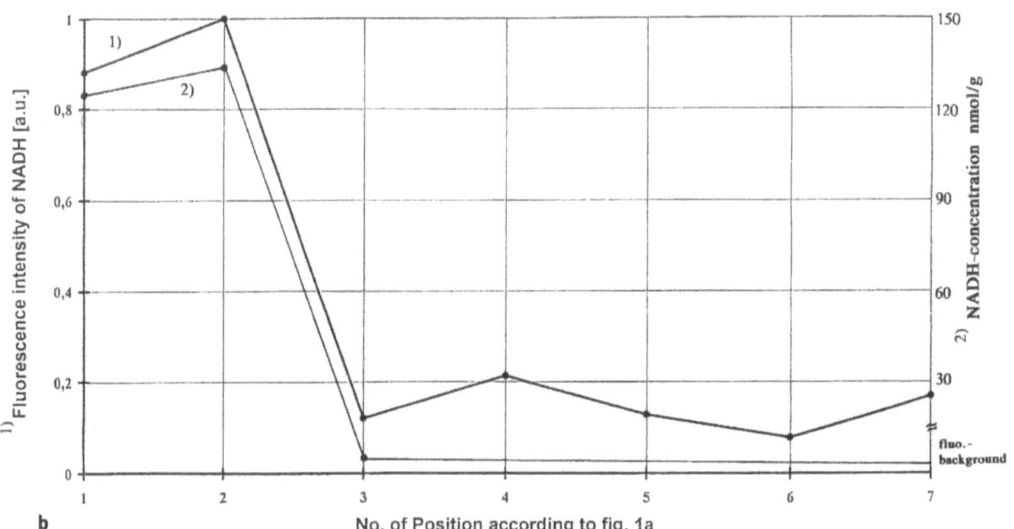

Fig. 1a, b. First *in vivo* fluorescence diagnosis of a malign kidney tumor (grade II) by imaging control with X-ray tomography. a Positions of fluorescence detection according to the CT image. b Corresponding intensities of the LIF signal (excitation wave length 337.1 nm, fluorescence wave length 460 nm). The peak of the signal lies in the border region of the tumor

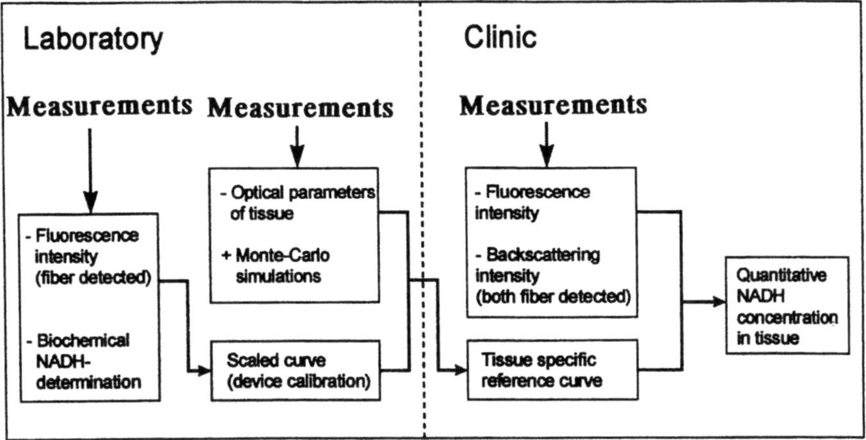

Fig. 2. Scheme of the rescaling procedure

proach for an underlying appropriate model is the diffusion approximation within the context of the transport theory for light in turbid media [18, 19]. Clinical fluorescence results often differ from this approximation because of the complex relationship of scattering, absorption and intrinsic fluorescence within the tissue. In particular, the optical parameters μ_a, μ_s, and the anisotropy factor g vary individually and locally in the target tissues. The main task for the evaluation algorithm is the appropriate consideration of these variations.

A recent method is based on the following two principles [20] (see also Fig. 2):

1. The calibration function of the experimental setup is determined by biochemical reference measurements.
2. Monte Carlo simulations using the measured optical parameters show a nearly linear functional relation between the optical tissue parameters μ_{ax} and the fluorescence intensity on the one hand and between μ_{ax} and the back-scattering intensity on the other (indices: a = absorption, s = scattering, x = excitation, f = fluorescence). The influence of the other tested parameters μ_{af}, μ_{sx}, and μ_{sf} on the fluorescence signal is one order of magnitude smaller and therefore negligible; the mean cosine of the scattering phase function g is assumed to be constant. Therefore, the two information channels, i.e. fluorescence and back scattering, together allow the determination of the actual, absolute concentration of NADH in the observed tissue region. While other methods are limited to the linear region [21], it is of course possible to also take a nonlinear functional relationship [22] between μ_{ax} and the Laser Induced Fluorescence (LIF) signal into account (see later in Fig. 6), assuming that these functions shows a monotone behavior. Due to physical reasoning like the energy conservation law, this presupposition is fulfilled for both absorption coefficients μ_{ax} and μ_{af} in general, whereas the question of the accuracy of this method depends mainly on the numerical value of the

gradient of the fluorescence intensity vs. μ_{ax} within the actual range of the parameters.

2
Materials and Methods

2.1
Experimental Setup

The principal experimental setup [23] for the observation of NADH fluorescence is shown in Fig. 3. The light source used is a N_2-laser with the following parameters: $\lambda = 337.1$ nm, pulse duration $\tau = 500$ ps, pulse energy E = 1 mWs, repetition rate = 50 Hz. The laser pulse hits a grating which deflects it into an optical fiber (Schott silica fiber, core diameter 360 μm, good transmission of UV radiation, length of 15 m). The fiber guides the pulse to the tissue. The NADH fluorescence of the biological probe provides information on the redox state (fluorescence intensity) and the state of binding (shape of the fluorescence decay). The fluorescence signal is returned to the grating via the same fiber and passed on to a Hamamatsu R 1450 photo multiplier with a spectral range from 300 to 650 nm, a rise time of 1.8 ns and a cathode sensitivity of 70 μA/lm. The separation of the NADH fluorescence from excitation light is achieved by the disperse properties of the grating and an interference filter. We used a Melles Griot filter from the 03 IFS series with a transmittance peak of 450 nm, FWHM

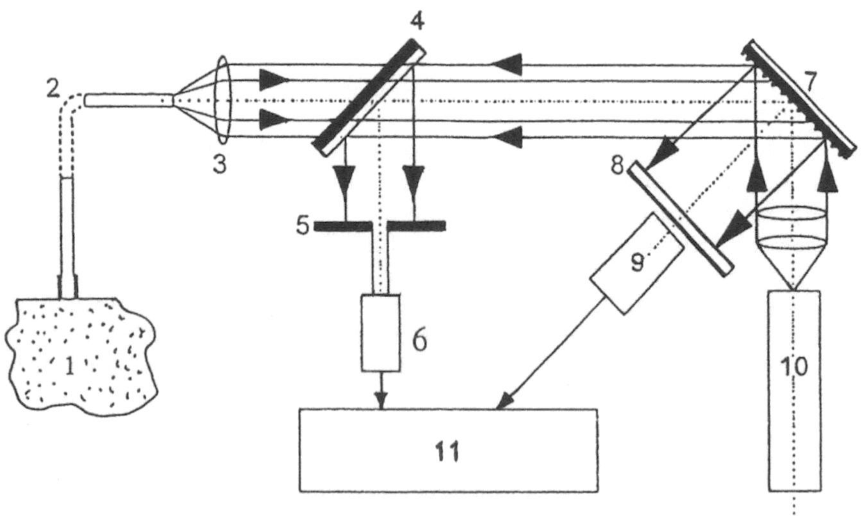

Fig. 3. Experimental setup of laser fluorescence spectroscopy for in vivo and in vitro measurements: *1* – biological probe, *2* – fiber, *3* – lens, *4* – semitransparent mirror, *5* – diaphragm, *6* – photodiode for the boxcar trigger and measurement of the excitation intensity, *7* – grid, *8* – interference filter, *9* – UV photomultiplier, *10* – N_2 laser, *11* – boxcar integrator

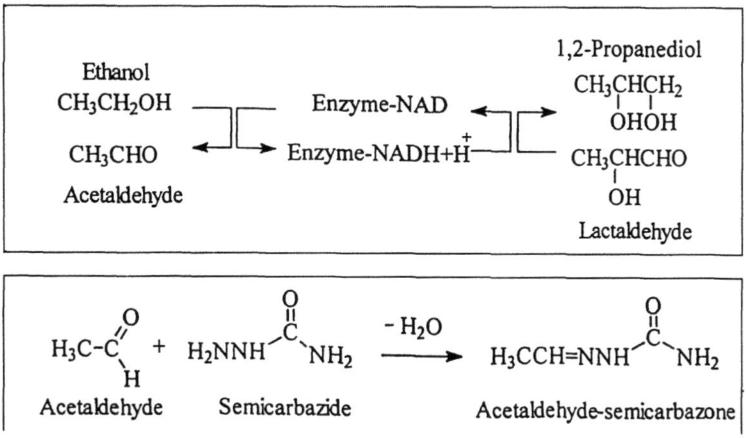

Fig. 4. Circle process and indicating reaction

40 ± 8 nm and a blocking of 0.01 %. A boxcar integrator (EG & G 4400) was used for data processing. The pure fluorescence response is received after the deconvolution with the excitation pulse.

The digital boxcar integrator allows a two-channel measurement. The intensity of the fluorescence light was detected in the first channel in a time-resolved manner. Simultaneously, the second channel served for the measurement of the intensity of the remitted part of the excitation light. By means of a corresponding look-up table (the graphical representation is given below) and the described evaluation procedure the intrinsic fluorescence was roughly calculated.

For the experimental setup concentration-dependent measurements of NADH were used to scale up the fluorescence values [24]. For this purpose, a well-known two-step analytical method was applied [25, 26]. First, in relation to the NADH quantity, acetaldehyde is generated during a biochemical circle process, in the second step the amount of acetaldehyde is determined by an indicating reaction (see Fig. 4).

Acetaldehyde is formed by the reaction of NADH in the presence of a surplus of ethanol and lactaldehyde. During the circle reaction the oxidation of ethanol is caused by NAD which is generated by LADH via NADH, whereas 1,2-propanediol is formed by the reduction of lactaldehyde. The circle reaction is limited by the concentration of the pyridinenucleotides NAD/NADH. The turnover number of the circle reaction corresponds to the variation in the enzyme concentration.

Using a stream of nitrogen, the gaseous reaction product acetaldehyde is absorbed into a solution of semicarbazide to yield the acetaldehyde-semicarbazone. This reaction product is then detected by UV measurements at 224 nm. In this way the NADH concentration can be directly quantified by the value of the optical absorption.

2.2
Monte Carlo Simulations

As target tissue we used human, cancerous liver parenchyma. At first, the optical parameters μ_s, μ_a, and the g-factor were determined by the double-integrating-sphere technique [27]. Then, these parameters were used in a well-known manner for Monte Carlo [29] simulations of the fluorescence intensity and the back-scattering intensity (see Fig. 5).

The simulation assumes full separation of parametric influence. All but one of the optical parameters involved were fixed to their mean values. In Fig. 6 the dependence of the fluorescence intensity on μ_{ax} is demonstrated for human, cancerous liver parenchyma. The variation is remarkably high, up to 80%. The dependence of the fluorescence intensity on the other optical tissue parameters, i.e. μ_{af}, μ_{sx}, and μ_{sf}, was one order of magnitude smaller and therefore neglected. The anisotropy factor g was assumed to be constant in a first approximation.

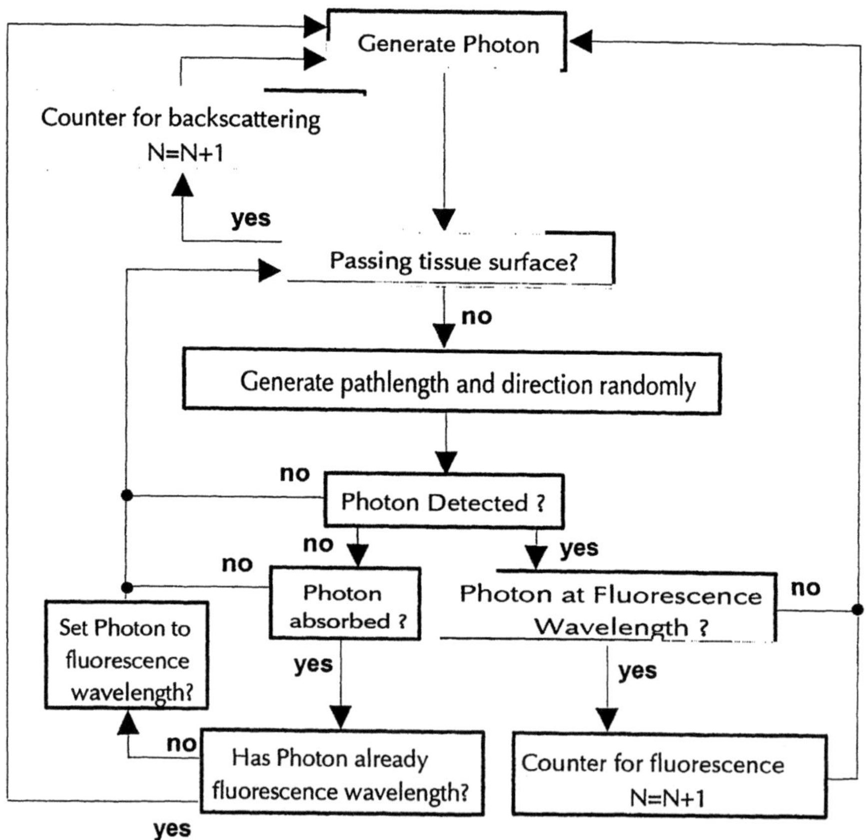

Fig. 5. Monte Carlo simulation flow chart

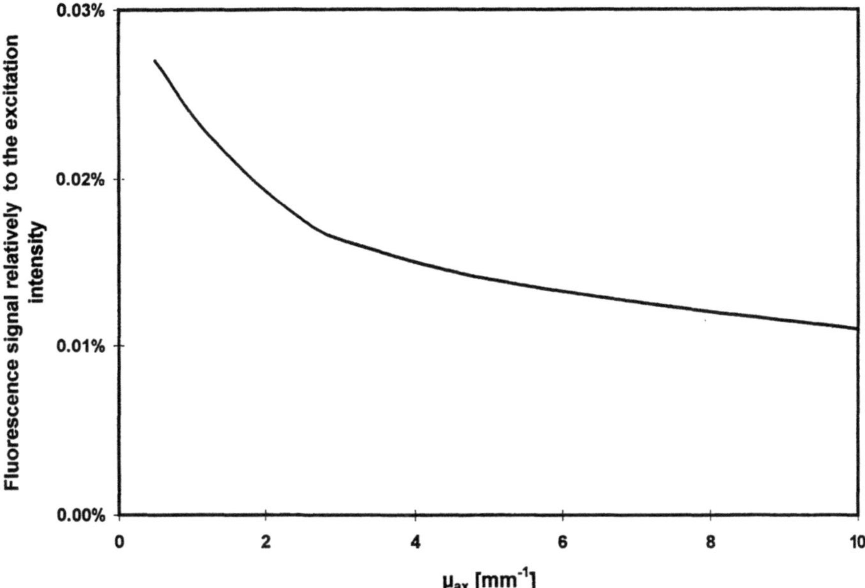

Fig. 6. Result of Monte Carlo simulations showing the decay of the fluorescence intensity detected by a fiber upon the absorption coefficient μ_{ax} (fiber diameter: 250 µm, numerical aperture of the fiber: 0.27, refractive index surrounding/fiber/tissue: 1.0/1.45/1.38, quantum yield for NADH: 0.02, $\mu_{sx} = 30.0$ mm^{-1}, $\mu_{af} = 2.0$ mm^{-1}, $\mu_{sf} = 30.0$ mm^{-1}, g = 0.9)

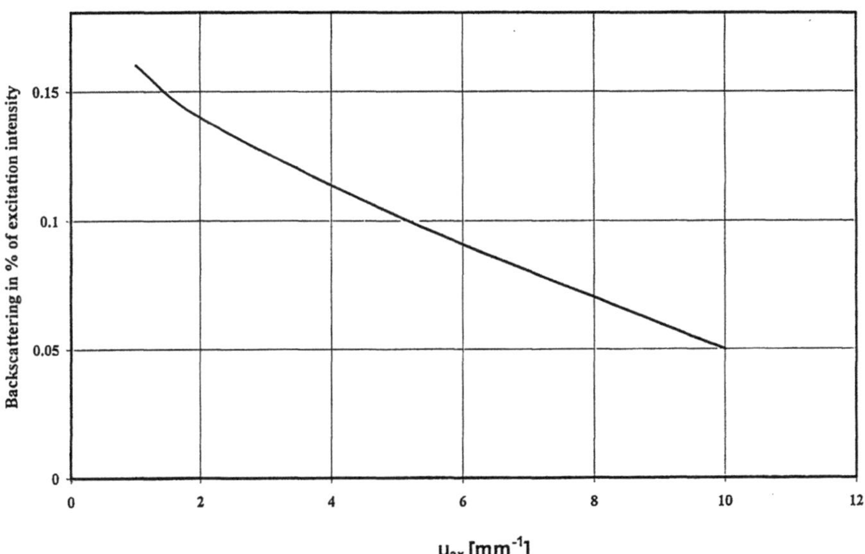

Fig. 7. Result of Monte Carlo simulations showing the dependency of the backscattering intensity detected by a fiber on the absorption coefficient μ_{ax}. The other parameters are taken from Fig. 6

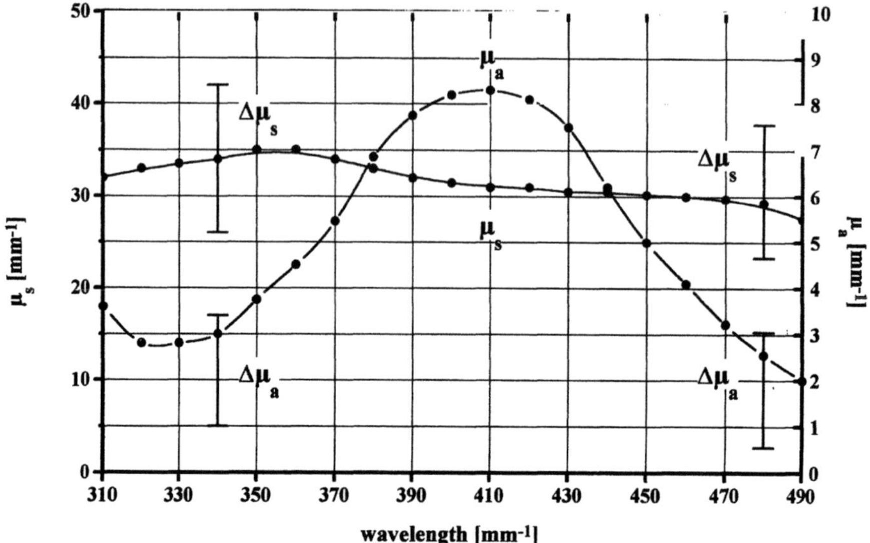

Fig. 8. Variation of the optical parameters for human liver tumor for wavelength between 310 nm and 490 nm

According to our Monte Carlo simulations the back-scattering intensity also depends linearly on μ_{ax} (see Fig. 7). Therefore a simultaneous measurement of the back-scattering intensity provides the data for a rescaling of the fluorescence intensity with respect to the deviations of the given μ_{ax} from its mean value (see Fig. 8). Furthermore, we can recognize that the scattering coefficient μ_{sx} is approximately constant in this spectral region.

2.3
Rescaling Procedure

As discussed above, the variations in the optical parameters have to be taken into account. From the Monte Carlo simulations we know that the necessary corrections are linear to the deviation of μ_{ax} from the mean value. Presuming our samples represent statistical variations in tissues we generated a three-parametric fitted curve which describes the fluorescence signal behavior of a tissue with the mean value of μ_{ax}. To get an analytical expression of the fitted curve we made the following assumptions and approximations:

1. The fluorescence intensity is essentially proportional to the concentration of NADH. This is especially true for small intensities I and small concentrations c, respectively:

$$\frac{dI}{dc} = k_1$$

2. Dealing with higher intensities and concentrations we also have to take into account the absorbency of the fluorescence radiation:

$$\frac{dI}{dc} = -k_2 I$$

Considering both aspects one gets:

$$\frac{dI}{dc} = k_1 - k_2 I .$$

Assuming there is no fluorescence intensity of NADH without the presence of NADH, i.e. $I(c = 0) = 0$, the solution to the above linear differential equation reads:

$$I(c) = K_1 (1 - e^{-K_2 c}) .$$

Nevertheless, even in the complete absence of NADH, there is a background fluorescence at the mentioned excitation wavelength because of other fluorophores being present in the tissue. We presume a constant background fluorescence intensity K_3 which is particularly independent of the excited NADH fluorescence:

$$I(c) = K_1 (1 - e^{-K_2 c}) + K_3 .$$

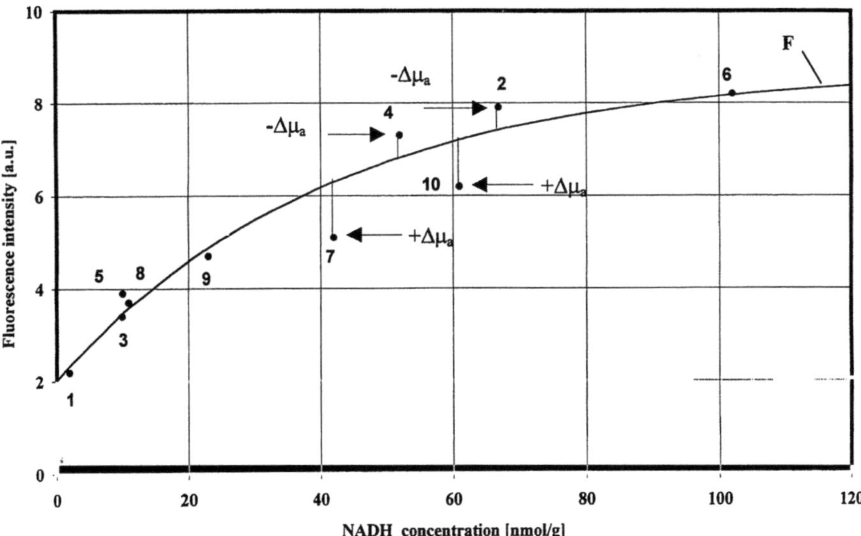

Fig. 9. Plot of the LIF signal vs. the NADH concentration for the 10 examples in Table 1 (*filled circles*). The *solid line* is the result of minimizing the sum of mean square deviations from an exponential function

Using the method of "minimizing the sum of the squares of the deviations" we achieved the best fit for the following parameters:

$K_1 = 6.73$
$K_2 = 0.024$
$K_3 = 2.03$

The resulting fitted curve is marked as a solid line in Fig. 9. The filled circles show the experimentally acquired data before the rescaling procedure was applied. We derived the rescaled values by proportionally correcting the experimental data according to the deviation of μ_{ax} from its mean value ($\mu_{ax} = 2.5$ mm^{-1}). Taking several starting points at random, the rescaling remained the same: 0.1 a. u. = 0.0027 nmol/g.

3
Results

As reported previously [27], the optical properties of six tissue samples were determined. Furthermore, the NADH concentrations in the same samples were measured using the biochemical method described above (see Table 1 for both results). Due to the small probe volumes an optical measurement was impossible for four other samples, while the fluorescence signal and the NADH concentration were measured. Nevertheless, these points can also be used for the determination of the rescaling curve F because the method of least squares is a statistical one. The probes were also imaged microscopically.

For a qualitative documentation of the NADH distribution in tissue slices of the liver tumor, the samples were colored with Nitroblue Tetrazolium Chloride (NBTC) stain. The microscopic enlargement shows the intensity of the color, which is equivalent to the NADH concentration. In Fig. 10a microscopical tis-

Table 1. List of the optical properties and the concentration of NADH measured by biochemical methods for tissue samples of a liver tumor

No.	Tissue type	Excitation wave length 337 nm			Fluorescence wave length 480 nm			NADH-concentration [nmol/g]
		μ_a [1/mm]	μ_s [1/mm]	g	μ_a [1/mm]	μ_s [1/mm]	g	
1	tumor necrotic	–	–	–	–	–	–	1.4
2	tumor border	1.0	24.7	0.87	0.4	24.2	0.94	66.2
3	tumor necrotic	1.7	42.7	0.91	1.2	30.9	0.94	7.2
4	liver metastasis	1.0	25.2	0.88	0.5	23.0	0.94	51.8
5	tumor necrotic	–	–	–	–	–	–	7.2
6	tumor border	–	–	–	–	–	–	102.2
7	tumor necrotic	7.0	32.7	0.81	2.9	35.9	0.95	41.8
8	tumor necrotic	–	–	–	–	–	–	8.6
9	liver metastasis	3.3	41.2	0.87	0.9	38.1	0.93	25.9
10	tumor partially necrotic	2.6	41.7	0.91	1.7	34.6	0.94	60.5

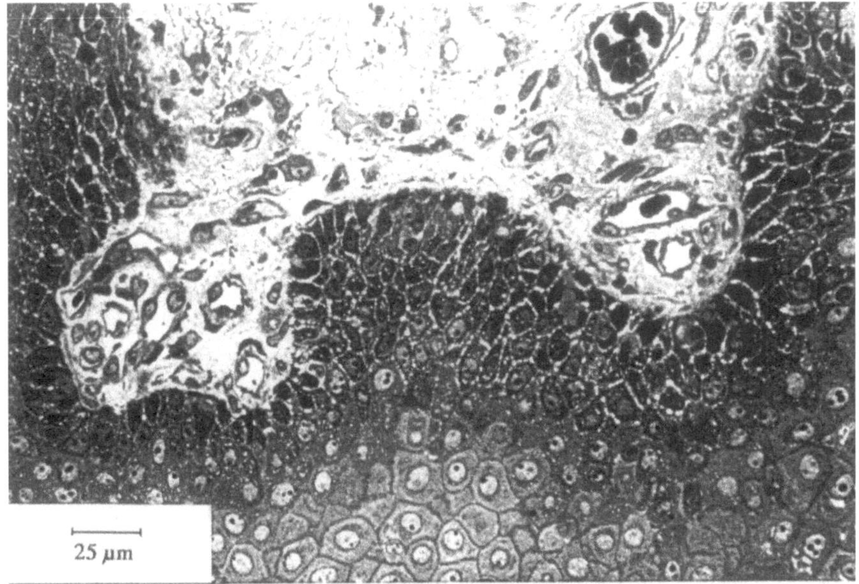

Fig. 10. Microscopic enlargement (20 times) of a tissue slice from the border region of a liver tumor using NBTC coloring. The color intensity is proportional to the NADH concentration

sue slice from the tumor border is shown. Normal tissue is dark blue in color. The border of the tumor contains a high concentration of NADH, whereas the necrotic area of the tumor contains only a small amount of NADH.

Furthermore, we can recognize some islands with relatively high metabolism. This fact corresponds with Table 1, which represents the variation of the NADH concentration within the tumor area.

The critical parameter for the intensity of fluorescence light from time-resolved measurements is the intensity of the peak at time t_m in Fig. 11. The correspondence between these intensities and the NADH concentrations was then established by the Monte Carlo method described above. Figures 6 and 7 show in the same manner the influence of a variation in an optical parameter on the intensity of fluorescence and back-scattering light. Most characteristic is the interdependence between the absorption coefficient of the excitation light and the fluorescence output in a fiber-based arrangement (see Fig. 6).

Curve F in Fig. 9 represents the scaling between the LIF signal and the concentration of NADH for mean optical parameters of the liver tumor. Furthermore, the values of the fluorescence signals for the samples from Table 1 and their NADH concentrations are shown. In general, these data points do not fit with the scaling curve due to the deviation of the respective values from the optical parameters, especially the absorption coefficient of the excitation light. Fig. 12 shows the correction procedure for LIF signals onto the true intrinsic concentration of the fluorophore. The back-scattering signal yields the information on the relevant deviation from the mean optical parameters and then

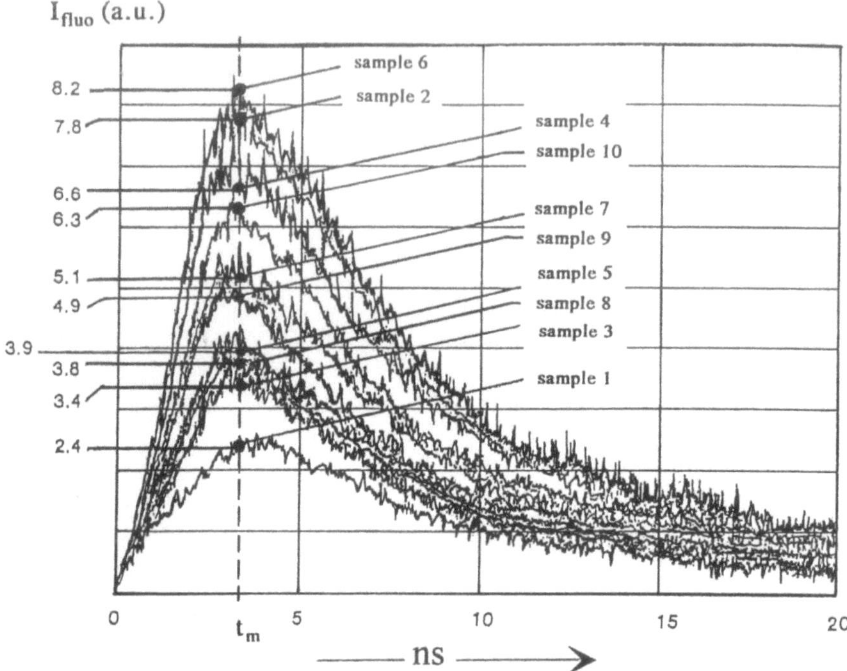

Fig. 11. Time-resolved fluorescence response from 10 tissue samples (according to Table 1). The intensity at time t_m is used for the rescaling fluorescence vs NADH concentration

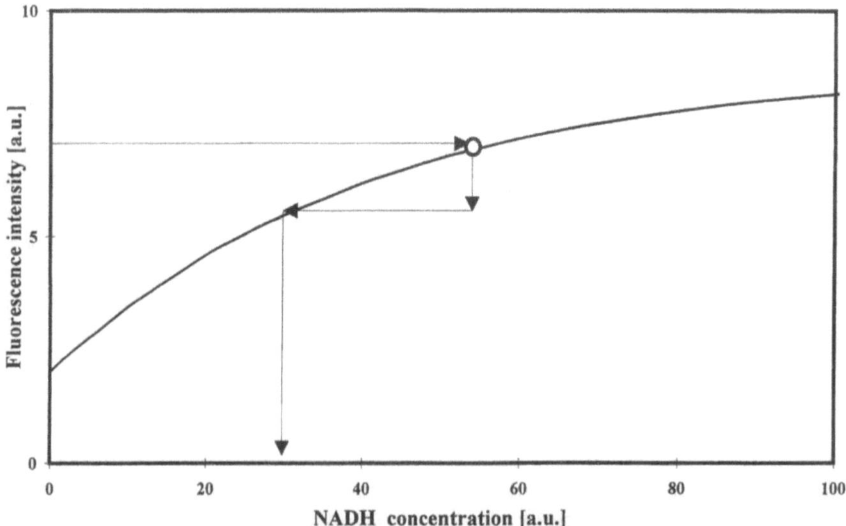

Fig. 12. Application of the rescaling curve (step 1 – measured fluorescence signal, step 2 – positive or negative correction according to the simultaneously measured back-scattering intensity, steps 3 and 4 – determination of the intrinsic NADH concentration)

the fluorescence signal can be corrected. Thereafter the initial measuring points fit the scaling curve within an error limit of about 10%. This good agreement confirms our assumptions and approximations described above.

4
Discussion

The limits of the procedure are determined from the ability to differentiate between different fluorophores and the sensitivity with regard to the error of the method and the various influences of tissue optics.

To determine the role of strange fluorescence – in this context the term "strange fluorescence" means fluorescence caused by the interaction of NADH with other compounds instead of being caused by pure NADH – samples of extracted connective tissue were investigated by spectroscopic measurements.

The compounds chondroitin sulfate A (CSA, Sigma) 0.1 g/L (physiological concentration: 0.2 g/L), a mixture of collagen type I and III (Sigma) 5 mg/mL and NADH (Sigma) 10^{-8} mol/mL, were separately dissolved in phosphate buffer at pH 7.4. For each solution the fluorescence was detected (excitation $\lambda = 337.1$ nm). The maxima of the fluorescence intensities show a divergent behavior compared to the NADH-fluorescence maximum ($\lambda = 460$ nm). For example, the mixture of collagen/chondroitin sulfate A shows a fluorescence maximum at 420 nm and at the same time the fluorescence intensity of this solution is increased by a factor of about three. Therefore it is necessary to discuss the spectral and time selectivity for this experimental setup [23]. The combination of both filtering techniques yields a relative contribution of 3–8% for the strange fluorescence in the time-resolved domain.

The investigations were restricted to ex vivo measurements. Hence, further sources of error, like motion artifacts or blood films between the probe and the fiber, are excluded.

The time response of the spectroscopic setup has also been described [23]. Although an optical fiber of about 15 m length was used for the temporal decoupling of the excitation and the fluorescence radiation, the autofluorescence of a UV quarz/quarz fiber could be neglected. The fiber has a numerical aperture of N. A. = 0.4. Our own Monte Carlo simulations show that, due to the small penetration depth of the UV radiation, a variation of the N. A. has relatively small influence on the fluorescence signal. Along with the influence of strange fluorescence we can estimate the total error of the described method to be about 10%.

It is obvious that the scaling curves for each type of tissue in Fig. 9 must be established individually. In contrast, the biological variability can be taken into account by the rescaling method (see Fig. 12). The problem of a quantitative fluorescence diagnosis is partly different for a fluorescence wavelength between 450 and 600 nm. In this case the variation of the scattering parameter also has an essential influence. To be able to establish a two-dimensional rescaling method analogous to Fig. 12, the optical parameters μ_a and μ_s' must then both be known. A simple method for the determination of these parameters is described [28].

5
Conclusion

Our results show the feasibility of a measuring method for the concentration of a fluorophore in turbid media by optical methods. It is necessary to measure the mean optical parameters of the sample and to establish a scaling curve in a statistically significant manner. Coincidental measurements of the fluorescence and back-scattering signal lead to the intrinsic fluorescence yield. The intrinsic fluorescence then has to be equated with the biochemical concentration values of NADH. Starting with these presuppositions, we can indeed measure the concentration to an accuracy of within 10 %. Consequently, the described method, together with our rescaling procedure, supplies full quantitative results thus opening a wide field of new applications.

The investigation of the aerobic cell metabolism is possible especially with NADH. For instance, the metabolic status of tissue regions during operations can be monitored instantaneously. In experimental medicine this method has been employed in various regions of the human body. It has proved particularly suitable for monitoring of minimally invasive procedures of laser-induced coagulation of liver metastases and for precise resection of squamous cell carcinomas in the oral region. In both cases the fluorescence relief in the tumor border region was evaluated before and after the surgical procedure. The results served as an indicator for the precisely and optimally performed coagulation and resection, respectively.

References

1. Chance B (1954) Spectrophotometry of intracellular respiratory pigments. Science 120:767
2. Duysens LNM, Amesz J (1957) Fluorescence spectrophotometry of reduced phosphopyridine nucleotide in intact cells in the near-ultraviolet and visible region. Biochim Biophys Acta 24:19
3. Grafni A, Brand L (1976) Fluorescence decay studies of reduced Nicotinamide Adenine Dinucleotide in solution and bound to liver alcohol dehydrogenase. Biochem 15:3165
4. Renault G, Raynal E, Sinet M et al. (1982) A laser fluorimeter for direct cardiac metabolism monitoring. Opt Las Technol 14:143
5. Alfano RR, Tata DB, Cordero J et al. (1984) Laser induced fluorescence spectroscopy from native cancerous and normal tissue. IEEE J Quantum Electron 20:1507
6. Ankerst J, Mont S, Svanberg K, Svanberg S (1984) Laser-induced fluorescence studies of hematoporphyrin derivative (HPD) in normal and tumor tissue of rat. Appl Spectrosc 38:890
7. Schneckenburger H, Feyh J, Götz A, Frenz M, Brendel W (1987) Quantitative in vivo measurement of the fluorescent components of Photofrin II (Tumor-selective accumulation). Photochemistry and Photobiology 46:765
8. Richards-Kortum R, Rava R, Fitzmaurice M. et al. (1989) A one-layer model of laser-induced fluorescence for diagnosis of disease in human tissue: Applications to atherosclerosis. IEEE J Biomed Eng 36:1222
9. Beuthan J, Zur Ch, Fink F, Müller G (1990) Laser fluorescence spectroscopic experiments for metabolism monitoring. Las Med Surg 6:72
10. Lohmann W, Nilles M, Bödeker RH (1991) In situ differentiation between nevi and malignant melanomas by fluorescence measurements. Naturwissenschaften 78:456

11. Lakowicz JR, Szmacinski H, Nowaczyk K et al. (1992) Fluorescence lifetime imaging of free and protein-bound NADH. Proc Natl Acad Sci 89:1271
12. van den Bergh H (1994) Photodynamic therapy and photodetection of early cancer in the upper aerodigestive tract, the tracheobronchial tree, the oesophagus and the urinary bladder. In: Amaldi U, Larsson B (eds) Hadrontherapy in Oncology. Elsevier, Amsterdam, pp 577–621
13. Sterenborg H, Thomsen S, Jacques SL et al. (1995) In vivo autofluorescence of an unpigmented melanoma in mice. Melanoma Research 5:211
14. Vo-Dinh T, Panjehpour M, Overholt B et al. (1995) In vivo cancer diagnosis of the esophagus using differential normalized fluorescence (DNF) indices. Las Surg Med 16:41
15. Bernarding J, Napiwotzki A, Kronfeldt H-D (1996) Zeitaufgelöste Fluoreszenz endogener Metabolite als Nachweismethode zur Tumorfrüherkennung, Teil 1: NADH und NADPH. Lasermed 12:27
16. Chance B, Legallais V, Schoener B (1962) Metabolically linked changes in fluorescence emission spectra of cortex of rat brain, kidney and adrenal. Nature 195:1073
17. Lehninger AL (1993) Principles of Biochemistry. Worth, New York
18. Ishimaru A (1978) Wave propagation and scattering in random media. Academic Press, Orlando, Vol 1
19. Wu J, Field MS, Rava RP (1993) Analytical model for extracting intrinsic fluorescence in turbid media. Appl Opt 32:3585
20. Beuthan J, Weber A, Minet O et al. (1994) Untersuchungen zur NADH-Konzentrationsbestimmung mittels optischer Biopsie. Lasermed 10:57
21. Durkin A, Richards-Kortum R (1996) Comparison of methods to determine chromophore concentrations from fluorescence spectra of turbid samples. Las Surg Med 19:75
22. Tinet E, Avrillier S, Ettori D et al. (1993) Monte Carlo evaluation of laser-induced fluorescence spectra modifications due to optical properties of the medium: Application to real spectra correction. Proc SPIE 2081:129
23. Beuthan J, Minet O, Müller G (1993) Observations of the fluorescence response of the coenzyme NADH in biological samples. Opt Lett 18:1098
24. Beuthan J, Zur Ch, Müller G et al. (1990) Ein experimenteller Beitrag zur Meßwertskalierung bei in vivo Laserfluoreszenzspektroskopie. Las Med Surg 6:127
25. Lowry P, Passeau A, Demoy J, Schulz R, Rock J (1961) The measurement of Pyridine Nucleotides by enzymatic cycling. J of Biol Chem 236:2747
26. Schulman S, Gupta I, Omachi A, Hoffman K, Marshall P (1974) A Nicotinamide-Adenine-Dinucleotide assay utilizing liver alcohol dehydrogenase. Anal Biochem 60:302
27. Roggan A, Minet O, Schröder C, Müller G (1993) Measurements of optical tissue properties using integrating sphere techniques In: Müller G et al. (eds) Medical optical tomography Bellingham. SPIE IS11, pp 149–165
28. Bocher T, Eberle H, Naber R, Minet O, Beuthan J et al. (1996) Extraction of optical tissue parameters in semi-infinite geometry using the PDW-approach. Proc SPIE 2925:66
29. Kalos MH, Whitlock PA (1986) Monte Carlo Methods. Wiley, New York, Vol I: Basics

Subject Index

1-anilino-naphthalene-8-sulfonate (ANS) 347
2,2′-azobisisobutyronitrile (AIBN) 374
2,7-oligo(pyrene)s 309
2,7-oligo(pyrenylene)s 313
3-pentanone 266
4-N,N-dimethylaminobenzonitrile (DMABN) 215
4-POBN 227
5- and 6-carboxyfluorescein 513
9,10-anthrylene dimers and trimers 314

absorption 92
absorption coefficient 6, 522
absorption cross section 199
activation energy 243
addition reactions 197
adenosine triphosphate 472
adiabatically 81
adrenocorticotropic hormone 517
AFM 126, 130
aggregation 113
aggregation number 111, 393, 399
AIBN 374, 375
all-anti conformation 81, 93, 97
allylamine 282
alpha-helix 474
amino acid derivatives 278
aminophthalimide 513
analyte 277, 289, 509, 513, 514
anisotropy 138, 139, 142, 147, 150, 349, 350, 439, 498
– coefficient 7
– decay 115, 142, 147, 150
– emission 349, 350
– time-resolved emission 498
– time-zero 139
anomalous emission 80, 89
– long radiative fluorescence lifetime 80
antenna molecules 483
anthracene 162
anthroate esters 185

anti (a_+, a_-) 92, 93
antibodies 54, 55, 57, 470, 509, 511, 513, 515
– labeled 54, 55
– molecules 54
– monoclonal 509
– mouse 515
– pig 515
– polyclonal 509
– rabbit 515
antibodybased sensor 288
anti-cancer drug 517
– α-(difluoromethyl)ornithine 517
antigen 55
antigen/antibody 102
antigen/hapten-antibody 511
antigen-binding sites 515
apparent size of the particle 111
argon ion 41
artificial receptors 277
Ascorbic acid 226
asymptotic proportionality factors 372
ATP 531
atrazine 278
autocorrelation 358
– orientational 358
autocorrelation curve 112, 210, 223
– Rhodamine 6G 112
– theoretical description 210
autocorrelation function 105, 209, 212
autofluorescence 424, 479, 484, 491, 497–499, 510
avalanche-photodiode 43
aza-15-crown-ether 167
azacrown-ether 163

β-adrenergic antagonists 278
β-barium borate (BBO) 62
Ba^{2+} 166
back scattering 539
background fluorescence 39, 491, 499
band-pass filters 43
BAPTA 534

barbiturates 278
bathochromic shift 62, 67
Beer-Lambert law 6
benzo-15-crown ether 163
benzyltriphosphonium chloride 278
Bessel functions 380
biexponential fit 55
bimolecular reactions 197
binding 282
binding principle 509
– competitive 509
biological 521
– bilayer 101
biological
biotin 513
bisanthraceno-crown ether 164
black level compensation 418
bleaching 193
bleaching channels 195
bleaching curves 208
bleaching rate constant 195
bleaching reaction 195
block copolymer 389, 391, 393
blood flow 17
blood serum/plasma 510
– matrices 510
bond-length isomerism 81
boxcar integrator 541
BP-HPC 115
B-Phycoerithrin 213, 214, 216
brain imaging 527
Brownian motion 46, 349, 350
bursts 39, 42, 45, 46, 48, 51
butyl-substituted 314
butylterrylene 311

Ca^{2+} 522, 527, 528, 531, 534
Ca^{2+}-binding 532
Ca^{2+}-dynamics 532
Ca^{2+}-sensor 532
Ca-green 528, 530
– confocal excitation 530
calcium channels 433
calcium pumps 433
cancerous liver parenchyma 542
capillary gel electrophoresis (CGE) 42
carbocyanine dye 42, 51
Carbostyryl-124 214, 216
carcinogenesis 479
carrier stream 511
cartenoids 226
Cascade Blue 513
catalytic reactions 288
CCD chip 522, 525
CCD-camera 43

cells 431
– living 431
cellular 530
– Ca^{2+}-signaling 530
central nervous system (CNS) 521
cerebellar slice 529
CFM 119, 126, 129
CH_3OH/O_2-mixtures 252, 254
chargecoupled device (CCD) 523
chelate complexes 161
– with metal ions 161
chemical dynamic 251
– calculations 251
chiral 97
chloramphenicol 291
chondroitin sulfate A 549
chromatography 517
– reversed phase 517
chromophore 48, 51, 63–65, 67, 68, 76
chromosomal subdomains 429
chromosomes 429, 465
cis-isomers 170
cis-trans isomerization 167
Cl^- 522
clinical 509
clusters 378
CNS cells 521
coherent anti-Stokes Raman spectroscopy
   (CARS) 254
collagen type I and III 549
collagen/chondroitin sulfate A 549
collected fluorescence intensity 104
collimated transmittance 7
collision model 243
collisional quenching 14, 263
combustion 241, 252
complexation 161, 162
conductivity 79
confocal 210, 523
– epifluorescence 523
– fluorescence 123
– fluorescence images 128
– fluorescence microscopy (CFM) 119, 126
– laser 529
– laser scanning microscopy 125
– microscopy 102, 127, 194, 204, 523, 525
– setup 42, 522
conformational changes 79
conformational constraint 81
conformational interactions 284
conformational isomerism 92
– ground state 92
conformational transition 101
conformers 92
– mixtures 92

conical intersection 94
conjugated linear chain 80
contrast agents 16
cooperative diffusion coefficients 371
coronene 304, 306, 308
– quinacridone 306
– alkylated hexa-*peri*-hexabenzo- 308
correlation function 441, 447
correlation time 209
COT 225
coumarin 166, 193, 221
Coumarin-102 214, 216
Coumarin-120 214, 216, 221, 222, 225, 233
Coumarin-307 214, 216, 222
Coumarin-39 214, 216
coupled electron 197
Cramérs inequality 40
critical amplitudes 378
critical exponents 372
critical micelle concentration (CMC) 101
critical phenomenon 376
critical point 326, 328, 331, 371, 372, 378
critical time 378
cross-linked polymer network 280
cross-linker 278, 374, 378
crown compounds 161
crown-containing benzoxazinone 169
crown-containing benzoxazinones 170
crown-containing dyes 161
crown-containing styryl dyes (CSD) 171
crown-ether 161–163
cubic lattice 378
Cyanin 5 214, 216
cyanine dyes 15
Cyanorhodamine B 42
Cystamine 225, 226
Cysteine 225, 226

DABCO 225, 227
dansyl glycine 282
dansyl monomers 280
– difunctional 280
– monofunctional 280
dansyl probes 280
dansyl-L-phenyl-alanine 278
Davydov splitting 73
DCFDA 534
deactivation 81
– radiative 81
– radiationless 81
decay time 292
degree of polymerization 371
delocalized excitation 96
deoxycholate 110
dephasing 59

desorbing time 383
desorption experiments 371, 376, 379
detection volume 209
detectors 524
detergent molecules 107
– number 107
Dexter mechanism 219
diagnosis 537
diagonally 29
diaza-18-crown-6 ether 163
diazepam 278
dications 163
dichlorodihydrofluorescein diacetate 534
dielectric relaxation 333
diffusion coefficient 113, 380
– collective cooperative 380
– intramicellar 113
diffusion constant 107, 111
diffusion equation 15
diffusion properties 102
– fluctuations 102
– fluorescent molecules 102
diffusion time 102, 111
digital CCD 524
digital imaging 206, 526
digitonine 110
dihydrorhodamine 123 534
dimers 314
dimethylnaphthalenes 164
diode laser 42, 44, 53
– lowcost 42
– pulsed 53
diphenylhexatriene 141
dipole oscillation 285
diseases 521, 532
distryrylbenzene 64, 67
DNA 39, 42, 48, 50, 150, 198, 422, 428, 465,
    466, 468, 479, 480, 483
– analysis 421
– base-sequence 483
– bending of 465
– cloned 428
– labeled 48
– lambda phage 422
– nucleotides 48
– sequences 479
– sequencing 39, 42, 50, 498
– strand 48
– surroundings 48
DNA-primer/DNA-target 102
donor-acceptor complex 463
Doppler profile 248
Down's syndrome 531
DPH-PC 115
dual emission 91

dual fluorescence 81, 163
dye laser chemistry 193
dynamic events 209
dynamic light scattering 392

EDMA 278
EGDM 373–375, 378
electron exchange energy transfer 197
electron transfer 179, 197
– photoinduced 179
electron-hole pair 67
electronic excitation 79
electroosmotic forces 49
electrostatic 284
ELIF 270, 272
ELISA 54, 55, 56
emission 69, 88, 113, 358
– anisotropy 358
– transition moment 113
– time-resolved 69
– spectroscopy 69
– localized states 88
endogenous 510, 537
– chromophors 537
– components 510
– sample 510
energy transfer 59, 64, 79, 188, 285, 335, 348,
    458, 463, 469, 481, 482, 503
– Förster-type 64
– singlet-singlet 335
– triplet-triplet 348
environment 282
environmental chemistry 509
enzyme 426
epifluorescence microscope 523
epitopes 54, 57
equilibrium geometry 94
ergodic hypothesis 331
ethylene glycol dimethacrylate (EGDM)
    282, 371
europium 468, 517
europium chelate 513
ex vivo measurements 549
excimer 162, 362, 389, 394, 396, 402, 405,
    411, 442
exciplex 162, 442
excitation 79, 80, 105, 522, 524
– cross section 105
– delocalization 79
– intensity 522
– localization 79
– trapping 80
excitation spectra 92
– oligosilanes 92
– low temperature 92

excited conformers 97
excited state potential energy surface 81
excited-state lifetimes 481
exciton 66, 68, 72, 76, 81
– Frenkel 72, 76
– Wannier 68
eximer laser induced fragmentation
    fluorescence (ELIF) 270
exonuclease enzyme 48

FCS 101, 102, 113, 115, 209, 210, 212
– liposomes 115
– principles 102
– size distributions 115
FFC 377
fiber guides 540
flow injection analysis 510, 516
fluctuation 102, 103, 193
Fluo-3 526
fluorescein 51, 213, 214, 216, 225, 482, 513
fluorescein-theophylline conjugate 516
fluorescence 12, 21, 25, 36, 39, 42, 42, 44, 45,
    47, 48, 50, 51, 54–56, 61, 63, 66, 67, 69, 71,
    74, 75, 122, 125, 132, 135, 180, 183,
    204–206, 224, 230, 241, 242, 246, 376, 540
– analysis 292
– anisotropy 113, 292
– autocorrelation traces 116
– bursts 45
– decay 54, 55, 66, 89
– decay time 67
– decay curves 71, 84
– depolarization 349, 350, 355
– fluctuations 102, 106, 209
– detection 42, 55
– Herzberg-Teller-like 75
– laser-induced 39, 241, 242
– lifetime 25, 42, 44, 47, 48, 50, 54, 55, 56
– near-infrared (NIR) 42
– non-exponential 66
– quantum efficiency 25
– quantum yields 42, 84, 103, 105
– quenching 162, 183
– saturation 201
– self-quenching 166
– spectroscopy 69, 74
– techniques 39
– time-dependent 45
– time-resolved 42, 55, 69, 74, 241
– up-conversion 61
– vacuum-ultraviolet 246
fluorescence correlation spectroscopy (FCS)
    51, 101, 194, 204, 206, 209
fluorescence immunoassays 511, 513
– heterogeneous 511, 513

fluorescence labels 510
- long-wavelength 510
fluorescence microscopy 149, 210, 417, 420,
    491, 504
- near-field 504
- quantitative 417
- signal-to-noise ratio 420
fluorescence probe 346–348, 360, 361
- micro-structural 348
- techniques 360
fluorescence resonance energy transfer
    (FRET) 285, 288, 457
fluorescence spectroscopy 132, 540
- experimental setup 540
fluorescent biosensors 278
fluorescent dyes 193
fluorescent labels 509, 513
fluorescent lipid probes 115
fluorescent markers 108
fluorescent probes 108
fluorescent proteins 513
fluoroionophores 162
fluorophore 180, 185
fractal dimension 333, 340, 363
Franck–Condon 64, 80
- forbidden fluorescence 80
Franck–Condon state 69, 167
free oxygen radicals 534
free volume 407
free-radical cross-linking copolymerization
    372
frequency domain 14, 18, 138
frequency doubled Nd/YAG 41
fundamental anisotropy 114
- global analysis 114
Fura Red 526
Fura-2 526–528

G *Streptococci* 515
gauche (g_+, g_-) 92
Gaussian broadening 88
Gaussian noise 27
gel 373, 376, 377, 378, 379, 380–383, 386
- formation 378
- fraction 371, 372, 377
gelation 371 373, 375, 379
globular domain 474
glutathione 227
glycoprotein molecules 51
glycoprotein mucine (MUC1) 54

H$^+$ 527
H$_2$O$_2$ 534
head-group mobility 449, 451
heart beating 431

heart cells 433
heart muscle 433
Henyey-Greenstein phase function 7
heptamethylenediammonium dication 164
herbicides 289
Herzberg–Teller coupling 74
heterodyne detection 19
heterogeneous 510
heterogeneous fluorescence immunoassays
    509
heterogeneous microenvironment 198
heterolytic dissociation 196
hexabenzocoronene 309
hexadecamethylheptasilane 82
- synthesis 82
high frequency vibrational modes 198
homogeneous assays 510
homogeneously broadened 63
HOMO-LUMO excitation 79
hormone/receptor 102
Huang–Rhys factor 68, 70
human protein 515
human serum 51, 53, 54
hydrodynamic radius 111, 107, 113
hydrogen bond strength 281
hydrophobic cavities 282
- imprints 282

illumination 525
- backside 525
imaging 3, 17, 25, 259, 435, 498, 528, 546
- cell cultures 528
- fluorescence lifetime 3
immunoassay 288, 510
immunoreactor 512, 513, 515
imprinted polymers 277, 279, 287, 291
- bulk polymerization 279
- characterization 279
in situ hybridization 428
Indo-1 526, 528
Indocyanine Green 17
inhibitor/enzyme 102
inhomogeneous broadening 60, 62, 63, 76
integrating-sphere 542
intensity distribution 105
intracellular changes 521
- Ca^{2+} 521
intracellular second messengers 532
intramicellar lateral diffusion 115
intramolecular eximer 163
ion binding 183
ion concentrations 421, 501
ion selective channels 432
ion sensitive probes 500
ionochromic effect 162

irradiance 205, 206
isomerization 91, 196, 198
isomers 80
– excited state 80

K^+ 166, 534
K^+-changes 522
Kohlrausch-Williams-Watts (KWW)
 function 66

labels 514, 415
ladder-type oligo(p-phenylene)s (LOPP)
 313
ladder-type polymer 317
LADH 541
lanthanide 481, 514
lanthanide chelates 457
lanthanide emission 459
laser 20
– titanium/sapphire 20
laser photolysis/laser-induced fluorescence
 (LP/LIF) 245
laser-induced fluorescence 193, 262, 537
laser-induced ignition processes 252
lateral diffusion coefficient 114
LIF 257, 259, 262, 538, 547
lifetime discrimination 510
– lanthanide luminescence 510
lifetime imaging 491
lifetime measurements 461, 492, 494
– frequency-domain 492, 494
– time-domain 492, 494
light cavity 103
light propagation 8
light scattering analysis 341
light-emitting diodes 73
light-emitting polymers 299
light-harvesting structures 213
lipid bilayer 361
lipid/protein 102
liposomes 102
liquid crystals 346, 362
Li-Tanaka equation 371
liver 17
liver tumor 537, 544, 547
living cells 101, 497, 498
LOC 249
localized excitation 91
logic gates 187
longitudinal osmotic modulus, M 380
loose bolt 183
LOPP 309
low-temperature flow cell 85
L-phenylalaninamide 286
LPPP 69, 71

Lucifer Yellow 513, 517
luminescence 54, 190
– background 54
– decay 54
– delayed 190
– biexponential fit 54
luminescent conjugated oligomers 299
luminophores 180, 299

macrocycle 161
macromolecules 150
malign kidney tumor 537
Marquardt-Levenberg algorithm 29
matrix-isolation 93
maximum likelihood estimator (MLE) 41
MCS 53
medical diagnosis 537
membrane 108, 143, 442, 443, 449, 451, 531
– DPH-labeled 143
– filtration 115
– mimetic systems 101
– permeable dye 531
– phospholipid 443
– potential 532
– probes 439
membrane/water interface 445, 446, 448
merocyanine 167
meso form 97
metabolism 532, 537, 547
metal ion 183
methanol 374
methanol/oxygen mixtures 253
methyl methacrylate (MMA) 371
methylated oligosilanes 79
Mg^{2+} 527
micelle 115, 361, 389, 393, 397, 405
– core 405
– structure 393
micelle-micelle interactions 396, 402
micellization 389, 398, 406
microcapillary 43, 49, 50, 57
microchannels 49
microenvironment 281
microfluorometry 521
micro-heterogeneity 280
– cross-linked polymers 280
microscopy 198, 494
– confocal 494
– conventional 494
microspectroscopic technique 101
microviscosity 498
millisecond lifetimes 514
mimic 101
mirror image (enantiomeric) 92
mitochondria 6

MLE 41
MMA 373, 374, 375
molecular association 471
molecular recognition 277
monoaza-crown ether 162
monomer 299, 402
mononucleotides 44, 45, 47, 48, 50
– fluorescently labeled 48
– labeled 44, 46–48
Monte Carlo 67
– master equation 72
– simulation 66, 68, 378, 539, 542, 544
Mowiol 227
multi-component assays 514
multi-photon photolysis 196, 203
multiple assays 514
multiple emissions 91
multiple emissive conformers 98
multiplex dyes 15, 42, 51
muscle 469, 472
– cell 469
– contraction 473
– mechanics 472

Na$^+$ 522, 527
NAD 541
NADH 537, 541, 544–547
naphthalene 372
NBTC 547
Near-Field Scanning Optical Microscopy
   (NSOM) 119, 122, 124–126, 129–134
near-IR fluorophors 3, 514
nervous system 421
network 280
neurodegenerative mechanism 532
neuronal somata 525
neurons 521, 525
NIR fluorophores 3, 514
NO concentration 259
non-linear crystal 61
n-Propyl-Gallate 227
nucleic acid hybridization 428
nucleotide 474

Octadecamethyloctasilane 82
– synthesis 82
oligo(p-phenylene)s 309
oligo(p-phenylenevinylene)s 309, 311
oligo(rylene)s 309, 311
oligomers 299
oligosilanes 80, 81
oligothiophene 60
one-photon-excitation (OPE) 230
optical antipode 286
– D-Phe-NH$_2$ 286

– L-Phe-NH$_2$ 286
optical sensor 289
optical tissue parameters 539, 542
OPV 64, 309, 313
organelles 530
– subcellular 530
organic field effect transistor 73
ortho (o_+, o_-) 92
osmotic bulk moduli 373, 380
osmotic pressure 373
Otto engines 259
oxygen concentration 219

π-systems 300
pacemaker 435, 436
– frequencies 436
particle territory 103, 104
partition coefficient 405
pathophysiological condition 522
pentarylene 311
peralkylated polymers 80
percolation model 371, 378
permethylated linear oligosilanes 79, 86,
   93
perylene 300, 306, 309, 319
– biperylenyl 301
– dicarboximide 300, 302
– dyes 301
– polyperylene 319
– tetracarboxdiimide 300, 302, 319
– tetraphenoxy-substituted 300, 319
pH changes 513
phenanthrene 372
phosphorescence 12
photobleaching 193, 194, 197, 199, 203–207,
   213, 219, 221, 223–225, 528
– experimental methods 206
– measure 106
– probability 194, 218
– rate constant 212
– reactions 194, 196
– simultaneous second- and first-order reaction 197
photodegradation processes 197
photodestruction 193, 235
photodynamic therapy 17
photoinduced electron transfer 162
photoisomerization 171
photoluminescence 59
photon migration 4, 8, 14–17
photons 418
– number of 418
photophysical parameters 79, 193, 209
photostability 193, 213
– organic fluorescent dyes 213

photostabilization 226
- additive compounds 226
phototoxicity 526
phycoerythrin 513
pigments 299
pinholes 523
pixel 36
plasticity 522
polarization 292
polarized fluorescence confocal images 127
polarons 59
poly(2-vinylnaphthalene) 348
poly(acenaphthylene) 348
poly-(methylmethacrylate) (PMMA) 372
poly(perylene-*co*-diethynylbenzene) 318
poly(p-phenylene) 313, 316
- ladder-type 316
poly(p-phenylene)s 315
poly(p-phenyleneethylinene) (PPE) 315
poly(p-phenylenevinylene) 315
- α, β-donor-acceptor-substituted 315
(poly)(p-phenylenevinylene)s 315
polyacetylene 59
polyarylenevinylenes 68
polydiacetylenes 62
polymer chain 280, 351, 355, 358
- diffusion of ionic reagents 280
- dynamics 358
- solvatation 280
polymer demixing 398
polymer fluorescence 79
polymer heterogeneity 280
polymer network 380
polymer side chain 354
polymer structure 280
- fluorescence 280
- analysis 280
polymeric networks 374
polymerization 279, 282, 299
polymethylene 163
Polyphenylenevinylene (PPV) 62
polysilanes 79
polystyrene 348, 358, 390
- end-capped 358
porous glass 334
Porphyrin 16, 126, 127, 129, 133, 185
porphyrin wheels 120, 126
potential energy surface 80, 95
- multidimensional 95
power law 372
power law analysis 335
PPBD 318
PPD 227
PPE 315
PPP 313, 315

PPPV 64, 67, 68
PPV 64, 67, 69, 311, 313, 315
- α, β-donor-acceptor-substituted 315
presynaptic membrane 528
ProLong 227
protein 101, 198, 467, 468, 515, 522
- kinase A 522
protein A 515, 516
protein A/G reactors 516
protein folding 360, 361
protein G 515, 517
protein/cell interactions 102
proton transfer 197
pseudo first-order reaction 197
pseudo-cyclic complexes 163
PtP 120, 127, 128-132
Purkinje cells 529
pyrene derivatives 371, 373

QCT 251
quantum efficiency 33, 522, 525
- CCDs 525
quantum yield of photobleaching 194
quantum yield of photoluminescence 315
quantum yields 80, 89, 93
- temperature dependence of 93
quasi-classical trajectory (QCT) 251
quaterrylenediimides 303
quinacridones 304, 306

racemic 93, 97
radical ion pair formation 197
radiative lifetime 80
radical copolymerization 374, 375
radioimmunoassays 509, 510
Raman scattering 39, 40, 45, 106
random walk 356
Rayleigh scattering 39, 40, 106, 256, 261, 262
Rb$^+$ ions 163
reaction cross sections 247
reactor 517
- protein A 517
rearrangement 196
receptor 185
receptor-binding studies 510
recognition 287, 288
recognition element 288
- enzyme-based 288
reconstruction 33, 34
reconstruction algorithm 31
reconstruction error 29
recoordination 175
recovery times 212
relaxed excited state 80
renormalization group 331, 335

rescaling procedure 544
resonance energy transfer 459, 473
resonance-Raman spectroscopy 201
Rh 123 221
Rh 6G 221
Rhodamine 123 214, 217, 221, 222, 534
Rhodamine 6G 40, 41, 102, 107, 204, 213,
   214, 216, 218, 221, 222
Rhodamine B 41, 234
rhodamines dyes 40, 41, 51, 193, 199, 201,
   213, 215, 221, 513
- photochemical properties 215
- triplet quantum yield 213
RNA 198
rotational correlation times 113, 145
rotational diffusion 209

S. aureus 515
saccharide 185, 190
sample photodecomposition 85
saturated polymers 79
saturation 205
scanning confocal microscopy 120, 529
scattering coefficient 6
scattering medium 31
SDS 110
self-assembling 161
self-diffusion 328
self-similarity 331, 333
sensitized emission 458, 466, 468, 479, 480,
   481
sensors 179, 187
sensory information 521
sexi-thiophene (6 T) 72
shear modulus G 373, 380
shear quench 341, 343
$Si_{10}Me_{22}$ 80, 81, 92, 98
$Si_{16}Me_{34}$ 80, 92
Si_3Me_8 81
Si_4Me_{10} 80, 81, 92
Si_5Me_{12} 80, 81, 86, 91, 92, 94, 97
Si_6Me_{14} 80, 81, 86, 91, 93, 97
Si_7Me_{16} 81, 86, 91-93, 97, 98
Si_8Me_{18} 93, 98
Si_9Me_{20} 98
sialic acid 278, 282, 284
signal transduction 101
signal-to-noise ratio 442
silicon chain 79, 81
single fluorescent molecules 209, 224, 417,
   422, 424
single handedness 93
single liposome 513
single photon counting (SPC) 46, 62, 119,
   125, 134

single-molecule detection 39, 42, 51, 54,
   193, 229
single-molecule spectroscopy 193
singlet oxygen 219, 235
SiSi bond 80
site bond percolation 378
size heterogeneity 115
Slowfade 227
SMD 194, 195, 209
sol 378
sol-gel phase transition 372, 373
solid-phase reactor 512, 516
solute-solvent exciplex 215
solution polymerization 375
solvation 440
solvatochromism 79, 162
solvent dependence 198
solvent dipolarity/polarizability 281
solvent uptake 379, 380, 381, 385
spatial resolution 522
spectral analysis 93
spectral reconstruction 441
spin statistics 219
S-propanolol 278
squid giant synapse 528
star-shaped oligo(phenylene)s 313
statistical mechanics 328, 329
steady-state concentrations 201
steady-state fluorescence (SSF) 371, 373
stereospecific [2 + 2]-cycloaddition 172
stilbene 304, 307
- hexa-*peri*-hexabenzocoronenes 304
- *N*-alkyl and *N*-aryl perylenes 307
- α, β-donor-acceptor-substituted stilbene
   307
Stokes shift 60, 64, 67, 89, 442, 451, 480, 499
- time-dependent 451
Stokes-Einstein relationship 113
Streak camera 59, 60, 61, 62, 76
stretched exponential 333, 355
subcellular organization 521
sub-picosecond 167
sulforhodamine 213, 513
sum frequency 62
superoxide generation 532, 534
supramolecular 161
swelling 285, 377, 379, 382
- curves 385
- experiments 379
- processes 371
- rate constant 380
switches 179, 187
swollen 382, 386
synchronous fluorescence 514
syringe pumps 511

tagged analyte molecules 40
target analyte 277
template 278, 284
terbium 476, 482
terrylene 303
– benzoylterrylene 303
terrylenebisdiimides 302
– benzoylterrylenebisdiimides 302
– quaterrylenebisdiimides 302
tetramethylrhodamine 213, 214, 221, 480,
   481, 482, 513
Texas red 40, 213, 214, 217, 513, 517
theophylline 278, 516
therapeutic strategies 521
thermochromism 79
Thesit 110
third-order hyperpolarizability 79
three dimension lattice 378
three-photon excitation 137
TICT 165, 185, 215, 347, 366
time-correlated single photon counting 40,
   44, 84, 120, 125
time-correlation function 113
time-resolved emission spectra 441
time-resolved fluorescence 55, 123, 127,
   138, 537
– clinical application 537
time-resolved fluorescence anisotropy 113
tissue 3, 8, 26, 36, 544
titanium/sapphire laser 41, 61
– frequency doubled 41
– Kerr lens mode-locked 61
TMR 217, 221, 222
toluene 374, 375, 378, 379, 381, 385
trans-cis photoisomerization 167
transient absorption spectra 167
trans-isomers 170
translational diffusion 106, 111, 209
– micelles 111

transmitter 528
trap emission 75
trimers 314
triplet 40, 41, 188, 215
– correlation time 212
– lifetime 40
– population 215
– quantum yield 40
triplet-triplet annihilation 197
trisomy 16 531
Triton X-100 110
tryptophan 145, 196
tumor 17, 547
Tween 80 110
twisted intramolecular charge transfer
   (TICT) 165, 185, 215, 347, 366
two emitting species 96
– Si_7Me_{16} 96
two-photon excitation 139, 197, 231, 496,
   526, 529, 530
two-step excitation 196
tyrosine 145, 196

up-conversion 59, 67, 68, 76
Urate and Vitamin E 227

vesicles 102, 111
vibronic splitting 63
vibronical coupling 215
vinylphenylboronic acid 282
viscosity 198

weight average degree of polymerization
   372, 378
Wellferon 517
Wigner-rule 219
wobbling-cone 345, 351

xerography 79

Springer
and the
environment

At Springer we firmly believe that an international science publisher has a special obligation to the environment, and our corporate policies consistently reflect this conviction.

We also expect our business partners – paper mills, printers, packaging manufacturers, etc. – to commit themselves to using materials and production processes that do not harm the environment. The paper in this book is made from low- or no-chlorine pulp and is acid free, in conformance with international standards for paper permanency.

Springer